AGRICULTURE
MADE EASY

For
Competitive Examinations

Dr. K. Sasikumar
I.F.S(P)

V. Saravanakumar
M.B.A

EIGHTH REVISED AND ENLARGED EDITION

2022

JAIN BROTHERS (New Delhi)

16/873, East Park Road, Karol Bagh, Near Dr. Joshi Hospital,
NEW DELHI-110005 Phones : 23618426, 45097398
E-mail : jain_bros@hotmail.com
Website : www.thejainbrothers.com

Published by :

Shri Sunil Kumar Jain
for
Jain Brothers (New Delhi)
16/873, East Park Road, Karol Bagh,
New Delhi-110 005
Phones : 011-23618426, 45097398

Branch Office at :

1. **M/s Sunil Book Service**
   Fatima Nagar, Opp. Andhra Bank,
   KAZIPET-506004 (A.P.)
   Ph. : 0870-2459617, Mob.: 9390115435
   E-mail : kpkulshrestha@yahoo.com
2. **M/s Jain Brothers**
   Vidya Vihar, Pilani (Rajasthan)
   Phone : 01596-242144

Available at:
Pusa Agricultural Book Service
Shishir Hostel, IARI, PUSA
New Delhi
Ph.: 011-25841048

First Published	: July, 2005
Second Revised Edition	: May, 2007
Third Revised Edition	: June, 2009
Fourth Revised Edition	: July, 2010
Fifth Revised Edition	: Oct., 2011
Sixth Revised Edition	: Jan., 2015
Seventh Revised Edition	: Sept., 2017
Eighth Revised Edition	: Sept., 2022

***Price :* ₹ 475.00**

ISBN : 978-81-8360-064-4

Laser Typesetting by :

Niyati *Graphics*
754/23 Khajoor Road,
Karol Bagh, New Delhi

Foreword

Agriculture has always been a popular optional for the UPSC civil services exam, Indian Forest Service exam etc, with the number of students also increasing with the years. There are also other exams like ICAR entrance test, ARS, Ph D. Entrance, Bank AO (Agri.) exams where agriculture is chosen as a subject. Despite the popularity and success of the subject, it remains that there is no one textbook that deals with Agriculture in its entirety, let alone with a focus on the syllabi of the competitive exams.

This has led to a situation were every aspirant re-invents the plough, studying several specialised textbooks, making synoptic notes, creating mnemonic codes and structuring the points in a form suitable for the purpose of exams **Dr. K. Sasikumar, IFS** and **Sri. V. Saravana Kumar, MBA** have written this book as a way of reaching out to student community to fill this critical gap. '***Agriculture made easy***' has comprehensively dealt with the subject of Agriculture. The ordering of the chapters closely follows the syllabi of UPSC IAS/IFS exams. Even as one without a background in Agriculture, I found the chapters on Agro-ecology, Food production trends, Recent advances in agriculture and Current issues in agriculture not found in any other book and particularly latter two chapters interesting and informative. The format of the book is pleasant, and easy to use as a reference manual and as well as an exam guide. I am sure this book will be of great use to the students.

S. NAGARAJAN I.A.S.
(Topper 2004)

Preface to the Eighth Edition

We feel pleasure to present Eighth revised and updated edition of the book. All data pertaining to various fields are updated with latest information. In many chapters new information has been added to make the text more useful to students preparing for competitive examinations.

We hope our sincere efforts will be useful for readers.

Suggestion are always welcome for improvements.

Date. 6-10-2022

K. Sasikumar
V. Saravanakumar

Preface to the Seventh Edition

Data pertaining to various fields are updated and new information are added in many chapters to make to text more useful for readers. Glossary of economic terms is also updated.

Authors will welcome any suggestion from readers to enrich the text

Date. 5-9-2017

K. Sasikumar
V. Saravanakumar

Preface to the Sixth Edition

In this revised edition two chapters 1. Agricultural Economics and 2. Agricultural Statistics are thoroughly updated and expanded to increase their utility to readers. A new chapter Overview of Indian Agriculture has been added to enhances the utility of text to readers. All tables and statistical information are updated. A glossary of common economical terms are added in the book.

Authors hope readers shall find the text more useful than earlier edition.

Suggestions are always welcome for improvement of text.

Date. 2-2-2015

K. Sasikumar
V. Saravanakumar

Preface to the Third Edition

In our preparation for various competitive exams, we found that there was no book that dealt with the subject in a comprehensive manner. Several specialised advanced level text books were available for each of the sub areas of agriculture. In the absence of a single book, it was a formidable task to study all these books, assimilate their contents, prepare synoptic notes suitable for revision on the eve of the exam. We have written this book to address the need of the student community for a good reference-cum-text book relevant to the realities of Indian examinations.

We have not attempted a treatise on agriculture. We have made the book suitable for preparation and revision for competitive exams, while at the same time maintaining academic rigour and factual accuracy.

Also, it has certain sections that one does not find in conventional textbooks. These include Agro-ecology, Food Production, Statistics, Recent advances and Current issues in agriculture.

In our effort, we have been greatly benefited by the experience and advice of seniors and peers of **Indian Agriculture Research Institute (IARI)**, Pusa, New Delhi, **too numerous** to mention, who have tread the treacherous path of the toughest competitive exams of India.

We thank Dr. Susheela Thirumaran. VC, Thiruvalluvar University Vellore (T.N.), Our teacher Dr. M.L. Choudary. Horticulture Commissioner G.O.I., Dr. S.K. Singh, Senior Scientist, Div. fruit and technology IARI, New Delhi. Dr. Krishna Moorthy PDF, Deputy Director (NIPCD), Mr. K. Balamurugan, I.F.S., B. Ravichandar, I.R.T.S., Dr. S. Bharatidasan, I.A.S., DR. K.S. Jayachandran I.R.S., Mr. S. Sundar, Executive Manager (TCL), Dr. V. Palanichamy Ph. D. (Horti), Dr. R. Thangapandian, Ph. D. (SS & AC), Dr. K. Karthikeyan Ph. D. (Ag. Engg.), Dr. K. Boomiraj Ph. D. (Envi. Sciences), Dr. S. Sarathbabu Ph D. (Patho), Dr. T. Karuppasamy Ph. D. (Horti), Dr. R. Gopi Ph. D. (Patho.), Dr. M. Anand Prakash IFS, Dr. Manish Kumar Ph. D. (Envi. Sciences), Dr. M.K. Kumar IFS (Ento.), Dr. R. Srini Vasan Ph. D. (Micro.), Mr. S. Jayakumar ACP, Dr. David Rajkumar Ph D. (Horti.), Dr. C.M. Madhavan Ph. D. (SS & AC), Dr. C.M. Senthil kumar Ph. D. (Ento.), Dr. Dhilip Kumar Ph. D (Breeding) and Dr. A. Prabhu Ph. D. (Ento)

I am grateful to Anita and Pradeep Kumar for their help in revision and proof reading.

In third edition of the book matter have been thoroughly revised with many new additions to enhance the utility of the book.

We hope that this book will help the readers in their endeavour. We invite any suggestions for making the book more useful to the readers.

K. Sasikumar
V. Saravanakumar

CONTENTS

Dedicated

to

Our Parents

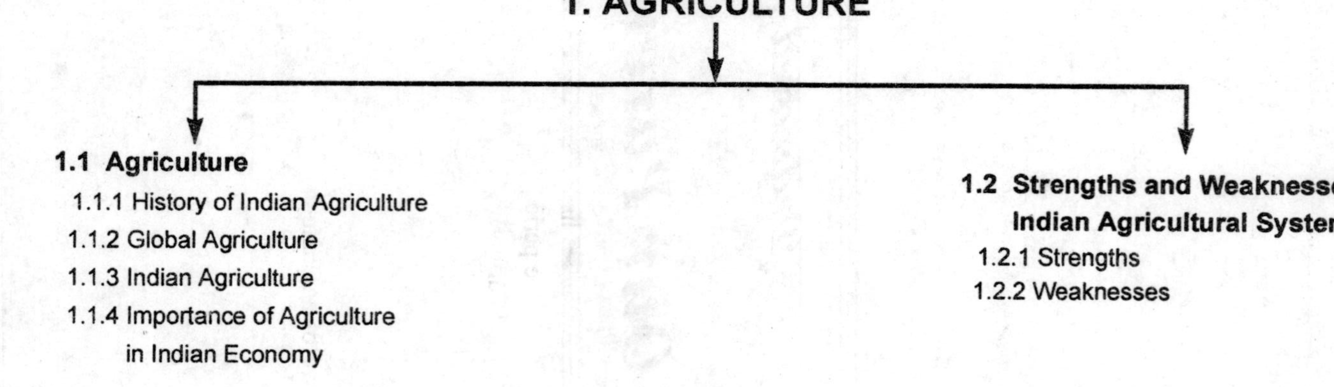
1. AGRICULTURE
1.1 Agriculture
1.1.1 History of Indian Agriculture
1.1.2 Global Agriculture
1.1.3 Indian Agriculture
1.1.4 Importance of Agriculture in Indian Economy
1.2 Strengths and Weaknesses of Indian Agricultural System
1.2.1 Strengths
1.2.2 Weaknesses

1. Agriculture

1.1 AGRICULTURE

The word agriculture comes from the Latin words *ager*-referring to the soil and *culture*-to its cultivation. Agriculture, in its widest sense can be defined as the *cultivation and production of crop plants or livestock products*. It is synonymous with farming : the field or field dependent production of food, fodder and industrial organic materials.

1.1.1 History of Indian Agriculture

The history of agriculture and civilisation go hand in hand as the food production made it possible for primitive man to settle down in selected spots leading to formation of society and initiation of civilisation.

Western Asia is considered to be the birth place of agricultural revolution where wild ancestors of wheat and barley and domesticated animals like goat, sheep, pig and cattle are found. The period from 7500-6500 B.C. was the period of discovery of agriculture.

The period from 3000-1700 B.C. marked the spread of agricultural revolution to Egypt and subsequently to Indus valley. Mohenjodaro to Harappa territory was the centre of agricultural revolution in Indus valley.

Agriculture was very important profession during vedic age 1500-1000 B.C. Use of iron implements, particularly iron ploughs became prevalent. Buddhist period 600 B.C. marks the importance of trees. It can be called as a period of *Arboriculture* and *Horticulture.*

During the first century of Christian era the most important development in agriculture was irrigated cultivation in agriculture. Irrigated cultivation of rice in South India cauvery river was the most important source of irrigation water. Cultivation of rice, finger millet, Sugarcane, Pepper and turmeric was quite common.

During British period the most important development in agriculture was cultivation of commercial crops like cotton, sugarcane and Indigo. The reason for introducing commercial crops in India by Britishers, they felt need of raw materials for their industrial growth and they got huge amount from European market by selling our commercial crops there.

After Independence during 1950's there was food crisis in India. India imported wheat from USA and rice from Myanmar. During mid 1960's green revolution was started in India. After green revolution India became a food surplus country.

1.1.2 Global Agriculture

Advancement of civilisation is closely related to agriculture which produces food to satisfy hunger. In the world nearly one billion people are living below poverty line. Therefore, the increased food production should aim at trebling food production in the next century.

For crop production the basic input is land. The FAO's analysis of growth patterns in crop output in developing countries shows that 63% of the growth in production come from higher yields and 15% from higher cropping intensity, only 22% is from land resource.

Environmental degradation is increasing at a pace that is impairing the productivity of land and undermining the welfare of millions of rural people.

Cereals are grown throughout the world to provide food for human consumption and fodder for livestock. They are grown in 73% of the world's arable land and contribute 74% of the global calorific production. Demand for food is growing with ever increasing population.

1.1.3 Indian Agriculture

In India, more than 70% of the population is depending on agriculture in one form or the other. The present population is about 1000 million which is expected to stabilize at about 1500 million by the middle of present century. This trend of population growth created alarming situation as the scope of increasing area under cultivation is limited.

1.1.4 Importance of Agriculture in Indian Economy

Importance of agriculture can be measured by the share of agriculture in national income and employment pattern etc. Knowing the importance of agriculture, agriculture sector can be related with the industrial sector.

Share of Agriculture in National Income

Agriculture GDP: The agriculture and allied sectors contributed approximately 13.9% of India's GDP (at constant 2004-05 prices) during 2013-14. Gross Domestic Product (GDP) of agriculture and allied sectors and its share in total GDP of the country during the last 4 years, including the current year, at 2004-05 prices is as shown in Table 1.1.

India GDP has shown robust growth since 1990-91. Which shows that non-agricultural sectors (Particularly service sectors) has grown at the expense of agriculture. This is a trend that takes place the fundamentals of Indian Economy closer to the developed economies

The share of agriculture in GDP increased to 19.9 % in 2020-21 from 17.8% in 2019-20.

India expected to grow by 6.4% in 2022 well below the 8.8% growth in 2021 as high inflationary pressures and uneven recovery of the labour market curbing private consumption and investment according UN report. War of Ukraine has impact the growth worldwide. Global economy is expected to grow by only 3.1% in 2022. Global inflation in expected to increase to 6.7% in 2022.

Indian Agriculture and Pattern of Employment

Agriculture sector is the biggest unorganized sector as labour force and about 39.8 crore people are employed in agriculture.

Very high proportion of working population in India is engaged in agriculture. According to India's census figure, 66 per cent of India's working population is engaged in agriculture. Where as in U.K. and U.S.A. 2 to 3 per cent, in France 7 per cent and in Australia 6 per cent of working population is engaged in agriculture.

As per the latest data from Ministry of Finance as on 2017, 49% people of India depend on agriculture from livelihood point of view.

According to CMIE data shows reverse migration to agriculture continues even in 2020-21.

The share of agriculture in total employment jumped to 39.4% in 2021 from 38% in 2019-20.

Importance of Agriculture for Industrial Development

Indian Agriculture has been source of supply of raw materials to our leading industry. Cotton, Jute, textile industry, Sugar, vanaspathi and plantation depend upon agriculture directly.

There are many other industries which depend on agriculture in indirect manner. Many of our small and cottage industries, handloom, oil industry, rice mills and sugar mills depend upon agriculture for their raw materials. They account for 50 per cent of income generated in manufacturing sector in India.

However in recent years the importance of food processing industry is being recognised both for generation of income and employment opportunities.

TABLE 1.1 (Rs. in Crore)

Gross Value Added from Agriculture & Allied Sector	2017-2018	2018-2019	2019-2020	2020-2021	2021-2022
1. At current Basic Price	2829826	3029925	3358364	3609494	3964392
2. At constant Basic Price (2011-12)	1840023	1878598	1882023	2048032	2115048

Role of Agriculture in the field of International Trade

Agriculture products like tea, oilseeds, tobacco and spices constitute the main items of exports of India. The proportion of agricultural goods which are exported may account for 50 per cent of our export and another 20 per cent of exports from manufactures with agricultural content.

- Rice is the largest exported agricultural product and contributed to more than 19% of the total agriculture export during 2021-22

In total 70 per cent of India's exports comes from agricultural sector. The increased exports pay for the imported oil, machinery, etc.

Even though export surpluses are available for these commodities along with competitive prices, the export is quiet often hampered by non price competition, viz Technical barriers to trade (TBT) and Sanitary and phyto sanitary measures (SPS) under WTO regulations.

These are not uniform in all countries. They vary from country to country. Developed countries use these to minimise import from developing countries.

To meet these challenges, Indian agricultural and agricultural processing sector have to adopt measures like Hazard analysis and critical control points (HACCP) and Total quality management (TQM).

Role of Agriculture in Economic Planning

Agriculture is main support for India's transport system. Transport systems like railway and roadways secure bulk of business from movement of agriculture goods and international trade is mostly in agriculture products. Many other economic activities like food grain business, dairying and fruit marketing seed and agricultural machinery etc. also depends upon agriculture. Economic condition of govt. to large extent depend upon prosperity of agriculture.

It is therefore clear that agriculture is back bone of Indian economy and prosperity of agriculture largely influence the Indian Economy.

Tenth plan of India has shown that agriculture sector is the prime moving force of the economy, Amartya Sen Nobel laureate has suggested measures on the same lines.

1.2 STRENGTHS AND WEAKNESSES OF INDIAN AGRICULTURAL SYSTEM

1.2.1 Strengths

- India has a vast diversity of basic natural resources in terms of land, water, climate, rainfall, flora and fauna. The country has a vast shoreline, excellent sunshine for about 280 days annually in most parts of the country and has a vast network of natural perennial river systems running over 1,32,000 kms.
- The country is recognized as one of the eight mega-centres of plant biodiversity having about 45,000 species of plants.
- It also has a complex animal genetic diversity consisting of cattle, buffaloes, sheep, goats, camels, horses, pigs and poultry. The Indian waters also have a wide range of fish species contributing to the wide biodiversity in aquatic resources.
- It has a large coastline with significant marine resources, a coastal belt that harbours some rare species of plants and animals, which could be used for genetic improvement.
- India is bestowed with excellent climate with all types of agroclimatic conditions prevalent anywhere in the world and making it suitable for the production of a variety of crops that includes food grains, fruits, vegetables, plantations and commercial crops.
- The country has the world's second largest irrigation system with an estimated potential of 156 million hectares making it suitable for the growth of many plants.
- There is a rich heritage of knowledge on farming systems and wisdom based upon indigenous technical built up through centuries of farming and agriculture.

- India has one of the largest and institutionally complex agricultural research systems in the world.
- The National Agricultural Research System (NARS) has been at the forefront to steer and guide technological breakthroughs that have ushered in the Green Revolution.
- The transfer of technology and extension system in the country are strong and extensive, with over 1,20,000 trained extension workers carrying the message of newer technology to the field.
- The country has well established and strong public and private sector industry for providing quality seeds of different crops.
- The input supply system for the supply of fertilizer, pesticides and the agro-chemicals is strong and is also supported by the government machinery, where needed.
- There is a good marketing network for agricultural produce with over 7000 regulated wholesale markets and 28,000 rural periodic markets.
- The country has an extensive network of educational institutions for imparting quality training in agricultural and allied areas. This is backed by a National Agricultural Research System (NARS).
- Highly trained technical and scientific manpower is available both at the Centre and in the States.
- Linkages with a number of external agencies, research and development organizations are strong and many bilateral and multilateral arrangements have been developed. This has helped bringing frontier technologies from many countries in the service of the farmer and have helped to improve agricultural productivity in general, and diversification of crops in particular.
- A fairly well developed industrial sector, capable of producing inputs needed for agricultural such as fertilizers, seeds, pesticides and agrochemicals, agricultural implements and machinery.
- An Exclusive Economic zone of 202 million sq. km of ocean, offering an economics scope for development of a marine fisheries and vast inland fisheries resources.

1.2.2 Weaknesses

- Even though the Green Revolution registered a remarkable increase in productivity of foodgrain crops, especially wheat and rice, the average yield continues to be low when compared to many other countries.
- The ultimate success of improved technology is measured by the stability in production, unaffected by the vagaries of season, or biotic and abiotic stresses. Indian agriculture is still heavily monsoon dependant and has to reach the level of stability required for continued growth.
- The Agro-based industry is very weak and thus farm-age prices of commodities have been low in the absence of facilities for even primary value addition at the farm and other production centers.
- Rural employment and income enhancement through value addition have become causalities in the process.
- Land holdings are small and scattered and the prevailing social systems of inheritance of family land holdings have further eroded their size.

- Market infrastructure, storage, transportation, credit support etc. are weak.
- Power availability to Indian farmers, which currently is only 1 KW/ha is inadequate to cope up with the demands of high intensity of cropping and high productivity.
- Despite the fact that the rate of return to investments in agricultural research and education are very high, investments in agricultural research and education in the public sector funding is low.
- Private sector investment in research is low, except in highly viable commercial sectors.
- Holistic thrust is lacking in the development of farm technologies that are appropriate to address the problems of rural poor.
- Indian farms are multi-dimensional and multi-enterprise oriented. Technology development, therefore has to address the farming system as a whole.
- Gross Capital Formation in the agriculture sector has not kept pace with the overall Gross capital Formation, limiting rural employment and development of rural industries.
- Inability of the system to organize demand-driven production system that calls for market interventions during gluts or shortages.
- Inadequate and skewed flow of farm credit effects production.
- India ranks 2nd in 2022 largest in agriculture product in the world and 9th largest in agriculture exports.
- Indian agriculture is now not a gamble on the monsoon.
- In terms of value of agriculture output. India is double than that of USA and E.U.
- Life expectancy in India in 2022 is 68 years.

Achievements in India Agriculture (2015-2016)

- India ranks first in Milk production.
- India ranks second worldwide in farm output. Agriculture and allied sectors like forestry and logging.
- India is largest producer in the world of milk, cashewnuts, coconut, tea, ginger, turmeric, blackpepper, okra, onions, brinjal etc.
- India is second largest producer of wheat, rice, sugar, groundnut, egg plant and inland fish.
- India has largest cattle production.
- India accounts for 10% of world fruit production with first rank in sappota and banana.
- India is the second largest producer of Fruits and Vegetables.
- India is third largest producer of tobbacco in the world.
- India ranks first in mango production in world.
- India ranks third in potato production in world.
- India ranks second in use of fertilizers in the world.
- India ranks third in numbers of tractors.
- India ranks 2nd in fish production.
- India ranks 2nd in vegetables production.
- India ranks 3rd in egg production.
- India ranks first in total pulses production.
- India is biggest exporter of cotton in the world.

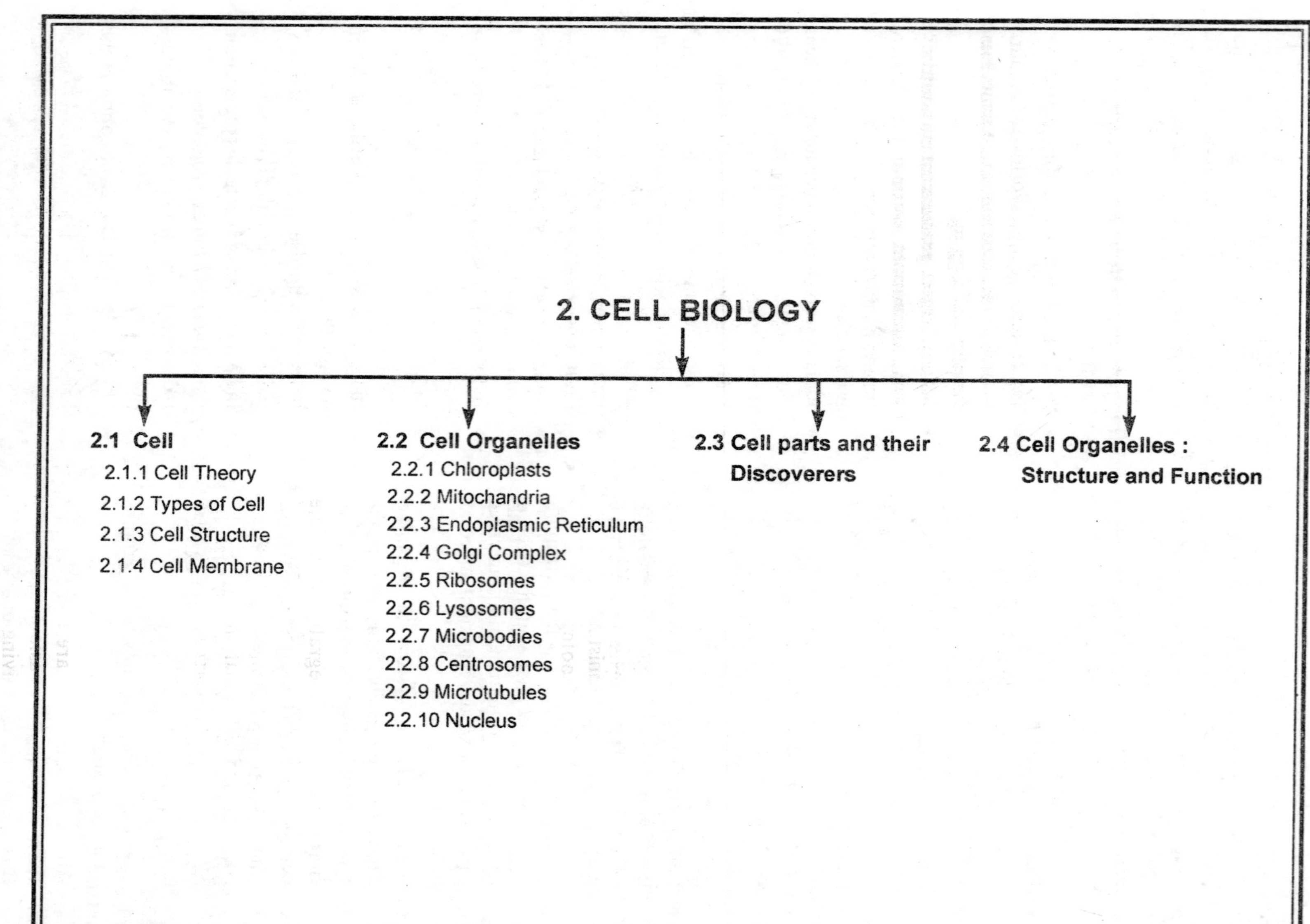

2. CELL BIOLOGY
2.1 Cell
2.1.1 Cell Theory
2.1.2 Types of Cell
2.1.3 Cell Structure
2.1.4 Cell Membrane
2.2 Cell Organelles
2.2.1 Chloroplasts
2.2.2 Mitochandria
2.2.3 Endoplasmic Reticulum
2.2.4 Golgi Complex
2.2.5 Ribosomes
2.2.6 Lysosomes
2.2.7 Microbodies
2.2.8 Centrosomes
2.2.9 Microtubules
2.2.10 Nucleus
2.3 Cell parts and their Discoverers
2.4 Cell Organelles : Structure and Function

2. Cell Biology

2.1 CELL

The first person to use the term *cell* was Robert Hooke (1635-1703). He used a simple kind of microscope to study thin slices of cork from the bark of a cork oak tree. He used the term cell in his publication Micrographia in 1665 and described his observations. A.V. Leeuwenhoek (1632-1723) was the first person to carefully study magnified cells. He is known as father of microbiology.

2.1.1 Cell Theory

The French scientist H.J. Dutrochet (1824), boiled the tissues in acid and separated the cells. On this basis he thought that all animals and plants are made of cells. This was the foundation of the *cell theory*. The German botanist *M.J. Schleiden* (1838) and the German zoologist *Theodor Schwann* (1839) found the essential similarity in the structure of plant and animal tissues. On this basis they gave an abstract generalization which is now known as cell theory.

The cell theory states that—

1. All plants and animals are made of cells. The cell is a unit of structure and function.
2. Each cell behaves independently but also functions as an integral part of the complete organism.
3. The new cells are formed by a process similar to crystal formation.

Virchow (1858), a German scientist stated that *the new cells orginate from the pre-existing cells only* (*Omnis cellula e cellula*). The main features of the cell theory as known today may be briefly summed up as follows :

1. All living organisms are composed of cells. Thus cell is the basic structural and functional unit of all living organisms.

   It is the smallest unit of an organism that is classified as living and is often called the building block of life.
2. All cells arise from the pre-existing cells of the similar kind. Thus continuity of life from one generation to another is through living cells.
3. The chemical composition and metabolism of all cells is basically alike.
4. The function of an organism as a whole is the outcome of the activities and interactions of the constituent cells.

Genetic material in all cells consist of nucleic acid

2.1.2 Types of Cells

On the basis of presence or absence of organised nucleus, two types of cells have been recognised. 1. *prokaryotic* and 2. *eukaryotic cells.*

Table 1 : Differences between Prokaryotic and Eukaryotic cells

Prokaryotes	*Eukaryotes*
1. **Protoplasm**	
Relatively rigid, usually non-vacuolate, relatively resistant to desiccation, osmotic shock and thermal denaturation.	Typical more fluid, generally vacuolate, more sensitive to desiccation, osmotic shock and thermal denaturation.
2. **Nucleus**	
True nucleus absent ; nuclear membrane and nucleolus not present, nucleoplasm undifferentiated from cytoplasm.	True nucleus present, nuclear membrane, nucleolus and nucleoplasm distinct.
3. **DNA**	
Scattered in the protoplasm, without histone proteins.	Organised with histone proteins to form distinct chromosomes.

Contd.....

Prokaryotes	*Eukaryotes*
4. **Organelles**	
Membrane bound organelles, like golgi bodies, plastids mitochondria, ER, etc., are absent.	Organelles like golgi bodies, plastids, mitochondria,., ER, etc., are present.
5. **Ribosomes**	
70 S type.	80 S type. (70S in Mit. Chl.)
6. **Mitotic apparatus**	
Absent.	Present.
7. **Cell division**	
Mostly amitotic.	Typically mitotic.
8. **Sexual process**	
Typical sexual process lacking, but genetic material is exchanged by parasexual processes	Typical sexual process (with alternating nuclear fusion and reduction division) often present.
9. **Examples**	
Cells of mycoplasma, bacteria, blue-green algae.	Cells of all the higher organisms, except the three group mentioned as prokaryotes.

2.1.3 Cell Structure

The cell can be divided into three major structural parts :

(1) **Cell wall,** (2) **Vacuoles** and (3) **Protoplasm.**

(1) ***Cell wall*** is in characteristic of plant cells only, being absent animal cells. It forms the outermost limiting layer of the plant cell and is freely permeable. Cell wall is composed of cellulose which lies away from plasma memberane.

(2) ***Vacuoles*** are spaces filled with cell sap or vacuolar sap. Vacuole is bound by vacuolar membrane or *tonoplast.* In unicelluclar organisms such as ameoba there is a food vucuole in which all food items are filled which is consumed by it.

(3) ***Protoplasm*** The living matter of cells is called protoplasm and consists of cytoplasm and the nucleus. Cytoplasm includes all the structures outside the nucleus. It is the fluid portion of protoplasm surrounding the nucleus.

Table 2 : Differences between Plant and Animal cells

Plant Cell	Animal Cell
1. **Cell wall**	
A cell wall made of cellulose which is present outside the cell membrane.	Cell wall is absent. The outermost is the cell membrane or plasma membrane.
2. **Chloroplasts**	
The photosynthetic cells are green due to the presence of chloroplasts.	Chloroplasts are absent
3. **Centrosome**	
Except a few plant cells, centrosome (centriole) is absent.	In most of the animal cells centrosome (centriole) is generally present.
4. **Vacuoles**	
Plant cells generally have a large, centrally located vacuole. The protoplasm and the nucleus are present near the cell wall.	Animal cells mostly lack vacuoles. If present, these are small and remain scattered throughout the cell.
5. **Cytokinesis**	
This occurs due to the formation of cell plate.	This is the result of formation of a furrow in the centre of the cell.
Golgi apparatus is single and large coiled.	Golgi apparatus is more than one is small sizes.

CELL WALL. The plant cells are bounded externally by a non-living, more or less firm covering called cell wall. Animal cells do not have a cell wall and thus their outermost covering is cell membrane. The cell wall does not act as a physiological boundary. Its main function is mechanical, *i.e.,* to provide protection and shape to the cell.

(I) Layers of the Cell Wall

Cell wall is derived from the living protoplast. It consists of *middle lamella, primary cell wall, secondary cell wall* and *tertiary cell wall.*

1. ***Middle lamella.*** It is the first formed layer, present between the two adjacent cells. Middle

Cell and its Parts

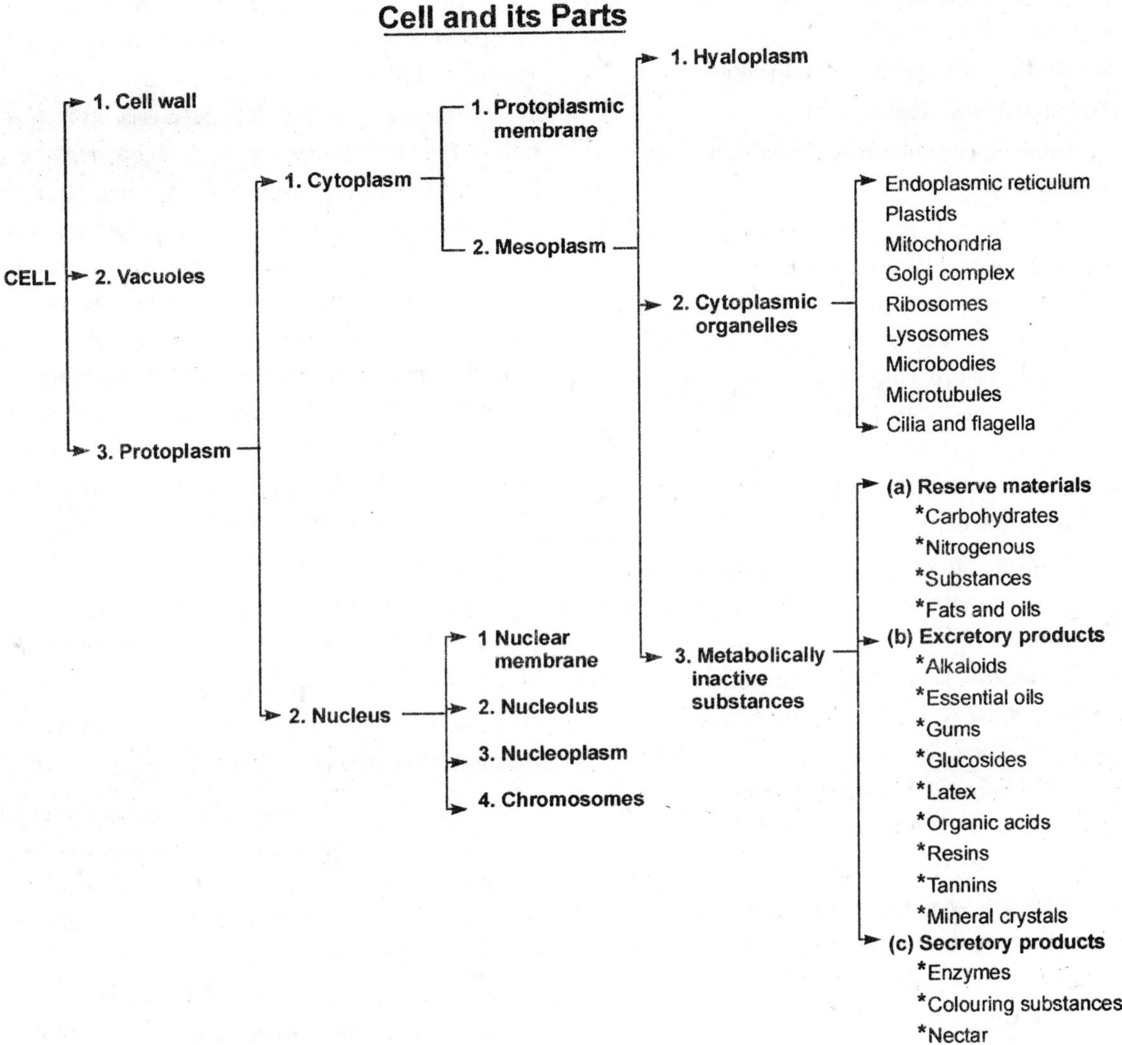

lamella is situated outside the primary cell wall. It is made of calcium and magnesium pectate. Middle lamella acts as a cement which holds the adjacent cells together.

2. ***Primary cell wall.*** This layer occurs inner to the middle lamella. The major components of this layer include cellulose, hemicellulose, pectic substances, lipids, proteins, mineral elements and water. This thin and elastic cell wall becomes thick and rigid after the completion of cell enlargement.

During the development of primary cell wall and the middle lamella, certain openings are left at places between the adjacent cells. These maintain cytoplasmic continuity between the neighbouring cells and are known as *plasmodesmata*. This term was used by *Strasburger* (1901) for the first time.

3. ***Secondary cell wall.*** It is situated inside the primary cell wall. It is thick being made of several layers of cellulose, hemicelluloses and polysaccharides. Secondary cell wall is deposited after the primary wall is fully formed, *i.e.*, after the cell has undergone elongation. It is thick and non-elastic, being mostly made of lignin.

4. ***Tertiary cell wall.*** Sometimes tertiary cell wall is deposited in a few cells. It is considered to be a dried residue of the protoplasm.

(II) Cell Wall Depositions

Some other substances found in the cell wall are as follows :

1. ***Lignin*** : It is a complex chemical substance, found in woody and hard tissues. The cells become woody as a result of lignification. Woody tissues are, hence, generally dead. Most of the vegetable fibres are lignified. Lignin deposition is generally not uniform and may result in annular, spiral, scalariform, reticulate or pitted patterns. These thickenings provide mechanical support to the cell.

2. ***Cutin.*** This waxy layers forms a thin or a thick cuticle on the stem and leaf. It is impervious to water.

3. ***Suberin.*** This fatty substance is present in the walls of the cork and is totally impermeable to water and gases.

4. ***Mucilage.*** This slimy substance absorbs water and stores it. Mucilage becomes hard when dry and viscous when moist.

5. ***Mineral substances.*** Silica, calcium oxalate, calcium carbonate, etc., are present in the cell wall in the form of crystals.

VACUOLES

A major part of the plant cell is occupied by a vacuole. Animal cells, on the other hand, have either very small vacuoles or these may be absent. A young dividing cell is completely filled with cell sap and has no vacuoles. But as the cell grows, many small vacuoles appear which ultimately coalesce to form a large central vacuole. Thus in a mature plant cell, cytoplasm occurs as a thin layer near the periphery of the cell surrounding the large central vacuole. Vacuole is delimited by a membrane, known as *vacuolar membrane* or , *tonoplast.* The vacuole is filled with *cell sap* or *vacuolar sap.*

Physiologically, vacuole is a very important part of the cell. It serves as a storage place and a place for depositing the cellular wastes. Vacuole may contain a variety of dissolved substances like sugars, salts, acids, alkaloids and pigment anthocyanin. Anthocyanins give purple or red colour to different organs. The tonoplast is a differentially permeable membrane of lipo-proteins. It maintains the cell structure and also the osmotic gradient. It also checks the outward flow of some ions, thus allowing their higher concentration in the vacuole than in the cytoplasm.

Plasmolysis

It is the process by which the cell memberane of cell will shrink along with the cell under the influence of hypertonic solution

PROTOPLASM

Physical Characters

Protoplasm is the only substance that shows the properties of life. It is a complex, jelly-like, gelatinous, granular, viscous, elastic and colourless substance. The protoplasm responds to various stimuli like temperature, light, chemicals, etc. It also acts as a semi-permeable membrane.

Composition of the Protoplasm

1. ***Water.*** Water forms most of the protoplasm. In aquatic plants the amount of water may be as high as 95% while in dry seeds it may be as low as 10%.

2. ***Proteins.*** Proteins form about 60-65% of the total dry weight. These are made of carbon, hydrogen, oxygen and nitrogen. In some proteins, sulphur and phosphorus are also found. The fundamental units of proteins are amino acids. There are some 20 essential amino acids

contributing to the protein structure. Each amino acid has an *amino group* - NH_2 and one or more acidic *carboxyl group* - COOH. During the formation of proteins, amino group of one amino acid joins the carboxyl group of another amino acid. This is called *peptide bond.* A simple protein consists of about 100 amino acid molecules.

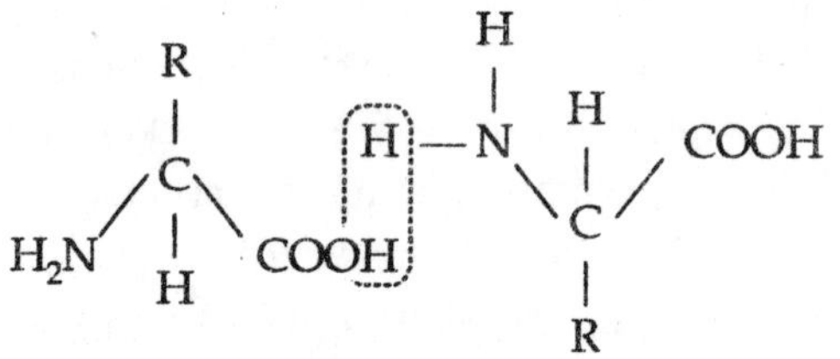

```
                              H
       R                      |
       |             [H]—N       H
       C                    \    |   COOH
      / \                    \   |  /
  H₂N  |  COOH]               C
       H                      |
                              R
```

Proteins are basic building blocks of protoplasm. Most of the enzymes found in plants and animals are proteins or they occur in conjugation with proteins.

3. *Lipids or fats.* Fats consist of carbon, hydrogen, oxygen, nitrogen and phosphorus. These are mostly insoluble in water. Lipids are of three types — true fats, waxes and phospholipids.

(*a*) *True fats.* These are made of glycerol and fatty acids.

(*b*) *Waxes.* These are lipids which are composed of fatty acids combined with long chain alcohols with 24 to 36 carbon atoms. These are found in combination with other substances as components of cell wall, *e.g.,* suberin and cutin.

(*c*) *Phospholipids.* They contain glycerol, fatty acids and atoms of phosphorus and nitrogen. In plants, these occur in least amounts among the three classes of lipids. These provide structural framework for various cell membranes and cellular organelles.

4. *Carbohydrates* : These are made of carbon, hydrogen and oxygen. Typically, the hydrogen and oxygen atoms are present in 2 : 1 ratio. The emperical formula of carbohydrates is CH_2O. On the basis of carbon atoms present, carbohydrates are divided into (*i*) monosaccharides, (*ii*) disaccharides and (*iii*) polysaccharides.

(*a*) *Monosaccharides or simple sugars.* They contain 3 to 7 carbon atoms and are accordingly named as *triose* (3 carbon), *tetrose* (4 carbon), *pentose* (5 carbon), *hexose* (6 carbon) and *heptose* (7 carbon) sugars. Common monosaccharides are sweet tasting, colourless solids and are soluble in water (*e.g.,* glucose).

(*b*) *Disaccharides.* Two monosaccharides join together to form disaccharides (double sugars). *Sucrose, maltose* and *lactose* are the common disaccharides occurring in protoplasm.

(*c*) *Polysaccharides.* They are composed of several monosaccharide units. The constituent monosaccharide units may be similar (homopolysaccharides; *e.g.,* starch, cellulose) or different (heteropolysaccharides; *e.g.,* pectin and chitins). Polysaccharides are tasteless and colourless amorphous powder which are little soluble in water, although some may form colloidal solution.

Carbohydrates constitute about 1% of the protoplasm. On oxidation they give energy. Cellulose, pectin, chitin, etc., are important parts of the cell structure.

5. *Inorganic substances* : The essential inorganic substances found in the protoplasm are in the form of solution. These occur in ionic form. Calcium, magnesium, potassium and iron are the main cations while nitrate (NO_3^-), sulphate (SO_4), phosphate (PO_4), chloride (Cl^-) and bicarbonate (HCO_3) are the major anions. Inorganic substances control pH of the cell and also influence numerous chemical reactions in the cell.

6. *Nucleic acids* : These are highly complex organic compounds, consisting of hydrogen, oxygen, carbon, nitrogen and phosphorus. These are of two types — *Ribose Nucleic Acid* (RNA) and *Deoxyribose Nucleic Acid* (DNA). Nucleic acid is made of many nucleotide units. Each nucleotide has three components — *nitrogenous bases* (purine or pyrimidine), *pentose sugar* (ribose or deoxyribose) and *phosphoric acid*.

7. *Other substances* : Besides the above major constituents, pigments, hormones, enzymes, alkaloides, latex, etc., are also found in the protoplasm.

2.1.4 Cell Membrane

Structure

It is typical bilayer of lipids in which protein molecules are embedded. It may also contain sterol of various kinds.

Cell membrane is not clearly visible under the light microscope but can be observed with electron microscope. Two types of cell membranes are recognised — *cytoplasmic membrane* which surrounds the whole protoplasm and *internal membrane* which surrounds various cellular organelles.

Following structures of cell membranes have been proposed by different workers.

1. *Trilamellar structure.* It was given by *Danielli* and *Davson* (1935). They suggested that cell membrane consists of a middle bimolecular layer of phospholipids, which is bounded on both sides by a protein layer (Fig. 1A). The lipid molecule of each monomolecular layer has its hydrophilic and close to the protein layer while its hydrophobic end is away from it.

2. *Unit membrane concept.* It was proposed by *J. David Robertson* (1959). He found that not only the cell membrane is trilamellar but most of the membranes of cellular organelles are also trilamellar (Fig. 1B). Thus all the membranes in the cell are now called *unit membranes.* According to this concept, unit membrane is made of three layers.

3. *Fluid mosaic model.* It was given by *Singer* and *Nicolson* (1972). This is one of the most accepted views. It differs from Robertson's model in the arrangement of proteins. According to this model, the proteins do not always cover the entire hydrophilic surface of the lipid bilayers but are of two types. These are : (*i*) *Peripheral* or *extrinsic* proteins. The peripheral proteins are present superficially on the surface and can be easily removed. (*ii*) *Integral* or *intrinsic* proteins of the integral proteins, some are completely

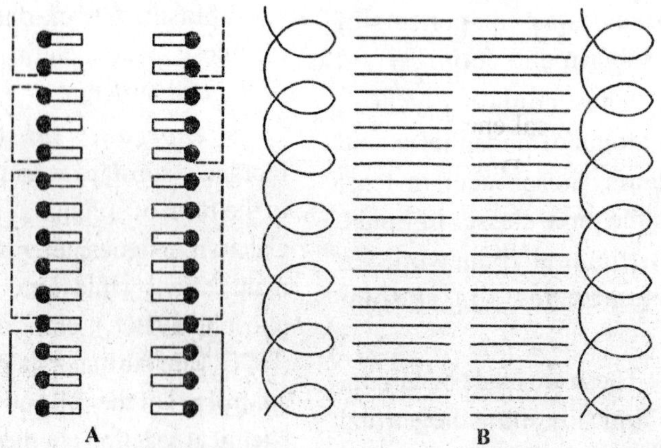

Fig. 1 A-B Cell membrane as proposed by [(A) Danielli and Davson, (B) Robertson]

embedded in the lipid layer while other are partly embedded in the lipid layer and partly projected on the surface. These are tightly held and can not be easily removed.

Lipids and integral proteins form a mosaic. The unit membrane is, hence, *quasi-fluid* (semi-fluid) in which lipids and integral proteins move within the bilamellar structure.

Chemical Composition

Cell membrane is mainly made of lipids and proteins. About 60-80% of the substances are proteins, some of which act as enzymes. Lipids constitute about 20-40% of the total substances. Besides, small amount of cellulose is also present. It is associated with proteins to give stability to the structure of cell membrane.

Functions

The primary function of the cell membrane is the regulation of passage of various substances in and out of the cell. Some of the important functions of the cell membrane are given below :

1. ***Permeability.*** Cell membrane allows the movement of small ions and molecules in and out of the cell. This membrane is *selectively premeable.*

2. ***Passive transport.*** Ions of different types move from their region of higher concentration to their region of lower concentration through cell membrane. It is called passive transport because it does not require chemical energy.

3. ***Active transport.*** During this process molecules move against the normal diffusion gradient. The process is called active because it requires chemical energy.

4. ***Exocytosis and Endocytosis.*** Macromolecules are known out of the cell through exocytosis and ingested through endocytosis. When the foreign substances ingested are fluids, the process is called *pinacocytosis* or *pinocytosis*, and if the substances ingested are solids, the process is known as *phagocytosis.*

5. **Cellular recognition and adhesion.** Two components of the cell membrane, glycoproteins and glycolipids, play an important role in selective adhesion of cells. It is due to this property of cell membrane that cells from different tissues like heart and liver do not adhere. Rejection or acceptance of transplanted tissues is determined by above mentioned components of the cell membrane. This occurs in animals only because the cell membranes of adjacent cell come in contact.

2.2 CELL ORGANELLES

2.2.1 Chloroplasts

The details of chloroplast are given below as they are of wide occurrence and great importance. These familiar type of plastids contain the photosynthetic pigment—the *chlorophyll.* Chloroplasts are present in all green coloured eukaryotic cells. Fungi lack chloroplasts and are, therefore, heterotrophic. In prokaryotic cells, like blue green algae and bacteria, organised chloroplasts are absent. Diameter of chloroplasts varies from 4-8 mm. It is about 2 mm thick.

Structure

Chloroplast is surrounded by two unit membranes. There is a space between two membranes, called *periplastidial space.* The internal structure of the plastid shows two distinct parts— (*i*) colourless ground substance called *stroma* and (*ii*) closed, flat, stack-like membrane system called grana.

Stroma is watery and proteinaceous ground substance. It contains starch grains, lipid droplets, DNA and free ribosomes. The dark reaction of photosynthesis takes place in the stroma.

Grana are densely packed stacks of membrane layers called *thylakoids*. Each thylakoid is bounded by a single membrane but because of the flatness of these structures they appear as double membrane layers or lamellae. Large number of grana are formed at frequent intervals by packed stacks of thylakoids. This is the place where light reaction of photosynthesis takes place. Two adjacent grana are joined with one another by lamellae called *intergranal lamellae* or *stroma lamellae* or *fret channel*. Each chloroplast contains about 40-60 grana.

Each lamella is made of a bimolecular layer of lipo-pigment sandwiched between two protein layers. The lipo-pigment layer contains chlorophyll and carotenoid pigments alongwith phospholipids. The chlorophyll molecules are so arranged in this layer that their hydrophilic porphyrin head is close to the outer layer of proteins. Amongst these lamellae, distinct groups, each consisting of about 230-300 chlorophyll molecules called *quantasomes*, are present. Quantasomes are the smallest morphologically distinct and photosynthetically functional units.

Chloroplasts also have 70S ribosomes which are either free or attached to intergranal lamellae. Pyrenoid is present in the chloroplasts of some algae and other plants. It consists of a protein core surrounded by peripheral starch plates.

Functions

The following are major functions of chloroplasts.

1. Absorption of light energy.
2. Production of $NADPH_2$ and evolution of oxygen through the process of Hill reaction.
3. Transfer of carbon dioxide obtained from the air to 5 carbon sugar (ribulose 1, 5 diphosphate).
4. Breaking of 6-carbon atom compound into two molecules of phosphoglyceric acid.
5. Hydrogenation of phosphoglyceric acid to form phosphoglyceraldehyde.

2.2.2 Mitochondria

Mitochondria are membrane bound cell organelles that contain the enzymes responsible for aerobic respiration of eukaryotic cell.

Distribution

Mitochondria are present in the living eukaryotic cells and absent from prokaryotic cells like bacteria and blue green algae. Mitochondria are lost secondarily in some highly specialised eukaryotic cells like mammalian RBC.

Structure

Mitochondria can be stained *in vivo* with *Janus Green B*. It reveals that mitochondria has two parts — an outer *envelope* and a *central cavity* filled with *matrix*. The envelope that surrounds the mitochondria is made of two unit membranes. It is called *perimitochondrial space*. The outer membrane is smooth but the inner membrane has many infoldings which extend into the matrix. These infoldings are called *cristae*.

Golgi appartus consists of a set memberane bounded, fluid-filled vesicles, vacuoles, and flattenened cisterinae. The single memberane structure is called cisternae.

The cristae provide an increased surface area within the mitochondrion for enzymatic activity. The inner surface of the inner membrane (*i.e.*, the one facing the matrix) is covered by small tennis racket-like particles with a head and a stalk. These particles have been variously named as *inner membrane sub-units*, *elementary particles or oxysomes*. *Parsons* (1963) called them *electron transport particles* (ETP).

Each particle consists of a head called F_1 sub-unit, approximately 100 Å in diameter and is attached to a base piece called *Fo* sub-unit. It is 35-50 Å in length F_1 sub-unit projects into the matrix. It is an integral protein of the inner membrane.

The inner membrane has all the enzymes required for electron transport. The F_1 - F_0 combination has special ATPase (ATP synthetase) for oxidative phosphorylation. ATP is called as the currency of the cells.

Functions

Mitochondria are known as *'power house of the cell'*. This is because of the formation and storage of energy rich compound ATP (adenosine triphosphate) as a result of oxidation. ATP is a readily available source of energy for numerous cellular reactions.

Mitochondria are concerned with oxidation and phosphorylation reactions of Krebs cycle during aerobic respiration. Mitochondria are absent in cells which do not respire aerobically. Besides oxidation of carbohydrates, mitochondria are also involved in the oxidation of proteins and fats.

2.2.3 Endoplasmic reticulum (ER)

The term ER was first used by Keith Porter (1953) to identify a fine reticulum (network) in endoplasm of the cell, ER (endoplastic reticulum) is a series of interconnected membranous tubules in the cytoplasm.

Distribution

All cells do not possess ER. It is generally absent in egg and embryonic cells. It is well developed in differentiated cells. In spermatocytes, though present, it is poorly developed. Generally ER is fully developed in cells actively engaged in protein and hormone synthesis. The size and shape of the ER changes with the type of cells in which it occurs.

Structure

ER is composed of three different types of structures. These are — *cisternae, vesicles* and *tubules.*

1. *Cisternae.* These are long, flat and unbranched plates or lamellae arranged in parallel rows.
2. *Vesicles.* These are usually round or ovoid sacs. They often occur isolated in the cytoplasm.
3. *Tubules.* They are irregularly branched tube-like structures. These structures are surrounded by thin unit membrane of 50-60 Å thickness and their lumen is filled with the secretory products of the cell.

Types of ER

Two types of endoplasmic reticulum have been recognized. These may be present in the same or different types of cells.

1. ***Agranular or smooth endoplasmic reticulum (SER).*** The surface of this type of ER is smooth, ribosomes being not attached. Smooth ER is present in cells which are actively engaged in steroid synthesis (*e.g.*, cholesterol, progesterone, testosteron, etc.), carbohydrate metabolism, pigment production, etc.
2. ***Granular or rough endoplasmic reticulum (RER).*** The rough endoplasmic reticulum have ribosomes attached throughout the surface. This type of ER is present in cells which show active protein synthesis.

Functions

Some lipids and proteins made by RER and SER functions as enzymes and hormones. ER is involved in following major functions.

1. ***Functions common to both smooth and rough ER.***
   (*i*) Forms skeletal framework.
   (*ii*) *Active transport* ER helps in the transfer of materials from cytoplasm to cytoplasm or b/w nucleus to cytoplasm.
   (*iii*) Metabolic activities due to enzymes and hormones formed by some lipids and proteins.
   (*iv*) Provides increased surface area.
   (*v*) Formation of new nuclear membrane during cell division.
   (*vi*) ER produces lipids and proteins which help in the formation of plasma mem brane by process called memberame biogenese
2. ***Functions of smooth ER.***
   (*i*) Lipid synthesis.
   (*ii*) Glycogen synthesis.
   (*iii*) Steroid synthesis like cholesterol, progesterone, testosterone, etc.
3. ***Functions of rough ER.***
   (*i*) It provides site for protein synthesis.
   (*ii*) It helps in transport of proteins.

2.2.4 Golgi Complex

Distribution

Golgi apparatus or complex as organelle present in all the eukaryotic cells but it is absent in prokaryotic cells. Among eukaryotes, Golgi complex is not found in the cells of fungi, male gametes of bryophytes and pteridophytes, mature sieve tubes and mature sperms and red blood cells of animals. It was first identified in 1898 by Italain physician Camillo Golgi.

Structure

It consists of three main parts—*cisternae, tubules* and *vesicles*. Cisternae, which form the central flattened or disc-shaped part of Golgi complex, are stacks of parallel double membranes or lamellae. In a plant cell 2 to 7 such cisternae constitute one Golgi apparatus.

Tubules arise from the periphery of the cisternae and form a highly branched anastomosing network.

Vesicles are of two types — (*i*) *rough vesicles* present at the ends of the cisternae and (*ii*) *smooth vesicles* usually present within the network or sometimes near the centre of the stack of cisternae.

Golgi apparatus forms an extensive intercommunicating membrane system in association with cell membrane, endoplasmic reticulum, lysosome and nuclear membrane. It is also called *endomembrane system*.

Golgi apparatus is very rich in phospholipids, proteins, various enzymes, carotenoids, fatty acids and vitamin C.

Functions

The following are some of the major functions of Golgi apparatus.

1. Synthesis of polysaccharides.
2. Formation of glycoproteins by combining carbohydrates with proteins.
3. Synthesis of pectins and other carbohydrates necessary for cell wall formation. Secretory vesicles of Golgi apparatus fuse and get incorporated into cell membrane. Similarly, in the plant cells, Golgi apparatus deposits pectic substances and cellulose microfibrils to form a cell plate during cell division.
4. It also secretes gum and mucilage.
5. It is associated with storage, condensation, packaging and transportation of various substances.
6. It is also active in the transformation of one type of membrane into another.
7. Secretory vesicles and lysosomes originate from Golgi apparatus.

2.2.5 Ribosomes

Ribosomes are small particles composed of rRNA and proteins.

Distribution

Ribosomes are present in both prokaryotic and eukaryotic cells. In eukaryotic cells, they are free as well as attached to the endoplasmic reticulum and nuclear membrane. Ribosomes are also present in mitochondria and chloroplasts. There may be as many as 5,00,000 ribosomes in an ordinary plant cell. In prokaryotic cells ribosomes are free.

Structure

These dense and rounded granules are perhaps the *smallest cell organelles* and can be seen only with the electron microscope.

Ribosomes are isolated by differential centrifugation, and its sedimentation coefficient is measured in *Svedberg Units* or *S* units. On the basis of sedimentation coefficient two types of ribosomes have been recognised.

1. 70 S ribosome. This type of ribosome is smaller of the two. 70S type ribosomes are found in all prokaryotic cells like bacteria and blue green algae and also in chloroplasts and mitochondria of eukaryotic cells. These consist of two sub-units—50S and 30 S,50 S sub-unit is larger and the smaller 30 S sub-unit is placed over it like a cap.

2. 80 S ribosome. This type of ribosome is larger in size with a sedimentation coefficient of 80 S. It has a molecular weight of 40×10^6 daltons. These ribosomes are found in the cells of animals and plants (eukaryotic cells). 80 S ribosome consists of two sub-units — 60 Sand 40 S. 60 S sub-unit is larger and the smaller 30 S sub-unit is placed over it like a cap.

Each sub-unit of ribosome is further composed of smaller sub-units (Fig. 2). For example, 50 S sub-unit of 70 S ribosome consists of core particles 40 S and split proteins — SP 50. Similarly 80 S ribosomes are also made of smaller sub-units. Core particles are further composed of RNA and core particles. The amount of RNA present in the ribosomes is about 70-75% of the total amount of RNA present in the cell.

The association of ribosomal sub-units is dependent upon the concentration of Mg^{++} ions. This association requires very low concentration of Mg^{++} (0·001 M). If the concentration is lesser, both the ribosomal sub-units remain separate. If the concentration of Mg^{++} ions is increased ten folds, two ribosomes get associated to form *dimer.* In bacterial cells, the two subunits occur freely in the cytoplasm.

During protein synthesis many ribosomes occur in a chain on common m-RNA strand and are called *polyribosomes* or *polysomes.*

Ribosomes (80S) remain attached to endoplasmic reticulum with their 60 S end by electrostatic reactions and also be the nascent polypeptide chain that grows from the ribosome and penetrates across the membrane. These could also be bound with ER by two transmembrane glycoproteins - *ribophorin I* and *ribophorin II.*

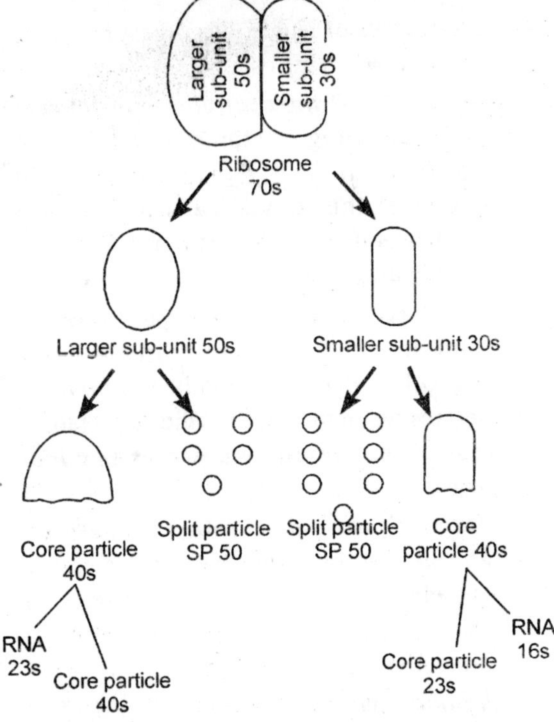

Fig. 2 Ribosome : Components of 70 S ribosome.

Functions

Ribosomes are the centres of protein synthesis.

2.2.6 Lysosomes

These were first reported by Belgian biochemist. Christian de Duve (1955). These are small spherical vesicles measuring about 500 nm and are involved in intracellular digestion.

Distribution

Lysosomes are common in animal cells but are also found in plant cells. These can be observed in secretory cells like pancreatic cells, spleen cells, leucocytes, liver cells, meristematic plant cells, etc.

Functions

The following are the main functions of lysosomes.

1. *Digestion of extracellular particles.* Lysosomes, pinocytes and phagocytes digest the external food particles. Lysosomes of leucocytes digest proteins, bacteria and viruses.
2. *Digestion of intracellular substances.* Lysosomes digest stored materials like proteins, fats and glycogen to provide energy to the starved cells. Lysosomes in plant cells hydrolyse stored material in germinating seeds.
3. *Autolysis.* Under certain pathological conditions lysosomes digest cell organelles. This is also called *autophagy* and results in the death of the cell. Hence, lysosomes are also known as *suicidal bags.*
4. *Extracellular digestion.* Sometimes enzymes from lysosomes are liberated outside the cell, *e.g.*, the lysosomes of sperm release enzymes which dissolve the protective coating of ovum.
5. *Mitotic division.* Lysosomes are said to initiate the process of mitosis.

2.2.7 Microbodies

In protozoans, fungi, some plants, liver and kidney, some small membrane bound organelles are found associated with ER, mitochondria and chloroplasts. These are called microbodies. Following are some of the common types of microbodies.

Peroxisomes

The term peroxisome was introduced by *Beaufaytt* and *Berther* in 1963. These are found in many plant and animal cells. Peroxisomes mainly occur in photosynthesizing cells of higher plants, mature pear fruit, algae, fungi, liverworts, mosses and cells of ferns.

It consists of a granular matrix enclosed by a unit membrane. A nucleoid that consists of parallel tubules or twisted strands, occupies the centre. The enzymatic contents of animal and plant peroxisomes differ. However, they contain some peroxodie producing enzymes like urate oxidase.

The exact function of peroxisomes is not known. However, glycolate metabolism is known to occur in them. Thus these are associated with photorespiration in plants (C_3 plants) and lipid metabolism in animal cells.

Sphaerosomes

They originate from endoplasmic reticulum by budding. Sphaerosomes are bound by a single membrane enclosing enzymatic proteins which synthesize fats and oils. Like lysosomes, sphaerosomes also contain phosphatases, esterases, ribonucleases and hydrolases. Thus, sphaerosomes do not appear to be different from lysosomes, except for their specific lipidic nature. The main function of spherosomes is to synthesize fats and their transportation.

Glyoxysomes

These occur in the cells of yeast, *Neurospora* and oil seeds of many higher plants. They have enzymes for fatty acid metabolism and gluconeogenesis (conversion of lipids of germinating seeds into carbohydrates).

2.2.8 Centrosome

Distribution

Centrioles are found in algae, fungi, bryophytes, ferns, gymnosperms, protozoans and most of the animals. However, these are absent from the cells of prokaryotes, diatoms, yeast and most of the higher plants (*e.g.*, conifiers and angiosperms).

Functions

Centrioles take part in the formation of basal bodies and mitotic spindle, and facilitate the movement of separating chromosomes during nuclear division.

2.2.9 Microtubules

Distribution

Microtubules are found in the cytoplasm of plant and animal cells, flagella, centriole, basal body, mitotic apparatus and meristematic plant cells, etc.

Functions

Following are the main functions of microtubules :

1. These are related with the movement of flagella and cilia.
2. During cell division they form spindle fibres which control the movement and alignment of chromosomes.

2.2.10 Nucleus

History

The credit for the discovery of nucleus goes to *Robert Brown* (1831). He observed the nucleus in the cells of an orchid. *J. Hammerling* (1953) conducted experiments on a green algae *Acetabularia* and proved hereditary role of nucleus.

Distribution

Nucleus is present in all the eukaryotic cells. Some cells contain nucleus when they are young but it degenerates when cells mature, *e.g.*, sieve tubes in plants and RBC in mammals. In prokaryotic cells 'true nucleus' is absent.

Generally, there is one nucleus in each cell, but this number may vary in different types of cells. Depending upon the number of nuclei, cells are called *uni–*, *bi–*or *multinucleate*. Multinucleate condition may arise either by fusion of many cells (*e.g.*, syncytium of coconut endosperm) or due to repeated nuclear divisions (*e.g.*, coenocyte of *Vaucheria*).

Parts of the nucleus

The nucleus can be easily distinguished into following four parts :

(1) **nuclear membrane,** (2) **nucleolus,** (3) **nucleoplasm,** (4) **chromatin**

1. ***Nuclear membrane.*** The nuclear membrane or *karyotheca* is the outer envelope of the nucleus. It occurs in all the eukaryotic nuclei except during later part of the cell division. In prokaryotic cells, nuclear membrane is always absent.

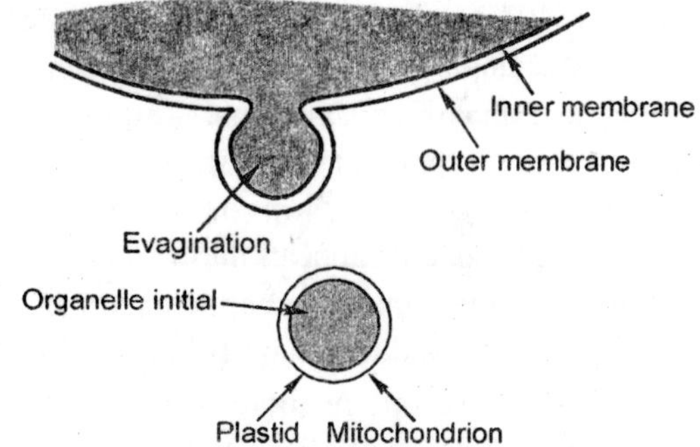

Fig. 3. Nuclear membrane (envelope) : Formation of organelles

Nuclear membrane separates nucleus from the cytoplasm. Membrane is nuclear composed of two layers of membrane. The space between the two membranes called perinuclear space, is about 75 Å. The outer unit membrane is continuous with endoplasmic reticulum. The nuclear membrane is interrupted by pores. These pores are usually octagonal and their diameter varies from 300-1,000 Å. These are called nuclear openings or pores. They maintain continuity between nucleo-cytoplasmic or regions. Nuclear pores, however, do not freely communicate with the cytoplasm. They are plugged by a cylinder of protein material called annulus. The passage of ions and small molecules through the pores is regulated by annulus.

The nuclear membrane breaks into small vesicles during prophase of cell division and gets dispersed in the ground substance. Later, during telophase, these vesicles either fuse once again to form a nuclear membrane or new vesicles are formed to produce a nuclear membrane.

Nuclear membrane allows a free exchange of ions. Nuclear pores permit mRNA to move out of the nucleus into the cytoplasm. Evagination of nuclear membrane results in the formation of organelle initials which may later transform into organelles like mitochondria and chloroplasts.

2. *Nucleolus.* This organelle was first seen by Fontana (1781) and was termed nucleolus by Bowman (1840). Nucleolus is a spherical, dense, colloidal and acidophilic body that remains attached to a special type of chromosome—nucleolar organising chromosome at a specific place called nucleolar organising region or secondary constriction. Nucleolus disappears during prophase of mitosis and meiosis and reappears during telophase in mitosis and telophase II in meiosis.

A cell may have one or more nucleoli. The size of the nucleolus also varies depending upon its synthetic activity. Nucleolus consists of three parts as given :

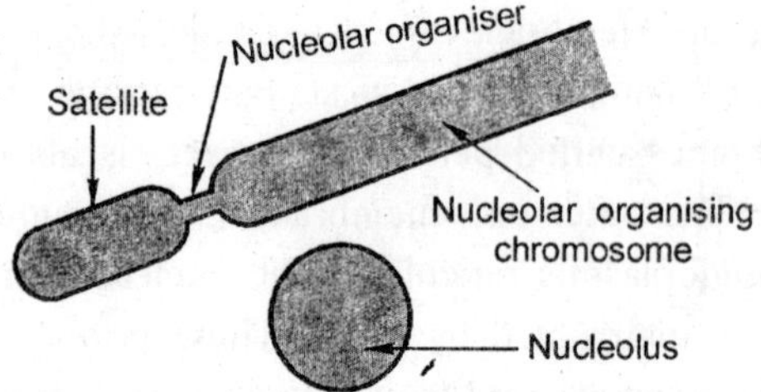

Fig. 4. Nucleolus : Nucleolar organising chromosome with nucleolus

(*a*) *Granular region.* This is made of ribonucleoprotein granules.

(*b*) *Fibrillar region.* This region, consisting of 50-80 Å long proteinaceous fibrils, is called nucleonema.

(*c*) *Amorphous matrix.* This, electrically less dense region, shows the presence of amorphous matrix and is called 'pars amorpha'.

3. *Nucleoplasm.* The nucleus is filled with a transparent, semi-solid, granular and acidophilic substance, known as nuclear sap, nucleoplasm or karyolymph. Nucleolus and chromatin fibres appear scattered in this sap.

Nucleoplasm is mainly made of nucleic acids, proteins, enzymes, lipids and minerals. The nucleic acids—DNA and RNA are found in the form of nucleoprotein. Proteins occur as basic proteins or histone proteins (*e.g.*, nucleoprotamines and nucleohistones) and acidic proteins or non-histone proteins (*e.g.*, phosphoprotein). Some of the important nuclear enzymes include DNA polymerase, RNA polymerase, nucleoside triphosphates, aldolases, enolases, etc. Lipids are found in small amounts. Minerals like phosphorus, potassium, sodium, calcium and magnesium are also present.

4. *Chromatin.* When the cell is not reproducing the DNA and its associated protein appear as a threadlike mass called *chromatin.* They are thin thread like intangled mass composed of DNA and proteins.

CHROMSOMES: are thread like structure which contain hereditary information in the form of genes. Genes are the hereditary units special functioning unit along with chromosomes, chromosome arrange into chromatin when cell is about to divide. Chromosomes and chromatin are really the same molecules, but they differ in structural arrangement.

Nucleoid: The part of bacterium or virus that contain nucleic acid and is analogous in function to nucleus of eukaryotic cell.

2.3 CELL PARTS AND THEIR DISCOVERERS

Part of the Cell	*Discoverer*
Cell Theory, (1839)	Scheleiden & Schwann
Living Cell, (1723)	Antony von Leeuwenhoek
Primary cell wall, (1665)	Robert Hooke
Plasmodesmata, (1901)	E. Strasburger
Protoplasm, (1840)	J.E. Purkinje
term sarcode, (1835)	Dujardin
Hyaloplasm, (1846)	Hugo von Mohl
Protoplast, (1880)	Hanstein
Cell membrane, (1855)	Nageli & Cramer
termed plasmalemma, (1931)	J.Q. Plowe
Unit membrane model, (1959)	Robertson
Fluid Mosaic model (1974)	Singer & Nicolson
Endoplasmic reticulum, (1945)	K.R. Porter
Plastid, (1865)	E. Haechel
term chloroplast, (1885)	A.F.W. Schimper
Mitochondrion, (1880)	A. Kolliker
named fila, (1882)	W. Flemming
named bioblast, (1894)	Altmann
named mitochondrion, (1867)	Benda
Observed in plants, (1904)	F. Meves
Golgi body, (1898)	Camillo Golgi* (NL 1906)
Ribosome, (1943)	
observed & termed microsome	A. Clande* (NL 1974)
discovery of RNA in ribosome, (1955)	*Palade (NL 1974)
term ribosome given, (1958)	Haguenau
Lysosomes, (1955)	Christian de Duve* (NL 1974)
observed in plants, (1964)	P.Matile
Peroxisomes, term given, (1963)	Beaufaytt & Berther
Microtubules, (1953)	De Robertis & Franchi
Centrosome, (1888)	T. Boveri
Nucleus, (1831)	Robert Brown
Nucleolus, (1781)	Fontana
described, (1838)	Schleiden
Chromosome, observed, (1875)	E. Strasburger
term chromosome given, (1888)	Waldeyer
Chromatin, (1879)	W. Flemming
Nucleosome, (1975)	Oudet, Gross-Bellard & Chambon
Nucleic acid,	
term nuclein given, (1869)	F. Miescher
DNA stained, (1924)	R. Feulgen & H. Rossenbeck
DNA, structure proposed (1953)	Walson & Crick* (NL 1962)
DNA synthesized, (1959)	A. Kornberg (NL 1959)*
RNA synthesized, (1959)	S. Ochoa (NL 1959)*
Clonal selection theory of antibody (1960)	F.M. Burnet and Medawar*
Three dimensional strucutre of DNA (1962)	Francis H.C. Crick and Watson*
DNA as a genetic material (1969)	Hershey & Max Delbrück

* Nobel Prize winners, NL = Nobel Laureate or Nobel Prize winner

2.4 CELL ORGANELLES : STRUCTURE AND FUNCTION

	Organelle Structure	Function
Cell wall	A rigid cell wall surrounding the cell,consisting of cellulose microfibrils running through a matrix of other complex polysaccharides,namely hemicelluloses and pectic substances. May be secondarily thickened in some cells.	Provides mechanical support and protection. It prevents osmotic bursting of the cell. It is totally permeable and hence pathway for movement of water and mineral salts. Various modifications, such as lignification, for specialised functions.
Middle lamella	Thin layer of pectic substances (calcium and magnesium pectates).	Cements neighbouring cell together.
Plasmadesmata	A fine cytoplasmic thread linking the cytoplasm of two neighbouring cells through a fine pore in the cell walls.	Enabels a continuous system of protoplasm, to be formed between neighbouring cells for transport of substances between cells.
Plasma membrane	Two layers of lipid (bilayer) sandwiched between two protein layers.	A differentially permeable barrier controlling exchange of water and minerals between the cell and its outside.
Endoplasmic reticulum	A system of flattened, membrane-bounded sacs called cisternae, forming tubes and sheets. It is continuous with the outer membrane of the nuclear envelope.	If ribosomes are found on its surface it is called **rough ER** and transports proteins made by the ribosomes through the cisternae. **Smooth ER**, (no ribosomes) is a site of lipid and steroid synthesis.
Chloroplast	Large plastid containing chlorophyll and carrying out photosynthesis. It is surrounded by an envelope of two membranes and contains a gel-like stroma through which runs a system of membranes that are stacked in places to form grana. It may store starch. The storma also contains ribosomes, a circular DNA molecule and lipid droplets.	It is the organelle in which photosynthesis takes place, producing sugars and other substances from carbon dioxide and water using light energy trapped by chlorophyll. Light energy is converted to chemical energy.

	Organelle Structure	Function
Mitochondria	Surrounded by two membranes, the inner being folded to form **cristae**. Contains a **matrix** with a few ribosomes and circular DNA molecule.	In aerobic respiration cristae are the sites of oxidative phosphorylation and electron transport, and the matrix is the site of Krebs cycle enzymes and fatty acid oxidation.
Golgi apparatus	A stack of flattened, membrane-bound sacs, called **cisternae**, continuously being formed at one end of the stack and budded off as vesicles at the other. Stacks may form discrete dictyosomes as in plant cells, or an extensive network as in many animals cells.	Processing in cisternae and transport in vesicles of many cell materials, such as enzymes from the ER. Often involved in secretion and lysosome formation.
Ribosomes	Smallest cell organelle with cell membrane. It consists of a large and a small subunit. They are made or roughly equal parts of protein and RNA. Ribosomes are found in mitochondria and chloroplasts in plants.	Sites of protein synthesis. They are either bound to the ER or lie free in the cytoplasm. They may form **polysomes** (polyribosomes), chain of ribosomes—along messenger RNA.
Lysosome	A simple spherical sac bounded by a single membrane and containing digestive (hydrolytic) enzymes. Contents appear homogeneous.	Many functions, all concerned with break down of structures or molecules like autophagy, autolysis, endocytosis and exocytosis. They are called as suicidal bags of the cell, they are made up of powerful digestive enzymes. So if there is any problem in cellcular metabolism. Then cells may be burst and eat themselves.
Microbodies	A roughly spherical organelle bounded by a single membrane. Its contents appear finely granular except for occasional crystailoid or filamentous deposits.	All contain catalase, an enzyme that breaks down hydrogen peroxide. All are associated with oxidation reactions. In plants, are the site of the glyoxylate cycle.
Centriole	Cylindrical structure without limiting membrane. It consists of 9 triplet fibres, each consistingof three secondary fibres or tubules. These are enclosed in amorphous matrix. Centrosome separates into a pair of centrioles. Mostly found in animals and a few plants.	Centrioles take part in formation of basal bodies and mitotic spindle. These also facilitate the movement of chromosomes during cell division.

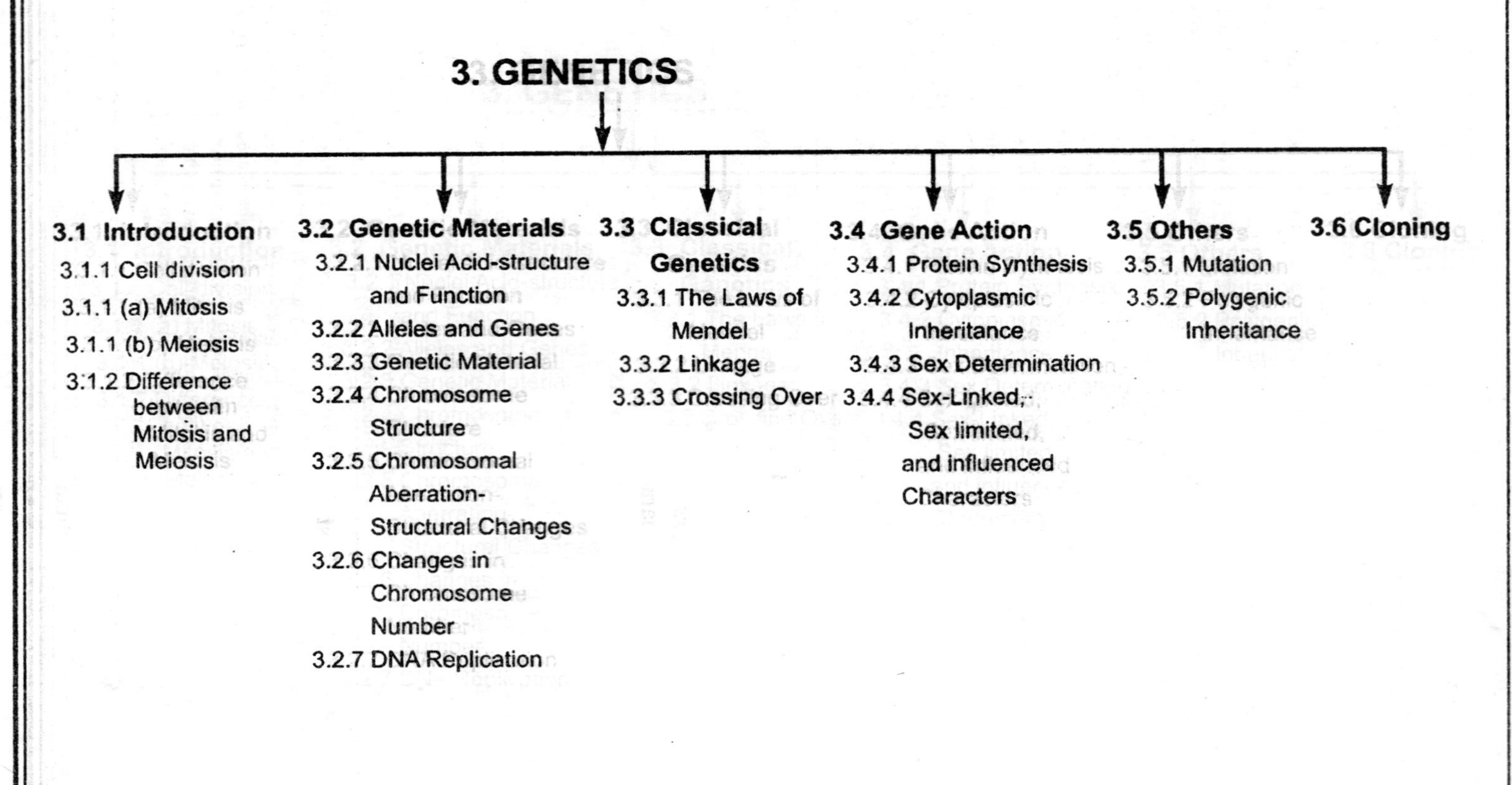
3. GENETICS
3.1 Introduction
3.1.1 Cell division
3.1.1 (a) Mitosis
3.1.1 (b) Meiosis
3.1.2 Difference between Mitosis and Meiosis
3.2 Genetic Materials
3.2.1 Nuclei Acid-structure and Function
3.2.2 Alleles and Genes
3.2.3 Genetic Material
3.2.4 Chromosome Structure
3.2.5 Chromosomal Aberration-Structural Changes
3.2.6 Changes in Chromosome Number
3.2.7 DNA Replication
3.3 Classical Genetics
3.3.1 The Laws of Mendel
3.3.2 Linkage
3.3.3 Crossing Over
3.4 Gene Action
3.4.1 Protein Synthesis
3.4.2 Cytoplasmic Inheritance
3.4.3 Sex Determination
3.4.4 Sex-Linked, Sex limited, and influenced Characters
3.5 Others
3.5.1 Mutation
3.5.2 Polygenic Inheritance
3.6 Cloning

Gregor Mendel
(1822–1884)

William Bateson
1861-1926

3. Genetics

3.1 INTRODUCTION TO GENETICS

Genetics is a science which unravels the inheritance of various character from one generation to another. The term first used by William Bateson in 1905.

Pre Mendelian Theory about Heredity

1. *Performation Theory*

Swammerdam and Bonnet (1770-1793) this theory states that a minute human called humanculus was already present in egg and sperm.

2. *Theory of Epigenesis*

Wolf (1738-94) egg is differentiated into various organs takes place only after fertilization from the zygote resulting various organs universally accepted. This concept view of origin of adult organ and tissue from zygote.

3. *Theory of Acquired characters*

Lamarck (1744-1829) new characters once acquired by an individual shall pass on to its progeny.

4. *Theory of Pangenesis*

Darwin (1809-1882) small, invisible copies of each body organs (gemmules) are transported by the blood tissue.

5. *Germ plasm Theory*

Weismann (1889). Body tissue in two parts germplasm and somatoplasms.

Contribution of Some Geneticist

1. **Mendel G.:** Father of Genetics.
2. **Bateson :** Linkage, Sex determination.
3. **Johannsen W.L. :** Pure line, Phenotype and Genotype.
4. **Morgan T.H. :** In 1911 Chromosomal theory of heredity. First geneticist to win Nobel Prize.
5. **Muller H.J :** X-ray : Physical mutant.
6. **George Beadle and Edward Tatum (1948)** showed that genes are acted by coding for catalytic proteins called *enzymes*.

   Beadle, George and Tatum : One gene one enzyme hypothesis (1944).
7. **Avery O.T ; Mac Leod, C.M. and Mc Carthy ;** Bacterial transformation.
8. **Watson J. D. and Crick :** Molecular Structure of DNA. (Double helix) (1953)
9. **Seymour Benzer (1955)** showed that mutant sites within gene could be mapped in relation to each other.

   Barbara McClintock : Jumping Gene/ Transposons. Nobel prize in 1983.
10. **Benzer, S :** Unit of functions cistron, Recon and muton. Pioneer of neurogenetics.
11. **Jacob and Monod :** Operon concept.
12. **Nirenberg M.W and Khorana HG :** Deciphers the genetic code. (1966)
13. **Ede Lman G :** Immunoglobulin structure, 1972 Nobel prize.
14. **Arber, Nathans, Smith and Mitchell :** Restriction endonuclease technology.
15. **Berg Paul, Gilbert Walter and Sanger:** Recombinant DNA technology, DNA sequencing technology. (Nobel Prize, 1980).
16. **Baltimore, David:** Nobel Prize for discovery of reverse transciptase.
17. **Babcob, E.B.:** US plant geneticist, pionred genetic analysis of genus, crepis.
18. **Jeppys Alec:** Developed DNA Finger-printing and DNA profiling techniques.
19. **Sturtevant, A:** Constructed first genetic map of a chromosome.
20. **Derald Langham:** American agricultural geneticst, Father of Seasame.

Organism Used in Genetics

1. **Bacteria :** E. coli
2. **Bacterio Phage :** λ Phages

3. **Insect :** Drosophila
4. **Fungi :** Neurospora crassa
5. **Plant :**
   (*i*) Corn / Maize (*ii*) Sweet pea
   (*iii*) Arabidopsis.
6. Nematode

3.1.1 Cell Division

Cell is a basic unit of structure and function in all living systems. The process of formation of new cells from the pre existing cells is referred to as cell division. The cell which undergoes division is known as mother cell and the new cells which are formed by the process of cell division are termed as daughter cells. In bacteria, cell division takes place by fission of pre existing cell. In eukaryotes there are two types of cell division viz mitosis and meiosis.

3.1.1 (a) Mitosis

It is the process by which a cell divides into two daughter cells, each having a nucleus with sets of chromosomes exactly identical to the parent cell. Cell cycle is divide into interphase, mitosis and cytokinesis.

1. *Interphase*

After the telophase stage of previous mitotic division and before the onset of prophase of the next one. Chromosomes are fully extended and uncoiled. Nuclei are surrounded by a nuclear envelope. It is the longest stage. Three substages –

(*i*) **G_1 Phase :** Between the beginning of interphase and the beginning of S phase. Protein and RNA Synthesis, Ribosome production in nucleoli, takes place during this phase.

(*ii*) **S Phase :** DNA Synthesis takes place as a result of which the chromosomes replicate. Production of two sisters chromatids from the single chromatid of each of the chromosomes in the nucleus. All the histones of the nucleus are synthesised. A DNA segment capable of undergoing replication is known as a *replicon*. Each chromosome has several replicons. Each replicon has a specific point the *initiation point* from where DNA replication is initiated. At the end of replication of all the replicons of an eucaryotic chromosome two sister chromatids are produced from the single chromatid of each chromosome. Since the two chromatids of a chromosome are produced by semi conservative replication of a single parent chromatid they are known as sister chromatid.

(*iii*) **G_2 Phase :** Period between termination of DNA synthesis and beginning of prophase of the next division. RNA and protein, particularly components of microtubules synthesis. Some mRNA and Ribosomes are produced.

RNA and protein synthesis are abundant during G_1 and G_2 depressed during S, prophase and upto early to mid metaphase and absent during late metaphase, anaphase and upto early telophase.

2. *Prophase*

The appearance of definite thread like structures in nuclei. Chromosomes become increasingly shorter and thicker due to increased condensation. Nucleous and nuclear membrane remain present. Spindle apparatus is not yet organised.

Chromosome condensation due to coiling of chromosomes

Relational Coiling :

- Plectonimic coiling (Mitotic prophase)
- Paranimic coiling (meiotic prophase)

(*i*) *Plectonimic Coiling* : Sister chromatids are twisted round each other, cannot separate from each other.

(*ii*) *Paranimic Coiling* : The coils of one of the sister chromatids are simply slipped into those of the other can easily separate from each other.

3. *Metaphase*

(*i*) Disappearance of Nucleous

(*ii*) Breakdown of nuclear envelope and distribution of its components into the ER

(*iii*) Appearance of spindle fibre

(*iv*) Arrangement of the chromosomes on a single plane called equatorial plate.

(*v*) The shortest and the thickest form of chromosomes

(*vi*) Absence of relational coiling between sister chromatids.

4. *Anaphase*

Anaphase begins when the centromeres of chromosomes appear to divide longitudinally so that the sister chromatids seperate from each other and ends with the reaching of the chromosomes to the opposite poles.

5. *Telophase*

(*i*) The chromosomes uncoil so that they become very long and thin

(*ii*) Nucleous reappears

(*iii*) Nuclear membrane is reorganised

(*iv*) Phragmoplast appears at the equatorial region of the cell.

Cytokinesis

It is completed by the end of telophase. At the equatorial plate, membranous elements of ER and products of golgi bodies organise into the phragmoplast gives rise to cell plate and subsequently to middle lamella. In plants the seperation of cytoplasm begins in the centre of the cell and gradually extends outwards on each side in a plane perpendicular to the axis of the spindle. The two daughter cells produced by mitosis contain one nucleus each, each nucleus has the same number of chromosomes as the parent cell.

Significance of Mitosis –

- Mitosis is responsible for development of a zygote into adult organism.
- It is essential for normal growth and development of living organisms.
- It is useful in maintaining the purity of types because it leads to production of identical daughter cells.

3.1.1 (b) Meiosis

Meiotic division of one somatic (2n) cell gives rise to 4 haploid (n) daughter cells.

The reduction in chromosome number of the daughter cells is mainly due to a single S phase during the pre-meiotic interphase followed by two successive nuclear divisions.

1. *Initial meiotic* : Meiosis occurs shortly after the zygotes are produced eg. fungi, yeast.
2. *Middle meiotic* : Meiosis produce spores which gives rise to the gametophyte generation. Gametes are produced through mitosis in the gametophyte eg. higher plants.
3. *Terminal meiotic* : Meiosis occurs at the end of the life cycle and it directly leads to the production of gametes eg. Animals.

Stages of Meiosis

Premeiotic Interphase : The interphase stage just prior to the entry of a cell into meiosis.

During the S phase of premeiotic interphase chromosome replication takes place. But approximately 0.3% of the total DNA present in the nucleus doesnot replicate during the S phase. This DNA replicates during the zygotene substage of prophase I.

A special type of histone specific to cells preparing for meiosis is synthesized during S phase.

Thus during meiosis chromosomes replicate only once, while the nuclei divide twice.

First Meiotic Division :

Events —

(*i*) pairing between homologous chromosomes.

(*ii*) Crossing over between them during prophase I.

(*iii*) Seperation of homologous chromosome and their migration to the opposite poles of a cell during anaphase I.

First Prophase :

Longest in duration comparatively mitotic prophase.

Main events —

1. Intimate pairing between homologous
2. Condensation of chromosomes
3. Crossing over between homologous chromosome
4. Movement of homologous chromosome away from each other (without seperation) so that chiasma become observable
5. Terminalization of chiasma
6. Movement of chromosomes during chromosome pairing and chiasma terminalization.

Substages :

(*i*) *Leptotene* :

(*a*) Increase in the nuclear volume

(*b*) Chromosome condensation

(*c*) RNA Synthesis

(*d*) Proteins required for condensation are synthesised.

(*ii*) *Zygotene*

(*a*) pairing between homologous chromosome.

(*b*) Completion of replication of the remaining 0.3% DNA of each nucleus. This DNA synthesis is referred to as Z-DNA synthesis (or) zygotene DNA synthesis.

(*c*) Synthesis of a specific nuclear protein.

(*d*) Development of the synaptenemal complex.

(*e*) Progressive condensation of chromosomes.

Pairing between homologous chromosome is often referred to a *synapsis*.

(*i*) May begin at both the ends of a homologous pair and proceed toward its centre

(*ii*) It may begin at the centromere and progress toward the tolomeres.

(*iii*) It may begin simultaneously at several places. Chromosome pairing occurs in two stages.

1st homologous chromosome move close to each other and stay about 1000 Å away from each other.

2nd Synaptenemal complex develops in this 1000 Å region between the two homologous chromosome.

Synaptenemal complex is composed of protein. It consists of 3 distinct elements —

1. dense central element
2. two dense lateral elements
3. two less dense transverse elements.

(*iii*) *Pachytene*

(*a*) Further condensation of chromosomes.

(*b*) As a consequence of synapsis between homologous chromosome the number of chromosomes is only half of the somatic chromosome number, that is only n. The haploid number of chromosome pairs commonly referred to as bivalents. Since each of the two chromosomes of a bivalent consists of two sister chomatids, each bivalent has four chromatids as strands. So, bivalents are said to be in four strand (or) tetrad stage.

(*c*) Chromosomes are easily recognisable during this stage of the nucleous more crossing over between homologous chromosomes.

(*iv*) *Diplotene*

(*a*) Homologous chromosome of each bivalent begin to move away from each other

(*b*) The two homologous of each bivalent appear to be attached with each other these attachments are known as chiasma.

(*c*) As diplotene progresses chiasmata slowly move find the ends of the homologous chromosome, this movement is referred to as *chiasma terminalization*.

(*v*) ***Diakinesis***

(*a*) Two homologous chromosome of each bivalent are now close to one or both the telomers only.

(*b*) Chromosomes become shorter and thicker.

(*c*) Nucleolus and nuclear envelope disappear

(*d*) Spindle apparatus is organised

Bivalents

(*i*) *Closed Ring Bivalent*. Metacentric and submetacentric chromosomes. Chiasma is formed in both the arms of the homologous chromosome.

(*ii*) *Open Ring Bivalent*. Chiasma is formed only one of the two arms of metacentric and sub metacentric

(*iii*) *Rod Shaped, Bivalent.* Acrocentric and Telocentric chromosomes chiasma in their long arms only.

3.1.2 Differences between Mitosis and Meiosis

Mitosis	*Meiosis*
1. Mitosis occurs in almost all organisms.	1. Occurs only in sexually reproducing organisms.
2. Mitosis takes place in all somatic cells whether haploid, diploid or polyploids.	2. Takes place in diploid cells such as zygote, spore mother cells, and gamete mother cells, also occurs in polyploid cells but never in haploid cells.
3. It completes in one sequence formation of two diploid nuclei from a diploid mother nucleus.	3. It completes in two sucessive nuclear divisions and results in the formation of four haploid nuclei.
4. Prophase -Shorter duration no sub phase.	4. Prophase-longer duration and five subphases.
5. Synopsis does not take place.	5. Synopsis in zygotene stage.
6. Crossing over, chiasmata formation absent.	6. Crossing over occurs during pachytene and chiasmata are formed.
7. Each chromosome consists of two chromatids.	7. Each bivalent consists of four chromatids.
8. Metaphase → the chromosomes are arranged at the equatorial plane. The centromers lies on the equatorial plane and arms remain free.	8. Bivalents are arranged at the equatorial plane during metaphase I. Arms of chromatids lie on the equatorial plane and centromers are away from it.
9. During Anaphase the chromatids separate and move to opposite poles.	9. Anaphase I homologous chromosome of each bivalent separate and move to opposite poles (disjunction).
10. Telophase daughter nuclei are organized.	10. Telophase I two daughter nuclei may be organized they are haploid.
11. The number of the chromosome in daughter nuclei is same as in the parent nucleus.	11. Daughter nuclei are hapioid and contain half the number of chromosome of the parent nucleus.
12. Chromosome of daughter nuclei are quantitatively and qualitatively similar.	12. Chromosome of daughter nuclei are qualitatively dissimilar due to crossing over.

Life cycles of a typical haploid and diploids indicating the stages when meiosis occur.

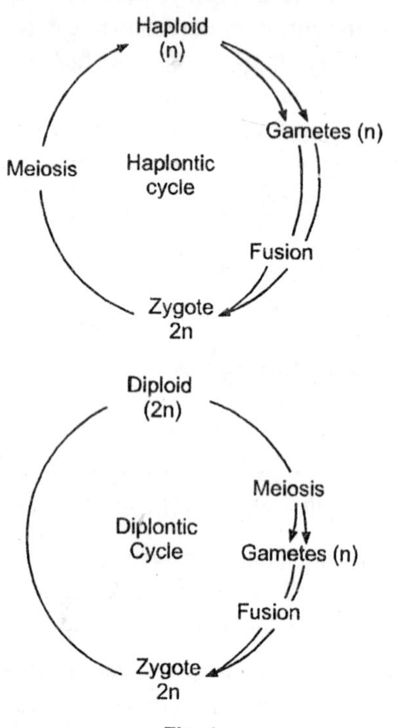

Fig. 1

3.2 GENETIC MATERIALS

3.2.1 Nucleic Acid-Structure and Function

Chemical analysis of chromosomes shows two nucleic acids — DNA (Deoxyribose Nucleic Acid) and RNA (Ribose Nucleic Acid). They were first discovered by a Swiss biochemist *Fredrick Miescher* in 1869. He used the term *nuclein* to denote nucleic acids because of their acidic properties.

Nucleic acids are macromolecules, composed of many (polymers) small units called nucleotides.

Nucleic acid = many nucleotides

Each nucleotide consists of phosphoric acid (phosphate), a five carbon sugar and a nitrogenous base.

Nucleotide = phosphoric acid + sugar + nitrogenous base

The sugar and base combination (without phosphoric acid) is called *nucleoside*.

phosphoric acid + sugar + nitrogenous base

Nucleoside (sugar + nitrogenous base)

Nucleotide (phosphoric acid + sugar + nitrogenous base)

As such nucleotides are phosphoric esters of nucleosides.

I. Phosphoric acid

The acidic nature of nucleic acids is due to the presence of phosphoric acid. Sugar of the nucleoside combines with phosphoric acid by a phosphodiester bond.

II. Sugar

It is a five carbon (pentose) sugar. There are two types of sugars — *ribose* and *deoxyribose sugars*. The nucleic acid containing ribose sugar is called *ribose nucleic acid* (RNA) and the other with *deoxyribose sugar* is called *deoxyribose nucleic acid* (DNA).

III. Nitrogenous bases

Each nucleic acid has four nitrogenous bases — two purines and two pyrimidines. The *purine bases* are adenine and guanine and the *pyrimidine bases* are thymine and cytosine. In RNA, uracil (pyrimidine) is present in place of thymine.

DNA : Structure

The full form of DNA: Deoxiribose Nucleic Acid.

The DNA is a double stranded helix made of many nucleotides. The nucleotides consist of deoxyribose sugar, phosphoric acid, purine bases (adenine and guanine) and pyrimidine bases (cytosine and thymine). The arrangement of these substances in DNA molecule long been a subject of curiosity. It was largely due to X-ray diffraction studies of *Wilkins* which provided basis for promising double helical structure of DNA. *Watson and Crick* (1953) finally described the structure of DNA and were awarded Nobel Prize in 1962 alongwith Wilkins.

Following are some of the characteristic features of this model.

1. Each nucleotide consists of sugar, phosphate and a nitrogenous base. Many such nucleotides are linked to form a polynucleotide chain or strand.
2. The adjacent nucleotides of the same strand are joined with each other by one phosphodiester bond between 5-carbon of sugar of one nucleotide and by another phosphodiester bond with 3-carbon of sugar of the next nucleotide.
3. Nitrogenous base is attached at 1-carbon of sugar. At this place purine is attached by its *9th*-position and pyrimidine by its *3rd*-position.
4. Thus a polynucleotide strand consists of sugar and phosphate forming its long axis.
5. The two polynucleotide strands are complementary to one another. If one strand has adenine, the other strand would have only thymine opposite to it. Similarly guanine and cytosine form the other complementary base pair. Thus, if the base sequence of one strand is CAT TAG GAC, the base sequence of another strand would be GTA ATC CTG.
6. The two strands are joined with one another by hydrogen bonds between their complementary nitrogenous bases. There are two hydrogen bonds between adenine and thymine and three hydrogen bonds between cytosine and guanine.
7. Two polynucleotide strands are helically coiled around a common axis to form a DNA-molecule. *The two strands are antiparallel, i.e.,* they run in opposite directions; the sugar molecule in one shows 5—P, 3—OH direction while the other shows 3-OH, 5-P direction.
8. The helical coiling of double strands is right handed. This DNA is called *B-DNA*.
9. Double stranded DNA molecule has a diameter of 20 Å, *i.e.*, the distance between two polynucleotide strands is 20 Å.
10. The helix makes one complete turn every 34 Å along its length.

RNA : Types and Structure

RNA or Ribose Nucleic Acid is present in all the living cells. It is found in cytoplasm as well as nucleus.

I. Types of RNA

RNA is generally involved in protein synthesis but in some viruses it also serves as a genetic material. The following are the types of RNA.

1. **Genetic RNA. H. Fraenkel-Conrat** (1957) showed that RNA present in TMV (Tobacco Mosaic Virus) acts as a genetic material. Since then it is known to be the genetic material of most of the plant viruses and some bacteriophages.
2. **Non-genetic RNA.** This type of RNA is present in those cells where DNA is the genetic material. Non-genetic RNA is synthesized on DNA template. It is of the following three types.

(*a*) ***Messenger RNA (m RNA).*** It carries the genetic information present in the DNA. It forms about 5-10% of the total RNA present in the cell.

(*b*) ***Transfer RNA (t RNA).*** It is also known as **soluble RNA** (s RNA). These are the smallest molecules which carry amino acids to the site of protein synthesis. It forms about 10-15% of the total cell RNA.

(*c*) ***Ribosomal RNA (r RNA).*** It is the most stable type of RNA and is found associated with ribosomes. It forms about 80% of the total cell RNA.

RNA is generally single stranded and is made of a chain of polynucleotides. The single strand is folded in such a way that the chain formed by sugar and phosphate is external while the complementary nitrogenous bases are projected inside and joined by hydrogen bonds as in DNA.

The differences between DNA and RNA are given in the table 1.

Table 1: DNA and RNA : A Comparison of their Structure, Reactions and Role in the Cell

DNA	RNA
1. Localization	
Primarily in nucleus also in mitochondria & chloroplasts	In cytoplasm, nucleolus and chromosomes
2. Pyrmidine bases	
Cytosine, Thymine	Cytosine, Uracil
3. Purine bases	
Adenine, Guanine	Adenine, Guanine
4. Pentose sugar	
Deoxyribose	Ribose
5. Strands	
Mostly double stranded[1]	Mostly single stranded[2]
6. Cytochemical reactions	
Feulgen	Basophilic dyes with Ribonuclease treatment
7. Hydrolysing enzyme	
Deoxyribonuclease (DNase)	Ribonuclease (RNase)
8. Role in cell	
Always genetic information	Synthesis of proteins, sometimes genetic
9. Replication	
self replicating	Formed from DNA. Self replication only in some viruses

3.2.2 Alleles and Genes

ALLELE

Alternative form of a gene is known as allele.

Main feature of Allele —

1. Govern same characters of an individuals.
2. A haploid cell has single copy of an allele, diploid two and polyploid more than two for a character.
3. All individual may have identical alleles.
4. They may be dominant and recessive type.

Test for Allelism

1. Recombination test
2. Complementation test.

Multiple Allele

1. Always belong to the same locus.
2. Control same characters.
3. No crossing over in the multiple allelism.
4. Wild type is always dominant.
5. Don't show complementation.

Example for multiple Allele —

1. Fur colour in rabbits.
2. Wing type in drosophilla
3. Eye colour in drosophila
4. ABO blood group in man.

Pseudoalleles

Closely linked and functionally related genes.

Characters

1. Govern different expression of the same characters.
2. Pseudo alleles occupy different position on the same locus.
3. Low frequency of Recombination by crossing over.
4. They exhibit cis-trans position effect.

Isoalleles

An allele which is similar in its phenotypic expression to that of other independently occuring allele is known as isoallele.

Types of Isoallele

1. Mutant Isoalleles.

2. Normal Isoalleles.

GENES

Smallest and Individually functional part of genetic material.

Properties of Gene

1. Form (alternative form allele).
2. Location (chromosome, linear, locus)
3. Status (several unit).
4. Number (each Diploid)
5. Sequence (specific sequence)
6. Expression - incomplete, complete.
7. Change in form – mutation.
8. Exchange of Gene – Translocation.
9. Composition – DNA Bacteriophage

Function

1. Control the expression of a specific character in an organism.
2. Based on gene control over characters. Oligogenic traits, polygenic traits and pleiotropic.
3. Gene interaction (when two or more gene govern a characters).
4. Linkage → two or more genes are inherited together.

Modern Concept of Gene

Genetical View

1. *Mendel Idea*
   - Factor for gene → responsible for transmission of characters from parents to their offspring.
   - Sutton and Boveri hypothesis → chromosomal theory of inheritence.
   - Morgan → linkage studies given that idea of genes are located on the chromosome in linear fashion.
2. *Modern Concepts*
   (*a*) Gene is divisible it was believed that gene is basic unit of structure which is indivisible by crossing over now observed that gene is divisible based on studies on Intragenic recombination like bar eye gene of drosophila.

   (*b*) Part of a gene can function :
   1. Recon → Region between recombination.
   2. Muton → Elements of gene rise to mutant.
   3. Cistron → Unit of function of gene.

Biochemical View

(*a*) *Earlier Concept* : Gene is sequence of nucleotides in DNA which control a single polypeptide chain.

(*b*) *Mordern Concept* : (*i*) fine structure of gene (Benzer) (*ii*) Split Genes.

The sequence of nucleotides were interrupted by intervening sequences such genes with interrupted sequence of nucleotides are referred as split genes or interrupted genes.

It have two types of sequences (*i*) Normal sequences (*ii*) Interrupted sequences.

(*c*) *Jumping Genes* : A gene keeps on changing its position within the chromosome and also between the chromosomes of the same gene. Such genes are known as jumping genes or transposons or transposable elements.

The transposable elements are of two types –

(*i*) insertion sequence

(*ii*) Transposons

(*d*) *Overlapping Genes* : Some nucleotide sequences (genes) can code for two or more proteins. The genes which code for more than one protein are known as overlapping genes.

(*e*) *Pseudo Genes* : There are some DNA sequences, especially in eukaryotes, which are non functional and defective copies of normal genes. Such DNA sequences of genes are known as pseudogenes.

Application of Genetics

1. *Taxonomy* : To study evolution, classification and identification.
2. *Agriculture* : Improvement of crop plants Yield, resistance to insects, disease, salinity, drought, frost, lodging and adaptability.
3. *In Medicine* : (*i*) Detection of Hereditary Disease (*ii*) Production of Antibiotics.
4. *Evolution* : Natural and Artifical selection have been responsible for evolution of various crop plants. (*i*) Polyploidy (*ii*) Mutation.

3.2.3 Genetic Material

1. Chemical Nature

Properties of the Genetic Material —

1. The Genetic material must be replicated with high fidelity.
2. Genetic material must be able to express itself.
3. Genetic material must be able to store the highly variable information.
4. The distribution of Genetic material must allow errors in a low frequency for the origin of new genetic variation.

Experiments of Griffith (1928)

His studies on Diploccocus pneumonea Different strains of Diploccocus from two types of colonies

1. Smooth 2. Rough.

Smooth colonies are enclosed in a polysaccharide capsule the strain are able to produce pneumonia, the virulent Rough colonies lack the polysaccharide capsule such strains are avirulent they can't produce pneumonia.

Virulent strains are classified into several types eg. II, III etc. on the basis of antigenic properties of the polysaccharides present in their capsule.

(*a*) Live IIIS (virulant) cells injected died (due to pneumonia).

(*b*) III S heat killed cells (virulent) → A live indicating that all the cells were killed by the heat treatment.

(*d*) Mix of heat killed IIIS and live IIR → died (pneumonia) since all the cells of heat killed IIIS culture were dead it was postulated that some of the cells IIR had changed into the IIIS type due to influence of dead IIIS cells present in the mixture. This phenomenon was called transformation and the component of IIIS cells which induced the conversion of IIR cells to IIIS was named transforming principle. Griffth demonstrated transformation but they did not hint at the identity of the transforming principle. It is now known that transformation is essentially a special type of recombination in which a segment from the transforming DNA replaces the homologous segment of the bacterial chromosome.

Experiments of Avery, Macleod and McCarty-1944

Avery and associates carried out experiments of Griffith in vitro (in test tubes).

A culture of live IIR cells : Rough colonies heat killed IIIS cells or DNA isolated from IIIS cells no colony.

Live IIR + Heat killed IIIS + Antibody IIR → IIIS colonies.

Live IIR + IIIS DNA + Anti IIR → IIIS colonies.

Anti IIR was used for inactivating IIR cells.

These findings show that DNA is the transforming principle.

In order to establish beyond any doubt that DNA is the transforming principle.

They treated IIIS DNA with RNA (the enzyme which digest RNA) or proteases (enzymes which dagrade proteins) before it was mixed with IIR cells.

In both the experiments some IIIS type colonies were obtained. This shown RNA and proteins present as impurities in IIIS DNA preparations were not responsible for transformation.

Then next they treated IIS DNA with DNA (enzyme digests DNA) before it was mixed with IIR cells. There is no colony. They established beyond any doubt that DNA is the transforming principle.

Experiments of Hershey and Chase (1952)

This led to the universal acceptance of DNA as the genetic material.

They studied the life cycle of T_2 phage of E. coli. They showed that only the DNA component of T_2 particle is transmitted to the progency. T_2 has a hexagonal head and a contractile tail. Head coat and tail are made up of protein, while the DNA is packed inside the head coat. Infection begins when the tail plate of a T_2 particle comes in contact with the cell wall of an E.coli cell.

They labelled T_2 DNA and protein in two separate experiments. DNA contains P but no S. Protein - S but no P. They labelled T_2 DNA → ^{32}P or ^{35}S labelled phage particles E.coli cells were grown for several generations on a culture medium containing ^{32}P or ^{35}S. The progeny phage particles thus obtained were labelled with either ^{32}P or ^{35}S.

In one experiment they mixed ^{32}P labelled T_2 particles with E-coli cells. The cells were than agitated in a blender to separate the empty phase particles (called ghosts) remaining outside the bacterial cells after infection. Most of the ^{32}P labelled form was present in the infected E-coli cells. The progency phage particles obtained after types of these E. coli cells also contained the ^{32}P labelled from this we come to know that DNA is transmitted from one generation to the next.

In the other experiment they used ^{32}S labelled T_2 particles obtained throrgh lysis of the infected E.coli cells was almost negligible. From this we come to know that proteins are not transmitted from one generation to the next.

RNA as Genetic Material (1957)

In TMV DNA is absent but these viruses are composed of RNA and protein.

The proteins and RNA of TMV can be separated when they are remixed they reassociate to produce TMV. In one experiment either RNA or proteins isolated from TMV were used for the infection of tobacco leaves. Mosaic symptoms developed only when RNA was used for infection and not the proteins were used.

They constructed two types of hybrid virus particles by mixing (1) RNA from strains A and proteins from strain B (2) Proteins from strain A and RNA from strain B.

When tobacco leaves were infected with hybrid TMV or the first type the disease symptoms of strain A developed the proteins also identical with those of strain A. Similarly when hybrid protein of the second type were used for infection of tobacco leaves symptoms of strain B. It is evident from these findings RNA (and not the protein) of TMV has the capacity to produce the disease and that the type of proteins present in the virus particle is determined by the RNA.

3.2.4 Chromosome Structure

Strasburger discovered chromosomes in 1886.

The darkly stained, rod shaped bodies visible under light microscope in a cell during metaphase stage of mitosis are referred as chromosomes.

Main features of eukaryotic chromosomes :

- Chromosomes are not visible during interphase under light microscope.
- Transmission of characters from generation to generation.
- It vary in shape, size and number in different species of plants and animals.
- It have property of self duplication, segregation and mutation.
- It composed of DNA, RNA and histones.

Chromosome Shape

It is usually observed during anaphase. It have three different shapes viz rod shape, S shape and V shape.

Chromosome Number

(*i*) *Haploid* : It refers to half of the somatic chromosome number of a species and is denoted by *n*.

(*ii*) *Diploid* : It refers to somatic chromosome number of a species and is denoted by 2*n*.

(*iii*) *Basic Number* : The genetic chromosome number of a true diploid species is called basic number.

Chromosome Morphology

- *Centromere* : The region of chromosome with which spindle fibres are attached during metaphase is known as centromere.
- *Chromatid* : One of the two distinct longitudinal subunits of a chromosome is called chromatid.
- *Secondary constriction* : The constricted or narrow region other than that of centromere is called secondary constriction.
- *Telomere* : The terminal region of chromosome on either side is known as telomere.
- *Chromomeres* : The linearly arranged bead like structures found on the chromosomes are known as chromomeres.
- *Matrix* : A mass of acromatic material in which chromonemta are embeded is called matrix.

Karyotype

Karyotype is a phenotypic appearance of chromosomes of a particular species. It is represented by a diagram which is known as idiogram.

Karyotype is of two types viz symmetrical and asymmetrical.

Special Types of Chromosomes

Lampbrush Chromosome : These are special types of chromosomes in which large number of loops are projected out from the chromatin axis giving a lampbrush appearance. They are formed in oocyte nuclei of both vertibrates and invertibrates and spermatocyte nuclei of drosophila during diplotene stage.

1. Extra ordinary length
2. Large number of loops
3. Lamp-brush appearance.

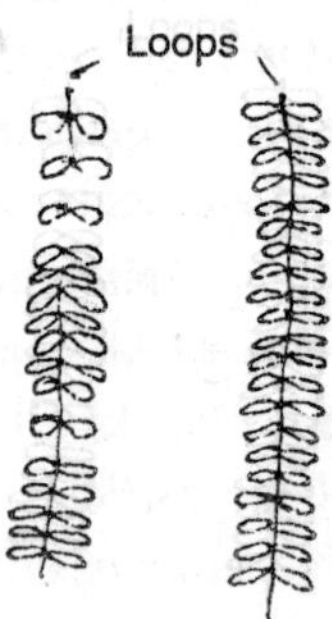

Fig. 2 Two homologous chromosomes.

Polytene or Giant Chromosomes : The multiple replicates of the same chromosome holding together in a parallel fashion resulting in very thick chromosome are known as polytene chromosomes.

These chromosomes have three main features:

1. Bands : The strips which are found in these chromosomes are known as bands.
2. Puffs : The smaller regions are known as chromosome puffs. These are the regions of genetic activity.
3. Giant Size .

B-Chromosomes :—

Some species possess extra chromosomes which are not members of normal chromosome complements. These are called as B chromosomes.

Classification —

on the basis of stability :

- Stable
- Unstable

on the basis of size

- Standard type
- Small type
- Very small type
- Large type

Behaviour at Mitosis and Meiosis

The meiotic behaviour of B chromosomes is studied during pachytene stage.

1. They do not pair with a chromosomes.
2. Lower degree of pairing is observed among B chromosomes.
3. When single B chromosome is present, it remains univalent during pachytene.

Chromosome Models

Chromatin fibres are the basic units of chromosome structure.

1. ***Folded Fibre Model***

Chromatin fibres are basic units of chromosome which are about 230 Å in diameters. A single chromatin fibre is found in each chromatid which consists of a single coiled double DNA helix. The folding of chromatin fibre in different ways results in the development of chromatin structure which is observed at metaphase. Two copies of chromatin fibre are formed from a single chromatin as a result of DNA replication during interphase. The replication of chromatin in the centromere region takes place where two chromatins have to separate out. Extensive folding of chromatin fibres leads to significant reduction in their length and increase in thickness and stainability.

2. ***Nucleosome – Solenoid Model***

Chromatin is composed of DNA, RNA, histones and other proteins. Chromatin fibres are 300 Å in diameter. The nucleosomes are sub units of chromatin and have bead like appearance. Each nucleosome is composed of a histone octamer and 146 bp of DNA. Each nucleosome consists of (1) a core particle and (2) linker or spacer DNA. The core particle has two copies each of H2A, H2B, H3 and H4 histone molecules. The core particle is about 110 Å in diameter and 60 Å in height one molecule of histone H_1 is connected with linker DNA. The super coiled nucleosome fibre is known as solenoid.

3.2.5 Chromosomal Aberration - Structural Changes

Any change which alters the normal structure of a chromosome is known as structural chromosomal aberration.

Types

Alter gene number in chromosomes → (*i*) Deletion (*ii*) Duplication.

Alter the sequence of genes in the Chromosome → (*i*) Translocation (*ii*) Inversion.

Deletion (Deficiency)

Loss of a portion of segment from a chromosome.

Observed → Drosophila, maize, tomato, wheat.

Depending upon location deletion two types.

1. *Terminal Deletion* : Loss of either terminal segment of a chromosome two types.
   (*i*) Heterozygous Deletion → Deletion occurs only in one chromosome of a homologous pair.
   (*ii*) Homozygous Deletion → Deletion occurs in both the chromosome of a pair.
2. *Interstitial Deletion* : Loss of a segment of chromosome from the intermediate portion or between telomere and centromere. The interstitial deletion generally does not involve centromere. In such deletion the break occurs at two places.

Detection :

Cytological Method — (*i*) Meiotic pairing (*ii*) Chromosome length.

In heterozygous Deletion : The pairing occurs between homologous segments.

In terminal Deletion : Normal chromosome remains unpaired at one end.

Interstitial Deletion : A loop is formed in the normal chromosome in the region of deletion. The loop confirms the presence of deletion.

Genetic Effects

1. *Fertility* : The pollen fertility is reduced in the presence of deletion. The pollens with deficient chromosomes are non functional.
2. *Viability* : A large deletion is lethal.
3. *Crossing over* : The crossing over is suppressed in the region of deficiency due to lack of corresponding segment in the area of deletion.
4. *Phenotype* : Deletion affects the phenotype. In the absence of dominant gene in the deletion region the recessive gene express. This results in the change in phenotype.
5. *Change in Karyotype* : The chromosomes with deletion can never revert to a normal condition. The gene number as well as the keryotype of the individual is changed.

Significance

It plays an important role in species formation and releasing variety through mutations.

Important cytological tools for mapping genes. Deletion mapping has been widely used in drosophila to locate various genes in polytene chromosomes.

Translocation

One way or reciprocal transfer of segments between non homologous chromosome is known as translocation.

Origin

Translocation originate through breakage and exchange of parts between non homologous chromosomes. When only one chromosome from each pair of two homologoues is involved, it gives rise to translocation hetrozygotes and when both chromosomes from each pair are involved it produces translocation homozygotes.

Detection

Translocation can be detected by cytological and genetic methods. Cytological methods includes study of pachytene configurations and metaphase configurations.

Table 2. Differences between translocation and crossing over

Translocation	*Crossing Over*
1. It involves in non-homologous chromosomes.	1. It involves in non-sister chromatids of homologous chromosome.
2. Change the linkage map.	2. Does not change the linkage map.
3. Breakage and reunion	3. Chiasma formation.
4. Pollen and ovule sterility	4. No sterility.

At Anaphase the chromosome disjoin (segregate) in three different ways —

1. *Alternate Disfunction :*

When two normal chromosome (N_1 and N_2) move towards one pole and two translocated chromosome (T_1 and T_2) to another pole, is known as alternate disfunction segregation. In such segregation all gametes receive full complement of genes and will give rise to viable individual.

2. *Adjacent one Segregation :*

The segregation of one normal chromosome with one translocated is called adjacent segregation. Such segregation occurs in open ring configuration. Here the chromosomes which go to one pole are non homologous ($T_1 + N_2$ and $T_2 + N_1$).

3. *Adjacent two Segregation*

Sometimes in open ring configuration, two homologous chromosome (T_1 N_1) *i.e.* one normal one translocated move to one pole and other homologous (T_2 N_2) move to another pole. Such disjunction is known as adjacent 2 segregation. Both adjacent types of segregation will produce gametes with duplication and deficiencies which may cause some sterility. Translocation can be detected by genetic methods based on pollen sterility, gene segregation and linkage studies.

Genetic Effects

1. *Sterility* : Translocations lead to duplication and deletion of genes. Thus translocations result in pollen sterility and ovule sterility. If there is a ring of four chromosomes 50% steility. If there is a ring of six chromosome → 75% sterility.
2. *Crossing over* : crossing over is generally suppressed due to competition in pairing.
3. *Karyotype* : Changes in chromosome number and karyotype. They may alter the size of chromosome as well as position of centromere.

Phenotype

In human down syndrome (Mangolism) can arise in the progeny of an individual heterozygous for a translocation involving chromosome number 21.

Significance

Translocation alter the chromosome size. chromosome number and karyotype and thus play an important role in the formation of species. Translocation are useful in locating the position of genes, centromere and other genetic markers on the chromosomes. They are useful tools in breeding programmes for transfer of desirable characters from one species to another.

Inversion

It refers to structural change in a chromosome in which a segment is oriented in a reverse order.

(*i*) *Paracentric Inversion* : The inversion in which centromere is not involved is called

Paracentric Inversion. In this type both breaks occur in one arm of the chromosome.

1 2 3|4 5|6   1 2 3 5 4 6

Only one chromosome of a homologous pair has inversion it is called inversion heterozygote. When both the members of a homologous pair have similar type of inversion it is called inversion homozygote. Meiosis is normal in inversion homozygotes. Crossing over within the inversion loop in a paracentric inversion heterozygote results in the formation of dicentric bridge and an accentric fragment after exchange. The other two chromosome remain normal.

The dicentric chromosome leads to formation of bridge at anaphase. The bridge is later on broken due to pull from both the poles, thus a centric segment is lost due to lack of movement. Thus out of four two are normal and two are deficient for some genes.

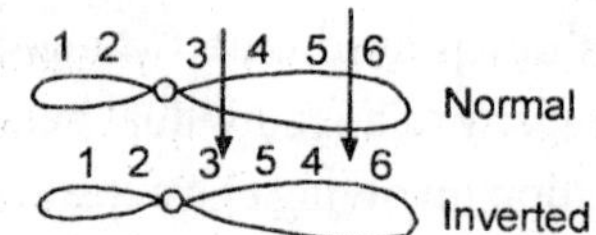

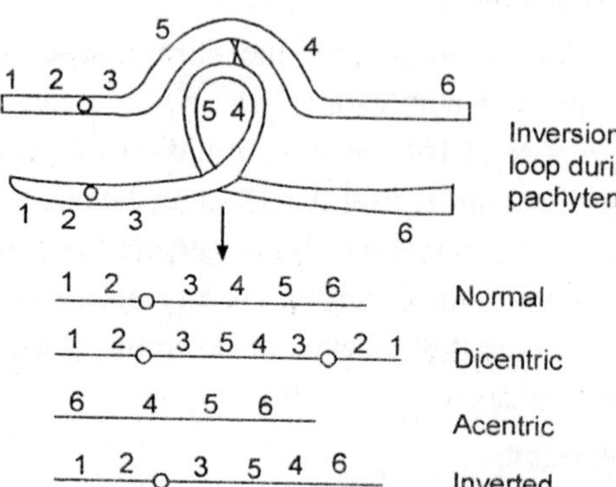

(*ii*) *Pericentric Inversion* : When centromere is involved in the inversion it is known as pericentric inversion. When a break occurs in each of the two arms of a chromosome the centromere is included in the detached segment resulting in a pericentric inversion.

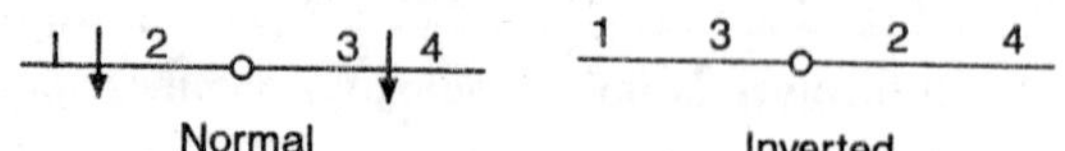

Crossing over within the inversion loop results in the formation of chromotids with duplication and deficiency. Out of four chromatids two are cross over produts and two are normal. One of the non cross over have original gene sequence and the other has inverted gene sequence.

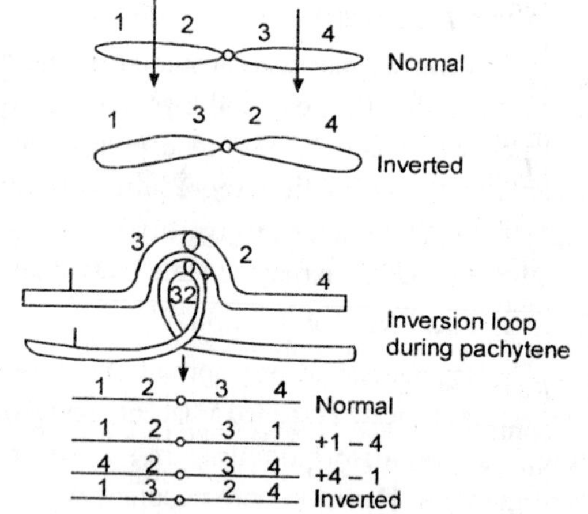

Origin

Inversion result when there are two breaks in a chromosome and the detached segment is reunited to the same chromosome in the reverse order.

Detection

Three cytological criteria (*i*) pachytene configuration (*ii*) anaphase configuration (*iii*) position of centromere are used for detection of inversion. Inversions can be detected in the meiotic nuclei by the presence of an inversion loop in the paired homologous during pachytene.

Genetic Effects

1. *Fertility* : The crossing over in the inversion loop leads to formation of chromosomes with duplication and deficiencies. Gametes with such chromosome are inviable and lead to 50% sterility.
2. *Crossing over* : Inversion heterozygotes often have pairing problems in the area of inversion. Thus competition for pairing reduces crossing over in the area of inversion.
3 *Gene order* : The Gene order is changed in the inverted segment of a chromosome. Inversion heterozygotes exhibit a linkage map with different gene order. In inverted chromosome there is no loss of genetic material provided crossing does not occur in the inversion loop.
4. *Karyotype* : Pericentric inversion sometimes results in change of karyotype by shifting the position of centromere in the inversion loop may lead to shift in the position of centromere.

Duplication

Occurance of a segment twice in the same chromosome. It results in addition of one or more genes to a chromosome. First reported in drosophila by Bridges in 1919.

Now it has been reported in maize and wheat.

Four types of duplication.

1. *Tandom Duplication*

Sequence of genes in the duplicated segment is similar to the sequence of genes in the original segment of a chromosome.

a b c [b c] d e f

2. *Reverse Tandom duplication*

The sequence of genes in the duplicated is reverse to the sequence of genes in the original segment of a chromosome.

a b c [c b] d e f

3. *Displaced Duplication*

When the duplication is found away from the original segment but on the same arm of the chromosome

a b c d e f g h i j k Normal

a [d e] b c d e f g h i j k Displaced

4. *Reverse Displaced duplication*

Duplication is also away from the original segment but found on the other arm of a chromosome.

These two types (3 & 4) are known as non-adjacent duplication because they are away from the segment which shows duplication.

Origin

Duplication originate due to unequal crossing over during meiosis. The homologous chromosome usually pair in such a way that all the identical loci match with each other in their positions. This facilitates equal crossing over between non-sister chromatids. Sometimes homologous chromosome pair in such a mis-aligned manner that the corresponding identical loci do not fall opposite to each other. Such situation leads to unequal crossing over between non sister-chromatids. This gives rise to two types of chromatids viz one with duplication other with deletion. When a gamete with duplication unites with normal ovule, it leads to formation of zygote with duplicate genes in a particular segment of a chromosome.

Detection

b c

a b c d e f g h i j

Duplication loop can be observed during pachytene stage when homologous chromosome pair. chromosome having duplicate segment are longer than normal chromosome. If a duplicate segment includes centromere it may be present as a small extra chromosome added to a normal chromosome complement duplication can also be detected by suppression of recessive characters. A single dominant gene in the duplicate region is enough to suppress the expression of two recessive alleles.

Significance

Duplications are less harmful than deletions. They do not reduce the viability of an individual. Duplications lead to addition of some genes in population.

3.2.6 Changes in Chromosome Number

Introduction

A basic or monoploid set of chromosomes of an individual is called genome. In a genome, each type of chromosome is represented only once. Most of the sexually reproducing plant species are diploid, *i.e.*, have two sets of chromosomes. Any change in the chromosome number from the diploid state is referred to as heteroploidy and the individuals having chromosome number other than diploid are called heteroploids. The heteroploidy is of two types *viz.*, I. euploidy and II. aneuploidy.

I. EUPLOIDY

The change in chromosome number which involves entire set is called euploidy. Euploidy includes, monoploids, diploids and polyploids.

Monoploids and Haploids

Monoploids contain a single chromosome set and are characteristically sterile. In a true diploid species, both monoploid and haploid chromosome number is the same ($n = x$). Thus a monoploid can be a haploid but all haploids cannot be monoploids.

Types of Haploids

Depending upon the origin, haploids are of two types *viz.*, euhaploids and aneuhaploids. Euhaploids develop from a euploid species and have complete chromosome set. Euhaploids are of two types, *viz.*, monohaploid – which develop from a normal diploid species, and polyhaploids – which develop from autopolyploid species. When a haploid develops from a tetraploid species, it is called dihaploid.

Aneuhaploids develop from aneuploid species and have either one additional or missing chromosome. Aneuhaploids include disomic haploids (n + 1), nullisomic haploids (n – 1), substitution haploids (n – 1 + 1), misdivision haploids etc. Mis-division haploids have an isochromosome which is produced by vertical division of centromere. Generally centromere divides longitudinally. Aneuhaploids are generally inviable.

Production

Now various methods are known by which haploids can be produced. These methods include (1) pollination with foreign pollen, (2) delayed pollination, (3) use of X-ray irradiated pollen for pollination, (4) temperature shock, (5) treatment with chemicals like colchicine, (6) interspecific and intergeneric crosses, and (7) anther and pollen culture.

Uses of Haploids

Haploids have several applications in plant breeding. They are used for (1) development of pure lines, (2) disease resistance, (3) development of inbred lines, and (4) production of aneuploids. These aspects are briefly described below.

Diploids

Normal diploids are known as disomics. They have regular bivalent pairing during meiosis. Diploids also have disomic genetics with two alleles at each locus. Pure line or inbred lines of diploids have homozygosity at each locus. Polyploids which behave like diploids are known as disomic polyploids like wheat, cotton, etc.

Polyploids

An organism or individual having more than two basic or monoploid sets of chromosomes is called polyploid and such condition is known as polyploidy. Polyploidy is of two types, viz, (1) autopolyploidy, and (2) allopolyploidy.

1. ***Autopolyploidy***

Polyploids which originate by multiplication of the chromosome of a single species are known as autopolyploids or autoploids and such situation is referred to as autopolyploidy. Autoploids include triploids (3x), tetraploids (4x), pentaploids (5x), hexaploids (6x), heptaploids (7x), octaploids (8x), and so on. Autoploids are also known as simple polyploids or single species polyploids.

Autotriploids

They have three sets of chromosomes of the same species. They can occur naturally or can be produced artificially by crossing between autotetraploid and diploid species. Triploids are generally highly sterile due to defective gamete formation. Triploids are useful only in those plant species which propagate asexually like banana, sugarcane, apple, etc.

Autoteraploids

They have four copies of the genome of same species. They may arise spontaneously or can be induced artificially by doubling the chromosomes of a diploid species with colchicine treatment. Tetraploids are usually very stable and fertile because pairing partners are available during meiosis. In such individuals diploid gametes (2n) are formed. Autotetraploids are usually larger and more vigorous than the diploid species. Rye, grapes, alfalfa, groundnut, potato and coffee are well known examples of autotetraploids.

2. ***Allopolyploidy***

A polyploid organism which originates by combining complete chromosome sets from two or more species is known as alloployploid or alloploid and such condition is referred to as allopolyploidy. Alloploids are also known as hybrid polypoids or bispecies or multispecies polyploids. An allopolyploid which arises by combining genomes of two diploid species is termed as allotetraploid or amphidiploid. Allopolyploidy can be developed by interspecific crosses and fertility is restored by chromosome doubling with colchicine treatment. Allopolyploids has played greater role in crop evolution than autopolyploidy, because allopolyploidy is found in about 50% of crop plants.

Natural Allopolyploids

Some important natural allopolyploid crops are wheat, cotton, tobacco, mustard, oats, etc. Interspecific crossing followed by chromosome doubling in nature have resulted in the origin of allopolyploids.

Induction of Polyploidy

Polyploidy is mainly induced by treatment with a chemical known as colchicine. This is an alkaloid which is obtained from the seeds of a plant known as **Colchicum autumnale,** which belongs to the family **Liliaceae.**

The colchicine induced polyploidy is known as colchiploidy. In plants, colchicine is applied to growing tips, meristematic cells, seeds and axillary buds in aqueous solution.

Colchicine induces polyploidy by inhibiting formation of spindle fibres. The chromosomes do not line up on the equatorial plate and divide without moving to the poles due to lack of spindle fibres. The nuclear membrane is formed around them and the cell enters interphase. Thus nucleus has double the chromosome number.

Effects of Polyploidy

1. Stems are thicker and stouter.

2. Leaves are fleshy, thicker, larger and darker green in colour.
3. Roots are stronger and longer.
4. Flowers, pollens and seeds are larger than diploids.
5. Maturity duration is longer and growth rate is slower than diploids.
6. Water content is higher than diploids, etc.

Applications in Crop Improvement

Polyploidy plays an important role in crop improvement. Both autopolyploidy and allopolyploidy are useful in several ways. However, allopolyploidy has wider applications than autopolyploidy. Applications of autopolyploidy and allopolyploidy in crop improvement are briefly presented below :

Autoploidy

Both triploids and tetraploids have been used in crop improvement. However, their applications have been limited to few species only. Autotriploids have been developed in sugarbeets and water melon only. The triploid sugarbeets have larger roots and higher sugar content than diploids. The triploid water melons are seedless or have rudimentary and soft seeds like cucumber.

Alloploidy

Alloploidy is useful in four principal ways, *viz.*, (1) in tracing the origin of natural allopolyploids, (2) in creating new species, (3) in interspecific gene transfer, (4) as a bridging species.

Limitations of Polyploidy

Polyploidy has several limitations. Some important limitations of polyploidy in crop improvement are briefly presented below :

1. *Limited Use.* The single species polyploidy has limited applications. It is generally useful in those crop species which propagate asexually like banana, potato, sugarcane, grapes, etc.
2. *Difficulty in Maintenance.* The maintenance of monoploids and triploids is not possible in case of sexually propagating crop species.
3. *Undesirable Characters.* In bispecies or multispecies polyploids characters are contributed by each of the parental species. These characters may be sometimes undesirable as in case of Raphanobrassica.
4. *Some other Defects.* Induced polyploids have several defects such as low fertility, genetic instability, slow growth rate, late maturity, etc.
5. Chances of developing new species through allopolyploidy are extremely low.

II. ANEUPLOIDY

Aneuploids are of three types, *viz.*, (1) monosomics, (2) nullisomics, and (3) polysomics. These are described below :

(1) Monosomics

An individual lacking one chromosome from a diploid set (2n-1) is called monosomic and such condition is known as monosomy. Monosomics may originate in three main ways, *viz.*, (*i*) from diploids, (*ii*) from nullisomics, and (*iii*) from trisomics as described below.

(*i*) ***From Diploids.*** Monosomics may originate spontaneously from diploids. Sometimes nondisjunction during meiosis gives rise to n-1 gamete. If this gamete is fertilized by a normal (n) gamete, a monosomic zygote (2n-1) is produced.

(*ii*) ***From Nullisomics.*** Nullisomics produce n-1 gametes. Union of such gamete with normal gamete gives rise to monosomics as shown below:

Nullisomic	Gametes	Union with normal gamete	Result
2n–2	n–1	n–1 + n	2n–1

(*iii*) ***From Trisomics.*** Trisomics (2n+1) also give rise to monosomics. Sometimes non disjunction of three homologous chromosomes in a trisomic during meiosis gives rise to n-1 gametes. Union of such gametes with normal one results in the development of monosomic zygote.

Trisomics	Gametes	Union with normal	Result
2n+1	$\frac{n-1}{n-1}$	n–1 + n	2n–1

(2) Nullisomics

An individual lacking one pair of chromosomes from a diploid set (2n-2) is called nullisomic and such situation is referred to as nullisomy.

(3) Polysomics

An individual having either single or one pair of extra chromosome in the diploid complement is known as polysomic and such condition is referred to as polysomy.

Applications in Crop Improvement

Aneuploids are useful in crop improvement in various ways. Some of the uses of aneuploids in plant breeding are briefly presented below :

1. Locating Genes. Aneuploids are useful tools for locating genes on a specific chromosome. Monosomics and nullisomics are used for this purpose.
2. Interspecific Gene Transfer. Monosomics are also used in transferring chromosomes with desirable genes from one species to another.
3. Aneuploids are used for developing alien addition and alien substitution lines in various crops.
4. Primary trisomics are useful in identification of chromosomes involved in translocations.

3.2.7 DNA Replication

The process by which a DNA molecule makes its identical copies is known as DNA replication. Three possible modes of DNA replication.

1. Dispersive Replication
2. Conservative Replication
3. Semi conservative Replication.

Semi Conservative Replication

Main features —

1. Seperation of two strands of parental DNA.
2. Complementary base pairing of the bases located in the single stranded regions.
3. Formation of phosphodiester linkages between the neighbouring deoxyribouncleotides.
4. This ensures that the base pairs of the new strands are complementary to the old strands.
5. The base sequence of a newly synthesized strand is dictated by the base sequence of old strand.

Evidence for Semi Conservative Replication

1. *Meselson and Stahl 1958*

They grew E. coli in ^{15}N for 14 cell generations till nitrogen in the bacterial DNA was only ^{15}N. This heavy DNA had more density than DNA with ^{14}N.

Then the bacteria were transferred to culture medium ^{14}N. They observed that such DNA was half dense indicating the presence of DNA hybrid (one ^{15}N strand + one ^{14}N strand). After second round of replication there would be 4 DNA molecules of these, two molecules would be hybrid (^{14}N - ^{15}N).

Methods of DNA Replication

1. *Initiation of Replication* : Replication begins at a replication origin.
2. *Unwinding of Helix* : Unwinding is brought by enzyme 'helicase.' This unwinding may result in formation of supercoils which are removed by enzyme DNA Gyrase (topoisomerase). Formation of replication fork (y shaped structure). Single strands are stabilised by a single strand binding protein (SSB).
3. *Formation of Primer Strand* : An enzyme primase initiates transcription of $3' \rightarrow 5'$ strands. This generates long primer RNA in $5' \rightarrow 3'$ direction.

   The free 3'-OH of this primer RNA provides the initation point for DNA polymerase for the sequential addition of deoxyribo nucleotides.

   DNA polymerase III catalyse DNA replication.
4. *Elongation of New Strand* : Synthesis of new strand occurs continously along the upper strand known as leading daughter strand - Synthesis of another daughter strand along lower parental strand takes place in the form of short pieces lagging daughter strand. Short pieces of DNA is called okasaki fragments. Discontinuous pieces of the lagging strand are joined together by the enzyme DNA ligase.

3.3 CLASSICAL GENETICS

3.3.1 The Laws of Mendel

Mendel's Predecessors :

1. The hybrids obtained by mating of two different variety of a species. The hybrids of some characters are identical to those of one of the two parents while some others are intermediate between those of two parents. The concept of dominance is present in these observation.
2. Characters of F_1 and F_2 progeny produced from reciprocal crosses are identical.
3. F_1 progeny are uniform in their characters. But F_2 generation show a large variation for different characteristics.
4. The appearance of the parental forms in F_2 phenotypes were intermediate in appearance.
5. Some phenotype in F_2 have entirely new character forms.

Reasons for their Failure

1. They studied the plant as a whole.
2. As a result of the above, the plants could not be classified into few clearcut classes.
3. The data from different generations were not kept accurately.
4. Complete control on pollination in the F_1 was lacking.
5. In many studies the F_1 was an interspecific hybrid exhibiting partial to considerable sterility.
6. The number of plants studied in F_2 was relatively small.
7. Most of the characters studied by the earlier workers were quantitative in nature.

Pea-advantages as an experimental material—

1. Several characters had a contrasting forms.
2. Pea-self pollination crop.
3. Flowers are relatively large. So, emasculation and pollination is very easy.
4. Duration single season every year one generation can be grown.
5. There is no problem in germination.

Table 3.7 pairs of contrasting characters of pea —

Characters	*Dominant Form*	*Recessive*
1. Length of stem	Tall	dwarf
2. Position of flowers	Axial	Terminal
3. Pod shape	Full	Constricted
4. Pod colour	Green	Yellow
5. Cotyledon colour	Yellow	Green
6. Seed coat colour	Grey	White
7. Seed shape	Round	Wrinkled

Reasons for Mendel's Success :

- An accurate and incisive analysis of the reasons for failure of earlier workers.
- He accurately diagnosed the weakness of their experimental materials, techniques and approaches and avoided them in his own studies.
- Studied the inheritance of only one pair of contrasting characters at a time.
- His selected pea variety had clearly different forms of one or more characters.
- He classified the plants of a population on the basis of contrasting characters, and kept an accurate record of the number of plants in each category for every generation.
- He carried out his experiments with great care and elaborateness.
- His knowledge of mathematics which helped him during the interpretation of his findings.
- He was able to formulate appropriate hypothesis.

Mendel was Lucky

1. The characters choosen by him was qualitative inheritance.
2. The contrasting forms of each of seven characters were governed by a single gene.
3. The seven characters studied by him, the genes for two were located in one chromosome, while three others were present in another chromosome.

Reasons for the Neglect of Mendel's Finding:

1. He used mathematical principles to explain a biological phenomenon this was not acceptable to biologists.
2. He studied contrasting pairs of characters exhibiting discontinuous variation many of his contemporaries (Darwin) pre occupied with characters exhibiting continuous variation.
3. The phenomenon of fertilization and the behaviour of chromosomes during cell divisions (mitosis and meiosis) were not known at that time.
4. Mendel did not publish his findings through further writing on the subject after his initial paper.
5. Mendel failed to domonstrate the validity of his conclusions in other species.

Highlights of the Mendel's Paper

1. Mendel concluded a complete fusion of male and female gametes, and equal contribution of the two parents to the development of characters of the hybrid (F_1).
2. He postulated the existence of genes, which are responsible for the development of various characters.
3. He clearly stated that genes were particulate.
4. He made clear distinction between the phenotype and genotype.

5. He gave the formula for determining the numbers of (*i*) different types of gametes produced by F_1 (*ii*) different genotypes in F_2 (*iii*) homozygous geno-types (*iv*) individuals in the perfect F_2 for segregation of *n* number of genes.
6. He introduced the concept of dominance and recessiveness.
7. He described laws of segregation and independent assortment.
8. He concluded that a large number of progeny would increase the precision of observation.
9. He stated that his explantations were based on two important assumptions (*i*) equal proportion of the different gametes produced by F_1 (*ii*) equal chance for each gamate thus produced to effect fertilization.

The Law of Segregation

Inheritance of Seed Shape in Pea

He crossed round with wrinkled. The seeds resulting from hybridization (F_1 seeds) were all round. The F_2 seeds (seeds produced by selfing of F_1 plants) were on an average 3 out of every 4 seeds were round and 1 out of every 4 seeds wrinkled. He allowed the F_2 plants to self pollinate. He found that all the plants from wrinkled -F_2 seeds produced only wrinkled F_3 seeds. But F_2 round seeds were of two types an average 1/3 only round F_3 seeds, 2/3 both round and wrinkled F_3 seeds in the ratio 3:1. He noted that the result from reciprocal crosses were identical.

The conclusion :

1. Both male and female parents make equal contribution to the development of characters in the progeny results.
2. $F_1 \rightarrow$ character of only one of the two parents is expressed (dominant character).
3. F_2 generation both dominant and recessive characters appear in 3:1 ratio.
4. The resessive character appears in F_2 unchanged and identical to that of the parent contributing this trait.
5. In F_2 1/3 of the individual with the dominant character are pure. 2/3 of them are hybrid.

The Law of Segregation —

This law is explained by making the following assumptions.

1. A character is produced by a specific gene
2. Each gene has two alternative forms (Alleles)
3. The two alleles of a gene govern the development of contrasting forms of the character governed by the gene.
4. Each somatic cell of an organism has two copies of each gene.

Definition

The two alleles of a gene present in the F_1 do not contaminate each other, they separate and pass into different gametes in their original form producing two different types of gametes in equal frequencies.

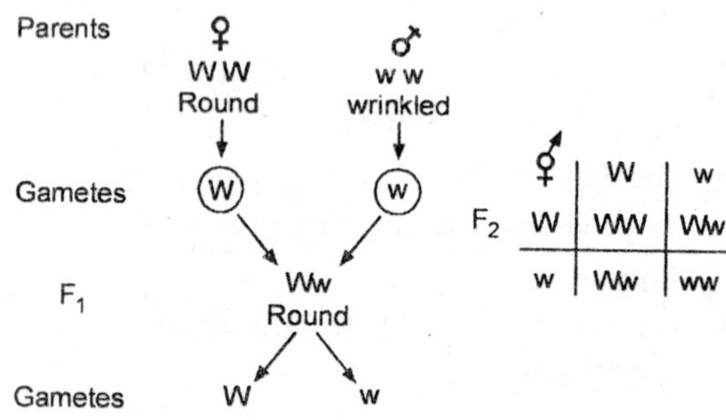

Essential features of this explanation of the 3:1 ratio in F_2

1. The presence of two copies of each gene in somatic cells.
2. Only one copy of each gene in gametes.
3. Lack of contamination or modification of each other by the two alleles of a gene during their stay together in the same cell of F_1 hybrids.
4. The separation of the two alleles of a gene and their transmission into separate gametes of F_1.
5. The production of two types of gametes with respect to the heterozygote gene by the F_1 in equal frequencies.
6. Random union between male and female gametes.

Law of Independent Assortment

This law states that when two pairs of gene enter in F_1 combination, both of them have their independent dominant effect.

Main features —

- It explains simultaneous inheritance of two plant characters.
- In F_1 when two gene controlling two different characters, come together, each gene exhibits independent dominant behaviour without affecting other gene.
- These gene pairs segregate during gamate formation independently.
- The alleles of one gene can freely combine with the alleles of another gene.
- Each of two gene pairs when considered separately, exhibits typical 3:1 ratio in F_2 generation.

Example —

When plants of garden pea with yellow round seeds are crossed with plants having green wrinkled seeds. We get yellow round seeds in F_1. Thus yellow colour of seed exhibits dominance over green and round seeds shape over wrinkled independently. The F_1 produces – Yellow round (YR) yellow wrinkled (yr), green round (YR), and green wrinkled (yr). Selfing of F_1 gives rise to all above four types of individuals in 9:3:3:1 ratio

Parents Yellow Round YYRR x Green wrinkled yyrr

↓

F_1 YyRr yellow round

F_2

	YR	Yr	yR	yr
yR	YYRR	YYRr	YyRR	YyRr
Yr	YYRr	YYrr	YyRr	Yyrr
yR	YyRR	YyRr	yyRR	yyRr
yr	YyRr	Yyrr	yyRr	yyrr

Independent assortment of two pairs of genes in garden pea.

3.3.2 Linkage

All the genes on a chromosome are said to be linked to one another and belong to the linkage group. The phenomenon of inheritance of linked genes in same linkage group is called *linkage*.

Features

- Two or more genes linked
- It may involve either dominant or recessive genes
- Linkage between closely located gene
- Higher percentage of parent in F_1
- It leads to desirable or undesirable trait
- Strength of linkage depends on the distance between the linked genes.
- Maximum number of linkage groups in an organism is equal to its haploid chromosome number.

Phases of Linkage

Coupling : Linkage between two or more either dominant (AB) or recessive (ab) allele is refered as coupling.

Repulsion : Linkage of dominant allele with recessive allele known as Repulsion.

Types of Linkage

1. *Based on crossing over —*
   (*a*) **Complete linkage :** Linkage in which there is no crossing over
   (*b*) **Incomplete linkage :** Some frequency of crossing over between linked genes.
2. *Based on genes involved*
   (*a*) Coupling Linkage
   (*b*) Repulsion Linkage
3. *Based on Chromosome involved*
   (*a*) **Autosomal linkage :** Linkage of genes which are located in other than sex chromosomes.
   (*b*) **X - chromosome linkage :** Linkage of genes which are located in sex chromosomes.

Detection of Linkage

1. Test Cross
2. Individual heterozygous at two loci in self pollinated linkage and pleiotropy.

The only way to distinguish between linkage and pleiotropy is to find out a cross over product between linked characters.

Significance of linkage in plant breeding

1. Effect on selection
2. Effect on Genetic variance
3. Effect on Genetic Correlation

Thomas H. Morgan (Nobel Lauerate) was first biologist to relate linkage to segregation of homologus chromosomes . He also found crossing over between homologus chromosomes during meiosis.

Table 4. Differences between crossing over and linkage

Crossing over	*Linkage*
1. Separation of linked genes	1. Keeps the genes together
2. Non-sister chromatids of homologous chromosomes involves.	2. Individual chromosome involves
3. Frequency of crossing over never exceed 50%.	3. It can never be more than haploid chromosome number
4. It increases variability.	4. It reduces variability
5. Provides equal frequency of parental and recombinant types in test cross progeny	5. Provides higher frequency of parental types than recombinant types in test cross progeny.

3.3.3 Crossing Over

Interchange of parts between non sister chromatids of homologus chromosomes during meiotic prophase.

Main features —

1. Take place during meiotic prophase.
2. It occurs between non sister chromatids
3. It takes place at four strand stage
4. It yields two recombinant types or crossover types and two parental types or non cross over types.
5. It leads to exchange of equal genes.
6. The frequency of recombinants can be worked out from the test cross progeny.

Chiasma and Crossing Over

The point of exchange of segments between non sister chromatids of homologous chromosomes during meiotic prophase is called chiasma. Here only crossing over takes place.

Chiasma Terminalization

The movement of a chiasma away from the centromere and towards the end of tetrads is called terminalization. Chiasma terminalization occurs between diplotene and metaphase I.

Three theories to explain the mechanism of chiasma terminalization.

1. Electrostatic hypothesis
2. Coiling hypothesis
3. Elastic chromosome repulsion.

Relationship between Crossing Over and Chiasma Formation

Two theories explain these relationship —

1. *Classical Theory* : This theory states that first chiasma is formed then crossing over takes place. The chiasma is formed at diplotene stage of meiosis and crossing over occurs between diplotene and anaphase.
2. *Chiasma type Theory* : Here first crossing over occurs and then chiasma is formed.

Molecular Mechanism of Crossing Over

1. *Copy Choice Theory* : The entire recombinant section arises from the newly synthesised section. The non sister chromatids when come in close contact they copy some section of each other resulting in recombination.
2. *Breakage and Reunion Theory* : The crossing over takes place due to breakage and reunion of non sister chromatids. The two segments of parental chromosomes which are present in recombinants arise from physical breaks in the parental chromosomes with subsequent exchanges of broken segments.

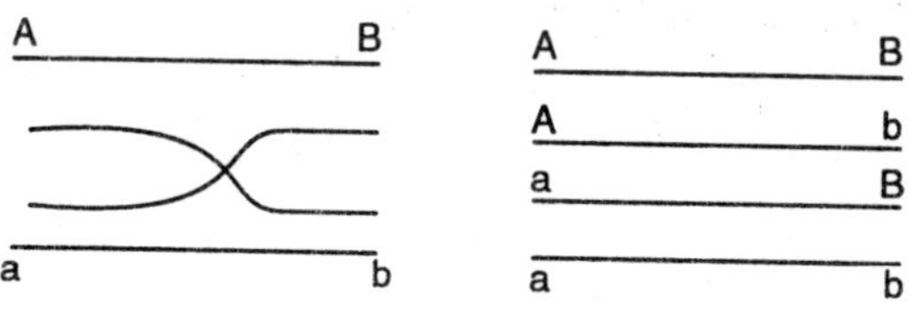

Interference (Muller)

Tendency of one crossover to reduce the chance of another crossover in its adjacent region.

Coefficient of interference (%) = 1 – Coefficient of Coincidence × 100.

Coincidence (Muller)

It explained the strength or degree of interference.

Coefficient of coincidence (%)

$$= \frac{\text{Observed double crossover}}{\text{Expected double crossover}}$$

Chromosome Mapping

A line diagram which depicts various genes present on a chromosome and recombination frequency between them. Such maps are also known as genetic maps or linkage maps. The mapping of chromosomes is done with the help of three point test cross. A three point test cross is a cross of trihybrid with its homozygous recessive parent.

Types of Crossing Over

Single Crossing Over : Formation of a single chiasma between non sister chromatids of homologous chromosomes.

Double Crossing Over : Formation of two chiasma between non sister chromatids of homologous chromosomes.

Multiple Crossing Over : Formation of more than two cross overs between non sister chromatids of homologous chromosomes.

Factors affecting crossing over are —

* Distance
* Temperature
* Sex
* Nutrition
* Chemicals
* Structural changes
* Centromere effect

Significance of crossing over —

- Creation of variability
- Locating genes
- Linkage Maps

3.4 GENE ACTION

3.4.1 Protein Synthesis

Protein synthesis depends on the sequence of amino acids in its composition. Union of several amino acids leads to the formation of polypeptide chain. Protein synthesis involves two main process (*i*) transcription (*ii*) Translation.

Transcription

The process of synthesis of messenger RNA from a DNA template is known as transcription.

$$\text{DNA} \xrightarrow[\text{Transcription}]{\text{Transcriptase}} \text{mRNA}$$

Mechanism of Transcription

The process of transcription takes place during interphase and continues upto prophase of cell division. The process of transcription require (*i*) DNA tempelate (*ii*) activated precursors (ATP, GTP, UTP and CTP) (*iii*) Divalent metal ions (Mg^{++} or Mn^{++}) (*iv*) RNA Polymerase.

The process of transcription consists of three main steps —

1. Initiation of transcription
2. Elongation of RNA chain
3. Termination of RNA chain.

1. *Initiation of Transcription*
   - Start at specific site in a cistron called promotor site.
   - The binding of RNA polymerase at initiation site results in separation of DNA double helix.
   - After opening of DNA strands, synthesis of mRNA begins on one of the two strands of DNA.
   - The RNA chain starts with triphosphate and the first base is always purine.
   - After synthesis of terminal triphosphate, nucleotides are removed by endonucleases.
2. *Elongation of RNA chain*
   - The elongation of RNA chain takes place by addition of activated ribonucloside triphosphates like ATP, GTP, UTP, and CTP to one strand of DNA molecule.
   - RNA synthesis takes place only on one strand of DNA.
   - The elongation of mRNA chain takes place with the help of core enzyme which moves along the DNA template.
3. *Termination of RNA Chain*
   - The chain is terminated either when DNA receives termination signal or an additional protein.
   - The terminator gene acts as a signal to end transcription.
   - RNA polymerase joins the nucleotides together to produce mRNA transcript.
   - When mRNA transcript is detached, the DNA template strand reforms H-bands with its complementary strand resulting in the formation of double helical structure.

Translation

The process of protein synthesis from information in mRNA is known as translation.

$$\text{DNA} \xrightarrow{\text{Transcription}} \text{RNA} \xrightarrow{\text{Translation}} \text{Protein}$$

Mechanism of Translation

The translation process requires mRNA, rRNA, ribosomes, 20 kinds of amino acids and their specific transfer RNA's.

It consists of five major steps —

(*i*) activation of aminoacids

(*ii*) transfer of amino acids to tRNA

(*iii*) Chain initation

(*iv*) Chain elongation

(*v*) Chain termination.

1. *Activation of Amino Acids*

There are 20 different amino acids which are used in protein synthesis. Each amino acid is activated in the presence of ATP by its own activating enzyme called aminoacyl synthetase to form aminoacyl adenylate.

2. *Transfer of Amino Acids to RNA*

The aminoacyl adenylate reacts with specific tRNA and gives rise to aminoacyl tRNA complex.

Thus synthetase has two active sites, one for recognizing the specific amino acid and the other for recognizing specific molecule.

3. *Chain Initiation*

- The 30s ribosome sub unit attached to mRNA to form an mRNA 30S complex.
- The amino acid tRNA complex binds to the initiation codon AUG on mRNA through its anticodon UAC to form 30S initiation complex.
- The larger ribosomal sub unit joins to the 30s initiation complex to form the complete initiation complex.
- The larger ribosomal subunit has two binding sites for tRNA *i.e.* acceptor site and peptide site.

4. *Chain Elongation*

- The second amino tRNA complex now occupies the acceptor site. There is enzymatic recognition of internal codons. This process requires EF-T_4, GTP and metallic ions mg^{++}.
- Formation of peptide bonds takes place. This process requires two metallic ions mg^{++} and k^{+}.

5. *Chain Termination*

- The termination codon provides signal to the ribosome for the attachment of release factors.
- The release factors interact with peptydyl transferase causing hydrolysis of the bonds between tRNA and the polypeptide chain.
- Hydrolysis of GTP results in the dissolution of the release factors from the ribosome.
- The released polypetide chain undergoes processing. The formyl group of starting amino acid methionine is removed. Proteins are synthesized from amino acids with the help of polysomes, three types of RNA, various enzymes and cofactors etc.

3.4.2 Cytoplasmic Inheritance

Most of the phenotypic characters are controlled by genes present in the chromosomes. But some characters expressed by factors present in the cytoplasm. These factors lying in the cytoplasm are called plasmagenes. The transmission of characters controlled by plasmagenes are called cytoplasmic inheritance.

The first case of cytoplasmic inheritance was reported by Correns in 1909 in four 'o' clck (*Mirabilis jalapa*) for leaf colour.

Important characteristics (or) features of cytoplasmic inheritance :

1. *Reciprocal differences*

Characters which are governed by Cytoplasmic Inheritance invariably exhibit marked differences in reciprocal crosses in F_1.

2. *Lack of Segregation*

In general F_2, F_3 and the subsequent generation do not show segregation for a Cytoplasmic

Inheritance trait. This is because the F_1 individuals generally receive plasma genes from one parent only.

3. *Irregular Segregation in Biparental Inheritance*

In some cases, plasma genes from both the parents are transmitted to the progeny, this is known as Biparental inheritance. Biparental inheritance gives rise to irregular segregation ratio in the F_1 generation of higher plants.

4. *Somatic Segregation*

Plasma genes generally show somatic segregation during mitosis.

5. *Association with organelle DNA*

Several plasma genes have been shown to be associated with CP-DNA not DNA.

6. *Mutagenesis*

Some Mutagens *e.g.* ethidium bromide, are highly specific Mutagens for plasma genes.

7. *Mappability*

It is very difficult to map cytoplasmic genes or prepare linkage map for such genes.

8. *Maternal Effects*

In case of cytoplasmic inheritance distinct maternal effects are observed. This is mainly due to more contribution of cytoplasm to the zygote by female parent than male parent.

Table 5. Difference between Mendelian Inheritance and Cytoplasmic Inheritance

Mendelian Inheritance	*Cytoplasmic Inheritance*
1. Governed by Nuclear Genes	1. Governed by plasma genes
2. Exhibits distinct segregation pattern	2. Does not exhibit distinct segregation
3. Reciprocal differences are not observed	3. Reciprocal differences are observed
4. Does not show maternal effects.	4. Exhibits maternal effects
5. Genes can be easily mapped on chromosomes	5. Mapping of plasma genes is very dificult.
6. Nuclear genes are associated with chromosomes	6. Plasma genes are associated with either chloroplast DNA or mitochandrial DNA.

Significance in Plant Breeding

- Studies of cytoplasmic inheritance have played key role in mapping of chloroplast and mitochondrial genome in yeasts, chlamydomonas, maize, humans etc.
- Cytoplasmic inheritance has been useful in explaining the role of various cytoplasmic organelles in the transmission of characters in different organisms.
- Development of cytoplasmic male sterility (CMS).

  CMS line have been developed in several crops like maize, pearl millet, sorghum, cotton etc.
- Availability of CMS lines has facilitated the production of hybrid seed in these crops at a cheaper cost than with hand emasculated pollination method.
- Mitochondria can be used in the manifestation of heterosis.
- Mutation of chloroplast DNA and mitochnodrial DNA leads to generation of new variants. Some of such variants are of special significance especially in ornamental plants.

3.4.3 Sex Determination

Sex determination is the process of sex differentiation, which utilises various genetical concepts to decide whether a particular individual will develop into male or female sex.

Mechanisms of Sex Determination

1. Chromosomal sex determination
2. Monogenic sex determination
3. Environmental sex determination

1. *Chromosomal Sex Determination*

(*a*) *Sex Determination by Allosomes*

Sex with similar type of sex chromosomes (xx) is known as homogametic sex and with dissimilar type of sex chromosomes (xy) as heterogametic sex.

There are four different systems of allosomal sex determination

1. xx – xy female – male system
2. xx – xo female – male system
3. xo – xx female – male system
4. zw – zz female – male system

(b) Diploid-Haploid System

In honey bees the females have diploid (2n = 32) chromosomes and males have haploid (n = 16) chromosomes unfertilized eggs develop parthenogenetically into haploid.

(c) Genic Balance System

Sex determining genes are present on both X-chromosomes as well as autosomes.

2. *Monogenic Sex Determination*

In some animals and plants, expression of sex influenced by a single gene.

3. *Environmental Sex Determination*

In horse tail plant development of sex depends on growing conditions. Plants grown under good conditions develop as females and those grown under stress conditions develop into males.

3.4.4 Sex Linked Characters

The characters for which genes are located on sex chromosomes are known as sex linked traits. Such genes are called sex linked genes, and linkage of such genes is referred to as sex linkage.

Main features —

1. *Location* : Sex linked genes are located on X chromosomes only.
2. *Number* : In diploid organism, each homogametic sex (xx or zz) has two copies of linked alleles.
3. *Expression* : In homogametic sex, a recessive gene can express. In heterogematic sex, a recessive allele can express.
4. *Transmission* : Sex linked genes are transmitted from female to male.
5. *Pattern of segregatin* : Inheritance of sex linked characters does not follow normal segregation pattern.

Sex Limited Characters

The characters which are expressed in one sex only are referred to as sex limited characters.

Features —

- They express in one sex only.
- Sex limited genes may be located either in sex chromosome or autosomes.
- Sex limited genes control the expression of primary and secondary sex characters.

Sex Influenced Genes

The dominance expression of some genes depends on the sex of individual. Such genes are known as sex influenced genes.

Features —

- Such genes are located in the autosomes
- They express more frequently in one sex than other.

- Expression of such characters appears to be governed by sex hormones.

3.5 OTHERS

3.5.1 Mutation

During the process of DNA replications errors ocassionaly occur in the polymerization of second strand. These errors are known as *mutations*. These can have an impact on the phenotype of an organism, especially if they occur within the protein coding sequence to gene. Error rates are usally very low- 1 error in every 10-100 million bases- due to the proofreading ability of DNA polymerases.

Processes that increase the rate of changes in DNA are called mutagenic. These chemicals promote errors in DNA replication, often by interferring with the structure of base pairing while UV radiation induces mutations by causing damage to DNA structure.

Mutation : Sudden heritable changes in the characters of origansm.

Mutants : Individual showing mutational changes.

Mutant allele : The allele producing the changed pheno-type variant.

Mutation is the ultimate source of all the Genetic variation.

Historical

Seth Wright (1791) : Male lamb with unusually short legs.

Hugo de vries (1900) : The term mutation was introduced sudden heritable changes observed by him is another contribution.

Morgan (1910) : White eye mutant of drosophila.

H.J. Muller (1927) : Mutagenic action of X-rays in drosophila.

Characteristics

- Generally mutant alleles are recessive.
- Random events.
- Most have harmful effects.
- Recurrent in nature.
- Occurs at very low frequencies.
- Mutable genes.
- Mutaor genes.
- Highly mutable sites within a gene - hot spots.
- Rate of induced mutation varies from one gene to the other.
- Most Mutant alleles are pleiotropic.
- Occurs in both forward and reverse directions.

Classification

- The direction of Mutation
  1. Forward 2. Reverse
- Cause of Mutation
  1. Spontaneous 2. Induced
- Dominance Relationship
  1. Dominance Mutation
  2. Recessive Mutation
  3. Co dominant (blood group)
  4. Partially dominant.
- Tissue of Origin
  1. Somatic Mutation (bud mutation)
  2. Germinal Mutation
- Effect on survival
  1. Lethal 2. Sublethal 3. Subvital
  4. Vital 5. Super vital

In crop improvement → Vital and Super Vital

- Type of Trait Affected
  1. Visible Mutation 2. Character Mutation
- Intensity of Character Expression
  1. Amorphic 2. Hypomorphic 3. Isoallelic
  4. Hyper morphic
- Quantum of Morphological Effect Produced
  1. Macro mutation 2. Micro Mutation
- Effect on the Expression of Neighbouring genes

1. Polar Mutation 2. Non polar Mutation

- Cytological Basis
  1. Chromosomal Mutation
  2. Gene Mutation
  3. Cytoplasmic Mutation
- Molecular basis
  1. Base substitution
  2 Deletion
  3. Addition
- Type of Amino Acid Replacement in the polypeptide
  1. Missense 2. Non sense 3. Frame shift

Change in base sequence of DNA molecule occur in two ways :

1. Base substitution
2. Base addition and deletion

Base Substitution

When one base in DNA molecule is replaced by another one.

2 types : (*i*) Transition (*ii*) Transversion

(*i*) *Transition* :

Purine is replaced by another purine *e.g.* Adenine is replaced by Guanine.

$$A \longleftrightarrow G$$

One base substitution affects the base sequence of only one codon as a result only amino acid is altered in the concerned protein.

(*ii*) *Transversion*

Pyrimidine (Thymine and Cytosine) is replaced by a purine (adenine and guanine) $A \longleftrightarrow T$ are base susbtitution affects the base sequence of only one codon as a result, only one amino acid is altered in the concerned protein.

Base Addition and Deletion

Base Addition

Insertion of one or more bases in a DNA molecule.

Base Deletioin

Loss of one or more bases in a DNA molecule.

Mutator Gene

Some genes increase the frequency of spontaneous mutation of other genes located in the same cell.

Anti Mutator Gene

Some genes decrease the frequency of spontaneous mutation of other genes located in the same cell.

MUTAGENS

Also, Refer to *Chapter* 4, *Aritcile No*. 4.3.4

Applications in Crop Improvement

1. *Development of Improved Varieties*
   Wheat - Sarbati Sonora, Pusa Lerma
   Rice – Jagannath
   Tomato – Pusa lal, Meeruti
   Cotton – MCU7, MCU10
   Sugarcane – Co8152, 8153
2. *Induction of Male Sterility*
   Genetic male sterility has been induced in durum wheat using radiations.
3. *Production of haploids.*
   Use of X rays radiated pollen in production of haploids in many crops.

3.5.2 Polygenic Inheritance

Characters or traits are governed by one or few genes. Such traits are called as qualitative or oligogenic characters. Some characters are controlled by several genes.They are known as quantitative characters or polygenic characters.

Features of Polygenic Traits

- Each polygenic character is controlled by several independent genes.
- Polygenic character exhibit continuous variation.
- Highly sensitive to environmental changes.
- Effect of individual gene is not easily detectable.
- Expression is governed by additive gene action.
- Transmission is generally low because of high amount of environmental variation.

Analysis of Polygenic Traits

- Measurement based on degree of variation
- Observation recorded in population.
- Studies with the help of mean, variance and covariance.

Assumption of Polygenic Traits

- Each of the contributing genes involved in the expression of a character produces an equal effect.
- Each contributing alleles has either cumulative or additive effect in the expression of character.
- There is no epistasis among genes at different loci.
- The linkage is in equilibrium.
- The environmental effects are absent.

Examples of Polygenic Traits

1. Kernel colour in wheat
2. Corolla length in Tobacco
3. Skin colour inheritance in man.

Transgressive Segregation

Appearance of transgressive segregants in F_2 is an important feature of polygenic inheritance. The segregants which fall outside the limits of both the parents are known as transgressive segregants.

Example :

parents AAbb × aaBB

↓

F_1 AaBb

F_2 AABB and aabb are transgressive segregants

Polygenic Variation

Three types —

1. *Phenotypic Variability*

It includes both genotypic and environmental variation.

2. *Genotypic Variability*

It is the inherent or genetic variability which remains unaltered by environmental conditions. It consists of additive, dominance and epistatic components.

3. *Environmental Variability*

It is non heritable variation. This uncontrolled variation is measured in terms of error mean variance.

Table 7. Differences between Additive Variance and Dominance Variance

Additive Variance	*Dominance Variance*
• Difference between homozygotes	• Deviation of Aa from the mean of AA and aa
• Genes show lack of dominance	• Genes show incomplete, complete or over dominance.
• Associated with homozygosity	• Associated with heterozygosity
• It is fixable	• Non fixable
• Selection is very effective	• Selection is ineffective
• It is the cause of transgressive segregation	• Cause of heterosis.

Table 8. Differences between Polygenic Traits and Oligogenic Traits

Polygenic Traits	*Oligogenic Traits*
• Governed by several genes	• Governed by few genes
• Effect of each gene is not detectable	• Effect of each gene is detectable
• Variation is continuous	• Variation is discontinuous
• Usually governed by additive genes	• Governed by non additive genes
• Highly influenced by environmental factors	• Not influenced by environmental factors.

Epistatic Variance

It is due to the deviations as a consequence of inter-allelic interaction, is interaction between alleles of two or more different genes or loci.

It is of three types —

1. Additive × Additive
2. Additive × Dominance
3. Dominance × Dominance

Significance of Polygenes

- It have great evolutionary significance
- Increase the crop yield
- Variation for speciation

3.6 CLONING

Cloning is the process of creating an identical copy of an original. Human and animal body is made of billion of cells. Each of these cells has a copy of all the genetic information needed by the body. This is known as *genetic code*. It means that each cell of our body has enough information to create another person exactly like us. For doing this we need an egg-cell– the single cell divides millions of times to give rise to the million of cell which form our body. In this process of cloning scientist extract the DNA that contains genetic code from an animal cell and implant it into on egg cell taken from another animal. Before implanting it, the DNA of the second animal is removed. In this way a carbon copy of the first animal is created and this is called *clone*.

3.7 CLONING IN PLANTS

Plant cloning is the act of producing indentical genetical plants from an original plant.

Cloning in plants have been in practice for thousand of years. For example, when we take a leaf cutting from a plant and grow into a new plant we are cloning the original plant, because the new plant has the same genetic makeup as the donor plant.

Recently scientists have been able to clone plants by taking pieces of specialized roots, breaking them up into root cells and using a technique to grow them into new plants, that are identical to the original plant from which the root pieces were taken. This method is called as *Plant Tissue Culture* widely used by horticulturists for grow prized orchids and rare flowers.

Most important modern method of plant cloning is meristem culture using the actively dividing tissue (meristems) of plants (shoots, roots, or axillary buds)

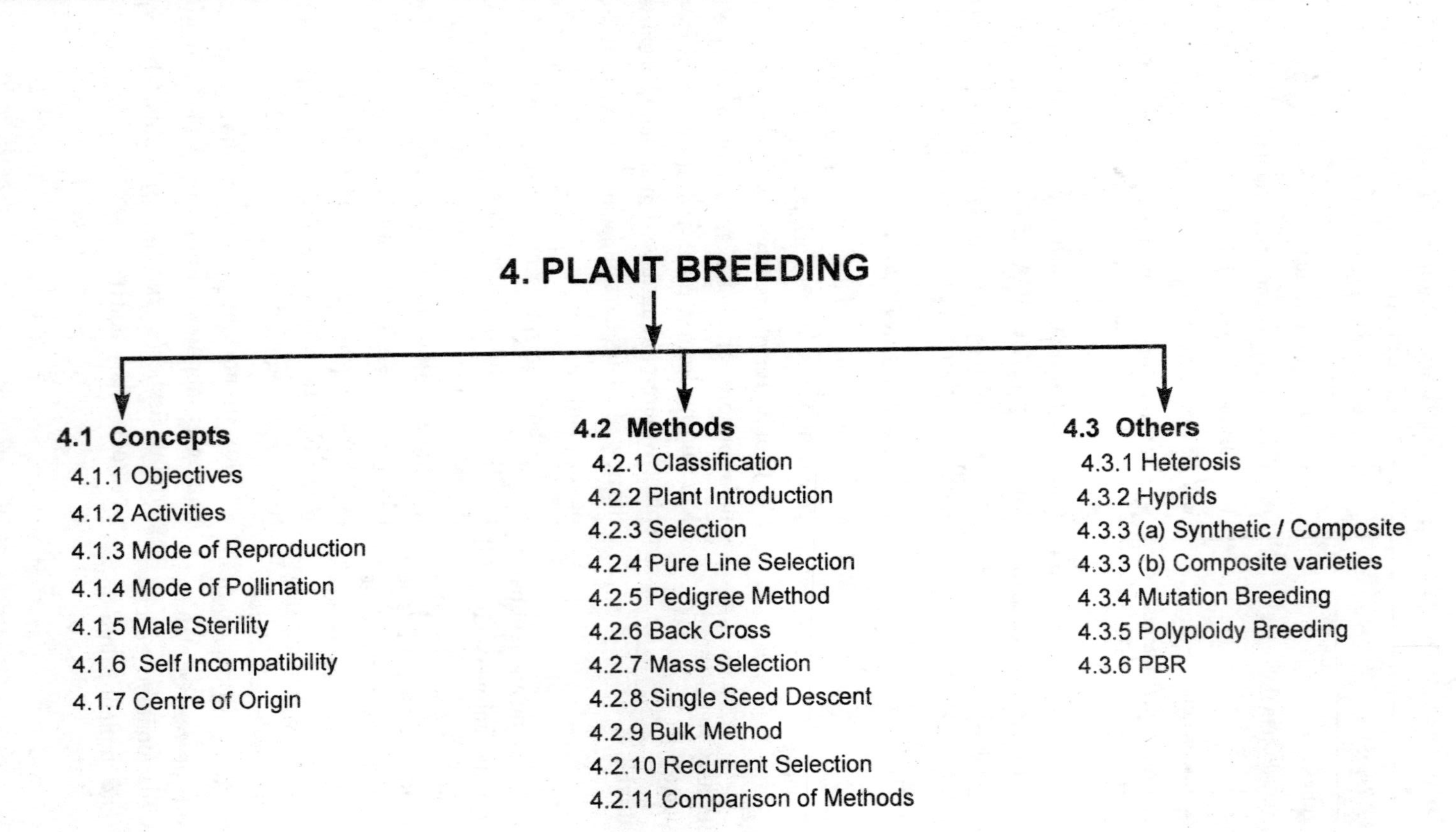
4. PLANT BREEDING
4.1 Concepts
4.1.1 Objectives
4.1.2 Activities
4.1.3 Mode of Reproduction
4.1.4 Mode of Pollination
4.1.5 Male Sterility
4.1.6 Self Incompatibility
4.1.7 Centre of Origin
4.2 Methods
4.2.1 Classification
4.2.2 Plant Introduction
4.2.3 Selection
4.2.4 Pure Line Selection
4.2.5 Pedigree Method
4.2.6 Back Cross
4.2.7 Mass Selection
4.2.8 Single Seed Descent
4.2.9 Bulk Method
4.2.10 Recurrent Selection
4.2.11 Comparison of Methods
4.3 Others
4.3.1 Heterosis
4.3.2 Hyprids
4.3.3 (a) Synthetic / Composite
4.3.3 (b) Composite varieties
4.3.4 Mutation Breeding
4.3.5 Polyploidy Breeding
4.3.6 PBR

4. Plant Breeding

4.1 CONCEPT OF PLANT BREEDING

Plant breeding aims at improving the characters of economic importance of a crop plant. The objective is to produce genotypes, which are better than the existing.

Plant breeding is the purpossful manipulation of plant species in order to create desired genotypes and phenotypes for specific purposes.

The sum of an organism's observable characteristics is the *phenotype*.

Genotypes are characteristics inherited from an organism's parents or genetic makeup of an organism.

4.1.1 Objectives of Plant Breeding

- Higher yields
- Improved Quality
- Disease and insect resistance
- Photoinsensitivity
- Synchronous Maturity
- Non shattering characteristics
- Varieties for new seasons
- Moisture stress and Salt tolerance

4.1.2 Activities in Plant Breeding

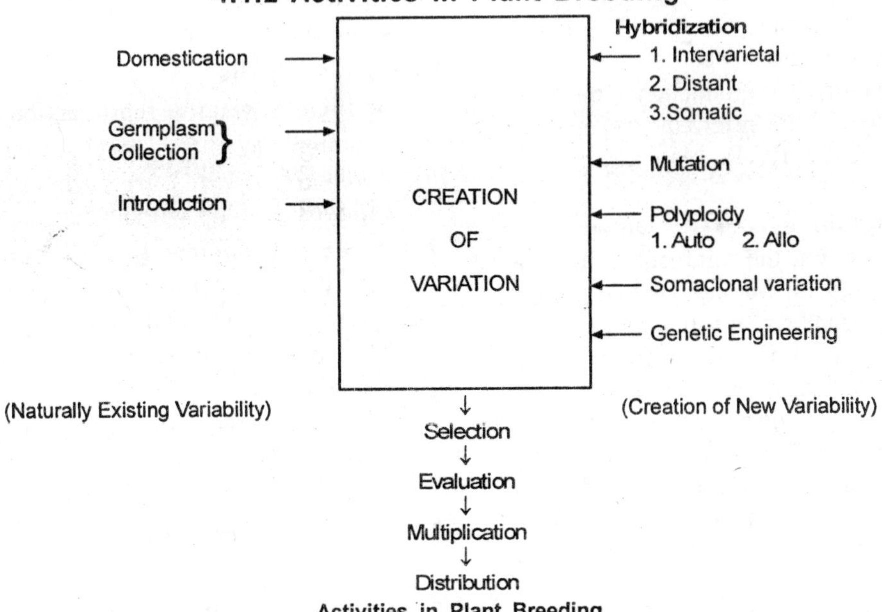

Activities in Plant Breeding

4.1.3 Mode of reproduction

(*a*) Sexual Reproduction

It means fusion of male and female gametes to form a zygote, which develops into an embryo.

In crop plants meiotic division of specific cells stamen and pistil yields microspores and megaspores respectively. This is followed by mitotic division of the spore nuclei to produce gametes like microspores and megaspores respectively.

Sporogenesis : Production of microspores and megaspores is known as sporogenesis.

Microspores are produced in anthers, while megaspores are produced in ovules.

1. ***Microsporogenesis*** **:** Microsporogenesis is the process of formation of pollen grains (microspores). During this process a microgametophyte developes inside a pollen grain. This development take place in three-called stages. In flowering plants angiosperms, microsprogenes process takes place with the involvement of microspore mother cell is present in the anther of the plant.

Each anther has four pollen sacs, which contain pollen mother cells. Each pollen mother cells undergoes meiosis to produce four haploid cells or microspores. This process is known as microsporogenesis.

2. ***Megasporogenesis*** **:** A single cell in each ovule differentiates into a megaspore mother cell. The megaspore mother cell undergoes meiosis to produce four haploid megaspores. Only one is functional megaspore per ovule.

Gametogenesis : The production of male and female gametes in the microspores and the megaspores respectively is known as gametogenesis.

1. ***Microgametogenesis***: During the maturation of pollen, the microspore nucleus divides mitotically to produce generative and vegetative or tube nucleus. When the pollen lands onto the stigma of a flower pollination occurs, shortly after pollination, the pollen germinates. The pollen tube enters the stigma and grows through the style. The generative nucleus undergoes a mitotic division to produce two male gametes. The pollen tube enters the ovule through a small pore, micropyle and discharges the two sperms into the embryosac.

2. ***Megagametogenesis*** **:** The nucleus of a functional megaspore divides mitotically to produce four or more nuclei. Now megaspore nucleus undergoes three mitotic divisions to produce eight nuclei. Three of these nuclei move to one pole and produce a egg cell and two synergid cells. Another three nuclei migrate to the opposite pole to give rise to antipodal cells. The two nuclei remaining in the centre is called polar nuclei. The megaspore thus develop into a mature embryo sac. The development of embryo sac from a megaspore is known as megagametogenesis.

Fertilization: Fusion of one of the spores with the egg cell, producing a diploid zygote is known as *fertilization*. The fusion of the remaining sperms with the secondary nucleus, leading to formation of a triploid primary endosperm nucleus, is termed as *triple fusion*. The zygote divides mitotically to produce a diploid embryo.

(*b*) Asexual reproduction

Asexual reproduction – Does not involve fusion of male and female gametes.

(*a*) *Vegetative reproduction:*

1. *Underground Stems*
   - Tubers • Bulb • Rhizomes
2. *Subaerial Stems*
   - Stolon • Suckers • Runners.
3. Bulbils – garlic
4. Artificial vegetative reproduction
   - grafting, layering • Tissue culture.

(*b*) *Apomixis* – Seeds are formed but the embryos develop without fertilization.

1. *Adventive embryony* – embryo by nucellus, integeument and chalaza :
   - There is no formation of embryosac.

   e.g. Mango, Citrus.
2. *Apospory* – non reduced embryo by vegetative cells other than embryosac.
   - Here embryosac is formed.

   e.g. Heiraceum, Ranunculus.
3. *Diplospory* – embryo formed by embryosac cells.

4.1.4 Mode of Pollination

SELF POLLINATION – Pollen fall on to the stigma of the same flower mechanisms promoting self pollination.

Mechanism promoting self pollination.

- *Cleistogamy* – Flowers do not open *e.g.* cereals.

- *Chasmogamy* – Flowers open, but only after pollination has taken place *e.g.* in cereals.
- Position of anthers in relation to stigmas ensure self pollination *e.g.* tomato and brinjal.
- Flowers open but the stamens and the stigma are hidden by other floral organs *e.g.* pea and soybean.

Cross Pollination :

Mechanism promoting cross pollination :

1. *Dicliny* : The flowers are either staminate (male) or pistillate (female).

(*a*) *Monoecy* – Staminate and pistillate flowers occur in the same plant, either in the same inflorescence *e.g.* castor, mango.

(*b*) *Dioecy* – The male and female flowers are present on different plants *e.g.* papaya, datepalm.

2. *Dichogamy* – Stamens and pistils of hermaphrodite flowers mature at different times.

(*a*) *Protogyny* – Pistils mature before stamens *e.g.* bajra.

(*b*) *Protandry* – Stamens mature before pistil *e.g.* maize.

3. Stigmas are covered with a waxy film *e.g.* alfalfa.

4. Self-Incompatibility.

5. Male sterility.

4.1.5 Male Sterility

Male sterility is a reproductive deficiency of some plants where male organs in hermaphrodite flowers are rendered defunctioning. It relates particularly to non-viable pollen grains which are formed through a chain of vital processes during microsporogenesis.

Characteristics Features

1. Better quality of hybrid seeds since human errors committed during detasselling are completely eliminated.
2. Larger quantity of hybrid seeds, since the production area no more remains a limiting factor, as also the manpower.
3. The possibilities of losses due to weather hazards at the time of detasselling are completely dissipated.
4. There is considerable wastage of plant energy in producing pollen grains in normal plants.
5. Male sterility can be utilized for developing mass reservoirs, particularly in outbreeders for the conservation of variability.
6. Male sterile lines can serve as tester genotypes for assessing the combining ability of a large number of stocks.
7. Male sterile plants produce seed only on cross pollination with functional pollen of other plants.
8. Presence of male sterility leads to heterozygocity in a species as it promotes out breeding and reduces homozygosity due to elimination of inbreeding.
9. It results from the action of nuclear genes or cytoplasmic genes or both Male sterility is caused due to pollen or anther abortion.
10. It occurs in nature from spontaneous mutations as well as can be induced artificially by chemical or physical mutagens.
12. It can be observed in all diploid species in crop plants both wild and cultivated if properly investigated.

Types of male sterility

It is of five types —

(*i*) Genetic male sterility

(*ii*) Cytoplasmic male sterility
(*iii*) Cytoplasmic genetic male sterility
(*iv*) Chemical induced male sterility
(*v*) Transgenic male sterility.

Genetic Male Sterility (GMS)

The pollen sterility is caused by nuclear genes is termed as genic or genetic male sterility.

Reported in barley, wheat, maize, cotton, sorghum, lucerne, cucurbits, tomato and sugar beet.

The characteristics of GMS are :

1. Male sterility genes are usually recessive.
2. In majority of cases, sterility is caused by single gene. In few cases two or more genes control male sterility.

Merits

- GMS usually recessive and monogenic. Hence fertility restoration in the hybrid and crossing plants are relatively easy.
- Can be used for the production of hybrid seed both in seed as well as vegetatively Propagated crops :
- Generally doesn't have undesirable agronomic characters.
- Requires less area and labour, because the breeder has to maintain only A and B lines.

Demerits

- Less stable
- 50% plants are fertile, which have to be removed every year. This increases the cost of hybrid seed.

Cytoplasmic Male Sterility

Features—

1. Plants carrying particular type of cytoplasm are male sterile, but will produce seed if pollinators are present. The F_1 seeds (cross seeds) produce only male sterile plants, because their cytoplasm is derived entirely from the female gamete.
2. It consists of only A line and B line. A is male sterile and B is male fertile. For all other characters A and B lines are similar or isogenic.
3. CMS can't be utilised for hybrid seed production without the use of restorer line because F_1 seeds produce only male sterile F_1 plants.
4. CMS can be used for development of hybrids in vegetatively propagated crops and ornamental crops, were grain is not the economic part.
5. CMS is not influenced by environmental factors such as low or high temperature.

Merits

- Highly stable and is not influenced by environmental conditions such as temperature and day length.
- CMS requires less area because breeder has to maintain only A and B lines.

Demerits

- Can't be used for development of hybrids in those crops where seed is the economic product.
- Can be used only in asexually propagated crops viz., sugar cane, potato and forage crops.
- CMS is solely governed by plasma genes Therefore, it is impossible to restore fertility in the hybrid.
- CMS line may have inferior agronomic performance.

Cytoplasmic genetic male sterility

When pollen sterility is controlled by both cytoplasmic and nuclear genes it is known as cytoplasmic genetic male sterility.

Features —

- Male sterility is controlled by the cytoplasm and nuclear gene.
- System includes A, B and R lines. A is male sterile line, B is similar to A in all features but its male fertile and R restores fertility in F_1 hybrid, hence its called restorer line (R). 'B' line is used to maintain the fertility and so its called maintainer line 'B' line.

Development of new male sterile and restorer lines :

New male sterile lines are developed by following the same procedure as that CMS. But the nuclear genotype of the pollinator strain must be rr in this transfer, otherwise fertility would be restored.

Restorer lines are developed by following the procedure :

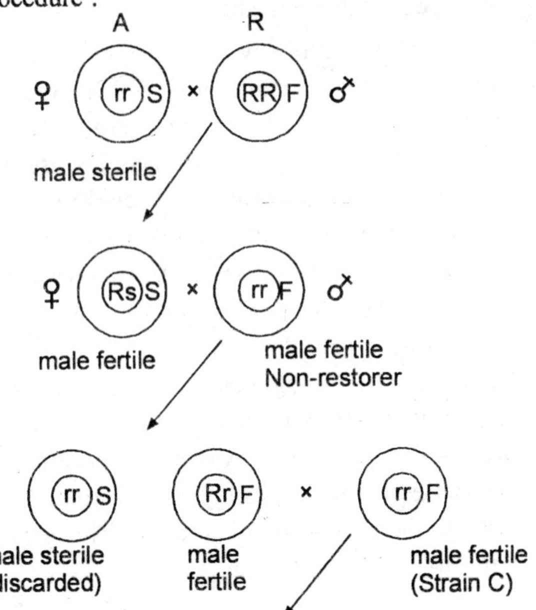

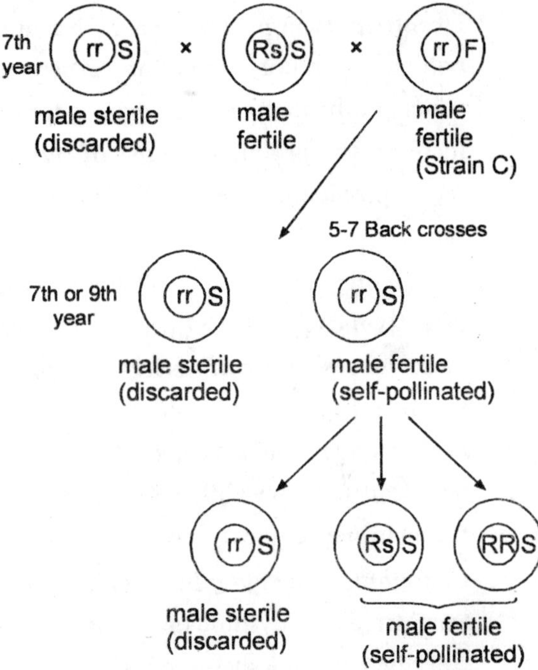

Utilization

- Commercially used to produce hybrid seed in maize, bajra, sunflower, cotton, jowar etc.
- Widely used for hybrid seed production in both seed as well as vegetatively propagated species.
- CGMS is highly stable and reliable as its not affected by environmental factors.

Limitations :

1. *Undesirable effects of cytoplasm :* Male strile cytoplasms have undesirable effects.
2. *Unsatisfactory fertility restoration :* Restoration of fertility is not satisfactory and so these sources can't be used in the hybrid seed production.

3. *Unsatisfactory pollination :* Natural pollination is not satisfactory except in wind. Pollinated species (maize) which increase the production cost of hybrid seed by reducing the production of the same.
4. *Modifying genes :* Reduce the effectiveness of CGMS. During backcrosses while transforming male sterile cytoplasms, the nuclear genetic background may also be disturbed, which may lead to some pollen production by the male sterile lines.
5. *Contribution of cytoplasm by sperm :* Sometimes, cytoplasm may be contributed by the sperm, which may lead to breakdown of male sterlity during long run.
6. *Non-availablity of a suitable restorer line :* In crops like wheat, polyploid nature of the crop and undesirable linkages with the restorer genes make it very difficult to develop a suitable restorer (R) line.

Chemically Induced Male Sterility

This type of male sterility can be induced by chemicals called male gametocides. The major difference between mutagen induced and gametocide induced male sterility is that the former is heritable and the latter is non-heritable.

(Eg) for male gametocides : (Chemical hybridizing agents (CHA))

Ethanol – Rice, sugarbeet, wheat.

GA_3 – Maize, onion, rice.

MH – Cucurbits, onion, tomato, wheat.

NAA – Cucurbits.

Sodium Methyl Arsenate – Rice.

Advantages

1. It can be used with any variety as per need and there is no need of seed for isolation or transfer of cytoplasmic or genetic male sterility.
2. Maintainer and restorer lines need not be developed.
3. Saves time, labour and money.
4. This method is being used in China to develop rice hybrids.

Disadvantages

1. Pollen abortion is incomplete and variable.
2. Treatments are effective only at specific stage of crop growth and effect is short lived.
3. Female fertility may also be reduced.
4. Production of undesirable side effects.
5. High cost of CHAs and their repeated application increases the cost of hybrid seed.

Transgenic Male Sterility

The male sterility which is induced by the technique of genetic engineering which is induced by the technique of genetic engineering is called TMS so the foreign gene (transgene) is used for the induction of male sterility. The gene responsible for male sterility is isolated from microorganisms. It comes under genetic male sterility is heritable.

The TMS has been induced in tobacco by transferring a gene from *B.* amylolique faciens. Two types of transgenes are used, one gene (Barnase) causes male sterility, hence this gene is integrated in '*A*' line. Another gene (Barstar) suppresses the male sterility gene (barnase) and leads to the restoration of fertility. This gene is transferred to restorer line.

When a cross is made between *A* and *B* line the F_1 is fertile. But this is an expensive method of obtaining male sterility. Maintenance of male sterility is a problem.

4.1.6 Self-Incompatibility (SI)

SI was first reported by Koelreuter. Here, pollen grains fail to germinate on the stigma of the flower that produced them. If some pollen grains do germinate, pollen tubes fail to enter the stigma. In many species, the pollen tubes fail to enter the stigma. In many species, the pollen tubes enters the style, but they grow too slow to effect fertilization before the flower drops. Sometimes fertilization is affected, but the embryos degenerate at a very early stage. All these conditions combinely referred as self-incompatibility.

Types

SI can be classified on the basis of

(A) flower morphology.

I. Heteromorphic System
- Distyly
- Tristyly

II. Homomorphic System
- Diallelic
- Poly allelic

(B) Genes involved
(*a*) mono-allelic
(*b*) Diallelic
(*c*) Poly allelic.

(C) Site of expression
(a) Stigmatic
(b) Stylar
(c) Ovarian.

(D) Pollen cytology
(a) Binucleate
(b) Trinucleate

I. Heteromorphic system : Here, flowers of different incompatibility groups are different in morphology.

(SI) in primula two types of flowers Pin and Thrum is called distyly.

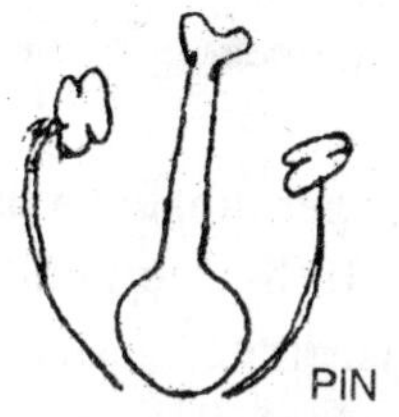

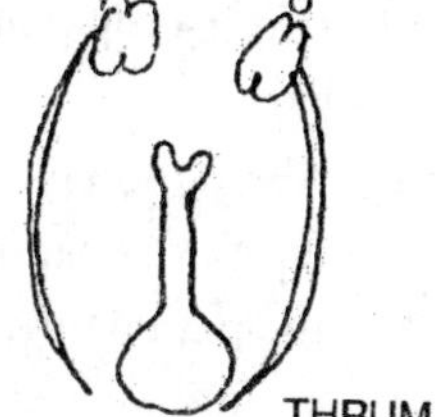

- Pin and thrum flowers produced on different plants. The only compatible mating is between pin and thrum flowers, which is governed by a single locus S; Ss produces thrum, while ss produces pin flowers.
- Here the incompatiblity reaction is governed by the genotype of the plant producing them.
- Allele '*S*' is dominant over '*s*' and incompatibility system is hetero-morphic-sporophytic.
- Results of four types of crosses in primula.

Pin × Pin	ss × ss	Incompatible
Thrum × Thrum	Ss × Ss	Incompatible
Thrum × Pin	Ss × ss	1 Thrum : 1 Pin
Pin × Thrum	sS × Ss	1 Thrum : 1 Pin

Tristyly

In tristyly anthers and style have three positions, short, medium and long. (Eg) members of lythraceae.

Three positions of style are genetically controlled by two genes (*Ss* and *Mm*). The '*S*' gives rise to short style, *sM* to medium style and *sm* to long style. The compatible matings in primula are long × medium, long × short and medium × short.

Homomorphic System

The incompatibility reaction of pollen may be controlled by the genotype of the plant in which its produced (sporophytic) by its own genotype (gametophytic control). This system is very much important in crop plants. Its can operate in many ways like

1. The pollen grains do not germinate on the stigma of same flower. If they germinate the pollen tube fails to penetrate the stigma as in a rye, cabbage and radish.
2. Pollen grains may germinate, but there is retardation of pollen tube growth.
3. There is slow rate of pollen tube growth and it rarely reaches the ovule.
4. Pollen tube growth may be normal, but it does not release the male gamete.

Gametophytic System (GS)

Described by *East and Mangelsdorf*

• Genotype of plant (Sporophyte)	$S_1 S_2$	$S_3 S_4$
Genotype of pollen and gametes	$S_1 S_2$ ↓ ↓	$S_3 S_4$ ↓ ↓
Incompatibility reaction of pollen grains	$S_1 S_2$	$S_3 S_4$
Incompatibility reaction of style	$S_1 S_2$	$S_3 S_4$

- The incompatibility reaction of pollen is determined by own genotype and not by the genotype of the plant on which it is produced. This is because the biochemical substance involved in the S_1 reaction of the pollen is produced after meiosis.

Sporophytic system (ss)

Discovered by *Hugher and Babcock* (1950).

In sporophytic system SI reaction is governed by single gene '*S*' with multiple alleles. In general, the number of '*S*' alleles is considerably larger in the gametophytic system than in the sponophytic system.

The incompatibility reaction of pollen is governed by the genotype of the plant on which the pollen is produced and not by the genotype of the pollen.

In sporophytic system, the 'S' alleles may exhibit dominance individual action (codominance) or competition.

- Polygenes (modifying genes) are known to increase as well as decrease the activities of 'S' alleles both in the GS and SS systems.

Mechanism of SI

1. *Stigmatic inhibition (pollen stigma interaction)* : Inhibition of pollen germination or pollen tube growth occurs on the stigma. Normally it is confined to sporophytic system. This type of inhibition is found in Radish, Cabbage, Cauliflower, Sunflower etc.
2. *Stylar inhibition* : Inhibition of pollen tube growth in the stylar region is a common feature in SI plants with binucleate pollen. This inhibition is restricted to GS. Pollen tubes usually swell and burst of the apex in the upper region of the incompatible style. (eg) Beta, Helianths, Fagopyrum.
3. *Ovarian inhibition (pollen tube style interaction)* : Self incompatibility reaction occurs only when the pollen tubes reacts the ovary. Such inhibition occurs in the ovary in the species those have hollow styles. (Eg) Lilium, Anona etc.

Elimination of SI

1. Doubling the chromosome number in case of single locus gametophytic system.
2. Isolation of self fertile (S_f) mutations.
3. Self-Incompatibility alleles may be transferred from related or same species through back cross.

Methods to Overcome SI

Temporary self fertility can be achieved in many ways so then the SI is fully functional in the selfed progeny. Such fertility is called Pseudo fertility and is brought about by

1. Bud pollination.
2. End of season pollination.
3. Mentor pollen technique (recognition pollen).
4. High temperature exposure to pistils – Tomato, Brassica.
5. Irradiation by x-rays/γ-rays – Solanaceae.
6. Grafting.
7. Double-pollination.
8. Other techniques such as.
   (*i*) Injecting styles with immuno suppressant.
   (*ii*) Treatment of pistil with phytohormones.
   (*iii*) Steel brush pollination.

Seed coat: Integuments of ovules harden as tough protective seed coats. The micropyle remains as a small pore in the seed coat. This facilitates entry of oxygen and water into the seed during germination.

Albuminous seed: Seeds which return a part of endosperm are called albuminous seeds *e.g.* wheat, maize, barley, castor, sunflower etc.

Non-albuminous seed: Seeds with no residual endosperm are called non-albuminous seeds. Endosperm is completely consumed during embryonic development in these seeds, *e.g.*, pea, groundnut etc.

Perisperm: In some seeds, remnants of nucellus are also persistent *e.g.*, black pepper and beet. This residual persistent nucellus is the perisperm.

Table 1. Development of fruits

Ovary	*Fruits*
Ovule	Seeds
Integuments	Tests (seed coat)
Nucleus	Perisperm
Two polar nuclei + sperm nucleus	Endosperm 3n
Egg nucleus + sperm nucleus	Zygote — embryo (diploid 2n)

4.1.7 Centres of Origin (Vavilov)

Name of the centre	*Areas covered*	*Plants*
1. Chinese centre	Mountaineous regions of central western China	Buck wheat, Soybean, Onion, Bamboo, Cherry, Citrus, Sugarcane Tea etc.
2. Indian centre		
(*a*) Main centre	Assam and Myanmar	Rice, Sugarcane, Mango, Orange, Jute, Coconut Pepper etc.
(*b*) Indo Malayan	Indo-China and Malayan Archipelago	Banana, Sugarcane, Clove, Coconut, Manila hemp, Pepper.
3. Central Asiatic	Kashmir, NWFP of Punjab, Afghanistan, Uzbekistan, Tian-Shan	Wheat, Pea, Beans, Cotton, Carnet, Garlic, Apple, Almond, Apricot.
4. Near Eastern	Interior of Asia-minor, Transcaucasia, Iran, highlands of Turkmenistan.	Cherry, Pomegranate, fig, Almond, Clove,
5. Meditrenanean	Borders of Mediterranean sea	Cabbage, Turnip, Clove, Mustards, Duram and emmer wheat, Pepper etc.
6. Abyssinian	Ethiopia, Somalia, Eritrea	Wheat and Barley, Sesame, Castor, Coffee, Okra.
7. South Mexican and Central American	South Mexico, Guatamala, Honduras, Costarica	Maize, Bean, Sweet-potato, Papaya, Tobacco, Upland cotton.
8. South American	High mountainous areas	Many potato spp, Tomato, Pumpkin,
(*a*) Peruvian-Bolivian		Egyptian cotton, Tobacco.
(*b*) Chile		Common potato
(*c*) Brazilian-Paraguian		Groundnut, Cassava, Pineapple, Rubber and Cashew nut.

Primary gene centre : Native wild relatives, dominant genes, diversing due to ancient cultivation.

Secondary gene centre : Devoid of relatives recessive characters, ecological diversity, farming practice, human migration patterns and internal biological dynamics of hybridisation, segregation-selection.

4.2 BREEDING METHODS

4.2.1 A generalized classification of selection procedure

1. Intra-population selection based on additive variance

(*a*) Based on sample size and degree of inbreeding

(*i*) Individual or pure-line or pedigree selction

(*ii*) Mass pedigree selection

(*iii*) Mass or bulk selection

(*b*) Based on family structure and relatives

(*i*) Individual selection

(*ii*) Family selection

(*iii*) Within-family selection

(*iv*) Combined selection

(*c*) Based on number of traits selected

(*i*) Tandem selection

(*ii*) Independent culling level

(*iii*) Index selection

2. Inter-population selection based on non-additive variance

1. Reciprocal recurrent selection
2. Recurrent selection for gca
3. Recurrent selection for sca

II. Stabilizing selection.

III. Disruptive selection.

4.2.2 Plant Introduction

Taking a genotype or a *group* of genotypes of plants into a new environment where they are not being grown before.

The materials are introduced from other countries or continents.

Movement of crop variety from one environment into another within the same country is also introduction.

e.g. within the country introduction are the wheat in West Bengal and rice in Punjab.

Classification

(i) Primary introduction

When the introduced variety is well suited to the new environment it is released for commercial cultivation without any alteration in the original genotype.

e.g. dwarf wheat variety Sonar 64, Lerma Ròjo dwarf rice variety - T.N. 1, IR8, IR-28.

(ii) Secondary Introduction

The introduced variety may be subjected to selection to isolate a superior variety. Alternatively it may be hybridized with local variety to transfer one or few characters from this variety to the local ones.

e.g. Kalyan Sona and Sonalika wheat variety selected from materials introduced from CIMMYT, Mexico.

Procedure for Plant Introduction

1. *Procurement of Germplasm*
   - Gift, Purchase or Collection
   - Propagules
2. *Quarantine*
   - Plant produce imported for consumption
   - Plant propagules imported for cultivation and Research
   - Quarantine procedure
3. *Cataloguing*
4. *Evaluation*
5. *Multiplication*
6. *Distribution*

Purpose or Objectives of Plant Introduction

- Provide an entirely new crop species *e.g.* Maize, Potato, Tobacco.
- Directly released as New varieties *e.g.* Sonara 64, Larma Rojo
- Utilization in crop improvement programmes *e.g.* Pusa Ruby tomato was derived from Meeruty × Sioux

- Saving a crop from a disease or pest coffee was introduced in South America from Africa to prevent losses from leaf rust.
- Utilisation in scientific studies.
- Used for Aesthetic value.

Merits :

- Provides entirely new crop plants.
- Provides superior variety.
- Introduction are the only feasible means of collecting germplasm and to protect variability from genetic erosion.
- Very quick and economical method of crop improvement.
- Crop species may be introduced in new disease from areas to protect them from damage *e.g.* coffee and rubber.

Demerits :

- Weeds entering through plant introduction *e.g.* phyllaries minor.
- *Diseases :* Late blight of potato from Europe, coffee rust from Srilanka.
- Bunchy top of banana from Srilanka.
- *Insect pest :* Potato tuber moth from Italy, wolly aphids of apple.
- *Ornamental turned weeds :* Water hyacinth, lantana camara.
- *Threat to ecological balance :* Eucalyptus spp introduced from Australia causes a rapid depletion of the sub soil water reserves.

4.2.3 Selection

Selection operates only genetic variability occuring in plants or crops.

1. *Mendelian variation :*

   All variations occur through gene mutation.

   macromutation ⇒ maize, Brassica

   micromutation ⇒ *e.g.* Rice, barley, tomato.

2. *Interspecific hybridization :*

   ⇒ less contribution to evolution → sterile progenies

   ⇒ introgressive hybridization ⇒ importance for evolution

   e.g. Tripsacum × primitive maize

   maize

3. *Polyploidy :*

   autopolyploidy ⇒ more vigour, robust

   1. auto triploids ⇒ Banana, water melon ($3x$)
   2. auto tetraploids ⇒ potato ($4x$)
   3. auto hexaploid ⇒ sweet potato

   Allopolyploidy ⇒ 50% of plants evolved

   Allohexaploid ⇒ wheat, triticale.

   Allotetraploid ⇒ cotton, tobacco

Germplasm

(*a*) Land Races. Naturally selected strains/ species

(*b*) Obsolete variety Systematic breeding but not cultivated now.

(*c*) Varieties in cultivation

(*d*) Wild forms and relatives

(*e*) Breeding lines.

4.2.4 Pureline selection

A large number of plants are selected from a self pollinated crop. The selected plants are harvested individually. The selected individual plants are grown in individual rows then evaluated and best progeny is selected.

Characters of pureline :

(1) All the individuals are homozygous.

(2) Variation in pureline is due to environment.

(3) Genetically become variable with time.

Procedure of Pureline Selection

Pureline selection consists of four major steps, *viz.* (1) selection of a heterogeneous population from which purelines have to be isolated, (2) Isolation of purelines by individual plant selection, (3) Testing of purelines in field trials, (4) Release of the best pureline as a variety. The year wise procedure is given as follows :

First year. An old variety or land race is used for pureline selection. Single plants are selected from the heterogeneous population keeping in view the objective of selection. The number of individual plants to be selected may vary form 200 to 1000 in various crops.

Second year. The progeny of each selected plant is grown separately in few rows and evaluated for the character under consideration. The top 15-20 progenies are selected and seed of all plants in each progeny is bulked which constitutes strains.

Third year. The strains constituted in second years are evaluated in field trials for 2-3 years for yield performance. In India, the selected entries (strains) are evaluated in All India Coordinated Crop Improvement Project. The best genotype is identified on the basis of yield performance.

Fourth to Tenth year. The best performing strain is released and notified as a variety. Then the breeder, foundation and certified seeds are produced. The production of certified seed takes two years after release of a variety. Thus the seed of new variety reaches the farmers in tenth year.

Merits :

(1) Uniform.

(2) Maximum improvement possible.

(3) Variety is easily identified in seed certification programme.

Demerits :

(1) Less adaptable.

(2) More time, labour and money.

(3) Upper limit and improvement is set by the genetic variation present in original population.

4.2.5 Pedigree method :

Individual plants are selected from F_2 and subsequent generations and their progenies are tested. During this process details about the plants selected in each generation is recorded in pedigree record.

Procedure steps :

Breeding Procedure

First parents are selected keeping in view the breeding objective. The cross is made between selected parents. The F_1 material is grown using wide spacings. The dominance behaviour for various characters is recorded. In F_2 also the material is grown using wide spacings. Individual plant selection is practised in F_2. The progeny of each selected plant is grown separately which constitutes F_3 generation. In F_3 and F_4 generations, selection is practised within and between families. From F_5 to F_8 between progeny selection is practised and superior progeny are identified and isolated in F_8. These progeny constitute strains. These strains are evaluated in replicated multilocation trials for a period of 3-5 years. Based on superior performance, the strain is released as a variety. Thus release of new variety by this method takes 14-15 years. The number of plants to be grown and selected in each generation is presented. This is a general procedure of pedigree method. The number of plants to be grown and selected in each generation is not fixed. It may slightly vary from crop to crop.

Merits :

- Gives maximum opportunity to the breeder to use his skill.
- Well suited for characters which are simply inherited.
- Transgressive segregants can be easily identified through records.

- Information about inheritance is precisely obtained.

Demerits :

- Maintenance of pedigree record is time consuming.
- The success in this method is largely dependent on skill of the breeder.
- Selection for yield in F_2 and F_3 is ineffective.

Achievements :

Rice	: Krishna, Ratna, Sabarmati, Padma, jaya, Bala, Kaveri, etc.
Wheat	: HD 2281, HD 2285, HD 2380, ND 2402, Janak.
Cotton	: LH 900, LH 1556, F 846, Vikas, Khandwa 3, Sharda, Abhadita, Sahana, Anjali, Surabhi, Suvin etc.
Pigeonpea	: T21, Prabhat
Greengram	: T2, T44, T51, Sheela, etc.

4.2.6 Backcross method

Backcross refers to crossing of F_1 with either of its parents when the F_1 is crossed with homozygous recessive parent, it is known as test cross. A system of breeding in which repeated backcrosses are made to transfer a specific character to a well adapted variety for which the variety is deficient is referred to as backcross breeding.

Main features :

Application – It is generally used to improve specific character of a well adapted variety for which it is deficient such as resistance to a specific disease.

Parent material – In this method two types of parents like recipient parent and donor parent involves. The parent which receives a desirable character is known as recipient parent. The parent which donates desirable character is known as donor parent.

Genetic constitution – The new variety resembles the parent variety in all the characters except for the character under transfer.

Number of backcross – 5 to 6 backcross are sufficient to retain genotype of original variety with new character.

Procedure :

Transfer of dominant gene

Suppose wilt resistance in cotton is controlled by a dominant gene (*RR*). The donor parent is a strain (*B*) from the germplasm. The resistance has to be transferred to an adapted variety (*A*) which is susceptible to wilt. The adapted variety (*A*) will be used as recurrent parent and strain (*B*) as donor parent. The F_1 will be wilt resistant but heterozygous (*Rr*). Backcrossing of F_1 (*Rr*) with susceptible variety (*rr*) will produce resistant and susceptible plants in equal number in BC_1 F_1 (1 *Rr* : 1 *rr*). The resistant cotton plant (*Rr*) can be identified by growing the material in wilt sick plot. The resistant plants (*Rr*) are then backcrossed to the adapted variety. Generally, 6-8 backcrosses are sufficient to obtain identical plants to adapted except for the added genes for wilt resistance. The wilt resistant plants are heterozygous (*Rr*). They are selfed for one genereation to obtain homozygous (*RR*) resistant plants. All the resistant true breeding plants are bulked and new variety is released. The variety developed in this way is identical to the adapted variety (*A*) expect for wilt resistance.

Transfer of recessive gene

Suppose wilt resistance in cotton is governed by a recessive gene (*rr*). In such case, a progeny of each backcross will segregate into two genotypes (*RR* and *Rr*) which cannot be identified. Therefore, it is necessary to self the population after each backcross obtain resistant homozygous recessive plants (*rr*). The resistant plants are identified by owing the F_2 material in wilt sick plot. The resistant plants are backcrossed with adapted variety. Here each backcross is followed by one selfing, whereas with dominant one continuous backcrosses are made.

Merits :

- It retains all desirable characters of a popular adapted variety.
- It is a useful method for transfer of oligogenic character like disease resistance.
- The male sterility and fertility restorer genes are transferred to various agronomic bases.
- It is the only breeding method which is used for interspecific gene transfer.

Demerits :

- The backcrosses have to be made for 6-8 generations.
- Sometimes undesirable character also transferred to the new variety.

Achievements :

Backcross method has been widely used for the development of disease resistant varieties in both self and cross pollinated species. It has also been used for interspecific gene transfer and development of multiline varieties in self pollinated species. Several varieties resistant to various diseases have been developed by this method in wheat, cotton and several other crops. In cotton varieties V797, Digvijay, Vijalpa and Kalyan which belong to *Gossypium herboceus* have been developed by backcross method.

4.2.7 Mass selection

Definition : A number of plants are selected on the basis of their phenotype and the open-pollinated seed from them is bulked together to raise the next generation.

Procedure :

First year : (*i*) Several plants selected on the basis of phenotype (*ii*) open pollinated seed from the selected plants harvested and bulked.

Second year : (*i*) Bulk seed from the selected plants grown (*ii*) Mass Selection may be repeated.

Third to fifth year : Main yield trials are carried out to determine the performance of superior mass selected strains in comparison with standard varieties as checks.

Sixth to eight year : Trials on the cultivator's holding.

Eighth year : The variety is produced and named.

Application :

- It cannot bring any new change in the phenotypes of the population but can only sort out the different types already existing it.
- It can be applied to any crop which posesses some amount of heterozygosity *i.e.* variability, which is the basis of mass selection.

In cross pollinated crops mass selection is more effective with following types of crop materials in which heterozygosity is found invariably :

- Wild varieties
- Local varieties growing on farmer's fields
- Introduced varieties
- Mixed varieties of a crop

Uses :

- Production of new varieties
- Purification of unimproved varieties
- Maintaining the purity of a variety

Achievements :

- First commercial types of hirsutum cotton in India *e.g.* Cambodias

- Variety of Groundnut such as TMV-1, TMV-2
- Maize – T13, Jaunpur, Darjeeling white round
- Bajra – Pusa moti
- Potato – K122

Demerits :

Varieties developed show variation and are not uniform as pureline varieties.

In the absence of progency test, it is not possible to determine whether the selected plants are homozygous.

Improvement is less than that of pureline selection.

4.2.8 Single seed descent method

A breeding procedure used with segregating populations of self pollinated species in which plants are advanced by single seeds from one generation to the next iş referred as single seed descent method.

Procedure :

Modified from Bulk method (one seed is selected)

Merits :

- This is a simple, convenient, less expensive and time saving method.
- Large number of crosses can be evaluated by this method, because less space and labour is required in each generation.
- This method is able to retain variability in a breeding population.

Demerits :

- It does not provide opportunity to select for superior plants till F_5 generation.
- The frequency of getting desirable genotypes in the advanced generation is reduced.
- The identity of superior F_2 plants cannot be maintained in this method.

4.2.9 Bulk method

In this method F_2 and subsequent generations are harvested as bulk to grow the next generation. The duration of bulking may be 6-7 generations. Selection can be made in each generation but harvesting is done as bulk. At the end of bulking period single plant selection is made and tested for yielding ability.

Procedure :

The bulk breeding method consists of four important steps, *viz.* (1) bulking period, (2) progeny selection and isolation of superior progeny, (3) multilocation trials of superior progeny, and (4) release of the best progeny as a variety. These steps are briefly discussed below :

Bulk period. The F_1 plants are grown and their F_2 seeds are harvested in bulk. The F_2 plants are raised from a sample of F_2 seeds and F_3 seeds are harvested in bulk. This process is repeated until the desired level of homozygosity is achieved. In general, bulk period is allowed upto F_5 generation. The material is subjected to biotic and abiotic stresses during bulking period to eliminate undersirable genotypes.

Progeny selection. In F_6, the material is space planted and individual plant selection is practised. The progeny of each selected plant is grown separately in F_7 and superior progeny are selected and isolated in F_7 and F_8.

Multilocation testing. The selected progeny constitute strains. In eighth year, preliminary yield trial is conducted. From 10th to 14th year multilocation testing is carried out and the best performing strain is identified on the basis of 3-4 years performance in the multilocation trials. The best strain is released and multiplied for seed distribution in the 15th year. Thus bulk method takes 15-16 years for release of new variety.

The bulk material should be tested under such environment which is expected to favour desirable genotypes. For example, screening for disease resistance should be carried out in the disease prone area. Moreover, the screening should be done under natural field condition rather than in green houses.

Merits :

- No pedigree record is maintained.
- If bulking period is longer natural selection operates and desirable genotypes selected.
- Chances of obtaining transgressive segregants more.

Demerits :

- Takes much longer time to develop a new variety.
- A large number of progenies are to be selected in each generation which requires much labour, time and space.
- We cannot get information on inheritance.

4.2.10 Recurrent selection

Recurrent selection is reselection generation after generation with inter-breeding of selects to provide for genetic recombination. It is used to improve the frequency of desirable alleles for a character in a breeding population.

Main features :

Application – It is more commonly used in cross pollinated species than in self pollinated species.

Base population – A heterozygous base population is required to start recurrent selection.

Important steps – (*i*) Selection of superior plants from base population (*ii*) selfing of selected plants (*iii*) growing progeny of selected plants in the next season (*iv*) intermating among progeny (*v*) bulking of crossed seed.

Use of end product – The population developed by recurrent selection can be used in producing homozygous inbreeds by selfing, in the production of hybrid varieties, in the production of synthetic varieties.

Impact – It is used to improve the frequency of desirable alleles for a character in a population.

Types of Recurrent Selection

1. Simple, Recurent Selection

A type of recurrent selection that does not include tester is referred to as simple recurrent selection. It is also known as phenotypic recurrent selection. This method is an extension of mass selection. Main features of simple recurrent selection are given below :

1. The tester is not used in this scheme.
2. It does not measure the combining ability.
3. The selection is based on phenotype or simple test.
4. This method is useful only for those characters which have high heritability.
5. This method requires only two seasons for the completion of one selection cycle.

2. Recurrent selection for GCA

First year. Superior plants for the character under improvement are selected from the base population. The selected plants are selfed and also crossed to a heterozygous tester having broad genetic base. The selfed seed is kept in cold storage.

Second year. The crossed seed is sown and the combining ability of the selected plants is evaluated and plants with good *gca* are identified.

Third year. The progeny of selected plants with good *gca* are grown from their selfed seed kept in cold store. These progeny are intermated in all possible combinations and their crossed seed is composited to form a new source population for further selection. This completes original selection cycle. In the same way another cycle can be completed in three years (4th to 6th year).

This is called first recurrent selection cycle. Many such cycles may be made to obtain desired results.

3. **Reciprocal Recurrent selection of SCA**

A form of recurrent selection that is used to improve the *sca* of a population for a character by using homozygous tester is referred to as recurrent selection for specific combining ability (RSSCA). It is also known as half sib recurrent selection with homozygous tester. This method was originally proposed by Hull in 1945.

Table 1 : Comparison of recurrent selection for *gca* and recurrent selection for *sca*

Particulars	*Recurrent selection for gca*	*Recurrent selection for sca*
Application	used to improve polygenic traits	Also used to improve polygenic traits
Basis of selection	Test cross performance	Test cross performance
Tester used	Heterozygous	Homozygous
Effectiveness	More effective with incomplete dominance	More effective with complete and overdominance
Condition of use	Used when additive gene action is important	used when non-additive gene action is important
Impact	It improves *gca* of a character	It improves *sca* of a character

4. **Reciprocal Recurrent Selection**

A form of recurrent selection that is used to improve both *gca* and *sca* of a population for a character using two heterozygous testers is known as reciprocal recurrent selection (RRS). It is also termed as recurrent reciprocal half sib selection.

First year. Several phenotypically superior plants are selected from population *A* and *B*. The pollen of some selected plants of *A* population is used to cross large number of randomly selected plants of population *B*. Similarly, pollen of some selected plants of *B* population is used to cross large number of plants of population *A*. All the plants of population *A* and *B* used as pollen parents in the crosses are selfed.

Second year. The progeny of test crosses made with pollen parents of *A* and *B* populations are evaluated in separate replicated trials. The superior progeny are identified.

Third year. The selfed seed of those *A* and *B* plants whose progeny were found among the progeny of *A* plants and also among the progeny of *B* plants. The crossed seeds of *A* block are composited in equal quantity to raise *A* generation. Similarly, crossed seeds of *B* block are bulked to raise B_1 generation. This completes original cycle of selection.

Fourth year. The A_1 and B_1 populations are grown from the composite crossed seeds of respective population obtained in third year. Then operations of first year are repeated.

Fifth year. The operations of second year are repeated.

Sixth year. The operations of third year are repeated.

The last three years constitute first cycle of reciprocal recurrent selection. Such selection cycles may be continued till the desired improvement is achieved.

Merits :

- It is an efficient breeding method for increasing the frequency of superior genes in a population.
- It helps in maintaining high genetic variability.

- Selection is made on the basis of test cross performance

Demerits :

- It is not used directly for the development of new varieties.
- It involves lot of selection, crossing and selfing work.
- It permits selfing which leads to loss of genetic variability.

4.3 OTHERS

4.3.1 Heterosis breeding

Heterosis hybrid vigour refers to the superiority of F_1 hybrids in one or more characters over its parents. Heterosis differs from luxuriance, as in the case of former, there is an increase of F_1 over parents in general vigour, yield and adaptation, while the later shows only an increase in vegetative growth and not the yield.

Heteroisis was first introduced by Shull in the year 1914.

Features of Heterosis

1. *Superiority over Parents* : Hybrid vigour/ Heterosis leads to Superiority in adaptation, yield, quality, disease resistance, maturity and general vigour over its parents.
2. *Confined to F_1* : Heterosis is confined only to the $F1$ generation of a cross. It declines and disappears in F_2 and subsequant generations of a cross as a consequence of segregation and recombination.
3. *Genetic Control* : Expression of heterosis is governed by nuclear genes. It results due to interaction between nuclear genes and cytoplasm.
4. *Reproducible* : Heterosis once identified can be easily reproduced in a definite environment.
5. *Association with SCA* : It has positive association. SCA is a measure of dominant variance and a significant amount of dominant variance is essential for heterosis breeding.
6. *Effect of heterozygocity* : Magnitude of heterosis associated with heterozygocity because the dominance variance is associated with heterozygocity.
7. *Conceals recessive Genes* : In case of heterosis, the deleterious effect of recessive genes are covered by the favourable effect of dominants.
8. *Low frequency* : The frequency of desirable heterotic combination is very low. All the F_1 crosses donot show desirable heterosis.

Fixation of heterosis :

1. *Vegetative propagation :* Which is based on mitosis. As a result genotype is transmitted intact to all its vegetative progeny. In crops like, sugar cane, potato, sweet potato, banana, where vegetative propagation is a rule, hybrids can be easily developed.
2. *Apomixis :* Here the seed develops without fertilization. If F_1 and its progeny produce only apomictic seeds having 2n embryos, the F_2 and subsequent generations will be identical in genotype to the F_1 and there by lead to the fixation of heterosis.
3. *Balanced lethal system :* In plants like, *Oenothera*, the homozygotes will die as they are lethal and only heterozygotes survive, which results in the fixation of heterosis.
4. *Polyploidy :* Heterosis can also be fixed by chromosome doubling or polyploidy especially in interspecific and intergeneric hybrids. For *e.g.* the heterosis in wheat-rye cross can be conserved in

amphidiploid hybrids through chromosome doubling.

Factors affecting heterosis :

I. *Mode of pollination :* Level of heterosis higher in cross pollination than self pollination crops.

II. *Genetic diversity of parents :* Higher heterosis associated with crosses of more distantly related parent and maximum heterosis occurs at optimum or intermediate level of parental diversity.

III. *Genetic base of parents :* If the genetic base is broad, then the heterosis will be higher.

IV. *Adaptability of parents :* If the parents are having wider adaptability, then there will be the occurrence of heterosis.

Advantages

- Higher yield potential.
- More uniform and attractive.
- Can be developed in cross pollination and self pollination crops.
- Can be possible to reconstitute the hybrid with same genotype.

Disadvantages

- Fresh seed to be produced every year.
- Hybrid seed is costlier.
- Cultivation require more inputs.
- Requires more technical skill.

4.3.2 Hybrids

Hybrids are the first generation from crosses between two purelines, inbreeds, open pollinated varieties of other populations that are genetically not similar.

Kinds of hybrids :

1. Single cross hybrids crossing two inbreeds or purelines, A × B.
2. Three way cross hybrid. A cross between a single cross hybrid and an inbreed (A × B) × C.
3. Double cross hybrid. Double cross involving the four inbreeds A, B, C and D.
4. Double top cross hybrid. Double cross hybrid cross with open pollinated variety.

Development of inbreeds :

1. By inbreeding , selfing etc.
2. Development of inbreeds from haploids.

Evaluation of inbreeds

1. Phenotypic evaluation. Based on phenotypic performance. Highly suitable for characters with high heritability.
2. Top cross test. The selected inbreeds will be crossed to a tester parent with wide genetic base. The cross progenies will be evaluated in replicated progeny rows.
3. Single cross evaluation. The developed inbreeds can be crossed and the single crosses can be estimated in replicated trial standing hybrids tested over years in different locations, then released.

Production of hybrid seed

Methods

1. Hand emasculation and dusting-cotton, tomato.
2. Use of male sterile lines

   (*a*) Cytoplasmic male sterility - ornamentals.

   (*b*) Genic male sterility - redgram, castor.

   (*c*) Cytoplasmic genic male sterility - jowar, bajra, rice.
3. Use of self incompatibility.

Merits :

- Hybrids from single cross varieties are more uniform than open pollinated varieties, synthetic and composite varieties.

- Hybrids produced in both self pollinated and cross pollinated crops.
- It exploits both self combining ability and general combining ability.

Demerits :

- Seeds are to be changed every year.
- Needs technical skill.
- In cross pollination crops require long isolation distance.

4.3.3 (*a*) Synthetic varieties

A synthetic variety is produced by crossing in all combinations a number of inbreeds that combine well with each other. Once synthesized, a synthetic is maintained by open pollination in isolation.

Steps involved in the production of synthetic varieties.

Procedure of developing synthetic cultivar from already available inbred lines

Year	*Main Breeding Activity*
First	Selection of inbred lines, crossing with a common tester, harvesting top crossed seed separately.
Second	Evaluation of top crosses in replicated trial using standard hybrid or open pollinated variety as check. Identification and selection of good general combining inbred lines based on top cross performance.
Third	Making all possible single crosses among inbred lines selected for good gca and harvesting crossed seed of single crosses separately.
Fourth and Fifth	Mixing seed of single crosses in equal quantity and seed multiplication by open pollination in isolation for one or two generations.
Sixth	Release as a new synthetic variety and distribution of seed to the farmers for commercial cultivation.

Merits :

- Farmer can maintain his synthetic variety for more seasons which is not possible in hybrids.
- Because of wider genetic base the synthetics are more stable over years and environments.
- Seed production is more skilled operation in hybrids and it is not so in synthetics.
- Synthetic varieties are good reservoir of genetic variability.
- Less costly compared to hybrids.

Demerits :

- Performance is little bit lower compared to hybrids because synthetics exploit only GCA while hybrids exploit both GCA and SCA.
- The performance may not be good when lines having low GCA are used.
- Synthetics can be produced and maintained only in cross pollinated crop species, while hybrid varieties can be produced both in self and cross pollinated crops.

Table 2 : Achievements

Name of crop	*Type of cross*	*Name of hybrids released*
(*a*) *Cross pollinated crops*		
Maize	Double and three way crosses	Ganga 2, Ganga 5, Ganga 11, Makka 1 and VL42
Pearlmillet	Single cross	HB 3, BJ 104, MBH 110, BK 560, MH 179
Sunflower	Single cross	BSH 1, HSFH 1, MSFH 10, KBSH 1 and KBSH 11
Castor	Single cross	GCH 2, GCH 3, Gauchi and GCH 4
Cucumber	Single cross	Pusa Sanyog
(*b*) *Often cross pollinated crops*		
Sorghum	Single cross	CSH1, CSH5, CSH6, CHS9, CSH10, CSNH
Cotton	Intraspecific	H 4, H 6, H 8, H 10, JKHY 1, JKHY 2, PKVHY 2, PKVHY 3, PKVHY 4
	Insterspecific	Varalaxmi, DCH 32, HB 224, TCHB 213, DHB 105
	Tetraploid hybrids	Sruthi etc.
	Interspecific diploid hybrids	DH 7, DH9, DDH2
Intra-arboreum hybrid	LDH 11	
Pigeonpea	Single cross	ICPH 8 from ICRISAT
(*c*) *Self pollinated crops*		
Tomato	Single cross	Hybrid Tomato
Brinjal	Single cross	Vijay, Pusa kranti, Arkasheel, Arkanavneet and Azad Kranti
Jute	Single cross	Hybrid C

All the varieties of sugarcane and potato are hybrids. Several promising hybrids have been developed in mango. The important mango hybrids include, Mahmood, Bahar, Prabha Shankar, Neeluddin, Swarna, jehangir, Mallika, Amrapalli and Ratna. Hybrids have also been developed in chilli and bottle guard.

In China, hybrid rice has been developed for commercial cultivation through the use of cytoplasmic genic male sterility. In India, efforts are being made to develop hybrids in rice. However, production of hybrid seeds is a major problem in wheat. In India, ICAR has launched a Hybrid Project entitled, "Promotion of research and development efforts on hybrids in selected crops" from 1989 to intensify hybrid research. The selected crops include rice, maize, sorghum, pearlmillet, pigeonpea, cotton, sunflower, rapeseed-mustard and castor. This project is aimed at developing high yielding superior quality hybrids in above nine crops.

Merits and Demerits

Merits

1. Hybrid varieties have higher yield potential as compared to synthetics, composites and open pollinated varieties.

2. Hybrid varieties are more uniform and attractive than synthetics, composities and open pollinated varieties.

3. Hybrid varieties can be developed in both self and cross pollinated species, whereas synthetics and composites are relevant to cross pollinated crops.

4. It is possible to reconstitute the hybrid with same genotype, which is not possible in case of composite varieties and open pollinated varieties.

Demerits

1. Fresh seed has to be produced every year, because in F_2 the hybrid produces various types due to segregation and recombination. Farmer has to purchase fresh seed every year.

2. The seed of hybrid is costlier than synthetics, composites and open pollinated varieties.

3. Cultivation of hybrids requires more input (fertilizer, water, plant protection, etc.) to exploit their full potential.

4. Production of hybrid requires more technical skill and areas as compared to synthetics, composites and open pollinated varieties.

4.3.3 (*b*) Composite Varieties

In cross pollinated crops, the mixture of genotypes from several sources that is maintained in bulk from one generation to the next is referred to as composite variety. Composites are constituted by seed mixture of advanced generation material of intervarietal or inter-racial cross.

Table 3 : Similarities and dissimilarities between synthetic and composite cultivars

Particulars	*Synthetic cultivars*	*Composite cultivars*
Similarities		
Equilibruim	Polymorphic Stable	Polymorphic Stable
Relevant to	Allogamous Species	Allogamous Species
Genetic Constitution	Heterogeneous, mixture of several homozygotes and heterozygotes	Hetrogeneous, mixture of several homozygotes and heterozygotes
Adaptation	Wide due to broad genetic base	Wide due to broad genetic base
Dissimilarities		
Base Material	Inbreds or clones	Open pollinated varieties or any other heterozygous source
Component genotypes	Two to eight	Two to many
Evaluation for gca	Always done	Usually not done
Seed Replacement	After 4-5 years	After 3-4 years
Exact Reconstitution	Possible	Not Possible

Procedure of developing composite variety

Year	*Main Breeding Activity*
First	Selection of base material, intermating in all possible combinations and harvesting crossed seed separately.
Second	Evaluation of F_1 crosses in replicated trial using hybrid or open pollinated variety as check.
Third to Fifth	Evaluation of F_2, F_3 and F_4 generations in replicated trials using standard check, identification of crosses exhibiting little or no inbreeding depression.
Sixth and Seventh	Mixing parental seed of superior crosses in equal quantity to constitute composite variety, seed multiplication by open pollination in isolation.
Eighth	Release of new composite variety, distribution of seed to the farmers for commercial cultivation.

Achievements

Several composite varieties have been developed in maize in India. The important varieties are Jawahar, Vijay, Kisan, Amber, Sona, Vikram, Rattan, Shakti, Vikas, Tarun, Renuka, Kanchan, Co 1, etc. In Brassica compestris variety toria an early maturing variety known as composite 1 has been released.

4.3.4 Mutation Breeding

The genetic improvement of crop plants for various economic characters through the use of induced mutations is referred to as mutation breeding. It is one of the special methods of crop improvement. Mutation breeding is commonly used in self pollinated and asexually propagated species.

Significance

1. When the desired variability is not found in the cultivated varieties or in the germplasm of cultivated species.

2. When a high yielding variety has oligogenic defect such as susceptibility to a disease, mutation breeding is the best course of line for a breeder. Because induced mutagenesis leads to improvement of such variety without much alteration in the genetic background.

3. When there is a tight linkage between desirable and undersirable characters, mutation breeding is the best way of overcoming such problems.

4. When in fruit crops, the improvement has to be made without change in the taste and colour of the fruits, it can best be achieved through mutation breeding, because mutation breeding causes little changes in the genetic background of parental material.

5. When a particular reaction is to be blocked, mutation breeding is the only way. For example in synthesis of morphine taken place in step-wise manner. If the biochemical path way is blocked at bane level, it will block the synthesis of morphine without affecting the conversion process of the bane into useful pharmaceutical product.

6. In those crops where sexuality is absent and generation of variability through recombination process is not possible.

7. In those species where generation cycle is very long, such as plantation crops, fruit trees and forest trees, mutation breeding is the short cut way of genetic improvement.

8. When attactive flower and foliage colours have to be developed in ornamental plants which fetch high price in the market, mutation breeding is the only short cut method.

Mutagens and their mode of action

Mutagens refer to a various physical or chemical agents which greatly enhance the frequency of mutations. Various radiations and chemicals are used as mutagens. Mutagens are of two types, *viz.* (1) physical mutagens, and (2) chemical mutagens.

Table 4 : Commonly used physical mutagens (radiations), their properties and mode of action

Type of radiation	*Main properties*	*Mode of action or changes caused*
1. X-rays	S.I., penetrating and non-particulate	Induce mutations by forming free radicals and ions. Cause addition, deletion, transitions and transversions.
2. Gamma rays	S.I., penetrating and non-particulate	Induce mutations by ejecting atoms from the tissues. Cause all types of changes as above.
3. Alpha Particles	D.I., Particulate, less penetrating and positively charged.	Act by ionization and excitation. Cause chromosomal and gene mutations.
4. Beta Rays Particles	S.I., particulate, more penetrating than alpha particles and negatively charged.	Act by ionization and excitation. Cause chromosomal and gene mutations.
5. Fast and Thermal Neutrons	D.I., particulate, neutral particles, highly penetrating.	Cause chromosomal breakage and gene mutations.
6. Ultra Violet Rays	Non-ionizing, low penetrating mutations.	Cause chromosomal breakage and gene

Note : Particulate refers to particle emitting property. DI = Densely ionizing, SI = Sparsely ionizing.

Table 5 : Some commonly used chemical mutagens and their mode of action

Group of mutagen	*Name of chemical*	*Mode of action*
1. Alkylating agents	Ethyl methane sulphonate	$AT \leftrightarrow GC$ Transitions
	Methyl methane Sulphonate	Transitions
	Ethyl ethane Sulphonate	$GC \leftrightarrow AT$ Transitions
	Ethylene Imines	Transitions
2. Base analogues	5 Bromo Uracil	$AT \leftrightarrow GC$ Transitions
	Amino Purine	$AT \leftrightarrow GC$ Transitions
3. Acridine Dyes	Acriflavin, Proflavin	Delection addition and Frameshifts
4. Others	Nitrous Acid	$AT \leftrightarrow GC$ Transitions
	Hydroxylamine	$GC \leftrightarrow AT$ Transitions
	Sodium Azide	Transitions

The mutation breeding process consists of four important steps. These are (1) choice of material, (2) choice of mutagen, (3) mutagen, treatment, and (4) handling of treated material.

Handling of treated material

General procedure for handling of mutagen treated material consists of five important steps as discussed below :

1. **M_1 Generation.** A large M_1 population is raised from treated seeds. Plants are grown using wider spacing for easy identification. Generally, the mutants are recessive. If some dominant mutations are obtained, they are selected in M_1 ; otherwise observations are recorded on chlorophyll sectoring and fertility. All the M_1 plants are selfed to avoid contamination from cross pollination. Each M_1 plant is harvested separately.

2. **M_2 Generation.** The M_2 generation is raised from the seed obtained from M_1 using wider spacings. The oligogenic mutants with distinct features are identified and selected. The seed of such mutants is harvested separately.

3. **M_3 Generation.** The M_3 progeny are raised from selected M_2 and evaluated for homozygosity. The homologous M_3 progeny are bulked together to conduct yield trail in M_4 generation.

4. **M_4 Generation.** The M_4 progeny are raised in replicated trail using local check for comparison.

5. **M_5-M_9 Generation.** Selected lines are tested in coordinated multilocation trials. The best performing line is released as a variety.

Applications in crop improvement

Induced mutations play an important role in crop improvement. The main uses of induced mutations are : (1) development of improved varieties, (2) induction of male sterility, (3) production of haploids, (4) creation of genetic variability, (5) overcoming self incompatibility, and (6) improvement in adaptation.

Table 6 : Achievements

Improved character	*Crops and mutant varieties*
Higher yield	High yielding mutation varieties have been developed in barley (DL 253), chickpea (Pusa 408, Pusa 413, Pusa 417), Cowpea (cowpea 88, V 16, V 37, V 38, V 240), Groundnut (Co 2, TG 17), Mungbean (ML 26-10-3). Oriental mustard (RLM 514). Okra (MDU 2), Pea (Hans). Ridge gourd (PKM 1), Tomato (PKM 1). Tossa jute (Mahadev TJ 40) and sorghum (Co 21).
Short stature Earliness	Barley (RDB 1), Rice (Hybrid mutant), Sesame (Kalika), Rice (Prabhavati), Rice (AU 1, IIT 48, IIT 60, Indira, K 84, Padmini, Sattari, CNM 20, CNM 25, CNM 6, CNM 31), Castor bean (Aruna) Cotton (MCU 7), Blackgram (Co 4), Mungbean (TAP 7), Pigeonpea (Co 5) and Sugarcane (Co 85035)
Stress resistance	Drought resistance in cotton (MA 9, MCU 10), Salt tolerance in Rice (Mohan) and water logging tolerance in Jute (Padma)

Advantages

Mutation breeding has several advantages. Some important advantages of mutation breeding are briefly presented below :

1. Induced mutagenesis is used for the induction of cytoplasmic male sterility. Ethidium bromide (EB) has been used for induction of cytoplasmic male sterility in pearl millet and barley.

2. Mutation breeding is a cheap and rapid method of developing new varieties as compared to backcross, pedigree and bulk breeding methods.

3. Mutation breeding is more effective for the improvement of oligogenic characters such as disease resistance than polygenic traits.

4. Mutation breeding is the simple, quick and best way when a new character is to be induced in vegetatively propagated crops.

Limitations

Induced mutations have some limitations which are briefly presented below :

1. Most of the mutations are deleterious and undesirable.

2. Identification of micro-mutations, which are more useful to a plant breeder is usually very difficult.

3. Since useful mutations are produced at a very low frequency (0·1%), a very large plant population has to be screened to identify and isolate desirable mutants.

4. Mutation breeding has limited scope for the genetic improvement of quantitative or polygenic characters.

POLYPLOIDY

4.3.5 Introduction

An organism or individual having more than two basic or monoploid sets of chromosomes is called polyploid and such condition is known as polyploidy. It is estimated that about one third species of flowering plants are polyploids. In wild species of grass family polyploidy has been reported upto 70%. In animals, polyploidy is rare because of its lethal effects. It is found only in those species of animals which develop parthenogenetically like aphids. Polyploidy is of two types, *viz.* (1) autopolyploidy, and (2) allopolyploidy.

Table 7 : Types of polyploidy

Types of polyploid	*Definition/brief description*
1. Autoploypoidy	Multiplication of the same genome
(*a*) Autotriploidy	Three copies of the same genome ($3x$)
(*b*) Autotetrapoloidy	Four copies of the same genome ($4x$)
(*c*) Autopentaploidy	Five copies of the same genome ($5x$)
(*d*) Autohexaploidy	Six copies of the same genome ($6x$)
2. Allopolyploidy	A condition in which complete genomes of two or more species combine together
(*a*) Allotetraploidy	Two copies each of two distinct genomes ($2x_1 + 2x_2$)
(*b*) Allohexaploidy	Two copies each of three distinct genomes ($2x_1 + 2x_2 + 2x_3$)
(*c*) Allooctaploidy	Two copies each of four distinct genomes ($2x_1 + 2x_2 + 2x_3 + 2x_4$)

Induction of polyploidy

Polyploidy is mainly induced by treatment with a chemical known as colchicine. This is an alkaloid which is obtained from the seeds of a plant known as Colchicum autumnale, which belongs to the family Liliaceae. Colchicine does not affect Colchicum from which it is extracted, because this plant has an anticolchicine substance.

Colchicine is applied in a very low concentration, because high concentration is highly toxic to the cells. For effective induction of polyploidy, usually concentrations of 0·01 to 0·5% are used in different plant species. The colchicine induced polyploidy is known as colchiploidy. In plants colchicine is applied to growing tips, meristematic cells, seeds and axillary buds in aqueous solution or mixed with lanolin. The duration of treatment varies from 24 hours to 96 hours depending upon the species of plants.

Effects of polyploidy

Polyploidy has marked effects on the morphology of plants. The distinct features of autoploids are increase in general vigour and size of various plant parts. Such features are generally referred to as gigantism. Autoploids have the following important features :

1. Stems are thicker and stouter.

2. Leaves are fleshy, thicker, larger and deeper green in colour.

3. Roots are stronger and longer.

4. Flowers, pollens and seeds are larger than diploids.

5. Maturity duration is longer and growth rate is slower than diploids.

6. Water contents are higher than diploids etc.

Applications in crop improvement

Polyploidy plays an important role in crop improvement. The autopolyploidy and allopolyploidy are useful in several ways. However, allopolyploidy has wider applications than autopolyploidy. Aplications of autopolyploidy and allopolypoloidy in crop improvement are briefly presented below :

Autotetraploids have been developed in forage crops like berseem, alfalfa and rye ; vegetable like radish, turnip and cabbage; and fruits like grapes.

Alloploidy

Alloploidy is useful in four principal ways, viz. (1) in tracing the origin of natural allopolyploids, (2) in creating new species, (3) in interspecific gene transfer, and (4) as a bridging species.

Limitations of Polyploidy

Polyploidy has several limitations. Some important limitations of polyploidy in crop improvement are briefly presented below :

1. **Limited use.** The single species polyploidy has limited applications. It is generally useful in those crop species which propagate asexually like banana, potato, sugarcane, grapes etc.

2. **Difficulty in maintenance.** The maintenance of monoploids and triploids is not possible in case of sexually propagating crop species.

3. **Undesirable characters.** In bispecies or multispecies polyploids characters are contributed by each of the parental species. These characters may be sometimes undesirable as in case of *Raphanobrassica.*

4. **Some other defects.** Induced polyploids have several defects such as low fertility, genetic instability, low growth rate, late maturity, etc.

5. Chances of developing new species through allopolyploidy are extremely low.

4.3.6 Plant Breeders' Rights

Plant Breeders' Rights (PBR) refer to the legal rights provided to an originating plant breeder or owner of a variety to regulate production and marketing of his/her variety. In other words, the legal rights which will allow plant breeders to get benefits of their products are referred to as plant breeders' rights.

PBRs are specialized patent like system for cultivated plants. PBRs are one of the most recent forms of Intellectual Property Law. The policy basis for PBR is the same as for patent, design and copyright laws. Features of PBR

1. PBRs are specialized plant like system for cultivated plant species.
2. PBRs consititute one of the recent forms of Intellectual Property Law.
3. PBRs provide legal rights to a plant breeder to get benefit of his/her innovation or variety.
4. PBRs differ from patents in the sense that the former allow farmers privilege and research exemption. Farmers' privilege is the right to hold material as seed source for subsequent seasons. The research exemption refers to the right to use protected material as the basis to develop a new variety or other research use. Patents donot provide such exemptions. Research exemption is also called breeders' privilege.
5. PBRs protect the variety but not the standard breeding procedures that are used for development of a variety.
6. PBRs are generally considered to provide less protection than patents.
7. Laws related to PBRs were first framed in 1961 by the union for the protection of new plant varieties (UPOV) which were further revised in 1972, 1978 and 1991 UPOV conventions were held in Geneva.

Main Features of UPOV

The first UPOV convention was held and signed in Paris in 1961, but it came into force in 1968 with its headquarters at Geneva. Since then many changes have taken place in UPOV, because it was revised in 1972, 1978 and 1991. The 1978 Act came into force in 1981 and the Act 1991 has not yet come into force. The main features of UPOV are briefly presented below :

1. Material to be protected

A protection right can be granted for varieties of all botanical genera and species. In other words, all new varieties, of cultivated species are protected under Plant Breeders' Rights ((PBR). The new variety must have a designation (name) as per the rules of International code of nomenclature.

2. Rights provided

UPOV provides legal rights to the original plant breeder or owner of a variety for commercial production, marketing and export of his variety. Thus original breeder gets benefit of his intellectual property or his invention.

3. Basic Requirements

There are four basic requirements for protection of a variety under PBR. These are : (1) novelty, (2) distinctness, (3) uniformity and (4) stability. The last three criteria were given by UPOV 1978 convention and the first criteria *i.e.*, novelty was added in UPOV convention 1991. Novelty refers to newness of a variety. The variety should be new one and it should not have been commercially cultivated for more than one year before granting protection under PBR. The second criteria is distinctness. The new variety must be distinguishable in one or more morphological, quality or other characters from previously available varieties. Thirdly, the variety should be uniform (homogeneous). In other words the variety should be pure and look uniform. Lastly, the variety must give stable performance in different generations. These four attributes are determined by grow out test.

4. Power of Authorization

The holder of PBR has power of authorization. He can authorise any other person for commercial production, marketing, export and import of his variety. Prior authorization of a breeder (owner) of a variety is required for commercial production, marketing, export and import of his variety. The use of a protected variety for further research does not require authorization as per UPOV 1978 Act. However, the use in research requires written permission as per UPOV Act 1991.

5. Farmers' Rights

It refers to legal rights that are provided to farmers to save, use, exchange, share or sell his farm produce of a variety. Here the sale is restricted to non-commercial sale. The 1978 Act is beneficial, because it protects farmers' rights. According to 1978 Act, farmers can use, reuse their produce as seed and have the right to dispose off their farm produces. However, 1991 Act has curtailed farmers' rights. Farmers' rights is also referred to as farmers' privilege.

6. Breeders' Privilege

It refers to legal rights that are provided to plant breeders to use protected plant material for further research. Breeders' privilege is also known as Research Exemption or Breeders' Exemption. The UPOV Act 1978 provides Breeders' Privilege. However, the Act 1991 has curtailed Breeders' Privilege to use the material of a protected variety for further research.

7. Period of Protection

The period of protection varies with plant species. For field crops, the minimum period of protection is 15 years as per 1978 Act and 20 years as per UPOV Act 1991. For forest trees, vines, fruit trees, ornamental trees and shrubs, the period of protection is 18 years as per UPOV Act 1978 and 25 years as per UPOV Act 1991. The UPOV Act 1991 is pro Plant Breeders.

Advantages of PBR

Plant Breeders' Rights have several advantages. Some important advantages of PBRs are briefly prsented below :

1. Incentive to breeders
2. Fast development of seed industry
3. Improvement in quality
4. Procurement of good material
5. Enrichment of genetic resources

Disadvantages of PBR

There are some disadvantages of PBR which are briefly presented below :

1. Exploitation
2. Encouragement of unhealthy practices
3. Increase in cost of seed
4. Reduction in genetic diversity
5. Ban on use of own seed

5. AGRICULTURAL BIOTECHNOLOGY

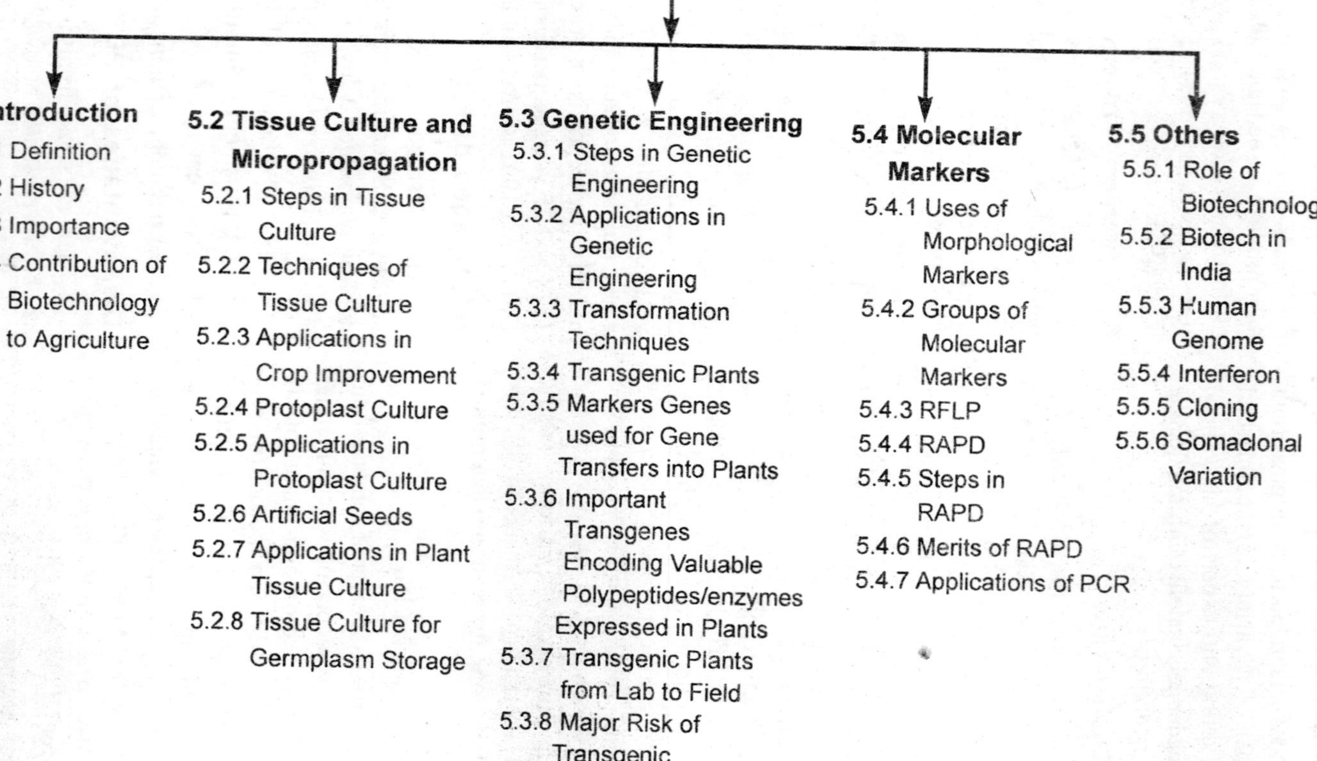

5. Agricultural Biotechnology

5.1 INTRODUCTION

5.1.1 Definition

The term biotechnology is composed of two words - bio (Greek-bios-means life) and technology (Greek technologia - means systematic treatment). In other words, it is the science of applied biological processes. Accordingly to the oxford dictionary, technology is the "Scientific study of the practical or industrial arts".

Spinks (1980) defined biotechnology as, "the application of biological organisms, systems or processes to manufacturing and service industries".

European Federation of Biotechnology (1981) defined biotechnology as, "the integrated use of biochemistry, microbiology and engineering sciences in order to achieve technological (industrial) application of the capabilities of micro-organisms, cultured tissue cells and parts thereof".

In the report Biotechnology: A Development plan for Canada (1981) biotechnology is defined as "the utilization of a biological process, be it microbial, plant or animal cells, and their constituents, to provide goods and services".

First food product produced through biotechnology was sold in 1970. By 2003 about 7 million farmers were utilizing biotech crops.

5.1.2 History

1. *Pre pastures Era* : Selection in plants and Fermentation.
2. *Pasture Era* : Microbes are cause of Fermentation, Enzyme convert sugar into alcohol.
3. *Third Era* : antibiotic (pencillin). Green revolution
4. *Modern Era* : DNA structure , Genetic Engineering
5. *Recent Era* : Transgenic, marker, tissue culture etc.

5.1.3 Importance of Biotechnology

1. To Increase Nutrient Content in crop.
2. To Increase resistance to Biotic and abiotic stress.
3. To introduce gene
4. To Create variation.
5. To minimize inputs and maximize output.

5.1.4 Contribution of Biotechnology to Agriculture

Technology	*Uses*
1. Embryo culture to rescue otherwise inviable hybrids, to recover haploid plants from interspecific hybrids, micropropagation of orchids etc.	The first two applications are the most remarkable.
2. Rapid clonal multiplication through meristem culture,. *e.g.*, of many fruit and forest trees, such as, teak.	Very high rates of muliplication; conventional rates very low.
3. Recovery of virus and other pathogen free stocks of clonal crops; meristem culture is generally combined with thermotherapy/cryotherapy.	Very useful in clonal crop; particularly for germplasm exchange.

Contd...

Technology	Uses
4. Germplasm conservation through storage in liquid nitrogen (at-196° C, cryo-preservation) or through slow growth.	Particularly useful in clonal crops, especially in those producing tubers, storage roots etc.
5. Rapid isolation of homozygous lines by chromosome doubling of haploids produced through anther culture/interspecific hybridization/ovary culture.	Very successful in variety development in China, *e.g.*, in rice and wheat.
6. Isolation of stable somaclonal variants with improved yield/yield traits/disease resistance/ resistance to cold, herbicides, metal toxicity, salt and other abiotic stresses.	Many examples of successful isolation; many variations are stable and heritable, often due to gene mutation, which may sometimes, be novel.
7. Gene transfers (genetic engineering) for insect resistance, protection against viruses, herbicide resistance, storage protein improvement etc.	Mainly using the Tiplasmid of *Agrobacterium*; also through particle gun, free DNA uptake; REVOLUTIONARY DEVELOPMENT in crop improvement.
8. Molecular markers, e.g., RFLPs and RAPDs, for linkage mapping and mapping of quantitative trait loci.	A powerful tool for indirect selection for quantitative traits; several other important applications.

In agriculture, rapid and economic clonal multiplication of fruit and forest trees, production of virus.-free stocks of clonal crop, creation of novel genetic variations through somaclonal variations, and transfer of novel and highly valuable genes (*e.g.*), for disease and insect resistance through genetic engineering have opened up exciting possibilities in crop production, protection and improvement.

5.2 TISSUE CULTURE & MICRO PROPAGATION

It bypasses sexual process in the development of new crop varieties.

It overcomes the barriers of crops incompatibility in distant crossed (interspecific and intergeneric crosses).

It helps in development of transgenic (genetically engineered) plants with resistance to herbicides, biotic and abiotic stresses and improvement in quality.

It is a rapid method of crop improvement. e.g. tetraploid plants can be obtained in one step through protoplast fusion.

Plant Tissue Culture

Plant tissue culture refers to growth of living plant tissues in a suitable culture medium (*in vitro*). In a broad sense, it is referred as cultivation of plant organs, tissues of cells in test tubes on an artificial media; often the techniques of plant cell and tissue culture are also called *in vitro* techniques.

"Culture medium" is a nutrient medium, which contains all essential micro and macronutrients, carbohydrates, vitamins and hormones, The p^H of the nutrient medium should be 5.5. However, the culture medium differs from species to species. In this technique, new plants are regenerated from tissues, cells and organs in test tubes in the nutrient medium. The organ includes any plant organ that has separate identity such as anther, ovule, embryo and bud.

The plant part, which is used for regeneration, is called '*explant*'. It may be a cell, a protoplast, a tissue or an organ. A mass of regenerated cells in culture medium is called '*callus*' and suspension of free cells of callus in a liquid medium is known as '*suspension culture*'. The regeneration capacity or ability of a plant cell to develop into a whole plant is known as '*totipotency*'

5.2.1 Steps in Tissue Culture

Tissue culture technique consists of four important steps :

1. Isolation of tissues
2. Regeneration and callus formation in the culture medium.
3. Embryogenesis
4. Organogenesis

Various steps involved in Tissue Culture Technique

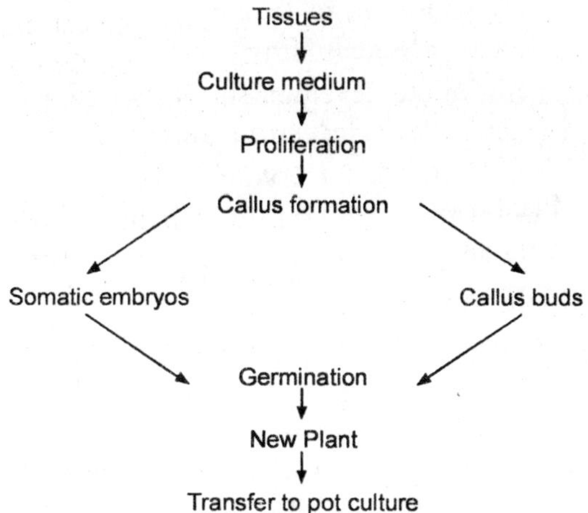

1. ***Isolation of tissues*** : Disease free tissues for regeneration can be isolated with the help of sterilized blade from any plant part *viz*. e.g. stem, apical bud, axillary bud etc.
2. ***Regeneration and callus formation :*** Tissues proliferate on the culture medium and give rise to a mass of cells called '*callus*'.

The callus is generally of two types :

1. Friable callus
2. Compact callus

The friable callus can be easily manipulated for suspension culture.

However, compact callus is not suitable for suspension culture.

5.2.2 Techniques of Tissue Culture

Tissue culture techniques are also referred to as *in vitro* techniques. In vitro techniques give rise to new plants by regeneration of tissues and organs in the nutrient medium. Various plant parts *viz.* meristems, embryo, anther, cells and ovules are used for regeneration. Depending upon the plant parts used as explants, plant tissue culture techniques are of five main types *viz.* meristem culture, embryo culture, anther culture, cell culture and ovule culture.

(A) Meristem Culture

Meristem culture refers to regeneration of whole plant from tissues of an actively dividing plant part such as stem tip, root tip or axillary bud. Generally, shoot apical meristem is used for regeneration. This technique is wisely used in vegetatively propagated plants such as sugarcane, potato, banana and several fruit trees and timber species.

The main applications of meristem culture in crop improvement are :

1. It is used for micro progaration (mass production of clonal progeny through tissue culture) in banana, strawberries, citrus etc.
2. Virus free plants can be obtained through this technique because meristematic cells are almost free from virus even in the virus infected plants.
3. The exchange of germplasm plantlets obtained by meristem culture is safe, because some material is free from insect and pathogens. This is useful in the

exchange of germplasm of asexually propagated plant species.

4. The germplasm can be conserved in the form of meristems of - 196° C for long term storage in the liquid nitrogen *i.e.*, meristems are suitable for cryopreservation.

(B) Embryo Culture

Regeneration of a whole plant from an embryo in the culture medium is called '*embryo culture*'. Embryos of appropriate stage are removed from the seed and are transferred to the culture medium. This technique is used when there is disharmony between embroy and endosperm. This technique is used to make distant crosses successful.

e.g.

1. Interspecific hybridization in Trifolium, *Lycopersicon*
2. Intergeneric hybrization in
   Hordeum & Secale
   Triticum & Secale
   Triticum & Aegilops etc.

Main Advantages are —

1. Useful in making interspecific and intergeneric (distant) crosses successfully
2. Useful in obtaining haploids *e.g.* barley and wheat.
3. In orchids, seeds are stored food and are unable to propagate. Hence embryo culture is useful
4. Long seed dormancy is noticed in *Prunus, Iris and Taxus* species. Embryo culture is useful in breaking the long dormancy and in obtaining seedlings.

(C) Anther Culture

Regeneration of whole plant from anther (or) pollen in the culture medium is called anther culture. The optimum stage differs from species to species. e.g. *brassica, tobacco, rice, barley, wheat, potato, tomato, triticale* etc.

Advantages are —

1. obtaining haploids
2. homozygous diploids are obtained simply by doubling chromosomes or obtained spontaneously from anther culture.
   (*a*) Mapping population
   (*b*) Breeding cycle is shortened

(D) Cell Culture

Regeneration of whole plant from callus and suspension cultures in the nutrient medium is known as *cell culture*. Protoplasts (naked cells or cells without cell wall) are also used for regeneration of whole plants. *e.g.*, wheat, rice, barley, maize, tobacco, fruit and timber trees.

Advantages are —

1. Useful in mass clonal propagation of different species.
2. Protoplast culture is useful in somatic hybridization and overcoming the barriers of cross incompatibility.
3. Useful in the development of transgenic (genetically engineered) plants.
4. Useful/suitable for cryopreservation *i.e.* preservation in liquid Nitrogen at –196 degree Celsius.

(E) Ovule Culture

Regeneration of whole plant from the ovule in the nutient medium is called *ovule culture*.

Two types of ovules :

Unfertilized → Haploids

Fertilized → Diploids

This technique is however used to a limited extent.

5.2.3 Applications In Crop Improvement

(A) Generation of Variability

Variation is created through invitro techniques. Depending upon the explants, variation induced by tissue cultural techniques is of three types:

1. **Gametoclonal variation:** Variation observed among plants, which are regenerated from gametic culture. (anther/ovule culture).
2. **Somaclonal variation:** The variation observed among plants obtained callus culture of somatic explants such as meristems.
3. **Protoclonal variation :** the variation observed among plants that are regenerated from callus cultures of protoplast.

Main features of the variation are —

1. Variation caused is genetic in origin and hence heritable.
2. Somaclonal variation occurs both in sexually and asexually propagated species. But, the frequency is very high in a sexually propagated species.
3. Occurs both in oligogenic and polygenic traits.
4. Somaclonal variation is heterozygous in origin (through mutation).
5. All three types of culture variations resulted due to chromosomal changes such as deletion, duplication, inversion and translocations.

Somaclonal variation is useful in isolation of diseases, herbicide tolerant and early maturing variants.

(B) Development of Hàploids

Made possible through anther and pollen Culture. In China, one variety in wheat (Jighua 1) and in rice (Guan 18) have been developed for commercial cultivation.

5.2.4 Protoplast Culture

Isolated plant protoplasts are cells from which the walls have been removed. Cell walls can be removed either by mechanical or enzymatic methods. Isolated protopalsts can be induced to divide in culture to form callus. From the callus whole plants can be regenerated, either through embryogenesis or organogensis.

Isolation of Plant Protoplasts

Plant protplasts are generally isolated from leaves by enzymatic method.

The Need for an Osmoticum : During isolation and culture, plant protoplasts require a suitable osmotic stabiliser. Such a substance is called *osmoticum*. In the absence of an osmoticum the protoplasts may burst. Several ionic and non-ionic solutes have been tested for adjusting osmotic potential of various solutions used in protoplast isolation and culture. The important osmotica are *sorbitol, mannitol, glucose or sucrose*.

Culture of the Protoplasts

Isolated protoplasts are *cultured invitro* to regenerate whole plants. The procedure is outlined below.

1. A culture medium (MS medium) with mannitol containing 1.5% agar in a petri dish is used. To this medium the protoplasts are embedded and incubated at 25°C in dim light.
2. Within 48 hours the naked protoplasts form cell walls.
3. Two to seven days after culture, cell division occures to form callus.
4. The callus is transferred to a medium without mannitol and auxin.
5. Within 3 to 4 weeks embryoids are formed.
6. These embryoids are dissected from the callus and transferred to filter-paper bridge in a culture tube containing liquid medium composed of :

    MS salts
    Kinetin- 0.2 mg/l
    Sucrose - 2.0% w/v
7. After a few weeks plantlets are produced. These are transferred to sterile soil and grown to maturity.

5.2.5 Applications in Protoplast Culture

Protoplast culture has wide applications. some of them are :

1. Cell Biology Studies

Isolated protoplast are ideal materials for the study of cell wall synthesis in plant cells.

2. Protplast fusion and Somatic Cell Hybridisation

Two protoplasts from the same source or different sources can fuse to form a hybrid protoplast. Whole plants can be regenerated *invitro* from this hybrid protoplast. This method is utilised to produce hybrids from two plants which show physical or chemical incompatibility in normal sexual crosses.

When protoplast isolated from two sources are allowed to fuse their cytoplasms mix. The nuclei of the fused protoplasts may fuse nuclei are called *heterokaryons or Heterokaryocytes*. Thus hybrid protoplasts are formed. These are now selected culture medium to regenerate hybrid plants. Thus isomatic cell fusion helps to obtain interspecific and intergeneric hybrids which are difficult to obtain by conventional plant breeding methods. This method of obtaining hybrids are called somatic hybrids.

Thus production of somatic hybrids is a multistep process:

(*i*) Islolation of protoplasts from two different speices
(*ii*) Fusion of protoplasts from two different species
(*iii*) Isolation of fused protoplasts. and
(*iv*) Regeneration of fertile hybrid plants from the fused protoplasts.
(v) Somatic hybrids have been produced in several crop plants like potato and tobacco.

3. Protoplast Fusion and Production of Cybrids

When protoplasts from two different species are allowed to fuse, their cytoplasms mix. Sometimes, the nucleus of one of the protoplasts completely disappears. The cytoplasms of the two parental protoplasts are still hybridized. This fusion product is called a *cybrid or cytoplasmic hybrid.* Some genetic factors are carried in the cytoplasm, e.g. factors for male sterility in some plants. Hence formation of cybrids has practical applications in plant breeding.

4. Production of Genetic Variants from Cell/ Protoplast Culture

Plants regenerated from cell cultures often show genetic variations. Since, such variation

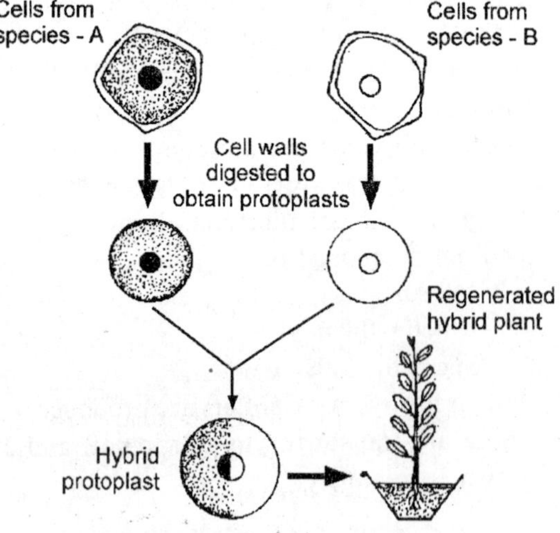

Fig. 1 : Scheme of somatic cell hybridisation.

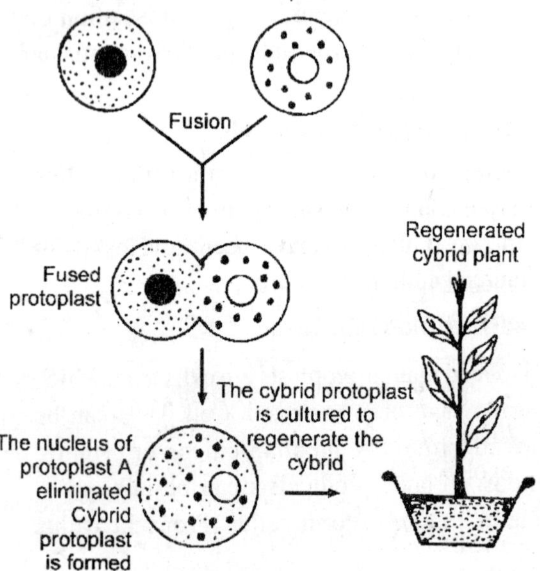

Fig. 2 : Production of cybrids

occures in cells of somatic origin, it is called *somaclonal variation*. Several kinds of mutations are formed inversion, duplication and deletion. Somaclonal variation in cell culture cannot be directed toward changing a particular kind of mutant. Beneficial genetic variations originating in cell cultures can be utilised in crop breeding for the development of improved varieties of crop plants. using this method genetic variants with traits such as resistance to herbicides, salt tolerance and resistance to toxins such as resistance to hebicides can be evolved. Somaclonal genetic variants have been evolved in several crop plants like wheat and sugarcane.

5. Gene-transfer into Higher plants by Co-cultivation of Plant Protoplasts with *Agrobacterium*

Transgenic plants can be evolved by introducing foreign genes into isolated plant cells/ protoplasts by co-cultivation with *Agrobacterium tumefaciens* that contains a co-integrate plasmid carrying the foreign gene. The transformed plant cells can be regenerated *invitro* to evolve transgenic plants.

6. Introduction of Foreign Genes into Isolated Plant cells/Protoplast by Microinjection, Electroporation and Microprojectile Bombardment etc.

Foriegn DNA can be introduced into isolated plant cells by microinjection, electroporation and microprojectile bombardment. Liposomes containing foreign genes can be fused with isolated protoplasts. The transformed cells/protoplasts are then selected by growing on suitable media and are then regenerated to produce transgenic plants.

Using these methods plants that are resistant to herbicides toxins, pests and pathogens have been evolved. Several transgenic plants that can produce compounds such as vaccines, interferons, plastics and thaumatin have been evolved.

7. Plant cell Cultures and Biosynthesis of Secondary Products

Plant cells can be grown as micro-organisms in culture media and many secondary matabolites have been synthesised from different plant species

The yield of these substances is more when compared with the yield from whole plants.

Table 1 : A few secondary products produced by plant cells in culture

Plant Product	*Plant Species*
Codeine(alkaloid) - analgesic	Papaver somniferum
Diosgenin (steroid) - antifertility agent	Dioscorea deltoidea
Dogoxin (Cardiac glycoside)	Digitalis lanata
Quinine (alkaloid) - Antimalarial agent	Cinchona ledgeriana
Pyrethrin (insecticide)	Chryanthemum cine rarialfolium
Thaumatin (Cholcone) non-nutritive Sweetener	Thaumotcoccus danielli
Ginseng - stimulant	Panax ginseng
Jasmine (perfume)	Jasminum sp.

8. Plant Cell Culture in Biotransformation

Plant cell culture can convert certain organic substances added to the cultures into useful products. This process is called *biotransformation*. Using this method useful substances that have pharmacological properties can be synthesised.

Hardening or Acclimatisation of Plantlets Regenerated Through Culture

The plantlets regenerated through tissue culture methods are transferred to soil for futher growth and development. Different species of plants differ in their capacity to adjust to the new environment. During this period the plantlet has to change from the heterotrophic to the autorophic state. Water loss from the regenerated plantlet is high. This is due to the following reasons:

1. The root system formed in the cultured plantlets is indadequate to maintain the plantlets in the soil.
2. The regenerated plantlets have thin leaves with reduced epicuticular wax on their leaves and stems.
3. The stomata do not respond efficiently to stress conditions. Due to these reasons the regenerated plants dehydrate more quickly than the normal plants. Hence the regenerated plantlets must be protected from dessication and hardened to attain some tolerance to moisture stress.

The regenerated plantlets have been grown under aseptic conditions. So, they will be easily attacked by soil pathogens. Such plants should be protected from soil borne pathogens so that they can grow and develop into healthy plants.

Hardening of regenerated plantlets is done by planting them in a sterile soil and covering the plants with a transparant hood. These plantlets are grown in shade for some time and then transferred to the field

5.2.6 Artificial Seeds

Somatic embryo (embryoids), shoot buds or any other plant material obtained as a result of *invitro* culture are covered (encapsulated) with a chemical membrane. Such encapsulated materials behave as seeds. These are called *artificial seeds* or *synthetic seeds*. The artificial covering acts as an artificial seed coat. Such seeds are bead like and can "germinate" and plantlets are also formed.

Several substances are used as artificial seed coats. Some of them are *agar, agarose, carrageenin, polyacrylamide, introcellulose, ethyl cellulose and sodium alginate*. Sodium alginate is most commonly used.

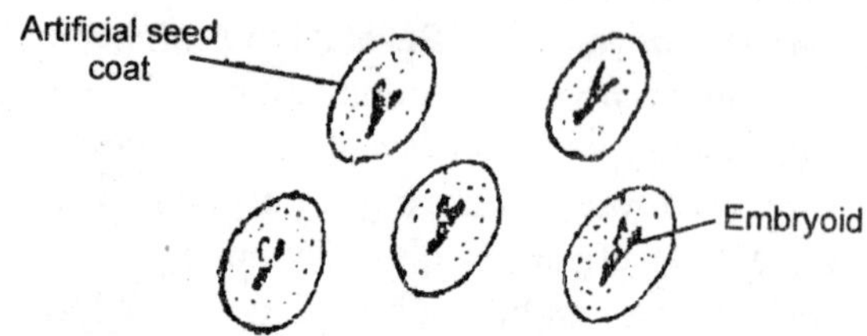

Fig. 3 : Artificial seeds

Advantages of Artificial Seeds

1. The size of the artificial seeds is smaller when compared to the natural seeds of the plant.
2. Storage and transportation of such seeds is easier.
3. Viability of seeds is 100%.
4. Artificial seeds can be made to germinate uniformly on a suitable substratum.
5. Such seeds do not show dormant.
6. The plant grower can grow the desired plant any time and this is not season dependent.
7. Large scale production of seeds from any kind of plant part is possible.

Disadvantages

1. The artificial seeds cannot be stored for longer time and it is temperature dependent.
2. The initial cost for the production of artificial seed is more than that for the natural seeds.
3. Production and germination of artificial seeds require aseptic conditions. Any deviation will affect the quality of the seeds and their subsequent development.

5.2.7 Applications in Plant Tissue Culture

Plant tissue culture has applications in agriculture, hoticulture, plant breeding and industry.

1. Rescue of hybrid embryos.
2. Culture of immature embryos to propagate orchids.
3. Culture of mature embryos to reduce seed dormancy.
4. Production homozygous lines of recessive genes through anther or pollen culture.
5. Production of disease free clones through nucellus culture and apical meristem culture.
6. Production of somatic hybrids.
7. Production cybrids.
8. Production of genetic variants from cell/ protoplast culture.
9. Creation of transgenic plants.
10. Biosynthesis of secondary products and biotransformation.
11. Micropropagation.

5.2.8 Tissue culture for Germplasm Storage

Invaluable plant species are stored to make them available for future breeding programmes. This is called *Germplasm storage*. Normally it is done by storing the seeds. It has certain limitaions: 1. The seeds may loose viability with the passage of time, 2. The seeds may be damaged by seed borne pathogens, 3. This method cannot be effectively used for vegetatively propagated crops.

Tissue culture is used as an alternative method for the storage of germplasm. Culture tissues can be indefinitely stored at very low temperatures in a frozen state. This method is called cryopreservation. When there is a need whole plants can be regenerated from the tissues under *cryopreservation*. When there is a need whole plants can be regenerated from the tissues are brought to a state of non-dividing and zero-metabolism. Commonly tissues are stored in liquid nitrogen at - 196°C. Certain substances are added to the culture media before freezing. These protect the tissues against ice damage. Such substances are called cryoprotectants, *e.g.*, Glycerol, Proline, Mannitol, Sucrose, Glucose, Polyethylene glycol. Cryopreserved tissues can be recultured to produce, whole plants or they can be sub-cultured. The viability of cryopreserved tissues depends upon the plant species.

5.3 GENETIC ENGINEERING

Genetic engineering is the construction of a recombinant DNA molecule which contains DNA fragments (genes) from different and unrelated organisms. Hence a gene of interest or a desired gene has to be identified and isolated. It is then joined with the DNA of another organism. The final step will be to allow this " desired gene" to express itself in an appropriate host or new environment and to produce the gene product. In several cases the gene product will be isolated.

5.3.1 Major Steps of Genetic Engineering

1. Isolation of the Desired Gene

Specific DNA of fragments or desired genes are to be identified. They are then isolated and purified for being transferred to an appropriate host. This DNA of interest is also called *passenger DNA, target DNA or donor DNA* . Isolation of DNA is done by any one of the following biochemical method.

(*a*) extraction of total cell DNA.

(*b*) synthesis of complementary DNA [cDNA]

2. Selection and Isolation of a Vector

If a desired gene is to be introduced into a host cell, a carrier molecule that can transport the gene into the host cell is required. Such a molecule is called a *cloning vehicle, carrier molecule or a vector*. A vector is a self replicating molecule of DNA to which the donor gene can be linked. The vector should also be able to propagate in the host organism or cell. Commonly plasmids and the DNA of viruses are used as vectors. A suitable vector is identified and then isolated. New techniques are now evolved to directly introduce the donor DNA into the host cell (Fig. 4).

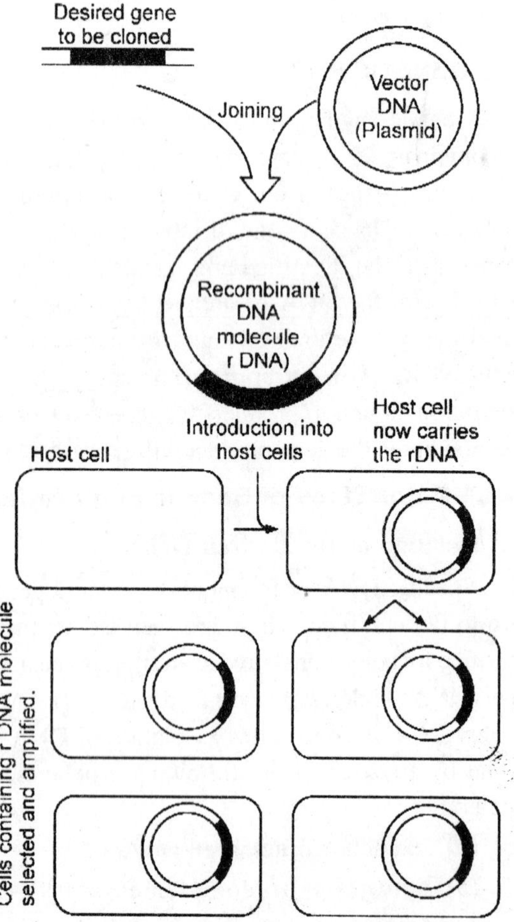

Fig. 4 : Outline of the synthesis and cloning of recombinant DNA molecules.

3. Construction of Recombinant Vectors

The donor DNA is inserted into the vector to produce a donor-vector hybrid DNA molecule. This hybrid is also called recombinant DNA a chimera or a recombinant vector as it contains DNA from unrelated sources. This recombinant DNA can now be introduced into a host cell.

4. Introductior of the Recombinant Vector into the Host Cell

The recombinant vector is introduced into a suitable host cell. Since most bacteria and eukaryotic cells can take up naked DNA molecules from the medium under suitable conditions they are used as host cells. Such host cells are also called target cells.

Inside the host cell the recombinant DNA is either linked to the host chromosome or maintained free in the cytoplam. If viruses are used as vectors, the synthetic viruses carrying the recombinant DNA infect the target cells. The genome of such sythetic virus contains the gene of interest. This gene is present in the place of a segment of viral DNA that is not essential for its replication. Inside the target cell the recombinant vector DNA may multiply. Thus numerous copies of the recombinant DNA are formed. This means, the desired gene it carries is also multiplied.

5. Selection and Multiplication of the Host Cells Carrying the Recombinant DNA Molecule

The cells which have successfully taken up the recombinant DNA molecule are selected. These cells are allowed to divide. After a large number of cell division, a colony of identical host cells is produced. At the cellular level it is a clone of cells. Each cell in clone carries one or more recombinant DNA molecules. Thus the desired gene carried by the recombinant DNA is *cloned*.

6. Expression of the Desired Gene

The desired gene can now express itself in the body of the host organism and produces the desired trait. In many cases of genetic engineering the desired gene will produce the gene product in the new environment which can be isolated and purified, e.g., synthesis of human insulin, synthesis of somatostatin etc.

5.3.2 Applications of Genetic Engineering

1. R-DNA technology in nitrogen fixation.
- Introduce nif-gene into several other microbe.
- To increase host range
- To transfer nif-genes from bacteria to plants.
2. Transgenic plants T_i with Herbicide Resistance.
- Atrazine resistant oil seed (*Brassica-napm*).
- Atrazine is detoxified by glutathione-s-transferase.
3. Transgenic Glyphosate tolerant plants glyphose is an active components of many herbicide.
4. Genetically Engineered plants that Resists/repel pests and pathogens.

   Bacillus thuringensis crop : cotton, maize, Brassica.
- NAPI-Nicogene natural insecticide effective against many pest from *Nicotiana alata*.
5. Genetically Engineered Protection Against viruses in transgenic Plants.
- The presence of one virus in plants can interfere with infection by another virus or strain. This is called cross protection.
- Coat protein.
6. Gnetically Engineered plants are used to produce Novel Compounds.
- Transgenic Banana with haptitis virus.
- GMO tobacco plants to produce interferon.
7. Plants that produce plastic :
- *Transgenic Arabidopsis thaliana* to produce polyhydroxy butyrate (PHB).
8. Transgenic plants to Improve Photosynthetic Efficiency.

   Modification of RUBISCO to favour up take CO_2 rather than O_2.
9. Transgenic plant to improve protein Quality.
- Transgenic soyabean with sulphur containing amino acid.
- Cereals with protein content .
10. To delay senescence.
- Tomato → "flavour savour" variety to delay ripening.
- Carnation - Antisence RNA technology for delay flower senescence.
11. Golden Rice.
- β-Carotin gene is inserted in rice to improve vitamin -A content.

5.3.3 Transformation Techniques

The uptake of foreign DNA or transgenes by plant cell is called transformation. A variety of technique have been used to introduce transgenes into plant cells.

1. Agro bacterium mediated.

(*a*) Co-culture with tissue Explants.

(*b*) In Planta transformation.

2. Agro infection : Introduction of a viral genome into plant by placing with in the T-DNA of a Ti Plasmid and using the Agro becterium containing this recombinant plasmid for co-culture with plant cell is called Agro infection.
3. Direct gene transfer. (Biolisitic method).

(*a*) Chemical methods [(Poly ethylene glycol) PEG].

(*b*) Electroporation.

(*c*) Particle gun delivery.

(*d*) Lipo fection.

(*e*) Micro injection.

(*f*) Macro injection.

(*g*) Pollen transformation.

(*h*) Delivery via growing pollen tubes.

(*i*) Laser induced.

(*j*) Fiber -mediated gene transfer.

5.3.4 Transgenic Plants

The technology of Gentic Engineering is the practice of altering or disturbing the gentic blue prints of living organism.

Transgenic for transfer the desirable qualities from one organism to another.

Vectors

R-DNA technology is based on the insertion of a DNA fragments (gene) into a suitable cloning vector and then producing into a suitable host to propagate the recombinant DNA.

1. ***Plasmids*** :

Circular, double standard, extrachromosomal self replicating DNA molecule.

Plasmids	*Size (nucleotide length in kb)*
PBR 345	0·7
PBR 322	4·363
Col E_1	6·36

Other Vector

Vector	*Capability(Kbp)*
1. Cosmid	45-48
2. BAC	15
3. PAC	15
4. Phagemid	15-60

Identification of transgenic plants by reporter gene.

Certain marker genes on which no selection pressure is imposed are need to identify the transformed cells/Plants. Such gene are called reporter gene.

Two types :

1. *Selectable marker* : Cell which persent survive under selective condition Eg. kanamycin, Neo mycine.
2. *Scorable marker* : Produce different phenotype which allow an easy identification of R-DNA cell Eg. SVS, β-galactriodase.

5.3.5 Marker Genes used for Gene Transfers into Plants

Table 2 : Some of the marker genes used for gene transfers into plants.

Gene	*Scorable/Selectable*	*Charateristic*
nos (Napoline synthase)	Scorable	Nopaline production
cat (Chlormaphemicol acetyi transfersase)	Scorable	Acetylation of chloramphenicol
lux (Luciferase)	Scorable	Phosphorescence
gusA (β-galactosidase)	Scorable	Blue colour
dhfr (dihydrofolate reductase)	Selectable	Methotrexate resistance
nptII (Neomycin phosphotranserase)	Selectable	Kanamycin, neomycin and G 418 resistance
hptIV (Hygromycin phosphotransferease)	Selectable	Hygromycin B resistance
bar (Phosphinothricin acety1-transferase)	Selectable	Phosphinothricin resistance

5.3.6 Important Transgenes Encoding Valuable Polypeptides/Enzymes Expressed in Plants

Polypeptide/Enzyme	*Source*	*Applications*
POLYPEPTIDES		
Enkephalin	Human	Opiate activity.
Hirudin	Synthetic	Thrombin inhibitor.
α-Trichosanthin	Chinese medicinal plant	Inhibition of HIV replication.
Epidermal growth factor	Human	Proliferation of specific cells.
Erythropoietin	Human	Regulation of erythrocyte levels.
Growth hormone	Trout	Growth stimulation.
Human serum albumin	Human	Plasma expander.
Interferon	Human	Anti-viral activity.
ENZYMES		
α-Amylase	*Bacillus licheniformis*	Liquefaction of starch.
(1-3, 1-4)-β-Glucanase	*Trichoderma reesei*; Hybrid of two Bacillus species	Brewing.
Manganese-dependent lignin peroxidase	*Phanerochaete chrysosporium*	Bleaching and pulping of paper.
Phytase	*Aspergillus niger*	Increased phosphate utilization from feed.
Xylanase	*Clostridium thermocellum*; *Cryptococcus albidus*	Animal feed, paper and pulp.

5.3.7 Transgenic Plants from Lab to Field

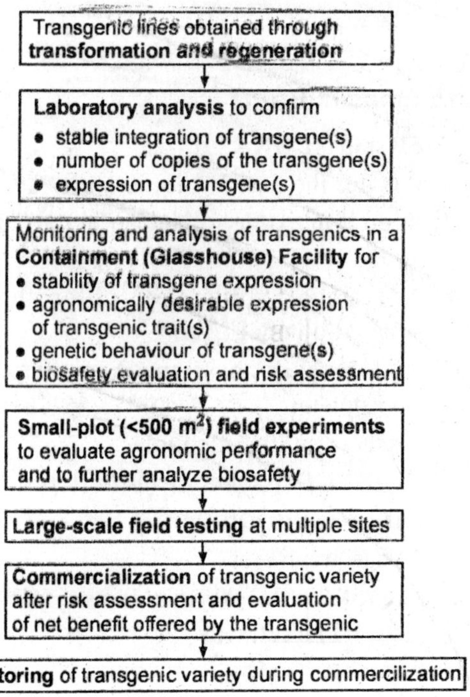

Transgenic lines obtained through transformation and regeneration

↓

Laboratory analysis to confirm
- stable integration of transgene(s)
- number of copies of the transgene(s)
- expression of transgene(s)

↓

Monitoring and analysis of transgenics in a **Containment (Glasshouse) Facility** for
- stability of transgene expression
- agronomically desirable expression of transgenic trait(s)
- genetic behaviour of transgene(s)
- biosafety evaluation and risk assessment

↓

Small-plot (<500 m^2) field experiments to evaluate agronomic performance and to further analyze biosafety

↓

Large-scale field testing at multiple sites

↓

Commercialization of transgenic variety after risk assessment and evaluation of net benefit offered by the transgenic

↓

Monitoring of transgenic variety during commercilization

5.3.8 Major Risk of Transgenics

1. Escape of Enginerred genes by gene flow or gene dispersal.
2. Non target effects or ecological effects.
3. Invasiveness or weediness of transgenics.
4. Creation of super weeds and super-virus.
5. Toxicity and allergenicity to human beings and animals.
6. Expression of undesirable phenotypic traits.
7. Erosion of Biological diversity.

Risk Management

1. First to outline the risk assessment.
2. Isolation zone.
3. Trap crop.
4. Refuge crop.
5. Male sterility.
6. Removal of flower from transgenic plants.

5.4 MOLECULAR MARKERS

Molecular Markers

Consisting of specific molecules which show easily detectable difference among different strains in the DNA sequences (polymorphism).

A linear map of different chromosomes of a species depicting the location of various molecular markers is known as molecular map.

5.4.1 Uses of Molecular Markers

1. Varietal and hybrid identification
2. Identification of MS line of sunflower red leaf petiole (anthocyanin) pigment.
3. Identification of TGMS line in rice. PTGMS lines with receptive purple leaf colours.
4. Identification of cultivars resistant to pests and diseases.

   Trichomeless trait in Cumbu - smut resistant

   Cotton-replant colour, hairiness of leaf glanded cotton offer resistance to various pests and disease.
5. Quality : Flax - varigated seed colour mutant used to identify fatty acid profile. Low linoleic acid varieties can be developed through this.

5.4.2 Groups of Molecular Markers

1. DNA hybridization based assays.
   - Restriction Fragment length Polymorphism (RFLP)
   - Fluorescent In situ hybridization (FISH)
2. PCR based techniques using arbitrary primers
   - Random Amplified polymer DNA (RAPD)
   - DNA Amplification Fingerprinting (DAF)
   - Amplified Fragment Length Polymorphism (AFLP)

5.4.3 RFLP

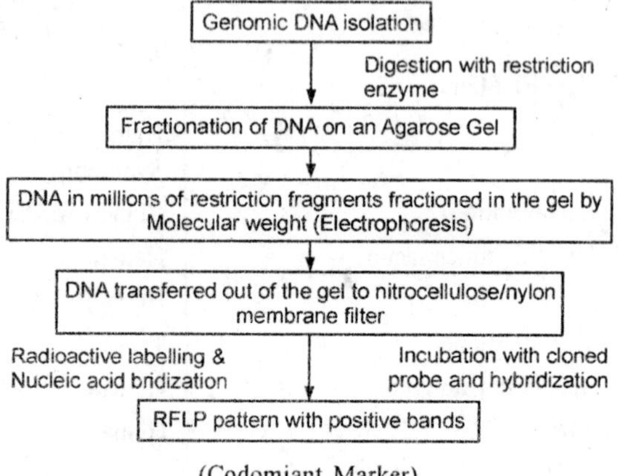

(Codomiant Marker)

5.4.4 RAPD

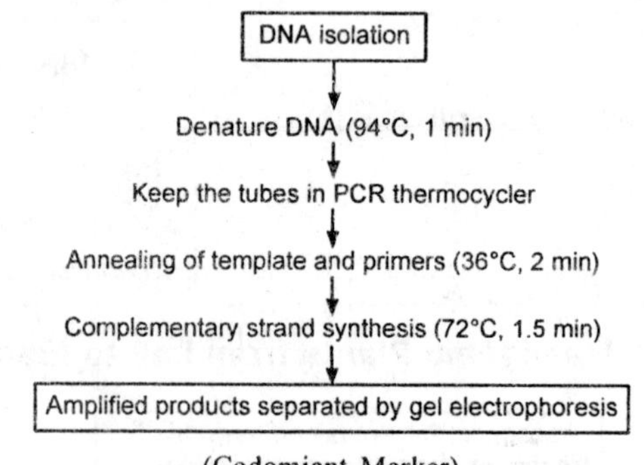

(Codomiant Marker)

5.4.5 Steps in RAPD

1. **Isolation of DNA**
   - Powder the leaf in Liquid Nitrogen.
   - Keep it in extraction buffer at 65°C for 1 hour.
   - Add equal volume of chloroform : isoamyl-alcohol (24:1).
   - Take aqueous phase and precipitate with isopropanol.
   - Wash with 70% alcohol.
   - Dry it and dissolve in TE buffer.
2. **Purification of DNA**
   - RNase treatment.
   - Phenol : chloroform : Isoamylalcohol (25:24:1) treatment

- Chloroform : Isoamylalcohol (24:1) treatment.
- Precipitate with isopropanol.
- Dry and dissolve in TE buffer.

3. **PCR Amplification**
   - Ten random primers will be used.
   - PCR product will be resolved on 0.8% agarose gol.
4. **Polymerase Chain Reaction (PCR)**

   In the year 1983 American biochemist Mullis, K.invented PCR and won Nobel Prize for same.

   Principle : Selective amplification of required region of DNA molecule.

   Steps :

   1. Denaturation of template DNA duplex by heating at 94°C.
   2. Annealing of oligonucleotide primers to the target sequences of separated DNA strands at 35-65°C.
   3. DNA Synthesis from the 3'-OH end of each primer by DNA polymerase.

5.4.6 Merits of RAPD

- High degree of polymorphism.
- Minimum quantity of DNA is required.
- Rapid analysis
- Inexpensive compared to other methods.

5.4.7 Applications of PCR

PCR can make use of very small amount of DNA. Even DNA in nanogram quantity can be easily handled. In view of this, PCR has many applications. Some of them are :

1. Remnants of extinct organisms is sufficient to recover their DNA and amplify specific segments using PCR. Using this technique, from the remnants of woolly mammoth which lived about 40,000 years ago specific DNA segments were recovered and amplified. Now it is proved that this animal is related to present day elephant.

2. PCR has revolutionised pre-natal detection of several genetic diseases. Only a very small sample of the foetus is sufficient to perform this work. Some of the diseases detected are : Phenylketonuria, cystic fibrosis, haemophillia A, beta thalassemia, sickle - cell anemia.

3. The sex of the *invitro* fertilised egg can be determined using PCR. The presence of the Y-chromosome can be detected at the 6-10 cell cleavage embryo without affecting the life of the embryo.

4. The PCR has wide and significant use in forensic science also. Small pieces of tissue, even a single hair left at the scene of the crime or one sperm lingering after a rape is enough to fix the identity of the criminal.

5. One can pick out a single antibody producing cell, amplify the genes encoding the antibody using PCR and clone them in bacteria to produce an anti-body producing bacteria. Thereafter, any amount of antibody can be made by growing such bacterial cells in a fermenter.

5.5 OTHERS

5.5.1 Role of Biotechnology

In Agriculture

Some of the areas where biotechnology has played a significant role are detailed below.

1. Production of pathogen - free plants: Plants traditionally being vegetatively propagated are systemically infected with viruses and other pathogens which greatly reduce yield and also quality of the marketed commodity. Virus diseases like potato leaf roll virus (PLRV) or potato virus Y (PVY) for example, can cause upto 95% reductions in the tuber yield of potato crops. Increase in yield upto 300% (Averaging 30%) has been reported following replacement of virus - infected stock with specific pathogen free plants.

Since majority of viruses infect plants in a systemic manner their elimination may be achieved through meristem tip culture.

2. Production of Disease Resistant Plants: This is one of the most useful applications of tissue culture in crop improvement. In potato, somaclones have been screened for both late and early blight resistance. In maize, somaclonal variation has induced resistance to race T of southern corn leaf blight. In sugarcane, resistance to diseases like fiji and downy mildew have been recovered. In lucerne, selection of cell lines and plants resistant to the toxin of *Fusarium oxysporium* has been accomplished.

3. Improvement of nutritional quality : One of the major sources of protein for human and animal consumption is constituted by the proteins contained in seeds of many plant species. The cereals and legumes which are major sources of storage seed proteins, contain limited amount of certain amino acids which are essential for human beings. Majority of these cereals are deficient in lysine whereas legumes are deficient in sulphur amino acids. A wide range of approaches have been employed for improving nutritional quality of various crop plants. Important among them are selecting cell lines resistant to amino acid or their analogues of lysine, tryptophan, proline and phenylalanine. Isolation of variants over producing specific amino acids in culture has been successful, but expression in the whole plant and especially in the seed has not yet reached the level required to make an impact on protein quality.

4. Selection for salt and drought tolerance: Continued efforts to increase intensity of cropping for increasing production from limited land resource by extending irrigation facilities have resulted in the gradual build-up of salt concentrations in the soil. This has resulted in loss of productivity on such soils.

Salt tolerant lines have been produced in crop plants such as tobacco, tomato, cereals. Salt tolerance has been incorporated into rice lines with improved plant type. Some of the improved rice cultivars *viz.*, IR 42, IR 43 and IR 52 are salt tolerant.

Development of cultivars tolerant drought can contribute significantly in agriculture economy. Tolerance to drought is a polygenic trait and involves highly complex osmo-regulatory functions. In tissue cultures, simulated drought conditions have been achieved through incorporation of non-penetrating osmotic solutes such as PEG (polyethylene glycol) and dextrans in the media.

5. Production of Genetically variable plants : The success of any crop improvement programme depends on the usable genetic variability in the base population cells in culture offer an excellent systems for inducing variations and regenerating pure mutant types. Genetic variation can be an option to lessen our reliance on cost intensive germ-plasm collection and conservation programmes. Variant producing capacity of cell culture can be augmented to a great extent by employing physical and chemical mutagens. Somaclonal variation has been extensively exploited for the improvement of a sexually propagated crops *viz.*, potato, sugarcane. From cell cultures, some superior cultivars have already been produced in sugarcane which are high yielding, drought resistant and temperature tolerant.

6. Biofertilization : Molecular nitrogen in the atmosphere is converted into Biologically converted forms by nitrogen fixing micro-organisms *e.g.* Rhizobium. The most sophisticated approach to biofertilization is to create plants that possess the genetic capacity for nitrogen fixation. Attempts are being made to transfer genes for nitrogen fixation (hifgenes) from bacteria to plants.

7. Rapid clonal propagation : Tissue culture has found its best commercial application in production of cloned plants at a very high rate as compared to conventional methods. It is important specially for initially building up of propagation stock of elite clones or individual plants which are otherwise slow to multiply. A number of agriculturally important plants have been clonally multiplied.

8. Germplasm storage : The primitive cultivars and wild relatives of crop plants constitute a pool of genetic diversity which is invaluable for further breeding programmes. There are over 20,000 plant species which are rare or threatened with degradation of their neutral habitats. The most economical form of storing germ plasm for seed propagated species is seeds. However, there are certain limitations of this method. Therefore, various methods of invitro storage of germplasm are thus of great practical significance for long term storage of germplasm. Presently, there are two approaches to invitro germ plasm storage. Slow growth technique and cryopreservation.

9. Biological control of Agricultural pests: Insects consume nearly one-third of human food supplies on earth. One of the major goals of biotechnology is to develop target specific biological pesticedes that will kill some specific pest but will not harm other species. Several insect-specific pathogens, developed through biotechnological processes, are being produced commercially for their use as microbial pestices. *Bacillus thuringienis inocula* have been used as bio-insecticides. Toxin produced by the bactenium kills the gypsy moth. The prevention of crown gall disease formation is achieved by spraying young plants or seeds avirulent species *Agrobacterium radiobactor* var, *radiobactor*. Microbial insecticides have a number of advantages over traditional chemical insecticides.

(*a*) they are specific for a small number of species of insects and do not kill plants, animals and beneficial insects.

(*b*) they are cheaper than organic pesticides and

(*c*) they do not leave any residue effect.

In Food

Can be effectively used to increase storage of food grains, increase nutrition value of food, enhance flavour etc. Fermented food through enzyme engineering (*beer biotech*).

In Health and Medicine

- Through human genetics, identification of defective genes, and gene therapy.
- In forensic medicine, identification of parents, criminals etc. by finger printing.
- Various vaccines are either produced or expected to produce eg: Rabies, Malaria, Hapatitis B, Cholera.
- Various vital ingredients like Insulin, Interferons can be produced. Also antibiotics like, penicillin, streptomycin etc.

Industry

Production of high fructose corn syrup – sweetening agent for soft drinks, enhance the production of alcohol, production of chemical from agri waste.

Environment

- Special Bacterial to eat away oil spilling by digesting Hydro Carbon.
- Bio-tech developed oragnisms to act as bio indicators to indicate level or pollution in environment.

5.5.2 Biotech-in-India

- To disseminate biotech for the socio-economic development of the country.
- Department of Biotech – 1986, entrusted with planning, promotion and Co-ordination major programmes undertaken by DBT.
- Development of vaccine and immunisation of especially vulnerable section.
- Increase oil seed production by tissue culture of coconut, oil plams etc.

- Cattle herd improvement by embryo technology.
- Development of bio-diagnostic kits for early detection of various diseases.
- Fertility control.
- In agri-development of crops-having high nutritional content, drought resistant, pest resistant etc.

5.5.3 Human Genome

- It refers to entire genetic makeup of human beings.
- Every human cell contains about a 1 lakh genes.
- Human genome project an international effort to identify and locate all the genes and their variants, understand their functioning and write down the sequence of nucleotides that comprise the genes.

Uses

1. Better understanding of genome might help in diagnosing many diseases at an early stage or even before birth.
2. Knowledge may help in replacing defective genes.

5.5.4 Interferon

These are natural cell signalling to challenges such as viruses, parasites and tumor cells proteins produced by the cells of the immune system in response.

Functions

- They interfere with the ability of a virus to infect healthy cells.
- They affect the cellular membrane of the virus-so that it cannot spread to other cells.
- They control virus multiplication, by inhibiting their protein synthesis.
- Now interferon is produced outside the body with the help of genetic engineering. There are three

5.5.5 Cloning

- Means making an identical copy physically as well as genetically.
- Clones are exact replica of plants or animals replicated asexually.
- Clones derive genetic matter from single parents.

Normally

Clones are not found in animals because offspring are sexually produced having genetic material from both parents.

With advent of genetic engineering an entire organism from a single somatic cell.

Recent Developments

Dr. Ian Wilmut of Roslin Institute, Scotland. At present Director of MRC centre of Regenerative Medicine at Unis of Edinburgh.

- He used nuclear transfer technology.
- He arrested the cell-cycle by starving it of a vital nutrient – calf-serum.
- In this state-cell was alive but no multiplication.

Advantages

1. To clone animals that can produce pharmacologically useful proteins such as clotting factors.
2. To make animals with human disease and then test therapies.
3. To make superior animals of uniform quality in large numbers.
4. Childless couple – baby.
5. Organ transplant – no question of rejection.
6. Xenotransplantation. (Animals to Humans)

Disadvantages

1. Propagation of few species-less genetic diversity – a single disease may eliminate the entire spectrum.
2. In Human Cloning even if same genetic characteristics – Social factor – man is a social animal.

5.5.6 Bioprospecting

- 30 million species of life on earth.
- only 1.5 m have been described
- It is estimated that 50,000 species are disappearing every year.
- Genetic wealth of the planet holds the key to resolution of our problems of poverty and health.

What is Bioprospecting ?

It implies using the genetic diversity in the Ecosystems of the earth for producing new drugs and new crops.

Bioprospectors often search for useful, substances in extremophiles, microorganisms plants fungi that grow in extreme environments such as rain, forest, deserts and hot springs.

5.5.7 Somaclonal Variation

Variation arised during tissue culture plants is known as somaclonal variation.

Causes

1. Change in chromosome number/ structure.
2. Gene mutation.
3. Plasmagene mutation.
4. Attention in gene expression.
5. Gene amplification.
6. Mitotic and crosing over.
7. Rearrangements in Cytoplsamic genes.

Applications

1. Resistant to disease variants can be isolated.
2. Stress Resistance.
3. Low-temperature mutant.
4. Mutants for Agronomic traits.
5. Mutants for Efficient Nutrient Utilization.

Achivements

1. 'Scarlet' variety in sweet potato.
2. Sigma – Alfalfa.

Advantages

1. High frequency mutagensis.
2. Variation is created.
3. Novelity.
4. Less time for variety release.

Limitations

1. Selected cell lines fail to regenerate.
2. Many of the selected clones are unstable.

5.5.8 Biotechnology and Indian Agriculture

We need to recognize that, just as the food requirements of today's population of nearly 6 billion people could not have been met by the technologies of the 1940's, we cannot assume that current practices will feed the population of 8 billion expected by 2020. New approaches are needed in addition to the continued improvement of existing methods of crop and animal husbandry and food processing. The strategic integration of biotechnology tools into India agricultural systems can revolutionise Indian farming and usher in a new era. Genetic engineering is clearly the most revolutionary tool to impact agricultural research since the discovery of genetics by Mendel. Prior to genetic engineering, the exchange of DNA material was possible only between individual organisms of the same species. With the advent of genetic engineering in 1972, scientists have been able to identify specific genes associated with desirable traits in one organism. For example, a gene from bacteria, virus or animals may be transferred into plants to produce genetically modified plants having changed characteristics. This method therefore, allows mixing of the genetic material from species that cannot otherwise, breed naturally.

Genetic engineering can be vital for an agrarian country like India. It can help in minimizing the crop damage through disease and pest resistant varieties, reducing the use of chemicals, enhancing stress tolerance in crop plants, thus permitting productive farming on unproductive lands, etc. One could even extend the growing season of crops and minimize losses due to environmental factors on the one hand and increase the shelf life of fruits and vegetables, on the other, thus minimizing losses due to food spoilage. We can expand the market vista and improve food quality. Biotechnology can also produce plants that possess healthy fats and oils, possess increased nutritive value, and create a whole range of higher value feeds. Biotechnology has even the power of producing biodegradable plastics, edible vaccines, etc.

It is only through the blending of the 'gene revolution' with our experience in the 'green revolution' that we can reach our goal of 'evergreen revolution' and also 'nutritional revolution'. The advantage of the gene revolution is that it is relatively scale neutral, benefiting big and small farmers alike, it is also environment friendly. Thus, it can be of great help to the smallest farmer with limited resources in increasing farm productivity through the availability of improved but powerful seed. It can also reduce a farmer's dependency on chemical inputs such as pesticides and fertilizers.

Modern biotechnology offers unlimited opportunity for enhancing genetic potential of crops and other commodities, management of biotic and abiotic stresses, bio-remediation and organic recycling. India is one of the main centres of agricultural biodiversity and its gene richness can greatly complement the developments in modern biotechnology. Likewise, new developments in GIS, remote sensing, and crop modelling, provide new opportunities for integrated management of natural resources. The revolution in informatics provides opportunities for sharing latest information for research and planning in a highly organized and efficient manner. With the globalization of economy, the opportunities for value addition and post harvest management are also immense. With clever blending of technologies, the India farmer can usher in new era of confidence and performance.

Biotechnology in India : In 1982, GOI has setup an official agency, National Biotechnology Board which functions under the Department of Science & Technology. In the year 1986 NBTB was replaced by a full fledged department, Department of Biotechnology under the ministry of S & T. This department is responsible for planning promotion and co-ordination of various activities of biotechnology in the country.

BIOTECHNOLOGY INFORMATION CENTRES IN INDIA

1. *Genetic Engineering*
   - Indian Institute of Science, Bangalore
   - Madurai Kamaraj University, Madurai
   - Bose Institute, Kolkata
   - Jawaharlal Nehru University, New Delhi
   - Indian Institute of Biotechnology, Bangalore.
2. *Animal Cell Culture and Virology*
   - Poona University, Pune
3. *Plant Tissue Culture, Photosynthesis and Plant Molecular Biology*
   - Indian Agricultural Research Institute, New Delhi
4. *Oncogenes, Reproduction Physiology, Cell Transformation, Nucleic Acid and Protein Sequences*
   - Centre for Cellular & Molecular Biology, Hyderabad
5. *Immunology*
   - National Institute of Immunology, New Delhi
6. *Enzyme Engineering, Immobilised Biocatalyst, Microbial Fermentation and Bio-process Engineering*
   - Institute of Microbial Technology, Chandigarh

BIRAC: Biotechnology Industry Research Assistance Council. (A Public sector enterprise set up by Department of Biotechnology).

SIBRI: Small Business Innovation Research Initiative Scheme launched in 2005. A scheme of Department of Biotechnology, Ministry of Science & Technology to boost Public-Private- Partnership (PPP) efforts in the country.

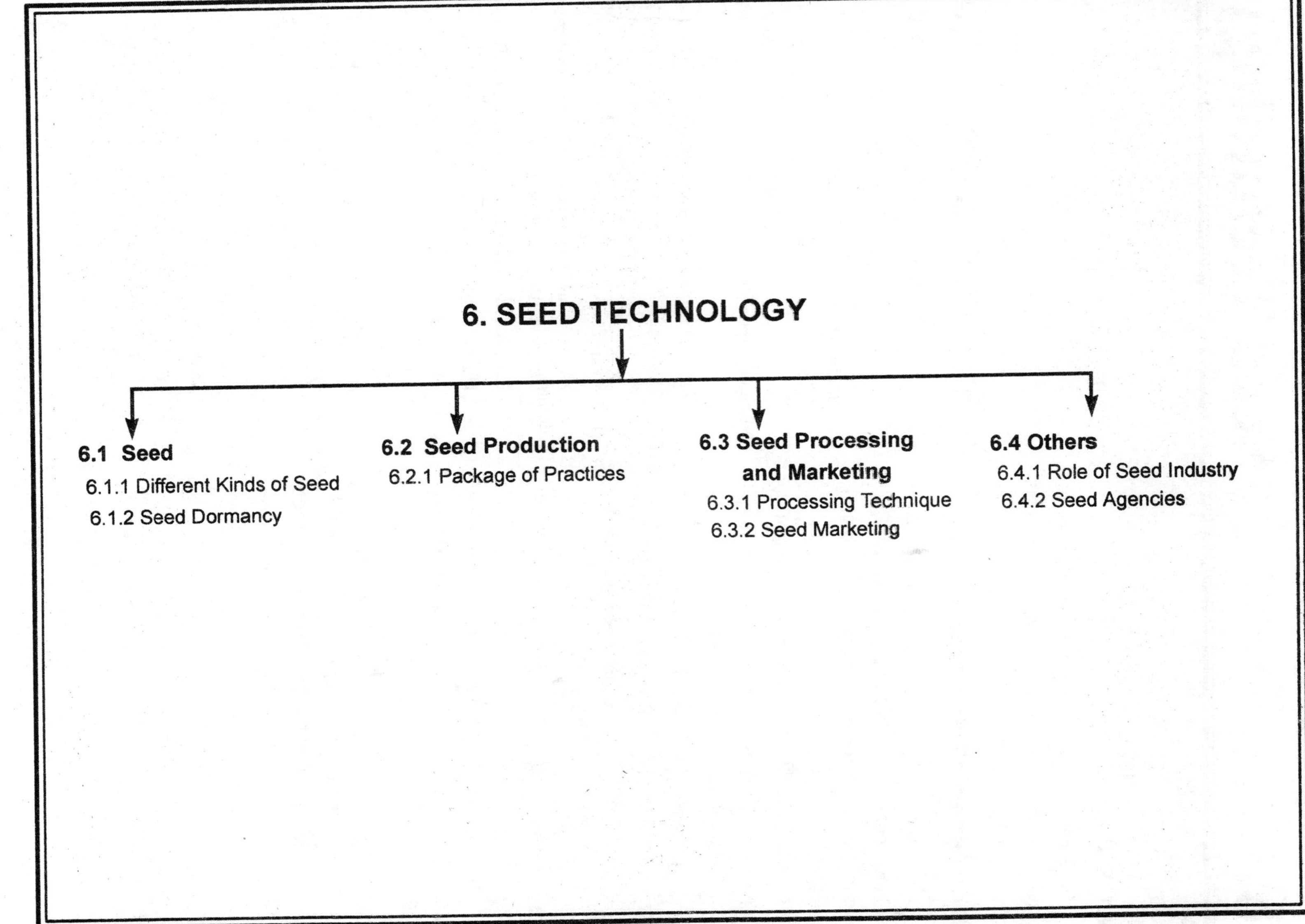
6. SEED TECHNOLOGY
6.1 Seed
6.1.1 Different Kinds of Seed
6.1.2 Seed Dormancy
6.2 Seed Production
6.2.1 Package of Practices
6.3 Seed Processing and Marketing
6.3.1 Processing Technique
6.3.2 Seed Marketing
6.4 Others
6.4.1 Role of Seed Industry
6.4.2 Seed Agencies

6. Seed Technology

6.1 SEED

A ripened ovule containing an embryo in arrested state of development usually with a food reserve and a protective coat.

6.1.1 Different Kinds of Seed

Multiplication of quality seed under supervision of breeder or seed certification agency to distribute quality seed of notified varieties for sowing purpose.

(*a*) *Nucleus Seed* : The true seed or vegetative propagules used for crop production and propagation are produced by the breeders. 100% gentically pure.

(*b*) *Breeder's Seed* : Progeny of Nucleus seed multiplied in larger area under the supervision of breeder.

(*c*) *Foundation seed* : Progeny of Breeder Seed.

It is handled by seed producing agencies in public and private sector under the supervision of Seed Certification Agencies (SCA).

SCA issues white colour certificate.

(*d*) *Certified Seed* : Progeny of Foundation Seed produced by registered seed growers under supervision of Seed Certification Agencies. SCA gives Blue colour certificate.

(*e*) Truthfuly labelled seed (TLS) minimum standard prescribed for the variety

(*i*) *Production of Nucleus Seed*

Basic Seed
↓
Sown in field with row to row and plant to plant spacing
↓
Plants are marked and observed
↓
Harvested

Breeding of Nucleus Seed —

Notification of variety
↓
Nucleus Seed (Bulk the seeds of selected rows)
↓
Breeders seed production plot
↓
Breeders seed

(*ii*) *Production of Breeder seed*

- Nucleus seed is sown with less seed rate in line with all requirements.
- Roguing is complete before flowering
- Harvesting, threshing and processing
- Monitored by a team of breeder, representative of crop co-ordinator, representative of National Seed Corporation Agency
- Quality of Processed Breeder seed is submitted to ICAR. Golden yellow certification is given.

(*iii*) *Production of Foundation and Certified Seed*

- Foundation and certified seed produced under supervision of seed certification Agency.

- The grower get Breeder seed from seed corporation.
- Grower harvests, threshes and dries the seeds, lastly seeds transported to seed processing plant.
- Seed is processed under the observation of seed inspector.
- Seed inspector submits a sample from the lot to seed testing lab..

6.1.2 Seed Dormancy

Condition in a viable seed by which it prevents germination even under optimal conditions provided for germination.

Types of Seed dormancy

Structural component of Seed : Seeds cannot germinate until maturity of embryo is complete.

Physiological : pH and osmotic pressure in the immediate environment of seed may act as inhibitors for germination.

Chemical : Inhibitors like phenols, Benzoic acids etc.

Breaking of Dormancy

Scarification : Physical or chemical treatment which weakens or softens seed coat.

Bitter gourd - Piercing with needle

Rubber - Removing the seed coat

Cotton - Soaking in conc H_2SO_4 for 20-30 min.

Stratification (Temperature treatment) :

- Pre chilling (0-5°C) for 3-10 days *e.g.* mustard.
- Pre heating (40-50°C) eg. Rice
- Hot water treatment - 80°C for 1 to 5 minutes.

Treatment with Growth Regulators and other chemicals

GA - For endogenous inhibitors

KNO_3 - For light requiring seeds - eg Tomato

Thiourea - For prechilling *e.g.* lettuce.

6.2 SEED PRODUCTION

6.2.1 Package of Practices

Agronomic Principles of Seed Production

- Selection of suitable Agro Climatic region
- Selection of seed plot
- Isolation Distance
- Selection of variety
- Seed treatment
- Seed rate
- Sowing method
- Roguing
- Weed control
- Pest and disease control
- Harvesting : (*i*) Time of harvesting
  (*ii*) Method of harvesting
- Drying of seeds
- Storage

6.3 SEED PROCESSING AND MARKETING

6.3.1 Processing Technique

1. **Seed Drying**

   Lowering of seed moisture for maintaining seed viability and vigour.

   (*a*) Sun drying (*b*) Air drying

2. **Pre conditioning**

   Includes all operations of shelling, debearding etc.

3. **Pre cleaning**

   Removal of extraneous matter like pieces of trash and stones etc.

4. **Basic cleaning**

   Removal of larger, smaller, lighter and thicker adulterants as compared to crop seed. It was done on the basis of weight, size and density using cleaner with air screen.

   Grader : Based on density and size

   Scalper : Inert matter of larger size than seeds.

   Aspirator : Lighter inert matter

5. Grading

Removal of smaller and shrivelled seeds from the healthy seeds. Specific gravity separators on the basis of resistance to air flow.

6. Seed treatment

Methods of seed treatment :

Physical : Cotton delinting, scarification
Chemical : Fungicide, insecticide etc.
Biological : Rhizobium

7. Packing

- Filling of seed bag to the exact weight
- Attaching labels regarding seed quality
- Storage

8. Seed Storage

Maintenance of high seed germination and vigour after harvest until planting is important.

- Stored in dry and cool place
- Only high quality seed should be stored.
- Proper sanitation should be there.
- Control on disease, insect and rodents.

6.3.2 Seed Marketing

It is a vital activity controlling the supply and price of seeds for the progress in crop production and productivity.

Marketing organisation —

Central Marketing Cell
↓
Regional offices

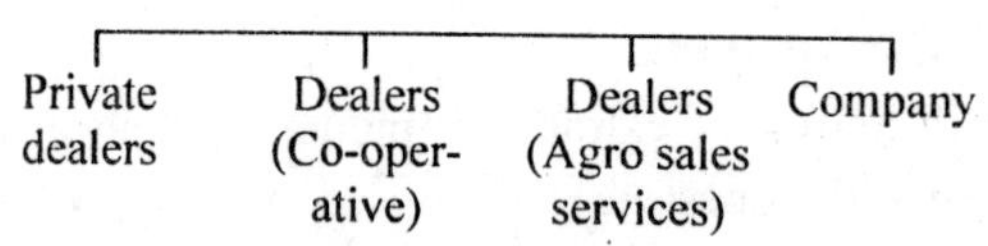

Marketing Chain

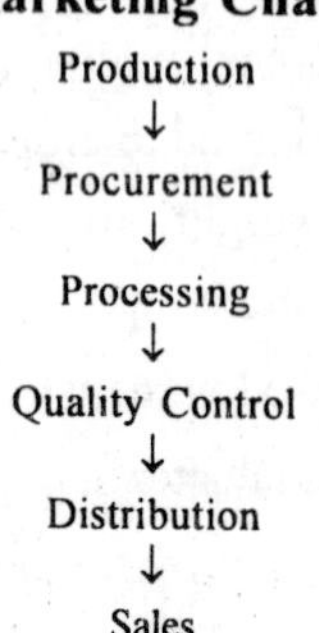

Problems with Marketing

- Irregular supply
- Adulteration
- Storage and warehouse inadequate
- Lack of financial facilities and market facilities
- Absence of proper grading and standardisation

6.4 OTHERS

6.4.1 Role of seed Industry

Seed sector is heavily dependent on research which has to provide a continuous supply of improvement seeds to keep pace with growing requirement. So far, more than 3,000 varieties have been notified under the Seeds Act, 1966 in different crops at National level. Today, the Indian seed programme boasts one of the biggest seed markets in the world, with annual sales at around US $920 million. Of this, domestic accounts for US $900 million and sales in the global market accounts for the remaining US $ 20 million.

The National Seeds Project (NSP) launched in 1975 with the assistance of World Bank had an objective of developing a broad-based decentralised network of seed production agencies throughout the country capable of meeting the requirements of different quantities of certified seed. That is, with adequate infrastructure for production, processing and storage of seeds. The National Seeds Projects I and II provided the basis for today's seed industry. Under NSP I and II, State Seed Corporations were set up in Andhra Pradesh, Bihar, Haryana, Karnataka, Maharashtra, Orissa, Punjab, Rajasthan and Uttar Pradesh.

The implementation of the New Policy for Seed Development in 1988, greatly encouraged foreign participation in the seed industry. The policy permitted private companies to import seeds of vegetables, flowers, and ornamental plants, subject to prior evaluation by the Indian Council of Agricultural Research (ICAR). Imports regarding seeds of coarse cereals, pulses, and oilseeds were also permitted but only under the condition that the Indian company importing the seed negotiate a technical cooperation agreement with the foreign supplier stipulating that parent lines, breeder seed, and seed production technology would be ceded to the Indian company within two years. Import duties on seed and seed processing equipment were lowered significantly.

NSP III was launched in March 1990 with the assistance of World Bank. The main objectives of the project were :

(*i*) To assist the farmers by ensuring timely and adequate availability of certified/ quality seeds of suitable varieties at reasonable prices.

(*ii*) To improve the working efficiency of the National and State level public sector seed companies so as to make them economically viable.

(*iii*) To provide facilities for the growth of private seed industry through adequate institutional finance.

Public sector

Like many agriculturally developed Asian nations, India too has a sizeable amount of public and private sector seed businesses. Giant public sector players include the National Seed Corporation (NSC), the State Farms Corporation of India (SFCI) and the thirteen State Seed Corporations (SSCs). NSC was the first public sector organisation, established in 1963, and remained virtually the only agency in seed production for about 13 years. Its role extended to several development programmes including training, quality control and extension activities in seeds. This was followed by the setting up of the SSCs under two consecutive plan periods, supported by the World Bank, and these largely adopted to role of the NSC in the Indian States. These corporations mainly engage in production and marketing of seed of high yielding and hybrid varieties developed by the public sector.

Private Sector

Although private seed companies such as Poacha and Sutton have been established since the pre-independence era, accelerated growth of the private sector began only after the introduction of the new seed policy in 1988 which ushered in a liberal business climate. Currently there are over 200 private seed companies, together with a few multinationals, and these tend to focus on low volume, high value crops with the principal effort being placed on creating hybrids for oilseeds, maize, cotton and vegetable crops. The private sector accounts for 70 per cent of the market turnover whereas the public sector has the greater share in volume sales.

6.4.2 SEED AGENCIES

State Seed Corporation (SSC)

- Production and supply of certified and foundation seeds of national and state varieties.
- Distribution and marketing of certified seeds within the state.

State Seed Certification Agencies

- Responsible for seed certification in the concerned state.
- Make field inspections and conduct seed tests.

State Farms Corporation of India (SFCI)

- Production of foundation and certified seed on behalf of NSC and SSC respectively.
- Taken up large multiplication programme of oilseeds and pulses.

Private Seed Companies

Mahyco, Rasi etc.

- 60% of seed in our country is produced by private seed companies.

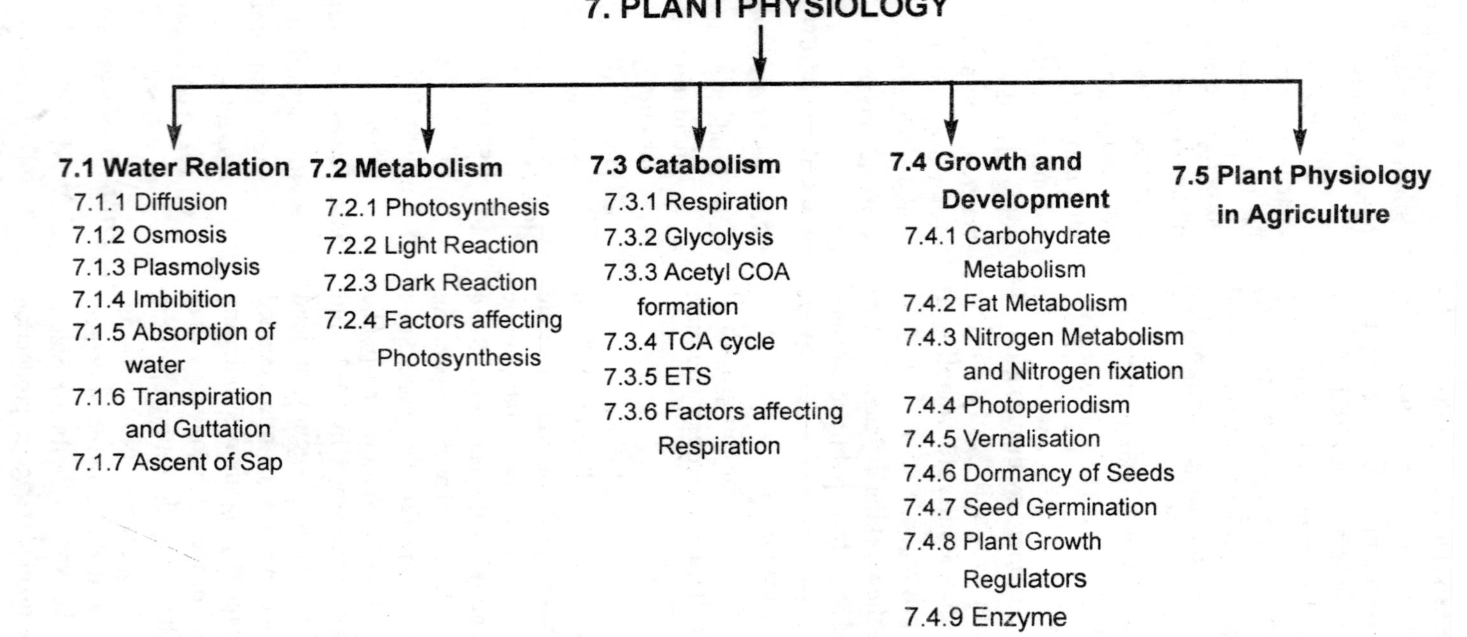
7. PLANT PHYSIOLOGY
7.1 Water Relation
7.1.1 Diffusion
7.1.2 Osmosis
7.1.3 Plasmolysis
7.1.4 Imbibition
7.1.5 Absorption of water
7.1.6 Transpiration and Guttation
7.1.7 Ascent of Sap
7.2 Metabolism
7.2.1 Photosynthesis
7.2.2 Light Reaction
7.2.3 Dark Reaction
7.2.4 Factors affecting Photosynthesis
7.3 Catabolism
7.3.1 Respiration
7.3.2 Glycolysis
7.3.3 Acetyl COA formation
7.3.4 TCA cycle
7.3.5 ETS
7.3.6 Factors affecting Respiration
7.4 Growth and Development
7.4.1 Carbohydrate Metabolism
7.4.2 Fat Metabolism
7.4.3 Nitrogen Metabolism and Nitrogen fixation
7.4.4 Photoperiodism
7.4.5 Vernalisation
7.4.6 Dormancy of Seeds
7.4.7 Seed Germination
7.4.8 Plant Growth Regulators
7.4.9 Enzyme
7.5 Plant Physiology in Agriculture

7. Plant Physiology

Plant physiology : Science that studies plant functions it helps to understand many thing that happen inside the plant.

7.1 WATER RELATION

7.1.1 Diffusion

Movement of particles or molecules from a region of higher concentration to a region of lower concentration is called as *diffusion.*

Significance

- It is an essential step in the exchange of gasses during respiration and photo synthesis.
- During passive salt uptake, the ions are absorbed by simple process of diffusion.
- In stomatal transpiration the diffusion of water vapours from the intercellular spaces into the outer atmosphere through open stomata.

7.1.2 Osmosis

If a solution and its pure solvent are separated by a semipermeable membrane the solvent molecules diffuse into the solution. This diffusion of solvent molecules into the solution through a semipermeable membrane is called as *osmosis*.

Significance

- Large Quantities of water are absorbed by roots from the soil by osmosis.
- Cell to cell movement of water and other substances dissolved in it involves this process.
- Opening and closing of stomata depend upon the turgor pressure of the guard cells.

7.1.3 Plasmolysis

If a plant cell is placed in a highly concentrated sugar or salt solution, water from the cell sap flows out due to exosmosis through the plasmamembrane outside the cell. The loss of water from the cell sap causes contraction or shrinkage of the protoplasm. Since the cell wall is firm and less elastic, it can not keep pace with the contraction of plasma membranes. Ultimatly the protoplasm separates from the cell wall and assumes spherical shape. This condition is called plasmolysis.

Plasmolysis can be of two types (*i*) Concave plasmolysis (*ii*) Convex plasmolysis.

Convex plasmolysis always irreversible while concave plasmolysis is usually reversible.

Significance

- It indicates the semi permeable nature of the plasma membrane.
- This phenomenon is utilized in addition of concentrated sugar solution to jams and jellies.
- It is also used in determining the osmotic pressure of the cell sap.

7.1.4 Imbibition

Certain substances like dry wood or dry seeds are placed in water they absorb the water quickly and swell up considerably so that their volume is increased. These substances are called as imbibants and the phenomenon as *imbibition.*

Term imbibition was coined by Sachs.

- Imbibition pressure is also called matrix potential $\rightarrow$ this potential is a measure of atmospheric pressure of the additional water molecules tendency for the matrix to adsorb.

Table. 1. Distinctions among osmotic pressure turgor pressure, wall pressure and diffusion pressure deficit

Osmotic Pressure (OP)	*Turgor pressure (TP)*	*Wall pressure (WP)*	*Diffusion Pressure deficit (DPD)*
1. It is a colligative property of the molecules particularly in a solution. The OP of a concentrated solution is more than that of a dilute solution.	1. It is the hydrostatic pressure generated within the cell against cellwall as a result of entry of water into it, due to osmosis.	1. It is equal and opposite pressure of the cell wall against the cell content to the turgor pressure.	1. When a cell is placed in pure water, the water enters into the cell as a result of the diffusion pressure deficit of cell sap.
2. If the two solutions are separated by a semi-permeable membrane, the water moves from the solution having less OP to the solution of more OP.	2. Turgor pressure of a cell is maximum when it is fully turgid and it is reduced as the cell loses water.	2. It is the resistance of the cell wall generated against the expanding protoplasm due to entry of water.	2. The pure water has more diffusion pressure. If some solutes are mixed into it then it reduces the diffusion pressure of water. This is called diffusion pressure deficit.
3. The OP is responsible for the movement of water across a semi permeable membrane.	3. The actual growth of a cell occurs when the TP exceeds wall pressure.	3. At a given time wall pressure equals to turgor pressure WP = TP.	3. If a plant cell is placed in water. The water enters into it due to OP. But this entry is restricted by the turgor pressure. Therefore, the actual pressure due to which water enters will be equal to OP minus TP. DPD = OP – TP.

Significance

- The first step in the absorption of water by the roots of higher plants is the imbibition of water by the cell walls of the root hairs.
- Imbitition of water is very essential for dry seeds before they start germination.
- Rate of imbibition increases with temperature increases.

7.1.5 Absorption of Water

Soil Water Relation

Plants absorb water mainly from the soil.

Five stages of water in the soil —

1. Gravitational water. 2. Capillary water.
3. Imbibed water. 4. Runaway water.
5. Chemically combined water.

Mechanism of water Absorption

1. Active absorption of water
2. Passive absorption of water

1. *Active Absorption of Water*

When the water is absorbed due to the activity of root itself, particularly root hairs, it is termed active absorbtion.

(*a*) *Osmotic Absorption*

The first step in the absorption of water is its imbibition on the cell wall surface. The water enters through the cell wall and comes in contact with differentially permeable plasma membrane and cytoplasm. The cell sap, filled inside the vacuole is now separated from external soil water by a differentially permeable membrane.

(*b*) *Non-osmotic absorption*

Sometimes, it has been observed that absorption of water takes place even when the osmotic pressure of cell sap. This type of absorption which is non osmotic and against the osmotic gradient requires the expenditure of metabolic energy probably through respiration.

2. *Passive Absorption of Water*

Passive absorption of water takes place when rate of transpiration is usually high. Rapid evaporation of water from the leaves during transpiration creates a tension in water in the xylem of leaves. This tension is transmitted to water in xylem of roots through the xylem of stem and the water rises upward to reach the transpiring surfaces.

Pathway of water across root cells —

There are three possible pathways for the movement of water across the cortex :

(*i*) Through a system of inter connecting cell walls and intercellular spaces.

(*ii*) Through the living cytoplasm via plasmodesmata.

(*iii*) Through the living cytoplasmic membranes, cytoplasm and vacuoles of the critical cells.

Table. 2. Difference between active and passive water absorption

Active water absorption	*Passive water absorption*
• It occurs mainly due to the activity of root hairs. Needs energy.	• It occurs mainly due to the activity of upper part of plant such as shoot and leaves.
• The water is absorbed by the osmotic and non osmotic mechanisms.	• The water is absorbed due to the process of active transpiration in the upper part.
• The osmotic process involves diffusion pressure dificit of the cells.	• The passive absorption occurs due to the tension created in xylem sap by transpiration pull.
• The water first enters into the cell sap and then passes from one cell to another (symplast movement).	• The water in passive absorption moves probably through the free spaces (apoplast movement)

Factors affecting the rate of water absorption

1. Available soil water : If water in the soil is present below the permanent wilting percentage the absorption is affected.
2. Temperature of the sòil : The low temperature of the soil reduces absorption of water.
3. Concentration of soil solution.
4. Aeration of Soil : Poorly aerated soil retards absorption of water.

7.1.6 Transpiration and Guttation

The loss of water in the form of vapours from the living tissues of aerial parts of the plant is termed as transpiration.

Types of Transpiration —

1. Stomatal transpiration
2. Cuticular transpiration
3. Lentricular transpiration

Factors affecting the rate of transpiration

External Factors —

1. *Atmospheric Humidity* : The rate of transpiration decreases with the increase in relative humidity of atmosphere.
2. *Temperature* : The rate of transpiration increases with the increase in atmospheric temperature.
3. *Light* : The rate of transpiration increases in light and decreases in dark.
4. *Atmospheric Pressure* : Lowering of atmospheric pressure permits more rapid diffusion of water vapours.
5. *Available soil water* : If the available water in the soil is not sufficient the rate of transpiration is decreased.

Internal Factors —

1. *Leaf area* : By reducing the size of leaves, plants decrease their rate of transpiration.
2. *Leaf structure* : Thick cuticle on the surface of plant parts, compact mesophyll cells in the leaves and reduction in number of stomata reduce the rate of transpiration.
3. *Age of plants* : Rate of transpiration is slow at the seedling stage and maximum at maturity state.

Advantages of transpiration —

- It controls the rate of absorption of water.
- It helps in the mineral absorption and ascent of sap.
- It regulates the plant temperature.

Disadvantages of Transpiration —

- Nearly 97% of the total water absorbed by the plants is lost in the form of vapours.
- Transpiration often results in water deficits which causes injury to the plants by desiccation.
- Rapid transpiration causes temporary wilting.
- Deciduous trees have to shed their leaves to check transpiration.

Guttation

The process of exudation of liquid drops from the edges of leaves is called Guttation.

Table. 3. Difference between Transpiration & Guttation

Transpiration	*Guttation*
• Occurs in day	• Night
• Water is given out in in the form of vapour	• In the form of liquid
• Occurs through stomata, lenticle or cuticle	• Occurs through hydathodes
• It is a controlled phenomenon	• It is a uncontrolled phenomenon

7.1.7 Ascent of Sap

The water is absorbed mainly by the roots and is moved upward to all the parts of plants via stem. Thus the upward movement of water through stem to tree top is called ascent of sap.

Path of Ascent of Sap

After absorption by root hair cells, the water moves through several root tissues such as cortex, endodermis, pericycle and finally enters the xylem tissue. Once the water it reaches the mesophyll tissues of the leaves. The bulk of water enters the mesophyll cells and finally evaporates and transpires through the stomata, only small amount of water is used in metabolism and growth.

Many theories have been put forward to explain the upward movement of water.

1. Root pressure theory
2. Vital theories
3. Physical force theories

1. Root pressure theory - stocking
2. Vital theories of ascent of sap - Godlewski, Bose and Westermeir
3. Physical force theories
   (*a*) Imbibition theory - **Uinger and Sachs**

(b) Capillary theory - **Boehm**

(c) Transpiration pull and cohesion of water theory. This theory was first proposed by Dixon and Joly in 1894.

The theory is based on the following features :

(i) Cohesive and Adhesive properties of water molecules to form a continuous water column in the xylem.

(ii) Transpiration pull exerted on this water column.

Water molecules remain joined to each other due to the presence of H-bonds between them.

Although H-bond is very weak but when they are present in enormous numbers as in case of water, a very strong mutual force of attraction or cohesive force develops between water molecules and hence they remain in the form of a continuous water column in the xylem.

The adhesive properties of water further ensure the continuity of water column in Xylem.

When transpiration takes place in leaves at the upper parts of the plant, water evaporates from the intercellular spaces of the leaves to the outer atmosphere through stomata. More water is into the intercellular spaces from the mesophyll cells. Due to all this tension is created in water in the xylem elements of the leaves. This tension is transmitted downward to water in xylem elements of the roots through the xylem of petiole and stem and the water is pulled upward in the form of continuous unbroken water column to reach the transpiring surfaces upto to the top of the plants.

Factors affecting Ascent of Sap

High temperature, low atmospheric humidity, high atmospheric pressure and wind velocity permit more rapid transpiration and thus these factors also promote upward transport of water. Deficiency of water in the soil also cause decrease in the ascent of sap indirectly by influencing the absorption of water. Stomatal opening and closing also regulates movement of water in plants.

7.2. METABOLISM

7.2.1 Photosynthesis

It is an anabolic process, where the carbon containing compounds are formed from CO_2 and water by illuminated green cells, water and oxygen being used to produce photosynthesis.

Estimates reveal that 90% world's photosynthesis is done by marine water and fresh algae.

Blue Print

1. Light Reaction (Grana)
   $H_2O \rightarrow ATP + NADPH_2 + O_2 \uparrow$
   1.a → Non cyclic Phosphorylation
   1.b → Cyclic Phosphorylation
   1.c → Pseudocyclic Phosphorylation
2. Dark Reaction (Stroma)
   $ATP + NADH_2 + CO_2 \rightarrow C_6H_{12}O_6$
   2.a → C_3 Cycle
   2.b → C_4 Cycle
   2.c → CAM

Photosynthetic Apparatus

The chloroplasts in green plants constitute the photosynthetic apparatus. The chloroplast is bounded by two membranes consisting of lipid bilayer and proteins. Internally the chloroplast is filled with a hydrophilic matrix called as stroma in which are embedded grana. Each granum consists of disk shaped grana lamella. In cross section these lamella are paired to form sac like structures and have been called as thylakoids.

Chlorophylls and other photosynthetic pigments are found in the form of protein pigment complexes mainly in thylakoid membranes of grana. The grana are sites of primary photochemical reaction. Dark reaction of photosynthesis occurs in stroma.

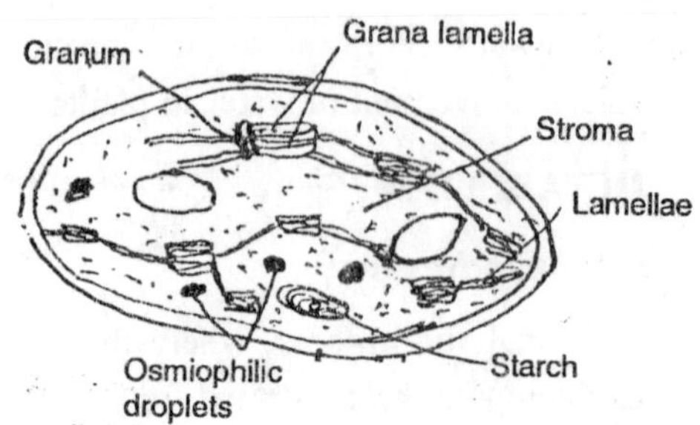

Fig. 1. Internal structure of chloroplast

Photosynthetic Pigments

Photosynthetic Pigments are of three types : 1. Chlorophylls 2. Carotenoids 3. Phycobillins.

Different pigments absorb light of different wavelengths.

(Refer Fig. 1 above) chlorophyll mostly occurus in the grana lambella. Seven types chlorophyll are known.

Red Drop and Emersons Effect

Proposed by Robert Emerson, (1943) while working on quantum yield of Photosynthesis in monochromatic light of different wavelengths in chlorella noticed a sharp decrease in quantum yield at wavelengths greater than 680 nm. Because this decrease in the quantum yield took place in red part of the spectrum and this phenomenon is called *red drop.*

Emerson and Lewis later found that, inefficient far-red light in chlorella, beyond 680 nm could be made fully efficient if supplemented with light of shorter wavelengths. The quantum yield from the two combine beams of light was found to be greater than the sum effect of both beams used separately. This enhancement of photo synthesis called as emersons enhancement effect.

Two Pigment System

Photosynthesis is driven by two pigment systems called pigment system I (PSI) and PSII. Wavelength of light shorter than 650 nm affect both the pigment systems and greater than 680 nm affect only PSI (700 nm).

(1) PSI Relatively Weak Fluorescent —

- Consists of chlorophylls, carotenoids, cytochrome 7, molecule of P700, Plastocyanin, cyt b, FRS (Ferredoxin reducing substance) and rich in chlorophyll-a, iron and copper.
- Controls, the process of producing a strong reductant to reduce NADP to NADPH + H^+
- Mainly found in stroma Lamellae

(II) PSII - Strongly Fluorescent —

- In addition to chlorophylls, carotenoids, moecule of P 680, Plastoquinone, four mn molecules bound to one or more proteins. Cyt b, cyt b_6, and chloride.
- Concerned with generation of strong oxidant and weak reductant coupled with the release of oxygen.

Production of Assimilatony Powers in Photo Synthesis

ATP and NADPH are the assimilatory Powers. The process of NADP in to NADPH + H^+ may be denoted as ETS in Photosynthesis or Reduction of NADP, while the process of formation ATP from ADP and inorganic phosphate using light energy called photo phosphorylation.

ATP molecule is a nucleoside triphosphate, consisting of a nitrogenous base (adenine) a ribose sugar and three serially bonded phosphate groups. ATP is referred as "energy currency" of the cell, as it provides readily releasable energy in the bond between the second and third phosphate groups.

7.2.2 Light Reaction

The light reaction and its concerned activities are found in thylakoids and grana.

$$6\ CO_2 \rightleftharpoons C_6H_{12}O_6 + 6O_2$$

The light reaction phase is a complicated process with several important events.

Events in Light Reaction —

(*i*) *Absorbtion of light energy by chloroplast Pigments :* Different chloroplast pigments absorb light in different region on the visible part of the spectrum.

(*ii*) *Transfer of light energy from Accessory Pigments to Chlrophyll-a* : All the

Photosynthetic Pigments except Chlorophyll-a, are called accessory pigments. Light energy absorbed by them is transferred by resonance to Chl-a which alone take part in the primary photo chemical reaction. Chl-a alone can absorb the light energy directly. In PSII, reaction centre is P680 and in PSI its P700.

(*iii*) *Activation of Chl-a molecule by Photons of Light* : When P680 or P700 forms chlorophyll - a molecule in two pigment systems receive a photon of light, it becomes an excited molecule having more energy than ground state energy and finally expels it energy along with an e^- and a^+ charge comes on the Chl-a which now becomes oxidised.

$$\text{Chl-a} \xrightarrow{\text{Light}} \text{Excited triplet state of chl-a} \longrightarrow (\text{Chl-a})^+ + e^-$$

(*iv*) *Photolysis of water and O_2 Evolution (Oxidation of water)* : Associated with PSII and catalysed by presence of Mn^{++} and Cl^- ions. PSII receives light and water molecules split into OH^- and H^+ ions (Photolysis of water). OH^- ions unite and form some more water molecules again and release O_2 and electrons.

$$2H_2O \longrightarrow 4H^+ + O_2 + 4\,e^-$$

(*v*) Transfer e^- from PS II to PS I, (Pc) Plasto cyanin is blue, cu containing protein art as Mobile carrier.

(*vi*) *Electron transport and the production of Assimilatory power ($NADPH + H^+ + ATP$)* : When the expelled e^- travelling through number of electron carriers either cycled back or is consumed in reducing $NADP^+$ to $NADPH + H^+$. The extra light energy liberated along with the e^- is utilised in the formation of ATP molecules at certain places during its transport, the process is called *photo-phosphorylation*.

There are 3 types of photophosphorylation, viz.,

1. *Non-cyclic photo phosphorylation/e^-transport. (z-scheme).*

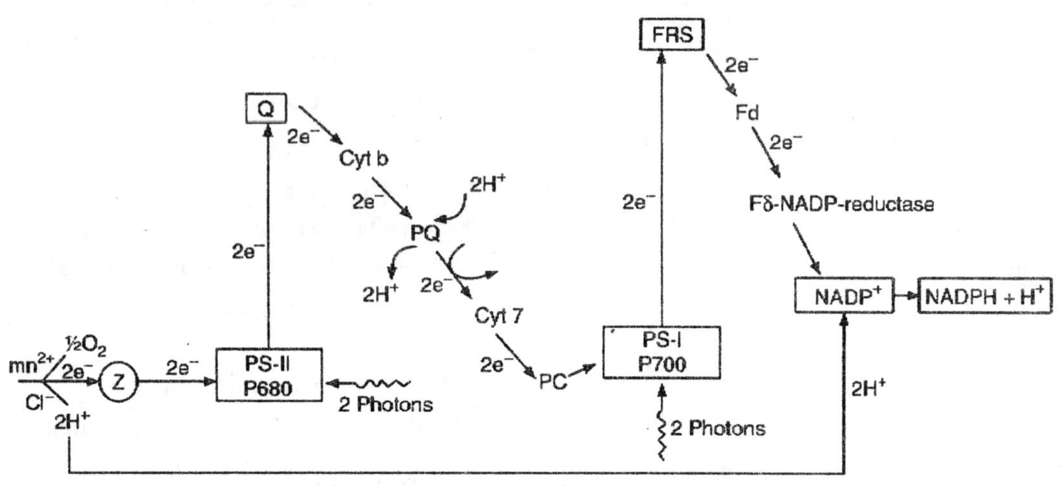

Fig. 2. Non cyclic Photo Phosphorylation

In non cyclic photo phosphorylation, the flow of electron is unidirectional, that is, electron denoted by PSII after passing through plasto quinones, Cyt b_6, Cyt F, plastocyanin and PSI eventually reaches ferredoxin which in turn denotes to reduce NADP. The reduced NADP is utilised for the reduction of Co_2. The electron does not complete the cycle. It starts from PSII reduction. So the ATP synthesis resulting from this type of non cyclic electron transport chain is known as non cyclic photophosphorylation. Water molecule is utilised as a source of electron in this system. In this process, two molecules of ATP are formed per two molecules of $NADP^+$ reduced.

During light reaction, the protons accumulate inside the thylakoid membrane resulting in a proton Gradient. The energy released by the protons when they diffuse across the thylakoid membrane into the stroma is used to produce ATP.

2. *Cyclic PhotoPhosphorylation*

Here ATP formed from PSI and wavelength of light greater than 680 nm is used. Here initially energy is provided from absorbed light Quantum instead of glucose. Electrons used here do not come from water. In the cyclic photophosphorylation light lifts the electron from P700 or feredoxin. The excited electron returns to P700 through two to three transfer steps to decreasing redox potentials. It is during such a downhill migration of the electron that enough energy is released for ATP Synthesis. In this process ATP can be formed between cytochrome b_6 and cytochrome and between ferredoxin and cytochrome b_6.

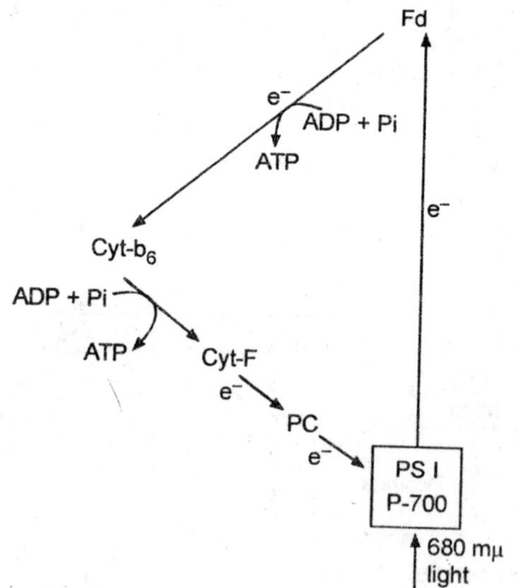

Fig. 3. Cyclic Phosphorylation

3. *Pseudocyclic Photophosphorylation*

Arnon and co workers demostrated, even in the absence of Co_2 and NADP, if chlorophyll molecules are illuminated it can produce ATP from ADP and Pi in presence of FMN and oxygen. The process otherwise is called oxygen dependent FMN catalylsed photophosphroylation, which involves reduction of FMN with the production of oxygen.

FMN is reoxidisable Hydrogen acceptor with the effect that the reduced FMN is be oxidised by oxygen. Thus the process can continue repeatedly to produce ATP.

Here the source of both hydrogen for the reduction of FMN and of oxygen for its oxidation in the water.

$$FMN + H_2O \xrightleftharpoons{\text{illuminated chloroplast}} FMNH_2 + \frac{1}{2}O_2$$

ADP + Pi → ATP

7.2.3 Dark Reaction

It occurs in Stroma —

The dark reaction process of photosynthesis has been named variously such as calvin cycle, Blackman reaction, carbon assimilation, Reductive pentose phosphate cycle etc.

It consists of —

(*a*) Synthesis of carbohydrate

(*b*) Regeneration of ribulose diphosphate

It can otherwise be called as Reductive Pentose Phosphate Pathway. (RPP) the cycle includes 3 phases.

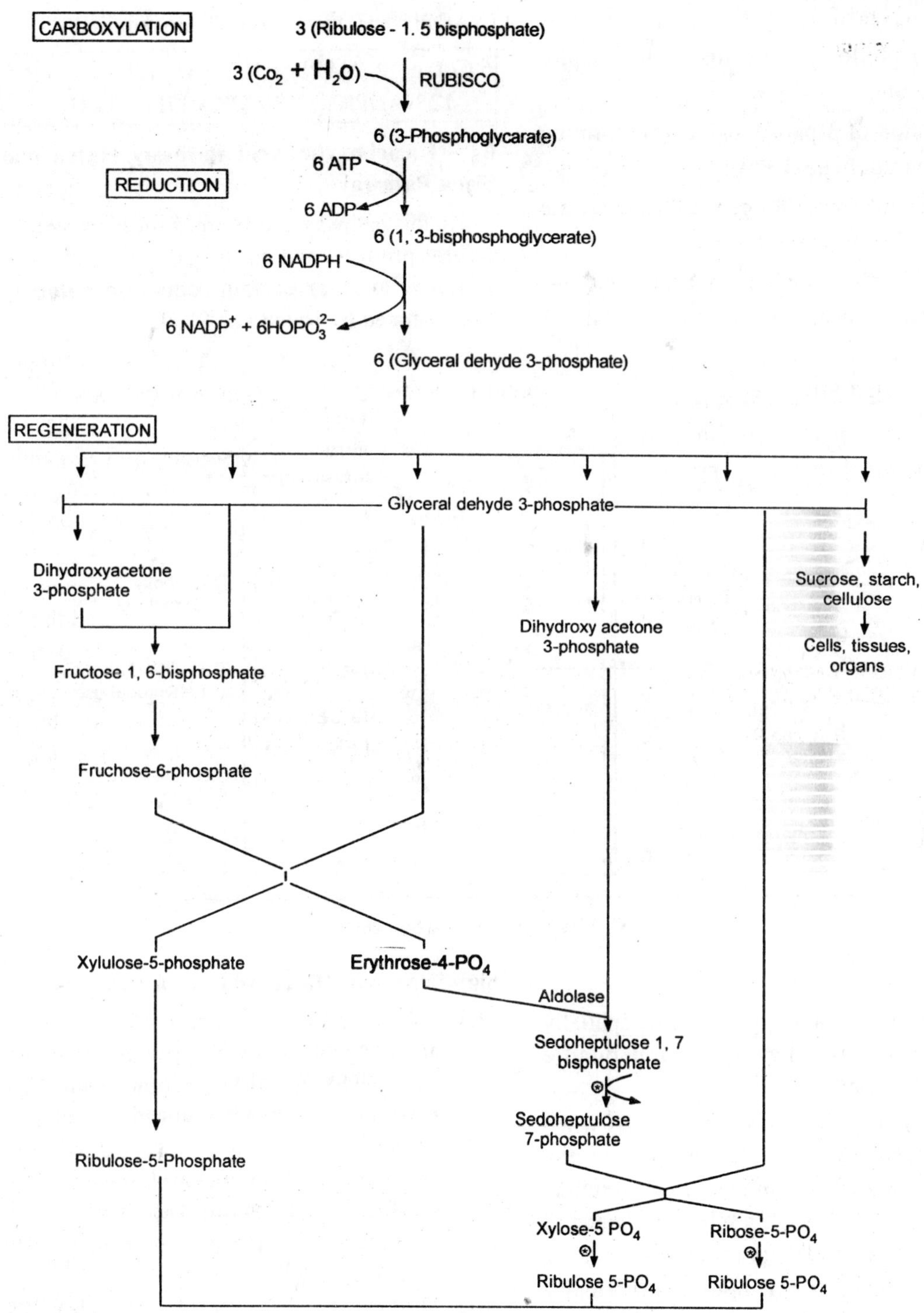

Fig. 4. Calvin Cycle

1. Carboxylation of the ribulose 1.5-bisposphate forming 3-phospho glycerate.
2. Redution of 3-phosphoglycerate forming Glyceraldehyde-3-PO_4.
3. Regeneration of ribulose 15-bisphosphate from Glyceraldehye 3-PO_4.

For the fixation of $6CO_2$ molecules into one hexose sugar molecule through calvin cycle, 12 ($NADPH + H^+$) and 18 ATPs are required.

$$6\ CO_2 + 12\ (NAPH + H^+) + 18\ ATP \rightarrow F\text{-}6\text{-}P + 12\ NADPH^+ + 18\ ADP + 17Pi + 11\ H_2O$$

C_4 - Dicarboxylic Acid Pathway Hatch and Slack Pathway

In this pathway, C_4-dicarboxylic acids are the earliest products and so its called C_4 path way, which is another carbon reduction pathway demonstrated by Hatch and Slack.

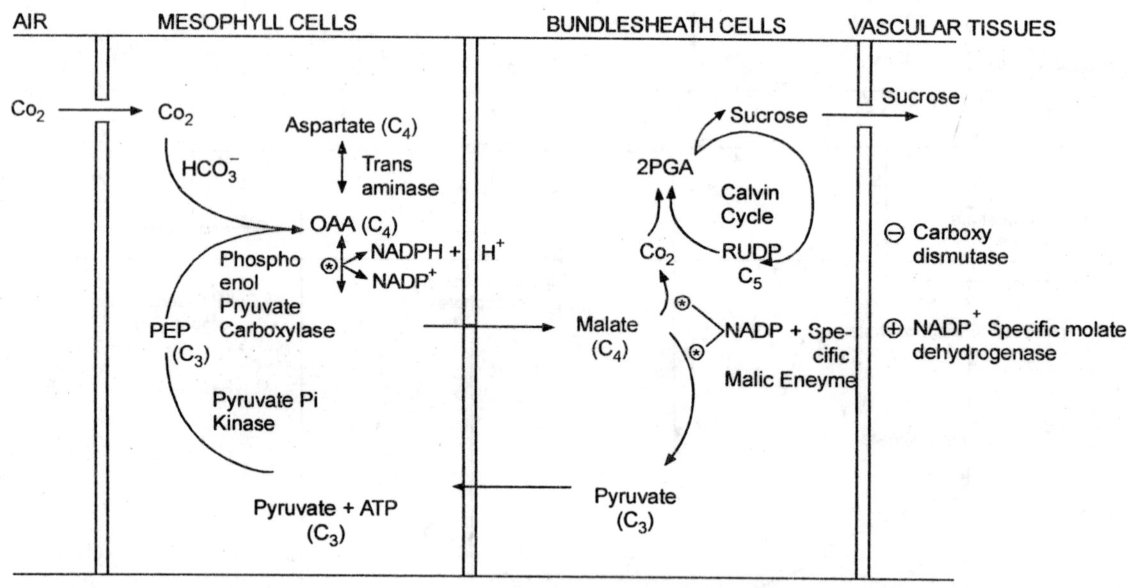

Fig. 5. C_4 Photosynthetic carbon cycle

Characteristics of C_4 Cycle

- The carbon fixation reactions are spatially seperated like mesophyll and bundle sheath cells.
- The cycle is less efficient in itself when compared to C_3 mode. Its because the fixation of 1 Co_2 mol in C_3 mode requires 2 NADPH + 3 ATP, while in C_4 mode, 2 NADPH + 5 ATP are required.
- But, C4 plants are photosynthetically efficient than C_3 plants because of the absence of Photorespiration in C_4 plants.

Significance of Hatch and Slack Pathway

(*i*) Its a modification of calvin cycle and is advantageous to plants growing in dense, stands of tropical vegetation where Co_2 concentration much reduced.

(*ii*) There has been a reduction of atmospheric Co_2 concentration since the evolution of photosynthesis which prompted C_4 plants to select this pathway.

(*iii*) Discovery of C_4 cycle, indicated the existence of yet undiscovered reaction other than the conventional calvin cycle.

Table. 4. Differences between C_3 and C_4 Plants

C_3 Plants	C_4 Plants
1. Wheat, oats, Barely, rice, cotton, Beans, spinach, sunflower etc.	1. Sugarcane, maize, sorghum and amarnathus.
2. 'C' Pathway in photosynthesis is only C_3 pathway (i.e.) calvin cycle only.	2. C_4 dicarboxylic acid pathway (Hatch and Slack pathway).
3. Phosphoglyceric acid [PGA] is the first stable product.	3. OAA (oxalo acetic acid) - I stable product.
4. RUBP-bisphosphate Primary CO_2 acceptor.	4. PEP is the Primary CO_2 acceptor.
5. Leaves have diffused mesophyll and only one type of chloroplasts.	5. Leaves have kranz anatomy with compact mesophyll around vascular bundles (Bundle sheath) and dimorphic chloroplasts present.
6. In each chloroplast two pigment systems are present.	6. In bundle sheath cells chloroplasts, PSII absent. So, dependent on mesophyll chloroplasts for the supply of NADPH + H^+.
7. Calvin cycle occurs in mesophyll chloroplast.	7. Calvin cycle occurs in bundle sheath cells chloroplasts.
8. CO_2 compensation point 50-150 PPm CO_2.	8. 0-10 PPm CO_2.
9. Photorespiration present and easily detectable.	9. Photorespiration is present only to a slight degree or absent.
10. The efficiency of CO_2 absorption at low concentration is far less. So they are less efficient.	10. Efficiency is higher and are more efficient.
11. Only C_3 cycle is found.	11. Both C_3 and C_4 cycles found.
12. Optimum temperature 10-25°C	12. Optimum temperature is 30-45°C.
13. 18 ATP for one glucose molecule	13. 30 ATP for one Glucose molecule.

Crassulacean Acid Metabolism (CAM Cycle)

Here the CO_2 *is concentrated at the site of Rubisco*, which is found in *crassulaceae, Euphorbiaceae, Orchidaceae* and *agavaceae*. This mechanism enables plants to improve water use efficiency and so, these plants have a competitive advantage in dry environment.

Unlike C_4 plants, the formation of C_4 acids, is both temporally and spatially seperated.

In CAM plants, stomata open at night and closed during most of the day. The distinctive diurnal fluctuation in acidity in plants showing CAM due to changes in amounts of vascular malic acid in mesophyll cells.

During night, malic acid synthesised using CO_2 which accumulates in the vacuole. During day this malic acid consumed resulting in decrease of acidity. Citric and isocitric acid also contribute to acidity, but their amount is negligible and these do not show consistent diurnal patterns of fluctuation as shown by malic acid.

Synthesis of Malate during night or Dark CO_2 Fixation

Large amounts of starch are consumed during acidification which indicates that carbohydrates are the source of malate synthesis

$$C_6H_{12}O_6 + 2\,CO_2 \longrightarrow \underset{\text{(malate)}}{2\,C_4H_6O_5}$$

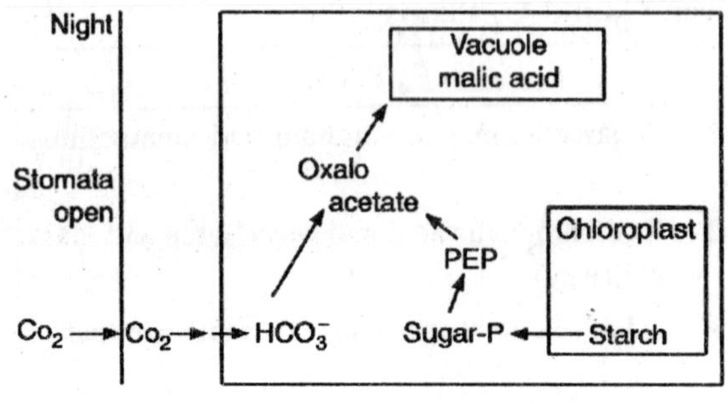

MESOPHYLL CELLS

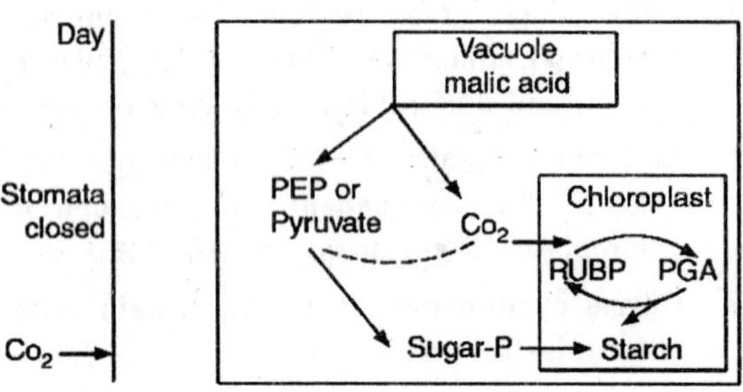

Fig. 6. CAM

Malate is synthesized during night in reaction in which some malate product derived from carbohydrate reserves *e.g.* pyruvate or PEP is carboxylated to produce malate either directly or first forming oxalo acetic acid which is then reduced to malate.

Consumption of malate in Light Deacidification

During the day, when acidified organs exposed to light, rapid consumption of malate occurs resulting in deacidification (due to releases Co_2 from malate). The malate may be decarboxylated in two ways.

(*i*) malate + $NADP^+ \underset{\text{enzyme}}{\overset{\text{NADP-malic}}{\rightleftharpoons}}$ Pyruvate + CO_2 + NADPH + H^+

(*ii*) (*a*) Malate + $NAD^+ \underset{\text{dehydrogenese}}{\overset{\text{Malate}}{\rightleftharpoons}}$ Oxaloacetate + NADH + H^+

(*b*) Oxalo acetate + ATP $\underset{\text{Kinase}}{\overset{\text{PEP-carboxy}}{\rightleftharpoons}}$ PEP + CO_2 + ADP

Pyruvate and PEP utilised for carbohydrates Synthesis during the day. Pyruvate first converted to PEP in the presence of Pyruvate Po_4 dikinase.

Pyruvate + ATP + Pi $\longrightarrow$ PEP + Amp + PPi

PEP in both types of CAM plants is converted into 3 PGA by release reactions of glycolysis. Then, 3-PGA is utilised in calvin cycle.

Photo Respiration

Otherwise known as photo respiratory carbon oxidation cycle (PCO). Rubisco, the most abundant protein in the world is having a distinct property, that it can catalyse both the carboxylation and oxygenation of RuBp. oxygenation is the primary reaction in a process known as photorespiration.

It's the process, in which, Co_2 is lost from cells that are simultaneously, fixing Co_2 by the calvin cycle. The cycle acts as a scavenger operation to recover the fixed Co_2 lost by the oxygenase reaction of rubisco. Three organelles are involved in the process viz., chloroplast peroxisomes and mitochondria.

- Photorespiration only in chlorophyllous tissues of plants. Glycolate is the chief metabolite and also its substrate. Other important metabolites are the amino acids glycine and serine.
- Like usual mitochondrial respiration, its also an oxidative process where oxidation of glycolate occurs with subsequent
- Glycolate cycle/C_2-cycle.

Factors Affecting

1. Co_2 compensation point
2. Co_2 concentration when high, photo respiration is also high and vice versa.
3. Inhibitors of Glycolic acid oxidase such as α-hydroxy-sulphonates inhibit the process of photorespiration.

Significance

1. It's a double wasteful process
2. Resulting in considerable decrease in photosynthetic productivity.
3. Unlike usual mitochondrial respiration, neither reduced co-enzymes are generated nor the oxdiation of glycolate coupled with the formation of ATP molecules.
4. Moreover, there is consumption of reduced coenzyme and ATP.
5. However, the knowledge on photo respiration is important. Therefore by manipulating different atmospheric conditions, use of inhibitiors of glycolic acid oxidase such as α-hydroxy sulphonates and through genetic control, the process can be regulated and photosynthetic productivity can be increased.

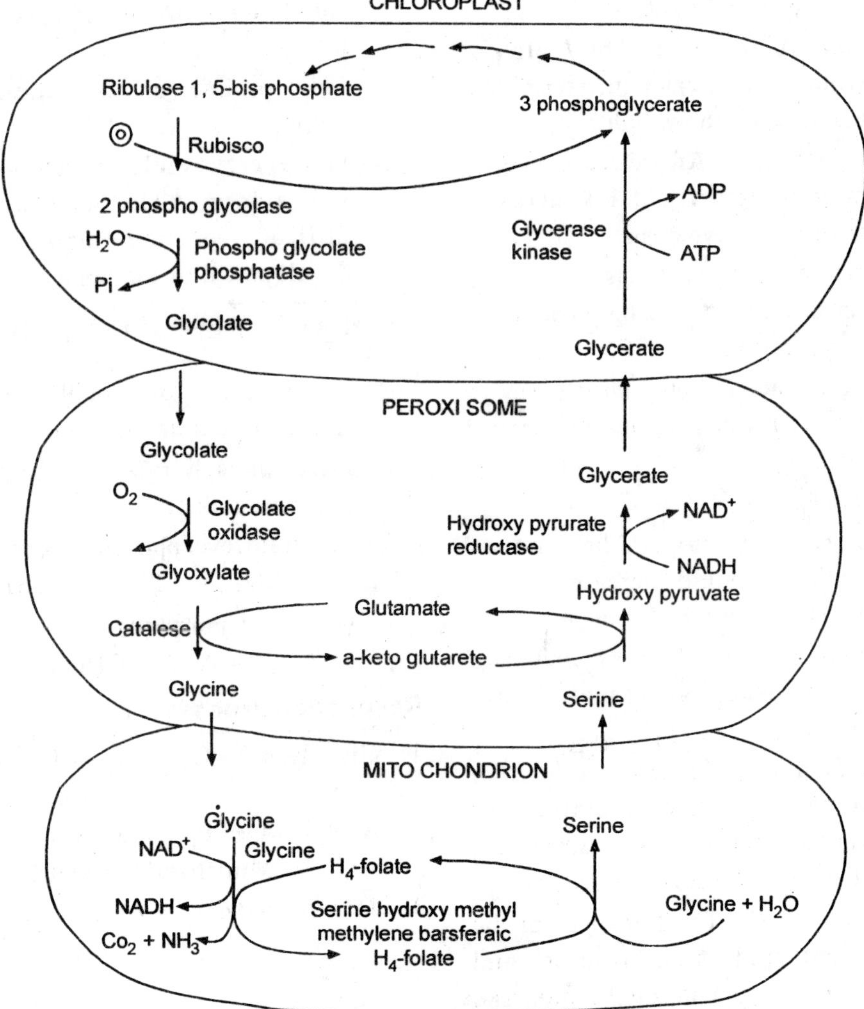

Fig. 7. Operation of the photorespiratory carbon oxidation cycle involves cooperative interaction among three organelles.

7.2.4 Factors Affecting Photosynthesis

External Factors —

1. *Light* : It affects the rate of phososynthesis in three ways :

(*a*) *Light Quality* : Due to heavy absorption of light rays in the red part of the spectrum of chlorophylls, maximum photosynthesis takes place in red light.

(*b*) *Light intensity* : The rate of photosynthesis is greater in intense light than in the diffused light.

(*c*) *Duration of light* : The rate of photosynthesis is greater in intermittent light than in continuous light.

2. *Carbon dioxide* : An increase in Co_2 concentration upto about 1% increases the rate of photo synthesis.

3. Temperature : Photosynthesis will stop in some plants at about freezing point and in some plants beyond 40-50°C.

4. Water : The rate of photosynthesis may go down if the plants are inadequately supplied with water.

Internal Factors —

1. *Chlorophyll content* : The rate of photosynthesis should increase with an increase in the chlorophyll content.

2. *Protoplasmic factors* : Proper hydration of the protoplasm is essential for protosynthesis.

3. *Accumulation of the end products of photosynthesis* : Accumulation of carbohydrates in the photosynthesizing cells retards the rate of photosynthesis.

4. *Anatomy of leaf* : The thickness of the cuticle and epidermis, structure and distribution of stomata, distribution and relative proportion of chlorophyllous and non chlorophyllous mesophyll tissue all influence the rate of photosynthesis.

7.3. CATABOLISM

7.3.1 Respiration

- It is a catabolic process
- Plant respiration (Aerobic) is the biological process by which reduced organic compounds are mobillized and subsequently oxidised in the controlled manner.
- Main product of respiration is energy output.
- Free energy is released and incorporated into a form ATP that can be readily utilized for the maintenance and development of the plant.

$$C_6H_{12}O_6 \xrightarrow{\text{oxidation}} CO_2\uparrow + H_2O + ATP$$

- Production of ATP is called phosphorylation (3types).

1. Photo phosphorylation : Light ⇒ ATP (light reaction)
2. Substrate level phosphorylation : Fructose 1, bis PO_4 → ATP → Glycolysis
3. Oxidative phosphorylation :
   ADP → ATP →ETS.

Respiration Process

1. Glycolysis ⇒ $C_6H_{12}O_6$ → Pyruvate (Cytosol/cytoplasm)
2. AcetylcoA ⇒ Pyruvate → Acetyl COA (Mitochandrain Matrix inner wall)
3. TCA/Kreb cycle ⇒ Acetyl COA → ATP, $FADH_2$, NADH (Matrix of Mitochandrion)

4. Electron transport system ⇒

$$\begin{matrix} FADH_2 \\ NADH \end{matrix} \begin{pmatrix} \text{Inner} \\ \text{Mitochandrion} \\ \text{Membrane} \end{pmatrix}$$

Energy released during respiration can be calculated on the basis of stages of respiration

1. Glycolysis — 2 ATP
2. Kreb's cycle — 2 ATP
3. ETS — 34 ATP (Total 38 ATP)

7.3.2 Glycolysis

- Greek word : Glykos – Sugar
  Lysis – Splitting
- The First stage of respiration, glucose (a six carbon molecule), is split into two 3 carbon molecules sugar are then oxidised and rearranged to yield 2 molecules of Pyruvate are partially oxidised and further oxidised through Link and Krebs cycle.
- Glycolysis occurs in all living organisms (Prokaryotes and Eukaryotes) .
- No oxygen is required to convert glucose to pyruvate and glycolytic mechanisms can become the primary mode of energy production in plant tissue when oxygen levels are low.
- In animal cells the enzymes of glycolytic pathway do not exist independently in the cytosol but associated in a supermolocular complex that may loosely bound to the surface of outer mitochandrion membrane.
- This complex is thought to facilitate the conversion of substrate to products in the multistep glycolytic process.
- In this process phosphofructokinase is one of the control points in both plants and animals.
- This type of ATP synthesis is referred to as substrate level phosphorylation because it involves the direct transfer of phosphat moiety from the substrate molecule to ADP to form ATP.

Alternative Glycolytic Pathway

- The synthesis of glucose through reversal of the glycolytic pathway known as gluconogenesis.
- It is not common in plants, but it does operate in seed of some plants (esp. oil seeds sunflower).
- Carbon stored in the form of oil (Triglycerols). After seed germinate, the oil, are converted into sugar by gluconogenesis.

Respiratory quotient

The molar ratio of carbon dioxide evolved per mole oxygen consumed in respiration is termed as the *respiratory ratio, respiratory coefficient* or *respiratory quotient (*RQ). If respiration proceeds according to the general equation given in the nature of the process, RQ should be equal near to 1 (unity). However due to different environmental factors and variations in the make-up of the substrate RQ may be deviate from this value. Actually, the quotient can vary from infinity (no oxygen taken in) to zero (no carbon dioxide given off). Some values found experimentally are presented in Table.

Table : Respiratory quotients in different plants (After Thomas et al., 1960)

Plant material	RQ
Leaves rich in carbohydrates	1.0
Darkened shoots of prickly pear (*Opuntia*)	0.03
Germinating starchy seeds	1.0
Germinating linseed (high fat)	0.64
Germinating buckwheat (high protein)	0.50
Germinating peas	1.5-2.4

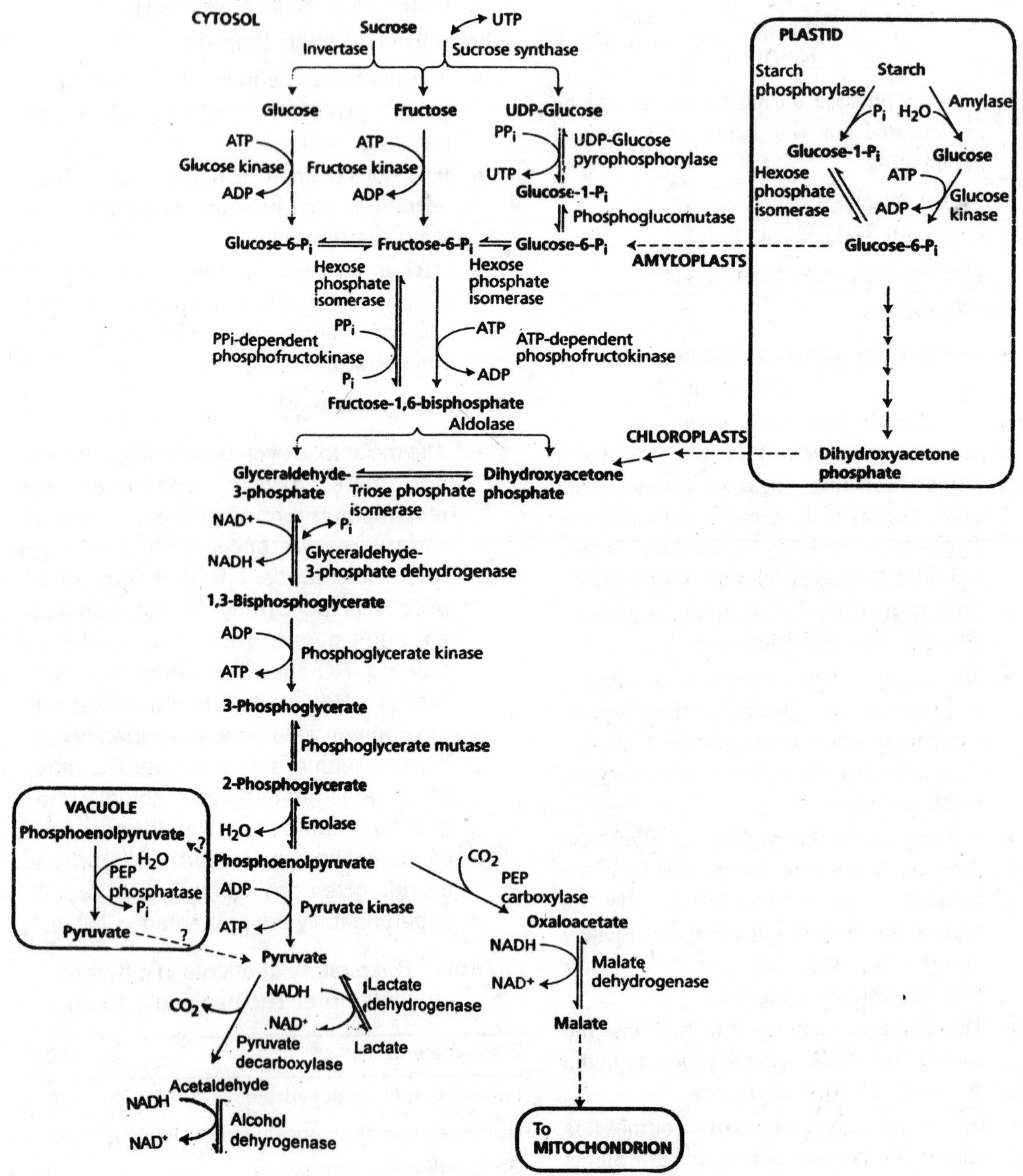

Fig. 8. Glycolysis

7.3.3 Acetyle COA Formation

- Inside the mitochandiral matrix, pyruvate is oxidatively decarboxylated by the enzyme pyruvate dehydrogenate to produce NADH, CO_2 and acetic acid.
- The acetic acid is linked via thioester bond to a sulphur-containing cofactor, co enzyme.

7.3.4 TCA/Kreb/Citric Acid Cycle

- It occurs in mitochondrial Matrix.
- 36 ATP released in Eukaryotes

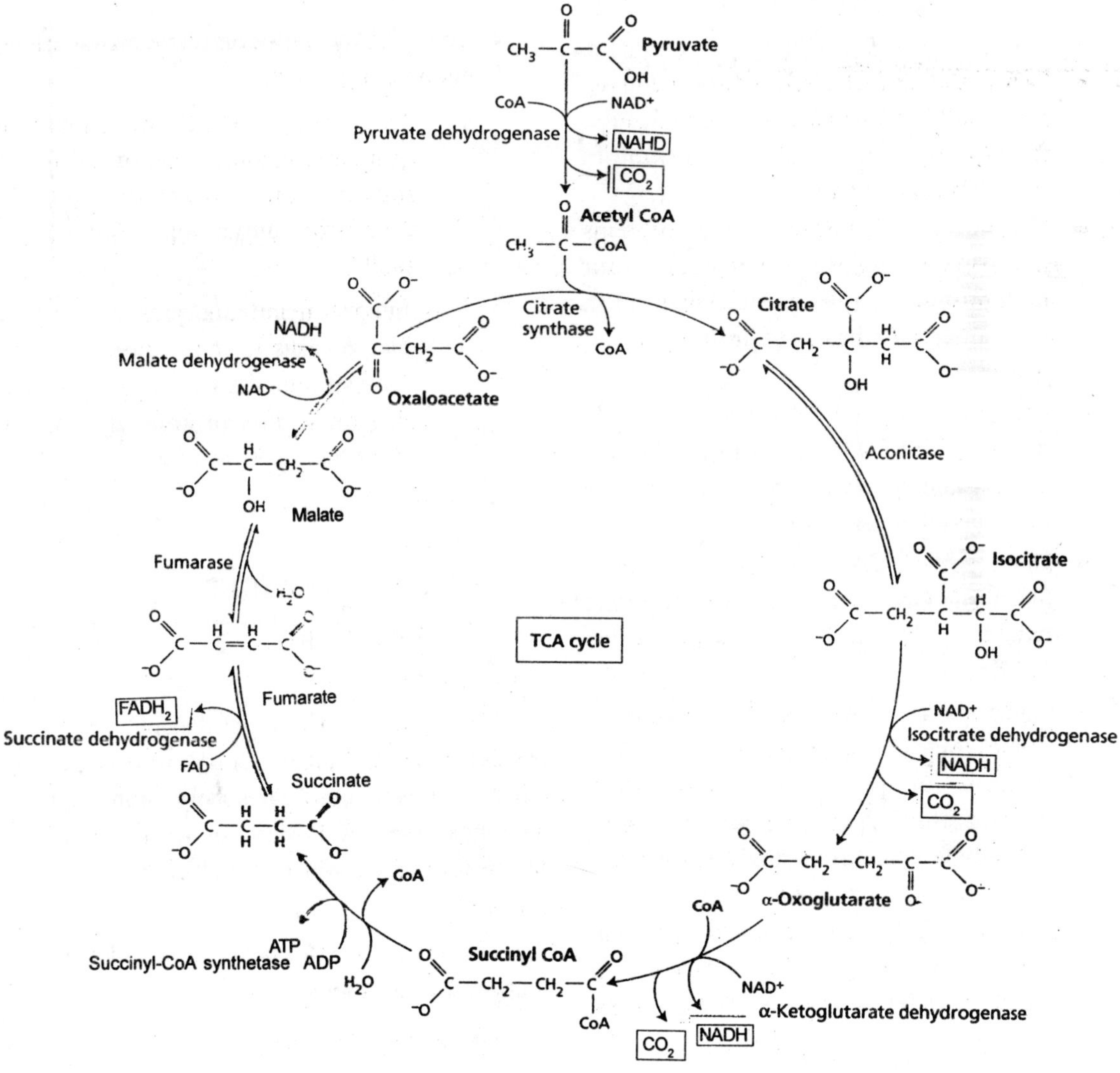

Fig. 9. TCA Cycle

7.3.5 Electron Transport System (ETS) (Oxidative phosphorylation)

- Here the high energy electron capture during the TCA cycle (In the form of $FADH_2$ and NADH) can be converted into ATP.
- This O_2 dependent process occurs on the inner mitochondrial membrane and invites series of electron carriers known as electron transport chain.
- Two molecules of NADH from cytosol, 8 molecules of NADH and Two molecules of $FADH_2$ from TCA - appear in mitochandrial matrix.
- The individual electron transport proteins are organised into a series of four multiprotein complexes and localised in the inner membrane of mitochondrial matrix.
- Complex I : TCA cycle NADH dehydrogenase, Flavin mono Nucleotide (FMN) and several sulphur proteins for the transfer of electrons.
- Complex II : TCA cycle enzyme succinyl dehydrogenase, electron derived from the oxidation of succinate are transfered via $FADH_2$.
- Complex III : Act as a ubiquinol : Cytochrome C Oxidoreductase, oxidising reduced ubiquinol and transfering electron through an iron-sulfur center, two b-type cytochrome and a membrane bound Cyt. C. Cytochrom C - is the only protein in ETC that is not an integral membrane properties, and it serves as a mobile carrier to transfer the electron from complex III to IV.
- Complex IV : Represents the cytochrom C oxidase. Which contain two copper centre (CUA and CUB) and Cyt a.
- Electron carriers within the mitochondrial inner-membrane allows for transfer of proton. The electron transport is associated with a transfer of protons from the mitochondrial matrix to the inner-membrane space.
- Complex V : This complex consist of two major components

  F_1 : Peripheral membrane protein complex contain atleast five different subunits and contains a catalytic centre for converting. ADP + Pi to ATP.

  F_2 : Integral membrane protein complex that consists of atlest three different polypeptides that form the channel through which protons are able to cross the inner membrane.
- Following compounds are known as *couplers.*

  2,4-dinitrophenol (DNP)

  2,5-Dinitrophenol

Uncouplers :

Uncouplers of oxidative phosphorylation in mitochondria inhibit the coupling between the electron transport and phosphorylation reactions and thus inhibit ATP synthesis without affecting the respiratory chain and ATP synthase (H(+)-ATPase).

Dinitrophenol. PMA break the membrane leads to free flow H^+ atoms.

So, no energy production.

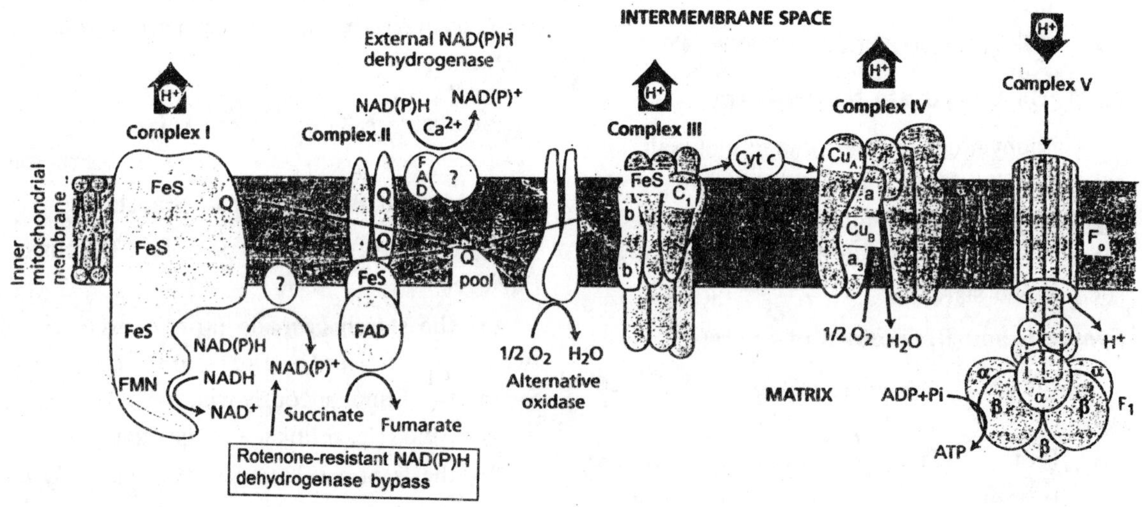

Fig. 10. Electron Transport system.

7.3.6 Factors affecting Respiration

External Factors —

1. *Oxygen*

O_2 is a important factor because of its role as substrate in the over all process.

- The diffusive movement of O_2 represents a limitation on plant respiration.

2. *Temperature*

- In the physiological temperature range, the respiration is temperature dependent.

3. *Injury*

Physical injury to plant tissues often stimulate O_2 uptake because of increases in respiration and in mitochandria oxygen consuming enzyme activity as catalysed by lipoxygenase, polyphenol oxidase and peroxidase.

4. *Water*

Proper hydration of respirating cell is essential for respiration.

5. *Mechanical Effects*

Gentle rubbing or bending of leaf blade shows increased rate of respiration.

6. *Light*

Indirect effect through synthesis of organic matter.

Internal Factors —

1. *Protoplasmic Factor*

The amount of protoplasm in the cells and its state of activity influence rate of respiration.

(*i*) High rate of young meristamatic cells which divide actively require more energy.

(*ii*) Rate of respiration is undoubtedly affected by internal structural relationship of various cell organells and occurance and distribution of enzymes.

2. *Concentration of Respiratory Factor or Substrate*

- Favourable and increased concentrate respirable food material brings about an increase in the rate of respiration.

- Under starvation condition, leaves are etiolated and slow downs the respiration.

7.4 GROWTH AND DEVELOPMENT

7.4.1. Carbohydrate Metabolism

Carbohydrates are the storage materials that are stored in plants which upon its breakdown yield energy required for the vital processes in plants. The prominent carbohydrates in plants are, sucrose and starch.

Synthesis and Breakdown of Sucrose

Synthesis : It may take place by three different ways :

1. From Glucose-1 phosphate and fructose in presence of enzyme sucrose phosphorylase. (eg.) in bacteria.

$$\text{Glucose-1-phosphate} + \text{Fructose} \xrightleftharpoons[\text{Phosphorylase}]{\text{Sucrose}} \text{Sucrose} + \text{inorganic phosphate.}$$

2. From UDPG (uridine Diphosphate Glucose) and Fructose in the presence of sucrose synthetase. (eg) in higher plants.

$$\text{UDPG} + \text{Fructose} \xrightleftharpoons[\text{Synthetase}]{\text{Sucrose}} \text{UDP} + \text{Sucrose.}$$

3. From UDPG and Fructose-6-phosphate in the presence of the enzyme sucrose phosphate synthetase (eg) Higher Plants.

$$\text{UDPG} + \text{Fructose} \xrightleftharpoons[\text{Synthetase}]{\text{Sucrose Phosphate}} \text{UDP} + \text{sucrose phosphate.}$$

Sucrose phosphate thus produced is hydrolysed in the presence of the enzyme phosphatase to yield sucrose.

$$\text{Sucrose phosphate} \xrightarrow[+ H_2O]{\text{Phosphatase}} \text{Sucrose} + \text{Phosphate}$$

Break Down : Sucrose is broken down or hydrolysed to yield glucose and fructose in the presence of enzyme invertase which is a reversible reaction.

$$\text{Sucrose} + H_2O \xrightarrow{\text{Invertase}} \text{Glucose} + \text{Fructose}$$

Synthesis and Breakdown of Starch

Synthesis of Amylose or α-1, 4 Glycosidic Linkages.

As the starch is made up of Amylose and Amylopectin, both the components are Synthesised Simultaneously with α-1, 4 Glycosidic linkages and α-1, 6 linkages respectively.

Synthesis of amylose may take place by the following ways.

1. Amylose can be synthesised in the presence of the enzyme starch phosphorylase from glucose-1-phosphate and an acceptor molecule consists of 3-20 glucose units joined together by α-1, 4 glycosidic linkages.

$$\text{n (Glucose-1-phosphate)} + \text{Acceptor} \xrightleftharpoons[\text{Phosphorylase}]{\text{Starch}} \text{Amylose} + \text{n (ip)}$$

2. Formation of α-1, 4 glycosidic linkages may also take place in the presence of the enzyme UDPG-transglycosylase (amylose synthetase) by the transfer of glucose from UDPG to an acceptor molecule with 2-4 or more glucose units joined together to the α-1, 4 glycosidic linkages or even a starch molecule.

$$\text{UDPG} + \text{Acceptor} \xrightleftharpoons{\text{UDPG trans glycosylase}} \text{UDP} + \alpha\text{-1, 4 Glucose acceptor.}$$

Synthesis of AmyloPectin or α-1, 6 Glycosidic Linkages

Transfer of small chains of glucose units joined together by α-1, 4 linkages to an acceptor molecule consisting of atleast four α-1, 4 linked glucose units in the presence of 'Q' enzyme. The α-1, 6 Glycosidic bond established between C-1 of the terminal glucose units of donor molecule and C-6 of one of the glucose units of the acceptor molecule.

Break Down by Means of Two Ways

1. By the enzyme Diastase

$$\text{Starch} \xrightarrow[+H_2O]{\text{Diastase}} \text{Glucose}$$

Diastase is a complex of many enzymes viz., *α-amylase*, *β-amylase*, *R-enzyme* and *maltase*.

2. *By starch phosphorylase*

$$\text{Starch} + \text{Phosphate} \xrightleftharpoons{\text{Starch Phosphorylase}} \text{Glucose-1-phosphate.}$$

$$\text{Glucose-1-phosphate} \xrightarrow{\text{Phoshpate}} \text{Glucose} + \text{Phosphate}$$

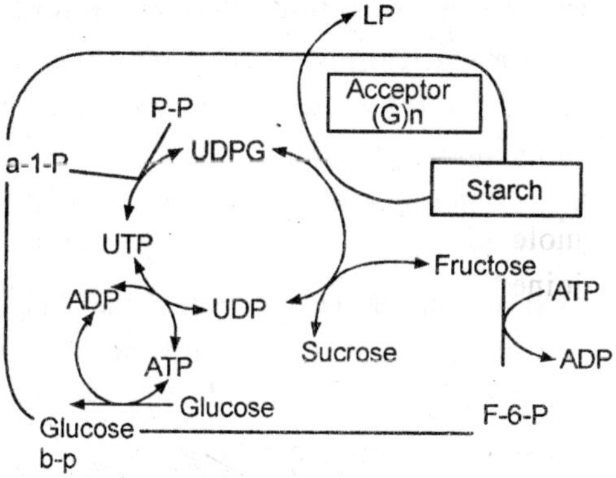

Fig. 11. Sugar/starch synthesis

7.4.2. Fat Metabolism

Lipids may be simple, compound or derived. Simple lipids are esters of fatty acids alcohol. If alcohol in the constitution is glycero they are called fats.

Fat metabolism consists of two parts

(*a*) Anabolism : Metabolic pathways that results is the synthesis of new, larger compounds, *e.g.* protein synthesis.

(*a*) Synthesis of fatty acid.

(*b*) Synthesis of glycerol.

(*c*) Condensition of fatty acids and glycerol into fats.

$$\text{Glycerol} + 3 \text{ Fatty acid COA} \xrightarrow[Mg^{++}]{ATP} \text{lipid} + 3\text{COA}$$

(*b*) Catabolism : Metabolic pathways that results in the breakdown of compounds, *e.g.*, glycolysis.

3 Path way

(*a*) β-oxidation (*b*) α-oxidation

(*c*) Peroxidation

β-oxidation

It is the catabolic process of fatty acid degradation in plants which is well established in saturated fatty acids not in unsaturated fatty acids. The place of occurence of this reaction is mitochondria and also in glyoxysomes which involves sequential removal of 2-C in the form of acetyl-COA from the carboxyl end of the fatty acid. This is called β-oxidation, because β-C of the fatty acid is oxidised during this process.

Reactions Involved

Fatty acid

Thio kinase ↓ ATP, CoAStd

Fatty acyl-COA

Acyl-COA Dehydrogenase ↓ FAD → $FADH_2$

Trans-α, β unsaturated fatty acyl-COA

Enoyl Hydrase ↓ $+H_2O$

β-hydroxy acyl-COA

β-Hydroxyacyl-COA dehydrogenese

↓ NAD^+ → $NADH + H^+$

β-keto fatty acyl-COA

β-keto acyl thiolase ↓ + CO ASH

Fatty acyl-COA + Acetyl COA

Here in this process, huge amount of energy is generated in the form of ATP molecules by the mitochandrial oxidation of fatty acids thro β-oxidation spiral and TCA cycle. Acetyl-COA may enter into TCA cycle or in case of germination of fatty seeds, they are converted to soluble sucrose through glyoxylic acid cycle.

Glyoxylate Cycle (or) Glyoxylic acid cycle.

In this cycle, the fats could be converted into sucrose during the germination of fatty seeds in plants. This cycle is to occurs in many other bacteria, yeasts, molds, and higher plants or is completed in Glyoxysomes, mitochondria and Cytosol.

Significance

- During Germination of fatty seeds, insoluble fats are converted into soluble sucrose, which is then available to the different growing regions by Glyoxylate cycle.
- Micro organisms that can grow in C_2H_5OH/acetate as a sole source use this cycle to produce longer carbon chains.
- It's an example of Gluconeogeneris process.

7.4.3 Nitrogen Metabolism and Nitrogen Fixation

Role of Nitrogen in Plants

- Proteins are the building blocks of protoplasm which are made up of Nitrogenous substances called aminoacids, which are synthesised when inorganic 'N' in the environment is converted to organic 'N' inside the plants.
- Proteins are the major substances in cell next only to water and mineral salts.
- 'N' is the constituent element of important organic compounds like chlorophylls, cytochromes, alkaloids, vitamins and Nucleic acids.
- Thus 'N' plays an important and fundamental role in metabolism, growth, reproduction and heredity.

Sources of 'N'

1. Atmospheric/molecular Nitrogen : Only some bacteria, some BGA, leguminous plants having root nodules can fix atmospheric 'N'.
2. Nitrates, Nitrites, Ammonia in soil (In organic 'N') : Nitrates is the chief form of 'N' taken up by the plants in the soil.
3. Amino acids (Organic Nitrogen) in Soil.
4. Ammonium Compounds of which NH_4^+ ions are absorbed directly by the roots of plants along with nitrate ions.
5. Organic Nitrogenous Compounds in the bodies of Insects.

'N' fixation in Plants : It can be achieved by the conversion of Nitrate in the soil to Ammonia by the plants which occurs as a two step process.

(*i*) Reduction of Nitrate to Nitrite

(*ii*) Reduction of Nitrite to Ammonia

The process of conversion is basically a reductive process, as the Nitrate (NO_3^-) which is a highly oxidised form gets converted into the reduced form (*i.e.*) Ammonia (NH_4) under the two major steps, there are many intermediate steps involved in the conversion, which are mediated by specific enzymes.

$$\underset{\text{Nitrate}}{NO_3^-} \xrightarrow{2e^-} \underset{\text{Nitrite}}{NO_2^-} \xrightarrow{2e^-} \underset{\text{Hypo Nitrite}}{H_2N_2O_2} \xrightarrow{2e^-}$$

$$\underset{\text{Hydroxyl Amine}}{NH_2OH} \xrightarrow{2e^-} \underset{\text{Ammonia}}{NH_3}$$

At each step R electrons are added and ultimately NO_3 in which 'N' has 5 positive charges converted to Ammonia (NH_3) with 3 negative charges. The electrons are supplied by NADH and NADPH.

I. *Reduction of Nitrate to Nitrite :*

Takes place in green leaves and roots and the enzyme is found in cytoplasm.

$$NO_3 + NADH + H^+ \xrightarrow[\text{reductase}]{\text{Nitrate}} NO_2 + NAD^+ + H_2O$$

The enzyme is the molybdoflavoprotein which contains FAD (Flavin Adenine Dinucleotide) as its Prosthetic Group associated with molybdenium.

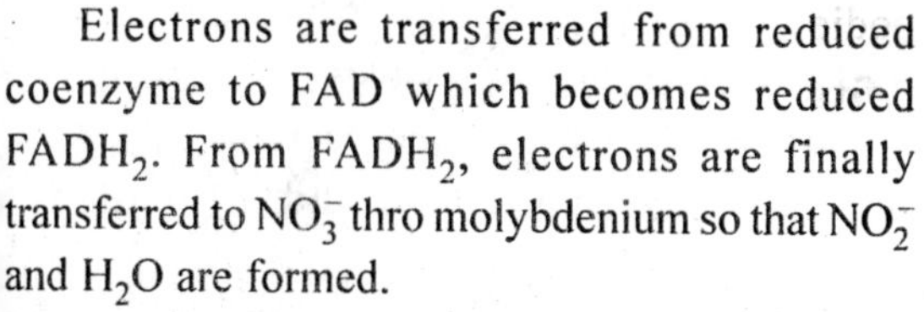

Fig. 12. Nitrogen reduction

Electrons are transferred from reduced coenzyme to FAD which becomes reduced $FADH_2$. From $FADH_2$, electrons are finally transferred to NO_3^- thro molybdenium so that NO_2^- and H_2O are formed.

II. *Reduction of Nitrite to Ammonia :*

$$NO_2^- \xrightarrow[OH^-]{2e^- + 2H^+} NOH \xrightarrow{2e^- + 2H^+}$$

$$NH_2OH \xrightarrow[H_2O]{2e^- + 2H^+} NH_4^+$$

The reduction takes place in the presence of Nitrite Reductase which catalyses by-reduction from NO_2^- to NH_4^+ with the formation of Nitroxyl/ Hypo Nitrite and Hydroxyl amine as inter mediates.

Electron donors for this reduction are, reduced ferredoxian and also reduced pyridine nucleotides.

As against the presence of Nitrate reductase in cytoplasm, this enzyme is found in chloroplasts. Sirohaem, an iron-porphyrin is known to be associated with this enzyme and probably mediates the transfer.

The involvement of two intermediate compounds in the process is doubtful because.

1. Hyponitrite is unstable
2. Hydroxyl amine is toxic.
3. These two, they never occur in free state in the cells. It's now generally believed that hyponitrite and hydroxyalamine may be formed at the surface of the enzyme and leave the surface only when they are completely reduced to further intermediate or ammonia.

There is yet another type of formation of Ammonia, through the Reduction of Hydroxylamine.

The reaction is catalysed by hydroxylamine reductase which requires manganese for its activity.

$$NH_2OH + NADH + H^+ \xrightarrow{mn} NH_3 + NAD + H_2O$$

The ammonia combines with organic acids to produce amino acids.

Biological Nitrogen Fixation

It can be defined as the conversion of molecular nitrogen of the atmosphere into inorganic nitrogenous compounds through the activity of some living organisms.

The process of Biological Nitrogen Fixation can be divided into two types.

Non Symbiotic Nitrogen Fixation

In case of non-symbiotic fixation, some of the components are necessary in the cell system of 'N' fixing bacteria.

1. Presence of hydrogenase and nitrogenase enzyme systems. Hydrogenase reversibly catalyses the reduction of hydrogen ion to molecular hydrogen while, nitrogenase reduces molecular nitrogen to NH_3 in the presence of molecular hydrogen.
2. Presence of Ferredoxin as an electron carrier.
3. Presence of pyruvate which acts as e^-donor and energy source.
4. Presence of co-factors like TPP, COA, ip and Mg^{2+} required in the metabolic utilization of pyruvate, that forms the energy [ATP).

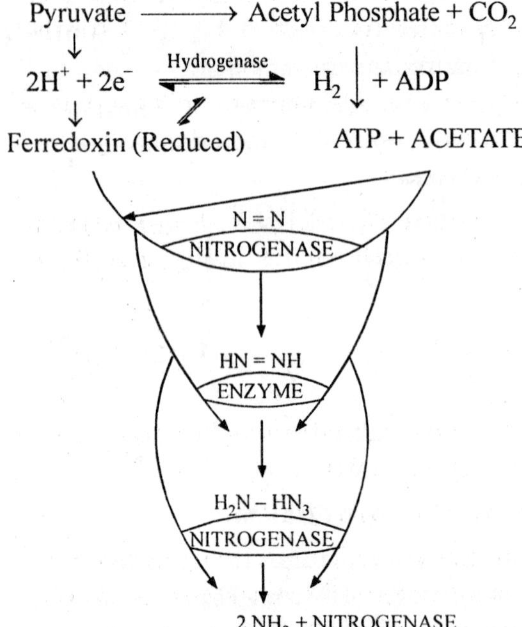

Fig. 13. NH_3 formation

Symbiotic Nitrogen Fixation

It is carried out by bacteria which is frequently found in the root nodules of leguminous plants. The most common bacterium is Rhizobium which has two types of strains.

1. Fast Grower (Eg) R. leguminosarum/ Bacillus radicicola and R. trifolii.
2. Slow growers (Eg) R. Japanicum and R. Lupinii.

Nodule Formation

It is the principal event in this process, which is a complicated process. In the initial stage, host cells have to recognise the correct strain of symbiotic bacteria where a critical specificity is involved in the association of various species of root nodule bacteria and leguminous plants and this step is controlled by specific protein found in the host cell called lectin. Lectins are the key proteins involved in the Rhizobium-legume symbiotic association and in the process calcium plays a critical role in modifying the ability of root to absorb rhizobia.

Rhizobia accumulate in the soil near the roots of the legumes probably due to the secretion of some growth factors by their roots. The bacteria now Penetrate the root through soft or injured root hairs whose tips become curved.

They now enter into the cells of the inner layers of the cortex through infection threads and cause cortical cells to multiply which ultimately result in the formation of nodules on the surface of the roots. The cells of the bacteria enlarge, assume pleomorphic shape and are called bacteriods.

The nodule contains a pink coloured leghaemoglobin, like true haemoglobin combine with O_2 and CO_2 gets readily oxidised to brown form with trivalent ion. The development of leghaemoglobin and Nitrogen fixing capacity are out come of symbiotic interaction. The leghaemoglobin is located in the space between bacteriods and membrane enclosing them or it's present outside the membrane and in the host cytoplasm.

Nitrogenase

The Process of 'N' fixation, whether its symbiotic or non-symbiotic involves the reduction of atmospheric 'N' to ammonia (NH_3) by the enzyme Nitrogenease. The enzyme is madeup of two protein components one containing iron and molybdenum and the other containing only iron called Fe-protein or azoferredoxin. The enzyme is extremely sensitive to oxygen. The development of leghaemoglobin pigment in nodule cells and heterocysts in BGA for protecting nitrogenase from oxygen inactivation. Thus the enzyme remains active under anaerobic conditions. The pigment leghaeomoglobin regulates the course of oxygen and provides suitable atmosphere to the enzyme. It combines rapidly with oxygen to avoid any inhibitory effect of it on enzyme, at the same time its able to make oxygen available to bacteriods for ATP production, required for 'N' fixation. Thus the process of 'N' fixation requires anaerobic conditions even though Rhizobium is aerobic.

Mechanism of Nitrogen Fixation

It's to be noted here, during the process of 'N' fixation, the free atmospheric nitrogen first bound to the enzyme surface and is not released until its completely reduced to ammonia. 'N' bound to the enzyme surface is reduced in step wise reactions before N-N bond is ruptured. Pathway of e-transport in nodule bacteriods.

Enzymes involved —

1. Glucose-6-PO4 dehydrogenase
2. NADPH-Ferredoxin oxido reductase
3. 'Fe' protein compund of Nitrogenase.
4. Fe-MO protein compund of Nitrogenase

The reduction of 'N' into NH_3 by nitrogenase in bacteroids depend upon the availability of ATP and reduced substrate capable of donating 'H^+ atoms to 'N' Glucose-6-PO_4 is the reduced substrate for the process and reduced NADP together with ferrodoxin function as electron carriers.

ATP interact with non-home iron (NHI) protein component of Nitrogenase and brings about confirmational change to convert it to a powerful reductant, which becomes capable of transferring electrons to reduce N_2 and NH_3.

Thus the reduction of one molecule of 'N' in to two molecules of ammonia requires twelve molecules of ATP because six electrons are required per molecule of 'N' reduced.

$$N_2 + 6H^+ + 6e^- + 12\ ATP \longrightarrow 2\ NH_3 + 12\ ADP + 2\ Pi$$

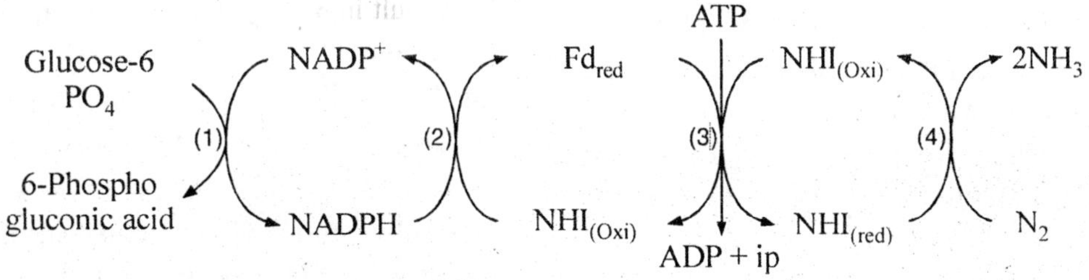

Glucose-6 PO_4 → 6-Phospho gluconic acid (1)
$NADP^+$ → NADPH
(2) Fd_{red}
ATP → ADP + ip (3)
$NHI_{(Oxi)}$; $NHI_{(red)}$
(4) N_2 → $2NH_3$

Fig. 14. Nitrogen fixation

7.4.4 PhotoPeriodism

The plants in order to flower require a certain day length (*i.e.*) relative day length and night is called photoperiod. The response of plants to the photo period expressed in the form of flowering is called as photoperiodism.

The observation of Garner and Allard (1920) that a mutant variety of tobacco, Maryland mammoth flowers when the relative length of the day was shorter than the length of the dark period, which led to the discovery of photoperiodism.

Depending upon the duration of photo period, they classified plants into three categories.

I. *Short Pay Plants* :

- Qualitative (flowering occurs only in short days)
- Quantitative (Flowering accelerated by short days)
- The plants which require a relatively short day light period (8-10 hrs) and a continuos dark period about 14-16 hrs for subsequent flowering.
- Examples are, maryland mammoth variety of Tobacco, rice, datura, onion, etc.

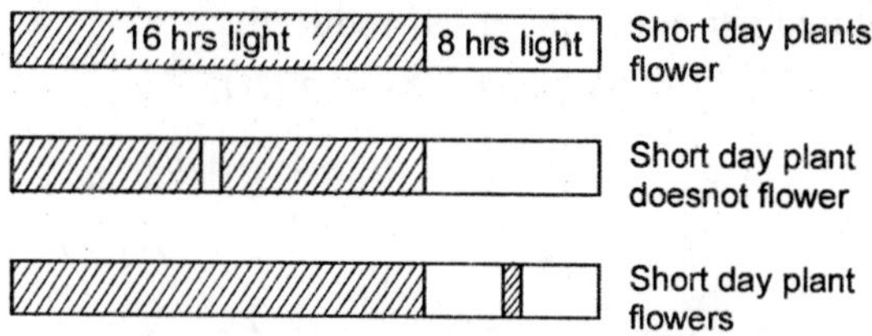

Fig. 15. Effect of night on short day plants

- In short day plants, dark period is critical and must be continuos. If this dark period is interrupted even with a brief exposure of red light, short day plant will not flower.
- Maximum inhibition of flowering with red light occurs at about the middle of the critical dark period.
- The inhibitory effect of redlight can be over come by a subsequent exposure with far-red light (730-735 nm).
- Interruption of the light period with redlight doesnot have inhibitory effect on flowering in short day plant.
- Prolongation of the continuos dark period initiates early flowering in short day plants.

II. *Long Day Plants (LDP)*

- Qualitative - flowers only in long days.
- Quantitative - Flowering accelerated by long days.
- These plants require a longer day light for a period of 14-16 hrs. in a 24 hrs cycle for subsequent flowering.
- Examples are, spinach, lettuce, radish alfalfa, sugar beet, opium poppy, oats and wheat.
- In the long day plants, the light period is critical.
- The best flowering of long day plant usually occurs in continuos light.
- For flowering they require either no dark period or a very short dark period.
- A flash of light in long dark periods can induce flowering in them even during short-day periods.
- Long day plants can flower even in short-day periods, if these short-day periods are accompained with still shorter dark periods.
- Too long dark periods inhibit flowering and so these are called short-night plants.
- A brief exposure in the dark period the prolongation of the light period stimulates flowering in long day plants.

III. ***Day Neutral Plants (Ambiphotoperiodic species)***

- The plants that flower in all photo periods ranging from 5 hrs to 24 hrs continuous exposure.
- Examples are tomato, cotton, cucumber, pea, sunflower maize and dandelion etc.
- Their flowering is not affected by the length of the day and they can flower even if the light period provided ranges from few hrs to continuos illumination.

7.4.5 Vernalisation

It can be defined as the method of inducing early flowering in plants by pretreatment of their seeds at very low temperatures. Lysonko gave the term vernalisation which means conversion of the winter variety into spring variety by low temperature or chilling treatment.

Soviet agronomist T. Lysenko in 1928 coined the term vernalisation.

Physiological Pre conditioning : The effect of the cold stimulus on plant is not immediately visible. It's expressed only at a certain later stage in the form of flowering. Thus like the photo periodism, the phenomenon is an excellent example of physiological pre conditioning.

Perception of the Cold Stimulus

The cold stimulus is perceived by the apical meristerms. And also all dividing cells including those in roots or leaves may be the potential sites of vernalization.

The perception of the cold stimulus results in the formation of a floral hormone, which is transmitted to other parts of the plant. It can also be possible, the cold stimulus may even by transmitted to another plant across a graft union and the floral hormone has been named as vernalin by melchers.

Conditions Necessary for Vernalization

1. Age of the Plant : In cereals, effective only at germinating seed.
2. Appropriate Low temperature and Duration of exposure - 1 - 6°C, one and a half months or more.
3. Oxygen - Vernalisation an aerobic process and requires metabolic energy provided by oxygen.
4. Water : In dry seeds, vernalisation is not possible.

Mechanism of Vernalization

Two theories are prevailing to deal about the mechanism of vernalization.

I. ***Phasic Development Theory***

According to this theory, the process of development of an annual seed plant consists of a series of phases which must occur in some predetermined sequence. An important condition is, commencement of any of these phases will take place only when the preceding phase has been completed.

Thermo Stage : First stage conditions required are –

1. Temperature (0-20°C)
2. Moisture
3. Aeration
4. Varied time

Photo Stage : When thermo stage is completed, the plant passes through next stage. There is pronounced effect of the relative length of day and night on the production of flowers, and this phenomenon is called photoperiodism.

Third Stage : It's quite necessary for seed germination as it has been connected with formation of sexual elements and is associated with garmetogeneris. The photoperiod requirement for the commencement of this stage is little shorter than required for the completion of photo stage.

II. ***Hormonal Theories***

Vernalization probably involves the formation of a floral hormone called as vernalin. The first hormonal theory proposed by long and melchers.

Process of Vernalisation

Germination of seeds and

cold tretment at 0-5°C

↓

Drying and after that

sowing of seeds

After the cold treatment, seedlings are allowed to dry for some time and then down. The drying period also should not be so long as the vernalize response decreases and it continues till seedlings completely devernalised.

Significance

1. Shortens the vegetative period of the plants.
2. Increases the cold resistance of the plants.
3. Increases resistance of plants to Fungal diseases.
4. Crops can be produced earlier (i.e.) crop can be harvested much earlier than the control crop.
5. Crops can be grown in the regions where they donot naturally reproduce.
6. Plant breeding work can be accelerated.
7. In colder countries like Russia, the harmful effects of the cold can be overcome and thus improving crop production.

7.4.6 Dormancy of Seeds

It may be defined as any phase in the lifecycle on plant in which active growth is temporarily suspended.

All the viable seeds have capacity to germinate if placed under suitable conditions necessary for germination. When seeds fail to germinate for some time even if placed under suitable conditions, that are ordinarily favourable for germination. During this period the growth of the seeds, remain suspended and they are said to be in dormant stage.

Factors Causing Dormancy of Seeds

1. ***Dormancy due to seed coat*** : It's of two types –

(*a*) *Seed coats impermeable to water* : Seeds of families leguminoceae, malvaceae, chenopodiaceae, and solanaceae have hard seeds coats, that are impermeable to water. The seeds remain dormant until the impermeable layer of testas decay by soil microorganisms.

(*b*) *Seedcoats impermeable to oxygen* : In cocklebur (xanthium), many grasses, compositae, the seeds coat is impermeable to oxygen. During the period of dormancy of seed coats gradually become more permeable to oxygen so that they may germinate afterwards.

2. ***Dormancy due to immature embryos*** : Some seeds are shed, before the embryo is matured and they will not grow, which require an "after ripening period" during which certain changes occur with in the seed. The changes include acidity, enzyme activity and respiratory rate etc. and once the embryo developement is complete, the seed then germinates without any special treatment.
3. ***Mechanically resistant seed coats*** : Seeds of certain weeds such as pigweed (Amaranthus), Sheperd's Purse (capsella), water plantain (Alismas) etc. remain dormant because their hard seed coat prevent any appreciable expansion of the embryo.
4. ***Dormancy due to chemical inhibitors*** : Seeds present in the juicy fruits such as orange, tomato donot germinate immediately. Para-ascorbic acid,

Phenolic acids, Dehydro acetic acid in the embryo, endosperm or other tissues of the seed fruit is also the cause of seed dormancy.

5. ***Need for after-ripening in dry storage*** : In many-plants (Eg.) Barley, oats, wheat, seeds though containing fully developed embryos are dormant, when they are harvested they require no special treatment to overcome this dormancy and germinate if kept under dry storage conditions at normal temperatures for few weeks to several months. This type of dormancy can be overcome by removing the seed coats or temperature treatments. (35-40°C storage).
6. ***Chilling or Low Temperature Treatment*** : Seeds of temperature species show dormancy which is over come by chilling (Eg) *Apple, Rose, Peach, Grapes*, etc., which will not germinate, if planted under moist conditions at 20°C, but will germinate if kept at 0-5°C for several weeks and then transferred to warmer conditions.
7. ***Dormancy due to light Sensitivity*** : The germination of many seeds affected by light. Such seeds are called photo blastic. The seeds in which germination is stimulated by light are called positively photoblastic whereas those in which germination is inhibited by light are negatively photoblastic.

Positive Photo blastic : Amaranthus, Sheperd's purse, , tobacco, and tomato etc.

Negative Photoblastic : Helleborus niger, Nigella damascena and Silene armenia.

Mechanism to Seed Dormancy

I. Regulation of dormancy is primarily hormonal.
II. Importance of interference with gaseous exchange by seed coats and other structure.

The hormonal theory visualises that the prevention of growth may be either (1) because of deficiency of some essential growth promoting substance. (endogenous gibberelins) (2) because of the presence of some inhibitory substances. (Dormins/abscisins).

Methods of Break Dormancy

1. *Scarification* : The process of rupturing or weakening the seedcoats by mechanical or other means and is employed, where the dormancy is due to impermeable seed coats. Scarification can be done either mechanically or chemically. Chemical scarification is accomplished by treatment of the seeds with organic solvents, such as alcohol, acetone, ether or xylol; and strong acids.
2. *Pressures* : When the dormant seeds are subjected to 2000 atm at 18°C for about 5-20 mins. and this works in cases, where the seed coats are impermeable to water.
3. *Low temperature* : When the dormancy is due to chilling requirement, it can be overcome, if the seeds are treated in moist medium at low temperature (5-10°C) for sufficient period of time, which is called stratification.
4. *Alternating Temperatures* : An alternation of low and high temperatures improves the germination of seeds.
5. *Light Dormancy of Positively Photoblastic* : Seeds can be broken by exposing them to red light or white light. This promotion of germination by redlight

and inhibition by far-red light probably involves physochrome.

6. *Germination Stimulating Compounds* : Such as gibberellins induce germination in tobacco, lettuce etc., and chemicals such as KNO_3, Thiourea and ethylene also involved.

Advantages of Seed Dormancy

1. In temperate zones, to overcome the severe colds.
2. In tropical zones, seed dormancy due to hard seed coat improves the chances of survival.
3. Dormancy of Cereal Seeds, increase the availability of them as food and in otherwise, they may not be used as food grains.

7.4.7 Seed Germination

The process of seed germination starts with the imbibition of water by seed coats and emergence of growing root tip of embryo. It ends when the embryo has developed into a seedling which is out of bounds of seed coats and has its own photosynthetic system.

Conditions for Germination

(*i*) Water (*ii*) Aeration (*iii*) Temperature (*iv*) Light.

(*i*) *Water* : It is absorbed initially by imbibition followed by absorbtion.

- Period required for soaking of seeds in water varies depending upon the nature of seed coat.
- Seeds with impermeable seed coats are to be soaked longer than those with permeable seed coats.
- Prolonged Soaking affects, as the metabolites leached out which are essential for germination (*e.g.*) Barley when soaked for 2-3 days.

(*ii*) *Aeration* : Oxygen requirement of different seeds vary considerably, but most seeds germinate at a lower oxygen level.

- High requirement of oxygen may be due to the presence of permeablity barrier. (*i.e.*) seed coat.
- Most seeds require active aerobic oxidation system for germination.

(*iii*) Temperature : The seeds of temperate plants require relatively low temperature as compared to tropical plants.

(*iv*) Light : The garmination of many seeds is affected by light. Such seeds are called ***photoblastic***. The seeds in which garmination is stimulated by light are called positively photoblastic, those in which garmination is inhibited by light are negatively photoblastic.

Physiological, Biochemical Changes During Germination

Water Uptake : Seed coat imbibes water thróugh seed coat, which is purely a physical process various hydrophilic groups such as $-NH_2$, $-OH$, $-COOH$ Proteins and CHO attract dipolar water molecules and swelling occurs. Due to imbibition of water. Seed coats become (*i*) more permeable to oxygen and water and (*ii*) less resistant to outward growth of the embryo.

Respiration **:** Uptake is accompanied by rapid increase in respiration rate of embryo. Initially its anaerobic, which is soon replaced by aerobic due to the availability of oxygen.

Mobilization of Reserve Materials **:** Occurs to provide,

(*i*) Building blocks for the development of embryo.

(*ii*) Energy for the biosynthetic processes.

(*iii*) Nucleic acids for control of protein synthesis and overall embryonic development changes in these components are as follows.

Nucleic Acids : In monocots, during imbibition stage, there is a rapid decrease of DNA and RNA content in the endosperm. Appreciable amount of RNA appears in the aleurone layer after 16 hrs. Due to more cell division the DNA content increased.

Carbohydrates : Starch are the important reserve materials of cereals in the endosperm. During germination starch is hydrolysed first into maltose in the presence of α-amylase and β-amylase and then maltose is converted into glucose by maltase. Glucose is absorbed by scutellum, converted into soluble sucrose and transported to the growing embroyonic axis.

Gibberellin is secreted during germination by embryonic axis into aleurone layer, causing denovo synthesis of α-amylase. This enzyme is not found in ungerminated seeds. GA induced synthesis of α-amylase is countered by ABA.

In contrast to α-amylase, β-amylase is already present in the seed in inactive form, which gets activated during germination. The activity of maltase is also regulated by GA.

Lipids : During germination, mobilization of lipids is brought about by hydrolysis of fats to fatty acids and glycerol by lipases and β-oxidation of fatty acids to acetyl-COA. Some of the acetyl COA is converted into sucrose via glyoxylate cycle and is transported to the growing embryonic axis.

Proteins : Some plants store proteins as reserve food in their seeds in the form of aleurone grains. Mobilization of these proteins involves the hydrolytic cleavage into amion acids by peptidases.

Inorganic Materials : Number of inorganic materials such as Phosphate, Ca, Mg, and Potassium are also stored in seeds in the form of Phytin.

7.4.9 Enzyme

Enzyme has been called the "agents of life". Since they control almost all life process. It is catalysts of life.

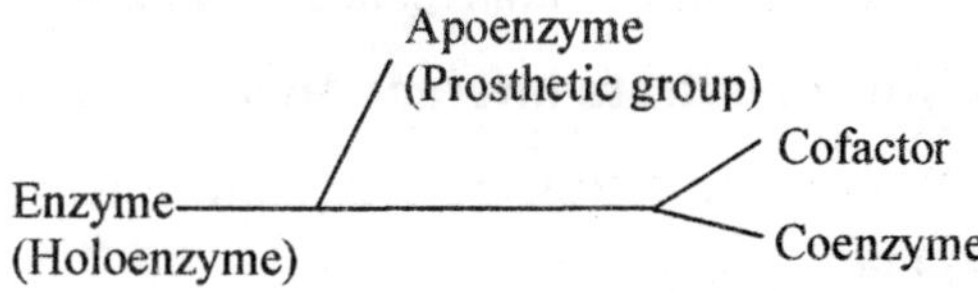

Terms co-factor is used to describe inorganic ions such as zinc, iron, copper and magnesium these ions chemically combine with enzyme and assist enzymes. If an organic molecule functions as co-factor it is called co-enzyme.

Cofactors : Fe^{+}, Mg^{2+}, Mn^{2+} (Inorganic in nature)

1. Cu – Cytochrome oxidase
2. Fe – Cytochrome oxidase, catalase, peroxidase
3. K^{+} – Pyruvate kinase
4. Mg^{2+} – Hexokiuase, Pyruvate kinase, glucose-6-phosphate
5. Mn^{2+} – Arginase, ribonuclease reductase.
6. Mo – Dinifrogenase
7. Ni – Urease
8. Se – Glutathione Peroxidase
9. Zn – Carboxianhydrase, alcohol dehydrogenase.

Coenzyme : Organic Group Transfer

1.	Biotin	CO_2
2.	Coenzyme A	Acyl
3.	Deoxy adenosolyl	H^{+}
4.	FAD	e^{-}

5. NAD H^+
6. Pyridoxy phosphate amino
7. TPP (Thymine pyrophosphate)

Basic equation :

For enzyme kinetic $V_0 = \frac{V_{\max(S)}}{\text{km} + [S]}$

- Small value of km – maximal catalytic efficiency at low substrate concentration.

Enzyme substrate Interaction :

Weak interaction between enzyme and substrate

Turnover Number :

The number of molecules of substrate that a single molecule of enzyme can react with in a given unit of time.

Characterstics of enzyme

1. Colloidal in nature
2. React with both acidic and alkaline substance
3. Show sensitivity
4. Inhibited
5. Activity can be accelerated
6. Size is greater than substrate
7. Specific in nature
8. pH sensitive

Effect of Inhibitors on Michalis-Menten reaction

Inhibitors	*Equation*	*Curve*	*Effect*
1. Normal	$V_0 = \frac{V_{\max(s)}}{\text{km} + [s]}$		
2. Competitive (action active site)	$V_0 = \frac{V_{\max(s)}}{\alpha\ \text{km} + [s]}$	Interspton $1/V_0$ axis at $1/V_{max}$	↑ km
3. Un-competitive (conformation in enzyme)	$V_0 = \frac{V_{\max(c)}}{\text{km} + \alpha'(s)}$	Identical shape	km–un affected V_{max}↓
4. Mixed (Both)	$V_0 = \frac{V_{\max(s)}}{\alpha\ \text{km} + \alpha'(s)}$	lift of $1/V_0$ axis	V_{max}↑ km ↑

Mode of enzyme action :

Michaelis and Menton (1913)

- Enzyme have active sites for the attachment of substrate molecule where as enzyme can form an intimate rehatible with the substrate.

Enzyme + substrate ⇌ enzyme substrate complete
⇓
End products + Enzyme

Factors affecting enzyme activity :

1. Temperature
2. pH
3. Anaesthetics
4. Substrate + Enzyme concentration.

7.5. PLANT PHYSIOLOGY IN AGRICULTURE

1. Role of Plant Physiology in the Improvement of Varieties

Plant breeders use the method of selection for crop improvement. Selection on the basis of only external characters is not fruitful. Therefore, its protein content, oil content, carbohydrate content and various other substances of nutritional importance must be tested before proper selection. Plant physiology thus plays an important and essential role in specifying the features for which the plant breeder must select the crop.

2. Role in Soil Fertility

Our knowledge of mineral nutrition of different crop plants provided extensive revisions in the application of fertilizers. It is now possible to recommend specific organic fertilizers or inorganic salts using knowledge of plant physiology.

Plant growth regulators

1. Role of Auxins

(*a*) Auxins are used to prevent premature fruit drop, when applied close to the time of harvest.

(*b*) Auxins (Particularly IBA) induce formation of roots on cuttings of many ornamental, horticultural and forest trees.

(*c*) Many auxins are used to promote sprouting, induce flowering and enhance yield in potatoes, peas, beans, etc.

2. Role of Gibberellins

(*a*) Gibberellins are used to break dormancy and induce uniform crop emergence, *e.g.*, seed potatoes.

(*b*) It is used to enhance the growth of vegetative parts of several floricultural species and used to increase flower size.

(*c*) Gibberellins are also used to produce parthenocarpic (seed-less) fruits.

(*d*) Cytokinins have less importance in agriculture. It is used in extending self-life of lettuce, tomatoes, apples, etc. It is also used to induce efficient root system in maize and rice.

3. Role of Ethylene

(*a*) Ethylene is a hormone which is well known to cause fruit ripening.

(*b*) Ethylene causes abscission.

(*c*) It is used to induce bulbing in onion.

(*d*) Ethylene increases tillering in rye, peas, etc.

3. Role of inhibitors or growth retardants

(*a*) A well known inhibitor CCC (2-chloroethyl-trimethyl ammonium chloride) is used to induce stiffening of plants, increase number of roots and litters, prevent lodging and increase yield in wheat, oat and several strains of Secale (rye).

(*b*) CCC is also used to increase bolting, flowering and increase yield in cotton.

(*c*) Many inhibitors, such as, CCC, phosphon, B-9, etc. are used in ornamental horiculture to retard the height of plants like chrysanthemum.

4. Role in Seed Production

Knowledge of plant physiology is used in production of high quality, viable and pure seeds from weed seeds.

5. Role in food technology

Plant physiology provides many techniques required for food preservation, food processing, such as drying, pickling and food storage.

7.4.8. Plant Growth Regulators

Contents	Auxin	ABA (ABSCISSIC ACID)
1. Definition	A compound that has a spectrum of biological activity similar to but not identical with IAA.	
2. History	kogl-named as Auxin.	Weering : • First identified in Acer pseudoplatanus (Abscisin-II) • from mature cotton fruit (Abscissin-I)
3. Site of Synthesis	Shoot, Root tip	Matured leaves
4. Transport	Polar, Active	Bipolar, Mainly through phloem
5. Mode of action at physiological level	1. Cell Elongation 2. Increase the extentibility of cell wall 3. Promote growth in stem 4. loosening of cell wall	• Inhibit seed germination
6. Natural Hormone	IAA, PAA,	Cis-ABA
7. Morphological effect	Tallness of Plant	Yellowing of leaves.
8. Physiological effect	1. Apical Dominance 2. Promote lateral and adventitious roots 3. Delay abscission 4. Vascular differentiation	1. Induce seed dormany 2. Accumulated in dormant bud 3. Inhibit α-amylase 4. Close stomata
9. Molecular effect	Stimulating cyclin dependent protein kinase (CDK) which is regulating cell cycle transitions from G_1 to S and from G_2 to mitosis	(ABA-response elements) DNA (↑ ABA) —Trans-cription→ mRNA → Protein → Stress Protein / adaptation to low temperature / Salt tolerance
11. Commercial use	1. Initiate Rooting in cutting (IBA) 2. Induce flowering in Dry material 4. Herbicide (2.4.-D) 5. Parthenocarpy eg. Brinjal	1. Used as stress harmone 2. Initiation of flower in short day plants.

7.4.8. Plant Growth Regulators (*contd...*)

Contents	GA	Cytokinine (CDK)
1. Definitioin	1. stimulate cell elongation and cell division or both.	1. Promote regulatory function
2. History	"Hori" observed, kurosawa - discovered.	2. Habenlandt - 1st observed
3. Transport	All shoot, leaves	3. Acropetal, through xylem and phloem.
4. Mode of action	Increase the number of cells, reduces the cell cycle	4. Increase cell number reduces the cell cycle.
5. Morphology action	1. Substitute for low temperature 2. Flowering in long day plants 3. Parthenocarpy 4. Fruit elongation 5. Breaking seed dormancy	1. Delay senescence 2. Weakens the apical dominance 3. Promote lateral bud development
6. Physiological action	1. Induce the activity of α-amylase 2. Break the genetic dwarf.	1. Mainly cell division 2. Shoot and root differention
7. Molecular action	1. Extensionability of cell wall 2. Stress relaxation	1. Increase the mitosis division 2. Increase in m-RNA synthesis in cytokinin
8. Commercial application	1. Parthenocarpy in grapes 2. Breaking seed dormancy 3. Substitute for kernalization	1. Mainly in Tissue culture 2. Uniform production of lateral branches 3. Delay senescence 4. Shoot elongation
9. Regulation	1. Carboxyl group $\Rightarrow$ gibberellin glycoside 2. Hydroxyl group $\Rightarrow$ gibberellin glycosyl ether	1. Zeatine 7-glucoside 2. Zeatine 9-glucoside

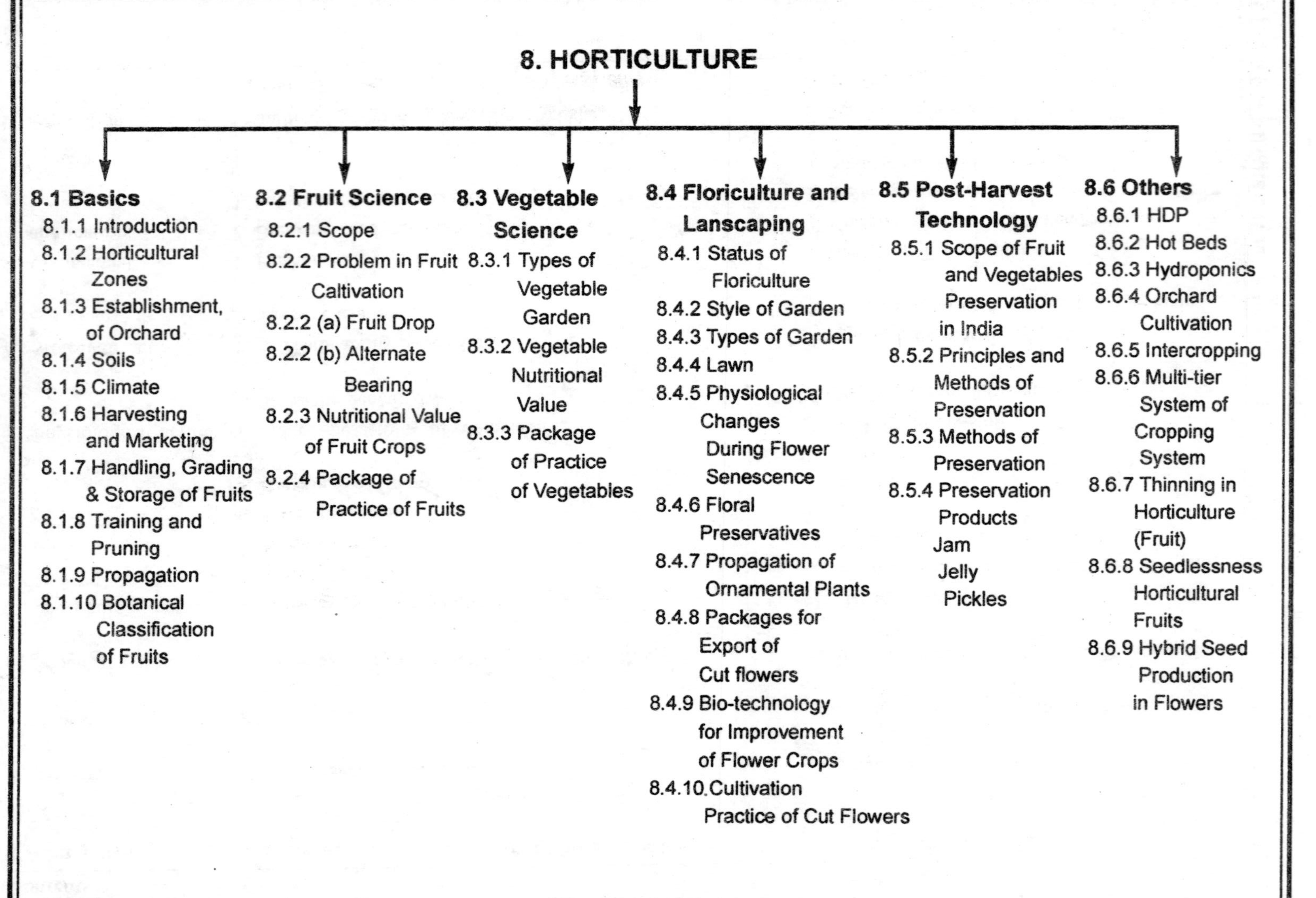
8. HORTICULTURE
8.1 Basics
8.1.1 Introduction
8.1.2 Horticultural Zones
8.1.3 Establishment, of Orchard
8.1.4 Soils
8.1.5 Climate
8.1.6 Harvesting and Marketing
8.1.7 Handling, Grading & Storage of Fruits
8.1.8 Training and Pruning
8.1.9 Propagation
8.1.10 Botanical Classification of Fruits
8.2 Fruit Science
8.2.1 Scope
8.2.2 Problem in Fruit Caltivation
8.2.2 (a) Fruit Drop
8.2.2 (b) Alternate Bearing
8.2.3 Nutritional Value of Fruit Crops
8.2.4 Package of Practice of Fruits
8.3 Vegetable Science
8.3.1 Types of Vegetable Garden
8.3.2 Vegetable Nutritional Value
8.3.3 Package of Practice of Vegetables
8.4 Floriculture and Lanscaping
8.4.1 Status of Floriculture
8.4.2 Style of Garden
8.4.3 Types of Garden
8.4.4 Lawn
8.4.5 Physiological Changes During Flower Senescence
8.4.6 Floral Preservatives
8.4.7 Propagation of Ornamental Plants
8.4.8 Packages for Export of Cut flowers
8.4.9 Bio-technology for Improvement of Flower Crops
8.4.10.Cultivation Practice of Cut Flowers
8.5 Post-Harvest Technology
8.5.1 Scope of Fruit and Vegetables Preservation in India
8.5.2 Principles and Methods of Preservation
8.5.3 Methods of Preservation
8.5.4 Preservation Products
Jam
Jelly
Pickles
8.6 Others
8.6.1 HDP
8.6.2 Hot Beds
8.6.3 Hydroponics
8.6.4 Orchard Cultivation
8.6.5 Intercropping
8.6.6 Multi-tier System of Cropping System
8.6.7 Thinning in Horticulture (Fruit)
8.6.8 Seedlessness Horticultural Fruits
8.6.9 Hybrid Seed Production in Flowers

8 Horticulture

8.1 BASIC HORTICULTURE

8.1.1 Introduction

* The term "Horticulture" first appeared in written language in 1631 in book written by Peter Laurenberg.
* Horticulture was first mentioned in book named the *New World of English Words*, written by Edward Phillips in London in 1658.
* The word *horticulture* is derived from the Latin word, ***hortus*** and ***culture*** meaning garden and to cultivate respectively.

Horticulture may be defined as the crop science which deals with the production, utilization and improvement of fruits, vegetables, ornamental plants and plantation crops including medicinal and aromatic plants.

- Horticulture crops generally require intensive cultivation.
- Horticulture produces are highly perishable.
- Horticulture crops are the rich sources of vitamins and minerals.
- Cultural operations such as propagation, training, pruning, harvesting and marketing are skilled operations and are specific to horticulture crops.

Horticulture consists of the following divisions :

- **Pomology** – study of fruit crops
- **Olericulture** – It is scientific study of vegetables.
- **Floriculture** – It is the art of growing, selling, designing and arranging flowers and foilage plants.
- **Spices** – Spices are those plants, the products of which are made use of as food adjuncts to add aroma and flavour *e.g.* pepper, cardamom, clove etc.
- **Condiments** – Condiments are also plants, products of which are used as food adjuncts to add taste only *e.g.* coriander, cumin.
- **Plantation crops** – Which are grown in extensive scale.
- **Arboriculture**– Study and selection; planting, care and removal of individual trees, shrubs, vines and other perennial woody plants.

Importance of horticulture

- Horticulture crops contribute to rational income.
- Horticulture crops generate more employment opportunities.
- These are highly remunerative and profitable than cereals.
- Fruits and vegetables are regarded as 'protective foods' since they supply minerals and vitamins.
- Fruits yield more calorific value.
- Many horticultural produces and their by products are the important raw materials for many industries.

8.1.2 Agro climatic zones for horticultural crops

1. **Temperate Northern region :**
   Jammu and Kashmir, Himachal Pradesh, Uttarakhand, West Bengal. Temperate fruits and cool season vegetables are the important horticultural crops.
2. **North western arid region :**
   Rajastan, Gujarat, parts of Punjab and Haryana. Arid fruit crops like Ber, Pomegranate, Aonla, Date Palm, Wood apple and spices like coriander and cumin.

3. **North Eastern sub tropical humid region:**
   Arunachal Pradesh, Assam, Manipur, Meghalaya, Mizoram, Nagaland and Tirupura.
   Fruits like Banana, Pine apple, Citrus, Jack fruit, plantation crops like tea etc.
4. **North central subtropical region :**
   U.P. Bihar, M.P. and parts of Maharastra. Mango, Sapota, Sweet Oranges and Guava.
5. **South central tropical region :**
   Tamilnadu, Andhra Pradesh, Karnataka and parts of Maharastra. Mango, Guava, Sapota, Pine apple and Turmeric.
6. **Coastal tropical humid region :**
   The entire region stretching along the Bay of Bengal in the East and Arbian sea in the West. Banana, Mango, Cashew and Coconut.
7. **Southern hilly region :**
   Comprising of Western ghat and Eastern ghat. Coffee, Tea, Cardamom, Pepper, Oranges and Pine apple etc.

8.1.3 Establishment of orchard :

Establishment of orchard is a long term investment and deserves a very critical planning. The selection of site, planting system and planting distance, choosing the varieties and the nursery plants have to be considered carefully to ensure maximum production.

Location and site :

Criteria for selection :

1. Location should be in a well established fruit growing region for getting benefits of others experience and selling produce through co-operatives.
2. It should be closer to a market.
3. The climate should be suitable for the chosen fruit crops.
4. Adequate water supply should be available round the year.
5. Before a grower establish an orchard, he must ensure the following factors.

(i) Suitability of soil, its fertility, the nature of sub soil and soil debris.
(*ii*) Site must have proper drainage and no water stagnation during rainy season.
(*iii*) Irrigation water must be of good quality.
(*iv*) These must be proper transport facilities either by road or rail within the reach.
(*v*) Suitability of climate must be ensured for immunity aganist dangers like cyclone, frost, hailstorms.
(*vi*) Whether there are seasonal gluts or over production in any particular period of the year.
(*vii*) Whether there is assured demand in the market for the fruits to be grown.
(*viii*) Availability of labour in the proximity.

Planning of an orchard :

A careful plan of the orchard is necessary for the most efficient economic management.

1. Optimum spacing to accommodate maximum number of trees.
2. Stores and office building in the orchard should be constructed at the centre for proper supervision.
3. Wells should be located at convenient places at the rate of one well for 2-4 ha.
4. Each kind of fruit should be assigned in a separate block.
5. Fruits ripening at the same time should be grouped together.
6. Pollinaters should be provided in deciduous fruits. Some variety, require pollen from other variety, to set fruits in them, otherwise, they will be barren. Such pollen donors are pollinators. Each 3rd tree in every 3rd row should be planted with a pollinater.
7. Irrigation channels should be laid along the gradients.
8. Roads should occupy minimum space for the economy.

9. Short growing trees should be allotted at the front and tall at the back for easy watching and to improve appearance.
10. Fruits attracting birds and animals should be close to the watchman.
11. A good fence is essential. Live fence is economic, cheap, should be drought resistant, easy to propagate from seed, quick growing, dense foliage, standing severe pruning and thorny plants, *e.g.* Agave, Prosopsis juliflora, Pithecolobium dulse.
12. Wind breaks, rows of tall trees planted close together around the orchard, are essential to resist velocity of wind.

Trees : Casuarina equisetifolia, ptenaspermum, polyalthia longifolia, eucalyptus globlus, Grevillea robusta, Azadiracta indica.

Laying out of orchards :

Any method of layout should aim at providing maximum no. of trees per ha., adequate space for proper development of the trees and ensuring convenience in orchard cultural practices.

Broadly 2 types :

(*a*) Vertical row planting pattern

(*b*) Alternate row planting pattern.

(*a*) *Vertical (square system, rectangular system)*: The trees set in a row is exactly perpendicular to those trees set in their adjacent rows. In the later planting pattern (*i.e.* Hexagonal, Quincunx, Triangular), the trees in the adjacent rows are not exactly vertical instead the trees in the even rows are midway between those in the odd rows.

The various layout systems used are :

1. *Square system :* Each corner of a square, whatever may be planting distance. This permits intercropping and cultivation in two directions.
2. *Rectangular system:* Any two rows are more than the distance between any two trees in a system. No equal distribution of space per tree.

(*b*) *Alternate row planting pattern :*

The various layout systems used are :

1. *Hexagonal system :* Planted in each corner of an equilateral triangle. 6 trees form a hexagon, 7th in the centre.

   This system provides equal spacing but difficult to layout. Accommodates 15% more trees than the square system.
2. *Diagonal or quincunx system :* Square system with one tree at the centre, the central filler must be of short duration.
3. *Triangular system :* Square system, difference being that those in the even numbered rows are midway between those in the odd rows instead of opposite to them.
4. *Contour system :* On the hills, planted along the contour across the slope. It particularly suits to land with undulated topography. Main purpose is to minimise land erosion and to conserve soil moisture. In South India tea is planted in contours either in single edge system or in double edge system.

8.1.4 Horticultural Soil

The soil act as a substratum and provides nutrients and water necessary for plant growth.

Kinds of Soil

Sandy soil is suitable for vegetables to obtain early crops and rapid root growth. It is the most common medium for propagating cuttings.

Loam of tall types are highly suitable for horticultural production it holds and releases nutrients for plant growth.

Soil fertility

The capacity of soil to nourish and sustain

plant growth is known as soil fertility. Plantation crops like coffee, cardamom and species need fertile soils rich in organic matter and humus content.

Soil depth :

Shallow soils with hard and compact subsoil layers like canker, heavy clay should be avoided for fruit crops as in such soils, the roots grow in the top layer of soil parallel to the surface such trees are easily uprooted by winds or premature death of trees may also occur.

Drainage :

Drainage depends upon the nature of sub soil. In a good sub soil, trees can stand drought better because of deeper root penetration on wet, heavy soils with improvable sub soil poor performance is noticed due to poor aeration and inadequate drainage.

8.1.5 Climate

Climate is one of the important complex factors which influences the production of horticultural crops.

Temperature

Optimum temperature range varies with crops and based on the optimum temperature requirement, horticultural plants are broadly classified into (*a*) temperate (*b*) subtropical (*c*) tropical plants. If a tropical plant is grown in subtropical zone their growth and performance may not be good.

Light

The intensity of light, quality and its duration affect the growth and development of horticultural plants.

Atmospheric humidity

Humidity affects the growth and development of horticultural crops. High humidity during flowering and fruiting promotes the incidence of pests and diseases.

Rainfall : Total rainfall and its distribution decide the growth and development of horticultural crops.

Air pollutants : The air pollutants like ozone and photo chemical oxidants such as sulphur dioxide reduce the assimilation rate, growth and development of plants leading to reduction in yield and quality of the produces.

Hailstorms : Occurrence of hails during pre-blooming or blooming period of apple, plum, peaches generally affect the fruit set.

8.1.6 Harvesting and Marketing :

Monetary returns of the fruit grower and the reputation of his fruit produce depends on the following factors :

1. Stage of picking of fruits.
2. Packing methods.
3. Storing and transporting.
4. Processing and delivery to the consumer in the form and at the time of the consumer desires to have the fruits.

Picking :

Picking refers to the harvesting of fruits from the trees at the right time. The time of picking depends upon :

(*i*) *Variety :* Certain variety only ripen in trees and certain variety will ripen even when they are harvested at 3/4th maturity.

(*ii*) *Purpose :* For which it is meant, e.g. tomatoes for table purposes can be harvested at red ripe stage, whereas for distant markets they may be harvested during breaker's (colour change) stage.

(*iii*) *Distance from the market :* The closer the market, ripe fruits may be picked, if distance is more mature green stage.

Methods to determine proper time :

1\. . ***Maturity index :***

Following are the rough but ready maturity tests of fruits employed to pick the well matured fruits.

(*a*) Colour changes : (mango, papaya).

(*b*) Increase in size.

(*c*) Softening of the tissue of the fruits *e.g.* figs, grapes.

(*d*) Case of detachment from the stalk *e.g.* sapota, annona.

(*e*) Shrivelling of fruit stalk, water melon.

(*f*) Time elapsed from the date of flowering to picking maturity.

(*g*) Sound by tapping jack and water melon produce hollow and dull sound at ripening stage.

2\. ***Accurate tests :***

(*a*) *Colour charts :* Charts prepared for different stages of maturity.

(*b*) *Penetrometer :* Instrument measuring the softness of tissues as an index of maturity.

(*c*) *Sugar/acid or Brix/acid ratio :* Based on the principle that acid content reduces and sugar increases on ripening.

The fruit growers should bestow more attention and considerable care during the picking season to reduce to a minimum level of careless handling of fruits by pickers.

1. Picking must commence from lower branches advancing towards the top to reduce dropping of fruits.
2. Dropping of the fruits from the tree should be avoided.
3. During picking, care must be taken to avoid any possible damages to the branches especially to the spurs as the subsequent cropping depends upon them.
4. Picking easily in the morning is best, fruits kept in shade. After picking fruits should be kept in coolest place available, which is well ventilated to arrest respiration and breakdown as much as possible.

Name of fruit	*Storage temperature (°C)*	*R.H. (%)*	*Approximate storage*
Apple	0-2.0	85-90	4-8 weeks
Grapes	0-2.0	80-85	2-4 weeks
Peaches	0-2.0	80-85	2-4 weeks
Banana	12.5-14.5	85-95	3 weeks
Mango	9-10.0	85-95	4 weeks
Orange	6-6.5	85-90	18 weeks
Papaya	4.5-6.0	80-85	5 weeks
Guava	9-10.0	85-90	3 weeks
Tomato (ripe)	0-4.5	85-90	10 days
Tomato (mature green)	10-15.0	85-90	30 days
Peas	0-05	90-95	2 weeks

Measures to reduce respiration

Temperature regulation

High humidity

Ventilation *i.e.* out door air.

5. There should not be any bruises while picking on it lowers the marketability of the fruits.
6. If picking is done in mid-day or hot weather, the fruits should be kept in a shed overnight to cool.

8.1.7 Handling, Grading and Storage of fruits :

Production of horticultural produce is limited by certain climatic condition and through proper storage of fruits, the availability, of produce is extended.

Factors influencing storage : Fruits and vegetables contain about 85 and 95% moisture content respectively. As these are living tissues after harvesting respiration goes on. So any storage procedure should aim at slowing down, regulating or preventing the increased respiration.

Temperature : The rate of respiration can be controlled by temperature regulation low temperature slows down.

Humidity : Maintaining high humidity during storage minimizes responsible.

Storage atmosphere : In addition to O_2 and CO_2 other gases may affect storage of fruits like ethylene by hastening ripening. These storage atmosphere can be improved by ventilation which mean introduction of outdoor air into the storage room.

Methods of storage :

1. *Air cooled storage :* Fruits placed in racks in insulation box. Keeping the ventilator open during night to admit the cool night air and during the day shut down. Thus storage rooms can be kept cool.
2. *Air cooled storage refrigerated with ice :* A few storages are filled with ice and salt and equipped with furs for quick cooling.
3. *Treating with wax emulsions :* In the absence of cold storage, fruits are dipped for a minute in a wax emulsion and dried completely for 10-15 minutes.
4. *Refrigerated storage :* 1. Prompt cooling of produce 2. Maintenance of optimum R.H. 3. An even holding temperature with 1.0-2.0 °C fluctuations 4. No need to depend upon the external conditions of air, their movements 5. Most efficient and scientific way of storing.

   Refrigerated storage involves the use of refrigerant (Freon NH_3-CH_3Cl).
5. *Controlled atmospheric storage (CAS) :* By reducing the O_2 supply available to the fruit and by increasing the amount of CO_2 around the fruit or by lowering the temperature, the respiration can be slowed down in fruits and vegetables. A combination of all these principles is involved in controlled atmospheric storage.
6. *Preservation by irradiation :* Potatoes, Onion, garlic and other crops lose their commercial value by sprouting or rooting during storage. These processes can be delayed by irradiating the produce with gamma.
7. *By use of chemicals :* Certain chemicals increase the shelf-life of fruits by delaying the ripening and senescence $KMnO_4$ an ethylene absorbent absorbs ethylene, from banana held sealed in polythene bags.
8. *Evaporative cooling (cool chambers) :* Evaporation of water produces a considerable cooling effect and this evaporative cooling occurs when air that is not saturated with water vapour, is blown across the wet surface.

   Cooling chamber works in the principle of evaporation cooling. This structure essentially contains a single layer of bricks as floor and a double layer of bricks as side wall with an interspace of 7.5 cm wide filled with river sand.

The top of the storage space is covered with piloted vetiver roots or gunny cloths in a bamboo structure.

Once the cool chambers are saturated with water, sprinkling of water once in the morning and once in the evening is enough to maintain temperature and humidity.

These storage chamber maintain a very high humidity of about 95% throughout the year. These chambers are ideal for storage of fresh fruits and vegetables for a short period.

Ripe mangoes can be stored upto 10 days as against 4 days in room temperature.

Similarly citrus fruits like grape fruit and orange can be stored upto 60-90 days as compared to 8-15 days in ambient room temperature.

8.1.8 Training and pruning

Certain fruit plants if left to grow of their own wild and do not bear abundantly unless trained or pruned to a specific form.

When a plant is tied, fastened, staked or supported over a trellis or pergola in a certain fashion or some of its parts are pruned with a view to giving the plants a framework, the operation is called 'training'

On the other hand 'pruning' refers to the removal of plant's parts such as buds, developed shoots and roots to maintain a desirable, form by controlling the direction and amount of growth.

Training	*Pruning*
1. It concerns orientation of plant.	1. It affects functions only
2. It determines plant's outline its branching and framework	2. Assists more in what the tree does in respect to fruiting
3. Generally concerned with first few years of the tree growth to determine its annual framework	3. This operation is carried out annually.

Objectives of training :

1. To admit more light and air to the centre of the tree and to expose maximum leaf surface to the sun.
2. To direct the growth of the tree so that varies cultural operations, such as spraying and harvesting can be done at the lowest cost.
3. To protect the tree from sun burn and damage.
4. To secure a balanced distribution of fruit bearing parts on the main limbs of the tree.

Method of leader training :

The method of leader training should be suited to the normal habit of the fruit tree the common system of training followed :

1. *Open centre :* The main storm is allowed to grow only upto a certain height and the leader stem is pruned to encourage scaffold branches production. This system is also known as vase-shaped system.
2. *Central leader :* A tree is trained to form a truck which extends from the surface of the soil to the top of the tree. This system of training is also known as closed centre one.
3. *Modified leader :* Intermediate open centre and central leader. This is developed by first training the trees to the leader type, by allowing the central axis to grow unhampered for the first 4 or 5 years. The central stem is then headed back and lateral branches are allowed to grow. So it possesses the advantages of both the system.

Trees trained with the help of a support can be :

1. *Power system* (Pandal, pergola system) in grapes, cucurbitaceous vegetables, vines spread over a cris cross network of wires.

2. *Espalier system :* Trained to grow flat on trellis or an horizontal wires by training the branches perpendicularly to the main stem.
3. *Kniffin system :* Two trellis of wire are strung supported by vertical parts.
4. *Telephone system :* Overhead trellis (3-4 wires).
5. *Tatura trellis :* Multi-layered wire trellis (v-shaped).

Trees trained do not require support but will be trained to a particular shape.

(*a*) *Head system* (grapes) vine is allowed to green single stem with the help of stakes.
(*b*) *Palmelle* central leader having 3-4 tiers of two opposite branches trained at an angle of 30 from the vertical.
(*c*) *Spindle brush :* Central stem with 15-20 semi-horizontal branches.
(*d*) *Dwarf pyramid :* Central leader tree and fully grown trees consist of central erect stem with 20-30 branches 45°-90°.
(*e*) *Head and spread system :* Variation of modified leader system. Primary branches are taken in different tiers, each tier being 60 cm above the lower tier.

Precaution should be taken to maintain the desired frame of the tree for long duration of time.

Pruning :

Commonly, trees are pruned annually in two ways:

Shoots or branches considered undesirable are removed entirely without leaving any stub, called 'thinning out'.

Removal of terminal portion of the shoots, branches or limb, leaving its basal portion contact is called 'beading back' thinning out. Involving large limbs as in old and diseased trees : bulk pruning.

Objectives :

(*i*) To remove surplus branches.
(*ii*) To open the trees so that the fruits will colour more satisfactorily.
(*iii*) To train it to some desirable form.
(*iv*) To remove the dead and diseased limbs.
(*v*) To remove the water sprouts.
(*vi*) To improve fruiting wood and to regulated production of floral bud.

Season of pruning :

Winter pruning is preferred to reduce bleeding summer pruning may have a dwarfing effect or an invigorating influence.

A light summer pruning may aid in coloration of fruit in certain variety.

Special pruning techniques :

1. Root pruning culture of dwarf fruit trees.
2. Ringing removal of a complete ring of bark from a branch to increase fruit bud formation in certain fruit crops.
3. *Notching :* Partial ringing of a branch above a dormant lateral bud. The practice of notching below a bud (nicking) is effective in starting a floral shoot.
4. *Smudging :* Practice of smoking the trees like mango, to produce off-season crop in Philippine.
5. *Bending :* Bending of branches for increasing fruit production in guava.
6. *Coppicing :* Complete removal of the trunk in trees like, Eucalyptus and Cinchona leaving 30-35 cm stump alone.
7. *Pollarding :* Removing growing point e.g. silver oak.
8. *Pinching* (In herbaceous plants removal of growing point).
9. *Disbudding* (removing unwanted flower bud).
10. *Alternate differential pruning*

8.1.9 Propagation of Horticultural Crops

I. Sexual propagation

Seed : Most annuals, biennials and many pernnial fruit plants; vegetables crops; plantation, aromatic and medicinal plants; ornamental, flowering and shade providing shrubs and trees.

II. Asexual propagation

(*i*) *Propagation by Apomicitic seedlings* : Citrus, mango, apple etc.

(*ii*) *Propagation by vegetative structures* :

A. Propagation by division

Division of stem tubers: Potato, caladium etc.

Division of tuberous roots: sweet potato, turmeric, dahlia, bigonia etc.

Division of rhizomes : Ginger, banana, blueberry, bamboo, iris, oxalis etc.

B. Propagation by separation

Bulbs : Onion, daffodil, Bird of Paradise, tulip, narcissus, hyacinth etc.

C. Propagation by sukers

Pineapple, banana, chrysanthemum, blackberry, raspberry etc.

D. Propagation by runners

Strawberry

E. Propagation by stolon

Doob grass.

F. Other vegetative structures

Bulbils : Lilies

Offsets : Pineapple, banana, cacti and succulent, tulip, iris, lily etc.

III. Propagation by cuttings

(*a*) *Stem cuttings* : Grape, fig, pomegranate, mulberry, kiwi fruit, olive, quince, rose, , honeysuckle wisteria, weeping willow, poplar etc.

Semi-hardwood cuttings : Mango, guava, jackfruit, camellia, lemon, rhododendron etc.

Softwood cutting : Juniper, hollyhock, azalea, lilac etc.

(*b*) *Leaf cuttings* : African violet, bigonia, bryophyllum etc.

(*c*) *Leaf bud cuttings* : Blackberry, Lemon, cammelia, hydrangea, kalanchoe etc.

(*d*) *Root cuttings* : Blackberry, fig, cherry, apple, pear etc.

IV. Propagation by layering

(*a*) *Tip layering:* trailing black berry, raspberry etc.

(*b*) *Simple layering* : Ornamental climbers, bougainvillea, grape, lemon etc.

(*c*) *Stooling/Mound layering* : Apple rootstocks, guava, mango, litchi, Gooseberry etc.

(*d*) *Serpentine layering* : Clematis, American grape etc.

(*e*) *Trench layering* : Apple rootstocks, cherry, plum etc.

V. Propagation by grafting

(*a*) *Whip grafting* : Temperate fruit plants

(*b*) *Cleft grafting* : Avocado, apple, pear, plum, camelia etc.

(*c*) *Bridge grafting* : For different fruit plants

(*d*) *Side grafting* : Mango, avocado, junipers etc.

(*e*) *Inarching* : Mango, sapota, guava, litchi etc.

(*f*) *Veneer grafting* : Mango

(*g*) *Epicotyl grafting* : Mango

VI. Propagation by budding

(*a*) *T-budding* : Citrus, plum, peach, cherry, ber, rose etc.

(*b*) *Patch budding* : Walnut, Indian rubber tree etc.

(*c*) *Chip budding* : Mango, grapes etc.

(*d*) *Ring budding* : peach, plum, ber, mulberry etc.

VII. Micropropagation

(*a*) *Shoot tip culture* : Citrus, papaya, grape, orchids, asparagus, lily, gladiolus, strawberry, carnation, chrysanthemum etc.

(*b*) Embryo culture : Citrus, rhododendron etc.

(*c*) *Single cell culture* : Carrot, sunflowers etc.

(*d*) *Embryo rescue* : Grape

8.1.10 Botanical clasification of fruits

Simple fruits			*Aggregate fruits*	*Multiple fruits*	*Accessory fruits*
1. Fleshy fruits	2. Dry fruits		Strawberry, custard apple	Pineapple, fig. etc.	Dewberries, black berries etc.
	A. Dehiscent fruits	B. Indehiscent fruits			
(*a*) *Berry* e.g. banana, grape, papaya, tomato, pepper, brinjal etc.	(*a*) *Capsule* e.g. iris, lily, gladiolus, azalea, rhododendron.	(*a*) *Achene* e.g. butter cup, sun flower etc.			
(*b*) *Hesperidium* e.g. orange, lime, lemon, grapefruit, etc.	(*b*) *Follicle* e.g. delphinium, marigold, larkspur, peony etc.	(*b*) *Caryopsis* e.g. wheat, rice, maize, barley, etc.			
(c) *Pepo* e.g. watermelon, muskmelon, cucumber, squashes pumpkins etc.	(c) *Siliqua* e.g. mustard, stocks, cabbage, radish, turnip, rapeseed, cauliflower, knol khol etc.	(c) *Nut* e.g. walnut, hazelnut, oak, cashewnut etc.			
(*d*) *Pome* e.g. apple, pear, quince, loquat etc. (*e*) *Drupe (stone)* e.g. mango, peach, cherries, apricot, almond, olive etc.	(*d*) *Pod or legume* e.g. peas and beans etc.	(*d*) *Samara* e.g. maple, elm, dioscorea etc.			

8.2 FRUIT SCIENCE

8.2.1 Scope of fruit growing :

Cultivation of fruits contributes to the health, happiness and prosperity of the people.

In spite of the fact that India is climatically favourable for production of a variety of tropical, sub-tropical and temperate fruits, our per capita availability of fruits is low (200 gm) in 2021 so in comparison to developed countries (250-450 gm) and developing countries like Hungary, Philippines, Pakistan (200-300 gm).

1. For that reason, there is a vast scope of growing fruit crops in our country because the total area under fruit crops is very less.
2. About majority of 'Indian population is vegetarian and production of fruits is far less.
3. With development of Agro-industries and recent Govt. initiatives of Mega Food Park storage facilities, improvements in transport facilities, preservation and canning will have great demand. Govt's promotion of Agricultural exports through Agri-export zones creation will ensure remunerative price to farmers.
4. The multiple cropping system in the orchards also generate more avenues for employment apart from giving additional income. The intensive horticultural practices may accommodate skilled and trained personnels where unemployment is a serious problem.
5. India processes only 2% of its produce as against 45% in Philippines. So there is a more distance to cover for catching up the world market with Govt. support in the value-added products.
6. The fruit production has a great possibility in trade and industrial development. A large number of ancillary industries and trade sectors are also developed.
7. With our rapid increasing population, the pressure on land is likely to continue and size of holding will remain small in the foreseeable future. Moreover fruit crops give much more production per unit of area and are certainly paying more than ordinary crops.
9. After achieving the food security when the Govt. is planning for agricultural diversification, at this moment fruits may well fit in crop diversification programmes.

8.2.2 Problems of fruit cultivation

While fruit cultivation is having lots of promise but still some of the problems dog up the idea of a successful fruit cultivation. Some of the national problems like malformation, alternate bearing and spongy tissue in mango still defining solution let no headway has been achieved in production of superior quality planting materials. There is lack of varieties resistant to biotic and abiotic stresses. Due emphasis needs attention of Integrated Pest Management. However, the most important problems being faced in the fruit growing, are :

(*i*) *High initial orchard investment :* Costs of plastics, expenditure on layout, the ordinary farmer having small land holding can't afford to establish an orchard without any capital support.

(*ii*) *Long juvenile period of fruit plants :* Stretching upto 10 years; fruit orchards do not ensure any return. Small and middle farmers cannot wait. for such a long gestation period without substantial income. It can be compensated by planting filler plants like Kinnow, Plum, Papaya, Guava and vegetables.

(*iii*) *Lack of high quality fruit plants :* A healthy nursery plant is the foundation of an orchard, use of poor planting material is the major constraint limiting higher production in most of the orchards.

(*iv*) *Faulty marketing system :* The marketing of fruits in India is the most sorry aspect of fruit industry. The lack of marketing facilities is a big hurdle in the expansion of horticulture. The fruit crop is auctioned by the growers to the middle man who can get better profit from this produce.

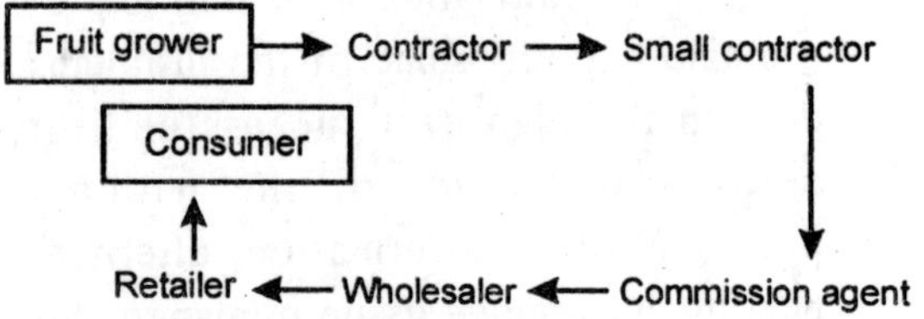

- very lengthy procedure and cost more to the consumer .

Solution. : (Cooperatives)

(*v*) *Lack of storage and transportation facilities :* The post-harvest losses account for 25% of the fruits. After harvest, the fruits need to be transferred to cooling units within half-an-hour. Due to poor cold storage and controlled atmospheric storage facilities the losses further increase. Inefficient transportation system aggravates the position.

(*vi*) *Lack of processing units :* When all over the world processing if fruits and vegetables. Processing units absorb the glit in the markets thus favouring the farmers income. But the processing industry may be located preferably established in the private sector with proper safeguards for the interests of the producer.

(*vii*) *Perishable nature of fruits :* Most of the horticultural produce is highly perishable because of high water content. Inspite of very heavy losses, very little work has been done on post harvest handling and storage systems.

(*viii*) Low purchasing power of people : The fruits are sold at very high cost in the markets low income and middle income group has low purchasing power and do not purchase, the fruits regularly.

- There is a strong need for developing the co-operative system of marketing, so that the fruit grows get decent return of their investment. Also the production of fruits has to be increased by putting more area under fruits with a view that the fruit should be available at reasonable price in the market.

8.2.2 (A) Fruit Drop :

- Fruit trees usually bear a large number of flowers and only a small percentage of which are enough to give a normal yield.
- When the fruit set is much more than the trees can carry to maturity these will be drop of fruits at various stages of fruit development.
- Such a drop will beneficial and natural to the trees and it will prevent exhaustion of the resources and breaking of branches by over bearing.

First drop : Flowers with aborted pistils drop off at this stage. Lack of pollination, lack of stigmatic receptivity defective flowers and some fertilized poor pollen transfer are causes.

2nd drop : Fortnight after first drop. Includes unfertilized and fertilized flowers.

3rd drop : Drop occurs when fruits are of *'minable size'* due to the formation of abscission layers on the young fruit stalks. The above drops which occur in 3 stages are beneficial to the fruit tree.

Pre-harvest drop : Loss to the grower. At this stage, half developed and 3/4th developed fruits are, shed due to many causes. This is a loss to the growers and is a serious problem.

Causes of fruit drop :

1. *Mechanical* : Wind and hailstorms cause fruit drop.
2. *Climatic factors* : High temperature, low humidity, very low temperature bost on the formation of abscision layers, consequently fruits drop.
3. *Physiological factors* : Abnormal fluctuations of soil moisture favour heavy fruit drop.
4. *Nutritional* : Lack of available N and several other nutritional factors may cause fruit drop.
5. *Cultural practices* : Deep digging and deep ploughing during the fruit development phase will damage the roots cause the fruits to shed.
6. *Pathological practices* : Incidence of parts and diseases will cause more shedding of fruits.
7. *Varietal factor* : Within a kind of fruit, the varieties differ among themselves in the extent of fruit drop.

Prevention :

- The pre-harvest drop may be reduced by controlling the causes to a certain extent.
- Proper and timely culture such as irrigation plant protection, provision of polliniser and wind breaks will help to prevent or reduce the amount of fruit drop.
- In the final stage of fruit growth, a rapid decrease in auxin content is correlated to segmentation of endosperm causing pre-harvest fruit drop. In high concern of auxin supplied exogenously may inhibit fruit drop.
- NAA (10-20 PPM) reduces drop of fruits in pome, fruits such as apples and pears.
- 2,4-D (less than 20 PPM) checks drop in citrus fruits.
- GA at 60 PPM reduces the drop in Jamun.
- Great care, must be exercised on the use of hormones as it is possible to cause much damage to the foliage, and new growth.

8.2.2 (B) Alternate bearing :

Alternate or biennial bearing is an age old horticultural problem and is also designated as irregular bearing.

An year of optimum or heavy fruiting is followed by an year of little or no fruiting. Essentially these tree, carries optimum load of crop in one year but it fails to flower and produce a satisfactory crop in the following year. This phenomenon is observed in most of the perennial crop in the following year with varying degrees.

Any perennial plant, which carries a heavy crop load in the current year is likely to bear less in the following year. However this phenomenon is distinctly expressed in horticultural crops like mango, olive, coffee, tamarind, apple, coconuts.

The heavy crop year is 'on year' and the less bearing is 'off year'.

If an 'on year' is regularly followed by an 'off year', then it is called "biennial or alternate bearing". Irregular bearing on the other hand refers to the rhythm of a heavy crop followed by more than one year of poor crops.

Disadvantages :

1. Compared to regular bearer, the average yield is less.
2. The quality of fruits may be inferior during 'on year'. Cropping compared to regular bearer.
3. More breakage of heavy loaded branches is likely on 'on year'.

Causes

1. *Genetic causes :* Certain trees which belong to the particular species or genus or family express this tendency always e.g., apple, pear, plum and apricot belong to Rosacea have a tendency. Within a species also cultivars differ on this tendency e.g. in mango, (Bangalora, Neelum, Mallika) are regular bearer but Mulgova, Dashari are alternate bearing.
2. *Bearing habit :* Terminal flower bed formation is often regarded as an explanation for alternate bearing in apple. The formation of spurs (which is a terminal bearer) is strongly stimulated in the 'off year', hence in the on year, the crop bears heavily.
3. *Age of the tree and size of the shoots :* In biennial apple, this tendency is found to be more, in young trees than in older trees. In mango, it has been found that bearing shoots of 8-10 months old branches produce flower buds.
4. *C/N ratio :* It plays important role in creating favourable conditions for the synthesis and action of the substances responsible for flowering.
5. *Hormonal balance :* Higher levels of auxin like substances and inhibitors (similar to ABA) and lower levels of GA_3 like substances are vital for a fruitfullness shoot in mango.

Remedial measures :

1. *Scientific management of orchards :* A proper schedule of numerus and fertilizer application, intercultural operations, irrigation at fruit set and fruit developmental stage etc. will help.
2. *Deblossoming :* Can be helpful only at full develop stage of panicles.
3. *Smudging and chemical regulation :* Practice of inducing flowering in mango by smudging is being followed in Java and Philippines.
4. *Pruning :* Only old trees may respond to pruning. It can be, helpful in reducing irregular bearing and not alternate bearing.
5. *Growing regular bearing variety :* Regular bearing varieties are to be preferred.

8.2.3 Nutritional value of fruit crops

Fruits being rich in vitamins and minerals, are known as protective food. Realizing the worth of fruits and vegetables in human health, Indian Council of Medical Reasearch (ICMR) recommends the use of 120 g fruits and 280 g vegetables per capita per day. Consumption of adequate amount of fruits helps in maintaining health and vigour of an individual.

Nutrient-wise richness of important fruits

Vitamin 'A' richness

Fruits (F)

Mango	4800 IU/100 g
Papaya	2020 IU/100 g

Vitamin 'B_1' (Thiamine) richness

Cashewnut	630 mg/100g
Walnut	450 mg/100g
Apricot (dry)mg/100g	

Vitamin 'B_2' (Riboflavin) richeness

Beal	1191 mg/100g
Papaya	250 mg/100g
Litchi 122.5 mg/100g	

Vitamin 'C' richness

Barbados cherry	1000-4000 mg/100g
Aonla	600 mg/100g
Guava	299 mg/100g
Lime	63 mg/100g
Sweet orange	5017 mg/100g
Lemon	39 mg/100g

Carbohydrate richness

Apricot (dry)	72.81%
Date (pind)	67.30%
Karonda (dry)	67.10%
Banana	36.40%
Bael	30.6%
Custard apple	23.90%
Cashewnut	22.37%
Jamun	19.7%
Jackfruit	18.9%
Tapioca	38.1%

Protein richness

Cashewnut	21.20%
Almond	20.8%
Walnut	15.60%

Fat richness

Walnut	64.50%
Almond	58.9.%
Cashewnut	46.90%
Avocado	22.80%

Fibre richness

Guava	6.90%
Kaintha	5.20%
Pomegranate	5.10%
Aonla	3.40%
Grape	3.00%
Walnut	2.60%

Calcium richness

Litchi	0.21%
Karonda	0.16%
Kaintha	0.13%

Phosphorus richness

Cashewnut	0.45%
Walnut	0.38%
Litchi	0.30%
Kaintha	0.11%

Iron richness

Dry Karonda	39.1%
Date (pind)	10.6%
Cashewnut	5.0%
Walnut	4.80%

8.2.3 Package of Practices of Fruits

	Mango	Banana
Soil	• Alluvial to laterite soils. • Slightly acidic soil with pH <7.5. • Black soils not suitable.	• Deep, well drained loamy soil. • Depth and drainage is very important.
Climate	• Tropical fruit 24-27 °C, 1400 MMSL.	• Humid tropics. • Frost is the limiting factor.
Propagation	• Veneer grafting (commercial) (Season : March-September-North India)	• Sword suckers (1.5-2.0 kg). • Throughout the year.
Season	• Monsoon season	• February-March (hills of South India).
Spacing	• (10 × 10) m; HDP : 2.5 × 2.5	• Tall var : (2.7 × 3.0)m. • Dwarf var : (1.8 × 1.8)m
Training pruning	• Hardly needs any pruning. • Periodical removal of dead and diseased branches.	• No training and pruning. • Desuckering (of unwanted suckers).
Flowering	• February-March (North India). • January-February (East India) • Duration 2-3 weeks	• After 8-9 months planting. • 10-12 leaves with pseudostem
Pollination & fruit growth & development	• Highly cross-pollinated crop. • Agent-house fly (Musca domeslica) < 1.0% fruit set. • Sigmoid growth curve.	• Vegetative parthenocarpic. • Double sigmoid growth curve.
PGR used	• Last week of March, 2nd week June. • NAA, GA, @ 25 PPM for fruit crop.	• Ethrel-ripening.
Harvesting	• Slight colour development-shoulders. • Specific gravity 1.0-1.1%.	• Skin changes from angular to round. • September-April.
Yield	• 10-15t/hac (10th year).	• Tall vars : 15-20t/ha.
Problems	• Mango malformation (less temperature). • Alternate bearing.	• Dwarf var : 30-40t/ha. • Panama wilt. • Bunchy top of banana.
Varieties	• Alphonso, amarapali, mallika, dashchari, neelum.	• Robusta, rasthali, poovan, dwarf cavendish.
Manures and fertilizers	• 1000 : 500 : 1000 g NPK/trees. • 10 years and above age.	• 100-200 g N • 100-150 g P • 200-300 g K

	Citrus	Grapes
Soil	• Loamy soil with pH 5.5-7.5. • Sensitive to salts and presence of kankar ($CaCO_3$).	• Sandy loam soils. • Moderatly tolerance to salinity and alkalinity.
Climate	• Subtropical fruit (frost free). • Light loving plant. • 5-8 m depending on species. • Monsoon season.	• Long, warm to hot, dry summer and cool. • Heat units for ripening in winters.
Propagation	• Budding.	• Hardwood cutting (23-46 cm length with atleast 4 buds.
Season	• February-March (major bloom). • (North India)	• January (1 year old rooted plants).
Spacing	8 × 8 m	• Bower system : (3 × 3)m.
Training pruning	• Pruning of young trees - training. • Bearing trees - no pruning.	• Training : Bower (S.I.), heat system (N.I.).
Flowering	• Honey bees - cross pollination. • Poor fruit set due to high drop.	• North India (January) : South India (March-April, September-October). • Time of bud break : March-June. • Full bloom : 6-8 weeks after bud break.
Pollination & fruit growth & development	• Sigmoid growth pattern (kinnow). • Linear pattern (sweet orange).	• Parthenocarpy (black cointh). • Stenospermocarry (per latten Pusa seedless). Self pollination is rule. • Double sigmoid growth curve.
PGR used	• 2,4-D is very effective for fruit drop. • (MH-preharvest drop)	• GA_3 @ 40 PPM - flowering and bearing set stage improves size and quality of berry ((Thompson seedless).
Harvesting	• TSS/acid ratio-good indices (14:1). • Generally green to pale yellow.	• TSS/acid ratio is 15:1 • Colour change and softening with taste.
Yield	• Sweet orange : 1000-1500 fruits/ tree.	• 15-20 tonnes/ha (Thompson seedless).
Problems	• Citrus decline (multiple factors). • Granulation	• Splitting of berries (excess moisture).
Varieties	• Sweet orange : Mosarabi, sathugudi. • Mandarins : Nagpur cantia, kinnow.	• Thompson seedless, Pusa seedless, Bhekri, Beauty seedless, per letter, cardiness.
Manures and fertilizers	• 1000:100:200 g NPK/tree.	• 1000 : 500 : 1500 g NPK/vine fully grown vine (3rd year).

Require-ments	*Sapota*	*Ber*	*Pomegranate*	*Apple*
Soil	Sandy loamy Red laterite	Marginal lands with high pH	Deep loamy	Slightly acidic soil
Climate	Tropical-warm and moist weather	Dry and hot climate	Subtropical fruit	Temperate
Propagation Season Spacing Training and pruning	Inarching early monsoon season 8 - 10 m Central leader system	Budding Aug-Sept (8 × 8) m square system Modified leader system	Rooting of Hard wood cuttings Monsoon season (5 × 5) m Training-allowed to bush form	Budding (shield/T budding) Monsoon season 5 – 8 m Modified leader system
Plant Growth Regulator	GA_3 @ 100 ppm, NAA @ 300 ppm	–	–	NAA @ 10 ppm – better retention of fruit
Harvesting	March-May	Feb-April	Skin turns and light yellow (Index of maturity)	Soluble solids increase
Yield Problems	100 fruits/tree Fruit drop for its cultivation	2 quintal/tree/year Excess moisture is harmful	100 fruits/tree/year Fruit cracking internal browning	30-50 kg/tree/year Apple wooly aphid,
Varieties	Cricket ball	Illaichi, Gola	Ganesh, Khandari	Red delicious, Red Gold

Require ments	*Guava*	*Papaya*	*Pineapple*	*Litchi*
Soil	Alluvial to laterite soil	Well drained sandy loam soil	Sandy and loamy soil	Acidic soils are good.
Climate	Tropical and sub tropical	warm, humid climate	Tropical and subtropical	Subtropical
Propagation	Stooling	Seeds	Slips	Air layering
Season	On set of monsoon	mid June-October	early winter	August-September
Flowering	April-May, Aug-Sept	Starts flowering and fruiting 5 months after planting	Irregular flowering	Jan-Feb.
Plant Growth Regulator	NAA @ 100 ppm, 2-4, D @ 30 ppm	Soakaing seeds with GA_3 to increase germination	NAA @ 200-300 ppm	NAA, 2-4, D used to increase the fruit size
Harvesting	Fruits turns greenish yellow	Greenish to yellowish green	Orange yellow	Smoothness of epicarp
Yield	150-200 kg/tree	40-50 t/ha	20-25 t/ha	80-150 kg fruit/tree/year
Disease	Guava wilt	viral disease	Irregular bearing	Fruit cracking
Manures & Fertilizers	350 : 280 :350 g NPK/tree	250 : 200 : 200 g NPK/plant	500 : 40 : 660 kg NPK/ha	60 kg FYM, CAN
Varieties	L-49, Allahabad safeda	Washington, Honey dew pusa dwarf	Queen, Mauritius	Rose Scented

8.3 VEGETABLE SCIENCE

8.3.1 Types of vegetable garden :

Vegetable gardens can be classified into *6* different types according to the purpose for which they have been developed.

These are (*i*) home or kitchen garden (*ii*) market gardens (*iii*) truck gardens, (*iv*) gardens for processing (*v*) gardens for seed production.

1. *Kitchen garden :* A home or kitchen garden refers to the raising of vegetable crops in the back yard of a house. With the growing demand for the valuable land, vegetables can be grown over the roof or in verandahs.

It not only supplies fresh vegetables but also saves some money as vegetables are now quite costly in the market. The fresh vegetables not only ensure a better balanced diet but there is also a delicate psychology behind the taste.

2. *Market garden :* This type of vegetable garden is intended to supply vegetable for the local market. Therefore they are confined to the immediate, vicinities of cities and towns. Such gardens are located within a radius of about 15-30 km from large towns and cities. Since these types of gardens are close to Urban, lands are generally costly. Hence intensive method of cultivation is followed. The gardener must be a versatile person as he is to grow a number of vegetables throughout the year to supply all kinds of vegetables to the market.

3. *Truck garden :* This refers to producing some special crops in relatively larger quantities for distant market. The location of this type of garden is determined by the soil and climatic factors suitable for raising those particular crops. Cultivation is more extensive. The commodities are usually marketed through middle-man. Growing of cabbage and cauliflower in Nilgiri hills and transported to distant places like Chennai, Trivandrum, Cochin are good example of truck gardens. With the development of a quick and easy transport system, the distinction between the market and truck gardens is continuously decreasing.

8.3.2 Vegetable nutritional value

Natritional value. Vegetables, being rich in vitamins and minerals, are known as protective food. Realizing the worth of fruits and vegetables in human health, Indian Council of Medical Reasearch (ICMR) recommends the use of 120 g fruits and 280 g vegetables per capita per day. Consupmption of adequate amount of vegetable helps in maintaining health and vigour of an individual.

Nutrient-wise richness of important vegetables

Vitamin 'A' richness

Beet leaf	9770 IU/100g
Spinach leaves	9300 IU/100 g
Colocasia leaves	10278 mg/100
Coriander leaves	6918 mg/100 g
Drumstick leaves	6780 mg/100 g
Fenugreek leaves	6450 IU/100 g
Mustard leaves	4370 IU/100 g

Vitamin 'B_1' (Thiamine) richness

Chillies	217mg/100g
Colocasia leaves	mg/100g
Tomato (red)	0.12 mg/100g

Vitamin 'B_2' (Riboflavin) richesness

Fenugreek leaves	0.31 mg/100g
Amaranthus	0.30 mg/100g

Vitamin 'C' richness

Drumstick leaves	220 mg/100g
Coriander leaves	135 mg/100g
Chilli	111 mg/100g
Tomato	31 mg/100g

Carbohydrate richness

Sweet patato	28.2%
Potato	22.6%
Curry leaves	18.7%

Protein richness

Peas	7.2 mg/100g
Cowpea	4.3 mg/100g
Indian Bean	3.8 mg/100g

Fat richness

Mustard	0.8mg/100g
Beet leaf	0.7mg/100g
Spinach	0.6mg/100g

Calcium richness

Agathi	1130mg/100g
Curry leaves	813mg/100g
Amaranthus	395mg/100g
Fenugreek leaves	395mg/100g
Radish leaves	265mg/100g
Coriander leaves	184mg/100g

Iron richness

Agathi	83.9mg/100g
Amaranthus tender	25.5mg/100g
Coriander leaves	18.5mg/100g

General Information about vegetables

Sex Forms

1. Cleistogamy : Lettuce
2. Chasmogamy : Tomato, capsicum
3. Protogyny : Cole crops
4. GMS : Tomato, Musk Melon, Water Melon, Cole crops, Onion
5. CMS : Carrot, Sweet Pepper., Cucumber
6. CGMS : Onion, Carrot, Raddish Tomato
7. Protandry : Onion, Carrot
8. Herkogamy : Lima Bean
9. Peas and Beans are self pollinated crop due to presence of 'Keel'

- **5 Sex Forms in Cucurbits**

1. *Hermaphrodite* : Satputia (Ridge Gourd), Cucumber - Original form (Primitve)
2. *Andromonoecious* : Musk Melon, Watermelons.
3. *Gynoecious* : Cucumber
4. *Monoecious* : Cucumber, Ash gourd, Bitter gourd
5. *Trimonoecious* : Cucumber, Musk Melon, Ridge Gourd

- *Broccoli* : Highest Cross Pollination (95%)
- *Spinach* : 5 Sex forms present

- **Pollination**

1. **Anemophily** : Spinach, Beet root, Palak, Amaranthus
2. **Entmophily** : Onion, Carrot, Raddish, Cole Crops.

- **Anthesis**

1. Bottle gourd : Evening 5 - 8 pm
2. Snake gourd : Evening 6 - 9 pm
3. Ridge gourd : 5 - 8 pm
4. Musk Melon : Morning 5.3 - 6.30 am
5. Cucumber : Morning
6. Sponge gourd : Morning
7. Bitter gourd : Morning 4.0 - 7.30 am
8. Pointed gourd : Evening 7 - 7.30 pm (Female), 7.30 - 8 pm (Male)

- **Fruit type**

1. Peas and Beans : Pod/Legume
2. Okra : Capsule
3. Sweet potato, Yams : Capsule
4. Berry : Tomato, Potato, Taro
5. Carrot : Shizocarp
6. Seed ball : Turnip, Beet root

8.3.3 Package of Practice of vegetables

Sl. No.	*Name of the vegetables*	*Important varieties*	*Method of planting*	*Seed rate/ha*	*Spacing in cm*	*Season of sowing*	*Manures and manuring (Kg/ha)*	*Duration of the crops (in days)*	*Yield (tons/ha)*	*Other parti-culars*
1.	Tomato	Co. 1, Co. 2, Pusa Ruby	Sown in nursery and transplanted 25 days after sowing	375 g	60 × 60	June-Nov, Nov -Apr, Feb-May	N-150 P-100 K-50 Hybrid Varieties require 250:250:250	135-150	25-50	harvest the half ripe fruits at 3 days interval
2.	Chillies	MDU-1. Pusa Jwala	Sown in nursery and transplanted 40-45 days after sowing	1.0 to 1.5 kg	30 × 30 or 45 × 30	June-July to Dec-Jan. Sep.-Oct. to Mar-Apr., Mar.-Apr., to Sep.-Oct.	N-70 P-35 K-35	210-240	Dry pods: 2-2.5 Green chillies 7.5-10.3	—
3.	Brinjal	Pusa Purple Long, Arka Navneet (F_1) Palur-1	Sown in the nursery and transplanted 45 days after sowing	375-500 g	75 × 60	May-Oct., Dec.-May	N-100 P-75 K-50	150-165	20-25	—
4.	Bhendi	Pusa Sawani, Punjab-Padmini	Seeds dibbled in rows	7.5 kg	60 × 30 March -April	July-Nov.,	N-40 P-50 K-30	90-110	7-10	Immature fruits are edible and should be harvested at 3 days interval
5.	Peas	Bonnevilla, Telephone, Gloriosa	Seeds sown in raised beds (hills) and in ridges (plains)	80-100 kg	40 × 10 in hills, 60 × 10 in plains	Hills: Feb-Mar to May-June Plains: Oct. Jan.	N-180 P-125 K-100	80 to 100	5-7.5	—
6.	French beans	Watex, Premier, Contender,	Sown in lines in the beds, sown on the side of the ridges (plains)	Hills 125 kg, Plains 62.5 kg	Hills: 50 × 20, Plains: 45 × 10	Hills: Mar.-May Plains: Oct-Jan.	N-50 P-75 K-75	90-100	4-5	Also grown as rainfed crop with the half recommended dose of N and should be done at correct maturity
7.	Bellary onion	Bellary red. N.53, Arka Niketan, Arka Pragati, Arka Kalyan	Sown in nursery and transplanted in the main field after 45 days	8-12 kg	30-45 × 8 - 10, transplan-ted on both the side of the ridges	May-June	N-75 P-150 K-75	145-150	15-18	harvest the bulbs when 75% of the tops fall.

Sl. No.	Name of the vegetables	Important varieties	Method of planting	Seed rate/ha	Spacing in cm	Season of sowing	Manures and manuring (Kg/ha)	Duration of the crops (in days)	Yield (tons/ha)	Other parti-culars
8.	Potato	Kufri Jothi, Kufri Muthu, Kufri Swarna, President	Healthy sprouted tubers of 40-50 g planted in furrows	3000 to 3750 kg	45 × 30	Summer:Mar.-Apr. to July-Aug. Autumn: Aug.-Dec. Irrigated: Feb.-June	N-100 P-125 K-50	100 to 140	20-25	Fresh tubers exhibit dormancy, broken by treating carbon-di sulphide @ 30g/100 kg of seeds. Harvest the crop when the foliage turns yellow and starts withering.
9.	Carrot	Pusa Kesar, half Long Danvers, Ooty-1	Seeds dibbled in ridges	3-4 kg	Hills: 25 × 5, Plains: 30 × 15	Throughout the year, Plains: Aug.-Sep.	N-180 P-125 K-100	Hills:120 Plains: 100	25 to 30	Thinning is essential
10.	Cabbage	September, Golden-Acre, Pusa Drum Head, Hero	Sown in nursery and transplanted 30-35 days after sowing	375 gm	Hills: 45 × 45 Plains: 60 × 45	Hills:Jan-feb, April, June-July. Plains: Aug.-Sep.	N-180 P-125 K-100	105-120	30 to 40	Harvest when the head is firm but tender.

8.4 FLORICULTURE AND LANDSCAPING

8.4.1 Current status of development in floriculture :

India is endowed with various agroclimatic conditions suitable for growing large number of flowers. India is known for growing the traditional flowers such as Jasmine, Marigold, Chrysnthemum tuberose etc.

Major cut flowers, which are grown in India, includes, rose, orchids, gladiolus, carnation, anthurium, gerbera etc.

Important flower growing states are Tamil Nadu, Karnataka, Maharastra, West Bengal, J & K, Sikkim and Andhra Pradesh.

The area under flower cultivation is reported to be around 73,536 ha during 1997-98. There has been tremendous appreciation of its economic value in addition to its aesthetic importance.

During the last 10 years period, the flower trade in India has been increased many times in view of the significant increase in floriculture activities.

Major issues

Major issues confronting floriculture industry, particularly in the areas of production technology, post harvest, product quality and infrastructure and measures taken to initiate address the issues are as follows :

Acknowledging the potential floriculture holds, as a diversified economic activity in terms of production, employment, income and exports, the GOI have initiated several measures for supporting floriculture.

1. Developing a comprehensive database.

2. Developing specific programmes for traditional flowers and stepping up research and extension efforts.

3. Programme for assistance to small and large nurseries.

4. Strengthening of infrastructure for post harvest management and marketing.

5. Export promotion and market development.

6. Establishment of wholesale market-cum-auction centres for flowers.

Flower cultivation is a good source to earn foreign exchange considering India's geographical advantage, it is in a better climatic region unlike the severe winters prevailing in the west.

But the trade in cut-flowers is negligible when compared to the Netherlands. This might be due to lack of enterprise, technical know-how, standard methods of growing and harvesting flowers of an internationally acceptable quality as well as the problems of packing and speedy and cheap transport.

Export potential :

In view of the immense potential for export of ornamentals from the country, recently the GOI and its agencies like Ministry of Commerce, APEDA, State Trading Corporation have given encouragement and support for export.

Future prospects :

1. Emergence of new markets Japan, Europe, S-E Asian countries like Singapore, Hongkong and India may emerge as consumers of floricultural products in the year to come.
2. Many small units may exit from the market due to uneconomic returns and high overheads.
3. India has several advantages like : (*i*) availability of skilled manpower, (*ii*) labour at competitive rates, (*iii*) presence of a reputed large tissue culture production facility which can propagate plant material at low cost etc.

Reasons for Lukewarm response to floriculture

1. In India most of the farmers still use traditional methods of agriculture with poor infrastructure base.
2. Cost of establishment is quite high.
3. Cost of planting material is quite high.
4. Industry requires a technologically advanced infrastructure to ensure quality.
5. Quality is the single best parameter for a good return.
6. The absolute necessity to ensure quality are specialized transport system, refrigerated trucks, cold storage system and regular international flights for timely transport for export.

Strategy

1. Industry and Government should work in close co-ordination and Government support is necessary in foreign countries to certify credit worthy buyers and help Indian companies to penetrate new markets.
2. A high degree of transparency and sharing of market information and business experience is required to faster the growth of industry and face international market challenges collectively.
3. Increased frequency of domestic carriers.
4. Loan procedures should be simplified since industry is capital intensive (lower rate of interest may assist).
5. Policy initiatives are necessary to revamp the present insurance system for perishables.
6. Group marketing approach to exploit economics of scale needs to be developed.

7. Investing in marketing information on the present level of area, production and locations of flowers and ornamental plants in the country.
8. Product positioning-identifying niche markets and product differentiation. Flowers in demand be given preference in growing.
9. Invest in R & D for development of priority products through a breeding programme. Research on improvement of varieties which suits the demand of international markets.
10. Specialise in regional production of products that have a specific local advantage. In India also there is a budding market gradually catching up the flower trends.
11. Small growers of flower and live plants for export may require assistance in marketing their products. Export houses and big industrial organisations may be induced to enter into the export trade of floriculture items, like those in Kenya, Srilanka.
12. Visualising the acute shortage of qualified and trained staff in floriculture, the Government should depute floriculture staffs for training abroad.

8.4.2 Style of Gardening

1. Formal Design

This design is very stiff and everything is done in a straight and narrow way in a symmetrical or a geometrical pattern. Everything is planted in straight lines. The hedges, edges and topiary are maintained in proper shape by regular training and pruning. Symmetry is always maintained in the garden by undertaking similar type of plantation and using similar garden technique and adornments.

2. Informal Design

In this design, the plants and features of the garden are arranged naturally without following any hard and fast rules.

3. Picturesque Design

The idea of picturesque design of garden was expounded by William Robinson in the last decade of nineteenth century. His idea was to naturalize plant in shrubbery. He also suggested that grass should not be mown and bulbous plants should be mown and bulbous plants should be grown scattered in the group to imitate wild scenery. He opined that passages should be opened in the woodland, trees, shrubs and bulbous plants should be planted in forest flora to have an effect of wild garden. His idea was to allow the creeper to climb over the tree naturally to imitate forest type of effect.

8.4.3 Types of Garden

1. Mughal Gardens

The gardens laid out during the regime of Mughal Emperor, Babar in India are known as Muagal Gardens, Babar (1494-1531 A.D.) was the first Mughal ruler to introduce this style in India. All other Mughal rulers like Akbar, Jahangir, Shah Jahan, Aurangzeb and Begum Nur Jahan shared their hand on developing Mughal gardens in India.

The main features of Mughal garden are :

(*i*) Site and Design,
(*ii*) Walls and Gates
(*iii*) Terraces.
(*iv*) Running water.
(*v*) Baradari
(*vi*) Tomb or Mosque.
(*vii*) Trees and flowers.

2. Persian Gardens

Persian garden style is one of the oldest garden styles. The garden styles were formal. Persians used crafted materials such as masonry, carved and pierced marble stones and highly polished stones. The gardens were laid out by cutting terraces on hill slopes. They tapped some natural spring to create straight water-course through the gardens and manipulated water course to undertake different movements along its run. If there no natural source of water, it was created artificially by diverting some river or stream.

3. French Garden

In according, the French style of garden were also very intricate and artificial. The new style of gardening now known as French style is largely due to contribution of Le-Notre who served in the Royal Garden in Louis-XIV from 1943 to 1700 A.D.

4 Italian Gardens

There is a similarity between Persian or Mughal styles with the Italian style of gardening. In all these styles of gardening, the similarity was the use of heavy masonry features. The Italian elites conceived that garden is just an extension of lavish palaces, as a glamorous outdoor hall for entertainment and showing of their wealth as well as status. Massive stairs, generally of marble, complete with balustrade to connect the different levels in the garden are the main features of Italian gardens. Decorative fountains generally in combination with stone sculpture are other important features of the Italian gardens

5. English Gardens

Due to favourable climatic conditions such as high annual rainfall, the natural ground cover in the England is grass. In consonance to this the famous British Garden architect Repton and Brown advocated the concept that the British Garden should look like the countryside. The gardens were created with the idea that the garden should merge with the countryside without any artificial barriers such as fences or hedges etc. During the eighteenth century, the British architects Repon and Brown alongwith kent brought through of natural in the garden by the middle of the sixteenth century. During the eighteenth century gardens were laid with more emphasis on architectural features. The main features of the garden during that period were curved paths, informal group of trees rivulet or streams, artificial waterfall and clipped hedges

6. Japanese Gardens

The plant materials were selected in such a manner that the garden would not undergo many major changes during the four seasons of year. During the period of Muromachi (1392-1573 A.D) both the flat garden (Hira-niwa) and hill gardens (Tsukyama-sansui) developed. Japanese gardens were developed in close association with nature. This has contributed a lot in maintaining its popularity upto now whereas other styles of gardens have fallen into disfavour. The most important teaching of Japanese garden is not worth a place visiting. It should be a place where mind find rest and relaxation. The immutability is achieved also because rather rocks, stepping stones, streams, waterfalls, bridges and stone lanterns contribute to the structure of the Japanese gardens. The Japanese garden emphasizes to capture natural scenery. The three important elements for this purpose are water, stone and plants . Low sculptured bushes and trained dwarfed trees are attraction of Japanese gardens.

8.4.4 Lawn

A lawn can be define as the green carpet for the landscape maintained by growing and mowing grasses. Thus lawn is an area maintained under thickly grown grass cover. It forms the heart of the garden and hold pivotal importance in adding beauty to the garden. No garden is considered complete without establishment of lawn. Lawn serves as a base for a flower bed, a border, a shrubbery or a specimen tree. Besides having monetary value, lawn possesses aesthetic value for reducing tension of the mind. It becomes very charming to enjoy the grandeur of the spacious lawn in the garden.

SUITABLE LAWN GRASSES

Doob or Bermuda grass : *Cynodon dactylon,* Manilla grass : *Zoysia matrella*

Korean grass : *Z. japonic*, Korean velvet grass : *Z. tenuifolia.*

PLANTING OF LAWN

Lawn grass is planted using one of the following methods suitably :

1. By Seed. When other planting material like cutting etc., are not available lawn is raised by seeds. The area over which grass planting needs to be taken is prepared well by tilling to a pulverized solid and adding sufficient quantity of FYM followed by proper levelling. About 500 g seeds per 200-300 square metre area germinates sufficiently to prepare thick lawn. The seeds of Doob being very light in weight are sown after mixing with double the quantity of finely sieved sand or ash. Mixing seed with sand or ash facilitates broadcasting of seed and, thus, ensures uniform germination of seed throughout the bed. After brodcasting the seed is covered with thin layer (1-2 cm thick) of finely sieved farmyard manure. The bed should be watered regularly using watering can fitted with fine nose of nozzles so as to keep the bed continuously moist. The seeds germinates in about 3-5 to weeks.

2. By Dibbling. Dibbling implies planting in hole. The bits of the mown grasses are used for dibbling. The cutting should be taken from grass growing in shade. The grasses growing in sun having close internode are used for dibbling Dibbling grass is done at a distance of 10 cm in the moist bed. Continuous moistening of the bed is maintained to facitlitate sprouting of grasses. The dibbled grasses start growth in about 5-6 weeks.

3. By Turf. This is quickest method of developing lawn. The turf is a piece of earth with grass thickly grown over it. The roll of turf with a width of 1-2 ft. are used for setting lawn. The pieces of turf are placed closely with each other. Before placing turf, the levelling of land is ensured. The joints of the turf are packed using sand.

The turf is made firm by beating it with wooden beater. The turf is water immediately and irrigation is continued till it sets with soil.

4. By Turf plastering. Mixture of bits of grasses along with mud and water is termed as turf plastering. The paste of grass is spread evenly on prepared ground. It is then covered with 1-2 cm thick layer of sieved manure or soil. Watering is done at regular intervals so that the grass make constant with soil and grow successfully., This method is not suitable in dry areas where grasses get dry after spreading.

MAINTENANCE OF LAWN

Weeding. Weed pose great problem in maintenance of lawn and infestation gives an ugly look to lawn. Motha *(Cyperus rotundus)* and doob *(Euphorbia thymefolia)* are major weed in lawn.

Weed is a problem in established lawn. Once established it becomes very hard to remove weed. It is advisable to go for regular weeding in order to remove weeds. It is very difficult to eradicate motha from lawn completely. Deep hoeing, removing stolon of the plant helps in minimizing infestation of the Doob. In established lawn the regular mowing can keep the growth of motha under check.

Liming. Frequent irrigation at close intervals and stagnation of water makes the soil of the lawn acidic. The growth of moss is indication of acidity of lawn. For ameliorating acidic situation lime @ 250 g per square metre is added in the soil.

Rolling. Rolling is practiced with the object of helping the grass in this anchorage to the soil. It is also worth in making the lawn level after weeding. The stone or pebbles also interfere with mowing blade. To press the pebbles rolling is beneficial. Rolling is avoided in wet soil as it makes the soil compact.

Mowing. Mowing is done to prevent excessive growth of grass higher the growth of grass higher is the frequency of mowing.

Sweeping. The removal of cut-over grasses from lawn is mown, the sweeping using broom is practiced. Sweeping is also done every morning by dragging hose pipe of irrigation to allow draining and mixing of dew drop with soil.

Irrigation. Doob is a shallow rooted crop and needs frequent and light irrigation. Now -a-days provisions are being made to irrigate the lawn using sprinklers. The frequency of irrigation depends upon soil and climatic conditions.

8.4.5 Physiological Changes During Flower Senescence

The senescence of flowers is marked by loss of dry matter, primarily due to hydrolysis of macromolecules such as starch, sugars, protein and nucleic acids. The starch is broken down into sugars which are transport to the flowers. The breakdown of proteins release ammonia in the cells, which causes blueing of petals in red cultivars of proteins rose. Senescence is also marketed by changes in the property of membrane such as in microviscosity and loss of phospholipids and permeability which results in leakage of pigments, sugars, mineral ions and total electrolytes, thereby leading to death of the tissues. It has been observed that flowers show two peaks of respiration; first when the bud starts opening and the second when the senescence takes place. This is followed by decrease in respiration of flowers. This decrease in respiration has been ascribed to the depletion of sugars or inability of mitochondria to utilize available sugars.

From an investigation at IARI, New Delhi on the post harvest life of cut flower of ten different cultivars, it has been observed that the cultivars which lasted for longer duration recorded comparatively lower rate of respiration and less electrolyte leakage during the course of senescence over the cultivars which lasted for shorter duration. While studying the changes in vascular morphology of the cut surface of the rose stems of the above cultivars, the freshly cut rose stems of all the cultivars showed clear vascular system; in case of short lasting cultivars, on the third day itself, globular bodies were observed with partial breakdown of vessel cells on the cut surface.

There is also clear evidence that synthesis of new proteins is also associated with petal senescence. The newly synthesized enzymes during senescence thus, appear to play an important role in senescence.

Factors Affecting Shelf Life

Important factors which determine the storage life of flowers are quality of the flowers; stage of harvest; temperature; relative humidity; light; ethylene; and pathogens.

The flowers to be stored should be of good quality. Flowers should be devoid of any physical injury and free from attack of pests and diseases. The injured and diseased tissues are potential sources of ethylene and are highly prone to infection with pathogens.

The flowers must be cut at an optimum harvesting stage. Some flower like gladioli, roses, lilies, narcissi and irises are generally stored at but stage as buds of these flowers open perfectly after storage.

Temperature management during storage is an important component which determine the storage life of flowers, temperature fluctuations in the cold room should be avoided. Relative humidity inside the cold room should preferably be maintained high (90-95). Low humidity causes excessive water loss from the flowers thereby stimulating their desiccation. Cold rooms should be cleaned at regular intervals to check the spread of pathogens.

8.4.6 Floral Preservatives

Term "Floral Preservative" is used for any chemical formulation which is used for extent the life of flowers. Besides increasing life, floral preservatives help to improve flower opening, flower size, shape and colour.

Types of Floral Preservatives

Floral preservatives can be applied in three ways *i.e.*, as (*i*) pulsing solution; (*ii*) bud opening solutions; and (*iii*) holding or vase solutions.

(*i*) *Pulsing solutions* : "Pulsing" refers to short duration (16-24 hr) pre-storage treatment. The effect of such a treatment lasts throughout the entire life of the flower. Sugar (sucrose) is one of the main components of the pulsing solution. Since pulsing involves short duration treatment, relatively higher levels of sucrose are used.

(*ii*) *Bud opening solution* : Immature buds of may flowers can be made to open in chemical solutions, referred to as "Bud opening solution". The components of bud opening solutions are essentially the same as those of pulsing solutions, but in case of bud opening solutions, lower concentration of sucrose are used.

(*iii*) *Holding or vase solutions* : Vase solutions are meant to hold flowers continuously, till the termination of their vase life. The level of sucrose in vase solutions is therefore, also kept very low (0.5 to 2 per cent), due to long duration for which flowers are kept in the solution. Many commercial vase preservatives are also available for use as holding solutions and should be used as per recommendations of the manufacturers.

Constituents of Floral Preservatives

The main constituents of floral preservatives are (*i*) Water (*ii*) Sugar; and (*iii*) Biocides. Besides, mineral nutrients , acidifying agents, anti-ethylene compounds and growth regulators are also used in floral preservatives. The brief account of these constituents is given herewith :

(*i*) *Water* : Water is the most important component of floral preservatives. It affects vase life in terms of its quality as well as quality. Sometimes tap water can be harmful due to its high pH (alkalinity), presence of total dissolved solutes (TDS) and toxic ions. High pH reduces vase life. Therefore, acidifying water to low pH 3.0-3.5 is advantageous because it decreases microbial growth and considerably improves water absorption. The longevity of Super Star roses was improved by the use of tap water over that of single and double distilled water.

(*ii*) *Sugar* : Sucrose is the most widely used sugar in floral preservatives. Sugar acts as additional food source also improves water balance of cut flowers. The optimum concentration of sugar required however, varies from species to species, Since sugar promotes microbial growth, it is usually combined with biocides before use.

(*iii*) *Biocides* : Biocides are chemical substances which are used to inhibit microbial growth in the vase water as well as on the stem surface. Important biocides used for treating cut flowers are 8-hydroxyquinoline citrate, silver nitrate, aluminium sulphate, citric acid, slow release chlorine compounds and quarternary ammonium compounds.

8-hydroxy quinone citrate (8-HQC) is very effective broad spectrum biocide. It acidifies water and also indices partial closure of stomata. This compound is used at the concentration of 200-600 ppm.

Silver nitrate ($AgNO_3$) is also very effective biocide; $AgNO_3$ at the concentration of 25 ppm completely inhibits the microbial growth.

8.4.7 Common house plants and their commercial propagation methods

Common name	*Scientific name*	*Commercial propagation method*
Aglaonema	*Aglaonema commutatum*	Division of basal shoots, nodal cuttings
Anthurium	*Anthurium magnificum*	Seeds
Aphelandra	*Aphelandra squarrosa*	Half ripened wood cuttings
Aralia	*Aralia filiciplia*	Division of crown
Araucaria	*Araucaria heterophylla*	Seeds
Asparagus	*Asparagus africuns*	Division of crown
Bird of paradise	*Strelitzia reginae*	Seeds
Calathea	*Calthea insignis*	Root cuttings
Coleus	*Coleus hybridus*	Herbaceous cuttings of slips
Croton	*Codiaeum spp.*	Air layering
Dieffenbachia	*Dieffenbachia exotica*	Terminal and nodal cuttings
Draecena	*Dracaena deremensis*	Cane pieces
English ivy	*Cissus spp.*	Cuttings
Indian rubber tree	*Ficus elastica*	Air layering
Iron plant	*Aspidistra elator*	Division of crown
Monstera	*Monstera deliciosa*	Air layering
Palms	*Different species*	Offsets
Pandanus	*Pandanus utilis*	Offsets and division of suckers
Peacock plant	*Calathea spp.*	Division of crown
Philodendron	*Philodendron scandens*	Cuttings
Pitcher plant	*Nepenthes khasiana*	Hard-wood cuttings
Podocarpus	*Podocarpus macrophyllus*	Shoot tip grafting
Sago palm	*Cycas revoluta*	Offsets
Sansevieria	*Sansevieria spp.*	Offsets
Spider plant	*Chlorophytum comosum*	Division of crown
Tradescantia	*Tradescantia spp.*	Cuttings
Yucca	*Yucca gloriosa*	Seeds and offsets
Zebrina	*Zebrina pendula*	Herbaceous cuttings

8.4.7 Important shrubs and trees and their commercial propagation technique

A. ORNAMENTAL SHRUBS		
Common Name	*Scientific name*	*Commercial propagation technique*
Acalypha	*Acalyha spp.*	Hard-wood cuttings
Allamanda	*Allamanda cathatica*	Air layering
Azalia	*Azalia spp.*	Hard-wood cuttings and layering
Bottle brush	*Callstemon spp.*	Hard-wood cuttings and layering
Brya	*Brya ebenus*	Hard-wood cuttings and layering
Camelia	*Cammelia japonica*	Hard-wood cuttings and layering
Cestrum	*Cestrum noctrunm*	Hard-wood cuttings
Cledendron	*Cledendron spp.*	Hard-wood cuttings
Clematis	*Clematis spp.*	Leaf bud cuttings
Coral tree	*Erythrina indica*	Hard-wood cuttings
Hibiscus	*Hibiscus rosa-shinensis*	Hard-wood cuttings and layering
Hydrangea	*Hydrangea spp.*	Softwood and leaf-bud cuttings
Ipomea	*Ipmea carnea*	Hard-wood cuttings
Ixora	*Ixora spp.*	Hard-wood cuttings and layering
Jatropha	*Jatropha multifida*	Hard-wood cuttings
Juniper	*Juniperus spp.*	Hard-wood cuttings
Kachnar	*Bauhinia acuminata*	Hard-wood cuttings

8.4.8 Packages for export of cut flowers:

Packaging for export of cut-flowers is as important as the quality of the flower. Only corrugated fibre board (CFB) packages are accepted in the trade. Different types of CFB for cut flowers.

Slotted box, telescopic boxes, rigid boxer, ready glazed boxes and folded type boxes.

An ideal package for export should be small in volume, air-tight non-water absorbent, of the required strength, should have minimal effect on quality and shelf/vase life of flowers.

The size of the packages especially the length of boxes depend upon the stem length of cut-flowers required in trade, e.g. Rose:100 cm.

The strengths of the containers should be enough to withstand a stack height of 3-5 times to protect the loads from damage due to shock and vibrations.

Ventilation holes are not necessary if the flowers are packed and transported in ideal conditions.

The flowers are to be immobilized and cushioned to avoid abscission due to vibration. This is achieved by lining the inside of boxes with thin polyethylene film.

Fine moist tissue paper spreading are also spread of each of the box to provide cushioning to the blooms.

8.4.9 Biotechnology for improvement of flower crops :

The major constraints in flower breeding are, very limited gene pool for improving agronomic traits, which compels breeders to opt for sexual crossing, consuming lot of time and energy.

There is no possibility of directed plant breeding through mutation manipulation, because if we increase the flower size other traits get affected.

Biotechnology can provide many scientific pathways including cell and tissue culture techniques and molecular approaches for the flower breeders.

Plant biotechnology is applied to manipulate diverse traits including flower colour, shape and fragrance, flower and leaf longevity, plant form and habit, resistance to diseases and pests and induction and timing of flowering.

With the advent of in vitro culture techniques, plants can be cloned in a miniature form (microplants) from tiny pieces of leaves, scales and apical buds. In some plants, cell suspension cultures can be used for plant regeneration.

It is estimated that there exists a global market of US $ 15 billion for tissue-cultured plants and products the demand is still growing. In India, approximate 76 companies have plans for setting tissue cultured plants and more than 30 are already in the market with their products.

The various approaches those are used for improvement in flower crops are

1. Micropropagation (Anthurium, Chrysanthemum, Carnation, Gerbera, Dieffenbachia, Gladiolus, Rose, Dahlia).
2. Meristem culture (Dahlia, Gladiolus, Lily, Rose, Gerbera, Kalanchoe, Dieffenbachia).
3. Embryo culture (Litium, Chrysanthemum, Carnation).
4. Somaclonal variation (Carnation, Chrysanthemum, Litium, Pelargonium).
5. Haploid culture (Pelargonium, Lilium, Gerbera).
6. Somatic embroyogenesis (Iris sp Anthurium, Geranium, Floribunda)

(1) Micropropagation in case of ornamental shrubs like, Jasmine can also be cloned through rapid proliferation of auxiliary buds. This technology is also used for production of disease free motherstock cultures, which are then used as a source of conventional cuttings.

(2) The culture of shoot meristems has proven useful in eliminating virus from plants like Dhalia, Gladiolus, Rose, Lilium sp.

(3) Embryo rescue by culturing seeds devoid of endosperm is being used for obtaining new types of orchids and their multiplication. It is also used to obtain interspecific hybrids of Lilium, chrysanthemum and carnation

(4) Regeneration of plants through somatic embryogenesis has been obtained in Irish, Anthurium,Floribunda rose.

(5) Somaclonal Variation where tissue culture results in plants that are physiologically and genetically different from their progenitors have been studied and utilized in number of important flowers crops carnation, chryesanthemum, Lilium, Pelargonium.

(6) The culture of anthers, ovules, microspores is of considerable value to flower breeders to produce haploid plants

(A) Bacillus thuringensis (Bt). Gene has been used in Chrysanthemum for resistance to insect pest *armyworm.*

• In Petunia, transgenic plants were obtained for resistance to Lepidopteran pest using potato protease inhibitor and bovine trypsin inhibitor.

• Increase in vaselife is an area of much concern to floriculturists, transgenic carnation having an artisense ACC oxidase, gene exhibited markedly low ethylene production and delayed petal senescence .

• *Many novel Plants* - blue anthurium, scented rose, chrysanthemum plants with altered flower morphology and flowers that do not fade will come into existence, flowers growing due to presence of bioluminescence genes (Lux Genes) from *vibrio fischeri.*

• Biotechnology offers immense prospects to improve flower crops using the advanced tools, thus increasing the market value of flowers.

8.4.10 Cultivation practice of cut flowers

Flower name /S.N./ Family	Varieties	Spacing (m)	Planting Time	Pruning Time	Propagation	Cultural practice	Harvesting stage	Temp.	Key feature
1. **Rose** *Rosa* *Xhybrida* Rosaceae	Dr. B.P. Pal, Raktima, First Red (GH), Nobleless (GH)	45×45	June	Oct.	T-Budding	Pruning	tight bud	0·5-2°C	By products Rose water aktar, rose oil
2. **Gladiolus** *Gladiolus* Sps Iridaceae	Suchitra, Shagun, Dhiraj, Neelambri	30 × 30	Sep. Nov.	Sep.	Corms and Cormels	de corming, thinning	two flororts open	4°C	No species is Native to India
3. **Chrysanthemum** *Dendrothema* *Sps* Compositae	Diana, Neelambri	30×15	Aug. Sep.	Sep.	Sukers, Stem cutting	de suckering, thinning, disbudding	halfly open	1°C	Sort day plant
4. **Carnation** *Dianthus* *Caryophylus* Caryophylaceae	Scarlet elegance, sam's pride	30×15	June July	Aug.	Stem cutting	pinching (Half, full, 1/2 full)	outer 2 whorls open	0 to 1°C	Long day plant
5. **Marigold** *Tagetus* *ereeta* Compositae	P. Naringi gainda, P. Basanti gainda	30×30	July Nov.	July	Seed	pinching	fully open flowers		male sterility used for F_1 Hybrid
6. **Jasmine** *J. sambac* *J. grandiflora* *J. auriculatum* Oleaceace	CO_1, CO_2 Surabhi	1 M × 1 M	June July	Sep.	Stem cutting	thinning pruning	Bud stage		Used for essential oil extration
7. **Orchids** Orchidaceae	dendrobium, cymbidium, cyanthea, vanillila				tissue culture, Pseudo bulbs. (kikus)	velamen root for adsorptions. of nutrition	fully open flower	5-7°C	lip and labelleum are ornamental

8.5 POSTHARVEST TECHNOLOGY

8.5.1 Scope of fruit and vegetable preservation in India :

Fruits and vegetables are an important supplement to the human diet as they provide the essential minerals, vitamins and fibre (roughage) required for maintaining health.

India with a great diversity in climatic conditions and soil types, is ideal for growing a large variety of fruits and vegetables, both indigenous and introduced.

Position of India - 2nd in vegetables (1st China) and 3rd in fruits.

However for various reasons, this abundance of production is not fully utilized and wastage due to spoilage is about 20-30%.

Most fruits and vegetables are seasonal and perishable in nature. In a good season there may be a glut but because of insufficient transport facilities, lack of good roads and poor availability of packaging materials, the surplus cannot be taken quickly enough to markets in urban areas.

Moreover the surplus often cannot be stored for sale in off-season because of inadequate local cold storage facilities. Thus, the cultivator do not get a good price for their produce.

Although, the R and D efforts on the development of post-harvest handling has helped in reducing the spoilage, considerable losses continue to occur.

Two approaches are possible for solving this problem.

(i) One is the creation/expansion of cold storage facilities in the fruit and vegetable producing regions, as also in the major urban consumption centres, to ensure supply of fruit and vegetable throughout year.

(ii) Another approach is to process the fruits and vegetable into various products which could be preserved for a long time and add value to the product.

With increasing in purchasing power of middle class, there is increasing demand for factory made jams, gullies, pickles.

In spite of all these, the fruit and vegetable preservation industry of present is able to utilise less than 2% of the total production for conversion into products like caned fruits, juices and their beverage squashes etc. as against 40% in developed countries.

Thus there is considerable scope for expansion of the industry, which in turn would give a fillip to development of horticulture, specially in hilly areas and through export of value-added products, earn more foreign exchange.

India is one of the cheapest producer of horticultural produce. If the Government addresses some of the gaps in processing industry, Indian economy will get a boost up.

(*i*) Most of the varieties grown in India are not suitable for processing.

(*ii*) Seeds of processble varieties can be imported and multiplied using tissue culture technique.

(*iii*) Lowering sales tax on agriculture produce may help the industry to grow.

(*iv*) In New Exim Policy of Union Government, setting of Agriculture Export Zones (AEZ) to identify each state's competitive advantage and to promote the cultivation and processing of fruit and vegetable accordingly.

8.5.2 Principles and methods of preservation :

Food preservation can be defined as the science which deals with the methods of prevention of decay or spoilage of food, thus allowing it to be stored in a fit condition for future use.

It is better if the following directions are kept in mind to control the spoilage.

1. Raw materials should be thoroughly examined and handled hyginically to avoid microbial spoilage.
2. Equipments must be cleaned every time before use.
3. Processing should take place as soon as possible after sealing of cans or bottles.
4. Use of contaminated coater should be avoided.
5. The finished products after canning or bottling should be stored in well-ventilated rooms in a cool and dry place.

Freshly prepared products are highly attractive in appearance and possess good taste and aroma, but deteriorate rapidly if kept for sometime. This is an account of several reasons.

(*i*) Fermentation caused by moulds, yeasts and bacteria.

(*ii*) Chemicals present in pulp/juice, may react with one another.

(*iii*) Air coming in contact with the product, may deteriorate.

(*v*) Traces of metal from the equipment may get into the product and spoil its taste and aroma.

In the preservation of foods by various methods, the following principles are involved :

1. Prevention or delay of microbial decomposition

   (*a*) By keeping out micro-organisms (asepsis).

   (*b*) By removal of micro-organisms e.g. by filtration.

   (*c*) By bindering the growth and activity of micro-organisms.

   (*d*) By killing the micro-organisms e.g. by heat or radiation.

2. Prevention or delay of self decomposition of food

   (*a*) By destruction or inactivation of enzymes e.g. by blanching.

   (*b*) By prevention or delay of chemical reactions e.g. prevention of oxidation by means of an antioxidant.

3. Prevention of damage by insects, animals, mechanical causes

To retain the natural taste and aroma of a product, it is necessary to preserve it soon after preparation, without allowing it to stand for any length of time. The product must be prevented from the damage of insect and animals and prepare transportation and packaging must be ensured to protect the food products from any mechanical problems.

8.5.3 Methods of Preservation :

1. *Asepsis* (Absence of infection) :

Asepsis means preventing the entry of Microorganisms. Maintaining of general cleanliness while, picking, grading, packing and transporting of fruits and vegetables increases their keeping quality and the products prepared from them will be of superior quality.

2. *Drying :*

Sun-drying or mechanical drying of fruits and vegetables involves complete removal of the moisture from them. They are then treated with sulphur fumes to maintain colour and also to avoid spoilage by Microorganisms. Dates, Grapes, Jack, Peaches, beans, bitter gourd, Onions, Potatoes etc. are preserved by this method.

3. *Freezing :*

When fruits and vegetables are brought suddenly to a very low temperature at which all the chemical reactions stop. But it requires rather expensive equipment and stripping and storing under refrigeration.

4. *Preservation by heating :*

This process consists basically the application of heat in varying degrees to the food in close containers for a sufficiently long time to sterilise the contents before they are hermetically sealed. This method of preservation by heat is also known as processing. Canned fruits and vegetables come under this category.

High temperature used for preservation : (*i*) Pasteurization temperature (below 100 °C) (*ii*) sterilization temperature (above 100 °C).

5. *Preservation by sugar :*

No organisms can grow on a substance which contains sugar concentration above 65%, as they get desiccated by osmosis. Jams, Jellies, marmalades, candies, crystallized fruits are preserved by this method.

6. *Preservation by salt :*

Products which contain about 15% salt makes it most unfavourable for the growth and multiplication of Microorganisms. Salt acts both by osmosis and as a poison on Microorganisms. Pickles of fruits, vegetables and other pickles products are preserved by this method..

7. *Preservation by chemicals :*

Chemicals like Benzoic acid SO_2 at specific concentration help to preserve the fruit juices and squashes. Benzoic acid in the form of 0.06-0.10% sodium benzoate is sufficient and is effective against yeasts and its action is effective in the presence of CO_2.

Potassium matabisulphite (KMS) when added liberates SO_2 which reacts with water to form sulphurous acid. The concentration of SO_2 recommended is *350 PPM.* It is more active against moulds, spores and bacteria.

8. *Preservation by fermentation :*

Fruit juices containing sugar are fermented to produce alcohol. These fermented substances keep off the spoilage organisms due to the presence of alcohol. 14% alcohol acts as preservative in wines. Wines, beer, vinegar, fermented drinks, fermented pickles etc. are, prepared by these processes.

9. *Preservation by irradiation :*

Sterilization of food by ionising radiations is a recently developed method of preservation which is yet to gain acceptance. In the irradiation of food, the radiation dose must be carefully controlled. The WHO and IAEA have recommended that radiation dose of upto 1 kilo rad is not hazardous. Radiation energy must be provided in such a manner that it reaches every particle of food to ensure adequate killing of all microorganisms.

Ionizing radiations can be used for sterilisation of food in hermetically sealed packs, reduction of the spoilage flora on perishable foods, control of infection in stored cereals, prevention of sprouting of potatoes, onions etc.

8.5.4 Preservation Products :

Jam : Jam is more or less a concentrated fruit pulp possessing a fruity heavy body form. It is rich in flavour. Pectin in the fruit gives it a good set.

It contains not less than 68% soluble solids and thus high concentration of sugar facilitate preservation.

Jam manufacturer can chose fresh fruit, frozen, chilled or cold stored fruits, fruits or fruit pulp preserved by heat, sulphite, fruits or fruit pulp or dried fruits.

Sound fresh fruits free of any damage and for better felling effect, green fruits should be mixed with the ripe fruits. Overripe fruit should not be used.

The fruit must be washed thoroughly to remove any adhering dust.

Leaves, stalks and other undesirable portions of the fruits must be removed.

Fruit is then subjected to preliminary treatment which varies from fruit to fruit, e.g. apples are peeled, cut into pieces and then pulbed. Strawberries are crushed between rollers but in stone, fruits pacts, they are Lye peeled, stones removed and pulp is cut into pieces.

Addition of sugar : Amount of sugar to be added depends upon the degree of ripeness and acidity of the fruit. To get a minimum of 68.5% of sugar in the jam, generally 55.0 g of sugar is added to every 450 g of fruit.

Addition of acid, colour, and flavour: Generally citric acid, tartaric acid are used to supplement the acidity of fruits for jam making because appropriate combination of Pectin, sugar and acid is essential to give a set to the jam. A good result is obtained if the pH of mixture is about 3.0.

Ordinarily, jams do not require addition of flavours of desired, may be added when jam boiling in hearing completes. Looking fruit and sugar mixture are boiled to concentrate the soluble solids to about 68.5% and also to bring about necessary degree of inversion of sugar.

Determination of soluble solids is easily carried out by using hand refractometer while, boiling is in progress.

Packing : Soon after the end point is reached the jam should be cooled in cooling pan to about 93 °C and filled into jars at this temperature either mechanically or by hand. The surface of the jam in the jars should be covered with a thin disc of waxed tissue paper and allowed to cool. The jars should be stored in a dry place.

Jams can be prepared from fruits like apple, tomato, papaya, strawberry etc.

Jelly

In Jelly making, pectin is the most essential constituent. This pectin is present in the cell walls of fruit. Precipitation of pectin while boiling causes jelly formation. Precipitation takes places, only when pectin, acid, sugar and water are in equilibrium.

Fruits for jelly : Many fruits are rich in pectin as well as acid and are thus well suited for jelly making. Apple, grape, guava, lemon, orange, plum are rich in pectin and acid.

Selection of fruits : Sufficiently ripe, but not overripe and should have good flavour. Fruits are washed thoroughly with water. Guava and apple, need not be peeled, while fruits like, orange, lemon require peeling before used.

Extraction of pectin :

Only a minimum quantity of water should be used to the fruit for a simple extraction of pectin the amount of water required depend or the kind of fruit.

To get the desired pectin requirements the fruit should be cooked. The cooking time also varies from fruit to fruit and the time of boiling of fruit required to extract pectin.

The pectin extract can be clarified by allowing to settle overnight and the supernatant pectin liquor separated as it may be filtered through a muslin cloth. Aluminium vessels are used for boiling.

Pectin requirement : Usually about 0.5-1.0% pectin of suitable quality in the extract is sufficient to produce a good jelly. If pectin is in excess of this a firm and tough jelly is formed and if it is less the jelly may fail to set.

Pectin, sugar, acid and water must be present approximately in the following proportion, pectin (1.0%), sugar (60-65%), fruit acid (1%) water (33-38%).

The sugar should be sprinkled on the fruit extract while it is boiling and should be thoroughly mixed by stirring to ensure complete dissolution. During boiling, the scum which rises to the top is removed.

Cooking of Jelly : The mixture is boiled for about 20 minutes and the end point is when the final Brix reaches 65°C. The end point can also be determined by flake test.

After training the end point, jelly is then cooled slightly and poured into hot and dry containers.

Pickles :

The preservations of food in common salt or vinegar is called pickling. Spices and oil may be added. Pickles are good appetizers and add to the partibility of a meal. They aid digestion by stimulating the flow of gastric juices.

Salt, vinegar and lactic acid are the three important ingredients used in pickling. These substances when used in sufficient quantities, act as preservatives either singly or collectively.

1. *Salt :* Vegetables don't foremast when covered with a large amount of salt.
2. *Vinegar :* This preservative as pickles, the final percentage should not be below 2%. To avoid dilution of vinegar by water from vegetables they can be kept in a strong vinegar of about (10% concentration).
3. *Lactic acid :* Sufficient quantity of lactic acid, prevents lactic acid bacteria, and any change in the composition of the material is prevented, provided it is air-tight.

Pickling process : Two stages

1. *Dry softening :* Vegetable is prepared, washed thoroughly and weighed. For every 100 kg of prepared vegetables 3 kg of salt is used. The vegetable is placed about a few cm and a small quantity of salt and another layer of vegetables thick layer is placed. This process of keeping the vegetables and sprinkling the salt is continued till 3-quarter of the barrel full. The top of the vegetable is sealed with a choose cloth and placed every a clean stone to give some weight to the vegetables placed in. The container is kept in a warm dry place and fermentation is allowed to proceed. In a short time juice from the vegetable forms the brine covering the whole mases. Fermentation then starts and is usually complete in about 8-10 days. This may be confirmed by tapping the containers gently.
2. *Fermentation in brine :* Soaking of the vegetable in salt solution of predetermined concentration for a certain length of time is called brining. This treatment is given only to vegetables like cucumber which do not contain sufficient juice, to form brine with dry salt.

Problems in pickle making :

1. Bitter taste, (due to use of strong vinegar or excess spice).
2. Dull and faded product (insufficient curing, inferior quality).
3. Shivelling (when fruit and vegetable directly placed in strong sugar, vinegar or salt.
4. *Swin formation :* White swin always form on the surface, vegetable in brine.
5. *Softness* and *stipperiness :* Inadequate, covering with brine.
6. *Blackening :* Fe in the brine equipment used in pickling.

8.6 OTHERS

8.6.1 High density planting system (HDP) :

Definition : Planting of fruit trees rather at a closer spacing than the recommended one using certain special techniques with the sole objective of obtaining maximum productivity per unit area without sacrificing quality is often referred to as HDP.

The introduction of HDP may well be one of the most important advances in fruit production all over the world.

First established in apple orchards in Europe.

In India HDP have been successfully demonstrated in banana, pineapple, papaya and more recently mango.

Advantages of HDP :

1. Efficient land use

2 Reduced labour costs

3. Early cropping and bigger yield for a long time

4. Improved fruit quality with efficient harvesting.

In India, where large number of fruit crops are grown, there is a need to develop crop-specific strategies for developing HDP with adequate emphasis on :

(*i*) Availability of suitable sites, land and other physical resource (*ii*) Selection and identification of dwarf and compact variety of fruits (*iii*) Selection and identification of suitable dwarfing rootstock (*iv*) The direction of training and pruning practices to develop ideal tree structure and canopy architecture for efficient use of light. (*v*) Use of plant growth retardants and mechanical devices.

Characteristics of HDP :

(*a*) Trees of HDP should have maximum number of fruiting branches and minimum number of structural branches.

(*b*) Trees are generally trained with a central leader surrounded by nearly horizontal fruiting branches.

(*c*) Pruned occur with canopy adjustment.

(*d*) The height should be according to the harvesting efficiency.

Planting system : Success of HDP depends upon planting system.

Generally rectangular planting with single row (within the row spacing is less), double row (wider spacing after every two rows) and triple row planting are followed.

In meadow orchards a bed of 10-15 rows are closely planted and separated by alleys of 2.5 m width between beds. This system is also called bed system.

Future strategies :

Fruit production in India is going through evolutionary phase exemplified by changes in cultivar, root stocks, cultural practices, handling and marketing. These changes are required to constrain in bringing into focus, the HDP to produce easily and sustained production of quality fruits.

In developing successful HDP of tropical and sub-tropical fruits, need-based, crop specific strategies are required to be formulated.

Latest technical skills in tree architecture and canopy management is required for which adequate manpower should be trained. The development and success of intensive orcharding need to be demonstrated for quicker adoption and practice by fruit growers.

8.6.2 Hot beds

The main use of hot beds is for starting plants to be grown in the main field and in certain cases to grow crops to maturity during non growing season .

Hot beds are usually located near the farm buildings and near a good water supply with south and Southeast exposures for sunlight.

Generally farmers use hot bed made of wood and cement . These beds may extend 30-45 cm above the surface of the ground on the backsides and 15-30 cm in the front, thus giving a slope.

Hot beds have different heating systems.

(*i*) Manure heated (*ii*) Flue heated (*iii*) Hot water heated hot beds (*iv*) Electrically heated hot beds.

(*i*) *Manure heated:* Fresh manure mixed with one post stress are placed in pit where hot bed actually is needed with optimum moisture condition for fermentation and steaming

(*ii*) *Flue heated*: Heated by hot air conducted through flues from a fire box located at one end of the bed.

(*iii*) *Hot water heated*: Most satisfactory method as the temperature can be controlled more easily, but is expensive.

(*iv*) *Electrically heated*: Electricity is passed through the specially constructed cable, consisting of electrically insulated resistance enclosed in a flexible metal sheath. Initial cost is high but beds are always ready for use and they are easily manipulated and readily removed.

8.6.3 Hydroponics

Intensive and profitable cropping in a greenhouse condition is possible after the advent of hydroponics . Hydro-water and pones -labour literally meaning water working .

Definition: It is science of growing plants without the use of soil, but use of an inert medium such as gravel, sand, peat, vermiculites, or saw dust to which is added a nutrient solution containing all the essential for normal growth and development.

Advantages

1. Much higher yields than those grown on the soil.
2. It can be used in places where ordinary agriculture is impossible.
3. Quality of the produce is always superior.
4. Freedom from soil borne diseases and weeds.
5. Less cost of management .

Essentially there are three methods by means of which plants may be grown without soil.

(1) Water Culture (Plant root in solution)

(2) Drip Culture (Sand culture, watered with nutrient solution)

(3) Gravel Culture (Most widely used)- It needs following accessories:

(*i*) Nutrient tank ; made of concrete, wood or metal

(*ii*) Benches : Gently sloping benches made of concrete or wood.

(*iii*) Medium: Crushed graphite of irregular shape

(*iv*) Solutions ; Different chemical solutions

Recent improvements in soil less cultures are Nutrient film Technique (NFT), Tube culture and Aeroponics .

(NFT) is a water culture, plants grown with roots in plastic film through which nutrient solution is continuously circulated.

Tube culture is developed NFT where the nutrient solution is pumped through PVC dress pipes through which the seedlings are set into the solution flowing along the bottom of PVC.

Aeroponics is growing of plants in an opaque trough or supporting containers in which the roots are suspended and bathed in a nutrient bath mist are suspended and bathed in a nutrient bath mist rather than solution .

8.6.4 Orchard Cultivation

Refers to the careful management of the orchard soil in such a way that the soil is maintained in a good condition suitable to the needs of the tree with least expenses.

A good system of orchard cultivation should ensure.

1. Weed control and saving in moisture.
2. Very little disturbance to soil and preventing soil erosion.
3. Reduced cost of cultivation.

Methods of soil management practices :

1. Clean culture (regular ploughing and removal of weeds).
2. Clean culture with cover crops : Plant a green manure crop before the monsoon and plough it into the soil towards the end of monsoon season e.g. sunhemp, cowpea, diancha.
3. Mulching : One of the important soil management practices adopted.

Crop residues like, straw dust, cotton stalks, leaves, saw dust, pine needles, air dust or polythene film spread in the tree barins.

Main objective is to conserve soil moisture end to control weed.

Advantages :

1. Keeps soil cool in day, warm at night hours.
2. Reduced surface run off.
3. Adds humus to soil.
4. Prevents soil erosion.
5. Fruits are protected and kept clean since they fall on mulches.
6. It allows the absorption of more rain water.
7. It reduces irrigation frequency.

4. *Sod :* Permanent cover of grass in orchard and no tillage. Useful in sloppy lands for preventing soil erosion. But they compete for soil moisture and available N.
5. *Sod mulch :* Similar to sod, but vegetation is cut frequently and the cut material is allowed to remain on the ground. Here moisture loss is not like sod. More of available N to the fruit trees should be applied in both the cases to compensate losses caused by vegetation.

8.6.5 Intercropping :

Efficient utilisation of land upto the fruit trees for several years or in case of long duration host crops like tapioca, turmeric, ginger and banana, some area remain nucleocapsid by the main crop for few months.

The practice of growing any economic crop in alley spaces of the fruit trees in the first few years is referred as intercropping.

Principles :

1. Intercrops should not occupy the area where the roots of the fruit trees are concentrated.
2. Soil fertility should be maintained or improved when intercrops are grown.
3. Water requirements of the intercrop should be coincide with those of the main fruit trees. The intercrop may require an irrigation at a time when it is detrimental to the trees.
4. Intercrops should be selected with reference to their effect on soil moisture.

Vegetable are the best intercrops compared to millets. The intercropping should be stopped when trees occupy the entire orchard space. Therefore green manuring or cover crop should be practised.

Some quick growing fruit trees could be grown as intercrop as in some places in temperate regions peach intercropped in apple.

In properly spaced mango orchards guava tree can be planted to bear fruits in 2-3 years and then removed. Such short-lived trees are known as fillers. Papayas, bananas or phalsa may be grown as fillers in orchards.

Crop	*Age*	*Intercrop*
Mango	upto 7 years	Legume vegetables, papaya (filler)
Grapes	Upto 8 months	Snake gourd, bitter gourd in pandal
Apple, pear	upto 5 years	Potato, cabbage
Banana	upto 4 months	Sunlemp, onion
Arecanut	upto 10 years	Pineapple
Coconut	upto 3 years	Banana, tapioca, vegetables

8.6.6 Multi-tier system of cropping :

Certain plants like coconut, arecanut are grown for about 50 years in a particular land. It takes nearly 4-7 years for the above trees to reach the bearing stage.

Adequate alley spaces (nearly 75%) are available in between these trees and being the palm trees, their root system will not also spread beyond one metre in diametre.

Hence these vacant spaces can be utilised profitably for raising other crops, thereby increasing the employment opportunity and profit. This is the chief objective of multitier system of cropping.

Inter cropping and mixed cropping involving jointly in it and is defined as a compatible comparison of crops having varying morphological frames and rooting habits, grown together in such a manner that their canopies intercept solar energy at varying heights and their roots forage, at different depth.

The main principle here is land, water and sunlight should be effectively used.

An ideal combination of crops for multitier cropping in coconut and arecanut plantations are given :

Tier	*Crop*
First (Top)	Coconut or arecanut
Second	Pepper trained over the trunk of coconut trees
Third	Cocoa or cloves planted at the centre of four arecanut or coconut trees.
Fourth (ground)	Pine apple, ginger and dwarf coffee

8.6.7 Thinning in horticulture (fruits) :

Fruiting is an exhaustive process to the tree especially if the crop is heavy. Therefore the chief goal is to permit the tree to mature as large a crop as possible and yet conserve sufficient nutrients and carbohydrates for good shoot and spur growth, leaf development and flowers but formation for the next year crop.

If the tree is permitted to mature an excessive crop, it becomes devitalised to the point where it not only produces an inferior produce but it becomes increasingly susceptible to diseases. So in order to ensure proper balance, thinning is practised in certain fruit crops.

Thinning is the removal of a part of flower buds, flowers or fruits (before it natural) with the object of :

(*i*) To increase the annual yield and marketable, fruits.

(*ii*) To improve the fruit size and colour.

(*iii*) It reduces the chance of large breakage.

(*iv*) If promotes tree vigour and ensures more regular bearing.

(*v*) It promotes more thorough spraying and dusting of fruits.

(*vi*) It ensures uniform ripening.

Thinning :

(*i*) Blossom thinning to prevent exhaustion.

(*ii*) Marble stage thinning to prevent over crowding.

(*iii*) Thinning after natural fruit drop of young fruits.

The amount of thinning of fruits depends upon 1. fruit set 2. response of the variety to thinning 3. nature of pruning gives 4. age of the tree 5. value of increased sized fruits in market.

Methods :

1. *Hand thinning :* Costly, advisable, hand thinning of pea sized berries is beneficial in grapes.

Manual thinning is commonly practised in date palm.

2. *Chemical thinning :* NAA 100 ppm when applied 2 to 4 weeks after petal fall stage in grapes reduces fruit set in Anab-e-sahi variety of grapes.

In mandarin, NAA, 600 ppm on marble sized stage is recommended to thin the overbearing fruits so as to increase the size and quality of fruit.

8.6.8 Seedlessness in horticultural fruits:

Craze for seedless fruits is increasing among the consumers. The seedlessness results in certain fruits due to "parthenocarpy".

'Parthenocarpy' which refers to the development of fruits without fertilization or even without the stimulus that comes from pollination.

If a parthenocarpic fruit develops even without the stimulus from the pollination, then this phenomenon is vegetative parthenocarpy *e.g.* banana with the stimulus of pollination, stimulative Guava and water melon.

Thompson seedless due to stimulative pollen-aborted embryo.

Need :

Greater preference among consumer owing to its quality and hence fetch higher prices than the seeded fruits *e.g.* seedless grapes, guava or oranges.

Besides the problem of unfruitfulness may not arise if a fruit develope parthenocorpically and the grower is assured of his crop *e.g.* banana.

Drawback : Fruits are usually small in size (grapes) and irregular in shape (guava).

Induction of seedlessness :

1. *Use of growth regulators.* Application of GA_3 7000-8000 PPM in lanolin paste on the cut end of the style of emasculated flowers of guava results in seedless.
2. By changing the ploidy level.

In Japan, it was domesticated that by developing a triploid water melon (3n = 33) by crossing tetraploid × diploid variety, seedlessness can be achieved.

[The seedlessness in naturally available seedless guava var is due to *autopolyploidy* (triploid) and not due to parthenocarpic fruit development]

8.6.9 Hybrid seed production in flowers :

During early 1940s, the significant revolutionary change in floriculture was the exploitation of hybrid vigour in Petunia, when the technique for producing all double F_1 hybrids was developed for the first time.

Advantages of F_1 hybrids :

The main advantage is the unique combination of appreciable *vigour and uniformity*. If selected judiciously, the *F1 hybrids* have other advantages too.

Dwarf and compact plants, with basal branching, extremely free-blooming, early flowering, doubleness, large flowers, long duration of flowering, tolerance to heat and humidity and disease resistance.

The triploid hybrid Nugget, F_1 hybrid of the African Marigold being sterile, the absence of fertilisation and seed set has the unique ability of holding the flowers fresh on the plants for a longer period.

Geranium hybrids raised from seeds do not carry disease unlike cutting or slips.

Hybrid breeding techniques :

The F_1 hybrid breeding process, though rewarding is long, costly and requires considerable skill.

The breeding cycle is accelerated by growing the breeding material out-of-season at any other suitable location.

The process of testing inbred lines, making test crosses and testing the resulting F_1 hybrids are repeated several times to find a hybrid good enough to introduce.

The inbred lines come from the standard open pollinated varieties or occasionally from wild related species.

The best plants from each inbred line are selected for test crosses. The promising inbreeds are crossed with a large number of plants from a different inbred line or lines.

The F_1 hybrids produced from these test crosses are then grown and observed in extensive greenhouse and field trials.

The year before a new hybrid is released, hundreds of samples are distributed to commercial plant growers. This provides additional information on performance and cultural method, under commercial conditions in each agro climatic regions.

There is a constant endeavour by flower breeders to produce new more attractive and brightly-coloured hybrids. With the advancements made in new bedding plant technology in USA such as automated systems, smaller containers, shorter crop time and energy stress, hybrid variety are tailored to meet the requirements of commercial floriculture.

Maintaining and improving hybrids :

Once a new hybrid is produced, the work is not over. The parent lines must be maintained and retested to make for sure some of the desirable, attributes of the hybrid are not lost. The uniformity in hybrids depends upon the genetic purity of parental lines.

Production of F_1 hybrid seeds :

Most of the hybrid seed is produced by hand in greenhouse. Both emasculation and pollination are done by hand which enables the breeder to have a coider choice of parents for crossing.

In flowers of family compositae where emasculation by hand are difficult, like Marigold, Zinnia, male sterility is used for hybrid seed production.

Collection of pollen with the help of a vacuum pump, dehiscence of anthers for pollen under VU lamp, storage of pollen at cool temperature and use of pollen gun for pollination are some of the useful innovations technique for hybrid-seed production.

Harvesting of seeds at proper ripening stage, cleaning, and grading of seeds are essential processes before sale.

At every stage of production care is taken to avoid selfing or natural cross pollination or both to ensure complete hybridity and genetic purity.

Harvesting, processing and storage of seeds are done carefully so that there is no mechanical mixture.

Work done in India :

Pronounced hybrid vigour in *Marigold, Antirrhinum, hollyhock, petunia.*

Indo-American hybrid seeds : Commercial hybrid seed production in *Petunia, Antirrhiunum, Geranium, Marigold and Gloxinia.*

The hybrid seed production being labour intensive has a great potential for employment in youths in rural and semi-urban areas.

It may also generate gainful income by setting up of ancillary industries dealing with equipments and other facilities needed like greenhouse, small seed processing machinery, seed packaging etc.

Post Harvest Processing and Cold Storage

Post Harvest Losses

- India-third largest producer of food -600m. tonnes
- Horticulture Production -311 million tonnes
- 30% of fresh produce wasted.
- Only 1% of production is processed.

What is Agro processing

- Involves transformation and preservation through physical or chemical alteration, storage, packing and distribution.
- The processing of raw traditional commodities such as cassava and banana into flour for the production of baked products. Processing of banana and potato for chips.
- Nature of processing and degree of transformation vary

Categories of Agro processing

Level-1	Level-2	Level-3	Level-4
• Cleaning • Grading • Storage	• Ginning • Milling • Cutting • Mixing	• Cooking • Pasteurization • Canning • Dehydration • Freezing • Weaving • Extraction • Assembly	• Chemical alteration • Texturization

Agro-processing activities are as under:

Paddy processing units, Dal mills, Oil extraction, Sago manufacture from Tapioca, Milk processing units, Vegetable processing, Canning of mango pulp, Extraction of oils/ essential ingredients from super critical extraction method, Chicken meat processing.

Benefits derived from Processing

Provides value addition, Provides choice to end user, Promotes retail boom.

Constraints of processing

- High prices
- Dominated by SSI units-to upgrade technology
- Lack of post harvest infrastructure-wastage.
- Difficulties in marketing
- Non availability of raw materials.

Quality of final products to meet international standards (TQM, HACCP, GMP, GHP).

SAMPADA: Umbrella scheme incorporating ongoing schemes of the ministry like Mega food 'ark, integrated cold chain.

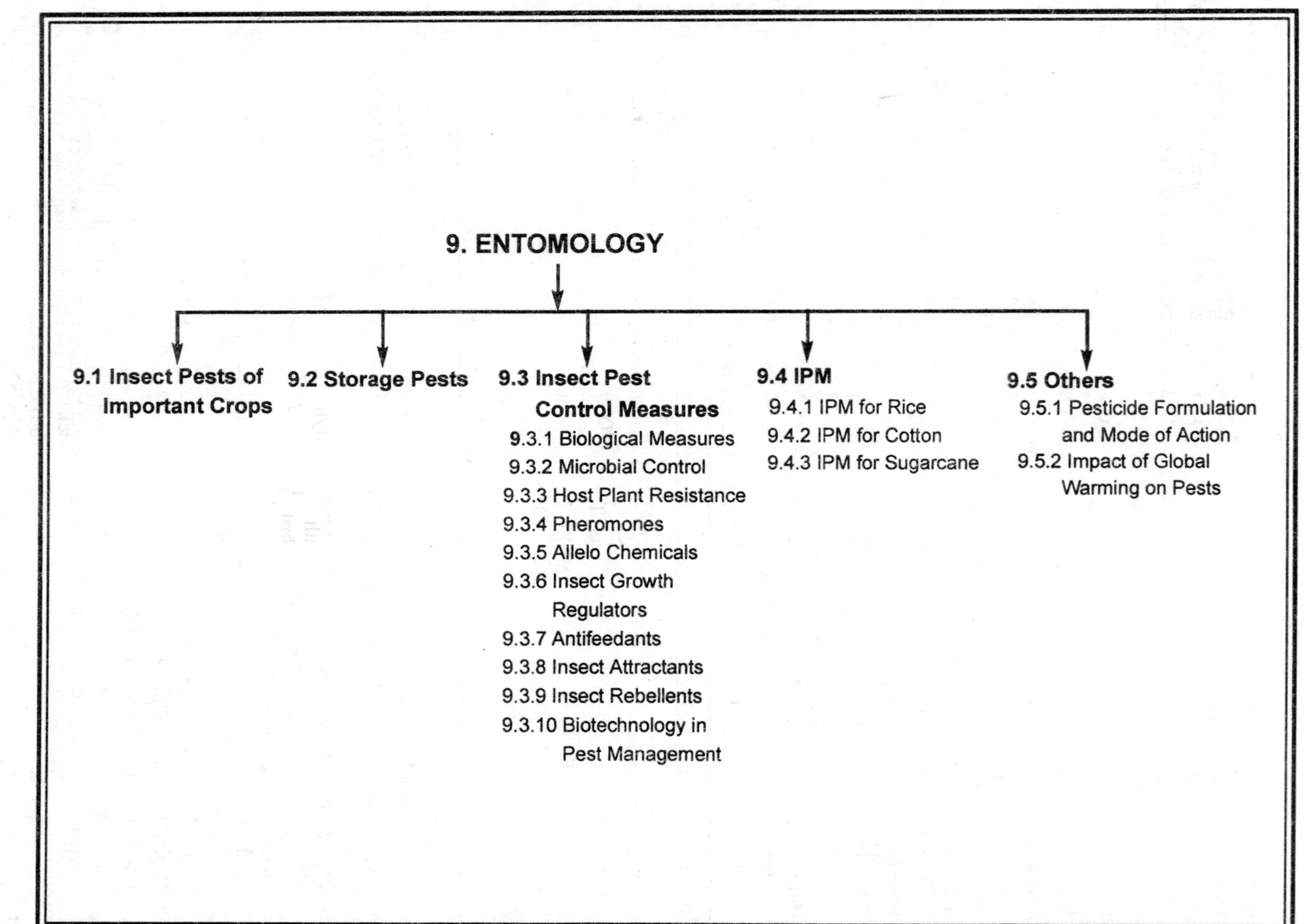
9. ENTOMOLOGY
9.1 Insect Pests of Important Crops
9.2 Storage Pests
9.3 Insect Pest Control Measures
9.3.1 Biological Measures
9.3.2 Microbial Control
9.3.3 Host Plant Resistance
9.3.4 Pheromones
9.3.5 Allelo Chemicals
9.3.6 Insect Growth Regulators
9.3.7 Antifeedants
9.3.8 Insect Attractants
9.3.9 Insect Rebellents
9.3.10 Biotechnology in Pest Management
9.4 IPM
9.4.1 IPM for Rice
9.4.2 IPM for Cotton
9.4.3 IPM for Sugarcane
9.5 Others
9.5.1 Pesticide Formulation and Mode of Action
9.5.2 Impact of Global Warming on Pests

9. Entomology

Entomology is made from Greek word *entomos* means cut in pieces, engraved or segmented hence insect and *logia* means scientific study. So entomology is scientific study of insects. Entomology is rooted in nearly all human cultures from pre-historic times but scientific study began only as recently as the 16th century.

9.1 INSECT PESTS OF IMPORTANT CROPS

Sugarcane

1. **Shoot borer** - *chilo infuscatellus*

   Nature of Damage : The Larva tunnels into the stem causing 'dead heart'. Its multiplication is favoured by high temperature and low humidity.

   Life History : The small pale grey brown moth with white hind wings. The egg, larval and pupal periods respectively 6, 25-30 and 6-8 days. The total life cycle is completed in 35 to 40 days.

   Control : Infested tillers collected and destroyed. Avoid late planting.

2. **Stem borer (or) internode borer** - *chilo sacchariphagus indicus*

   Nature of Damage : It occurs on sugarcane in its growing phase. It affects juice quality.

   Life History :

   Fgg (5–6 days) – Larva (37–53 days) – Bores (nodal region) – Pupa (in leaf sheath (8–10 days) – Adult (Pale brown)

   Control : Release of egg parasite *Trichogramma australicum*.

3. **Top Borer** – *Scripophaga excerptalis*

   Nature of Damage : It occurs in later stages of growth. Its infestation, is detected by presence of a reddish brown charred 'dead heart'.

   Life History :

   Egg (7–9 days) – Larva (35–50 days) – Pupa (inside stem) (8–10 days) – Adult (Whitish in colour with buff coloured anal tuft)

   Life cycle 50-70 days.

   Control :

   - Egg masses collected and destroyed.
   - Soil application of carbofuran at 2 kg a.i/ ha.

4. **Sugarcane Leaf Hopper** - *Pyrilla Perpusilla*

   Nature of Damage : It remains colonies on the lower surface of leaves and both the adults and nymphs suck up the sap causing yellowing and drying of leaves.

   Life History :

   Egg – Nymph (50–60 days) – Adult (brownish)

   Control : Endosulfon 0.07% spray.

Cotton

1. **Cotton Aphid** - *Aphis Gossypii*

   Nature of Damage : It infest tender shoots and the undersurface of leaves severe infestation results in curling of leaves, stunted growth and gradual drying and death of young plants.

   Life History :

   Nymphs (7–9 days) – Adults (greenish brown)

   Control :

   - Methylparathion 0.025 %
   - Phosalone 0.05%

2. **Cotton Leaf Hopper** - *Amrasca Devastans*

   Nature of Damage : The nymphs and adults suck the plant sap from the undersurface of leaves. The leaves show symptoms of 'hopper burnt' such as yellowing, curling and bronzing.

Life History :

Eggs (4–10 days) – Nymphs (7–20 days) – Adult

Control :

- Methyl parathion 0.025 %
- Phosalone 0.05%

3. Cotton white fly - *Bemisia tabaci*

Nature of Damage : The nymphs are found on the undersurface of leaves and drain of sap due to sucking in cases of severe infestation result in premature defoliation and development of sooty mould on honey dew excretion.

Life History :

Eggs (3–5 days) – Nymphs (9–14 days) – Pupa (2–8 days) – Adult

Control : Spray applications of phosalone 0.05% + neem oil 0.3%.

4. Spotted boll worm - *Earias insulana*

Nature of Damage : The larva causes drying of the shoot due to its feeding by boring into it. Heavy shedding of early formed flower buds due to this pest.

Life History :

Eggs (under surface of leaves (3 days) – Larva (10-12 days) – Pupa (7-10 days) – Adult.

Total life cycle - 20-22 days.

Control :

- The infested portions of buds and bolls should be removed and destroyed.
- Synthetic pyrethroids like permethrin @ 100-150 g ai/ha and cypermethrin @ 80 g ai/ha.

5. Pink boll worm - *Platyedra gossypiella*

Nature of Damage : The larva enters the developing boll through the tip portion and the entrance hole gets closed up as the boll matures. The infested flower buds shed prematurely. A rosette shaped bloom ware found.

Life History :

Egg (4-25 days) – Pink larva (20-35 days) – Pupa (6-20 days) – Small dark brown moth

Control :

- Attacked parts collected and destroyed
- Fumigation of seeds with methyl bromide at 1.5 kg/100 cum for 24 hrs.

Rice

1. Rice stem borer - *Scirpophaga incertula*

Nature of Damage : The larva feeds inside the stem causing drying of the central shoot or 'dead heart' in young plant and drying of the panicle or 'white ear' in older plant.

Life History :

Eggs (5-8 days) – Larva (dark brown head (35-40 days) – Pupa (inside the stem (6-10 days) – Adult moth

Total life cycle 50-70 days.

Control :

- Removal and destruction of rice stubbles from field.
- Collection and destruction of egg masses.
- Moths may collected by setting light traps.

2. Swarming Caterpillar - *Spodoptera mauritia*

Nature of Damage : The leaves are completely eaten away leaving behind the midribs of leaves. They feed mostly during night.

Life History :

Egg – Larva (20 – 25 days) – Pupa (10 – 15 days) – Adult (dark colour moth)

Control :

- Flooding the nursery makes the hiding larvae to come to the surface
- Spray application of methyl parathion 0.05%.

3. **Rice Gall Midge** - *orseolia oryzae*

Nature of Damage : It is popularly known as 'Silver shoot' or 'onion shoot' because of the formation of hollow pink, dirty white cylindrical tubes bearing at their tips a green reduced leaf blade.

Life History :

Eggs (3-4 days) – Maggots (15-20 days) – Pupa (2-8 days) – Adult (Yellowish brown fly

Control : Spray application at 0.5/kg ai/ha are chlorfenvinphos and dicrotophos.

4. **Rice mealy bug** - *Heterococcus rehi*

Nature of damage : Its infestation causes round patches in a normal crop and plants have distorted panicles with chaffy grains.

Life History :

Eggs – Nymphs (17-35 days)

Control :

- The grass on the field bunds should be removed.
- The infested plants should be removed and destroyed.

5. **Rice leaf roller** - *Cnaphalocrocis medinalis*

Nature of Damage : The larva rolls the leaf blade by fastening its edges. In a severely infested field the whole crops give a sickly appearance with white patches.

Life History :

Eggs (4-7 days) – Larva (15-25 days) – Pupa (6-8 days) – Adult (Brownish orange colour moth)

Control :

- Removal of grass
- Periodical spray application of dicrotophos and chlorpyriphos at 0.5 kg ai/ha.

Wheat

Termites : Odentetermes Obesus

Nature of Damage : Attack young seedling, wither and die.

Control : Partially decomposed manure should be avoided apply Endosulfan 25 kg/ha.

Aphids : Macrosiphum miscanihi.

Nature of Damage : Nymphs and adults suck sap from tender shoots, immature grain and result in poor yield.

Control : Avoid heavy manure, and succulent crop, 0·02% dimethoate.

Stemborer : Sesamia inference

Nature of Damage : dead heart

Control : Pullout and destroy dead hearts, dust 0·05% sulfur.

Ground Nut

1. **Red hairy cater pillar** - *Amascta albistriga*

Nature of Damage : The larvae feed the under surface of tender leaf lets and they also feed nature leaves leaving behind the petiole and mid ribs of leaves.

Life History :

Eggs (2–3 days) – Larva (40–50 days) – Pupa – Adult

Control :

- Pupae collected and destroyed during summer ploughing.
- Putting light traps to attract moths.
- Larval parasites like Apanteles flavipes and Apanteles bosei.

2. **Leaf minor** - *stomoperyx subrecivella*

Nature of Damage : The larva mines into tender leaf lots and feeds on the tissue. In a severely infested crop the whole field presents a burnt up appearance.

Life History :

Eggs (3 days) – Larva (14 days) – Pupa – Adults

Control:

- Putting light traps to attract adults.
- Spraying Fenitrothion 0.025% or phosalone 0.05%.

3. **Stem borer** - *sphenoptera perotetti*

Nature of Damage : The grubs of the dark brown shining beetle bore into the stem of groundnut plants and cause wilting and drying of plants.

Control : The infested plants should be collected and destroyed.

Coconut

1. **Rhinoceros beetle** - *oryctes rhinoceros*

Nature of Damage : The adult beetle cut across the leaf in the leaf lets. Frequent infestation results in stunting of trees and death of growing point in young trees.

Life History :

Eggs (8-15 days) – Larval grub (90-180 days) – Pupa (10-25 days) -- Adults

Life cycle - 3 to 8 months.

Control:

- Decaying trunk of trees should be burnt
- Castor cake @ 1 kg soaked in water in small mud pots attracts beetles.

2. **Black headed cater pillar** - *opisina arenosella*

Nature of Damage : Larvae live on the undersurface of leaf lots within galleries of silk material and feed by scraping the green matter.

Life History :

Eggs (1-5 days) – Larva (40 days) – Pupa (12 days) – Adult (Greyish white small moth)

Control:

- For young trees spray carbaryl 0.1%.
- Releasing parasitoid such as *perisierola* and *Trichospilus pupivora.*

3. **Red Palm weevil** - *Rhyncophocus ferrugineus*

Nature of Damage : It attack the central shoot shows sign of wilting. In the grown up trees the crown region is infested.

Life History :

Eggs (2–5 days) – Grub (35 - 75days) – Pupa (12–35 days) – Adult (reddish brown weevil)

Control:

- Damaged palms should be destroyed.
- The infested portion should be scooped out.

PESTS OF VEGETABLES

Brinjal

1. **Shoot and fruit borer** - *Leucinodes orbonalis*

Nature of Damage : The larva bores into tender shoots in the early stage and causes 'dead hearts'. It also bores into flower buds and developing fruits.

Life History : The moth has brownish and red markings on the whitish forewings. The female lays eggs on tender shoots and developing fruits. The egg, larval and pupal periods respectively 5, 15 and 6-8 days.

Control : The damage portions should be removed.

Periodical spray application of phosalone .07%.

2. **Stem borer** - *Euzophera perticella*

Nature of Damage : The larva bores into the stem and covers the hole with excreta. The plants become stunted growth and the infestation is generally noticeable in the later stage of crop growth life history.

The moth has greyish brown fore wings and transverse lines in the middle wings. The yellowish larva with a red head. Life cycle ranges from 26 to 55 days.

Control : The attacked plants should be removed periodical application of carbaryl 0.1%.

3. **Spotted leaf beetle** - *Epilachina vigintioctopunctata*

The grubs and adults scrape the leaves and feed.

Life History : The female lays eggs in group of 10 to 20 on the under surface of leaves. The egg period is 2-4 days. The yellowish spiny grubs become full grown in 10-20 days. The total life history takes 15-40 days.

Control : Spray application of carbaryl 0.1% malathion 0.1%.

4. **The Grey weevil** - *Myllocerus subfasciatus*

Nature of Damage : The adult beetles feed on leaves of brinjal, the grubs feed on roots and cause wilting and death plants.

Life History : The weevil lays eggs in the soil. The incubation period is about a week. The grubs become full grown in 2 months. The adults emerge in 10-20 days.

Control : The adult beetles should be collected and destroyed or killed by dust application of carbaryl.

Cucurbits

1. **Snake gourd semilooper** - *Plusia peponis*

Life History : The brownish moth has shiny brown fore wings. The caterpillar is greenish with white longitudinal lines. The pupa is greenish but turns dark brown before the emergence of the adult moth. Life history occupies around 1 month.

Control : The larvae when found in small numbers may be hand picked and destroyed. Spray application of parathion .025% controls the pest.

2. **Snake gourd stem weevil** - *Baris trichosanthis*

The small black weevil feeds on the green leaves. It lays eggs in nodes. The incubation period is 5-6 days. The grubs bore into the stem and cause damage.

3. **Fruit flies** -*Dacus cucurbitae*

Nature of Damage : It is slightly bigger with brown marking on wings. The maggots feed on the contents of fruits and cause premature dropping of fruits. The insects attack the fruits of snake gourd, bitter gourd etc.

Life History : It lays eggs on flowers or tender fruits. The maggots that hatch out from the eggs in 2-9 days feed on the seeds of the fruit. The larval and pupal periods occupy respectively 3 and 5 days. The total life history occupy 12 to 35 days.

4. **Pumpkin beetles** - *Raphidopalpa foveicollis*

Nature of Damage : The beetles bite holes on leaves and also feed on flowers. The roots as well as the stem and fruits that come in contact with the soil are damaged by the grubs.

Life History : The brownish eggs are laid in the soil. The egg period is 5 to 8 days. The grubs become full grown in 13-25 days and pupae in the soil. The pupal period from 7 to 17 days.

COLE CROPS

Cabbage

1. **The diamond back moth** - *Plutella Xylostela*

The larvae bite leaves and cause serious damage.

Life History : The moth is greyish brown with narrow wings having pale white markings anteriorly which form diamond like white patches. It lays eggs on the undersurface of leaves. The larva is greenish. The period from egg to adult is 25 to 35 days.

2. **The Cabbage borer** - *Hellula undalis*

Nature of Damage : The larvae web the leaves or bore into stem and cause damage.

Life History : The moth lays eggs on the leaves and hatch in about 4 days. The brownish larva has a black head. It becomes full grown in about 9 days and pupates in the soil. Pupal period is about 6 days.

Radish

1. **Leaf webber** - *Crocidolomia binotalis*

Nature of Damage : The larvae remain undersurface of leaves in webs and feed on them. They also attack flower buds and pods.

Life History : The small moth with light brownish fore wings lays eggs in masses. The incubation period ranges from 5-15 days. The larval becomes full grown in 20-25 days. Pupates emerges as adult in 14 to 40 days.

2. **Painted bug** - *Bagrada Cruciferarum*

Nature of Damage : The nymphs and adults suck the sap from leaves, shoots and pods.

Life History : The bug is small, black in colour with red and yellow lines. It lays barrel shaped brownish eggs. The egg and nymphal periods take respectively 1 and 3 weeks.

9.2 STORAGE PESTS

Excess moisture and mould, insects and rodents are known to be responsible for damage and loss to stored products.

Internal Feeders

The larvae feed entirely within the kernels or grain or stored material.

1. **Rice weevil** - *Sitophilus oryzae*

The weevil is dark brown. The female makes a small hole on the grain, deposits an egg. The grub feeds inside the grain, pupates there itself and emerges through an irregular hole made on the grain.

2. **The lesser grain borer or paddy borer beetle** - *Rhizopertha dominica*

The dark brown beetle has its head bent under the thorax and the posterior abdominal end blunt. The grubs develop within kernels.

3. **The Tobacco Beetle** - *Lasioderma Serricorne*

The brown round beetle has its thorax and head bent downward. The creamy white oval eggs are laid on the surface of stored material. The whitish hairy grubs feed on stored tobacco, ginger, turmeric and chillies.

4. **Pulse Beetle** - *Callosobruchus Chinensis*

The beetle is small, dark brown. White elongate eggs are laid on the pulse. The grubs feed on the inner contents of pulses and pupate inside them. The egg, larval and pupal periods are respectively 3, 12 and 4 days.

5. **Angoumois grain moth** - *Sitotroga cerealella*

The caterpillars feed on the internal contents of grains and pupate inside the grains. It inflicts severe damage to unhusked paddy.

6. **Potato tuber moth**- *Phthorimaea operculella*

In potato, the larva attacks all vegetative plant parts of the crop. Typical symptoms of leaf damage are mines caused by larvae feeding in the mesophyll, without damaging the upper and lower epidermis.

External Feeders

The larvae and adults feed on the grains from outside.

1. **The Rice Moth** - *Corcyra Cephalonica*

The adult moth is greyish brown. The caterpillars web together the grains and feed within. It attacks broken grains and flour especially the milled products.

2. **The Indian meal moth** - *Plodia interpunctella*

The moth has brown fore wings with a white band. It lays about 300 eggs in clusters and the life cycle is completed in about 5-6 weeks.

3. **Fig Moth** - *Cadra Cautella*

The small moth has greyish fore wings. The

caterpillars web together with the grains and feed on them.

4. **The khapra beetle** - *Trogoderma granarium*

   The reddish brown beetle. The yellowish brown grubs are clothed with long hairs. It attacks stored wheat.

5. **Red Flour better**- *Tribolium castaneum*

   Grubs and adult.

Secondary Pests and Scavengers

The larval or adults feed on broken grains which are out of condition and damp.

1. **Saw toothed grain beetle** - *oryzaephilus surinamensis*

   The dark brown beetle has a row of saw like sharp teeth on each side of the prothorax. It feeds by scarving of grain surface or burrowing holes in them.

2. **Long headed flour beetle** - *Latheticus oryzae*

   It occurs as a secondary infestation in stored wheat. The beetle is light reddish.

3. **Flat grain beetle** - *cryptolestes minutus*

Measures for control of stored products

Preventive methods :

- Before infestation grains should be cleaned and made free from insects.
- Disinfestation is done by dusting BHC 10%.
- Material should be dried well in the sun.
- The grain bags should be stacked in such a way as to allows proper ventilation and sufficient moving space for periodical inspection.
- Malathion at 15 ml of 50% EC or 4.5 lit of water may be sprayed as a thin film on bags.

Curative Methods :

- The stocks of grains should be periodically examined and if insects are noticed to appear the stock should be removed sundried and cleaned.
- Heavily infested material should be fumigated with hydrogen phosphide and methyl bromide.
- Seed material may be mixed with insecticides like DDT malathion.
- For tobacco beetle aluminium phosphide fumigation at 20-30 tablets/1000 cft for 100 hrs.

9.3 INSECT PEST CONTROL MEASURES

9.3.1 Biological Control

"The study and utilization of Parasitoids, predators and pathogens and other natural mechanisms for the regulation of pest population densities".

It is an important component of Integrated Pest Management (IPM).

Biological control is associated with scientist Paul DeBach (1914-1993).

First report of biological control of use of an insect species to control pest comes from *Nanfang Comu Zhuang* book attributed to Western Jin Dynasty scholor and botanist *Ji Han*.

In biocontrol, pest population is reduced to a level, that no longer of economic or health concern, yet leaves sufficient pest to allow survival of the control organisms. The biocontrol organism maintain its own population and prevents the pest from reaching to damaging level.

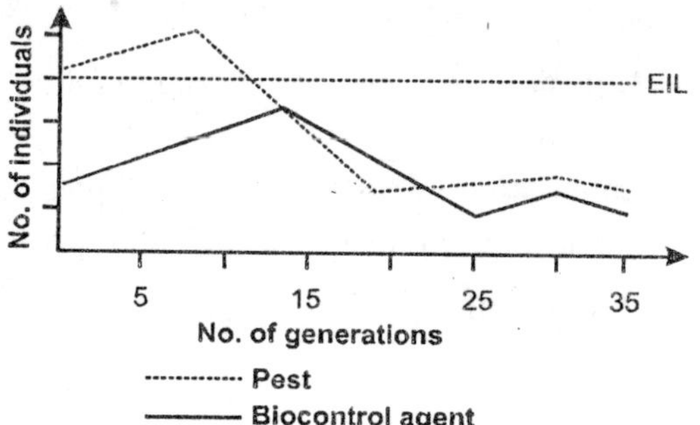

Factors governing Biological Control

- *Tolerance limit of injury:* It is successful in crops with high tolerance limit.
- *Crop value* : It is successful in crops with high economic value.

- Crop duration : Long duration crops are highly suitable *e.g.* Sugarcane, fruit trees.
- Indigenous or exotic nature of pest : *e.g.* For cotton pests identify indigenous natural enemy by survey.
- Level of control provided by the natural enemy on different stages of the pest (Egg-larval-pupal).
- Nature of pest complex (seasonal pest or perennial pest)
- Availability of selective insecticides.
- Insecticide resistant strains of natural enemies.
- Identical situation for successful control.

Qualities of an Effective Natural Enemy

- Host searching capacity
- Host specificity
- Adaptability
- Dispersal ability
- Amenability to culturing
- Ability to withstand competition
- Ability to out number the pest
- Survival capacity

Steps Involved in Classical Biological Control

- Conservation and encouragement of indigenous natural enemies
- Importation of exotic natural enemies
- Augmentation (mass rearing and release)

9.3.2 Microbial Control

Definition

"The control of pests by the use of microorganisms such as viruses, bacteria, protozoa, fungi, rickettsia and nematodes or their by-products" is known as microbial control. These microbes can be grouped as **ingested microbes** (bacteria, viruses, rickettsia and protozoa) which enter insect body along with food (like stomach insecticide) and **penetrating microbes** (nematode and fungi) that enter by penetrating the integument (like contact insecticide).

Desirable Attributes

- Pathogen should be highly virulent, able to cause disease in short period and spread from one insect to another.
- Host specificity.
- Cost effective and economical.
- Harmless to other forms of life (Safe to non-target organisms and man).
- Rapid prevention of pest feeding.

Viruses

Viruses are submicroscopic, obligate, intracellular, pathogenic entities. These are pathogenic to arthropods belongs to atleast eleven families.

Nuclear Polyhedrosis Virus (NPV)

Important features of NPV are —

- Occluded (rod shaped) singly or in groups in polyhedral (many sides) inclusion bodies.
- Site of multiplication is cell nucleus of epidermis, fat body, blood cells and trachea.
- Tree top disease: Diseased larvae not able to feed, move to the top of the tree or plant and hang inverterdly and die. (*e.g.*) NPV of *Spodoptera, Helicoverpa,*

Mode of Entry

The virus should be ingested to produce the disease. Due to alkaline gut juice, the virions are liberated from the polyhedral coat which attack nuclei of cells of tissues *viz.*, fat body tracheal matrix, haemocytes, sarcolemma of muscles, neurilemma and nerve cells of anglion and brain.

Symptoms

Insects become dull in colour, feeding rate is reduced and larvae become pinkish white especially in the ventral side due to accumulation of polyhedra. In advanced stage larvae become flaccid, the skin becomes very fragile and eventually rupture. Diseased larvae hang upside down from the plants. This is called tree top disease.

Incubation period

4-6 days depending upon the stage of the infection, weather conditions and dose of virus.

9.3.3 HPR in Pest Management

Methods of developing resistant varieties

Screening of available germplasm

Available germplasm collections are sown at a time in a single row in a location where there is a moderate to heavy incidence and evaluated by comparing with susceptible and resistant check varieties.

Selective Screening under Natural Infestation

Select promising lines from general screening and screen under natural condition in a single row or 2-3 rows in replicated trials.

Selecting Screening under Artificial Condition

To test true resistance, the selected varieties are screened under artificial condition in which insects are bomborded over the plant. Results are compared with a resistant and susceptible check. Breeders start screening of lines or cultivars from F_2 to F_6 stages for yield and resistance. If found suitable they will be forwarded to multilocation trial (MLT) and for Adaptive Research Trials (ART). If a line or cultivar is succeeding in all stages, it will be released as a variety.

Resistant Varieties

Some examples of resistant varieties are given below :

Rice —

Yellow stem borer	: Vikas, Ratna,
Brown plant hopper	: IR 64, IR 36

Sugarcane —

Early stem borer	: CO 421, CO 661
Top shoot borer	: CO 745, CO 6515

Cotton —

Bollworm	: Abhadita
Spotted bollworm	: Hopi, Deltapine

Compatibility of HPR with other Pest Management Practices

HPR easily combines with other tactics of pest management, due to following reasons.

Resistant host will make the pests weak, or will suppress the population, so that less quantity of chemicals or botanicals or lesser rounds of chemicals or botanicals will control the pest.

Resistant host will make the pests weak or will suppress the population, so that efficiency of predators, parasitoids or microbes (biological agents) on these population is increased.

e.g..,

- American root stocks of Grape vine resistant to aphid *Phylloxera vitifolia.*
- Rice varieties IR 60, IR 36 are multiple resistance sources, PY3, IR 26 are BPH resistant and MDU3 gall midge resistant.

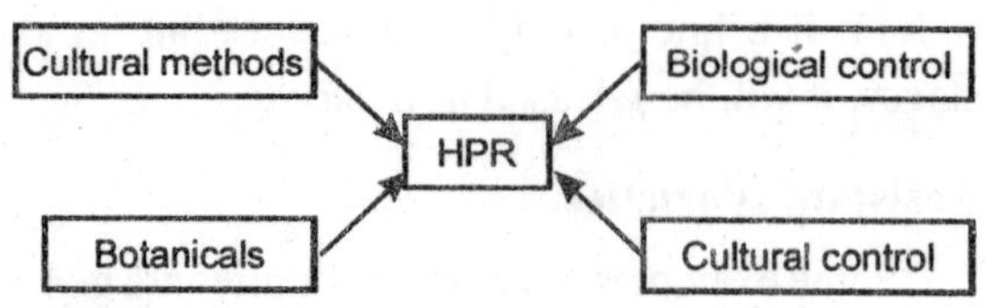

Merits

- Specificity
- Cumulative effect
- Ecofriendly
- Easily adoptable
- Effectiveness
- Compatibility

Demerits

- Time consuming
- Genetic limitations
- High initial cost

9.3.4 Pheromones

Semiochemicals are chemical substances that mediate communication between organisms. Semiochemicals may be classified into Pheromones (interspecific semiochemicals) and Allelochemicals (intraspecific semiochemicals).

Pheromones are chemicals secreted into the external environmental by an animal which has a specific reaction in a receiving individual of the same species. Pheromones are volatile in nature and they aid in communication among insects.

Based on the responses elicited pheromone can be classified into 2 groups.

Primer Pheromones : They trigger off a chain of physiological changes in the recipient without any immediate change in the behaviour. They act through gustatory (taste) sensilla. (*e.g.*) reproduction in social insects like ants, bees, wasps and termites are mediated by primer pheromones. These pheromones are not of much practical value in IPM.

Releaser Pheromones : These pheromones produce an immediate change in the behaviour of the recipient. Releaser phermones may be further subdivided based on their biological activity into

1. Sex pheromones
2. Aggregation pheromones
3. Alarm pheromones

Releaser pheromones act through (smell) sensilla and directly act on the central nervous system of the recipient and modify their behaviour. They can be successfully used in pest management programs.

1. Sex Pheromones

Sex pheromones are released by one sex only and trigger behaviour patterns in the other sex that facilitate in mating. They are most commonly released by females but may be released by males also. In over 150 species of insects, females have been found to release sex pheromones and about 50 species males produce. *Aphrodisiacs* are substances that aid in courtship of the insects after the two sexes are brought together. In many cases males produce aphrodisiacs.

Major differences between male and female produced pheromones are listed below :

Properties	Female sex pheromone	Male sex pheromone
Range	Acts at a long range. Attracts males from long distance.	Acts at a short distance
Action elicited the other sex	Attracts and excites males to copulate	Lowers females resistance to mating
Importance	More important	Less important

Insects orders producing sex pheromones : Lepidoptera, Orthoptera, Dictyoptera, Diptera, Coleoptera, Hymenoptera, Neuroptera and Mecoptera. In Lepidoptera, sex pheromonal system is highly evolved.

Pheromone producing glands : In Lepidoptera they are produced by reversible glands at the tip of the abdomen of the females. The posture shown during pheromone release is called '*calling position*'. Aphrodisiac glands of male insects are present as scent brushes or hair pencils at the tip of the abdomen (e.g.. Male butterfly of *Danaus* sp.).

Pheromone reception : Female sex pheromones are usually received by olfactory sensilla on male antennae and males search upwind, following the odour corridor of the females. In pheromone perceiving insects, the antennae of male moths are larger and greatly branched than female moths to accommodate numerous olfactory sensilla.

Chemical nature of sex pheromones : In general pheromones have a large number of carbon atoms 10 to 20 and high molecular weight 180 to 300 Dalton. Narrow specificity and high potency are two attributes which depend on long chain carbon atoms and high molecular weight. But since pheromones are volatile their molecular weights cannot be very high as they cannot be carried by wind.

2. Aggregation Pheromones

Pheromones which induce aggregation or congregation of insects for protection, reproduction and feeding or combinations of these are called aggregation pheromones. They are released by either male or female which attract both male and female of the same species. Aggregation pheromones are known mostly in Coleoptera and Dictyoptera.

Females of bark beetle *Dendroctonus frontalis* produce aggregation pheromone which attract both sexes. Similarly males of Phloem beetle *lps confusus* incorporate their pheromone in faecal matter which attract both male and female to the infested tree. Some common names of aggregation pheromones are :

Pheromone	*Insect*
Frontalin	– *Dendroctonus frontalis*
Ipsenol	– *lps confusus*
Periplanone	– *Periplanata americana*

Aggregation pheromones of following insects are used for monitoring or mass trapping the pests in India.

Coconut Rhinoceros beetle : *Oryctes rhinoceros*

Red palm weevil : *Rhynchophorus ferrugenius*

3. Alarm pheromones

Alarm pheromones are chemical substances released by insects to warn members of the same species about the presence of or attack by an enemy. This warning elicits different behaviour in different insects as listed below :

Dispersion or escape in aphids and bugs

Aggression in ants and soldier termites

Attraction in wasps and worker bees

Alarm pheromones have been reported in Hemiptera, lsoptera and Hymenoptera.

Organs producing alarm pheromones in various insects are listed below.

Aphids	– Cornicles
Termites	– Cephalic glands
Worker bees	– Sting glands
Ants	– Anal, mandibular and Dufours's gland

The chemical nature of alarm pheromones are terpenese (aphids), aldehydes (hemipteran) and formic acid (ants). Alarm pheromones can be used in IPM for controlling aphids. When alarm pheromone of aphids is sprayed on plants, the aphids try to escape, fall down from the plants and get killed.

9.3.5 Allelochemicals

Allelochemicals are intraspecific semiochemicals. They mediate the communication between two different species of organisms or insects. Allelochemicals may be classified into :

1. Allomones
2. Synamones
3. Kairomones

Allomones : Allomone is a chemical or mixture of chemicals released by one organism that induces a response in another organism which is advantageous to the releaser. For example the defensive secretions of insects and plants that are poisonous or repugnant to attacking predators. Ants release a defensive allomone called citral from its mandibular glands.

Kairomones : Kairomone is a chemical or mixture of chemicals released by one organism that induces a response in another organism which is advantageous to the recipient. The releaser turns out to be lower and the recipient (parasite or predator) a gainer.

Synamones : Synamone is a chemical or mixture of chemicals released by one organism that induces a response in another organism which is advantageous to both the releaser and the recipient. It encourages mutualistic relation between organisms. *e.g.*. Termites and protozoans.

9.3.6 Insect Growth Regulators

Insect Growth Regulators (IGRs) are compounds which interfere with the growth, development and metamorphosis of insects. IGRs include synthetic analogues of insect hormones such as acdysoids and juvenoids and non-hormonal compounds such as precocenes (Anti JH) and chitin synthesis inhibitors.

IGRs used in Pest Management

Ecdysoids : These compounds are synthetic analogues of natural ecdysone. When applied on insects, kill them by formation of defective cuticle. The development processes are accelerated by passing several normal events resulting in integument lacking scales or wax layer.

Juvenoids : They are synthetic analogues of Juvenile Hormone (JH). They are most promising as hormonal insecticides.

Juvenoids have anti-metamorphic effect on immature stages of insect. They retain *status quo* in insects (larva remains larva) and extra mouldings take place producing super larva, larval-pupal and pupal-adult intermediates which cause death of insects Juvenoids are larvicidal and ovicidal in action and they disrupt diapause and inhibit embryogenesis in insects.

Anti JH : They act by destroying corpora allata and preventing JH synthesis. When treated on immature stages of insect, they skip one or two larval instars and turn into tiny precocious adults. They can neither mate, nor oviposit and die soon. *e.g.*. EMD, and PB (Piperonyl Butoxide).

Chitin synthesis inhibitors : Benzoyl phenyl ureas have the ability of inhibiting chitin synthesis *in vivo* by blocking the activity of the enzyme chitin synthetase. Two important compounds in this category are diflubenzuron (Dimilin) and Penfluron. The effects they produce on insects include.

- Displacement of mandibles and labrum
- Adult fails to emerge from pupal skin leading to death
- Ovicidal effect.

IGRs from Neem : Leaf and seed extracts of neem which contains azadirachtin as the active ingredient, when applied topically causes growth inhibition, malformation and mortality in insects.

Hormone mimics from other living organisms : Ecdysoids from plants have been reported from plants like mulberry, ferns and conifers. Juvenoids have been reported from yeast, fungi, bacteria, protozoans, higher animals and plants.

Advantages of using IGRs

- Effective in minute quantities and so are economical.
- Target specific and so safe to natural enemies
- Bio-degradable, non-persistent and non-polluting
- Non-toxic to humans, animals and plants

Disadvantages

- Kills only certain stages of pest
- Slow mode of action
- Since they are chemicals, possibility of build-up of resistance
- Unstable in the environment.

9.3.7 Antifeedants

Antifeedants are chemicals that inhibit feeding in insects when applied on the foliage without impairing their appetite and gustatory receptors or driving them away from the food. They are also called gustatory repellents, feeding deterrents and rejectants. Since insects do not feed on treated surface they die due to starvation.

Groups of Antifeedants

Organotins : They are compounds containing tin. Triphenyl tin acetate is an important antifeedants effective against cotton leaf worm, Colarado potato beetle, caterpillars and grass hoppers.

Carbamates : At sublethal doses thiocarbamates and phenyl carbamates act as antifeedants of leaf feeding insects like caterpillars and Colarado potato beetle. Baygon is a systemic antifeedant against cotton boll weevil.

Botanicals : Antifeedants from non-host plants of the pest can be used for their control. The following antifeedants are produced from plants.

(*a*) *Pyrethrum* : Extracted from non-host plants of the pest can be used for their control. The following antifeedants are produced from plants.

(*b*) *Neem* : Extracted from leaves and fruits of neem *Azadiracta indica* is an antifeedant against many chewing pests and desert locust in particular.

(*c*) *Solanum alkaloids* : Leptine, tomatine and solanine are alkaloids extracted from Solanum plants and are antifeedants to leaf hoppers.

Miscellaneous compounds : Compounds like copper stearate, copper resinate, mercuric chloride and phosphon are good antifeedants.

Mode of action : Antifeedants inhibit the gustatory (taste) receptors of the mouth region. Lacking the right gustatory stimulus the insect fails to recognize the treated leaf as food. The insect slowly dies due to starvation.

Advantages

- Affect plant feeders, but safe to natural enemies
- Pest not immediately killed, so natural enemies can feed on them
- No phytotoxicity or pollution

Disadvantages

- Only chewing insects killed and not sucking insects
- Not effective as sole control measure, can be included in IPM

9.3.8 Insect Attractants

Chemicals that cause insects to make oriented movements towards their source are called insect attractants. They influence both gustatory (taste) and olfactory (smell) receptors.

Types of Attractants

Pheromones : Pheromones are chemical secreted into the external environment by an animal which elicit a specific reaction in a receiving individual of the same species.

Food lures : Chemical present in plants that attract insect for feeding. They stimulate olfactory receptors.

Use of Attractants in IPM

Insect attractants are used in three ways in pest management.

(*a*) Sampling and monitoring pest population
(*b*) Luring pests to insecticide coated traps or poison baits
(*c*) In distracting insects from normal mating, aggregation, feeding or oviposition. The female insects if lured to wrong plants for egg laying, the emerging larva will starve to death.

Examples of Poison Baits

- *For biting insects* : Moistened Bran + Molasses + insecticides
- *For sucking insects* : Sugar solution + insecticide
- *For fruitflies* : Trimedlure/Cuelure/ Methyl eugenol + insecticides
- *For cockroaches* : Sweet syrup + white or yellow phosphorus
- *For sweet loving ants* : Thallous sulphate + sugar + honey + glycerine + water

Advantage : They are specific to target insects. Natural enemies not affected. But they cannot be relied as the sole method of control and can only be included in IPM as a component.

9.3.9 Insect Repellents

Chemicals that induce oriented movements in insects away from their source are called repellents. They prevent insect damage to plants or animals by rendering them unattractive, unpalatable or offensive.

Types of Repellents

1. ***Physical repellents :***

Produce repellent by physical means.

Contact stimuli repellents : Substances like wax or oil when applied on leaf surface changes physical texture of leaf which are disagreeable to insects.

Auditory repellents : Amplified sound is helpful in repelling mosquitoes.

Barrier repellents : Tar bands on trees and mosquito nets are examples.

Visual repellents : Yellow light acts as visual repellents to some insects.

Feeding repellents : Antifeedants are feeding repellents. They inhibit feeding.

2. ***Chemical repellents :***

Repellents of plant origin : Essentials oils of Citronella, Camphor and cedarwood act as repellents. Commercial mosquito repellent 'Odomos' uses citronella oil extracted from lemongrass.

Pyrethrum extracted from *Chrysanthemum* is a good repellent and has been used against tsetse fly.

Synthetic repellents : Repellents synthetically produced

Important synthetic repellents are listed below:

Insects	*Repellents*
Mosquito, blood suckers	Dimethyl pthalate
Mites (chigges)	Benzyl benzoate
Phytophagous insects	Bordeaux mixture
Wood feeders	Pentachlorophenol
Fabric eaters	Naphthalene or mothballs
Bees	Smoke

Uses of Repellents

- Applied on body toward off insects
- Used as fumigants in enclosed area.
- Used as sprays on domestic animals.

- Used to drive away insects from their breeding place.

9.3.10 Biotechnology in Pest Management

Molecular biology techniques are used for the management of insect pests. The following are some of the strategies.

Wide hybridization : This technique involves transfer of genes from one species to other by conventional breeding. The genes for resistance are transferred from a different species. *e.g.* WBPH resistant gene has been transferred to *Oryzae sativa* from *O. officinalis.*

Somaclonal variability : This variation is observed in tissue culture derived progeny. *e.g.* Somaclonal variants of sorghum resistant to *Spodoptera litura* has been evolved.

Transgenic plants : Transgenic plants are plants which possess one or more additional genes. This is achieved by cloning additional genes into the plant genome by genetic engineering techniques. Transgenic plants have been produced by addition of one or more following genes.

- Bt endotoxin from *Bacillus thuringiensis*
- Protease inhibitors
- α-Amylase inhibitors
- Lectins
- Enzymes

Bt endotoxin Gene

The gram positive bacteria *Bacillus thuringiensis* produces a crystal toxin called δ (delta) endotoxin. The δ endotoxin is a stomach poison and kills the lepidopteran insects if consumed. The gene (DNA fragment) responsible for producing δ endotoxin is isolated from Bt and cloned into plants like cotton, potato, maize, etc. to produce Transgenic cotton, etc.

Transgenic Bt plants	*Target insect pests*
1. Cotton	Bollworms, *S. litura*
2. Maize	European corn borer
3. Rice	Leaf folder, stem borer
4. Tobacco, Tomato	Cut worms
5. Potato, Egg plant	Colarado potato beetle

Protease Inhibitors (PI) Gene

Insects have proteases in their gut which are enzymes helping in digestion of protein. Protease inhibitors are substances inhibit the proteases and affect digestion in insects. The protease inhibitor gene are isolated from one plant and cloned into another to produce transgenic plants.

Transgenic apple, rice, tobacco contains PI. gene.

Cowpea trypsin inhibitor (Cp TI) is a PI isolated from cowpea and cloned into tobacco. This transgenic tobacco is resistant to *Heliothis virescens.*

α-Amylase inhibitors gene

α-Amylase is a digestive enzyme present in insects for digestion of carbohydrate. α-Amylase inhibitor, affect digestion in insects. Transgenic tobacco and tomato expressing α-amylase inhibitor have been produced which are resistant to Lepidopteran pests.

Lectins genes

Lectins are proteins that bind to carbohydrates. When insect feed on lectins, it binds to chitin in peritrophic membrane of midgut and prevents uptake of nutrients.

e.g. Transgenic tobacco containing pea lectin gene is resistant to *H. virescens.*

Enzymes genes

Chitinase enzyme gene, and cholesterol oxidase gene have been cloned into plants and these show insecticidal properties.

Pyramiding genes

Engineering transgenic crops with more than one gene to get multi-mechanistic resistance is called pyramiding of genes. *e.g.* The Cp Ti gene and pea lectin gene were cloned to produce a transgenic tobacco.

Potential Advantages of Biotechnology in IPM

- Slow Development of resistance against transgenic Bt, PI, lectins
- All plant parts express toxin and so no need for insecticide spray
- No need for continuous monitoring
- No environmental pollution, safe to natural enemies, non-target organism.

9.4 INTEGRATED PEST MANAGEMENT (IPM)

Definition

"Integrated pest control is a pest management system that, in the context of associated environment and population dynamics of the pest species, utilizes all suitable techniques and methods in as compatible manner as possible and maintains pest populations at levels below those causing economic injury".

Principles of strategies

Monitoring insect pests and natural enemies : It is a fundamental tool in IPM for taking management decisions.

Concepts of injury levels : Following the concepts of EIL and ETL to reduce the usage of insecticides and their impacts on environment.

Integration of tactics : Proper choice of compatible tactics and blending them so that each component complement the other.

Relationship between different components

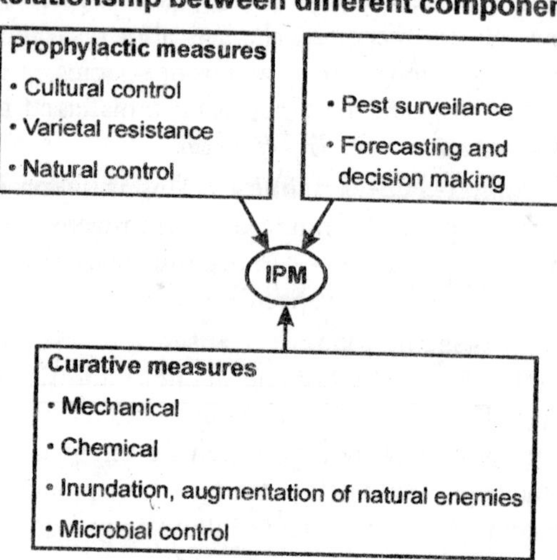

Approaches in IPM when compared to normal method

Parameter	*Normal method*	*IPM*
• Goal	Reduce losses due to pests	Reduce costs of production, ecological and social approach
• Diversity	Low to medium	High
• Ecosystem stability	Uncertain	Striving towards stability
• Spatial scale	Single farm	Entire region
• Time scale	Immediate	Long term/Single season
• Target	Single pest or closely related groups of pests	Several pests around crops and natural enemies
• Basis for control	Past experience or presence of pest	Economic Threshold Level (ETL)
• Principal method	Cultural, mechanical pesticides	All components
• Effect on environmental quality	Highly detrimental	Moderately detrimental/Negligible

Essential Requirements for IPM

- Proper identification of insect pests
- Life history and behaviour of the pest
- Natural regulating factors
- Need for control measures
- Timing of control measures
- Selection of suitable control measures
- Farmer's participation
- Government support

Advantages of IPM Compared to Pesticide-based Plant Protection Programme

- Proper
- Sustainability
- Economics
- Health
- Environment quality
- Social and political stability
- Local knowledge
- Export of agricultural commodities

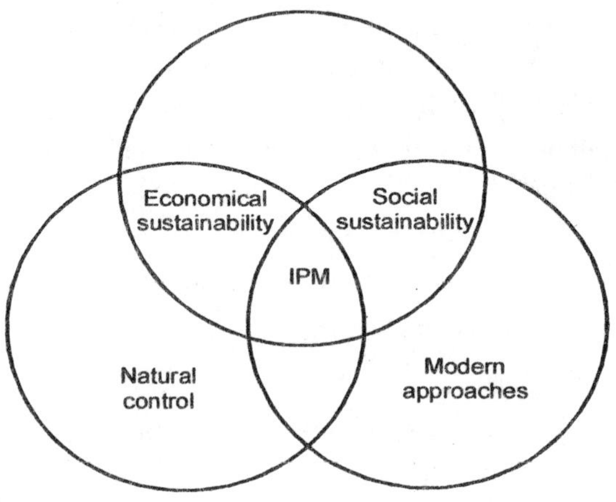

IPM for Rice, Cotton, Sugarcane

9.4.1 IPM for Rice

Observing ETL concept —

Insect	*ETL*
Stemborer	10% dead heart
Gall midge	10% silver shoot
Whorl maggot	25% damaged leaves
Leaf folder	10% leaf damage at vegetative stage
GLH	5 per hill at vegetable stage
	10 per hill at flowering stage
BPH	1 per tiller
	2 per tiller when spider is present at 1 per hill
Earhead bug	5 bugs per 100 panicle at flowering stage
	10 bugs per 100 panicle at milky stage

- Avoiding use of excess nitrogen which induces BPH and leaf folder.
- Alternate wetting and drying for BPH.
- Passing the rope on the crop and draining of water for case worm.
- Using light traps to monitor BPH, GLH and stem borer.
- Providing rogue spacing at every 2 m interval to take up the plant protection operation.
- Avoiding resurgence causing insecticides like synthetic pyrethroids, methyl parathion, and quinalphos for BPH.
- Avoiding shade for leaf folder.
- Use of Neem blended urea (1:5) which helps in the slow release of nutrient.
- Use of egg parasitoid *Trichogramma japonicum* for stem borer.
- Use of egg parasitoid *T. chilonis* and bacterial agent *Bacillus thuringensis* for leaf folder.
- Use of Neem oil for Ear head bug.
- Use of resistant varieties.

  eg. PY3, CO 42 for BPH

9.4.2 IPM for Cotton

Observing ETL concept —

Pests	*ETL*
Leaf hopper	1-2 per cent
White fly	5-10 per leaf
Boll worms	10% shoots, squares or bolls damaged
Stem weevil	10% infested plants
Spodoptera	8 egg masses per 100 m row

- Using light trap to monitor hoppers, *Spodoptera* and boll worms.
- Using pheromone traps to monitor and also for mass trapping of boll worms.
- Collection and destruction of infested plant parts, squares and bolls for boll worms.
- Growing trap crop (e.g..) Castor for *Spodoptera.*
- Manual removal of egg masses (e.g..) *Spodoptera.*
- Hand picking of larvae of boll worm.
- Avoiding late sowing of crop to escape from whitefly damage.
- Following crop rotation with unrelated crops to check the life cycle of important pests like boll worms, stem weevil, leaf hopper, etc.
- Removal of alternate weed hosts.
- Judicious use of Nitrogen and water for hoppers and whiteflies management.
- Use of yellow sticky traps for whiteflies.
- Avoiding early stage direct application of pesticides.
- Use of Neem based formulations.
- Use of predators like *Chrysoperla camea*
- Growing trap crop.
- Use of NPV virus.
- Use of egg parasitoid *Trichogramma* spp. for boll worms.

9.4.3 IPM for Sugarcane

- Trash mulching along the ridges to modify the moisture content of soil which checks the build up of early shoot borer.
- Detrashing on 105th day and 210th day for leaf hopper.
- Avoiding excess nitrogen for leaf hopper.
- Draining excess water for scales and mealy bugs.
- Use of predators like Coccinellid beetles for scales and mealy bugs.
- Use of egg parasitoid *Trichogramma* spp. for early shoot borer and inter node borer.
- Use of Granulosis virus and egg parasitoid *Sturmiopsis inferens* for Inter node borer.
- Use of egg parasitoid *Isotima javensis* for top borer.

9.5 OTHERS

9.5.1 Pesticide Formulations

Insecticides are available in different formulations after an insecticide has been commercially manufactured, formulators are added in various additives and prepare the insecticides in various forms such as dusts, granules, emulsifiable concentrates and solution etc.

3 types of formulations —

1. Solid 2. Liquid 3. Gaseous.

1. Solid Formulations

(*a*) ***Dusts*** :

- Insecticides mixed with dust carriers called dusts.
- Toxicity of an insecticide increase as the particle size decrease.
- It should be applied in the morning hrs on wet foliage.

(*b*) ***Granules*** :

- Dusts having 0.25 - 2.4 mm particle size.

- Used as dressing on or in the soil.

(*c*) ***Insecticide - fertilizer Mixture***

- Granular insecticides added to fertilizers.
- Applied as regular fertilization time to provide nutrition to plants as well as to control the soil insects.

2. Liquid Formulations

(*a*) *Solutions* :

- Pesticide is in a single liquid phase either in water or in an oil.
- Most organic insecticides are insoluble in water and need an organic solvent to dissolve.

(*b*) *Emulsifiable Concentrates (EC)*

- Commercially sold as EC to which a certain amount of water is added to make aqueous suspension suitable for spraying on plants.
- Clean solution and yields an emulsion when diluted with screen.

  (*e.g.*) Alkaline Soaps, Organic amines, etc.

(*c*) Suspension : Active material is suspended as solid particles.

(*d*) Foam Spray : Chemical additive which when sprayed then special foam nozzles converses the pesticide.

3. Gaseous Formulations

Fumigants :

- Compound volatile at ordinary temperature and sufficiently toxic.
- Advantage of using fumigant, places not easily accesible to other chemicals can easily be reacted due to the penetration and dispersal effect of the gas.
- Used to control Pests in storage bins, buildings, and soil insects, nematodes.

9.5.2 Impact of Global Warming on Pests

What is global warming ?

Warmth from sun heats the surface of the earth. Earth absorbs most of the energy but reflects back some energy in the form of infra red radiations. Greenhouse gases *e.g.* CO_2. Methane, CFC (Chloro Fluoro Carbon). Nitrous oxide present in atmosphere transmit the infrared radiation and reflects back to earth. This reflected energy falls on earth and also at lower atmosphere and keeps it warmer *i.e.,* heats the earth's surface. This is called global warming or green house effect.

Effect of Global Warming on World and Agriculture

- Increases overall temperature on earth.
- Causes changes in climate tremendously
- Results in melting of ice in polar region
- Increases sea level resulting in submerging of coastal areas
- Flooding and intense down pours
- Drought in warmer regions

Impact of Global Warming on Pest Status

- Due to change in climate, temperature and water availability, the farmers may change the type of crops grown.
- Due to increase in temperature, there can be outbreak of certain insect pests and diseases.
- In forest areas there will be a shift in tree species and also pest species.
- In agriculture lands since cropping pattern is changed, new crops are introduced to suit the climate and new pests are also introduced.
- When water availability is less, crops will be raised as rainfed. It will be difficult to take up control measures without water.

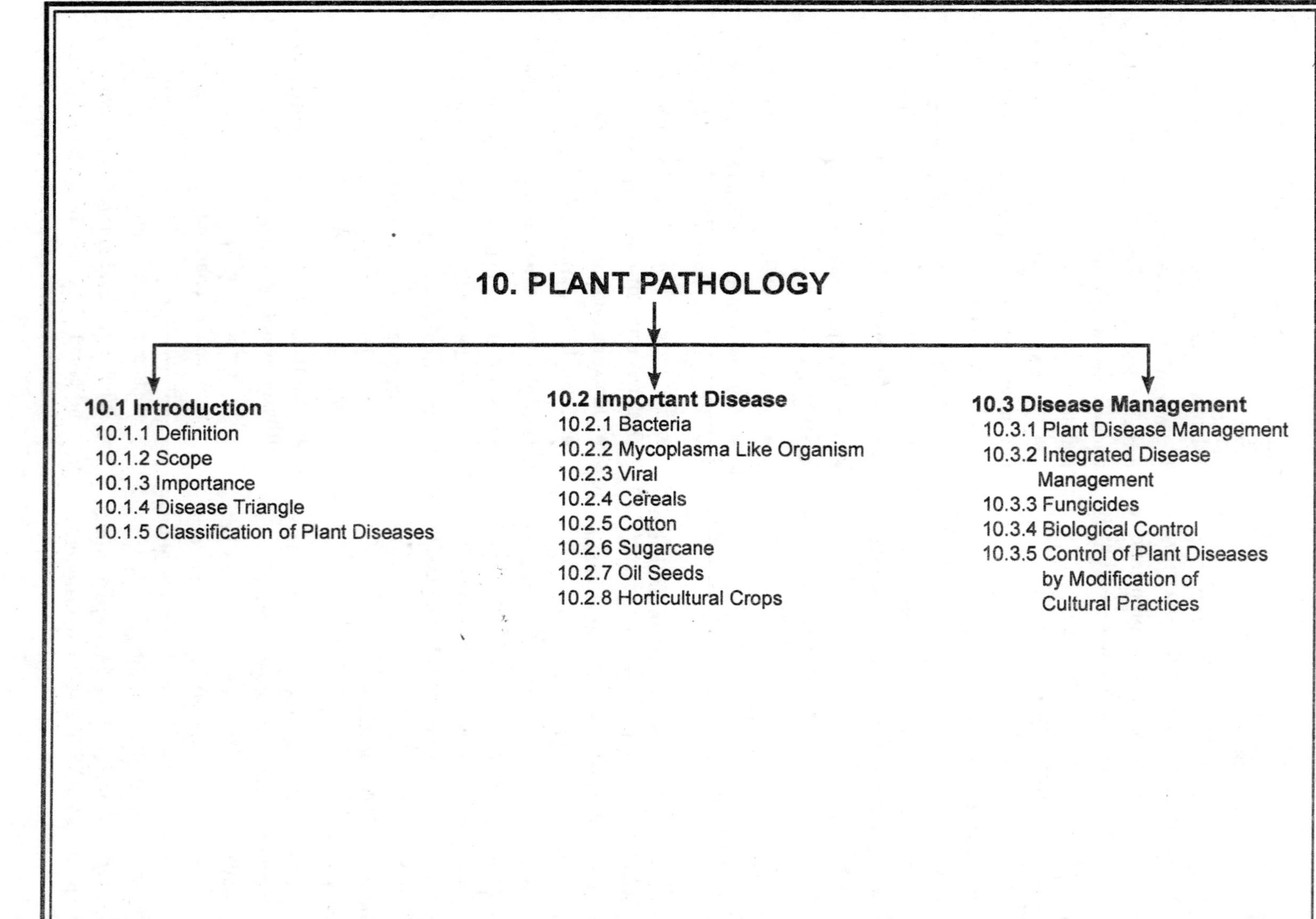

10. PLANT PATHOLOGY
10.1 Introduction
10.1.1 Definition
10.1.2 Scope
10.1.3 Importance
10.1.4 Disease Triangle
10.1.5 Classification of Plant Diseases
10.2 Important Disease
10.2.1 Bacteria
10.2.2 Mycoplasma Like Organism
10.2.3 Viral
10.2.4 Cereals
10.2.5 Cotton
10.2.6 Sugarcane
10.2.7 Oil Seeds
10.2.8 Horticultural Crops
10.3 Disease Management
10.3.1 Plant Disease Management
10.3.2 Integrated Disease Management
10.3.3 Fungicides
10.3.4 Biological Control
10.3.5 Control of Plant Diseases by Modification of Cultural Practices

10. Plant Pathology

10.1 INTRODUCTION

10.1.1 Definition

Plant pathology or phytopathology consists of three Greek words (1) *phyton* = plant, (2) *pathos* = ailments or diseases and (3) *logus* = discourse or knowledge. Thus, the term plant pathology or phytopathology means knowledge of plant disease.

10.1.2 Scope of Plant Pathology

Scope and responsibilities of plant pathology is unlimited. Its ultimate goal is to prevent and control plant diseases of economic importance. Responsibilities of the science of plant pathology may be summarized as under ;

1. Survey of fields, orchards and areas in order to find out prevalence of diseases and their incidence.
2. Recording new diseases of economic importance if any, with their identification and extent of incidence.
3. Assessment of losses caused by different diseases of economic importance.
4. Study of etiology, symptoms, predisposing factors and recurrence of such diseases.
5. Find out suitable and economic methods of management of economically important plant diseases.
6. Assisting in breeding of disease resistant varieties.
7. To train the extension workers and subject matter specialists in order to bridge the gap between pathologists and farmers for better crop production.

10.1.3 Importance of Plant Pathology in Agriculture

1. Plant pathology has advanced the knowledge to protect the crop from losses due to diseases.
2. The science of plant pathology has contributed disease free certified seed production.
3. Most of the diseases with known disease cycle can now be avoided by the modification of cultural practices.
4. With the knowledge of mode of disease spread, many diseases of economic importance can now be checked, minimized or controlled.
5. Crop improvement and varietal resistance have been achieved against many diseases through the joint effort of breeder and plant pathologist.
6. Plant pathology has made possible to restrict the spread of plant diseases from one place to other and one country to other through suitable measures and quarantine legislation.
7. With the knowledge of plant pathology various prophylatic measures are adopted for successful management of diseases. These measures are seed treatment in seed borne diseases soil treatment and crop rotations.
8. Diseases can be avoided in cold storage by the application of plant pathological

measures as per recommendations made for different diseases for protection of fruits and vegetables in storage.

9. Plant pathology has made possible to recognise, discard or utilize toxic substances by producing toxic substances or by competition or by parasitism. Organism exerting such lethal or damaging effect on the other is called antagonist.

10.1.4 Disease Triangle

Plant disease are caused by the interaction among susceptible host x abundance and virulence of pathogen x favourable environmental conditions for desirable time. These three factors constitute disease triangle, which are essential for disease incidence and its development. Each side of the triangle represents one of the three components. The size of each side is proportional to the sum total of the characteristics of the each component favouring disease.

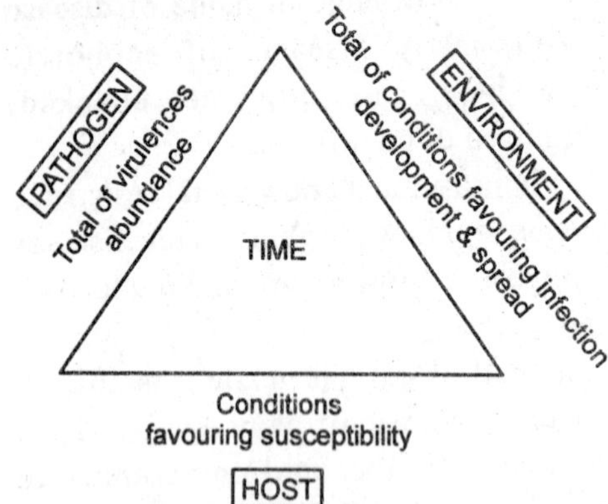

10.1.5 Classification of Plant Diseases

Plant disease have been classified into three groups on the basis of the occurrence of the pathogens.

***(i)* Endemic diseases.** Endemic diseases is one which is more or less constantly present in an area or locality. Such diseases are observed year to year in moderate to severe form. The causal agent of the diseases remains well established in the field or in the locality. By virtue of its ability of survival in weed, the disease is seen almost every year. Wilt of pigeon pea, red rot of sugar cane are observed as endemic diseases in Eastern U.P., Onion smut in Andhra pradesh and flag smut of wheat is Punjab are endemic diseases observed over there every year in moderate or severe form depending upon presence or predisposing factors.

***(ii)* Epidemic or Epiphytotic diseases.** Word epidemic is popularly used for animal and human diseases occurring in severe form. Therefore, in case of plant diseases the word epiphytotic should be used which means the occurrence of a disease widely and severely but periodically. The casual agent of such diseases may or may not remain present constantly in the locality. It is because the environmental conditions which become favourable occasionally.

A disease may be endemic or sporadic but may take form of epiphytotic at occasions. Late blight of potato-occurred in epiphytotic form in 1913 in the district of Rangpur now in Pakistan.

***(iii)* Sporadic diseases.** Sporadic diseases occur at very irregular intervals and locations. Green ear disease of bajra and black rust of wheat may be cited as examples of sporadic diseases which may appear in a locality or a district after one or more than one year or every year in sporadic form.

***(iv)* Pandemic diseases.** Diseases which occur all over the world and cause mass mortality, are called pandemic diseases. *Example* - late blight of potato.

***(v)* Classification according to Causal Agents.** According to causal agents the diseases may be classified into two main groups :

1. Parasitic or biotic diseases and
2. Non-parasitic diseases

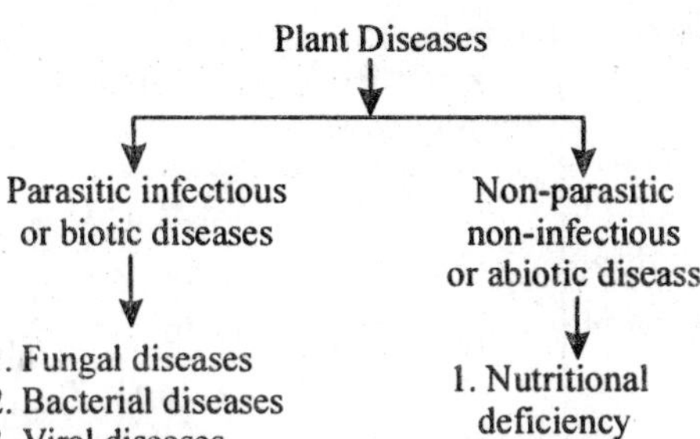

(*a*) *Parasitic or infectious diseases.* These diseases are caused by fungi, bacteria, viruses, nematodes and phanerogamic plant parasites. These diseases are usually contagious in nature and there is easy and efficient dispersal of the pathogens causing these diseases.

(*b*) *Non-parasitic or non-infectious diseases.* These diseases are not infected by biotic pathogens like fungi, bacteria, virus and nematodes etc. but are infected by the nutritional deficiencies or excess, by low and high temperature, by low and extreme light and by air pollutants etc. These diseases are not infectious in nature.

(*vi*) Classification According to nature of Diseases. Diseases classified according to the nature may be of following types :

(*a*) *Seed borne diseases.* In such diseases the inoculum is found associated with the seed internally or externally and are called as internally seed borne or externally seed borne diseases.

(*b*) *Soil borne diseases.* These diseases are incited by the inoculum present in the soil in the form of oospores, sclerotia and mycelia etc.

(*c*) *Air borne diseases.* These diseases are incited by the inoculum present in the air in the form of conidia, chlamydospores and sporangia etc.

(*vii*) Classification According to Symptoms. Symptoms of the diseases may also be considered as a base of classification and according to the symptoms, the diseases, may be grouped as rust diseases, leaf spot diseases, smut diseases, downy mildew diseases etc.

(*viii*) Classification According to the Host According to the host diseases may be grouped as cereal diseases and plantation diseases etc.

(*ix*) Classification According to the Nature of Infection. According to the nature of infections, the diseases may be classified into tow groups namely, (1) the systemic diseases such as green ear diseases of bajra and loose smut of wheat and, (2) Localized diseases such as leaf spot diseases.

10.2 IMPORTANT DISEASES

10.2.1 Bacteria

Table 1 : Important diseases caused by bacteria

Name of Diseases	*Diagnostic symptoms*	*Management*
Yellow ear rot or Tundu disease of wheat *C.O. Corynebacteriumtritici*	1. Twisting of the leaves. 2. Exduation of blight yellow sticky slime covering the entire ear charactreristic. 3. The slime envelops, the glume stems, and leaf sheaths. 4. Plant growth is checked and stem is distorted. 5. Distoration of ear head and rotting of spikelets.	1. Seeds free from nematode galls should be sown in nematode free drained field. 2. Deep summer ploughing. 3. In standing crop disease plants should be located, up rooted carefully and destroyed. 4. Suitable crop rotation of 2-3 years should be adopted.
Bacterial blight or Angular leaf spot or Black arm of cotton *C.O Xanthomonas exnopodis malvacearum*	1. Symptoms on all the parts of the plants above the ground level at different stages of growth. 2. The shape of the cotyledons are distorted causes them to dry and wither, 3. On the leaves similar spots appear on the under surface.	1. Removal and destruction of the alternate host and diseased plant debris. 2. Crop rotation. 3. Early irrigation. 4. Seed treatment. 5. Use of resistant varieties are recommended to control the diseases.
Bacterial brown rot of potato *pseudomonas solanacearum*	1. Stunting. yellowing of foliage, wilting and finally collapse of palnt. 2. Brown ring is formed in the tubers due to discolouration of vascular bundles.	1. Disease could be minimized by selection of healthy seeds. 2. Crop rotation, proper drainage and tuber treatment with0.02%. streptocycline for 30 minutes.
Citrus canker C.O. *Xanthomonas compestris pv citri*	1. The disease affects all aerial parts of the plant. 2. Round watery raised. yellowish brown spots appear on the leaves which later become rough and corky and surrounded by yellow halo. 3. Cankers on fruits are the same as in leaves except that yellow hallow is absent.	1. Use of disease free nursery stock for planting in new orchards. 2. In old orchard pruning of the affected twigs and spraying with 1 % copper oxychloride at periodical intervals.

10.2.2 Mycoplasma-like Organism (MLOs)

Japanese workers in 1967 showed that mulberry dwarf disease was incited by a mycoplasma like organism.

Table 2 : Plant diseases caused by mycoplasma like organisms (MLO's)

Name of Desease	*Vector*
Citrus greening (RLB)	Psyllid
Clover phyllody	Leaf hopper
Legume little leaf	Leaf hopper
Papaya bunchy top	Leaf hopper
Potato witches broom	Leaf hopper
Rice yellow dwarf	Leaf hopper
Tomato big bud	Leaf hopper
Sugarcane white leaf	Leaf hopper

Rickettsia-like Organisms (RLOs)

These disease causing agent have been recognised in recent years. They have distinct cell walls and belong to the group of bacteria. They are readily seen in the electron microscope and are usually transmitted by leaf hopper. Unlike mycoplasma-like organism, they are susceptible to pencillin. Rickettsia like organisms usually occur in xylem and phloem vessels.

10.2.3 Transmission of Plant Viruses

The chief methods of transmission of plant viruses are:

1. ***Seeds and pollen grains transmission***: Certain viruses like bean mosaic and curly top virus of beet sugar are transmitted by seed and pollen grains.
2. ***Transmission by vegetative propogation***: One of the chief methods of transmission of viral diseases especially of potato, raspberry, turnips, bulb plants, fruit trees and many ornamentals. The diseased vegetative parts such as tubers, bulbs, root, off shoots, buds and scions which are used for propagation will contain the virus present with parent plant. The new plants so raised by above mentioned methods are usually infected.
3. ***Transmission by mechanical means*** : Many viral diseases especially mosaic viruses are mechanically transmitted by the following methods:
   (*a*) By contact of infected leaves with healthy leaves brought about by wind.
   (*b*) By rubbing the juice of the diseased plant over the surface of leaves of healthy plants.
   (*c*) By grafting infected buds to healthy plants.
   (*d*) Knife and other mechanical contaminated implements may spread the disease.
   (*e*) Roots of diseased plant may come in contact of roots of healthy plants and thus spread the disease.
   (*f*) Dodder (Cusutta) may serve as a bridge between the healthy and diseased plants resulting in viral infection.
4. ***Soil transmission*** : Soil borne viruses are potato mosaic virus and oat mosaic etc. they are transmitted thorugh the soil.
5. ***Insect transmission*** : About 150 species of inscets are known to transmit from one plant to other. Insects capable of virus transmission are known as insect (carries). Some of them insect vectors which play a major role in dissemination viruses are the aphids, leaf hoppers, beetles, scale insects, thrips and white files the insect vectors are sucking insects.

Table 3 : Some insect vectors of plant virus diseases.

Crop	*Disease*	*Insect vectors*
Citrus	Veinal necrosis and twig decline	*Toxopetra citricidus*
Banana	Bunchy top	*Aphis gossypii*
Tomato	Leaf curl	*Bemisia tobaci*
Sugarcane	Grassy shoot	*Longuinuis sacchari*
Sesamum and sunhump	Phyllody	*Deltocephalus species*
Chinese-sarson	Mosaic	*Aphis gossypii, Mysus persicae,*

6. ***Transmission by fungi***

Gorgon (1985) found that the diseased lettuce was infected by Olpidium. He later discovered that the fungus act as a vector of the big vein mosaic. The virus acquired by the fungus remains in oospore. Similarly tobacco mosaic virus has been reported by Teakle (1960) to enter roots of its host by the zoospores of Olpidium brassicae.

7. ***Nematodes***

Some soil borne viruses are transmitted by nematodes. Examples are tobacco rattle, pea early browning and Colocasia mosaic etc.

Viruses have a great potential for high genetic variability due to their rapid replication and generation of large population in short time.

Detection of plant viruses:

1. ELISA (Enzyme-linked immunosorbent assay).
2. Molecular Technique (hybridization and DNA amplification).
3. High throughput sequence.
4. Multiplex PCR or RTPCR.
5. Multiplex LAMP
6. Lateral Flow Assay (LFA)
7. Oxford Nanopore Technologies.

Management of virus diseases

1. **Elimination of source of viral infection:**
   (*a*) Perennial weed hosts, annual weed hosts, unrelated crops, volunteer crops and approach crops.
   (*b*) Plant residue.
   (*c*) Rogueing
   (*d*) Shoot tip culture
2. **Avoidance of vectors :**
   (*a*) Breaking the life cycle of hosts, vectors and viruses
   (*b*) Creating physical barriers
   (*c*) Early sowing
   (*d*) Crop rotation
   (*e*) intercropping
3. **Use of resistant varieties**
4. **Insecticidal sprays or neem oil sprays.**

In case of non-persistent viruses such as Potato Virus-Y, use of insecticides to control vectors can reduce the vector population and disease incidence but cannot check the entry of vectors and spread of the disease.

Table 4 : Important diseases caused by viruses

Name of the Diseases	*Diagnostic Symptoms*	*Management*
Yellow dwrf of rice C.O. *Mycoplasma like organism (Phytoplasma)*	1. Loss of chlorophyll of new leaves. 2. affected plants may die immediately after seedling stage. 3. Severly infected plants may be devoid of grain formation.	1. Disease plant must be uprooted. 2. Grasses as alternate hosts nearby the field must be destroyed. 3. Leaf hoppers must be destroyed to check the spread of the disease.
Bunchy top of banana *caused by Banana virus I*	1. Leavess are typically branched together forming a dense rossette at the apex, 2. Green steaks appear on the secondary veins under side of the lamena, 3. Older leaves show symptoms of and curling.	1. Diseased plant material should be prohibited by law. 2. Diseased plant should be uprooted and burnt.
Mosaic of Papaya C.O. *Papaya mosic virus*	1. Profuse mottling and puckering of leaves especially the young ones. 2. Degeneration and marked reduction in growth. 3. First infection occures on top leaves, 4. Leaves are often modified into tendiril-like structures called shoe strings.	1. Spray 1% groundnut oil help to prevent infection.
Grossy shoot of sugarcane C.O. MLO now called as *Phytoplasma*	1. Production of numerous thin tillers from the base of affected shoots giving a dense or crowded bunch of tillers bearing pale yellow or chlorotic leaves. 2. Leaves remain thin, narrow, reduced. 3. Retarded elongation of internodes.	1. Use of disease free sets for sowing. 2. precaution against transmission by machenical means. 3. Vectors management by insecticides, 4. Use of resistant verieties.
Leaf roll of potato C.O. Potato leaf roll *virus*	1. Rolled leaf lets are rigid stiff thick and leathery, 2. Phloem necrosis also take place, 3. Tubers remain small in size.	1. Obtain certified seed, 2. Rogue out the sick plants, 3. Detop in the third of forth week of December, 4. Sparying of crop with Metasystox (0.1%) to control the insect vectors.

10.2.4 Cereals

RICE

Blast-pyricularia oryzae

Symptoms

The typical symptoms appear on leaves, leaf sheath, nodes and even the glumes are also attacked. On the leaves, the lesions start as small water soaked bluish green flecks, soon enlarge and form characteristic spindle shaped spots with grey centre and dark brown margin.

In infected nodes, irregular black areas encircle the nodes can be noticed.

Pathogen

Mycelium of the fungus, is hyaline to olivaceous, septate and highly branched. Conidia are produced in clusters on long septate, olivaceous slender conidiophores.

Favourable Conditions

Application of excessive doses of nitrogenous fertilizers, intermittent drizzles, cloudy weather, high relative humidity (93-99 per cent), low night temperature (between 15-20 °C), more number of rainy days, longer duration of dew, cloudy weather, slow wind movement and availability of collateral hosts.

Mode of Spread and Survival

The disease spreads primarily through airborne conidia since spores of the fungus present throughout the year. Mycelium and conidia in the infected straw and seeds are important sources of inoculum.

Management

Remove and destroy the weed hosts in the field bunds and channels. Treat the seeds with Captan of Thiram or Carbendazim or Carboxin or Tricyclazole at 2g/kg. Seed treatment with biocontrol agent *Trichoderma viride* @ 4g/kg.

Brown Spot or Sesame leaf spot

Helminthosporium oryzae

Symptoms

The fungus attacks the crop from seedling in nursery to milk stage in main field. Symptoms appear as lesions (spots) on the coleoptile, leaf blade, leaf sheath, and glume, being most prominent on the leaf blade and glumes. The disease appears first as minute brown dots, later becoming cylindrical or oval to circular.

Pathogen

H. oryzae produces greyish-brown to dark brown septate mycelium. Conidiophores may arise singly or in small groups.

Favourable Conditions

Temperature of 25-30 °C with relative humidity above 80 per cent are highly favourable. Excess of nitrogen aggravates the disease incidence.

Mode of Spread and Survival

The infected seeds are the most common source of primary infection. The conidia present on infected grain and mycelium in the infected tissue may viable for 2 to 3 years. The fungus may survive in the soil for 28 months at 30 °C and 5 months at 35 °C. Airborne conidia infect the plants both in nursery and in main field.

Management

Removal of collateral hosts and infected debris in the field. Crop rotation, adjustment of planting time and proper fertilization are suggested. Use of slow release nitrogenous fertilizers are suggested.

Sheath rot

Sarocladium oryzae

Symptoms

Initial symptoms are noticed only on the upper most leaf sheath enclosing young panicles. The flag leaf sheath show oblong or irregular greyish brown spot. They enlarge and develop grey centre and brown margins covering major portions of the leaf sheath.

Pathogen

The fungus produces whitish, sparsely branched and septate mycelium. Conidiophore is slightly thicker than the vegetative hyphae. Conidia are hyaline, smooth, single celled and cylindrical in shape.

Favourable Conditions

Closer planting, high doses of nitrogen, high humidity and temperature around 25-30 °C. Injuries made by leaf folder, brown plant hopper and mites increase infection.

Mode of Spread and Survival

Mainly through air-borne conidia and also seed-borne.

Management

Apply recommended doses of fertilizers. Adopt optimum spacing. Spray Carbendazim 250 g or Edifenphos 1 lit or mancozeb 1 kg or Chlorothalonil 1 kg/ha at boot leaf stage and 15 days later.

Sheath blight

Rhizoctonia solani

Symptoms

The fungus affects the crop from tillering to heading stage. Initial symptoms are noticed on leaf sheaths near water level. On the leaf sheath oval or elliptical or irregular greenish grey spots are formed. As the spots enlarge, the centre becomes greyish white with an irregular blackish brown or purple brown border.

Pathogen

The fungus produces usually long cells of septate mycelium which are hyaline when young, yellowish brown when old. It produces large number of globose sclerotia, which are initially white, later turn to brown or purplish brown.

Favourable Conditions

High relative humidity (96-97 per cent), high temperature (30-32 °C), closer planting and heavy doses of nitrogenous fertilizers.

Mode of Spread and Survival

The pathogen can survive as sclerotia or mycelium in dry soil for about 20 months but for 5-8 months in moist soil.

Management

Avoid excess doses of fertilizers. Adopt optimum spacing. Eliminate weed hosts. Apply organic amendments. Avoid flow of irrigation water from infected fields to healthy fields. Deep ploughing in summer and burning of stubbles.

Stem rot

Sclerotium oryzae

Symptoms

Small black lesions are formed on the outer leaf sheath and they enlarge and reach the inner leaf sheath also. The affected tissues rot and abundant sclerotia are seen in the rotting tissues. The culm collapses and plants lodge.

Pathogen

White to greyish hyphae produce smooth, spherical black and shiny sclerotia, visible to naked eyes as black masses.

Favourable Conditions

Infestation of leaf hoppers and stem borer and high doses of nitrogenous fertilizers.

Mode of Spread and Survival

The sclerotia survive in stubbles and straw are carried through irrigation water.

Management

Use of recommended doses of fertilizer. Deep ploughing in summer and burning of stubbles. Avoid flow of irrigation water from infected fields to healthy fields.

Bacterial leaf blight

Xanthomonas oryzae p.v. oryzae

Symptoms

The disease is usually noticed at the time of heading but in severe cases occur earlier also. Seedlings in the nursery show circular, yellow spots in the margin, later enlarge, coalesce and cause drying of foliage.

In grown up plants water soaked, translucent lesions appear usually near the leaf margin. The lesions enlarge both in length and width with a wavy margin and turn straw yellow within a few days, covering the entire leaf.

Pathogen

The bacterium is strict aerobe, gram negative, non spore forming.

Favourable Conditions

Clipping of tip of the seedling at the time of transplanting, heavy rain, heavy dew, flooding, deep irrigation water, severe wind, temperature of 25-30 °C and application of excessive nitrogen, especially late top dressing.

Mode of Spread and Survival

The infected seeds as a source of inoculum may not be important since the bacteria decrease rapidly and die in the course of seed soaking.

Management

Burn the stubbles. Use optimum dose of fertilizers. Avoid clipping of tip of seedling at the time of transplanting. Avoid flooded conditions. Remove weed hosts. Grow resistant cultivars like IR 20 and TKM 6.

Maize (Zea mays)

Downy mildew/Crazy top

Peronosclerospora sorghi

Symptoms

The most characteristic symptom is the development of chlorotic streaks appear on the leaves and the plants exhibit a stunted and bushy appearance due to the shortening of the internodes. White downy growth can be seen not only on the lower surface of leaf but also on the chlorotic streaks.

Pathogen

The fungus grows as white downy growth on both surface of the leaves, consist of sporangiophores and sporangia.

Favourable Conditions

Low temperature (21-23 °C), high relative humidity (90 per cent), water logging condition and light drizzling. Young plants are highly susceptible.

Mode of Spread and Survival

The primary source of infection is through the oospores present in the soil and also the dormant mycelium present in the infected maize seeds. Secondary spread is through air-borne conidia. The oospores survive in the soil as well as in the infected plant debris.

Management

Deep ploughing. Crop rotation with pulses. Adjust the time of sowing. Rogue out infected plants. Treat the seeds with Metalaxyl compound (Apron 35SD) at 6 g/kg.

Leaf blight

Helminthosporium maydis

Symptoms

The fungus affects the maize plant at young stage. Small yellowish round to oval spots are seen on the leaves. The spots gradually increase in area into bigger elliptical spots and it is greyish brown colour in the centre with dark brown margins.

Pathogen

Conidiophores are in group, geniculate, mid dark brown, pale near the apex and smooth. Conidia are distinctly curved, fusiform, pale to mid dark golden brown with 5-11 pseudosepta.

Favourable Conditions

Optimum temperature for the germination of

conidia is 18 to 27 °C. Infection takes place early in the wet season.

Mode of Spread and Survival

It is a seed-borne fungus.

Management

Treat the seeds with Captan or Thiram at 4 g/kg. Spray Mancozeb 1.25 kg/ha.

Rust

Puccinia sorghi

Symptoms

Circular to oval, elongated, brown powdery pustules are scattered over both surface of the leaves. As the plant matures, the pustules become brown to black owing to the replacement of red uredospores by black teliospores.

Pathogen

Uredospores are globose or elliptical finely echinulate, yellowish brown with 4 germ pores. Teliospores are brownish black, or dark brown, oblong to ellipsoidal, rounded to flattened at the apex.

Favourable Conditions

Cool temperature and high relative humidity.

Management

Remove the alternate hosts. Spray Mancozeb at 1.25 kg/ha.

Head smut

Sphacelotheca reiliana

Symptoms

Symptoms are usually noticed on the cob and tassel. Large smut sori replace the tassel and the ear. Sometimes the tassel is partially or wholly converted into smut sorus. Under such conditions the individual spikelets on tassels may form shoot like growths, or the entire tassel may develop into leaf like structures.

Pathogen

Smut spores (chlamydospores) are produced in large numbers which are reddish brown to black, thick walled, finely spined, spherical or slightly irregular.

Favourable Conditions

Low temperature favours more infection and this fungus also infects the sorghum crop and *vice versa*.

Mode of Spread and Survival

The smut spores retains its viability for two years. The fungus is externally seed-borne and soil-borne. The major source of infection is through soil-borne chlamydospores.

Management

Field sanitation. Crop rotation with pulses. Treat the seeds with Captan or Thiram at 4 g/kg.

Charcoal rot

Rhizoctonia bataticola

Symptoms

The fungus attacks roots of seedlings and young plants. The affected plants first exhibit the wilting symptoms. The stalk of the infected plants can be recognized by grayish streaks and become weak.

Pathogen

The fungus produces large number of sclerotia which are round and black in colour. Sometimes, it produces pycnidia on the stems or stalks.

Favourable Conditions

High temperature and low soil moisture.

Mode of Spread and Survival

The fungus has a wide host range, attacking sorghum, cumbu, ragi and pulses. It survives for more than 16 years in the infected plant debris. The primary source of infection is through soil-borne sclerotia.

Management

Long crop rotation with crops that are not natural host of the fungus. Irrigate the crops at the time of earhead emergence to maturity. Treat the seeds with Carbendazim or Captan at 2 g/kg.

WHEAT (Triticum aestivum)

Leaf or brown or orange rust	Yellow or stripe rust
Puccinia recondita	*Puccinia striiformis*
• Mainly seen on leaves, rarely on leaf sheath. Rust pustules (Uredia) apper on leaves at early stage of crop and at maturity they turn to brown or orange (Telia).	• Mainly occur on leaves than the leaf sheaths and stem. Bright yellow pustules (Uredia) appear on leaves at early stage of crop and at maturity, pustules are arranged in linear rows along with dull black lesions (Telia).
• Brown, spherical, minutely echinulated with 7-10 germpores.	• Yellow, spherical to oval, minutely echinulated with 6-16 germpores.
• Brown, two celled, smooth, thick walled with rounded apex	• Dark brown, two celled, thickwalled and flat at the top.
• Alternate host: Thalictrum sp Uredospores and dormant mycelium persist on crop debris and weed host.	• Alternate host: Unknown. Uredospores and dormant mycelium survive on stubbles and straws and also on weed hosts and self-sown wheat crops.
• Wind-borne uredospores from hills	• Wind borne uredospores from hills.
• Warm weather (15°C) and heavy rain during January.	• Very low temperature (11°C) and heavy rainfall during November-December.
• Mixed cropping with suitable crops. Avoid excess dose of nitrogenous fertilizers. Dust Sulphur at 35-40 kg/ha. Spray Zineb at 2.5 kg/ha. Grow resistant varieties like Lerma Roja, Safed Lerma and Sonalika.	• Mixed cropping with suitable crops. Avoid excess dose of nitrogenous fertilizers. Dust Sulphur at 35-40kg/ha. Spray Zineb at 2.5kg/ha. Grow resistant varieties like Lerma Rajo, Safed Lerma, Sonalika and Chotilerma.

SMUTS

Disease	*Loose Smut*	*Flag Smut*
Causal organism	*Ustilago nuda*	*Urocystis tritici*
Symptoms	Symptom is visible only after emergence of ear. Diseased ear heads emerge out of the boot leaf earlier than healthy ones. The smutted spikelet is covered by a silvery membrane which contain black, dry powdery mass of spores.	The symptom can be seen on stem, culm and leaves from late seedling stage to maturity. The seedling infection leads to twisting and drooping of leaves followed by withering. Grey sori occur on leaf blade and sheath. The sorus contains black powdery mass of spores.

Contd...

Spores	Spores germinate to produce promycelium with sporidia. Sporidia attack the ovary through stigma and remain dormant in embryo.	Aggregated sporeballs, consisting 1-6 bright lobose, brown, smooth walled spores surrounded by a layer of flat sterile cells.
Mode of Spread and Survival	Internally seedborne as dormant mycelium (Embryo infection)	Seed and soil borne. Smut sopres are viable for more than 10 years.
Favourable conditions	Temperature of 18-24°C. High Humidity (60-85 per cent) during flowering	Temperature of 18-24°C. Relative humidity 65 percent and above.
Management	*Hot water treatment* : Soak the seeds in water at 26-30°C for 5hrs to induce dormant mycelium to grow. Then immerse the seeds in hot water at 54°C for 10 min. to kill the mycelium.	Treat the seeds with Sulphur or Carboxin at 2g/kg. Grow resistant varieties like Pusa 44 and WG 377.

BUNT

Smooth spored bunt	**Karnal bunt**
Tilletia foetida	*Neovassia indica*
The fungus attacks seedling of 8-10 days old and become systemic and grows along the tip of shoot. At the time of flowering, hyphae concentrate in the inflorescence and spikelets and transforming the ovary into smut sorus of dark green colour with masses of chlamydospores.	The infection occurs only at the flowering stage. Only a few spikelets are converted into dark coloured sori.
Reticulate, globose and smooth walled. No resting period. Produce primary sporidia upon germination.	Globose and smooth walled. Long resting period. Germainate to produce primary sporidia which are needle shaped and then secondary sporidia which are sickle shaped.
Externally seed-borne	Soil- borne and air-borne.
Temperature of 18-20°C. high soil moisture	Temperature of 15-20°C, High humidity, cloudy weather during flowering.
Treat the seeds with Carbendazim at 2g/kg. Grow the crop during high temperature periods. Adopt shallow sowing. Grow resistant varieties like Kalyan sona, HD2021, HD4513 and HD4519.	Deep ploughing during summer. Avoid continuous cropping of wheat in the same field. Grow resistant varieties like HD1907, H1358, HP753, and L176.

10.2.5 Cotton

Wilt

Fusarium oxysporum f.sp. vasinfectum

Symptoms

The disease affects the crop at all stages. The earliest symptoms appear on the seedlings in the cotyledons which turn yellow and then brown. The base of petiole shows brown ring, followed by wilting and drying of the seedlings. In young and grown up plants, the first symptom is yellowing of edges of leaves and area around the veins *i.e.* discolouration starts from the margin and spreads towards the midrib.

Pathogen

Macroconidia are 1 to 5 septate, hyaline, thin-walled, falcate with tapering ends.

Favourable Conditions

Soil temperature of 20-30 °C, hot and dry periods followed by rains, heavy black soils with an alkaline reaction.

Mode of Spread and Survival

The fungus can survive in soil as saprophyte for many years and chlamydospores act as resting spores. The pathogen is both externally and internally seed-borne.

Management

Remove and burn the infected plant debris in the soil after deep summer ploughing during June-July. Apply increased doses of potash with a balanced dose of nitrogenous and phosphatic fertilizers.

Root rot

Rhizoctonia bataticola

Symptoms

The fungus causes three types of symptoms *viz.*, seedling disease, sore-shin and root rot. Germinating seedling and seedlings of one to two weeks old are attacked by the fungus at the hypocotyl and cause black lesions, girdling of stem and death of the seedling, causing large gap in the field. In sore-shin stage (4 to 6 weeks old plants), dark reddish-brown cankers are formed on the stems near the soil surface, later turning dark black and plant breaks at the collar region leading to drying of the leaves and subsequently the entire plant.

Pathogen

The fungal hyphae are septate and fairly thick and produce black, irregular sclerotia.

Favourable Conditions

Dry weather following heavy rains, high soil temperature (35-39 °C), low soil moisture (15-20 %)

Mode of Spread and Survival

The disease is mainly soil-borne and the pathogen can survive in the soil as sclerotia for several years. The spread is through sclerotia which are disseminated by irrigation water, implements, heavy winds and other cultural operations.

Management

Apply farm yard manure at 10 t/ha or neem cake at 2.5 t/ha. Adjust the sowing time, early sowing (First week of April) or late sowing (Last week of June) so that crop escapes the high soil temperature conditions. Adopt intercropping with sorghum.

Anthracnose

Colletotrichum capsici

Symptoms

The fungus infects the seedlings and produces small reddish circular spots on the cotyledons and primary leaves. The lesions develop on the collar region, stem may be girdled, causing seedling to wilt and die. In mature plants, the fungus attacks the stem, leading to stem splitting and shredding of bark. The most common symptom is boll spotting.

Pathogen

The pathogen forms large number of acervuli on the infected parts. The conidiophores are slightly curved, short, and club shaped.

Favourable Conditions

Prolonged rainfall at the time of boll formation and close planting.

Mode of Spread and Survival

The pathogen survives as dormant mycelium in the seed or as conidia on the surface of seed for about a year. The pathogen also perpetuates on the rotten bolls and other plant debris in the soil.

Management

Remove and burn the infected plant debris and bolls in the soil. Rogue out the reservoir weed hosts.

Boll rot

Symptoms

Initially, the disease appears as small brown or black dots which later enlarge to cover the entire bolls. Infection spreads to inner tissues and rotting of seeds and lint occur. The bolls never burst open and fall off prematurely.

Favourable Conditions

Heavy rainfall during the square and boll formation stage, punctures caused by the insects, especially red cotton bug *Dysdercus cingulatus*, close spacing and excessive nitrogen application.

Mode of Spread and Survival

The fungi survive in the infected bolls in the soil. The insects mainly help in the spread of the disease. The fungi make their entry only through the insect punctures.

Management

Adopt optimum spacing. Apply the recommended doses of fertilizers. Carbendazim 1 kg from 45 day at 15 days interval.

Leaf blight

Alternaria macrospora

Symptoms

The disease may occur in all stages but more severe when plants are 45-60 days old. Small, pale to brown, irregular or round spots. Each spot has a central necrotic lesion surrounded by concentric rings.

Pathogen

The fungus produces dark brown, short, 1-8 septate, irregularly bend conidiophores with a single conidium at the apex.

Favourable Conditions

High humidity, intermittent rains and moderate temperature of 25-28 °C.

Mode of Spread and Survival

The pathogen survives in the dead leaves as dormant mycelium. The pathogen primarily spreads through irrigation water. The secondary spread is mainly by air-borne conidia.

Management

Remove and destroy the infected plant residues. Spray Mancozeb or Copper oxychloride at 2 kg/ha at the initiation of the disease.

Bacterial blight

Xanthomonas campestris p.v. *malvacearum*

Symptoms

The bacterium attacks all stages from seed to harvest.

Pathogen

The bacterium is a short rod with a single polar flagellum. It is gram negative, non-spore forming.

Favourable Conditions

Optimum soil temperature of 28 °C, high atmospheric temperature of 30-40 °C, relative humidity of 85 per cent, early sowing, delayed thinning, poor tillage, late irrigation and potassium deficiency in soil.

Mode of Spread and Survival

The bacterium survives on infected, dried plant debris in soil for several years. The bacterium is also seed-borne and remains in the form of slimy mass on the fuzz of seed coat. The primary infection starts mainly from the seed-borne bacterium. The secondary spread of the bacteria may be through wind, wind blown rain splash, irrigation water, insects and other implements.

Management

Remove and destroy the infected plant debris. Rogue out the volunteer cotton plants and weed hosts. Follow crop rotation with non-host crops. Early thinning and early earthing up with potash.

10.2.6 Sugarcane (Saccharum officinarum)

Red rot

Colletotrichum falcatum

Symptoms

Typical symptoms of red rot are observed in the internodes of a stalk by splitting it longitudinally. These include the reddening of the internal tissues which are usually elongated at right angles to the long axis of the stalk. The presence of cross-wise white patches is the important diagnostic character of the disease. The diseased cane also emits acidic-sour smell.

Pathogen

The fungus produces thin, hyaline, septate, profusely branched hyphae containing oil droplets. The fungus produces black, minute velvety acervuli with long, bristle-like, septate setae. Monoculturing of sugarcane, successive ratoon cropping, water logged conditions and injuries caused by insects.

Mode of Spread and Survival

The fungus is sett-borne. The fungus also persists in the soil on the diseased clumps and stubbles as chlamydospores and dormant mycelium. The primary infection is mainly from infected setts. Secondary spread in the field may be through irrigation water and cultivation tools.

Management

Adopt crop rotation by including rice and green manure crops. Select the setts from the disease free fields or disease free area. Avoid ratooning of the diseased crop.

Smut

Ustilago scitaminea

Symptoms

The affected plants are stunted and the central shoot is converted into a long whip-like, dusty black structure. The length of the whip varies from few inches to several feet. In early stages, this structure is covered by a thin, white papery membrane. The whip may be straight or slightly curved. On maturity it ruptures and millions of tiny black smut spores (teliospores) are liberated and disseminated by the wind. Affected plants are usually thin, stiff and remain at acute angle.

Pathogen

The fungal hyphae are primarily intercellular and collect as a dense mass between the vascular bundles of host cell and produce tiny black spores.

Favourable Conditions

Monoculturing of sugarcane, continuous ratooning and dry weather during tillering stage.

Mode of Spread and Survival

Teliospores may survive in the soil for long periods, upto 10 years. The spores and sporidia are also present in the infected plant materials in the soil. The smut spores and dormant mycelium also present in or on the infected setts. The primary spread of the disease is through diseased. The secondary spread in the field is mainly through the smut spores developed in the whips, aided by air currents.

Management

Plant healthy setts taken from disease free area. Remove and destroy the smutted clump. Discourage ratooning of the diseased crops having more than 10 per cent infection. Follow crop rotation with green manure crops or dry fallowing.

Sett rot or Pineapple disease

Ceratocystis paradoxa

The disease primarily affects the setts usually two to three weeks after planting. The fungus is soil-borne and enters through cut ends and proliferates rapidly in the parenchymatous tissues. The affected tissues first develop a reddish colour which turns to brownish black in the later stages.

Pathogen

The fungus produces both macroconidia and microconidia. Conidiophores are linear, thinwalled with short cells at the base and a long terminal cell.

Favourable Conditions

Heavy clay soils, temperature of 25-30 °C and prolonged rainfall after planting.

Mode of Spread and Survival

The fungus survives as conidia and chlamydospores in the soil and in the infected. buried cane tissues. The inoculum moves from field to field through wind-borne conidia or irrigation or rain water.

Management

Soak the setts in 0.05% Carbendazim or 0.25% Emisan for 15 minutes. Provide adequate drainage during rainy seasons.

Rust

Puccinia erianthi

Symptoms

Minute, elongated, yellow spots (uredia). The pustules turn to brown on maturity. Late in the season, dark brown to black telia appear on the lower surface of leaves.

Pathogen

The mycelium is hyaline, branched and septate.

Favourable Conditions

Temperature of 30 °C, relative humidity between 70 and 90 per cent, high wind velocity and continuous cloudiness.

Mode of Spread and Survival

The fungus survives on collateral hosts like *Erianthus fulvus* and *Saccharum spontaneum*. The uredospores also survive in the infected stubbles in the soil. The disease is mainly spread through air-borne uredospores.

Management

Remove the collateral hosts. Spray Tridemorph 1 kg/ha.

Gummosis

Xanthomonas campestris p.v. *vasculorum*

Symptoms

The bacterium produces two distinct types of symptoms. On the mature leaves longitudinal stripes or streaks appear around the affected veins near the tip. Initially these stripes are pale yellow in colour, later turn to brown. The infected canes are stunted with short internodes, giving a bushy appearance.

Pathogen

The bacterium is a short rod, Gram negative, non spore forming with a single polar flagellum. It is a facultative anaerobe and it produces yellow slimy growth.

Mode of Spread and Survival

The bacterium remains viable in the soil as well as in infected canes. The primary transmission is through naturally affected diseased setts or through soil-borne contamination. The secondary spread may be through wind splashed rain, harvesting implements, animals and insects.

Management

Remove and burn the affected clumps and the stubbles in the field. Select setts from disease free areas. Avoid growing collateral hosts like maize, sorghum and cumbu near the sugarcane fields.

Grassy shoot

Mycoplasma Like Organism (MLO)

Symptoms

The disease symptoms are usually seen two months after planting. The disease is characterised by the production of numerous lanky tillers from the base of the affected shoots. Leaves become pale yellow to completely chlorotic, thin and narrow. The plants appear bushy and 'grass like' due to reduction in the length of internodes and premature and continuous tillering.

Pathogen

MLO is found to present in the sieve cells of infected plants.

Mode of Spread and Survival

The pathogen is transmitted by aphids *viz.*, *Rhopalosiphum maidis*, *Melanaphis sacchari*. In addition, leaf hopper, *Proutista moesta* also involves in the transmission.

Management

Select setts from middle of the healthy fields or from the commercial seed nursery. Remove and burn the infected clumps periodically. Treat the setts before planting as follows : Hot Water Treatment (HWT) at 50 °C for 2½ hr. Hot Air Treatment (HAT) at 54 °C for 8 hr or Aerated Steam Therapy (AST) at 50 °C for 1 hr or Moist Hot Air Treatment (MHAT) at 54 °C for 2 hr.

10.2.7 Oil Seeds

GROUND NUT (*Arachis hypogaea*)

Rust (*Puccinia arachidis*)

Symptoms

The disease attacks all aerial parts of the plant. The disease is usually found when the plants are about 6 weeks old. Uredosori appear on the lower surface of leaves. The epidermis ruptures and exposes a powdery mass of uredospores. Corresponding to the sori, small, necrotic, brown spots appear on the upper surface of leaves. The rust pustules may be seen on petioles and stem.

Favourable Conditions

High relative humidity (above 85 per cent), heavy rainfall and low temperature (20-25 °C).

Mode of Spread and Survival

The pathogen survives as uredospores on groundnut plant. The fungus also survives in infected plant debris in soil. The spread is mainly through wind-borne inoculum of uredospores.

Management

Avoid monoculturing of groundnut. Spray Mancozeb 1 kg. or Wettable sulphur 2.5 kg. or Tridemorph 500 ml or Chlorothalonil 1 kg/ha.

Collar rot or seedling blight or crown rot

Aspergillus niger and A. pulverulentum

Symptoms

The fungus is both seed-borne and soil-borne. So the infection can be seen at any stage from sowing onwards.

Crown rot : The infection when occurs in adult plants show crown rot symptoms. Large lesions develop on the stem below the soil and wilting of plant.

Favourable Conditions

Deep sowing of seeds, high soil temperature (30-35 °C) and low soil moisture.

Mode of Spread and Survival

The pathogens survive in plant debris in the soil, not necessarily from a groundnut crop. Soil-borne conidia cause disease carry over from season to season. The other primary source is the contaminated seeds. The fungi are carried on the seed surface or under the testa.

Management

Select good quality seeds. Treat the seeds with Carbendazim 2 g or Thiram 4g/kg. Avoid deep sowing of seeds. Destroy the crop debris by burning.

Root rot

Macrophomina phaseolina

Symptoms

In the early stages of infection reddish brown lession appears on the stem just above the soil level. The leaves and branches show drooping, leading to death of the whole plant. The decaying stems are covered by whitish mycelial growth. The rotten tissues contain large number of black or dark brown, thick walled sclerotia. When infection spreads to underground roots, the sclerotia are formed externally in the rotten tissue. Pod infection leads to blackening of the shells and sclerotia can be seen inside the shells.

Favourable Conditions

Prolonged rainy season at seedling stage and low lying areas.

Mode of Spread and Survival

The fungus remains dormant as sclerotia for a long period in the soil and in infected plant debris. The primary infection is through soil-borne and seed-borne sclerotia. The secondary spread of sclerotia is aided by irrigation water, human agency, implements, cattle etc.

Management

Treat the seeds with thiram 4 g or carbondazim

GINGELLY (*Sesamum indicum*)

Root rot or stem rot or charcoal rot

Macrophomina phaseolina

Symptoms

The disease symptom starts as yellowing of lower leaves, followed by drooping and defoliation. The stem portion near the ground level shows dark brown lesions and bark at the collar region shows shredding. The sudden death of plants is seen in patches. In the grow-up plants, the stems portion can be easily pulled out leaving the rotten root portion in the soil. The infection when spreads to pods, they open prematurely and immature seeds become shrivelled and black in colour.

Favourable Condition

Day temperature of 30 °C and above and prolonged drought followed by copius irrigation.

Mode of Spread and Survival

The fungus remains dormant as sclerotia in soil as well as in infected plant debris in soil. The infected plant debris also carry pycnidia. The fungus primarily spreads through infected seeds which carry sclerotia and pycnidia. The fungus also spreads through soil-borne sclerotia.

Management

Apply farm yard manure or green leaf manure at 10 t/ha or neem cake 250 kg/ha.

Powdery mildew

Erysiphe cichoracearum

Symptoms

Initially greyish-white powdery growth appears on the upper surface of leaves. When several spots coalesce, the entire leaf surface may be covered with powdery coating. In severe cases, the infection may be seen on the flowers and young capsules, leading to premature shedding. The severely affected leaves may be twisted and malformed.

Favourable Conditions

Dry humid weather and low relative humidity.

Mode of Spread and Survival

The fungus is an obligate parasite and disease penetrates through cleistothecia in the infected plant debris in soil. The ascospores from the cleistothecia cause primary infection.

Management

Remove the infected plant debris and destroy. Spray Wettable sulphur at 2.5 kg/ha or dust Sulphur at 25 kg/ha.

Leaf blight

Alternaria sesami

Symptoms

Initially small, circular, reddish brown spots appear on leaves which enlarge later and cover area with concentric rings. The lower surface of the spots are greyish brown in colour. Dark brown lesions can also be seen on petioles, stem and capsules.

Favourable Conditions

Low temperature (20-25 °C), high relative humidity and cloudy weather.

Mode of Spread and Survival

The fungus is seed-borne and also soil-borne as it remains dormant in the infected plant debris.

Management

Treat the seeds with Thiram or Carbendazim at 2 g/kg. Spray Mancozeb at 1 kg/ha.

Wilt

Fusarium oxysporum f.sp. sesami

Symptoms

The disease appears as yellowing, dropping and withering of leaves. The plants gradually wither, show wilting symptom leading to drying. The infected portions of fungus produces macroconida, microconidia and chlamydospores.

Mode of Spread and Survival

The fungus survives in the soil in the infected plant debris. It is also seed-borne and primary infection occurs through infected seeds or through chlamydospores in soil. The secondary infection may be caused by conidia disseminated by rain splash and irrigation water.

Management

Treat the seeds with Thiram or Carbendazim at 2g/kg. Apply heavy doses of green leaf manure or farm yard manure.

Bacterial leaf spot

Pseudomonas sesami

Symptoms

The disease appears as water-soaked yellow specks on the upper surface of the leaves. They enlarge and become angular as restricted by veins and veinlets. The colour of spot may be dark brown with shiny oozes of bacterial masses.

Mode of Spread and Survival

The bacterium remains viable in the infected plant tissues. It is internally seed-borne and secondary spread through rain-splash and storms.

Management

Keep the field free of infected plant debris. Spray with Streptomycin sulphate.

SUNFLOWER (*Helianthus annuus*)

Leaf blight

Alternaria helianthi

Symptoms

The fungus produces brown spots on the leaves, but the spots can also be seen on the stem, sepals and petals. The lesions on the leaves are dark brown with pale margin surrounded by a yellow halo. The spots later enlarge in size with concentric rings and become irregular in shape.

The fungus produces cylindrical conidiophores, which are pale grey-yellow coloured, straight or curved, geniculate, simple or branched, septate and bear single conidium.

Favourable Conditions

Rainy weather, cool winter climate and late sown crops are highly susceptible.

Mode of Spread and Survival

The fungus survives in the infected host tissues

and weed hosts. The fungus is also seed-borne. The spread is mainly through wind blown conidia.

Management

Remove and destroy infected plant debris. Rogue out weeds at periodical intervals. Sow the crop early in the season (June sowing). Treat the seeds with Thiram or Carbendazim at 2g/kg.

Rust

Puccinia helianthi

Symptoms

Small, reddish brown pustules (uredia) covered with rusty dust appear on the lower surface of bottom leaves. Infection later spreads to other leaves and even to the green parts of the head. In severe infection, when numerous pustules appear on leaves, they become yellow and dry. The black coloured telia are also seen among uredia on the lower surface.

Mode of Spread and Survival

The pathogen survives in the sunflower plants and in infected plant debris in the soil as teliospores. The disease spreads by wind-borne uredospores from infected crop.

Management

Remove and burn the infected plant debris in the field. Remove the volunteer sunflower plants. Spray Mancozeb at 1 kg/ha.

Root rot or charcoal rot

Rhizoctonia bataticola

Symptoms

The fungus is seed-borne and primarily causes seedling blight and collar rot in the initial stages. The grown-up plants also show symptoms after flowering stage. The infected plants show dropping of leaves and death occurs in patches. The bark of the lower stem and roots shreds and associated with a large number of sclerotia. Dark coloured, minute pycnidia also develop on the lower portion of the stem.

Mode of Spread and Survival

The pathogen survives in soil and in infected crop residues as sclerotia and pycnidia. The pathogen is seed-borne and it serves as primary source of infection. Wind-borne conidia cause secondary spread. The soil borne sclerotia also spreads through rain splash, irrigation water and implements.

Management

Treat the seeds with Carbendazim at 2 g or Thiram at 4g/kg or treat the seeds with *Trichoderma viride* at 4g/kg.

Head rot

Rhizopus sp.

Symptoms

The affected heads show water soaked lesions on the lower surface, which later turns brown. The discolouration may extends to stalk from head. The affected portions of the head become soft and pulpy and insects are also seen associated with the purified tissues. The larvae and insects which attack the head pave way for the entry of the fungus which attacks the inner part of the head and the developing seeds. The seeds are converted into a black powdery mass. The head finally withers and droops down with heavy fungal mycelial nets.

Favourable Conditions

Prolonged rainy weather at flowering and damages caused by insects and caterpillars.

Mode of Spread and Survival

The fungus survives as a saprophyte in host debris and other crop residues. The disease is spread by wind blown spores.

Management

Treat the seeds with Thiram or Carbendazim at 2g/kg. Control the caterpillars feeding on the heads. Spray the head with Mancozeb at 1 kg/ha during intermittent rainy season and repeat after 10 days, if the humid weather persists.

CASTOR (*Ricinus communis*)

Seedling blight

Phytophthora parasitica

Symptoms

The disease appears as circular, dull green patch on both the surface of the cotyledonary leaves. It later spreads gradually and cause rotting. The infection moves to stem and cause withering and death of seedling. In mature plants, the infection initially appears on the young leaves and spreads to petiole and stem causing black discolouration and severe defoliation.

Favourable Conditions

Continuous rainy weather, low temperature (20-25°C) and low lying and ill drained soils.

Mode of Spread and Survival

The fungus remains in the soil as chlamydospores and oospores which act as primary source of infection. The fungus also survives on other hosts like potato, tomato, brinjal, sesamum etc. The secondary spread takes place through wind-borne sporangia.

Management

Remove and destroy infected plant residues. Avoid low-lying and ill drained fields for sowing. Treat with Thiram or Captan at 4g/kg.

Rust

Melampsora ricini

Symptoms

Minute, orange-yellow coloured, raised pustules appear with powdery masses on the lower surface of the leaves and the corresponding areas on the upper surface of the leaves are yellow. Often the pustules are grouped in concentric rings and coalesce together to form drying of leaves.

Pathogen

The fungus produces only uredosori in castor plants and other stages of the fungus are unknown. Uredospores are of two kinds, one is thick walled and other is thin walled. They are elliptical to round, orange-yellow coloured and finely warty.

Mode of Spread and Survival

The fungus survives in the self sown castor crops in the off season. It can also survive on other species of *Ricinus*. The fungus also attacks *Euphorbia obtusifolia*, *E. geniculata*, and *E. marginata*. The infection spreads through air-borne uredospores.

Management

Rogue out the self-sown castor crops and other weed hosts. Spray Mancozeb at 1 kg/ha or dust Sulphur at 25 kg/ha.

Leaf blight

Alternaria ricini

Symptoms

All the aerial parts of plants viz., leaves, stems, inflorescences and capsules are liable to be attached by the fungus. Irregular brown spots with concentric rings form initially on the leaves and covered with fungal growth. When the spots coalesce to form big patches, premature defoliation occurs. The stem inflorescence and capsules are also shown dark brown lesions with concentric rings. On the capsules, initially brown sunken spots appear, enlarge rapidly and cover the whole pod. The capsules crack and seeds are also get infected.

Favourable Conditions

High atmospheric humidity (85-90 per cent) and low temperature (16-20°C).

Mode of Spread and Survival

The fungus also survives on hosts like *Jatropha pandurifolia and Bridelia hamiltoniana*. The pathogen is externally and internally seed-borne and causes primary infection. The secondary infection is through air-borne conidia.

Management

Treat the seeds with Captan or Thiram at 4g/kg. Remove the reservoir hosts periodically. Spray Mancozeb at 1 kg/ha.

10.2.8 Horticultural Crops

BANANA (*Musa spp.*)

Panama disease/Fusarium wilt/Vascular wilt

Fusarium oxysporum f.sp. cubense

Symptoms

The infected plants show characteristic yellowing of leaf blades developing as a band along the margin and spreading towards midrib. The leaf wilts and the petiole buckles. The leaf hangs between the pseudostem while the middle of lamina is still green. All the leaves collapse whereas the petioles join the pseudostem and die.

Mode of Spread and Survival

The pathogen is soil-borne. It survives in soil as chlamydospores for longer periods. The primary spread of the disease is through infected rhizomes and secondary spread is through irrigation water.

Management

Infected plants should be removed and destroyed. After removal and destruction of affected plants with the rhizome, the pits have to be treated with lime at 1 to 2 kg/pit.

Anthracnose/Fruit rot

Gloeosporium musarum

Symptoms

The fungus attacks the young banana fruits usually at the distal end. The skin of the fruit turns black and shrivels and becomes covered with characteristic pink acervuli. Finally the whole finger is affected. Later the disease spreads and affects the whole bunch. The disease results in premature ripening and shrivelling of the fruits which are covered with pink spore masses.

Mode of Spread and Survival

The spread of the disease is by air-borne conidia and numerous insects which frequently visit banana flowers also spread the disease.

Management

In the field the distal bud should be removed when all the hands have opened to prevent infection. Infected materials should be burned.

Black spot or black tip

Deightoniella torulosa

Symptoms

The fungus causes three kinds of symptoms *vis.*, black spot on leaves, black tip or black end on fruits and speckle (fruit spot). The earliest symptoms on leaves are round pin point, black spots on the main veins of the lamina, usually close to the leaf margin. These increase in size, become lenticular in the direction of the veins and are usually separated from the normal green leaf tissue by a narrow bright yellow peripheral band.

Mode of Spread and Survival

The conidia are spread by air.

Management

Affected leaves and fruits should not be left on the garden and they should be collected and destroyed along with dead and dried sheaths. Removal and destruction of trashes before heavy rain reduces the level of infection.

Bunchy top

Musa virus

Symptoms

The affected leaf shows green streaks on the secondary veins on the under side of lamina and on the midrib and petiole. The powdery bloom covers the midrib and petiole. If this is rubbed off the dark green streaks can be clearly seen. The streaks vary from a series of dark green dots to a continuous dark green line with a ragged edge.

Mode of Spread and Survival

It is not transmitted mechanically. The major vector is banana aphid *viz.*, *Pentalonia nigronervosa*. It is transmitted in a persistent manner.

Management

The disease management can be attained by the following ways :

1. Selection of healthy suckers for planting.
2. Removal of affected trees and suckers at periodical intervals.
3. Control of aphid vector by suitable chemical methods.

MANGO (*Mangifera indica*)

Malformation- *Fusarium moniliforme var- Subglutinans*

Symptoms

Three distinct types of symptoms are produced. They are bunchy top of seedlings (BT), vegetative malformation (MV) and floral malformation (MF). *Bunchy top phase* (BT) appears on young plants in the nursery beds when they are 4 to 5 months old. The characteristic symptoms of the disease is the formation of a bunch of thickened small shootlets bearing small rudimentary leaves or occasionally several bunches arising from a leaf axil at the top or lower down the main shoot. These shootlets are much thicker than main axis from which they arise. The shoot remains short and stunted. The growth of the plant is stopped and it gives an appearance of bunchy top. *Vegetative malformation* (MV) induces excessive vegetative branches of limited growth in young seedlings. They are swollen and are with very short internodes forming bunches of various sizes. *Malformation of inflorescence* (MF) shows variation in the panicle formation.

Mode of Spread and Survival

Diseased propagation materials help in the spread of the disease.

Epidemiology

The disease is serious in the northwest region where the temperature is between 10°C and 15°C during December-January before flowering.

Management

The disease incidence is reduced by spraying with NAA at 100 to 200 ppm during October. Pruning of diseased parts along with the basal 15 to 20 cm apparently healthy portions, followed by spraying of carbendazim 0.1 per cent or captafol 0.2 per cent effectively control the disease.

Powdery mildew

Oidium mangiferae

Symptoms

It usually occurs during December-March. A whitish bloom of fungal growth covers the stalks of the panicle, flowers and tender fruits. The affected flowers and fruits drop prematurely reducing the setting of fruits. At higher altitudes, the infection extends to the young leaves and twigs. Many of these are covered by the white powdery fungal growth and may exhibit distorted growth.

Mode of Spread and Survival

During off season, the pathogen has been reported in intact green malformed panicles mostly hidden under dense foliage. During flowering period (January-March), the conducive environmental conditions activate the dormant mycelium in necrotic leaves. Abundant conidia are produced and blown over to the new flushes through wind on young panicles which provide sufficient spore load for initiating the disease.

Epidemiology

Rains or heavy mist in the morning accompanied by cool nights during the flowering period favour the disease.

Management

Three or four dustings with sulphur powder at fortnightly interval controls the disease.

Anthracnose

Colletotrichum gloeosporioides

Symptoms

The fungus attacks tender shoots and foliage.

Brown or dark circular or irregular spots are formed on the leaves and such leaves are crinkled. The affected portions dry up and fall off and leave ragged margins. Often these leaves are shed leaving the twigs bare. The infection spreads to the green twigs and forms dark brown lesions on them. Young branches die-back. Older twigs and branches are infected through wounds and are killed.

Mode of Spread and Survival

Inoculum remains on dried leaves, defoliated branches, mummified flowers and flower brackets and they serve as primary inoculum. Secondary spread is through air-borne conidia. The fungus can enter the pores of green fruits.

Epidemiology

Under field conditions, the disease can be found on the fallen leaves and the blighted peduncles frequently remain *in situ* for many weeks. They produce spores under favourable moisture conditions and serve as foci of infection for the succeeding bloom. Even after they had fallen to the ground, they may continue to be the sources of infection for few weeks.

Management

Diseased twigs, leaves and fruits which fall on the ground in the orchard should be collected and all infected twigs should be pruned and burnt. Spraying of Bordeaux mixture 0.6 per cent in the young plants during February, April and September controls the disease.

Sooty mould

Capnodium ramosum

Symptoms

The fungi produce mycelium which is usually superficial and dark. They grow on the flowers, both tender and old leaves, stems and fruits. They grow and thrive on the sugary secretions of the plant hoppers. Black encrustations are formed on the surfaces of different parts of the plant.

Mode of Spread and Survival

Diseased leaves serve as primary inoculum.

Epidemiology

Disease is severe in old and dense orchards where light intensity is low. Trees exposed to eastern side has less incidence while the trees in centre of the orchard has more incidence. High infestation with plant hoppers and the sugary substances secreted by them favour development of sooty mould.

Management

Both the insects and the sooty moulds are to be simultaneously controlled in the eradication process. The insects are killed by spraying with carbaryl or phosphamidon 0.03 per cent.

Red rust

Cephaleuros virescens

Symptoms

The disease is characterised by red rusty spots on the leaves and young twigs. The spots are initially circular, slightly elevated and later coalesce to form irregular spot.

Epidemiology

The disease is more common on close plantations. The zoospores cause initial infection.

Management

It is controlled by spraying with Bordeaux mixture 1 per cent or copper oxychloride 0.1 per cent or lime sulphur.

Anthracnose/Wither tip/Die-back

Colletotrichum gloeosporioides

Symptoms

The plant is affected at all stages. The weakened or injured twigs and branches are generally affected. The branches begin to wither from the tip downwards. The drying back gradually progresses downwards with the leaves turning yellow, withering and drooping and gum formation on the affected stem.

Mode of Spread and Survival

The fungus, survives in a dormant condition in the dead twigs and branches.

Epidemiology

It causes decline of acid lime trees in North India. Deficiency of nitrogen and unfavourable soil conditions make the plant weak and susceptible to the disease.

Management

Dried twigs should be pruned. The cut ends should be protected by painting with Bordeaux paste or any other copper fungicide. Such trees may be sprayed thrice with carbendazim 0.1 per cent or captafol 0.2 per cent after pruning. The trees should be adequately manured with urea.

Powdery mildew

Acrosporium tingitaninum

Symptoms

All the aerial plant parts including young and actively growing leaves and twigs are affected. On the upper surface of the leaves white powdery patches of fungal growth are seen. The patches extend and cover the whole leaf. Petioles and stems are also covered by whitish fungal growth.

Mode of Spread and Survival

The fungus spreads through wind-borne conidia. The water shoots under the canopy are first infected.

Epidemiology

The disease appears usually from October to march and in higher elevations. Damp mornings with few hours of sun shine are the most favourable conditions for the onset of disease.

Management

The affected plant parts should be removed and destroyed carefully. Water shoots should be pruned to reduce infection. Dusting finely powdered sulphur gives effective control.

Scab/Verucosis

Elisnoe fawcetti

Symptoms

The disease attacks leaves, twigs and fruits. The lesions on the leaves in early stages consist of small, semi-translucent dots which become sharply defined pustular elevations usually on the under side, flat or somewhat depressed at the centre.

Mode of Spread and Survival

The pathogen perpetuates and survives in off-season as ascospores. The fungus spreads through conidia. It attacks sour orange, rough lemon and pummelo.

Epidemiology

The fungus infects tissues only when the surface is wet but prefers 16 to 20 °C temperature.

Management

The diseased leaves, twigs and fruits should be collected and destroyed. Spraying with Bordeaux mixture 1.0 per cent is quite effective.

GRAPEVINE (*Vitis Vinifera*)

Powdery mildew-*uncinula necator*

Symptoms

The fungus attacks all the stages of the crop growth. The characteristic symptom of this disease is the appearance of white powdery patches on affected parts. On the leaves the disease starts as small, white patches which later become larger in size and powdery in appearance. Sometimes the entire leaf is covered with dusty white fungal growth..

Mode of Spread and Survival

It survives as dormant mycelium and as

cleistothecia on the shoots and buds from season to season. The disease spreads by the air-borne conidia.

Epidemiology

The disease occurs in severe form during October-November in North India and February-June in South India. Disease is favoured by warm weather and retarded by sunshine.

Management

Overcrowding and dense growth of the vines should be avoided by proper pruning. Effective control can be achieved by spraying wettable sulphur 0.2 per cent or dinocap 0.7 per cent or carbendazim 0.1 per cent or sulphur dusting at berry development stage with a prophylactic treatment.

Downy mildew

Plasmopara viticola

Symptoms

The disease is usually first observed as small translucent, pale yellow spots with indefinite borders on the upper surface of leaves. On the under surface of leaves and directly under the spots, a downy growth of the fungus appears.

Mode of Spread and Survival

The pathogen survives on the infected leaves and vines as oospores and dormant mycelium. The secondary spread is through wind-borne sporangia and zoospores which are found on the new flush.

Epidemiology

Disease development is favoured during rainy season when there is a heavy dew, relative humidity is above 80 per cent and temperature is between 23°C and 27°C.

Management

Sanitation is very important in the management of this disease. Removing and burning of all diseased leaves, shoots, flowers and berries that may contain hibernating oospores help in preventing epidemics. Vines should be planted with proper spacing and should be trained in such a manner that leaves do not remain near the ground surface.

Anthracnose/Bird's eye spot

Elsinoe ampelina

Symptoms

The fungus attacks shoots, tendrils, petioles, leaves, veins and stems. Numerous spots occur on the young shoots. These spots may unite to girdle the stem and cause death of the tips. Spots on petioles and leaves cause them to curl or become distorted.

Mode of Spread and Survival

The pathogen survives as mycelium in the cankers on the stem and on the infected twigs. Secondary spread is through conidia which are carried by wind and rain water.

Epidemiology

The disease is severe during July-August and November-December months. Infection in new sprouts takes place during rainy season.

Management

The diseased leaves and twigs should be pruned and burnt. Spraying pruned the canes and leaves with a mixture of ferrous sulphate 2.5 kg + 0.5% sulphuric acid in 4.5 litres of water controls the disease.

Bacterial canker

Xanthomonas campestris pv. viticola

Symptoms

The disease starts as small water-soaked spots surrounded by yellowish halo at lower surface of the leaf. These spots enlarge in size and become dark brown and angular. Sometimes these spots coalesce to form larger patches. Leaves also show vein infection. Infected leaves after drying remain attached to the stem. Lesions are brown to black, elongated and cankerous on petioles and canes.

Mode of Spread and Survival

The alternate hosts are neem and mango. The bacterium survives in the infected, dry leaves upto 65 days. Secondary infection takes place through wind splashed rain. Disease spreads to distant places by diseased cuttings.

Epidemiology

Temperature range of 25 to 30 °C is favourable for the disease development. Free water from dew, irrigation or rain on leaf are more important for pathogenesis.

Management

Regular inspection of vineyard, destruction of infected plant materials, use of disease-free cuttings and late-October pruning are recommended for its management.

GUAVA *(Psidium guajava)*

Fusarium wilt

Fusarium oxysporum

Symptoms

The disease is characterised by yellowing and browning of leaves, discolouration of the stem and death of the branches along one side. Sometimes the infection girdles the stem and the whole plant may wilt. Leaves die and the twig barks split.

Mode of Spread and Survival

The fungus first colonises on the surface of the roots and enters the stem tissues at the basal portions near the ground level. It multiplies in vascular region and affects the cortical cells.

Epidemiology

Higher disease incidence is noticed during the monsoon period. The disease appears in August and increases sharply during September-October. It is severe in alkaline soils.

Management

Dry branches should be cut off and wilted plants uprooted. Soil should be treated with lime or gypsum to make the soil pH 6.0 to 6.5. Balanced nutrition of host reduces severity of the disease when organic nitrogen is supplied.

Fruit canker/Scab/Grey blight

Pestalotiopsis psidii

Symptoms

Infection generally occurs on green fruits. Minute, brown or rust-coloured, unbroken, circular, scabby lesions of 2 to 4 mm dia, appear on the fruit which later tear the epidermis open in a circinate manner. The margin of the affected area becomes raised.

Mode of Spread and Survival

Infected leaves and fruits serve as primary inoculum. Secondary spread is through the wind-borne conidia.

Epidemiology

The fungus is capable of growing at temperature between 20 and 25 °C. Mycelial growth with intensive sporulation takes place at 5.5°C.

Management

Spread of the disease can be checked by three or four spraying with Bordeaux mixture 1.0 per cent or copper oxychloride 0.2 per cent.

Anthracnose/Die-back/Fruit spot/Twig blight

Gloeosporium psidii

Symptoms

Severity of the disease may show die-back of

main branches resulting in death of plants. The most characteristic symptoms appear during the rainy season as small pin-head sized spots on the unripe fruits. They gradually enlarge to form sunken and circular, dark brown to black spots.

Mode of Spread and Survival

The pathogen remains dormant for about three months in the young infected fruits. It becomes active and incites rot when the fruit begins to ripe. In moist weather, acervuli appear as black dots scattered throughout the dead parts of the twigs. From the twigs, the fungus penetrates the petioles and attacks the young leaves, which become distorted with dead areas at margins or tips.

Epidemiology

The cool season (January-March) and the hot, dry weather (April-June) prevent the spread of infection. In moist weather, acervuli are produced in abundance on dead twigs and pinkish spore masses are seen. Conidia initiate fresh infection.

Management

Spraying the trees with Bordeaux mixture 0.6 per cent or copper oxychloride 0.2 per cent before the onset of monsoon reduces the disease incidence.

APPLE (*Pyrus malus*)

Collar rot

Phytophthora cactorum

Symptoms

Important symptoms of collar rot are production of rough patches on the stem around the collar or crown region of the tree at soil level which develop into cankers. Rapid extension of the cankered area, both in lateral and vertical directions, results in girdling of the tree.

Mode of Spread and Survival

The pathogen survives as oospores in the soil.

Epidemiology

A soil temperature of 12°C to 20°C with pH of 5 to 6 is found to be the best for the survival of the fungus.

Management

Grafting of rootstocks 30 cm above the ground level reduces initial infection. The disease can be controlled by painting wounds with copper paints and drenching the area around the trees with fungicides like mancozeb or metalaxyl.

Stem cankers

a. Black canker *sphaeropsis malorum*

Symptoms

It is also known as *black rot, die-back* or *smoky blight canker*. The disease appears in three phases, *i.e.*, stem canker, leaf spot (frog eye leaf spot) and fruit rot (black rot). Among them the canker phase is the most destructive. In this, reddish brown sunken lesions develop on trunk and branches. The lesions turn smoky with alternate rings. The wood below is stained reddish brown.

Mode of Spread and Survival

The pathogen survives winter in the form of dormant mycelium and fruiting structure in cankered.

Epidemiology

High humidity and a temperature range of 20 to 22 °C favour the disease development.

Scab-*venturia inequalis*

Symptoms

The most striking symptoms of scab are commonly observed on leaves, fruits and rarely on 1 to 3 years old shoots. In severe cases, petioles and blossoms also exhibit scab symptoms. On leaves, lesions first develop on the lower side. Most of these lesions do not have a definite margin in contrast to those on the upper side of the leaf. Lesions appear as olivaceous spots which turn dark brown to black and become velvety.

Mode of Spread and Survival

Pseudothecia formed in autumn and winter mature in spring to produce ascospores, the chief inoculum for primary infection. The secondary spread is through conidia.

Management

Clean cultivation reduces primary infection. Fallen leaves should be collected and destroyed along with pruned materials. Control programmes should have aim of elimination/reduction of pseudothecial formation and secondary infections.

Powdery mildew *Podosphaera leucotricha*

Symptoms

The disease appears soon after the buds develop into new leaves and shoots. Sometimes buds are so heavily infected in the previous season that they are killed before developing into leaves. The disease may be seen on both surfaces of leaves in severed infections.

Mode of Spread and Survival

The fungus overwinters in the form of mycelium in diseased vegetative buds and fruits. Secondary spread is through wind-borne conidia.

Epidemiology

The conidial germination takes place when air is saturated and the temperature is between 10 and 25 °C.

Management

Pruning after shedding of leaves, thinning out and cutting back of laterals, removal and destruction of all diseased parts are important in the control of powdery mildew.

10.3.1 Plant Disease Management

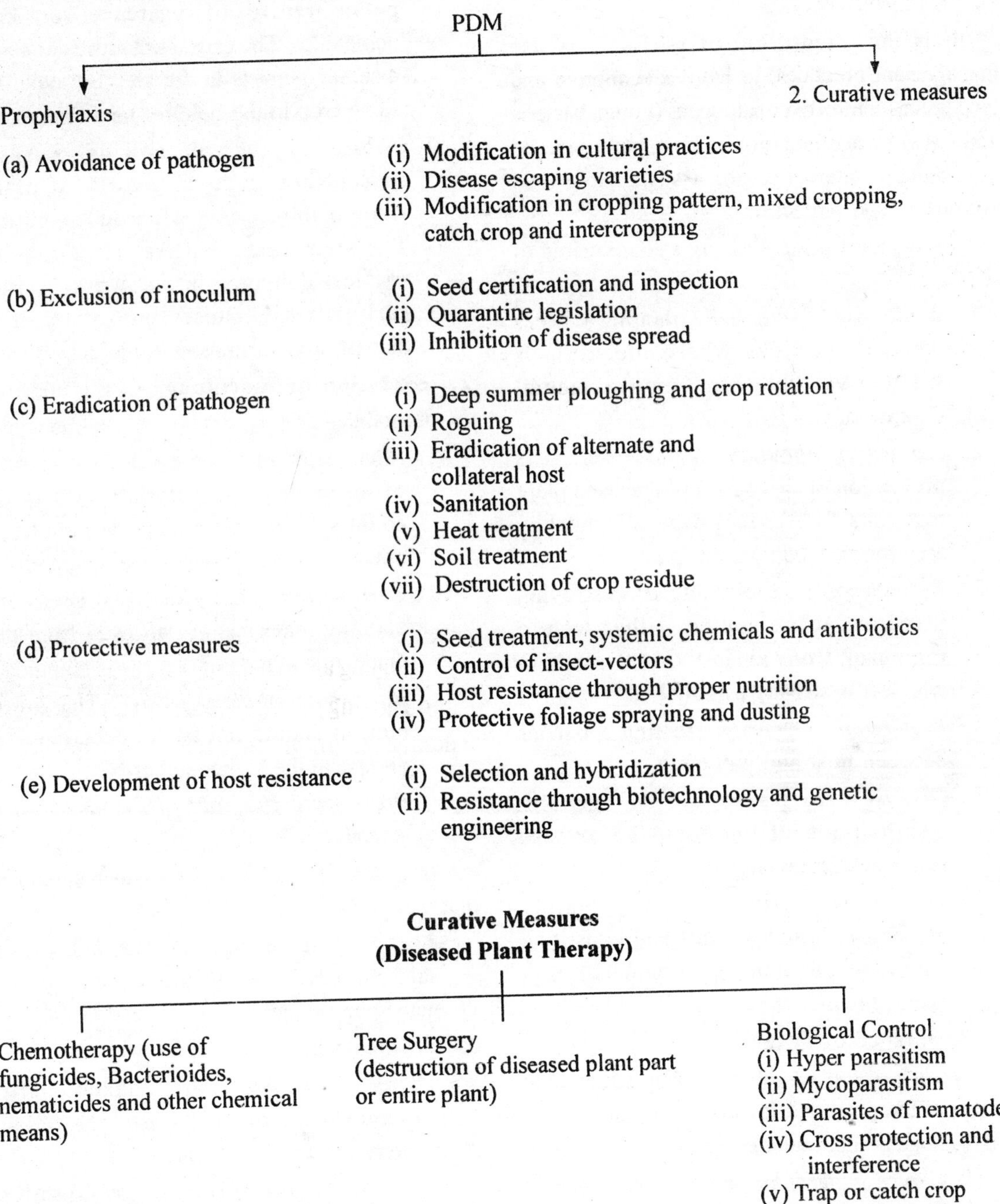

10.3.2 Integrated Disease Management (IDM)

It is the integration of various disease management practices, to have a economic and long lasting control over pathogens (Fungi, bacteria and virus) by manipulation of host, pathogen and environment interaction for a sustainable crop production and protection.

The general principles (or) components of IDM are :

1. *Avoidance of pathogen* : planting at times when or in areas where inoculum is ineffective due to environmental condition.
2. *Exclusion inoculum (legislation)*: It is prevention of movement of diseased plant materials from infected area to healthy area through legislation.
3. *Eradication* : Eliminating or destroying inoculum at the source, either from a region or from an individual plant in which it is already established.
4. *Protection* : It means creating a barrier between host and pathogen.
5. *Disease resistance* : by genetic manipulation of the host, the disease resistance is developed.

   The above principles are mainly preventing (prophylactic) and constitute the major procedures for plant disease management.

1. Avoidance of Pathogen

Avoid host contact with pathogen. Susceptible stages of the plant and favourable conditions for the pathogens should not coincide.

(*i*) *Choice geographic area* : many bacteria diseases are severe in wet areas. Smut and ergot diseases of bajra are severe in wet regions. So avoid planting bajra in wet areas.

(*ii*) *Selection of the field* : Becterial unit of potato, red rot of sugarcane, root knot nematodes. The causal organisms of above diseases persists in the soil. So avoid the same crop in the field for next time.

(*iii*) *Choice of time of sowing* : Avoid susceptible stage of growth of plant, when pathogen is in vigorous condition. The stem rust of wheat (P. Graminis trictici) damages late sown crop then early sown because time of onset of disease and formation coincide.

2. Exclusion of Inoculum

It includes :

(*i*) *Plant quarantine* : It prevents the movement of diseased plant materials with in the country (domestic quarantine) and from other countries (International quarantine).

(*ii*) *Certification* : Only certified seeds and planting materials are allowed from one place to another.

(*iii*) *Notification of plant disease* : The grower himself should notify the occurrence of disease to the higher authority

(*iv*) *Preventing the sale of diseased plant materials* :

It is done to prevent the further spread of disease.

e.g. : Potato seed tubers are not allowed for sale from Nilgiri's to prevent the golden nematode spread.

3. Eradication

The measures are taken after establishment of the disease. The various measures are

(*i*) *Eradication of alternate host* : Alternate host serves for the pathogens to complete its life cycle by eradicating alternate host, we can control several pathogens *e.g.* : To control wheat stem rust eradicate barbery plants. Brinjal serves as alternate host for cumbu rust

(*ii*) *Eradication over wintering host* : Several crops serve as over wintering host for pathogens during off-season.

e.g. : Eradicate Brachiaria mutica and Panicum ripens to control rice blast pathogen (Pyricularia oryzae) Bean yellow mosaic over winters sweet clover. Little leaf of brinjal-pathogen over winters Datura and vinca. Eradicate all over witntering hosts.

(*iii*) *Eradication of diseased plants* : By removing the diseased plant parts the inoculum potential of the pathogen is reduced

(*iv*) *Eradication of diseased plant parts* : Remove bleeding patches in the stem of coconut to control Ceratocystis parodoxa (stem bleeding of cocount) by knife and then sterilisze it with fire and apply Bordeaux paste.

(*v*) *Sanitation* : Burning of crop residue-we can control RTV, sheath blight of paddy. In sugarcane smut (*Ustilago scitaminea*) collect the infected whips and burn it.

Avoid smoking in the tobacco field to prevent the spread of TMV. By rouging RTV in the rice field can be controlled.

(*vi*) *Sterilization* : soil borne pathogens can be controlled by soil sterilization.

(*vii*) *Cultural practices.*

(*a*) *Crop rotation* : Mono cropping is the main cause for the disease severity in the fields. The crop rotation is effective when the pathogen is soil invader (i.e. pathogen remain in the soil for short period as long as host is present)

e.g. : Fusarial pathogens and Damping off caused by Pythium, etc.

(*b*) *Wider spacing:* closer spacing enhances disease development due to higher humidity. By wider spacing we can control rice blast, sheath blight of rice and bacterial blight of rice.

(*c*) *Avoid growing of susceptible varieties*: By this we can reduce the inoculum potential .

e.g. : IR-50 Susceptible to rice blast, Co-43, IR-20 are susceptible to sheath blight of rice.

(*d*) *Raising raised nursery bed* : To avoid damphing-off in chillies, tobacco caused by Pythium use raised nursery beds for seedling raising.

(*e*) *Mixed cropping* : Cotton root rot can be controlled by mixed cropping with Phaseolus aconitifolius.

(*f*) *Trap Cropping* : Striga (partial root parasite) is controlled by trap crop like sudan grass. Orabanche (total root parasite) in tobacco field is reduced by chilli (trap crop). The alkaloid present in sudan grass and chilli reduce the germination of Striga and Orabanche

(*g*) *Adjusting the date of sowing* : Sugary disease of cholam and cumbu is severe during winter (Nov-Feb). So adjust date of sowing that flowering should not coincide with winter periods.

- Rice blast is severe in winter months. So do not plant IR-50, TKM 9 rice varieties in winter (Samba season).

(*h*) *Adjusting soil pH* :

- Alkaline condition favours Streptomyces scabies (potato scab). Apply gypsum to reduce the pH and to control this disease.

- Acid condition favours *Plasmodiophorum brassicae* (club root of cabbage). So apply lime to increase the pH to control this disease.

(*i*) *Adjusting irrigation* : High soil moisture favours damphing-off, high soil temperature favours dry root caused by Macrophomina phaseolina (Dry root rot of sorghum, pulses) in summer month's. Give frequent irrigations to control this disease. By reducing soil moisture we can control damphing-off.

(*j*) *Adjusting fertilizer application* : High 'N' favours blast, sheath blight and bacterial blight of rice, So minimise the disease incidence, apply more potash.

- Heavy application of FYM reduces soil borne pathogens, FYM favours multiplication of antagonistic fungi.

(*k*) *Deep Ploughing* : summer ploughing expose soil borne sclerotia of Rhizoctonia solani, Sclerotium rolfsi, under direct sunlight. Which is used to control above pathogens.

(*l*) *Trenching* : Movement of disease is prevented from infected area to healthy area

e.g.. For coconut wilt make trenches 3-4 feet away from base of the trunk to prevent the further spread.

4. Protection

It includes physical, biological and chemical methods of control.

(*i*) *Physical method* : It includes

(*a*) Hot water treatment

(*b*) Cold storage

(*c*) sun drying

(*d*) store in air tight container

(*e*) Irradiation.

(*a*) *Hot water treatment*: This adopted for the control of internally seed borne diseases. *Jensen (1886)* evolved it for the control of loose smut of wheat. Soak the seeds in hot water at 54°C for 10-15 min. or 4-5 hrs in cold water.

- *For Helminthosporium* : leaf spot and bacterial leaf blight of paddy, soak the seeds in hot water at 50°C for 10 min.
- *Sugarcane mosaic and smut* : Sett treatment in 55°C for hours.

(*b*) *Cold water treatment of cold storage* : Fruits and vegetables are stored at 1-5°C in cold storage to prevent the storage diseases.

(*c*) *Store in air tight containers* : Soak the seeds in cold water for 4 hours and put it in air tight containers for 48 hours. It creates anaerobic condition to kill the spores.

(*d*) *Irradiation* : Fruits and vegetables are irradiated in UV rays or X-rays to prevent the post harvest or storage diseases.

10.3.3 Fungicide

A good fungicide should not only be toxic to the parasite or inhibit the germination of its spore without causing phytotoxicity, but should have following qualities also:

1. High field performance
2. Inherent fungitoxicity
3. Availability of the active constituents
4. Good coverage of host surface
5. Low phytotoxicity
6. Stability in storage
7. Stability after dilution to spray strength
8. Low toxicity to human beings and cattles
9. Should stick to the surface after drying

10. Good wetting property
11. Compatibility with pesticides, nematicides, herbicides, viricides and fertilizers.

Fungicides and their Methods of Applications

According to their chemical nature and formulations the fungicides may be divided into following categories:

1. *Spray materials* : Materials available in solutions the fungicides may be divided fungicides are used as spray which provide a protective covering over the surface.
2. *Dust treatment materials* : Insoluble powders are dusted over the surface.
3. *Seed treatments materials* : seed treatment with fungicides are done with dry powder dips or slurries.
4. *Soil treatment fungicides* : Fungicides to control soil borne diseases in certain cases are used as soil treatment.
   There are following methods of soil treatments:

- Drenching of soil with solution or suspension.
- Soil fumigations such as chloropicrin, methyl bromide.
- Furrow application.
- Broadcasting of dusts, powders and granules.

5. *Treatment of tubers, bulbs and seed-materials* : potato scab and black scurf disease are controlled by treating the seed materials in solutions of mercuric chloride, farmaldehyde etc. Tubers are also dip with mercury fungicides like aretan, emison etc.
6. *Treatment of tree wounds or tree surgery*: Cut-wounds are resulted from pruning or by cutting of diseased part of a plant. Such wounds may require wound dressing so as to avoid infection of cut parts. Therefore cut wounds are sterilized first by swabbing of 0.5 to 1.0 p.c. of sodium hypochloride or 70 p.c. ethyl alcohol or with 1: 1000 solution of mercuric chloride. Wound parts so sterilized are dressed with mixture of lanolin 10 parts, rosin 2 parts and gum 2 part. Bordeaux paste or Chaubattia paste are also used as wound dressing.
7. *Disinfection of ware houses* : Floors and walls of warehouses should be disinfected with copper sulphate solution.
8. *Control of insect-vectors* : Most of the virus diseases are transmitted by insects known as vectors. Such insect vectors should be controlled by use of suitable insecticides or through biological control in order to avoid transmission of viral diseases.

Classification of Fungicides

According to Chemical nature	Mode of action	According to general uses.
1. Sulphur fungicide	1. Protectant	1. Seed protectant
2. Copper fungicide	2. Therapentant	2. Soil fumigants
3. Mercurial fungicide	3. Eradicant	3. Foliage protectant
4. Systemic fungicide		4. Antibiotics
		5. Miscellaneous fungicides

Systemic Fungicides

systemic fungicides are those which are absorbed by the plants and remain present in protoplasms of different systems of the plant, making it to be artificially immune to parasitic infection. Some of the systemic fungicides are Plantvax, vetavax and benlate etc.

Common Chemicals used as Fungicides

Sulphur fungicides —

1. *Inorganic sulphur* : Sulphur dust, wettable sulphur, lime sulphur etc.
2. *Organic sulphur* : Such as zinc dimethyl dithiocarbamate used as spray.

Copper Fungicides : Blitox-50, Gupramar, Fytolan and Copper Sandoz 53.

Organo-Mercurial Fungicide : Ceresan and emison etc. are commercial names and are used as seed dressers to control externally seed borne diseases of different crops.

Non-Mercurial Organic Fungicides : Dethane-Z-78 (Zinc ethylin bisdithio-carbamate) is usually used as spray to control leaf spot and blight diseases of fruits, vegetables and flowers.

10.3.4 Biological Control

Garret (1965) explained biological control as methods of controlling a pathogen by any other organism and thus to reduce the infectivity of the pathogen is known as biological control. There are following important methods practiced as biological control.

1. **Biological control through hyperparasitism**

(*a*) Bacteriophages or bacterial viruses

(*b*) Mycoparasitism

(*c*) Parasites of nematodes

2. **Through cross protection and interference**
3. **Through trap and catch crop**
4. **Through antagonistic plants**
5. **Entomogenous fungi**

2. **Control through cross protection and interference :** When any plant is infected by any mild strain of a virus the same plant escapes to be infected by the virulent stains of the same virus, such reaction is known as cross protection. If any plant is susceptible to two different strains of a virus-one being mild and other virulent. The plant can be saved to be infected by virulent strain by vaccinating the plant with mild stream virus
3. **Disease control through trap or catch crop :** There are some crop plants which are free from infection of some nematodes and secrete such biochemical substances which initiate hatching of eggs of nematodes and disrupt the life cycle of nematodes resulting in their death. Such crop is called trap or catch crop.
4. **Disease control through antagonistic plants :** some plants like asparagus and marigolds are antagonistic to some nematodes as their roots secrete some toxic substances lethal to nematodes. In order to minimize the population of nematodes these antagonistic plants should be grown with the major crop.
5. **Entomogenous fungi :** Insecticides are not always effective to control some of the insect-pests. Besides, they also kill predators along with the pests. Keeping in view the demerits of insecticides, biological control of pests with entomogenous fungi may be suggested.

10.3.5 Control of Plant Diseases by Modification of Cultural Practices

Cultural practices are modified in such a way that the growth of the plant is not affected and the pathogen is directly or indirectly affected due to unsuitable environment conditions created by the modification of cultural practices. Some of the important practices my be summarized as under:

1. **Crop rotation :** The elimination of pathogen may be brought through crop

rotation which may result due to the absence of suitable host to the pathogen till the period of its survival or due to toxic effects of root exudates of the intervening crops.

2. **Mixed cropping :** Growing a susceptible crop mixed with a non-susceptible crop have several advantages such as reduction of disease susceptible host surface and also acts as barrier in disease spread in the field especially for vectors.

3. **Spacing and method of sowing :** Spacing between plants sometimes helps in reducing disease incidence due to reduced atmospheric humidity. Pathogens which flourish in high humidity are discouraged. Similarly regulation of depth of planting may help the young plants to escape infection.

4. **Alternation of date of planting and irrigation :** Some diseases are very destructive when susceptible stage of the host and predisposing factor for pathogenesis coincide. Therefore, alternation of date of planting and irrigation frequencies may be helpful in some diseases to escape the disease.

5. **Drainage :** Accumulation of water in the field helps survival of inoculum, therefore proper drainage is essential to keep the field free from disease organisms as proper drainage makes the plants vigorous and resistant.

6. **Organic soil amendments and fertilizer application :** Decomposition of organic matter have a sanitary effect in the soil by inhibition and arrestation of pathogens by producing a number of toxic substances lethal to most of the pathogens. Similarly judicial application N, P and K, help in better and quicker development of roots which reduces the disease incidence.

7. **Destruction of crop residues and roguing :** Roguing of diseased plants or parts and eradication of alternate, collateral and weeds should be done so as to control most of the soil borne pathogens.

8. **High ridging :** High ridging upto 10-15 cms. above the roots have been found to be beneficial against late blight of potato.

9. **Catch crop and trap cropping :** Catch crop and trap crop have been found effective to control some of the diseases incited by nematodes. Under such circumstances these crops should be cultivated in nematode manifested field.

   Catch crop is fast growing crop that is grown between successive plantings of a main crop or between two rows of regular crops in the same season.

   Trap crop is planted to attract insect pests from another crop especially the one in which the pests fail to survive or reproduce.

10. **Adjustment of soil reaction** is essential to control some of the nematode and bacterial diseases.

LIST OF DISEASES

Diseases of Cereals

I. Rice

1. Blast - Pyricularia oryzae
2. Brown spot - Helminthosporium oryzae
3. Bakanae/Foot rot/Fusarium blight – Fusarium moniliforme.
4. Stem rot – Sclerotium oryzae
5. Sheath blight – Rhizoctonia solani
6. Sheath rot – Sarocladium Oryzae
7. Udbatta disease – Ephelis oryzae
8. False/Green smut – Ustilaginoidea virens
9. Stackburn disease – Trichoconis padwickii
10. Bunt/Kernal smut/Black smut – Tilletia barcylayana.
11. Bacterial blight – Xanthomonas oryzae PV oryzae.
12. Bacterial leaf streak – Xanthomonas Compestris P.V. Oryzicola
13. Yellow dwarf – Mycoplasma like organism (MLO)

II. Wheat

1. Black/stem rust – Puccinia graminis tritici
2. Leaf, Brown/Orange rust – Puccinia recondita
3. Yellow/stripe rust – Puccinia striiformis
4. Loose smut – Ustilago tritici
5. Flag smut – Urocystis tritici
6. Stinking smut – Tilletia caries
7. Karnal bunt – Neovassia indica
8. Foot rot – Pythium graminicolum
9. Leaf blight – Alternaria triticina
10. Leaf blotch – Septoria tritici

III. Sorghum

1. Leaf spot – Cercospora sorghi
2. Anthracnose – Colletotrichum graminicolum
3. Leaf blight – Helminthosporium turcicum
4. Smut :
   (*i*) Covered/Kernal/short/grain – Sphacelotheca sorghi.
   (*ii*) Loose – Sphacelotheca cruenta
   (*iii*) Head – Sphacelotheca reiliana
   (*iv*) Long – Tolyposporium ehrenbergii
5. Downy mildew – Pernosclerospora sorghii
6. Charcoalrot – Macrophomina phaeseolina
7. Ergot/sugary disease – sphacelia sorghi
8. Phenerogamics parasite – striga asiatica.

IV. Pearl Millet

1. Downy mildew – Scelerospora graminicola
2. Rust – Puccinia penniseti
3. Smut – Tolyposporium pencillariae
4. Ergot – claviceps fusiformis

V. Maize

1. Smut – Ustilago maydis
2. Brown spot – Physoderma maydis
3. Rust – Puccinia sorghi
4. Leaf blight – Helminthosporium maydis
5. Downy mildew – Peronosclerospora sorghi

Sugarcane

Fungal Diseases :

1. Red Rot - Colletotrichum falcatum
2. Smut – Ustilago Scitaminea
3. Settrot (or) Pineapple disease-ceratocystis paradoxa

4. Wilt – Cephalosporium sacchari.
5. Rust – Puccinia erianthi
6. Downy Mildew – Peronosclerospora sacchari
7. Damping off – Pythium aphanidermatum
8. Eye spot – Helminthosporium sacchari
9. Ringspot – Leptosphaeria sacchari

Bacterial Diseases of Sugarcane

1. Gummosis – Xanthomonas campestris P.V. Vasculorum
2. Red Stripe – Pseudomonas rubrilineans
3. Leaf Scald – Xanthomonas albilineans

Viral Diseases of Sugarcane

1. Mosaic – Sugarcane mosaic virus
2. Spike

M.L.O. Diseases of sugarcane

1. Grassy shoot
2. White leaf

Oil Seeds

Groundnut

Fungal Diseases —

1. Early leaf spot : Cercospora arachidicola
2. Late leaf spot : Phaeoisariopsis Personata.
3. Rust – Puccinia arachidis.
4. Collar rot/seedling blight/crown Rot-Aspergillus niger and Aspergillus pulverulentum.
5. Root rot – Macrophomina phaseolina.
6. Stem rot – Corticium rolfsii
7. Wilt – Fusarium oxysporum
8. Anthracnose – Colletotrichum dematium
9. Grey mould – Botrytis cinerea.
10. Yellow Mould – Aspergillus flavus

Bacterial Disease of Groundnut

1. Bacterial wilt – Pseudomonas solanacearum

Viral Diseases of Groundnut

1. Rosette virus
2. Peanut spotted wilt/Bud Necrosis/ Groundnut ring Mosaic/Tomato spotted wilt virus.

Gingelly

1. Root rot/stem rot/charcoal rot- Macrophomina phaseolina
2. Leaf Blight – Alternaria sesami
3. Leaf spot – cercospora sesami
4. Wilt – Fusarium oxysporum F.sp. sesami.
5. Stem blight – Phytophthora parasitica var. sesami
6. Powdery mildew – Erysiphe cichoracearum.
7. Anthracnose – Colletotricum. sp.

Castor

1. Seedling Blight – Phytophthora parasitica
2. Rust – Melampsora ricini
3. Leaf Blight – Alternaria ricini
4. Brown leaf spot – cercospora ricinella.
5. Powdery mildew – Leveillula taurica
6. Stem rot – Macrophomina phaseolina
7. Bacterial leaf spot – Xanthomonas Campesteis P.V. ricinicola.

Fibre Crops

Safflower

1. Rust – Puccinia carthami
2. Leaf spot – cercospora carthami
3. Leaf Blight – Alternaria Carthami

Coconut

1. Bud rot – Phytophthora palmivora
2. Basal stem rot/Thanjavur wilt/Bole rot-Ganoderma lucidum.
3. Stem bleeding – Ceratocystis paradoxa.
4. Grey leaf blight – Pestalotia palmarum

M.L.O. Disease of Coconut

1. Kerala wilt/Root wilt disease.
2. Tatipaka Disease.

Cotton

1. Wilt – Fusarium oxysporum f.sp. vasin fectum.
2. Verticillium wilt – Verticillium dahaliae
3. Root Rot – Rhizoctonia-bataticola
4. Anthracnose – Colletotrichum capsici
5. Boll rot – Fusarium moniliforme.
6. Leaf Blight – Alternaria macrospora
7. Leaf spot – cercospora gossypina
8. Myrothecium leaf spot – Myrothecium roridum
9. Rust – Phakopsora desmium
10. Sooty mould – capnodium sp.
11. Bertical blight – Xanthomonas campestris P.V malvacerum

Diseases of Tabacco and Fruit Crops

Tobacco

1. Damping off – *Pythium aphanidermatum.*
2. Black shank – *Phytophthora parasitica* var. *nicotianae*
3. Anthracnose – *Colletotrichum tabacum*
4. Powdery mildew – *Erysiphe cichoracearum* var. nicotianae.
5. Frog eye spot – *Cercospora nicotianae.*
6. Brown spot – *Alternaria longipes.*
7. Tabacco mosaic – Tobacco mosaic virus.

Mango

1. Powdery mildew – *Oidium mangiferae*
2. Anthracnose – *Colletotrichum gloeosporioides.*
3. Grey blight – *Pestalotiopsis mangiferae.*
4. Stem end rot – *Diplodia natalensis*
5. Black mould – *Aspergillus niger.*
6. Sooty mould – *Capnodium ramosum.*

Banana

1. Panama wilt/Fusarium wilt – *Fusarium oxysporum* f. sp *cubense.*
2. Anthracnose – *Gloeosporium musarum.*
3. Sigatoka leaf spot – *Mycosphaerella musicola*
4. Cigar-end rot – *Verticillium theobromae.*
5. Crown rot – *Botryodiplodia theobromae.*
6. Bunchy top – Musa virus.

Citrus

1. Gummosis – *Phytophthora palnivora.*
2. Anthracnose – *Colletotrichum gloesporioides.*
3. Bacterial canker – *Xanthomonas axonopodis pv. citri.*
4. Powdery mildew – *Acrosporium tingitanium.*
5. Scab – *Elsinoe fawcetti.*
6. Tristeza or quick decline – Citrus tristezavirus.
7. Citrus Greening – *Rickettsia-like organsim.*

Guava

1. Wilt – *Fusarium oxysporum f. sp psidii.*
2. Anthracnose : *Gloeosporium psidii.*
3. Fruit canker – *Pestalotiopsis psidii.*

Papaya

1. Damping off – *pythium spp.*
2. Stem rot – Pythium aphanidermatum.
3. Powdery mildew – *Oidium caricae.*
4. Anthracnose – *Colletotrichum papayae.*
5. Mosaic – Papaya mosaic vinis.

Grapes

1. Downy mildew – *Plasmopara viticola.*
2. Powdery mildew – *Uncinula necator.*
3. Anthracnose – *Elsinoe ampelina.*

Ber

1. Powdrey mildew – *Oidium erysiphoides.* f.sp. zizyphi.

Pineapple

1. Mealybug wilt – Pineapple wilt virus.
2. Heart rot – *Phytophthora, Cinnamomi* and *P. parasitica.*

Pomegranate

1. Leaf spot – *Cercospora punicae.*

Sapota

1. Leaf spot – *Pestalotiopsis versicolor.*
2. Sooty mould – *Capnodium* sp.

Jack fruit

1. Fruit rot – *Rhizopus artocarpi*
2. Leaf spot – *Collelotrichum gloeosporioides.*

Apple

1. Apple scab – *Venturia inaequalis*
2. Powdery mildew – *Podosphaera leucotricha.*
3. Fire blight – *Erwinia amylovora*

Pear

1. Fire blight – *Erwinia anylovera.*
2. Pseudomonas blight – *Pseudomnonas syringae.*

Peach

1. Peach leaf curl – *Taphrina deformans.*
2. Bacterial canker – *Pseudomonas syringae.*

Plum

1. Wilt – Verticillium alboatrum.
2. Leaf curl – Taphrina deformans

VEGETABLES

Tomato

1. Bacterial wilt– *Pseudomonas solanacearum*
2. Damping off & root rot – *Pythium ultimatum, Rhizoctonia solani, Pythium parasitica*
3. Early blight – *Alternaria solani*
4. Fusarium wilt – *Fusarium oxysporium lycopersci*
5. Tomato leaf curl – *Leaf curl virus*
6. Spotted wilt (TSWV) – *Tomato spotted wilt virus*
7. Mosaic – *Tobacco mosaic virus*
8. Fruit cracking – When rain follows a **long dry spell** & high day temperature (concentric & radial are followed by low temperature with high Relative Humidity very common). Deficiency of **boron**

Brinjal

1. Damping off – *Pythium sps, Rhizoctonia solani, Sclerotium rolfsii*
2. Fruit Rot – *Phytophthora nicotianae*
3. Little Leaf of Brinjal – *Phytoplasma* (MLO's)
4. Alternaria leaf Spot – *Alternaria melongenae, Alternaria solani*
5. Collar rot – *Sclerotium rolfsii*
6. Chilli Tobacco – TMV
7. Stem borer – *Euzophera perticella*
8. Brown leaf hopper – *Cestius phycitis*
9. Calyx withering – Effected fruits have higher calcium and nitrate content than normal fruits
10. Poor fruit set – Out of 4 types of flowers [long style, medium styled, short and pseudo short styled] fruit set does not occur in short and pseudo short styled flowers.

Chilli & Capsicum

1. Damping off – *Pythium* sps
2. Anthracnose / Fruit rot / Dieback of pepper– *Colletotrichum capsici*
3. Powdery mildew – *Leveillula taurica*
4. Fusarium wilt – *Fusarium oxysporum f.sp.capsici*
5. Bacterial leaf spot – *Xanthomonas campestris pv. vesicatoria*
6. Chilli mosaic – Chilli mosaic virus
7. Leaf curl – Chilli leaf curl virus

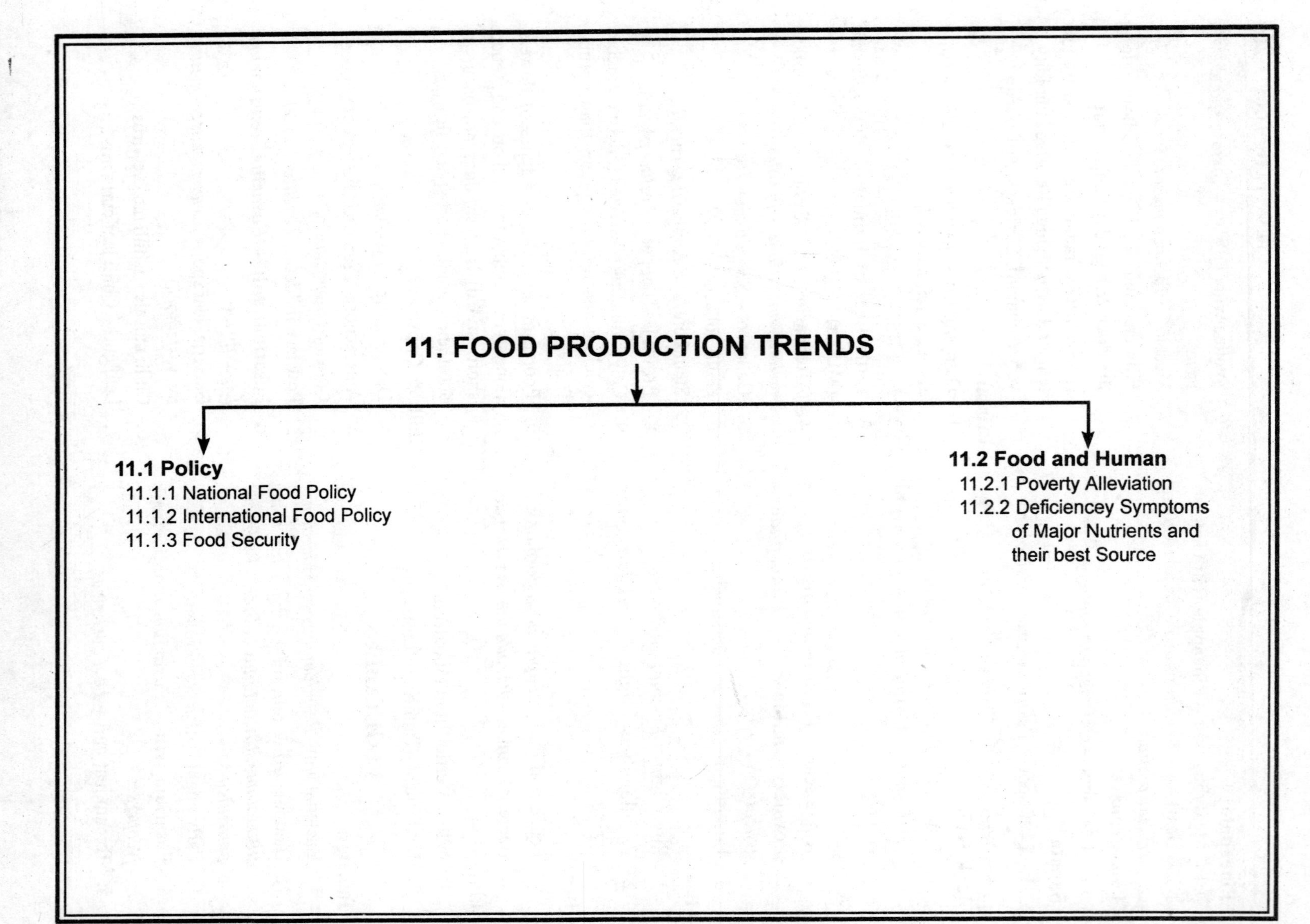
11. FOOD PRODUCTION TRENDS
11.1 Policy
11.1.1 National Food Policy
11.1.2 International Food Policy
11.1.3 Food Security
11.2 Food and Human
11.2.1 Poverty Alleviation
11.2.2 Deficiencey Symptoms of Major Nutrients and their best Source

11. Food Production Trends

11.1 POLICY

11.1.1 National Food Policy

Food policy has five aspects as follows :

1. **Measures to increase output of food grain.**
   - In view of shrinking land and water resources, the singular option for India to meet the future challenge food, fiber and other needs is through increase in productivity.
   - The only way to achieve this goal in a sustainable manner is through the scientific and need based integrated nutrient management.
   - The micronutrient deficiency in crops is growing rapidly both in extent and intensity with the green revolution affecting human nutrition as well. Zinc, Copper, Iron, Manganese have emerged as major micronutrient deficiency in the soil.
   - Availability of necessary micronutrient carriers. There is need to promote increased use of organic manners, which will meet some of the micronutrient requirements.
   - The question of subsidy on fertilizer should be viewed from a national perspective of food security nutritional security and national independence.
   - A gradual change is preferable to sudden ad-hoc change in subsidy.
   - There is need a critical review of pricing policy of fertilizers, agricultural commodities and of subsidy by a panel of experts, agronomists, soil scientists, plant scientists, agricultural economists, fertilizer industry and farmers representatives.
   - It also refers (*i*) to increase the technological measure (*i.e.*) adopt of improved technology in production and (*ii*) improvement in land reform.
2. **Measures Concerning Distribution of food grains**

It includes improvement in organization of food zone and creation of buffer stock in state trading. To improve food situation, it is necessary to ensure regularity and certainty in food supplies. Accordingly it becomes essential to create sufficient buffer stock in the years of excess supply to meet their extra demand during the lean years.

Hence Govt. set up Food Corporation of India in January 1965 to undertake purchase, handling, transport, storage and distribution of food grains on behalf of government. The storage capacity of corporation currently is 17.26 m.t.

3. **Procurement of food grains and regularization of public distribution**
4. **Procurement from whole sales**
5. **Import of food grains**

11.1.2 International food policy

According to the report of the International Food Policy Research Institute (1992), Washington, food security is basically defined as access by all people, at all times, to the food, needed for a healthy life. The food security concept thus addresses people's risk of not having access to the required food (*i.e.,* food insecurity). These risks can broadly encompass the major demands of crop production and productivity improvement, efficient public distribution systems, employment generation for poverty alleviation; and last but not the least, household food security ensuring enough food to meet adequate dietary intake of all its members.

The latest estimates of the FAO in its Bulletin published on the **state of food in security in the world (FAO, 2000)** have indicated that roughly about 826 million people are undernourished–of them, 729 million people in the developing world, and 34 million in the developed world. The report, in analyzing region-wise depth of hunger regime, has critically observed that **'to develop lasting solution to end hunger, it is important to know not only how many people are hungry but how hungry they are'.** The daily diets of the 826 million chronically hungry people in the world lack an average intake shortage of 100-400 kilocalories, diminishing their ability to lead an active life. The greater the depth of this hunger, the greater is the vulnerability to the nutrition–related health risk. However, expressing optimism, FAO projection to 2015 AD suggests that *'due to progressively declining population growth and increase in productivity and income level, more people will escape the prison of hunger'*. The food security, thus in its true sense of the word, as derived from the Latin word 'Secures', means free from care and anxiety and hence implies not only access to but right to food.

11.1.3 Food security

One of the finest Indian success stories of post-independent era has been the Green Revolution of sixties, which salvaged the country from being a chronic importer of foodgrains into an exporter. Since independence, while the population has increased 3 times, foodgrains production increased 4 times from 50.8 million tonnes during 1950-51 to 200 million tonnes by 200 AD. In the post-green revolution era, besides foodgrains production, significant achievements have been made in the field of milk, building up of strong agriculture R&D systems, extension services and human resource development programmes etc.

Attainment of food security is going to be the biggest challenge for the country from the very beginning of new millennium. According to the World Watch Institute, Washington, the average annual increase in foodgrains production, on a global basis, was 30 million tonnes between 1950 and 1954, which dropped down to 12 million tonnes during 1984-92 caloric gap in human nutrition. To meet the demand of our population @ 1.8% and average annual growth rate of 4.5%, according to some recent estimates, the country will need about 260-264 million metric tonnes of foodgrains, 130-150 million metric tonnes of milk, 151-193 million metric tonnes of vegetables, 80-106 million metric tonnes of fruits, 10-14 million metric tonnes of fish and 12 million metric tonnes of edible oils to provide adequate nutrition to 1.35 billion people by 2020 AD.

Between now and 2020 AD world population will increase by about 40%, to a total of about 8 billion people. Much of this increase will occur in developing countries, including India, where urban population will grow at the rate of more than triple as that of now. According to some estimates, India's population is projected to grow @ 1.62% annually during Ninth Plan and will gradually decline to 1.5% per annum during millennium ending 2021 AD. The country's population will thus increase from 934 million in 1996 to 1012 million in 2021 AD and 1350 million in 2011 AD, an increase of about 45% over 1996.

Against the population projected above, the long-term projection made by Dr. Swaminathan and Sinha in 1970 is that the ultimate foodgrains production potential of the country by 2050 AD is estimated at 450 million tonnes of foodgrains on the basis of total photosynthetic abilities of foodgrain crops around that period.

Based on the estimates made by the Ministry of Agriculture in its Draft Document '*Indian Agriculture : Vision 2020 AD,* the demand for foodgrains is estimated at 324 million tonnes to meet the foodgrains demands of 1350 million people.

During 1999-2000, foodgrains production touched 200 million tonnes mark; together with a very significant achievement in horticulture sector. India becoming the largest producer of fruits and second largest producer of vegetables in the world. To add further, India has emerged the largest producer of milk with 78 million tonnes production level before entering into twenty-first century, leaving behind USA and the European Union.

A strategy for food security based largely on self-sufficiency in food production has the advantage of promoting both productivity and purchasing power among small peasants and agricultural labourers. In general, policies for improving household food security, forming an integrate component of food security, should include:

(*i*) development strategies and macro-economic policies that would create conditions for growth with equity;

(*ii*) accelerating growth in the food and agricultural sectors which provide direct sources of food and income with which to buy food;

(*iii*) promoting rural development that focuses on the poor;

(*iv*) improving access to land and other natural resources;

(*v*) providing cheap credit for poor households;

(*vi*) increasing employment opportunities;

(*vii*) introducing income transfer scheme, including provision of public distribution of subsidised cheap food;

(*viii*) stablizing food supplies and food prices;

(*ix*) improving emergency-preparedness-planning for providing food and during natural disasters like drought, flood, earthquakes etc.

The right to food for freedom from hunger

The United Nations have legitimately considered the access of around 826 million hungry people of the developing world to adequate food as a universal human right and collective responsibility, as are approximately reflected from the set milestones and the understandings reached from the conclusions of the following international conventions held during past half-a-century.

- The Universal Declaration of Human Rights (1948), which recognized that *'everyone has the right to a standard of living adequate for the health and well-being of himself and his family, including food...'.*
- The International Convention on Economic, Social and Cultural Rights (1966), which stressed the 'right of everyone adequate food' and specified that to be free from hunger is a fundamental right of everyone.
- The World Food Conference (1974), convened in Rome by the United Nations, which reaffirmed that 'every man, women and child has the inalienable right to be free from hunger and malnutrition in order to develop fully and maintain their physical and mental faculties'.
- The World Food Security Compact (1984), which stated that 'food security was the common responsibility of humankind, requiring a moral commitment and international co-operation'.

- The International Conference on Nutrition (1992), convened by FAO and the World Health Organization, which examined in detail the problems of hunger, malnutrition and diet-related diseases.
- The World Food Summit (1996), prepared and hosted by FAO, which aimed to secure a political commitment to establishing food security within an agreed frame-work of action, including targets for the reduction of hunger.
- The State of Food Insecurity in the World (2000) draws on FAO's on-going work of monitoring the nutritional status of populations world-wide and analyzing their degree of food insecurity and vulnerability. This work represents part of FAO's contribution to the Food Insecurity and Vulnerability Information and Mapping System (FIVIMS) initiative, which is being established at the global and national levels.

Realignment of Research Priorities to Address Food Security

In the words of Dr. Amartya Sen '*Most countries in the world have achieved economic prosperity and eliminated economic insecurity through industrialization and along with that, by putting agriculture on a higher technological basis*'. The entire research and development programmes aiming at improvement of production and productivity to achieve self-sufficiency, from the First to the Ninth Plan, spanning five decades, can be grouped into four phases. According to *ICAR : Vision 2020.*

- Conservation, planned enhancement and utilization of agro-biodiversity.
- Enhancing productivity through evolution of high-yielding hybrids and varieties.
- Research on diversification, quality improvement, post-harvest technology, value-addition and export-oriented commodities.
- Sustaining enhanced productivity of irrigated agriculture and judicious development and use of energy, especially renewable sources of energy.
- Characterization and development of sustainable land-use models for rainfed agriculture in high-rainfall areas.
- Development of the Integrated Pest Management (IPM) and Integrated Nutrient Management System (INMS) approaches and systems for sustainable agriculture.
- Fostering excellence in relevant basic and strategic research.
- Generating research and technologies geared to promote equity among regions, sectors of society and gender.
- Strengthening social science, policy planning, agri-business and research monitoring mechanisms.
- Strengthening Agricultural Research Information System (ARIS).
- Promoting the Agricultural Human Resource Development (AHRD).
- Linking scientists with the farmers through Institute-Village-Linkage Programme (IVLP) as an innovative technology transfer model.
- Institutionalization and strengthening linkages/partnerships with CGIAR and under national and international agencies and research and development establishment, non-government organizations (NGOs), farmer organizations, private sector etc.
- Explore alternative or non-conventional plant resources for diversification of agriculture so as to enhance the food, fodder, fibre and fuelwood production.
- Provide solutions which are relevant, technically sound, economically viable,

socio-culturally acceptable, eco-friendly and systems-responsive.

- Enhance the use of biofertilizers and biopesticides and biodegradable chemicals in agriculture.
- Strengthen the production base by harnessing modern scientific knowledge and translate it in the form of crop plants which are high-yielding, resistant to biotic and abiotic stresses, and efficient in input use.
- Improve skill and productivity of farm women with accent on social engineering and empowerment.
- State-of-the-art technology for post-havest management and agro-processing.

11.2 FOOD AND HUMAN

11.2.1 Poverty Alleviation

The concept of food security is an integral part of the wider social security systems, which again is related to entitlement, capability and functioning of capability. The recent World Bank report on '*India : Policy to reduce poverty and accelerate sustainable development' provides a detailed account of how Indian economy's potential vulnerable sections can be uplifted through an integrated approach linking sustainable development with poverty alleviation programmes requiring sound micro-economic analysis and policy support together with open trade relations and enhanced access to human and physical capital.*

In India, about 244 million people are estimated to be below the poverty-lines in rural areas. This gives an indication about the magnitude of the problem of rural poverty and deprivation; and hence the compulsion of the Indian Government to ensure food security at local, individual and household levels through poverty reduction routes. The Ministry of Rural Development is the nodal agency at the apex level in so far as alleviation of poverty of rural India is concerned and the District Level Rural Development Agency (DRDA) co-ordinates various programmes at local levels. The schemes for alleviation of rural poverty which are being implemented under the Ministry of Rural Development can be broadly grouped into four categories, *viz.* wage employment scheme, self-employment scheme, area development scheme and social welfare scheme. Prominent among such schemes that directly or indirectly provide food security through the process of economic upliftment and/or employment and empowerment, especially of the vulnerable sections of the rural poor, are :

(*i*) Jawahar Rozgar Yojana,
(*ii*) Employment Assurance Scheme,
(*iii*) Swarnajayanti Gram Swarozgar Yojana,
(*iv*) Training of Rural Youth for Self-Employment,
(*v*) Development of Women and Children in Rural Areas,
(*vi*) National Social Assistance Programme,
(*vii*) Annapurna Scheme,
(*viii*) Jawahar Gram Sambridhi Yojana,
(*ix*) Pre-poor Initiative Programme (co-sponsored by the UNDP),
(*x*) Drought-prone Area Programme,
(*xi*) Desert Development Programme, and
(*xii*) Integrated Wasteland Development Programme.

Besides above national level programmes, Ministry of Rural Development have forged inter-institutional linkages through programme initiatives on cross-cutting food security issues affecting Asia and Africa with two international organizations, *viz.* the Centre of Integrated Rural Development for Asia and Pacific (CIRDAP) region and the Afro-Asian Rural Reconstruction (AARRO), which has been recently been rechristened as Afro-Asian Rural Development Organization (AARDO).

11.2.1 Deficiency symptoms of major micronutrients and their best source of availability

Major micronutrients	*Best source*	*Deficiency symptoms*
Iron	Leafy and other green vegetable, ragi, jaggery, meat, fish, poultry, dried fruits, fresh fruit juice, jamun, phalsa, karonda, plum.	Anaemia, reduction of haemoglobin concentration, poor absorption of nutrients, haemorrhage, loss of blood through illness.
Iodine	Iodised salt, sea foods, sea weeds	Obesity, lethargy, goitre (swelling of thyroid glands), dwarfism and crotinism or deformity in children.
Calcium	Leafy green vegetables, curry leaves, coriander leaves, green beans, banana, apple milk and milk products, ragi, guava, wood apple, seetaphal, tamarind, citrus fruits.	Bone deformities in children, stunted growth, poor digestion, poor muscular control, back and leg pain, softening of bones in women (Osteomalacia)
Phosphorus	Meat, poultry, fish, eggs, vegetables.	Poor appetite, fragile bones.
Potassium	Green vegetables, fruits, milk.	Lethargy, loss of appetite, weak bones, slow blood clotting.
Selenium	Sea food, bread, nuts.	Slow regulation of thyroid hormone, less protection against artery clogging.
Zinc	Pumpkin seeds, eggs, oysters and other sea foods.	A constituent of more than 300 enzymes so its deficiency adversely affects tissue repair, digestion, fertility, vision, cell reproduction and protein synthesis.
Manganese	Nuts, seeds, pineapple.	Less protection from osteoporosis.
Magnesium	Nuts, grains, fish, meat, leafy vegetables.	Lethargy, B vitamins are inactivated and imbalance of insulin levels occur.
Vitamins		
Vitamin A	Leafy vegetables, yellow fruits and vegetables, sweet potato, liver, eggs, cod liver oil, milk, butter cream.	Rough and dry skin, susceptibility to disease, night blindness, glare blindness, retarded growth, slow reproduction.
Vitamin E	Oil, nuts, seeds, egg yolk, leafy vegetables.	Retarded vitamin A activity, prone to heart diseases, enhanced ageing, slow blood clotting.
Vitamin C	Broccoli, citrus fruits, guava, cashew apple, cherry, drumstick leaves, brussel sprouts, leafy greens, bittergourd, chillies, raw vegetables, sprouted grains.	haemorrage, weak bones, teeth and gums, lack of blood leading to anaemia, slow healing of wounds, scurvy in severe cases, prone to cancerous cells, premature ageing.
Vitamin D	Sunlight, irradiated (UV) foods, fish liver oil.	Rickets, poor teeth, pigeon chest, protruding forehead, fat belly, bow leggs, slow absorption of calcium.

Major micronutrients	*Best source*	*Deficiency symptoms*
Thiamine (B1)	Pulses, groundnut bran, whole grain cereals, wood apple, asparagus, apple, banana, apricot, cabbage.	Poor appetite, constipation, faulty assimilation of food, depression, irritability, restlessness, berberi in severe cases.
Riboflavin (B_2)	Leafy vegetables, plum, litchi, banana, cowpea, pulses, papaya, milk and milk products, liver, meat, eggs, whole grain cereals.	Pale lips, splitting at the corner of the lips, eye strain, itching and burning eyes, sensitivity to light and other eye troubles.
Niacin (B_5)	Groundnut, meat, prawn, pulses, green vegetables, peach, cherry, banana, strawberry.	Poor growth, lethargy, poor appetite.
Folic acid	Beans, meat, leafy greens, mango, raisins.	Disturbed levels of homocysteine, prone to heart diseases, birth defects, improper production of haemoglobin.

11.2.2 Amino Acids

S. No.	*Essential amino acids*	*Non essential amino acids*
1.	Methionine	Alanine
2.	Arginine	Asparatic acid
3.	Threonine	Glutamic acid
4.	Tryptophan	Cysteine
5.	Valine	Glycine
6.	Isoleucine	Proline
7.	Phenylalanine	Serine
8.	Lysine	Tyrosine
9.	Histidine – essential only for children	

(*Source* : Notes ANGRAU, Hyderabad, 2012)

11.2.3 Important Proteins in Cereals

S. No.	*Cereals*	*Prolamines*	*Glutelins*	*Globulins*
1.	Wheat	Gliadin	Glutenin	*
2.	Rice	Gliadin	Oryzenin	*
3.	Maize (corn)	Zein	Traces	*
4.	Barley	Hordein	*	*
5.	Millets	Panccin	*	*
6.	Rye	Gliadin	Glutenin	Endestin
7.	Oats	*	Evenin	Evenalin

(*Source* : Khader V. ICAR. 2001) * = Minor

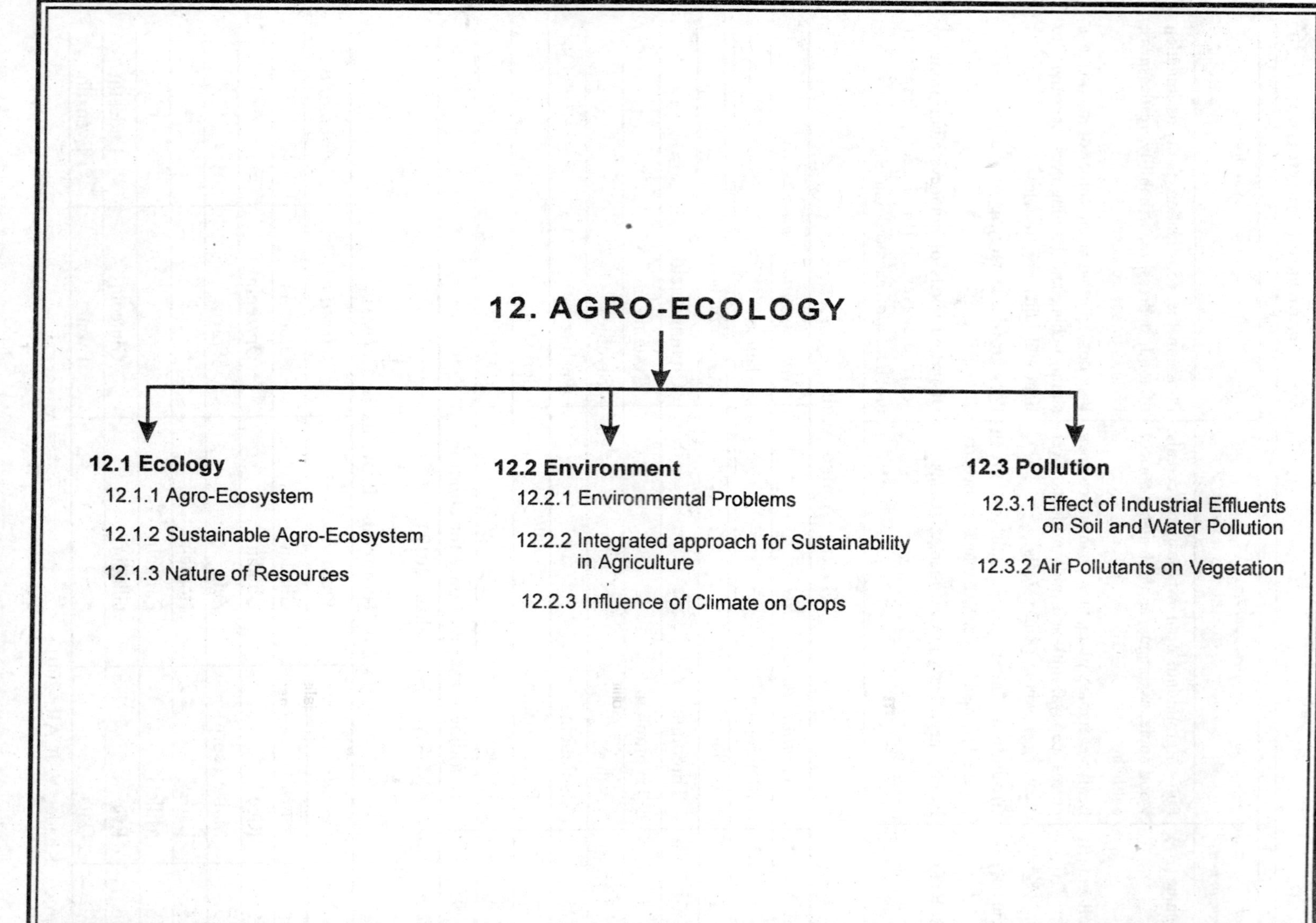
12. AGRO-ECOLOGY
12.1 Ecology
12.1.1 Agro-Ecosystem
12.1.2 Sustainable Agro-Ecosystem
12.1.3 Nature of Resources
12.2 Environment
12.2.1 Environmental Problems
12.2.2 Integrated approach for Sustainability in Agriculture
12.2.3 Influence of Climate on Crops
12.3 Pollution
12.3.1 Effect of Industrial Effluents on Soil and Water Pollution
12.3.2 Air Pollutants on Vegetation

12. Agro-Ecology

12.1 ECOLOGY

The science of relationship between organisms and their environment.

12.1.1 Agro-Ecosystem

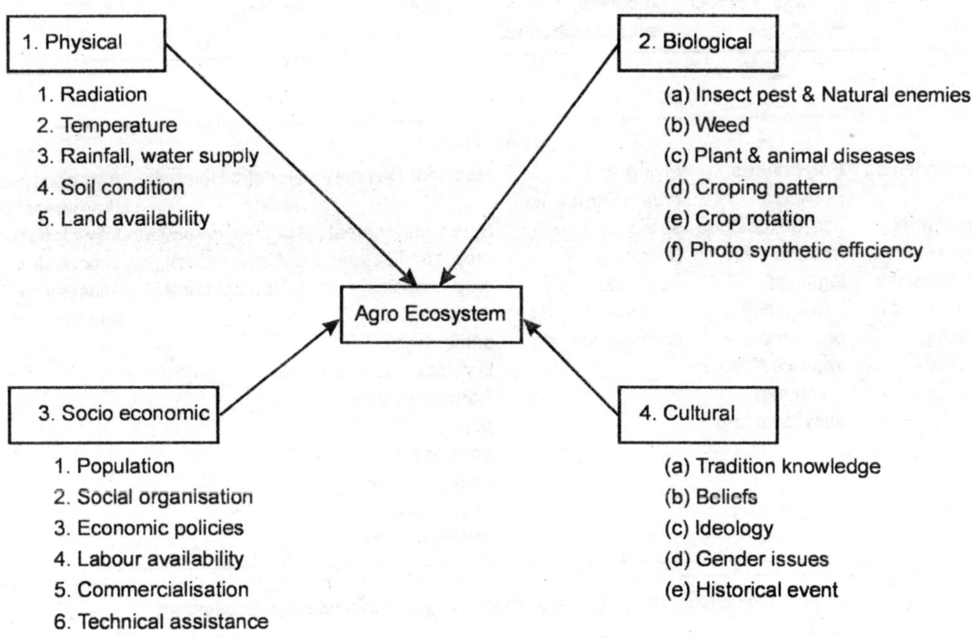

Fig. Agro ecosystem determinants

12.1.2 Sustainable Agro-Ecosystem

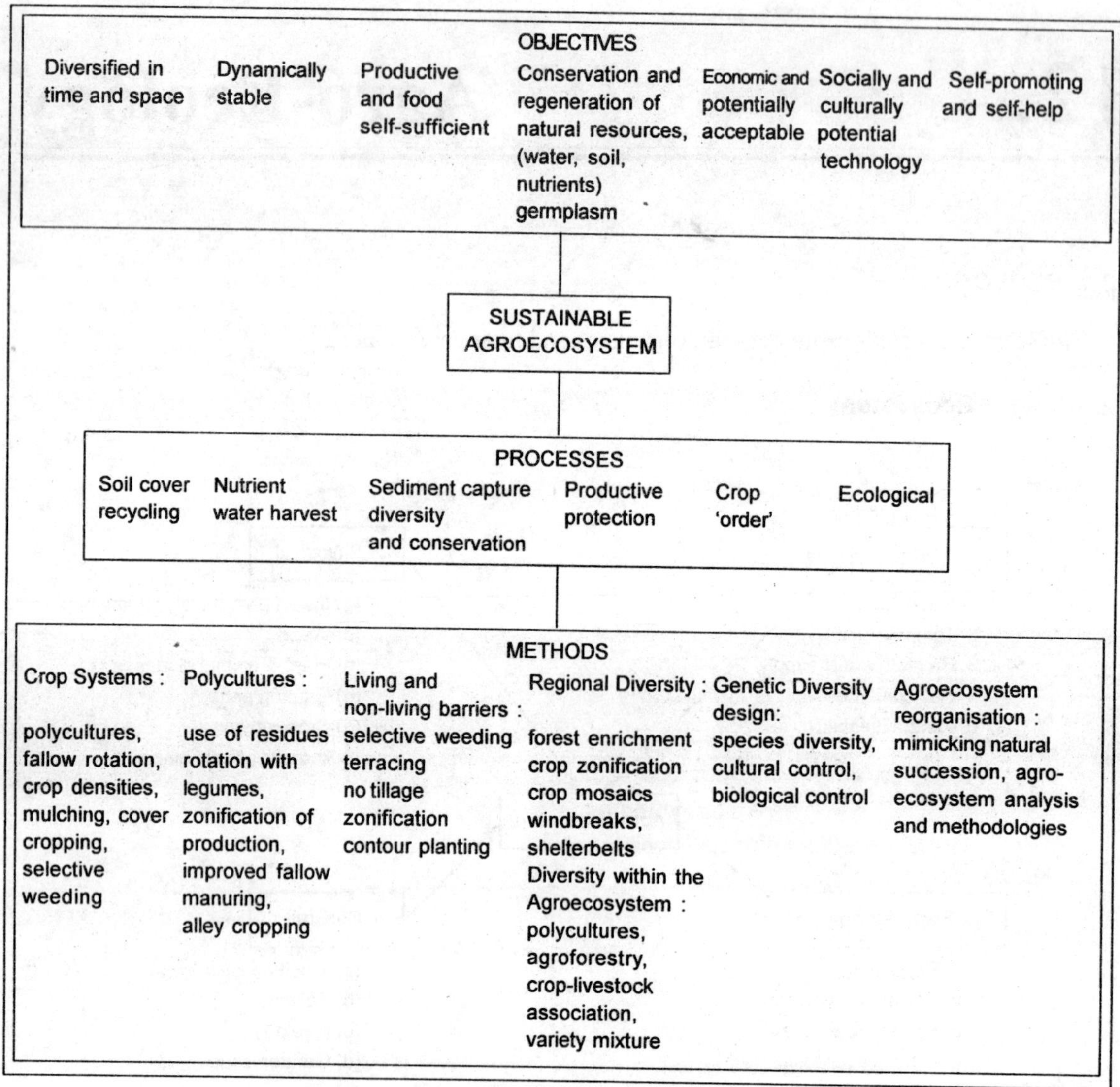

Objectives, processes and methods in sustainable agroecosystem

12.1.3 Natural resource

Natural resource is something that is useful and valuable in the condition we find it in its raw or modified state, it may be an input or raw material in the production process.

Natural resources, are those resources that are not man made, including all the earth's natural elements and environmental forces.

Natural resources fall into four categories as follows

1. Basic natural resources such as land, water and air.

2. Natural resource commodities such as timber and fish.
3. Environmental amenities such as clean air and scenic views.
4. Environmental pollution, soil erosion, ground water recharge and species generation.

Attributes of a Resource

A resource is multi attribute in nature. It has four attribute as follows

1. **Quantity :** A resource available in nature may be infinite or finite in quantity.

   (*e.g.*) Coal, iron ore and petroleum products.
2. **Quality :** A resource will be different in quality also.

   (*e.g.*) Recovery of iron from iron ore will be high or low depending upon the quality of ore excaved from different parts of the world.
3. **Time :** It refers to relative abundance and scarcity of resources at different point of time.

   (*e.g.*) A country becomes rich in resources by discovering new resources or by finding new uses for already known resources.
4. **Space :** It refers to availability of different resources in different parts of the world. While availability is not uniform in all the regions, demand for resources is more or less same for all the people.

Importance of natural resources

Natural resource includes land, water, fisheries, mineral resources, forest, marine resources, climate, and topography. Some of resources are known to man. But man utilises small portion of land resources and other surface resources. There are large amount of undiscovered resources under the surface. Importance should be given to explore natural resources in surface and undersurface.

Classification of Resources

Based on the attributes, resources can be classified as

1. ***Exhaustible Resource or Stock Resource***

These resources are finite in quantity and hence its exploitation and use should be in an optimum way. Spacio-temporal (space and time) dimensions of stock resource are important for its optimum use.

Examples

A. *Non renewable resource*

Coal, oil, iron, other metal, old growth timber and soils.

B. *Renewable resources*

Water stored in tanks, solar energy stored in cells.

2. ***Inexhaustible Resource or Flow resource***

These resources can be exploited continuously without diminishing the stock. It must be utilised when it is available.

A. *Non renewable resource*

Wind, flowing river water.

B. *Renewable resources*

Young-growth timber.

3. ***Biological Resources***

These resources represent complex category of resources. It involves solar energy, hydrological resources and soil nutrients.

(*e.g.*) Crops, Forest and Animals.

Man derives value from a complex system of solar, atmospheric, geological, hydrological and biological resources.

There are several factors that influence the scarcity of natural resources as follows

1. ***India's Population Growth and Demography*** **:** It is a fundamental determinant of raising demand for natural resources. Population growth increases pressure on the resources required for subsistence land for producing crops and grazing animals.

2. ***Income Growth*** : As people become wealthier, their consumption increases and its form changes. Rich people consume more energy in the form of transportation, heating, and air conditioning. They seek improved housing that uses more space and they buy more manufactured goods. Raising income mean fewer people depend directly on natural resources for their subsistence needs.

3. ***Market Forces*** : Pressure on natural resources also increases demands in national and international markets.

4. ***Technology*** : Improved technologies to harvest and harness natural resources have increased demand for them over the last century.

(*e.g.*) Better transportation infrastructure made forest areas more accessible to distinct markets; improvements in pumps and power sources made it economical to tap scarce ground water supplies.

Improved agricultural technologies and production practices have stimulated vast increases in food production on India's limited area of agricultural land.

Mitigating Scarcity of Resources

All nations and regions are not uniformly endowed with natural resources have to mitigate resource scarcity for economic development. The various possibilities :

1. ***Substitution*** : Scarce resource can be substituted with resource available in plenty.

(*e.g.*) Aluminium can be used for copper.

2. ***Technological change*** : Technological improvements mitigate resource scarcity in a variety of ways.

(*a*) Better efficiency in resource use.

(*e.g.*) Better technology helps to recover more metal from its ore.

3. ***Trade*** : Import of resources mitigates resource scarcity. It depends on infrastructural facilities like ports, roads.

4. ***Discovery*** : Discovery of new resources by better exploitation techniques.

(*e.g.*) Discovery of new fields or mines.

5. ***Recycling*** : Recycling of once used resources is a very important method of mitigating resource scarcity.

(*e.g.*) Waste paper, metal.

Guiding Principles of Resource Development

The principle objective of resource development is to maximise national output. For this purpose, there should be optimum utilisation of resources not only for short period but also for sustained long period. The following principles have been evolved for resource development:

- Economic use of resources to achieve minimum waste.
- Sustained use of economic resources through conservation of renewable resources and economic use of exhaustible resources.
- To increase multipurpose use of resources.
- Integrated planning in use of natural resources.
- Minimising transport cost by locating industries nearer to source of raw material.
- Optimum utilization of electrical power and other resources.
- To employ modern technology to explore possibility of bringing more natural resources.

12.2 ENVIRONMENT

12.2.1 Environmental Problems

There are three types of environmental problems.

(I) Water related problem (II) Soil related problem and (III) Technology related problem

(I) Water Related Problem

The problems are as follows :

(*i*) ***Increasing Water Table*** : In some of the areas, ground water level increases to ground surface. Water table less than 1.5 m below the ground level is considered to be harmful for agricultural operation.

(*ii*) ***Loss of Irrigation Water*** : The improper water management techniques result in large extent of loss of irrigation water. This is mainly because canal system and its branches are not properly lined. The seepage through improper lines is estimated the heavy loss of irrigation water.

(*iii*) ***Water Logging*** : Productivity in water logged areas decreases considerably.

(*iv*) ***Decreasing Water Quality*** : Quality of water to the agricultural sector decreases. This factor should be taken care of while programming major and minor irrigation systems.

(II) Soil Related Problem

Soil degradation is another problem in agriculture.

(*i*) **Salinization** : Seepage through unlined irrigation stream, improper land planning is main cause leading to soil salinity.

(*ii*) **Wind and water erosion of soil** : Erosion of soil due to wind and water are also given importance.

(*iii*) ***Increasing toxicity in soil***

(III) Technology Related Problem

This problem are more due to application of chemical fertilizer and pesticides.

(*i*) High quality application of fertilizer and pesticide.

(*ii*) Increase in chemical content of soil and related degradation.

(*iii*) Higher toxicity affecting root zone of plants.

Overdoses of chemicals and pesticides affect the soil structure and make it difficult for further cultivation.

The knowledge on proper methods of application should be given to the farmers through better extension methods.

Remedial Measures

To increase the agricultural productivity, proper remedial measures are to be taken for the above said three problems. Since the area under cultivation is a serious limiting factor in India, proper technology measures should be taken to increase the productivity.

12.2.2 Integrated Approach for Sustainability in Agriculture

1. *Resources*

Problem

(1) Salinity (2) Cultivable waste (3) Flood (4) Drainage (5) Improper distribution of irrigation water.

Remedy

(1) Soil reclamation (2) Quality improvement (3) Land management (4) Ground water management and (5) Policy measures.

2. *Cropping System*

Problem

Monocropping affect the fertility of the soil.

Remedy

Crop substitution for higher valued crops, increasing cropping intensity, policy measures relating to adding the mixed farming concepts to improve the productivity and income level of farms.

3. ***Inputs***

Problem

Inefficient distribution system of seed, chemical and fertilizer.

Remedy

Privatization, bulk purchase and storage in proper places and policy measure to strengthen their infrastructure to facilitate proper distribution of inputs in time and correct place.

4. ***Technology***

Problem

Recent technologies are costly.

Remedies

Cost effective technology measure should be formulated to facilitate the poor farmer to adopt the technology. Policy for reorienting the measures regarding the cost-effective technologies are as follows :

- Incentives for higher productivity through technical change.
- Remunerative prices, subsidy on inputs.
- Manpower improvement in training and management skills.
- Infrastructure administration management.
- Farmer responsible, farmer friendly administrators are needed.

Ecological Approach for Environmental Problems

The key ecological factors are land, water, vegetation, air, sea and wild life. The threats to them include overuse and pollution leading to soil erosion, salination, fertility decline, and degradation of plant quality, desertification, and poisoning via polluted water and airborne chemicals and radiation. There are limits to ecological carrying and self-regenerating capacity. They can be increased and but not yet surpassed.

To despair, ignore or reject an ecological message of doom is not likely to protect ecology. But the decision lies with thousands of farmers who must understand and act right way and in time.

12.2.3 Influence of Climate on Crops

Climate is the most dominating factor influencing the suitability of a crop to a particular region. The yield potential of a crop mainly depends on climate. More than 50% of variation in yield of a crop is due to climatic differences. The type of vegetation or crops determined by climate.

Solar Radiation

(*i*) ***Thermal effect of Radiation*** : More than 70% of the solar radiation absorbed by plants is converted into heat. This heat energy is utilized for transpiration for convective heat exchange with the surroundings. These changes determine the temperature of leaves and of other plant parts.

(*ii*) ***Photosynthetic effect of Radiation*** : A portion of solar radiation upto 28% of energy is used in photosynthesis.

(*iii*) ***Photoperiodic effect*** : It plays an important role as regulator and controller of growth and development.

Temperature

Cardinal Temperature

For each species of plants, there are maximum and minimum limits of temperature as at which growth is negligible optimum temperature at which growth is maximum. Most of the crop plants grow at 15 °C to 30 °C.

Influence on Growth Substances

At optimum temperature, the activity of the growth promoters are high and activity of the growth regulator is low with the result that growth rate is increased. At high or low temperature, the balance of growth substances change and affect growth.

Influence of Temperature on Development

Temperature has great influence on developmental rates like rate of germination, leaf initiation, tillering, flowering. Spikelet initiation and grain filling. All the development processes proceed at a faster rate at higher temperature.

Low Temperature Stress

Cell division and cell elongation.

Photosynthesis.

Water transport.

Vegetative growth.

Reproductive growth.

Rainfall

Rainfall analysis helps in taking decisions on time of sowing. Scheduling of irrigation, time of harvesting etc.

Amount of Rainfall

Rainfall may also be in excess of the optimum and thereby cause reduced yields which may appear paradescical to semi arid climates. When the rainfall is concentrated in 4-5 months of the year, there may be periods when the rate of precipitation exceeds the intake rate of soil.

Intensity of Rainfall

It influences erosion of soil.

Relative Humidity

Relative humidity has considerable influence on evapotranspiration and hence on the water requirement of crops.

Photosynthesis

When relative humidity is low, transpiration increases causing water deficit in plant. Water deficit cause partial or full closure of stomata and increase mesophyll resistance blocking entry of CO_2.

Pests

The incidence of insect pests and diseases is higher under high humidity conditions.

Grain Yield

Very high or very low relative humidity do not conducive for high grain yield.

Impact of changing environment on cropping pattern as indicators of environment

Factors Responsible for Environment Change

- Human population
- Urbanisation and industrialisation
- Fossil fuel utilization — over exploitation
- Deforestation
- Mismanagement in agriculture
- Emission of green house gases

Impact in India

(*i*) *Increasing temperature*
- Tree line shifting towards the higher altitude (Apple in Shimla).
- Delaying of sowing of wheat crop.
- Rabi crop in north India production is under threatened (wheat, rapeseed, chick pea).

(*ii*) *Precipitation*
- All mathematical models projection is increase in precipitation in India.
- But the intensity will be more *i.e.* less rainy. days which cause floods and drought.
- Which cause production decline.

(*iii*) *CO_2 level increase*
- Favourable for C_3 plants upto some extend.
- But increase the C_3 weed competition to C_4 plants.
- Increase the production and reduce the quality of the produce.

Finally due to climate change the wheat production in north India will be severely affected and the people should change their food habit towards rice.

12.3 POLLUTION

12.3.1 Effect of Industrial Effluents on Soil and Water Pollution

(*i*) **Major industrial effluents**

(*a*) Sugar factory (*b*) Tannery
(*c*) Distillery (*d*) Textile and dye
(*e*) Paper mill

(*ii*) **Effect on soil health**

(*i*) *Chemical Properties:* CEC, EC, pH, Heavy metal accumulation increase alkalinity, salinity and acidity of the soil.

(*ii*) *Physical properties :* Decrease the porosity and aeration, affect the structure and infiltration.

(*iii*) *Biological properties :* Reduce the population of microorganism and fauna of the soil.

(*iv*) *Effect on water* : Surface water pollution

Soil Pollution

Soil is a thin layer of organic and inorganic materials that covers the earth's rocky surface soil pollutant is any factor which deteriorates the quality, texture and mineral content of the soil or which disturbs the biological balance of the organisms in the soil.

Causes

- Indiscriminate use of fertilizers, pesticides, insecticides and herbicides
- Pollution due to urbanization
- Dumping large quantities of solid waste
- Deforestation and soil erosion

Effects

- Reduced soil fertility, nitrogen fixation and crop yield.
- Increased salinity and erosion.
- Bio-magnification
- Leaching and water pollution.
- Ecological imbalance and reduced vegetation

Ground Water Pollution

Sources:

1. Natural sources
2. Septic systems
3. Hazardous waste disposal
4. Petroleum products
5. Solid waste
6. Surface impoundment
7. Use of agricultural chemicals
8. Injection wells.

A. *Surface water pollution by direct discharge of effluents.*

- Heavy metal accumulation.
- Eutrophication.
- Reduce the productivity of the system.
- Biomagnification.

B. *Ground water pollution by infiltration*

- Reduce the water quality by increasing salinity alkalinity and heavy metal.
- Reduce the plant productivity by affecting its metabolism.
- Not fit for drinking and irrigation.
- Major pollutants fluoride, arsenic and nitrate.

12.3.2 Air Pollutants on Vegetation

Pollutant	*Injury symptoms*	*Age of leaf affected the most*	*Leaf part affected*
SO_2	Chlorosis, inter-venial bleaching, tip and marginal necrosis; similar to drought, insect and chilling injury	Middle aged most sensitive; older least sensitive	Mesophyll cells
O_3	Fleck, stipple, bleached and necrotic spotting, pigmentation; browning of conifer needle tips	Oldest most sensitive;	Palisade or spongy parenchyma in leaves
PAN	Glazing, silvering or bronzing of lower leaf surface	youngest most sensitive	Spongy parenchyma tissue
NO_2	Irrigular, white or brown collapsed lesions on near leaf margin	Middle-aged leaves most sensitive	Mesophyll cells
HF	Tip and marginal leaf burning, dwarfing, leaf abscission, narrow brown-red band, separate necrotic tissues from green tissues; fungal, cold and high thermal injuries make similar marking	youngest leaves most sensitive cells	Epidermis and mesophyll

Pollutant	*Injury symptoms*	*Age of leaf affected the most*	*Leaf part affected*
C_2H_4	Sepal withering, leaf abnormality; flower shedding, failure of proper leaf opening; excess water stress shows similar symptoms	Young leaves recover but older do not recover fully	All leaf part
Cl_2	Intervenial bleaching, marginal and tip burning , leaf abscission	Mature leaves more sensitive	Epidermis and mesophyll
NH_3	Cooked green appearance, becoming brown after drying	Mature leaves most sensitive	Complete leaf tissue
HCl	Leaf margin necrosis on broad leaves, acid type necrotic lesion, tip burning in fir	Oldest leaves most sensitive cells	Epidermis and mesophyll
H_2S	Basal and marginal scorching	Youngest leaves most sensitive	All leaves
H_2SO_4	Necrotic spots on upper (Acid rain) conducive for leaf damage	All plant leaf surface, high humidity more	All leaf parts tissue

Names of sensitive and tolerant plant species to various types of air pollution

Pollutant	*Sensitive plants*	*Tolerant plants*
Sulphur dioxide	Alfalfa, Barley, cotton, wheat, soybean, clover, rye, pine, safflower, apple, squash, tomato	Cantaloupe, celery, corn, oak
Hydrogen sulphide	Chinese apricot, gladiolus, grape. pine, sweet corn, peach, plum.	Aspergus, cherry, cotton, pear, squash, strawberry, tomato, wheat
Peroxy acetyl nitrate	Annual bluegrass, lettuce, spinach, beans, beet, pepper, tobacco, clover, mint,	Broccoli, chrystanthemum, corn, cotton, sorghum, cabbage, onion, cauliflower, carrot, cucumber, strawberry
Ozone	Oak, potato, radish, soybean, tobaco, tomato, white pine	Beet, geranium, gladiolus, maple, mint, pepper, rice
Ethylene	Cotton, cowpea, orchid, tomato	Grasses, lettuce

13. AGRONOMY

13.1 Farming System
- 13.1.1 Cropping Systems for Different Agro-Climatic Zones of India
- 13.1.2 Cropping Pattern in India
- 13.1.3 Cropping Systems
- 13.1.4 Impact of Cropping Patterns
- 13.1.5 Package of Practices
  - (5a) Cereals and Millets
  - (5b) Pulses
  - (5c) Oil Seeds
  - (5d) Fibre
  - (5e) Fodder

13.2 Weed Science
- 13.2.1 Weed
- 13.2.2 Characteristics of Weed
- 13.2.3 Propagation
- 13.2.4 Dispersal
- 13.2.5 Classification
- 13.2.6 Allelopathy
- 13.2.7 Activity and Selectivity Herbicide
- 13.2.8 Crop-Weed Competition
- 13.2.9 IWM
  - (9a) Principle
  - (9b) Biologic
  - (9c) Recent Advances in Weed Science

13.3 Dry Land
- 13.3.1 Dryland Farming
- 13.3.2 Soil-Moisture Conservation
- 13.3.3 Drought/Adoptic
- 13.3.4 Drought Management Technique for Dryland
- 13.3.5 Integrated Dry land Technology
- 13.3.6 ALU
- 13.3.7 Water shed.
- 13.3.8 Modern Concepts in dryland farming

13.4 Irrigation
- 13.4.1 Water use Efficiency
- 13.4.2 Scheduling of Irrigation
- 13.4.3 Methods of Irrigation
- 13.4.4 WUE
- 13.4.5 Drainage
- 13.4.6 Quality of Irrigation Water
- 13.4.7 Guide for Selection Methods of Irrigation

13.5 Agro Forestry
- 13.5.1 Social Forestry
- 13.5.2 Agroforestry
- 13.5.3 Propagation of Trees

13.6 Others
- 13.6.1 Shifting Cultivation
- 13.6.2 Taungaya
- 13.6.3 Puddling
- 13.6.4 Ratooning
- 13.6.5 Cardinal Temperature
- 13.6.6 Flue Curing
- 13.6.7 Harvest Index
- 13.6.8 MPTS

13. Agronomy

DEFINITION

Agronomy is the branch of agriculture that deals with scientific field crop production and soil management. Agronomic experiments focus on variety of factors relating to crop plants, including yield, diseases, cultivation and sensitivity to factors such as climate and soil.

Pietro de' Crescenzi is the Father of Agronomy.

13.1 FARMING SYSTEM

13.1.1 Efficient Cropping Systems for different Agro-climatic Zones of India

	Agro-climatic region	*Soil type*	*Cropping systems*
1.	Western Himalayas (cold-arid) Jammu and Kashmir, Ladakh Plateau	Shallow, sandy to loamy	Rice-Potato-Wheat
2.	Western Kachchh and part of Kathiawar	Desert and saline soils	Sorghum-wheat-millet, Maize-Green manure Maize-wheat-sorghum
3.	Deccan plateau (hot-arid) Andhra Pradesh, Karnataka	Red and black soils	Pigeon pea + sorghum/ groundnut cotton-millet-sorghum
4.	Northern plains and central highlands (semi-arid) Gujarat, Haryana, Madya Pradesh, Punjab, Uttar Pradesh, Rajasthan	Alluvium derived soils	Maize-green manure Rice-wheat-Maize-wheat Cotton-wheat
5.	Central highlands Gujarat plains Kathiawar penisula (hot semi-arid)	Medium deep black soils	Groundnut-wheat Rice-sugarcane Rice-wheat
6.	Deccan plateau (hot-semi-arid)	Shallow and Medium	Pearl millet-potato-cotton

Contd.........

	Agro-climatic region	*Soil type*	*Cropping systems*
7.	Deccan plateau and Eastern ghats (hot-semi-arid)	Red and black soils	Sorghum-safflower cotton-millet Pigeon pea-sunflower
8.	Eastern ghats, Tamil Nadu uplands and Deccan plateau (hot semi-arid)	Red loamy soils	Rice-Rice-pulses cotton-groundnut Groundnut-sorghum
9.	Northern plain (hot sub humid)	Alluvium derived soil	Rice-wheat, Maize-wheat, Maize-mustard-sugarcane
10.	Central High lands (hot sub-humid)	Black and red soils	Sorghum-wheat Maize-wheat
11.	Eastern plateau (hot sub-humid) soils	Red and yellow soils	Rice-wheat-green manure Rice-pigeon pea
12.	Eastern plateau and Eastern ghats (hot sub-humid)	Red and lateritic soils	Rice-Rice-Wheat Groundnut-sun flower
13.	Eastern plain (hot sub-humid)	Alluvium derived soil	Rice-wheat, Rice-lentil Rice-chickpea, jute-rice
14.	Western Himalayas (warm sub-humid)	Brown, forest and podzoilic soils	Maize-wheat, finger millet-rice-mustard
15.	Bengal and Assam plains (hot sub humid)	Alluvium derived soils	Rice-wheat-rice Rice-potato-sesame
16.	Eastern Himalayas (warm humid)	Brown and red hill soils	Rice-rice, rice-wheat jute-rice
17.	North-eastern Hills (warm humid)	Red and lateritic soils	Rice-rice, rice-wheat jute-rice
18.	Eastern coastal plains (hot sub humid to semiarid)	Coastal alluvium	Rice-rice, Rice-groundnut, Jute-rice
19.	Western ghats and coastal plains (hot humid)	Red lateritic alluvium derived soils	Rice-rice, Rice-sorghum, Plantation crops
20.	Islands (hot humid)	Red loamy and sandy soils	Rice-rice, Plantation crops

13.1.2 Cropping Patterns in India

Rice Based Cropping Patterns

- Found in heavy rainfall areas (or) supplemental irrigation available areas.
- 46.4% areas in Monsoon belt – U.P., Bihar, M.P..
- 53.6% area in Multiseason belt – Assam, WB, Orissa, A.P., T.N, Kerala, Karnataka.
- 30 – rice based cropping patterns in India.
- Bihar, WB, Orissa – Summer rice is emerging second crop competing jute.
- Introduction of Rice-non traditional areas – Punjab Haryana, Western U.P. Rice – wheat due to introduction of short duration of Rice has replaced maize in Punjab.

Wheat Based Cropping Patterns

- 19 - wheat based cropping patterns in India.
- Introduction of photoinsensitive varieties grow wheat from himalayas to kanyakumari.
- Traditional areas – north west India Wheat – wheat rotation replaced by Moong – wheat ; redgram – wheat.

Reasons : Availability of short duration varieties of moong and redgram.

Kharif Sorghum based Cropping Patterns

- Kharif sorghum – 11 m. ha Maharashtra, M.P., Karnataka, Gujarat.
- 17 major cropping patterns.
- Maharastra – Cotton, Pulses, Groundnut and small millets are alternatives.
- M.P. – Cotton, Pulses and small millets – are alternatives.
- Rajasthan – Wheat, Cotton, bajra, Maize are alternatives.
- A.P. – Cotton, Pulses, Groundnut are alternatives.

Rabi sorghum based Cropping Patterns

- 5 m. ha – Maharashtra, A.P, Karnataka
- 13 cropping patterns – Maximum in Maharashtra with bajra, Pulses, Oilseeds are alternatives.

Pearl millet based Cropping Patterns

- arid and semiarid tropics – 12 m. ha – Rajasthan and Gujarat.
- 20 major cropping patterns.
- along with pulses, Groundnut, oilseeds, kharif sorghum – Rajasthan.
- Cotton, tobacco – Gujarat.

Maize based Cropping Patterns

- 6 m. ha – U.P, Bihar, Rajasthan, M.P. and Punjab.
- 12 cropping patterns.
- U.P, Bihar → Rice (Kharif) – Wheat (rabi).

Ragi based Cropping Patterns

- 2.5 m. ha – Karnataka, T.N., A.P. Gujarat, Maharashtra, Hilly areas of U.P., H.P.
- Bajra, Sorghum, Maize, Pulses, and Oilseeds are alternative.

Cotton based Cropping Patterns

- 7.5 m. ha – Maharashtra, Gujarat, M.P.
- 16 cropping patterns.
- Cultivation extended - new command areas in Karnataka
- New short duration varieties – possible two crops of cotton in year.

Gram and other Pulses based Cropping Patterns

- Short duration varieties of green gram, Bengal gram, red gram, Cowpea made them suitable multiple cropping.

  eg. Rice – Wheat, Moong (summer)

  Moong – Wheat

  Redgram – Wheat

- Short duration Redgram – grown in rabi season - a non traditional season in Karnataka, A.P. Tamilnadu.

Groundnut based Cropping Patterns

- 70 m. ha – Gujarat, A.P., T.N., Karnataka
- Grown mainly in rainfed conditions – 9 cropping patterns.
- Kharif sorghum, cotton, Pulses are alternative – A.P. and T.N.
- Kharif and rabi sorghum and small millets are alternative – Maharashtra.
- Sorghum, Cotton, tobacco sugarcane – Karnataka.

13.1.3 Cropping System

It represents cropping patterns used on farm and their interaction with farm resources, farm enterprises and available technology to determine the farm makeup.

Cropping Pattern

Proportion of area under various crops at a point of time in a unit area.

Types of Cropping Systems

1. ***Monocropping (or) Monoculture***

Growing of only one crop on a piece of land year after year.

Reasons :

(*i*) Climatological and socio economic conditions.

(*ii*) Specilization of farmer growing particular crop.

2. ***Multiple Cropping***

- Intensification of cropping for both time and space dimensions.
- Growing two (or) more crops on the same piece of land in one calendar year.

Types :

(*a*) *Intercropping*

- Growing two (or) more dissimilar crops simultaneously on the same piece of land in a definite row pattern.
- Cropping intensity in space dimension is achieved.

Intercropping is of two types —

(*i*) Additive Series

(*ii*) Replacement Series

(*i*) *Additive Series* :

- one crop sown as a base crop.
- another crop introduced into base crop by adjusting (or) changing crop geometry.

(*ii*) *Replacement Series* :

- Both the crops are component crops.

Requirements – Successful Intercropping

(*i*) The time of peak nutrients demands of component crops should not overlap.

e.g. Maize + green gram.

Green gram – peak demand period – around 35 DAS. (Days after sowing)

Maize – Peak demand period – around 50 DAS.

(*ii*) Competition for light (solar radiation), nutrients, air space are minimum among component crops.

(*iii*) Complementary should exist between the component crops.

On the basis of above knowledge, suitable combinations are.

Principal Crop	*Intercrop*
Sugarcane	Wheat, Cowpea, Soybean, Moong, Sunflower
Sorghum	Cowpea, Soybean, Moong, Urd, Arhar
Maize	Cowpea, Soybean, Urd, Arhar. Castor
Bajra	Cowpea, Soybean, Urd
Cotton	Soybean, groundnut
Potato	Wheat, radish

Advantages

(i) To raise productivity per unit area.
(ii) Better utilization of growth resources – nutrient, water, light, air.
(iii) Additional income from component crop.
(iv) Quick growing component crops, suppress harmful weeds by thriving in the interspace of base crop.
(v) Gainful utilization of labour – increasing more man days employment potential.
(vi) Less incidence of insect pests and disease attack.
(vii) Less erosion losses.

Disadvantages

(i) Fertilizer management is difficult because nutrients requirements is different for different crops.
(ii) Difficult in harvest – different seeding time for different crops.
(iii) Certain circumstances (*e.g.* allelopathy) – one suppress growth for another crop.

(b) *Mixed Cropping*

- Growing two or more dissimilar crops, simultaneously intermingled without any row pattern.

(c) *Sequential Cropping*

- Growing of two or more crops in sequence on the same piece of land in a farming year.
- Succeeding crop is planted after preceding crop harvested

(d) *Relay Cropping*

- Planting of succeeding crop before the harvest of preceding crop.

(e) *Ratoon Cropping (or) ratooning*

- To raising a crop with regrowth coming out of roots or stalks after harvest of the crop.

  e.g. – sugarcane, fodder, grasses.

(f) *Multistoried Cropping/Multilevel Cropping*

- System of growing crops of different heights together at the same time on same piece of land.
- Using land, water, space most efficiently or economically.

  e.g. Coconut + pepper + grasses,
  Coffee + banana + arhar.

3. ***Crop Rotation***

- Recurrent succession of crops on the same piece of land either in a year (or) longer period of time, to get maximum profit with least investment and without impairing soil fertility.
- Crops should be rotated on at least a three to four year cycle.

Principles for Successful Rotation —

- Crop with tap roots and fibrous root system.
- Leguminous crops with non-leguminous crop.
- Selection of a crop based on demand of people of that area.
- Selection of crop should be problem based.

(a) Sloping lands – alternate cropping of erosion promoting crops with erosion resisting crops.
(b) Dryland (or) partially irrigated area – crops which can tolerate drought *e.g.* oilseeds and pulses.
(c) Low lying (or) flood prone area – crops which can tolerate water stagnation *e.g.* Rice, jute.

Advantages

- Improvements in soil structure.
- Proper utilization of all the resources and inputs.
- Helps in controlling insect pests, diseases and weeds.
- Agricultural operation done timely for all crops.
- Soil fertility is maintained.
- Get better price for produce.

How to Rotate Crops:

Crops should be rotated on at least a three to four year cycle. They should be rotated every year. So a crop of corn planted this year is not planted in the same field for the next two or three years. Ideally, altogether different crops should be used each year as insects and disease that affects one crop will also likely affect similar crops, i.e. cabbage and broccoli are of the same family and should not follow each other. Crops are changed year by year in a planned sequence.

Planning Considerations Patterns, though not agreed upon by all, emerge when deciding the best rotation plan. Legumes are generally a beneficial preceding crop. Potatoes yield best after corn. Some preceding crops (peas, oats, barley) increase the incidence of scab on potatoes. Corn and beans are not greatly influenced by the preceding crop. Carrots, beets and cabbages are generally detrimental to subsequent crops.

When legumes are used in a crop rotation, the nitrogen formed by fixation should be taken into account when determining the nutrients needed for future crops. Soil fertility levels should be regularly monitored and fertility maintained within the acceptable range for all crops in the rotation.

When planning out a rotation, divide the crops into their families. This follows the principle of not growing out the same crop or one in the same family. Beets, chard and spinach are in the same family for example. Also look at how much space the crop will require. Radish requires much less than corn.

In addition to rotating crops, many farmers rotate their livestock among different sections of pasture. This facilitates the dispersal of manure in the fields as well as prevents over-grazing of any one section. Overgrazing of pasture can lead to a depletion of vegetation and consequent soil erosion.

A few notes:

- Crops must be suited to your soils.
- Rotations that include small grains or meadow provide better erosion control.
- Small grains and meadow can always be used to replace any row crop or low residue crop to gain better erosion control.
- Corn (grains) can always be used to replace soybeans or any other low residue crop in the rotation to gain better erosion control.
- For crop rotations, which include hay (meadow), the rotation can be lengthened by maintaining the existing hay stand for additional years.
- Avoid planting a grass after a grass if possible.

13.1.4 Impact of Cropping Patterns

Factors responsible for Changes in Cropping Patterns

- Development of high yielding and short duration varieties in main crops.
- Development of production technology to grow them.
- Incentives and innovations provided by central and state govts.
- Spread of irrigation.
- Development of fertilizer production and technology.
- Govt's special programmes - enhancing production of particular crop.
- Changes in the socio-economic conditions of the people and rapid industrialisation in the country.
- Transformation of subsistence agriculture into market oriented economy.

Negative impacts of – Changes in Cropping Patterns

- Introduction of HYV and short duration varieties. Consequently led to intensive cropping and resulted in decline in fertility and deficiency of Zn, Mn in Punjab due to introduction of paddy-wheat rotation cropping.
- Cause more incidence of insect pests and weeds.
- More use of fertilizers – environmental pollution and salinity.
- Increased use of irrigation water – salinity and water logging.
- Indiscriminate use of pesticides – adverse effects on environment and microorganisms.
- Resistance of insects and pathogens towards.
- HYV – rice, wheat pushed pulses and oilseeds to marginal areas.

(a) Package of Practice for Cereals and Millets

S.N.	Crop	Sowing Time	Varieties	Spacing and Seed rate	Manures and Fertilizers	Water Management	Weed Management	Yield
1.	Rice	IR-8, IR-28, TKM-6, Jaya, Ratna Basmati - 370	Kharif-Jun-July Rabi - Mid Oct. Mid Nov.	(K) 20 × 15 cm (R) 15 × 15 cm Broad Casting 100 kg/ha Nursery 30-40 kg/ha	FYM - 10-15 t/ha NPK - 120-60-60 kg/ha	Critical stage – (CS) Seedling, Tillering, Flowering, grain formation	Pre-planting– Basalin @ 1 kg/ha Pre-emr-2kg ai/ha Post-emr - 3 kg ai/ha	40-50 q/ha
2.	Wheat	Sonalika, Kalyansona Arjuna	Rabi-Mid Oct- Mid Nov.	100 kg/ha	FYM - 10-15 t/ha NPK - 80-40-40 kg/ha	CS - Crown Root initiation, Tillering, flowering.	Phalario-Isoproturon @ 7·5 kg ai/ha 30 DAS	40-50 kg/ha (Irrigated) 20-25 q/ha (Rainfed)
3.	Barley	Kailash, Ratna, Jyoti	Rabi-Mid Oct. – Mid Nov.	20 × 100 cm 100 kg/ha	FYM – 10-15 t/ha NPK – 60-30-30 kg/ha	2-3 Irrigation	Phalario-Isoproturon @ 7·5 kg ai/ha 30 DAS	30-35 q/ha
4.	Maize	Ganga safed-2 Vijay, Vikram	Kharif – Mid June - July Rabi – mid. Oct.	50 × 20 cm Hybid - 20-25 kg/ha Composite – 20 kg/ha	FYM – 10-15 t/ha NPK – 100-40-40 kg/ha	CS – Seedling, Tasseling, silking, grain formation	Atrazine @ 1 kg ai/ha (pre-emr)	40-50 q/ha (Hybrid) 30-40 q/ha (Composite)
5.	Sorghum	CSH-1 – 6 CSV-1 – 7	Kharif June-July Rabi–June-Oct.	50 × 15 cm 15 kg/ha	FYM – 10-15 t/ha NPK–100–50–40 kg/ha	CS – Flowering, grain filling	Atrazine @ 1 kg ai/ha (pre-emr)	50 ql/ha (irrig.) 25 q/ha (RF)
6.	Pearl Millet	HB-5, PHB - 10	June-July	50 × 15 cm 5 kg/ha	FYM – 15 cm NPK – 80-40-30 kg/ha	CS – Tillering, Flowering, grain, filling	Atrazine @ 1 kg ai/ha (pre-emr)	35 q/ha (Irrig.) 15 q/ha (RF)

(b) Package of Practice for Pulses

S.N.	Crop kg/ha	Varieties	Sowing Time	Spacing and Seed rate	Fertilizers kg/ha	Water Management	Weed Management	Yield
1.	Gram	PUSA 209, Gausar, Ujjain 21	IInd fortnight of Oct.	80-100 kg/ha	N - 15-20 kg/ha P_2O_5 - 50-60 kg/ha	Flowering Grain Development	Pre-planting Basalin 1 kg ai/ha Pre-emergence Tribunil 2.5 kg ai/ha	15-20 q/ha
2.	Red gram	UPAS 120, ICPL 151	June-July	60 × 15 cm 10-15 kg/ha	N – P – K 20 : 50 : 30	Flowering, Grain Development	Pre-planting Basalin Pre-emergence Tribunil 2.5 kg ai/ha	20 -30 q/ha 1 kg ai/ha
Development								
3.	Green gram/ mung	T-44, Vamban-1	Kh.-June-July Sum.-Mar.-Apr.	30×45 cm 12-15 kg/ha	20 : 40 : 0	Flowering, Grain Development	Pre-planting Basalin 1 kg ai/ha Pre-emergence Tribunil 2.5 kg ai/ha	12-15 q/ha
4.	Urd/Black gram	T-9, T-27, Pragati	Kh.-June-July Sum.-Mar-Apr.	15 kg/ha	20 : 40 : 0	Flowering, Grain Development	Pre-planting Basalin 1 kg ai/ha Pre-emergence Tribunil 2.5 kg ai/ha	12-15 q/ha
5.	Cowpea	Pusa phalguni,	Kh.-June-July Sum. Mar-Apr.		20 : 40 : 0	Flowering, Grain Development	Pre-planting Basalin 1 kg ai/ha Pre-emergence Tribunil 2.5 kg ai/ha	10 q/ha
6.	Horse gram	CO-1, HPK - 1 & 2	Kh.-June-July Sum.-Mar-Apr.	40 kg/ha	20 : 40 : 0	Flowering, Grain Development	Pre-planting Basalin 1 kg ai/ha Pre-emergence Tribunil 2.5 kg ai/ha	7-8 q/ha
7.	Lantil	L 9-12, Pusa 1		20 × 30 cm 40 kg/ha	20 : 40 : 0	Flowering, Grain Development	Pre-planting Basalin 1 kg ai/ha Pre-emergence Tribunil 2.5 kg ai/ha	18-20 q/ha

(c) Packages of Practices for Oil Seeds

S.N.	Crop	Sowing Time	Varieties	Spacing and Seed rate	Manures and Fertilizers kg/ha	Water Management	Weed Management	Yield
1.	Groundnut	Kh.-Jun-July Ra.-Nov. Sum.-Mar-Apr	TMV - 1, 3, 4 AK-12-24	Bunch-40×15 cm Sprouting - 60 × 20 cm, 80-100 kg/ha	20 : 40 : 15	Branching, Flowering	Pre-planting Basalin 1 kg ai/ha Pre. em. Pendimat- halin 0.75 ai/ha	30-35 q/ha
2.	Rapeseed/ Mustard	Toria-Sept.-Oct Mus. - Oct.	Vasuna, Prakash, Pusa basari	Toria - 30×10 cm Mus. - 45×10 cm 5-6 kg/ha	60 : 40 : 80	Pre-flowering Pod filling	Pre. plt. Fluchloralin 1 kg ai/ha Isoproturon 1 kg ai/ha 30 DAS	Toria-20 q/ha Mus. 25 q/ha
3.	Soybean	Mid June - 1 week July	Bragy, Clask, Shilaject	45 × 10 cm 80 kg/ha	20 : 60 : 40	Seedling, Flowering, pod formation	Pre. plt. Fluchloralin 1 kg ai/ha Pre. em. Metribuzin 1 kg ai/ha	30 - 35 q/ha
4.	Sesamum	Kh.-June-July Rabi Oct. - Nov.	T-12, 13	45 × 15 cm 3-4 kg/ha	Fym - 15-20 t/ha 30 : 60 : 30	Seedling, Flowering, Pod Development	Pre plt. Basalin 1 kg ai/ha	8-10 q/ha
5.	Sunflower		CO1 BSH 1 KBSH 1	8-10 kg/ha	60 - 40 - 20	2 weeks before flowering to 2 weeks after flowering		20-25 q/ha
6.	Castor	June - July	HC-6 Aruna T-3	90 × 40 cm 8-10 kg/ha	40 - 40 - 20			400 - 950 kg seeds/ha

(d) Packages of Practices of Fibers

S.N.	Crop	Variety	Sowing Time	Seed rate and spacing	Fertilizers kg/ha	Water Management	Weed Management	Yield
1.	Cotton	Hybrid-4, Varalaxmi, Suria, Sujata	May-June	50 × 25 cm American cotton 20 kg/ha Indian cotton 15 kg/ha	NPK - 80:40:20 kg/ha	CS – Seedling, branching, flowering Boll formation	Diuron @ 1 kg ai/ha (pre-plant)	15-20 q/ha of seed cotton
2.	Jute	RC-212, 321 RO-632, 878	April–May	30 × 10 cm 10 kg/ha	NPK - 30:20:20	1 - pre sowing Irrigation 3 - post-sowing Irrigation	2-3 weeding at 20 days interval	20-25 q/ha of fibre
3.	Sun hemp	Kanpur-12 ST 55 Bellary	kh - in North rabi - in South	30 cm (Row) 15 kg/ha				14 q/ha
4.	Flax	F895 F 896	Rabi	20 × 2 cm 40 kg /hac		1-2 Irrigation		

(e) Packages of Practices for Fodder Crops

S.N.	Crop	Variety	Sowing Time	Seed rate and spacing	Fertilizers kg/ha	Water Management	Yield (q /ha)
1.	Berseem	Pusa giant, T 526	Rabi-Oct.	25-35 kg/ha	15-20 t/ha Organic Manure	10-12 Irrigation	600 - 800 q/ha
2.	Lucerne	Rarrbler	Rabi - Mid Oct. - Nov.	30 cm (Row) 15 kg/ha		Irrigate the crop after every cut Winter – 4 days Interval Summer - 10 days Interval	500-600 q/ha
							2000 q/ha
3.	Napier	Pusa giant, Pusa Napier 1	July	90 × 30 cm 12500 cuttings/ha	2500-300 q Org. Manure	Sensitive to water logging	

13.2 WEED SCIENCE

13.2.1 Weed

Plants out of place. Weeds are unwanted and undesirable plants which interfere in the utilization of land and water resources and thus adversely affect human welfare.

- Relative weeds-crops plants in a crop
- Off type plants-rogue-local wheat variety in Sonalika variety.

13.2.2 Characteristics of Weeds

- Prolific seed producer
- Prolonged viability
- Dormancy
- Capable of reproducing by underground portions.
- Dispersed through various agents.
- Inherent hardiness against heat, cold, soil abnormalities
- Capable of escaping the destruction caused by man and animal
- Capacity to self regenerate

13.2.3 Propagation of Weeds

1. Asexual – Algae, Ferns
2. Vegetative-Rhizomes, Runners, Stolons, Suckers, Offsets, Tubers, Bulbs, Bulbils, Stems and Roots.

13.2.4 Dispersal/Dissemination of Weeds

- Reproductive unit may fall near the mother plant
- A portion of mother plant may be carried away as contaminant with agricultural crop seeds
- Dispersed to short\long distances by dispersing agents

1. **Wind :**
    - Weed seeds and fruits have specific organs to float in air
    - Pappus hairs (dandelion)
    - Whole plant may be detached and rolled-wild safflower
    - Wind swings entire plant and cause mature capsule to disperse the seeds.
2. **Water :**
    - May drift either as whole plant or plant fragments.
3. **Birds and animals :**
    - Fruits are eaten and seeds are dispersed
    - Lantana seeds by Myna, Chinese turtle dove.
    - Farm animals carry seeds and fruits on their skin and hair.
4. **Man :**
    - Farm machinery, carts and automobiles
5. **Manure :**
    - Farm yard manure.

13.2.5 Classification of Weeds

1. **Based on Duration :**

    (*a*) Annuals : Weeds present along with annual crops. Duration is short.

    - Multiply by seeds and produce large number of seeds/plants.
    - Prevention of flowering or removal of weeds before flowering is important.

    e.g. Amaranthus, Argemone, Cenchrus.

    (*b*) Biennials : *Alternanthera echinata, Eichorium intybus.*

    - Life cycle two years
    - I year- vegetative; II year – produce flowers and seeds.

    (*c*) Perennials : *Cyanodon dactylon and Cyperus rotundus*

    - Grow more than two years.
    - Propagate vegetatively in addition to seeds.

2. **Based on cotyledons ;**
   Monocots – Dalapon and Phrchotaralin
   Dicots – 2,4 D, MCPA.
3. **Morphological characters :**
   Grasses : *Echinochloa, Cyanodan.*
   Sedges : Cyperaceae – *C. rotundus; C. deformis.*
   Broad leaf weeds : all dicot weeds

13.2.6 Allelopathy

Definition

The phenomenon of one plant having detrimental effect on another through the production of chemical compounds.

- It depends upon a chemical compound being added to the environment by an allelopathic agent.
- Differs from competition in that competition leads to removal or reduction of some factors (water, nutrients and light) from the environment that is required by some other plant sharing the habitat.

Effect

- May be direct or indirect harmful effect.
- Inhibit seed germination or reduce the growth of other plant species.
- Phenolic acids, terpenoids, flavonoids.
- Released as vapour, as leachings from the foliage, as exudates from the roots or decomposition of dead plant residues.
- It is the mechanism through which weeds affect crop growth.
- Significant role in maintaining the balance.

Types

True allelopahy : The direct or indirect harmful effect on the other crops through the release of toxic substance as such from the plant.

Functional Allelopathy : When precursor is released which is converted into active substances by some microorganisms called functional allelopahy.

Based on Inhibition —

Auto inhibitors : Toxic substances released from one species affect the other species.

Allo inhibitors : Toxic substances released from one species affect the same species.

I. Effect of weeds on crops : *e.g.* wheat.
   Seed exudates *Avena fatua* affect the germination and seedling growth of maize.
   Leaves and inflorescence, *Amaranthus spinosus* affect the vegetative growth.
II. Crop plants on weeds : Wheat, oats and peas *c . album* affect biomass production and leaf Surface.

13.2.7 Activity and Selectivity of Herbicides

Activity refers to the ability of herbicides to control weeds, where as selectivity is the ability of herbicide to control weed without affecting the economic plant in a mixed population.

Herbicide activity is related to phytotoxic effects of a chemical on the plant growth and development.

Herbicide selectivity refers to the phenomenon where in a chemical kills target plant species in a mixed population without harming or slightly affecting other plants.

Factors affecting herbicide activity and selectivity

(*a*) *Formulation, time, method and rate of application* :

Formulation affect solubility, volatility and specific gravity and phytotoxicity of a herbicide.

Variation in the placement of a herbicide can affect its activity.

A herbicide is selective at a lower rate may become non selective when applied at higher rates.

(*b*) *Plant Morphology* :

Horizontally arranged leaves intercept more of the spray than upright leaves. Presence of leaf hairs, thick cuticle prevents the absorption.

(*c*) *Cultivation Practices* :

Tillage practices bring out the dormant weeds to the top soil exposing them to sun resulting in germination, thereby increasing the susceptibility to herbicides.

(*d*) *Absorption* :

The differential absorption from one plant species to the other which determines herbicide activity and selectivity.

(*e*) *Translocation* :

Herbicide enters the plant must be transported to the site of action and disturb metabolic activities of plant.

(*f*) *Metabolism* :

Any herbicide which enters the plant undergoes metabolism. It is well known that tolerant plant inactivates a herbicide in one or many path ways.

(*g*) *Environmental Factors* :

The response of a plant to an applied herbicide depends upon the environmental stress occurs.

13.2.8 Crop-Weed Competition

Weed species absorb and accumulate nutrients faster than the crop plants. Application of high levels of fertilizers may increase the impact of the weed while weed growth is stimulated more than crop growth.

Competition for Light

Weeds grow faster and shade the crop plants if not checked. Weeds deplete photosynthetically active radiation resulting in reduction of photosynthesis and shortening the life of lower leaves. For *e.g. Amarnathus hybridus* reduced the photon flux density reaching the cotton canopy by 90% at midday.

Competition for Water

The amount of water required to produce an unit amount of dry matter is known as the transpiration ratio. The transpiration ratio of weeds is high as compared to crops.

Critical Period of Crop-Weed Competition

The critical period of crop weed competition is the period from the time of sowing upto which the crop is to be maintained in a weed free environment to get higher yield.

Importance of critical period of crop-weed competition in weed management

Weeds generally germinate either before or at the same time as the crop, offer serious competition to the crop plants since they got opportunity to establish and accumulate dry matter faster than crop plant *e.g.* In maize the first 2-3 weeks of it emergence, weeds put forth 15-18% of their growth while maize attained only 2-3%. Hence it is desirable to identify the critical period of crop-weed competition for all crops to get higher economic returns.

13.2.9 Integrated Weed Management (IWM)

"A weed population management system that uses all suitable techniques in a compatible manner to reduce weed population and maintain them at levels below those causing economic injury.

Need for IWM

- No single control measure can be effective in all cropping situations.
- Usage of same control method leads to build up of tolerant weeds.
- It is uneconomical to eradicate all weeds that some may provide food and shelter for insect predators and predators that reduce other pests.

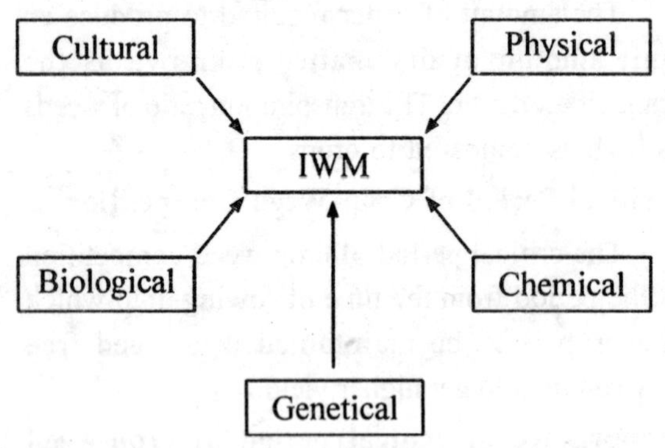

Cultural Methods —

(*i*) Field Sanitation – ploughing, burning

(*ii*) Sowing/Planting Method – weed seed free

(*iii*) Variety Selection – Selecting done

(*iv*) Planting Density/Spacing – closer

(*v*) Irrigation and Drainage – submergence alternate wet and dry

(*vi*) Cropping Systems – crop rotation, intercropping, mixed cropping

Physical Methods —

(*i*) Hand weeding – rice, tomato

(*ii*) Hand hoeing – onion, finger millet

(*iii*) Mowing –cutting at ground level

(*iv*) Dredging and chaining – aquatic weeds

(*v*) Burning

(*vi*) Mulching

(*vii*) Intercultivation

Biological Control Measures —

Insects – *zygogramma (parthenium sp.)*

Snails – *Mariaga sp.*

Fungi – Water hyacinth.

Chemical Methods —

Selective herbicides – Atrazine, 2,4 – D, Butachlor.

Non selective herbicides – Paraquat, Diquat.

Genetical Methods —

Manipulation of herbicide resistant gene in cultivated crop that facilitate application of herbicides that are non selective. E.g. tobacco, soyabean and maize etc.

Advantages of IWM

- All types of weeds can be controlled.
- Yield reduction is prevented.
- Evolution of tolerant weed is arrested.
- Ecologically sound and environment friendly.
- Avoiding herbicides reduces the pesticide load in the environment.
- Prevent perennial weed shifts.
- Economically feasible.

13.2.9 (*a*) Principles of Weed Control

1. Prevention
2. Eradication
3. Control
   - Mechanical
   - Cultural
   - Biological
   - Chemical

1. Prevention

- All measures to deny the entry and establishment of new weeds in an area
- Use of clean crop seeds
- Care on the grains fed to cattle
- Prevent movement of cattle
- Clean farm implements
- Quarantine

2. Eradication

- Complete elimination of particular weed species from an area.
- Killing and destroying the viability of organs of reproduction.
- Eradication of perennial weeds cannot be accomplished in one season.

3. **Control :**

(*a*) ***Mechanical or physical method*** :

(*i*) *Hand pulling* :

- In home gardens where hoe cannot reach.
- Best for annuals and biennials weeds.
- Useful in lawns after irrigating it.
- Time consuming and costlier.

(*ii*) *Hoeing* :

- Used in wide row crops.
- Complete destruction of all top growth.
- Hand rotary weeder and wheel hoe.

(*iii*) *Tillage* :

- Weed control is one of the principles of the tillage.

(*iv*) *Mowing* :

- The object is to prevent seed production and remove unsighted weed growth.
- Used along roadsides ,waste places and pastures.

(*v*) *Flooding* :

- In rice fields.
- Water is stagnated 6-10 inches for 3-8 weeks.

(*vi*) *Burning* :

- To destroy top of weeds.
- Used in cotton to destroy weeds in rows.

(*vii*) *Smothering with non living materials* :

- Use of artificial Mulch- straw, hay, plastic film, etc.
- Main purpose –to cut the light supply.
- 2-4 inch thick mulch layer is required.

(*b*) ***Cultural method*** : Not directly related to control weeds but helps to increase the yield.

(*i*) *Crop stimulation* :

- Modification of cropping conditions and soil leads to vigorous growth of crops which can compete better with weeds.
- Addition of sulphur, gypsum and altering the pH favourable to crops.
- Banding or seed dressing of fertilizers and manures.
- Foliar spray of fertilizers.

(*ii*) *Crop rotation* :

- Wild oat –using peas/gram as break crop for 2-3 years.
- Fodder –planting grain crops instead of lucene.

(*iii*) *Competitive or Smother cropping* : (cowpea, millets, lucerne)

- They germinate quickly and form larger foliage and deep roots which exert their influence against weed seedlings.

(*iv*) *Summer fallowing* :

- Air temperature of 40-45°C.
- Not recommended for control of perennial weeds on light soils due to fear of erosion.

(*c*) ***Biological control*** :

- Control by living organisms called natural enemies which are encouraged or disseminated by man.
- Maintenance of balance between the target weed and its natural enemies.

Criteria for selection of biological agents :

- Host specific.
- Able to withstand new environment.

- Should be free of its parasites and predators.
- High mobility.
- Faster reproduction.

Kinds of bioagents :

(*a*) Herbivorous fish-carp fish for aquatic weeds.

(*b*) Snails

(*c*) Mites

(*d*) Fungi-Cephalosporium sp. on *Acacia glauca*

Merits :

- Self regulatory.
- No residue and environment is safe.
- Useful when larger area is infested by a single weed.

Demerits :

- Slow process and complete control is not possible.

(*d*) ***Chemical control*** :

Inorganic herbicide — Sodium chlorate, sodium arsenate, copper sulphate, copper nitrate.

Organic herbicide — Organic acids, diesel oil and salts.

Selectivity :

Selective herbicide — Simazine, Atrazine and 2,4 – D.

Non selective herbicide – Paraquat and Diquat

Translocation :

Systemic — Simazine, 2,4 – D.

Contact — Paraquat and Diquat.

13.2.9 (*b*) Biological Weed Control

Use of living organisms *viz.* insects, disease organisms, herbivorous fish, snail or even competitive plants for the control of weed is called biological control.

Merits

1. Least harm to the environment
2. No residual effects
3. Relatively cheaper and comparatively long lasting effect
4. Will not affect non-targeted plants and safer in usage.

Demerits

1. Multiplication is costlier
2. Control is very slow
3. Success of controlling is very limited
4. Very few host specific bio-agents are available at present

Outstanding and feasible examples of biological weed control

(A) *Specific bio-agents*

Weed Control by Bio-agents

Sl. No.	*Weed species*	*Agent*
1.	*Salvinia molesta* (water fern)	Curculinoid weevil - *Crytobagus singularis*
2.	*Tribulus terrestris*	*Microlarinus lypriformis, M lareynil*
3.	*Opuntia dilleni*	Cochineal scale insect - *Dactylopius tomentosus*
4.	*Lantana camara*	*Crocidosema lantana, Teleonemia scrupulosa*
5.	*Parthenium hysterophorus*	*Zygrogramma bicolarata*

Bio-herbicides : The bio-herbicide approach involves spray of specific fungal spores or its fermentation products against the target weed. These preparations are called 'Mycoherbicides'.

Non-specific agents : White amur feeds several weed species Lemna, Hydrilla, Potamogeton. Besides silver carp, common carp also feed large amount of algae.

Mycoherbicides and target weed

Sl. No.	*Product*	*Content*	*Target weed*
1.	Devine	*Phytophthora palmivora*	Strangle vine Morrenia odorate in citrus
3.	Bipolaris	*Bipolaris sorghicola*	Jhonson grass-Sorghum halepens
4.	Biolophos hygroscopius	*Steptomyces general vegetation*	Non-specific

Competitive plants : The weed parthenium is suppressed by growing *Cassia sericea*. The plant leachates of Cassia have 'Kolinei' which accumulate in the soil and interfere with the germination of parthenium.

Paragrass - *Brachiaria mutica* has been found highly competitive to *Typha* sp. in ditches.

Biological control of weeds in crop fields in India

Eg. Control of *Ludwigia parviflora* in rice by steel blue beetle (*Haltica cynea*)

Control of *Cyperus rotundus by Bactra verutana.*

13.2.10 Recent Advances in Weed Science

Adoption of integrated weed management practices, appropriate cultivation technologies, sound crop rotation and crop diversification check the spread of weed seeds. However, crop weed interactions are complex and needs application of recent advancements for the development of future weed management strategies.

Weeds of Wetland, Garden Land and Dryland

Wetland weeds	*Gardenland weeds*	*Dryland weeds*
I. Sedges	***I. Sedges***	***I. Sedges***
Cyperus diformis	*Cyperus rotundus*	*Cyperus rotundus*
Cyperus iria		
Cyperus rotundus	***II. Grasses***	***II. Grasses***
	Cynodon dactylon	*Cynodon dactylon*
	Echinochloa colonum	
II. Grasses	*Panicum repens*	*Eleusine indica*
Echinochloa crusgalli	*Chloris barbata*	*Panicum sp.*
		III. Broad leaved weeds
Panicum sp.	***III. Broad leaved weeds :***	*Tridax procumbens*
III. Broad leaved weeds :	*Acalypha indica*	*Striga lutea*
Ammania	*Amaranthus viridis*	
baccifera	*Boerhaavia diffusa*	

A. Modelling crop-weed interactions

Competition between crop and weed for resources of growth (light, water and nutrients) is a critical process in agricultural ecosystems. The mechanisms of competition are not simple. Models of weed invasion, population growth and control will be useful for organising biological information on weeds and for developing weed control strategies.

B. Impact of Herbicide Use

The potential weed resistance to herbicides effect on non-target organisms and herbicide residues in the ecosystems needs to be monitored.

Herbicide resistance in weeds : The weeds may develop partial resistance, cross resistance and multiple resistance to the applied herbicide.

Herbicides in Agro-ecosystem : The persistence is very important factor because it determines the degree of weed control, the risk of carry over effect on the succeeding crop and the residue in soil and ground water. Various management techniques, viz., use of optimum dose of herbicides, application of FYM, crop rotation, and leaching the soil have been developed to minimise the residue hazards in soils.

C. Biological Control of Weeds

The idea of biological control originated for dealing with noxious weeds on non-crop lands, and later it was extended to water bodies. But now, some success has been achieved in biological control of weeds in crop fields.

Biological control approach makes use of the invasive plant's naturally occurring enemies to help reduce the invasive plant's impact on agriculture and the environment. It simply aims to reunite weeds with their natural enemies and achieve sustainable weed control. These natural enemies of weeds are referred to as biological control agents.

D. Biotechnological approaches

Bioherbicide which can be derived from fungi are Myco herbicides *viz.*, **'Devine'** containing a formulation of soil borne fungus *Phytophthora palmivora* and **'collego'** containing spores of *Colletorichum gloeosporoides* are commercialised bioherbicides

Use of Bioherbicides is another way of controlling weeds without environmental hazards posed by synthetic herbicides. These are made up of micro-organisms (*e.g.* bacteria, viruses and fungi) and certain insects (*e.g.* parasitic wasts, painted lady butterfly) that can target very specific weeds. The microbes possess imasire agents that can attack the defensegenes of the weeds, thereby killing it. The better understanding of the genes of both micro-organisms and plants allowed scientists to isolate microbes (pathogens) whose genes match particular weeds and are effective in causing a fatal disease in those weeds. Bioherbicides deliver more of these pathogens to the fields. They are sent when the weeds are more susceptible.

The benefit of using bioherbicides is that it survive in the environment long enough for the next growing season where these will be more weeds to infect. It is cheaper compared to syntheticpesticides.

- Molecular markers in weed science
- Herbicide resistant crops

E. Allellopathy in weed management

Although the allelopathic suppression of crops by weeds has been well documented, the research on allelopathic effect of crops on weed has been very limited release of phytotoxic ferulic acid from wheat crop residues suppressed *Ipomea purpurea* and *Sida spinosa.*

13.3 DRY LAND

13.3.1 Dry Land Farming

Dry land farming is defined as farming dependent exclusively on rainfall which is often deficient, erratic and unreliable. There are no supplemental irrigation sources, annual rainfall is low, usually less than 750mm.

Characteristic Features

- Farming dependent solely on rainfall
- Potential evapotranspiration is more than precipitation.
- Soil moisture storage is often inadequate for crop growth.

Differences between Dry Farming and Rainfed Farming

Dry Farming	*Rainfed Farming*
Rainfall is less than 750 mm.	Rain fall more than 750 mm.
Rainfall distribution is erratic.	Rainfall is well distributed.
PET exceeds rainfall.	Rainfall is more than Potential Evapo Traspiration (PET).
Arid and semiarid climate.	Humid and temperate climate.
Crops are subjected to severe moisture stress.	Crops do not face moisture stress.
Soil moisture conservation is important.	Runoff and erosion control are important.

- Cropping is confined to short seasons only.
- Droughts are frequent.
- Crop yields are not stable and fluctuate widely from year to year.
- Farming is mostly of the subsistence type.

13.3.2 Soil Moisture Conservation

It is a method of using and managing land based on the capabilities of the land itself, involving the application of the best practices leading to profitable crop production.

Prerequisites for soil moisture conservation are—

1. Physiographic. 2. Rainfall.
3. Flood, Drought.

Needs/Objectives

1. Efficient collection of rain water.
2. Reducing percolation losses and Evaporation losses.
3. Preventing soil erosion.
4. Water harvesting at the time of availability.
5. Efficient use of stored moisture.
6. Absence of supplementary irrigation source.

Types of Soil Moisture Conservation

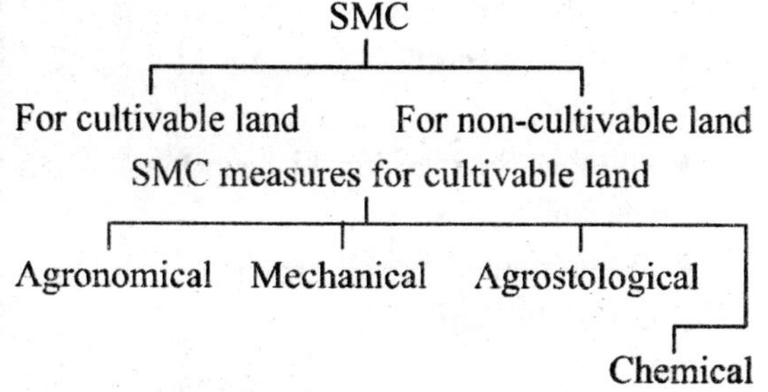

1. ***Agronomical/Cultural***
   - Mulching
   - Crop Management
   - Conservation tillage

(I) Mulching

- Covering the soil surface with any material.
- Materials- crop stubble, straw, coir pith, groundnut shell, husk.

Types of Mulching —

1. Vertical-Trench - 40 × 15 cm at 2-4 m internal across slope - filled with organic matter.
2. Live mulch : Plant can grow by inter cropping.
   e.g. - sorghum and cowpea.
   Sorghum + Sword bean.
3. Stover mulch/straw mulch.
   - cumbu, sorghum straw, sugar cane trash.
4. Stubble mulch.
   - Stirring of soil - Crop residues protect from erosion by minimum disturbance.
   - even up to 3% slopes.
6. Pebble mulch.
   - Useful in dry land fruit tree culture on base of tree increase infiltration, reduce runoff loss.

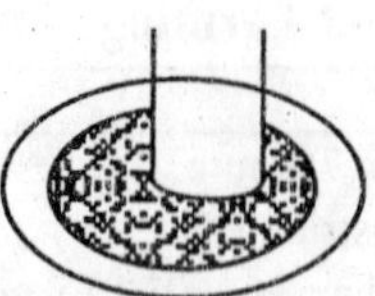

(II) Crop Management

- Multiple cropping
- Agroforestry
- Crop alternation
- Strip cropping

1. Multiple cropping :
   - Avoiding fallow
   - Crop rotation
   - Smothering crop mixture
   - Mix with shallow and deep rooted crop (*e.g.*) maize + Red gram
2. Agroforestry :
   - Agri Silvi culture
   - Silvi pastural
   - Agro silvipastural
3. Crop alternation :
   - Shallow rooted crops with deep rooted crops
4. Strip cropping :
   - Crops are planted in narrow strips across the slope altered with strips of erosion resistant crops and erosion permissible crops.
   - 2 types

(*i*) *Contour strip cropping* :

- Inter tilled row crop + growing erosion resistant crop - right angle to flow of water

(*ii*) Field strip cropping :

- Uniform strip against slope but without contour.

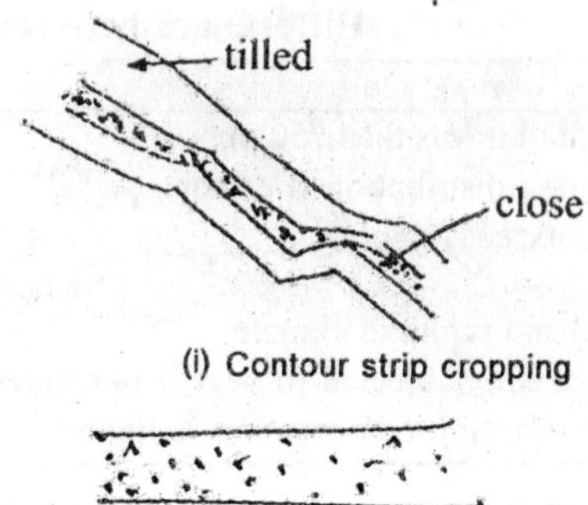

(i) Contour strip cropping

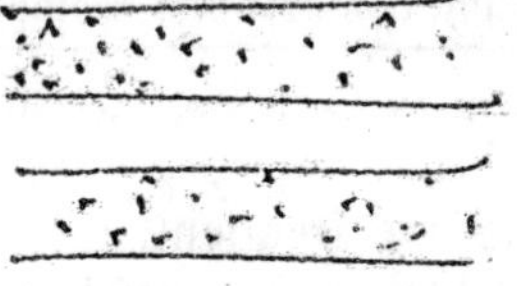

(ii) Field strip cropping

(III) Conservation tillage

1. Off season/summer tillage
2. Contour ploughing/contour tillage
3. Minimum tillage

2. *Mechanical Measures*

(*a*) Basin listing

(*b*) Bunding

(*c*) Ridges and furrows

(*d*) Broad bed Furrows

(*e*) Terracing

I. Basin Listing

- Small depressions of 10-15 cm depth at vegetative interval by basin listing.
- Usually before sowing.
- Basin listing provides maximum time for waater infiltration into the soil.

II. Bunding

- across slope at suitable intervals

III. Ridges and Furrows

- 30-40 cm width, 15-20 cm ht - across the slope
- suitable for medium to deep black and red soil
- Wide spaced crops grown i.e. maize, cotton, tomato

3 Types bunding

Contour bunding

- 1 m basal width 0.5 m ht. along contour
- Suitable for red soil with 1% slope
- Not for heavy black soil
- Permanent

Graded bunding

- 20-40 cm basal width, 15 cm ht.
- across + both direction
- deep red soil upto 1% slope
- not for black
- reshaped when required

Compartmental bunding

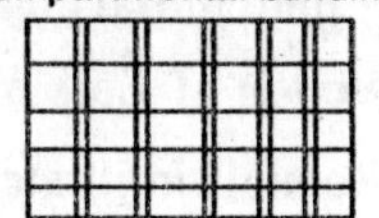

- 15 cm width 15 cm ht.
- along + across both
- red, black soil 0.5 - 1% slope
- Suitable for broardcasted crop

IV. Broad bed Furrow

- beds with 1.5 m width, 15 cm height
- Furrows of 30 cm height
- Suitable for heavy black soil
- Slope upto 0.6%

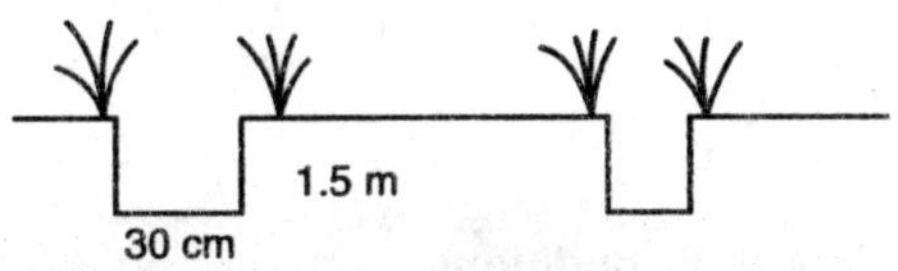

Broad bed furrow

Advantages —

- accomodate wide range of crops
- Suitable for intercropping
- Runoff reduced
- Sowing can be done by seed drill

V. Terracing

(*a*) Bench terracing

- 16-33% slope

Cut
Fill
to rain
Cut
bench portion
Shoulder bund

Bench terracing

- series of level strips (or) plat forms across slope
- intensive cropping possible

(*b*) Zing terracing

- 3-10% slope
- medium to deep soil of heavy rainfall areas
- field divided - 2 : 1 [donor; receiving area]

3. *Agrostological/Biological*

(*i*) Pasture/Grass land

(*ii*) Strip cropping with grasses

(*iii*) Ley farming

- fodder grass + legume in rotation

4. *Chemical*

Soil Moisture Conservation measures for Non cultivable lands

1. Agroforestry
2. Pasture
3. Vegetative barriers
5. Mechanical measures
   - Bunding
   - Terracing

Advantages —

- Prevent direct 'splash' erosion by raindrop
- Increase infiltration
- Fit easily in any farming system
- Less expensive
- Less management

Disadvantages —

- identification very difficult
- high initial cost

Reduction of Loss of Stored Soil Moisture

Rainfall infiltrates into the soil and permits downward and laterally gets stored in soil profile. Part of it percolates down to ground water. Stored water is absorbed by plants and weeds. It is lost from the soil surface as evaporation and from crop and weed canopy as transpiration.

Reduction of Evaporation Loss

Evaporation happens to maintain soil thermal regime and is governed by

- Soil moisture content
- Vegetative cover on surface
- Soil type
- Temperature gradient between soil and atmosphere

Measures to control evaporation loss :

- Shallow surface tillage
- Mulching
- Use of anti evaporant chemicals

The rate of transpiration is governed by:

- Soil moisture potential
- Atmospheric water demand
- Plant canopy characters such as leaf area, stomatal resistance

13.3.3 Drought/Adoptic

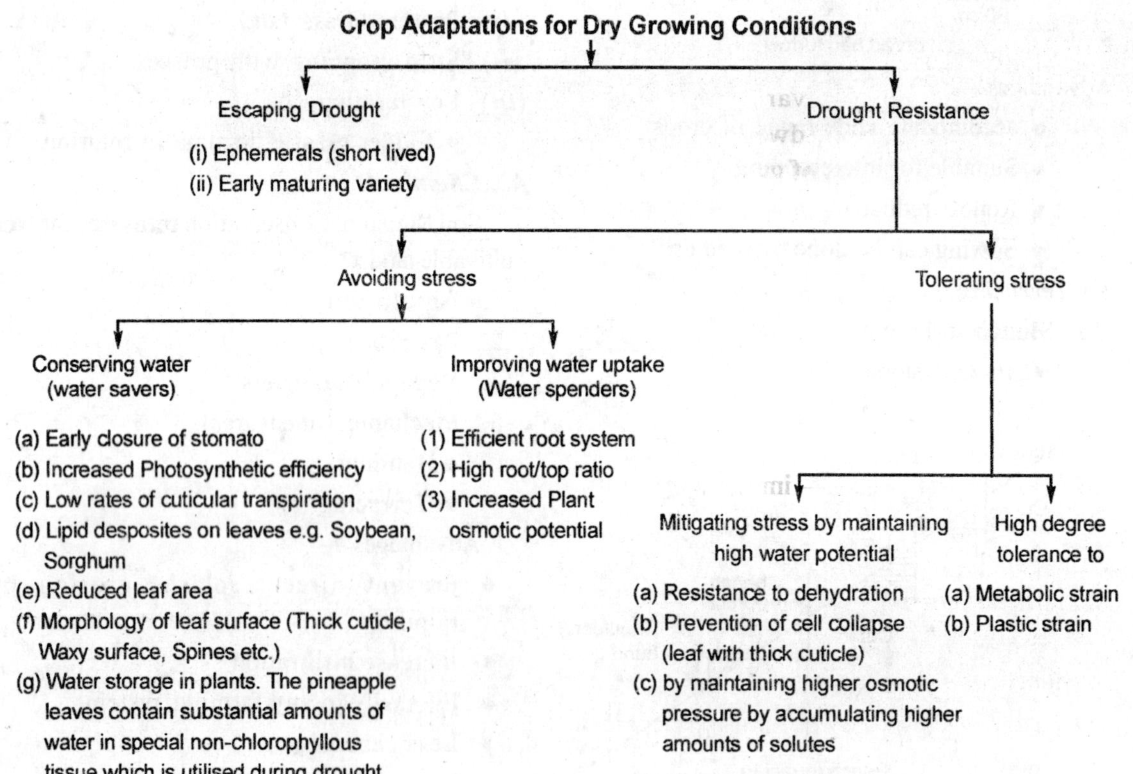

13.3.4 Drought Management Technique for Dryland Development

Dry farming; — Farming practiced in rainfall areas of 400-700 mm/annum.

- Source of water only Rainfall.

Rainfed farming ; > 700 mm Rainfall per annum.

Impact of Drought

(*i*) shortage of food production— failure of crop.

(*ii*) shortage of fodder

(*iii*) shortage of drinking water leads to migration.

(*iv*) shortage of fuel for cooking leads to deforestation.

(*v*) Death of cattle & Humans

Management of Drought

1. ***Low monitory system***
   - Moisture conservation — intercropping.
   - Choice of crops & varieties — Early maturity, deep root, dwarf shoot, thick leaves. more number of grain per pod
   - Optimum plant population — Reduction in evaporation.
   - Pest management
   - Mulching of plant material
2. ***High cost inputs***
   - Fertilizer use — for maximise the fertilizer use efficiency.
   - Use of tractor power — improve infiltration & better root development.
3. ***Water harvesting*** — Farm ponds, perculation ponds. checkdams, retaining wall etc.

   Benefits —

   (*i*) Ground water available through percolation ponds.

   (*ii*) It creates droughtless condition.

   (*iii*) Less cost.

   (*iv*) Involvement of community participation.
4. ***Transfer of Technology***

   Lab to land programmes through K.V.K.
5. ***Miscellaneous Efforts***
   - Financial assistance in the form of subsidy.
   - Low interest loans for the constructing of farm ponds.
   - Improving the socio-economic status of poor by Govt.
   - Small scale & cottage industries in the rural areas.
   - More emphasis on meterological research (computer & satellites)
   - Comprehensive price policy by Govt.

Infrastructure Requirement to Meet Drought

(*i*) Seed banks - facility to the farmers.

(*ii*) Fodder banks

(*iii*) Farm ponds.

13.3.5 Integrated Dry Land Technology

A single technology in isolation will not give desired results. Adoption of all related technologies as an integrated dryland technology alone will provide a synergistic effect and improve the crop productivity on dry regions. The various components of such an integrated dry land technology are the following :

- Insitu soil moisture conservation
- Choice of suitable crops and crop substitution.
- Selection of high yielding drought tolerant varieties
- Tillage to conserve moisture
- Establishment of optimum population
- Soil fertility management

- Crop protection against weeds, pests and diseases

Transpiration loss can be reduced by the use of antitranspirants and by some cultural methods.

Antitranspirants

Antitranspirants are substances or chemicals applied on plant foliage to control rate of transpiration. Based on their mechanism of action, antitranspirants are classified into various types.

- Stomatal closing type e.g. phenyl mercury acetate
- Film forming type e.g. paraffin and wax emulsions
- Reflectant type *e.g.* kaolinite
- Growth retardant type *e.g.* cycocel (CCC)

Cultural methods of transpiration control

- Weed control
- Shelter belts
- Alley cropping

13.3.6 Alternative Land Use Systems for Dry Lands

Uncertain rainfall, poor soil conditions and low level of management has made annual cropping of field crops a non-remunerative enterprise in many pockets of drylands. Sometime, cropping has been given up altogether and lands remain fallow and become wastelands overgrown with unwanted vegetation. To arrest this trend and to bring back the land under economically useful vegetation, alternate land use systems such as grasslands/pastures, agroforestry and horticulture are recommended. This has become necessary for the following reasons.

(*a*) Annual field crop production is nonviable and uneconomical in many years.

(*b*) Yield of field crops is low and fluctuate.

(*c*) Continued use of the eroded and degraded lands under the present system of annual cropping may ecologically degrade the lands.

(*d*) Alternate land use systems such as grasslands and tree culture are less risky, more productive and remunerative in these marginal lands.

The choice of an alternate land use system depends on the land capability. Most of the lands under dry farming tracts come under the land capability classes of III and above.

Land Capability Class	**Alternate Land Use Recommended**
Class II	Dry land horticulture
Class III & IV	Agroforestry or lay farming
Class V	Pastures or Silvipasture or tree farming
Class VI	Range lands/ wood lots

Ley farming offers the following advantages

Ley farming is an agricultural system where the field is alternately seeded for grain and left fallow other name of the method is alternate husbandry. In this method, field is alternately used for grain or other cash crops for a number of years and laid down to ley *i.e.* left fallow, used for growing hay or used for pasture for another number of years. After that period it is again ploughed and used for cash crops.

- Fodder for cattle
- Low risk system
- Soil moisture conservation
- Enrichment of soil fertility
- Control of Perennial weeds

Choice of trees for drylands

Trees suitable for drylands have the following characters —

- Multipurpose tree species
- Adaptable to wide variations in soil and climate
- Withstand severe pruning
- Rapid growth

e.g.

Black soil – *Azadirachta indica*

Red soil – *Albizzia lebbeck, Acacia nilotica*

Dryland Agriculture

The criteria for selection of fruit trees for drylands are :

- Drought tolerance
- Adaptability to varying soil conditions
- Quick regeneration after pruning
- Rapid recovery after stress is removed

e.g.

Black soil – Ber, Sapota, Jamun, Amla

Red soil – Mango, Pomegranate, Amla

Silviculture System for Drylands

- Staggered planting - Trees are grown scattered in the field with annual crops *E.g.* neem + pulses/sorghum
- Border trees - Trees can be grown along farm boundaries. *e.g.* neem, tamarind
- Agrisilviculture

Alley Cropping

Annual field crops are grown in alleys formed by hedge rows of trees and shrubs. e.g. *Leucaena* as hedge with sorghum, maize as intercrop.

Benefits

- Green fodder from hedge rows during dry season and dry fodder from annual crops during rainy season.
- Hedge rows check runoff and erosion when formed along contour or across slope.
- Pruning from hedgerows can be used as fodder, fuel wood or for mulching.

13.3.7 Water Shed

Water shed is a manageable hydrological unit that makes a harmonious use of the prevailing climate, soils, water, locally available material and human resources towards stepping up crop yields.

It is also referred as catchment area or drainage basin.

Common shape of water shed as:

1. The common shapes of watershed are square, rectangular, triangular and polygon.
2. The watershed shape is generally expressed by the terms, form factor and compactness coefficient.
3. There are two types of watershed's shape, in which one is fan shape and other is fern shape. The fan shape watershed produce peak rate of runoff very early, than the fern shape.

Size of Water Shed

- Macro watershed : > 50,000 ha
- Sub-watershed : 10,000 to 50,000 ha
- Milli-watershed : 1000 to 10000 ha
- Micro watershed : 100 to 1000 ha
- Mini watershed : 1-100 ha

Principles of Water Shed Management

1. utilizing the land based on its capability.
2. minimizing silting up of tanks, reservoirs and lower fertile lands.
3. conservation of rainwater.
4. Safe diversion of surface runoff to storage structures.
5. water harvesting for supplemented irrigation.

Objectives

1. Production of food, fodder, fuel, fruit, fish combine on a sustained basis.
2. Over exploitation of resources to be maintained.
3. Water storage at convenient locations for different purposes.
4. Erosion control, eco system safety, economic stability.
5. Recharge of ground water.
6. Reduction of drought.

Steps in Water Shed Management

1. *Basic information* :
   - Statistics of population and livestock.
   - Pattern of land ownership.
   - Topography, cropping system and yield and land capability for farming.
   - Data on rainfall, erosion problems and ground water.
   - Information on existing water sources like tanks, well etc.
   - Service facilities like schools, market, health and veterinary facilities.
2. *Developmental components* :
   - Soil conservation measures.
   - Run off harvest in storage structures and its recycling for protective irrigation.

- Ground water recharge and development
- Efficient use of available water through proper field layouts, land shaping, leveling, lining of water courses and life saving

Main Components of Watershed

Management Programme :

1. Soil and water conservation –
   - Permanent measures-Bunds, Terrace
   - Semi permanent measures-Inter bunds Treatment
   - Temporary measures-Vegetative barrier
2. Water harvesting.
3. Crop management –
   - Optimum seeding time
   - Fertilizers schedules and balanced use of plant nutrients
   - Weed management and package of practices for aberrant weather
   - Contingent crop planning

Benefits of Water Shed Management

1. Increased soil health
2. Decrease in silting
3. Increase in natural resource

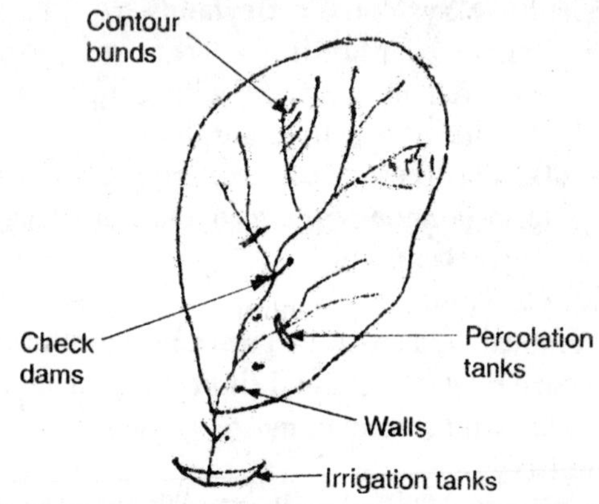

Water shed development for moisture conservation and water management

13.3.8 Modern Concept in Dryland Farming

Aspect of Cultivation	*Traditional Practice*	*Modern Concept*
1. Choice of crop and variety	Local varieties drought tolerant long duration poor yield potential dependence on traditional cropping system.	Improved varieties drought tolerant short or medium duration-high yield potential crop substitution. Sole cropping intercropping. Intercropping with suitable geometry sequential cropping.
2. Tillage	Immediately before sowing summer tillage and off season tillage bullock power.	Off season tillage deep tillage depending on soil mechanisation for early completion.
3. Insitu moisture conservation	No systematic efforts	Land shaping into various configurations depending on rainfall soil crop, slope.
4. Erosion control	Absent	Mechanical cultural and biological methods of erosion control.
5. Sowing		
(*a*) Time of sowing	After rains delay between rains and sowing	Early sowing-premonsoon sowing depending on soil, crop and rainfall.
(*b*) Seed Treatment	Nil	Seed hardening-biofertiliser inoculation.

13.3.8 Modern Concept in Dryland Farming

Aspect of Cultivation	*Traditional Practice*	*Modern Concept*
(*c*) Sowing method	Broadcasting sowing behind plough local seed drill	Mechanised seed drill bullock drawn or tractor drawn.
(*d*) Density	Low	Optimisation through sowing, seed treatment, proper depth and method of sowing and correct crop geometry.
6. Soil fertility managment	Limited or no use of organic manures and fertilisers	Integrated nutrient management through organic manures, optimal nutrient level time and method of application. Enriched FYM, biofertilser.
7. Crop protection		
(*a*) Weed control	Manual weeding	Integrated weed management-use of efficient tools-interculture.
(*b*) Pest and disease control	Absent in food crops limited in commercial crops.	Integrated pest management through resistant varieties early sowing, seed treatment, intercropping trap crops, summer tillage and dust formulations of chemicals if necessary.
8. Contingency planning and correction	Limited	Recommendations based on rainfall aberration.
9. Landuse and farming system	Mostly subsistance farming with food and fodder crops mostly arable cropping.	Land use system depending on land capability and suitability-integrated farming system.
10. Development approach	Planning for each farms as independent unit.	Development based on watershed approach

13.4 IRRIGATION

13.4.1 Water use Efficiency

The water utilized by the crop is evaluated in terms of water use efficiency.

It can be classified into :

1. Crop water use efficiency

It is the ratio of crop yield to the amount of water used by the crop for evapotranspiration.

2. Field water use efficiency

It is ratio of crop yield to the total amount of water used in the field.

3. Physiological water use efficiency

It is calculated in terms of the amount of fixed per unit of water transpired.

Types of Efficiency of water :

1. Application efficiency

$$= \frac{\text{water directly available to the crop}}{\text{water received at field inlet}} \times 100$$

2. Conveyance Efficiency

$$= \frac{\text{water received at inlet to a block of fields}}{\text{water released at project head works}} \times 100$$

3. Project Efficiency

$$= \frac{\text{water made directly available to the crop}}{\text{water released at head works}}$$

13.4.2 Scheduling of Irrigation

Definition

It's the process of determining when to irrigate and how much water to apply or frequency with which water is to be applied based on needs of the crop and nature of the soil.

Objectives

- Proper scheduling is essential for efficient use of irrigation water.
- It Leads to saving water and energy.
- Higher crop yield, efficient use of inputs and lower production costs.

Approaches to Divide when to Irrigate —

I. ***Soil approach*** : Methods involve determination of water content of the soil and finding the deficit in available at which it's proposed to irrigate.

(*a*) Gravimetric approach

(*b*) Feel and appearance

(*c*) Tensiometers

(*d*) Electrical resistance

(*e*) Water budget technology (ET, Effective Precipitation and soil water contents).

II. *Plant approach* : Monitoring plants is the most direct method of determining when to irrigate.

(*a*) *Appearance and growth* : Indicators which include leaf and shoot wilting, leaf colour, drooping of leaves, rolling of leaves etc., not often these are very effective and confusing.

(*b*) *Leaf temperature* : When partial or full stomatal closure occurs due to reduced transpiration, because of reduced availability of water and increase in heat temperature.

(*c*) *Leaf water potential* : Which indicates the plants need for water. Here the pressure used to measure the leaf moisture potential. This method is not extensively used since considerable time, care and training are required, no obtain reliable results.

(*d*) *Stomatal resistance* : Leaf resistance to vapour diffusion into the atmosphere is primarily governed by the degrees of stomatal closure. Its therefore an index to the need for water.

III. *Meteorological Approach* :

(*a*) *IW/Epan ratio* : Ratio between a fixed amount of irrigation water and cumulative pan evaporation (EPan) from an open pan evaporimeter minus effective rainfall since previous irrigation is used as a basis for scheduling irrigation to crops.

- Smaller IW/CPE ratio - Irrigation @ large intervals.
- Larger ratio - Irrigation @ shorter intervals.

IV. *Rough Methods for Farmers :*

(*a*) *Soil-cum-sand miniplot technique* : Based on plant as an indicator of water need. It's to reduce artificially available water holding capacity, of the soil in root zone depth in the miniplot by mixing sand with it. The field is irrigated when the plants in the mini-plot show moisture deficiency symptoms by wilting.

(*b*) *Can Evaporometer method* : Evaporation from one litre can has a correlation with evopo transpiration from a crop. When the water level in the can has dropped to an amount equal to the amount of water to be applied in each irrigation, its time to irrigate.

How much Water to Apply ?

The amount of water to be applied to the plants can be determined by the following methods :

Net Irrigation Requirement

It's the depth of irrigation water, required to bring the soil moisture in the effective root zone depth to field capacity.

$$\text{NIR} = \frac{(M_{FC} - M_b U)}{100} \times D_b D$$

where,

M_{FC} - Moisture content @ field capacity (%)

M_{bi} - Moisture content before irrigation (%)

D_b - Bulk density of the soil (g/cc)

D - Depth of soil (cm)

Gross Irrigation Requirement

$$\text{GIR} = \frac{\text{Net irrigation requirement}}{\text{Field efficiency of System}}$$

Irrigation Interval

It refers to the number of days between two irrigations in the period of highest consumptive use of the crop growth without rain fall. It's the function of soil, crop and water.

$$\text{Design interval} = \frac{MF_C - M_{bi}}{P_{CO}}$$

P_w = peak period moisture use rate of the crop (cm).

Irrigation Period

It's the number of days that can be allowed for applying the irrigation to a given design area during the peak consumptive use period of the crop being irrigated.

$$I_p = \frac{I_n}{ET\ \text{(crop) Peak}}$$

I_n = net amount of moisture (cm)

ET = peak period water use rate of the crop.

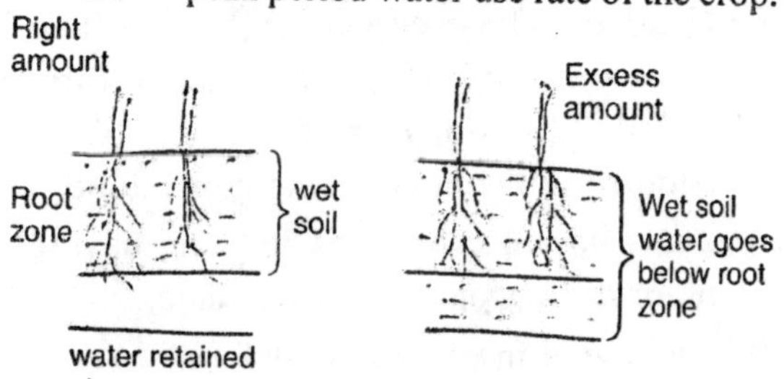

Irrigation requirement = crop water requirement – Effective rain fall.

It can be expressed in mm/day or mm/month.

Kharif-Rabi ratio or Crop ratio—It is the ratio of area to be irrigated in kharif season to that in the Rabi season is called kharif-Rabi ratio. Generally, it is 1:2.

Paleo irrigation—When soil is very dry, it is moistened by water to create favorable conditions for sowing of crops. This is known as paleo irrigation.

Kor-watering—The first watering given to crop, when crop is a few centimeters high called as kor-watering.

13.4.3 Methods of Irrigation

Definition

Application of irrigation water to cropped field by different types of lay outs are called Methods of Irrigation.

Objectives

- Necessary to save the water by proper methods to arrest run-off loss, perculation loss and evaporation losses etc.
- To optimise the crop water need.

Factors Influencing Irrigation Methods

1. ***Soil Type*** : Physical properties such as texture, structure, porosity, infiltration rate influence the selection of method.

2. ***Soil Depth*** : If the soil is shallow, holds less water, needs levelling and forming bunds to hold maximum water to increase irrigation interval.

If soil is deep, holds more water and needs longer irrigation interval.

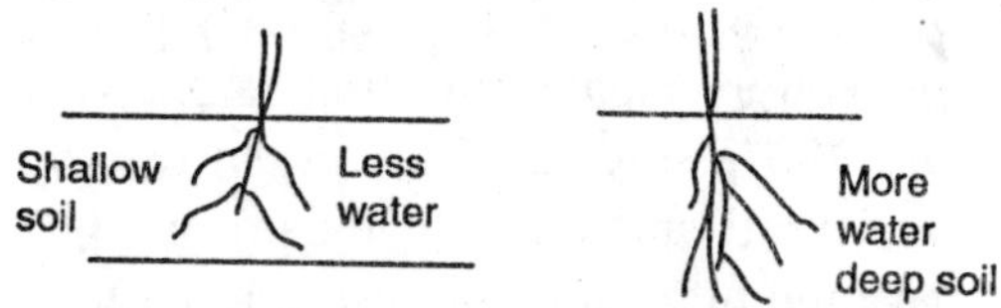

3. ***Topography of Land*** : Topography decides the method of irrigation. If the land is more slopy, basin method can't be used. Instead strip method can be used. In undulating topography, sprinkler/ Drip irrigation method can be used.

4. ***Climate*** : Rainfall, temperature, humidity, wind velocity, radiation etc., influence the irrigation methods. For *e.g.* heavy wind affects sprinkler irrigation and temperature affects surface method of irrigation by high evaporation loss.

5. ***Water source*** : Flow velocity, quantity, quality of available water, decide the methods of irrigation methods.

6. *Crops to be grown* : Value of the plant and geometry of the crop to be cultivated are main criteria.

For e.g., for high value/cash crop or wide spaced crop, sprinkler/drip can be used.

Classification of Irrigation Methods

The methods can be classified as—

1. Surface method or gravity method
2. Sub-surface/sub-irrigation
3. Pressurized or micro irrigation
   (*a*) Drip-irrigation
   (*b*) Sprinkler irrigation

I. *Surface/Gravity Irrigation* : Common method of irrigation practiced all over the world. Water is applied directly to the surface by providing some checks so the water flow.

1. *Border Strip Method* : Field is divided in to long parallel strips by providing small parallel earthen bunds or levees or dykes along both the side of the strips. The end of the strip may or may not be closed, based on the length of the strips. The application efficiency of this system is 75-85%.

- Suitable to soils with moderately low to moderately high infiltration rate, fields with 0-5% slope and not suited to very sandy soil and very clayey soil as they have too high and low infiltration rate.
- For closely spaced crops like pulses, wheat, barley, alfalfa, and ragi.

It can be divided into —

(*i*) Graded Borders, with slope 0.1 to 0.5% in the longitudinal directions and there are very little slope across the strip.

(*ii*) Level border, where there is no slope.

2. *Check basin method* : Adopted for levelled land surfaces. Also known as basin method. Land is divided into small basins. Area of basin surrounded by earthen bunds or levees or dykes.

- Suitable in fine textured soil.
- Suitable for crops like cereals, millets, pulses and oil seeds.

Disadvantages —

- Needs high degree of levelling.
- Not suitable for coarse textured soil.

It can be further classified into,

(*i*) Rectangular or square or Irregular basin.

(*ii*) Ring basis - mostly for orchard crop. These are circular basins formed around individual trees.

3. *Furrow method of Irrigation* : Small, evenly spaced shallow furrows or channels are formed in beds.

Suitable for crops like, cotton, maize, sugar cane, potato, beet root, onion and vegetable crops.

It can be classified into —

(*a*) Corrugated furrow
(*b*) Graded furrow
(*c*) Level furrow
(*d*) Contour furrow

(*a*) *Corrugated furrow* : Small furrow made by plough. 60 mm of bottom width and a depth of 100-150 mm.

(*b*) *Graded* : Open ended which facilitate more time for infiltration and uniform distribution along the furrows. This is adopted in the land having more slope.

(*c*) *Level furrow* : Adopted for land having no or little slope or levelled lands.

(*d*) *Contour furrow* : Furrows formed along the contours.

4. *Surge Irrigation* : It's the application of water into the furrows intermittently in a series of relatively short on and off times of irrigation cycle. This method gives more uniform soil moisture distribution and storage in the crop root zone. The irrigation efficiency is 85-90%.

II. *Sub-surface Irrigation* : Water is applied below the ground surface through net work of pipes or some devices. Main aim is to reduce evaporation loss and to maintain an artificial water table near the root zone of the crop.

Mainly suitable for high temperature area, where ET losses are very high.

III. *Pressurized Irrigation Methods* : Under this method we have both sprinkler as well as drip irrigation methods, where water is applied thro network of pipelines by means of pressure services.

(*A*) *Sprinkler Irrigation System* :This is the method of applying irrigation water above the soil surface in the form of spray or droplets, similar to natural rainfall.

It's made by pumping water under pressure through network of pipelines and allowing to eject out by means of small orifices or nozzles or holes.

It can be classified into —

1. *Rotating head system* : Special device to sprinkle the water called sprinkler head. Which consists of small nozzles and metal ring or water ejected through the nozzle strike the metal ring which changes its direction by the help of the spring attached to it.
2. *Perforated pipe system* : Lateral pipes, small holes are made, based on the nature of the crops to distribute water uniformly.

Suitability and advantages :

- Highly suitable for sandy soil, where infiltration rate is more.
- Lands having undulating topography or steep slopes where levelling is economically not advisable.
- Stream size is very small, where surface flow is low.
- Controls canopy temperature.
- Wastage of land or basin, ridges and furrows and irrigation channels are reduced.
- Almost suitable for all crops except rice which needs stagnation of water.

Disadvantages :

- Distribution efficiency decrease in heavy windy areas.
- With saline water, causes clogging of pores.
- Continuous power supply required to operate system to maintain pressure.
- Costly to install and maintain.
- Uniformity of application is difficult.

13.4.4 Drip or Trickle Irrigation

Water is applied through network of pipelines and allowed to fall drop by drop at crop root zone by a special device called emitters or drippers. The main principle is to apply the water at crop root zone based on the daily ET demand of the crop without any stress. Highest efficiency of irrigation is achieved by drip irrigation.

Components

Advantages :

- Application of water in slow rate, facilitates the easy infiltration into the soil.
- Root zone is always maintains with field capacity level.
- No seepage or percolation/evaporation losses.
- Restricted weed growth due to limited wetting zone.
- Fertigation, chemigation and herbigation possible.
- Salt content near the root zone is reduced.

- Saline water can also be used.
- Adopted for any type of topography.
- More area can be maintained with little quantity of water.
- Used for widely spaced crops like cotton, sugar cane, tomato, brinjal, coconut and orchard crops.

Disadvantages :

- Clogging in emitters due to salt content of water and other impurities like, dust, etc.
- Damage of pipelines by rodents.
- Not economical for closely spaced crops which require more number of pipes and drippers per unit area.
- Proper maintenance and periodical cleaning of drippers and pipelines (with 1% HCl) are very important to maintain the system efficiently.

13.4.5 Drainage

Definition

It's the process of removal of excess water or gravitational water from the surface and the sub surface of farm lands with a view to avoid water logging and creates favourable soil conditions for optimum plant growth.

Need or Objectives

- It raises the soil temperature, that is, keeps the soil relatively warm.
- To have optimum soil water balance.
- It has greater necessity due to heavy precipitation.
- Excess salts are leached down. Areas of salinity and alkalinity where annual evaporation exceeds rain fall and capillary rise of ground water occurs.
- Low lying flat areas surrounded by hills.

Characteristics of Good Drainage

- Should be permanent.
- Must have adequate capacity to drain the area completely.
- Should have minimum interference with cultural operations.
- Should have minimum loss of cultivable area.
- Should intercept or collect water and remove it quickly with in shorter period.

Methods of Drainage

1. Surface method
2. Sub-surface method

1. *Surface Method*

It's to remove the excess water from the surface of the soil profile. It can be done by developing slope in the land.

It's suitable for, slowly permeable clay and shallow soil and for regions of high intensity rainfall. The land with less than 1.5% slope is also needed a good drainage.

It can be made by land smoothing, making field ditches.

The Method can be further classified into,

(*a*) Lift drainage

(*b*) Gravity drainage

(*c*) Field surface drainage

(*d*) Ditch drainage

(*a*) *Lift drainage* : It's to drain the excess water from low lying area or areas having water due to embankment. It can be achieved by the use of scoops or pump.

(*b*) *Gravity drainage* : Water is allowed to drain from the areas of higher elevation to lower reaches through, the regulated gravity flow through, the outlet of various types.

(*c*) *Field surface drainage* : Excess water received from the rain or irrigation is drained through this method. Irrigated basins or furrows are connected with the drainage under lower elevation, which is connected to the main out let to the form pond used for water harvesting.

(*d*) *Ditch drainage* : Ditches of different dimension are constructed at distances inside the soil up to the depth of ditch. Such ditches are interceptors or relief drains. This method can be adopted for nurseries, seed beds and rainfed crops.

2. *Sub-surface Drainage System*

Sub surface drains are under ground artificial channels through which excess water may flow to a suitable outlet. The main purpose is to lower the ground water level below the root zone of the crop. It's mostly needed to medium textures soil, high value crop and high soil productivity.

Types of Sub-surface drainage —

(*a*) Tile drainage

(*b*) Vertical drainage

(*c*) well drainage/Drainage wells

(*a*) *Tile drainage* : Continuous fine of tiles laid at a specific depth and grade so that the excess water enters through the tiles and flow out by gravity. The tile drains are made up of clay and concrete.

(*b*) *Vertical drainage* : Disposal through well into porous layers of earth. Such a layer must be capable of taking large volume of water rapidly. These layers found in river bed.

(*c*) *Drainage wells* : These wells are used for the drainage of agricultural lands especially in irrigated areas.

Systems of Drainage

1. Random
2. Parallel
3. Herringbone
4. Grid iron
5. Interceptor

1. ***Random*** : Lines are laid more or less @ random to drain these wet areas. This is used where the wet areas are scattered and isolated from each other.
2. ***Parallel*** : Poorly drained soils of homogenous texture and mild slopes are drained through this system. It consists of the drains laid out in long parallel laterals which discharge into a common main drain.
3. ***Herring bone*** : Here, the mains are in a narrow depression and the laterals enter the main from both sides @ an angle of 45°, like the bones of a fish.
4. ***Grid Iron*** : Similar to herring bone, but the laterals enter the main only from one side @ right angles. This is adopted in flat regularly shaped fields. This is an efficient drainage system.
5. ***Interceptor*** : Ditches of different dimensions are constructed at distances to drain the excess water accumulated on the surface and inside the soil upto the depth of ditch. Such ditches are interceptors or relief drains. Its an effective and efficient method, which is adopted for nurseries, seed beds and rain fed crops.

13.4.6 Quality of Irrigation Water

I. *Suitability mainly depends on*

- Amount of type of salts present in the water

II. *Main constituents are,*

- Calcium, Magnesium, Sodium as cations.
- Chloride, Sulphate and Bicarbonate

- Other ion → like Bo, Mo

III. *Parameters which divide the quality of Irrigation water*

- Total salt concentration (Salinity)
- Sodium adsorption ratio (Sodicity)
- Bicarbonate and Boron content

1. Total concentration of Soluble Salts

- Total dissolved solids
- Electrical Conductivity

2. Concentration of Na

- SAR

$$SAR = \frac{(Na^+)}{\sqrt{\frac{Ca^+ + mg^+}{2}}}$$

- SSP

$$SSP = \frac{N_a^{2+}}{\sqrt{Ca^{2+} + mg^{2+} + Na^+ + k^+}} \times 100$$

or

$$\frac{\text{Soluble Sodium Concentration}}{\text{Total Soluble Cation Concentration}}$$

IV. *Quality parameters for irrigation water*

Class	*EC (dsm–1)*	*SAR*	*RSC (me q/l)*
Good	<2.0	< 10	<2.5
Normal	2-4	< 10	<2.5
Sodic	< 4	> 10	> 2.5
Poor	> 4	> 10	> 2.5

3. Magnesium calcium ratio : The adsorption of sodium by soils and clay minerals is more at higher Mg:Ca ratio.

4. Bicarbonate content : The water with *RSC* > 2.5 meg/lit is not suitable for irrigation, water containing 1.25-2.5 meg/lit of *RSC* is marginal and that containing less than 1.25 meg/lit is safe.

5. Boron concentration : The boron concentration beyond 2 ppm in irrigation water becomes toxic for most of the plants.

V. *Management of Saline water for Irrigation*

(*a*) *Irrigation practices*

1. Salt tolerant crops – Barley, Sugarbeet, Cotton, Mustard.
2. Semi-tolerant – Sorghum, rice, Castor.
3. FYM/green manure.
4. Application of fertilizer
5. Planting seed on the side of a ridge.

(*b*) *Irrigation management*

1. Giving heavy – Pre-sowing Irrigation
2. Frequent shallow Irrigation
3. Drip Irrigation
4. Furrow Irrigation

(*c*) *Improving Water Quality*

1. Adding chemical to precipitate harmful constituents
2. Good water + Saline water

(*d*) *Soil Management*

1. Irrigation and drainage
2. Mulching
3. Gypsum application

Drainage coefficient

Rate of water removal which is used in drainage design to obtain the desired protection of crops from excess surface and sub-surface water. It can either be expressed as the depth of water in centimeteres (inches) to be removed in a specific time or flow rate per unit of area or flow rate per unit time varying with the size of the area.

13.4.7 Guide For Selecting Method Of Irrigation

Irrigation Method	Topography	Crops	Remarks
Widely spaced borders	Land slopes capable of being graded to less than 1% slope and preferably 0.2%	Alfalfa and other deep rooted closed growing crops like cotton, sunflower etc.	The most desirable surface method for irrigating close-growing crops where topographical conditions are favourable. Even grade in the direction of irrigation is required on flat land and is desirable but not essential on slopes of more than 0.5%. Cross slope is permissible when confined to differences in elevation between border strips of 6-9.
Closely spaced borders	Land slopes capable of being graded to 4% slope or less and preferably less than 1%.	Pastures or grasses	Especially adapted to shallow soils underlain by claypan or soils that have a lower water intake rate. Even grade in the direction of irrigation is desirable but not essential. Sharp grade changes and reverse grades should be smoothed out. Cross slope is permissible when confined to differences is elevation between borders of 6-9.
Check basin and cross furrows	Land slopes capable of being graded to 0.2% slope or less	Groundnut, Soybean, Rice, Wheat, etc.,	This method is especially designed to obtain adequate distribution and penetration of moisture in soils with low water intake rates.
Corrugations	Land slopes capable of being graded slopes between 0.5% and 1.2%.	Alfalfa pasture and grain Row crops and fruit trees.	This method is especially adapted to steep land and small irrigation stream. An even grade in the direction of irrigation is desirable but not essential. Sharp grade changes and reverse grades should at least be smoothed out.
Graded contour furrows	Variable land slopes of 2-25% but prefeably less	Row crops and fruit trees.	Especially adapted to row crops on steep land, though hazardous due to possible erosion from heavy rainfall. Unsuitable for rodent-in-fested fields or soils that crack excessively. Actuall grade in the direction of irrigation 0.-1.5%.

Irrigation Method	Topography	Crops	Remarks
Contor ditches		Hay, pasture and grain	Especially adapted to foothill conditions. Requires little or no surface grading.
Rectangular checks (levess)	Land slopes capable of being graded so single or multiple tree basins will be levelled within 6cm	Orchards	Especially adapted to soils that have either a relatively high or low water intake rate.
Contour levee	Slightly irregular land slopes of less than 1%	Friut, rice, grain and forage crops	Reduces the need to grade land. Frequently employed to avoid altogether the necessity of grading.
Sub-irrigation	Smooth-flat	Shallow rooted crops such as potatoes or grasses	Requires a water table, very permeable subsoil conditions and precise levelling. Very few areas adapted to this method
Contour bench terrace	Sloping land - best for slopes under 3% but useful to 6%	Particularly suited to cultivated crops	Considerable loss of productive land due to bunds. Requires expensive drop structures for water erosion control.
Localized (drip, trickle, etc)	Any topographic condition for row crop farming	Wide spaced row crops or fruit trees.	Perforated pipe on the soil surface drips water at base of individual vegetable plants or around fruit trees. Has been successfully used in Israel with saline irrigation water. Still it is in infant stage
Sprinkler	Undulating 1-35% slope	All crops	High operation and maintenance costs. Good for rough or very sandy land in areas of high production and good markets. Good method where power cost are low. May be the only practical methods in areas of steep or rough topography. Good for high rainfall areas where only a small supplemental water supply is needed.

13.5 AGRO FORESTRY

13.5.1 Social Forestry

Meaning

Practice of forestry on lands outside the conventional forest area for the benefit of rural and urban communities.

Since various social objectives are met out by such type of forestry that is why it is called social forestry.

Social forestry is also described as forestry for the people by the people and of the people.

Objectives

1. Fuel wood, timber, bamboo, fodder and minor forest produce.
2. Employment
3. Cottage Industries
4. Using land
5. Soil and water conservation
6. Aesthetic value
7. Protecting agriculture

Scope

1. Forest area - 33% target.
2. Wood requirement
3. Employment
4. Soil and water conservation
5. Farmers - own timber, fuel etc.

Components

I. Farm forestry (private land)

II. Extension forestry (public land)

(*a*) mixed forestry

(*b*) shelter belts

(*c*) linear strip plantation

III. Afforestation of degraded forests.

IV. Recreational forestry

Trees

1. Jamun 2. Neem 3. Prosopis 4. Tamarindus 5. Pongamia

13.5.2 Agroforestry

Meaning

Suitable land management_system which increases the yield of the land combines the crop and forests plants and/or animals simultaneously or sequentially on the same unit of land.

Scope

1. Alternative to shifting cultivation
2. Sand dune stabilization
3. Employment opportunities
4. Rural development
5. Utilization of sub-marginal land

Benefits

1. Rational use of land
2. Complete use of land
3. Increase in production of agricultural crops.
4. Increase fuel wood.
5. Increase industrial wood
6. Pasture/fodder
7. Additional employment
8. Solving fuel, fodder and food problem

Components

A. Agri-silviculture system

B. Silvi pastural system

C. Agro-sylvi-pastural system

D. Multipurpose forest tree production system

Characteristics of Tree Species

1. Self-pruning
2. Artificial pruning
3. Nitrogen fixing
4. Permit-sunlight
5. High protein content
6. Different root zone
7. No allelopathy effect
8. Litter composition
9. Shade tolerant

Example —

1. *Albizzia, Eucalyptus* – Paddy, Millet.
2. *Poplar, Acacia* – Wheat, Tapioca.
3. *Cocos* – Maize, mustard.
4. *Leucaena, Gliricidia* – Potato, pulses.

13.5.3 Propagation of various types of Forestry Plantations

Plants are raised by two ways —

(*i*) By seeds

(*ii*) By vegetative parts of the plant

Direct sowing :

Any artificial plantation can be raised either by sowing seeds directly at plantation site or in the nursery for raising seedlings, and when in nursery they obtain palntable size, are planted at permanent site.

Advantages

- No need of nursery.
- Saving of time, man and material.
- Plant grow freely without any disturbances.

Disadvantages

- More seed is needed.
- At permanent site effective supervision is not possible for large area.
- At permanent site, it becomes very difficult to provide good soil conditions for good germination . Tree species raised by direct sowing.

e.g. Acacial nilotica, Albizia lebbeck, Prosopis alba. Tree species raised by transplanting of seedlings.

Propagation by vegetative means

Plant species which do not set seeds or their seeds or not viable or by nature they have the capability of making union when they are layered or planted by their stem cuttings and a few by root cuttings. Tree species propagated by stem cuttings are, *Acacia melanoxylon, Bombex wiba, cestrum diurnum, Erythria indica, punica granatum.*

13.6 OTHERS

13.6.1 Shifting Cultivation

Definition

Shifting cultivation is a primitive method of cultivation which is usually followed in hills by tribal. People in which forest land is cleared by cutting down all the trees , shrubs , herbs and climbers near the ground level , felled materials is dry and set on fire , sowing is followed in cleared area by dibbling small holes .

Other names : Land rotation, juming (North Eastern region) podu (A . P & Orissa)

Shifting Cultivation in India

1. N . E hill region
2. Andhra Pradesh
3. Madhya Pradesh
4. Orissa
5. Bihar
6. Karnataka

Features

1. Cropping is done for few years on same Unit of land , after clearing and burning.
2. Depletion of soil productivity causes cultivator move to other area for cultivating same crop.
3. Here crop is fixed burnt land is rotated
4. It causes soil erosion
5. Less productivity
6. Seldom use of manures and fertilizers
7. The size of plot from 1.0 to 2.5 ha per family having members 3-5

Demerits

1. Faulty land use (social evil)
2. Primitive method of cultivation (less productivity)
3. Change in soil microbial population

4. Loss nutrient through leaching, run off and percolation
5. C : N ratio reduced
6. Causes flood in adjoining area
7. Affect the flora and fauna

Ways for improving degraded areas :

- Efficient land use management
- Preventing loss of soil fertility by making contour bunds, levelling, terracing and good drainage system
- Soil management by growing cover crop, strip crops, mixed crop and erosion checking crops
- Adoption of production based crop management by growing HYV, weed and water management, fertilizer application and plant protection
- Problems created by this can easily be overcome by adoption of suitable agro forestry models

13.6.2 Taungya System

Taungya word derived from Burmese word, consisting of "Taung" means hill and "Ya" means cultivation i.e. cultivation in the hills.

Definition

Taungya is also a shifting cultivation in which cultivators are allowed to plant forest tree sp and to raise agriculture crops in between rows of forest trees simultaneously and cleared for about 3-5 years. Thereafter shift to another patch of land.

Aim : Wood and food production.

History :

(1) First started in Burma in 1850 (teak plantation).

(2) Java in 1856.

(3) In India this was introduced by Brandis in1863.

(4) In India first plantation raised in North Bengal in1863

Types of Taungya

(1) Departmental taungya

(2) Leased taungya (3) Village taungya

Crops grown in taungya system in India

State	*Tree crop*	*Associated agricultural crop*
Uttar Pradesh	Sal tree	Maize, Paddy
Andhra Pradesh	Cashew, Teak	Hill Paddy, Ground Nut
Kerela	Teak, Eucaly-ptus	Paddy, Ginger
Assam	Sal tree	Paddy
Tamil Nadu	Teak, Bamboo	Millet, Pulses
West Bengal	Teak, Sal tree	Paddy, Maize

Merits

1. Suppression of weed and climbers.
2. Regeneration of forest species in cheap.
3. Employment opportunity to rural people.
4. Full fertilization of available land for production of food crops.
5. Increase forest wealth.

Demerits

1. Erosion and soil fertility loss.
2. Exploitation of human labour.
3. Danger of epidemics.
4. Forest trees are not cared.
5. Cultivators may claim legal right on land.
6. Insecure land tenure.

13.6.3 Puddling

Introduction

Rice growth and yield are higher when grown under submerged conditions. Maintaining standing water throughout the crop period is not possible without puddling.

Definition

Manipulation of soil under water with tools and implements to break down the soil segregates is called puddling.

Objective

- To provide optimum seed bed for planting rice.
- To create impervious layer below the surface to reduce deep percolation losses of water.
- Weed control.

Method of Operation

- 5-10 cm of water is applied and first ploughing is carried out.
- After 3-4 days, another 5cm of water is applied and second ploughing is carried out.
- Within 3-4 days, another 4-5cm of water is given and third ploughing is done in both direction.

Formation of Impervious Layer

- Unlike other tillage operations, puddling aims at destroying soil structure. The individual soil particles viz., sand, silt and clay are separated during this operation. The impervious layer are formed over compacted soil.
- The sand particle reach the bottom, over which silt particle settle and finally clay particles fill the pores.

Implements Used

Depending upon the availability of equipment and nature of the land, different implements are used.

1. Wet land puddler
2. Cage wheel
3. Helical blade puddler
4. Green manure trampler
5. Rotovator

Advantage

- Avoid percolation loss of water.
- Minimization of leaching loss of nutrients.
- Uniform and soft seed bed preparation.
- Incorporation of weeds, stubbles, green manure and green leaf manure is possible.
- To some extent, perennial weeds can be managed.

13.6.4 Ratooning

Cultivation of crop regrowth after harvest.

Advantages

- Reduced cost of production through saving in land preparation and seed material.
- Reduced crop cycle period for ratoon crop.
- Better utilization of growing season in monsoonal climates.
- Less input (irrigation, fertilizer) requirement than main crop.

Disadvantage

- Lower yield than plant crop.
- Insects, diseases problem.
- Greater cost per unit produced.

Ratooning in sugarcane

- Ratoon provide cheaper raw material.
- Better utilization of resources.
- Give fairly high sucrose in juice.

Good Management

- Harvesting at proper time.
- Stubble shaving.
- Judicious fertilizer application.

Varieties co-740, co-6304

13.6.5 Cardinal Temperature

For each phase in the growth of cultivar, there is a temperature range within which growth and development is optimum. These temperature ranges are called cardinal temperature.

- Minimum – below which there is insufficient heat for biological activity.
- Optimum – at which the rate of metabolic processes are at their maximum. 15-30°C

- Maximum – above which growth ceases.

Cool season crops

Wheat, barley, potato, oats.

- grow best in cool weather period and grown in winter season (November to February)
- Temperate crops
- Cardinal temperature
  - Maximum 30 - 38°C
  - Minimum 0 - 5°C
  - Optimum 21 - 30°C

Warm season crops

Rice, sorghum, maize, sugarcane

- Grown in monsoon season and summer season.
- Tropical crops.
- Cardinal temperature
  - Maximum 45 - 50°C
  - Minimum 15 - 20°C
  - Optimum 30 - 38°C

Table showing cardinal temperature for germination of seeds of crops is shown in Appendix.

13.6.6 Flue Curing

Introduction

Curing is a process in tobacco by which harvested leaves are made ready for marketing. The sole object of curing is dry the green leaf to get final product required colour, texture and aroma.

Curing

Curing is a process of drying, decomposition of chlorophyll till green colour disappear, hydrolysis of starch into sugar and fermentation of sugar.

Flue curing is an improved method of curing, in which drying of leaves under artificial condition without contact of green leaf with smoke.

Feature —

- Drying of leaf under controlled conditions.
- Starch get converted into sugar.
- Green tobacco becomes bright, aromatic and fine textured.

Variety Suitable for Flue Curing

- Virginia Gold
- Harison Special

Production Technique of Flue Cured Tobacco

1. Raised with low nitrogen supply
2. Harvest at an advanced stage when rich in starch, low in Nicotine content.
3. The leaves are harvested by priming 3-4 leaves at a time.

Method of Flue Curing

Harvested green leaves are strung on sticks

Stacked into a specially constructed flue cured pan.

Artificial heating with gradual increase in temperature till leaves dry.

Stages in Flue Curing

1. Yellowing of leaf 2. Fixing of colour 3. Drying of leaf

1. *Yellowing of Leaf*
   - Slow heating is to give optimum humidity.
   - Retain Moisture
   - At end, leaf attain lescion yellow colour
   - Starch converted to sugar
   - completed within 1 to 1½ day
2. *Fixing of Colour*
   - Increase temperature gradually with decrease in humidity.
   - Completed within 1 to 1½ day.
3. *Drying of Leaf*
   - Drying the midrib of leaf
   - Temperature of pan increase @5°C for 1 hr. till reaches to 170-180°F for one day.
   - After drying leaves leaf in pan overnight with ventilation.

13.6.7 Harvest Index

Harvest Index defined as the ratio of economic yield to biological yield.

$$HI = \frac{\text{Biological yield}}{\text{Economic yield}}$$

HI - can also be expressed as %.

Biological Yield : The total dry matter produced by a crop.

Economic Yield: Fraction of the biological yield which is useful for man.

$B.Y \times K = E.Y.,\ K =$ constant

[Coefficient of effectiveness of Harvest Index]

Importance

A balance between the productive parts of the plant and reserves which form economic yield is essential. A considerable Increase in the yield of the economic product is usually dependent on an increase in total dry matter produced.

Factors Affecting Economic Yield

- Source (photosynthetic organs)
- Sink (storage organs)
- Capacity to translocate assimilates from one part to other.
- Water
- Nutrient uptake/Transport
- Plant population • Radiation

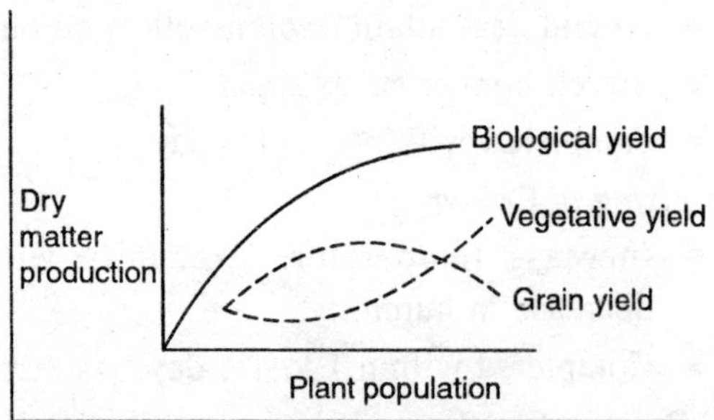

Recent Views/Approach

Yield is limited by photosynthetic rate and attention is paid to selection of plants with more erect leaves (or) higher photosynthetic rate or Calvin cycle plants lacking photorespiration.

Constraints in Pulse Production

Area 24 mn ha, productivity 550 kg/ha. Pulse requirement - 80 gm/day. Availabiilty ⇒ 40-50 g/day.

Factors

1. Unfavourable weather conditions
2. Abnormal soil conditions
3. Variable constraints
   (*a*) lack of high yielding varieties
   (*b*) High susceptibility to disease and insect
   (*c*) Flower drop
   (*d*) lack of short duration variety
   (*e*) poor response to inputs
   (*f*) instability in performance
4. Seed constraints
   - Poor quality • High price
   - Delay in supply
5. Agronomic constraints
   - Improper showing time
   - Low seed rate
   - Defective method of sowing
   - Inadequate interculture
   - Insufficient irrigation
6. Plant protective constraints
7. Inadequate extension facilities
8. Socio economic constraints
9. Inadequate research infrastructure
10. Miscellaneous
    - loss during storage
    - loss during processing

13.6.8 Multi Purpose Trees and Shrubs (M.P.T.S)

In agroforestry systems, it is necessary and economical if selected trees and shrubs are having their various uses so that each and every part, of the trees are fully utilized for domestic purpose, industrial purpose etc. Besides, multipurpose trees and shrubs help in protecting environment.

Domestic and Industrial Uses

1. Fuel wood for cooking foods and warming houses.
2. Timber, posts and poles
3. Paper cardboards, and construction boards.
4. Vegetables, fruits and nuts, seeds and firewood.
5. Edible leaves and shoots for sauces, curries, salads and beverages.
6. Fodder for livestock
7. Host plant for silk worms and lac insects.
8. Green manure for incorporating into soil.
9. Medicines and Pharmaceuticals.
10. Extractives such as a gums, resins, rubber and dyes.
11. Oil (edible), and non-edible for illumination, lubrication and soap making
12. Cake obtained after oil extraction for the use of animal feed, if edible and non-edible types are for the use as manure.
13. The fruit of anola, bahera are used as soap substitutes.
14. Essential oil, perfumes are for cosmetics.

Environmental Protection and Other Uses

1. Shade and support for planations crops.
2. Aid to the soil improvement by controlling soil erosion and sand dunes stabilization.
3. Improvement in soil structure, chemical properties, water balance and weed control.
4. Useful in hedges, windbreaks, shelterbelts, and fire breaks.
5. Ornamental, aesthetic or recreational value.

Table 1 : A few multipurpose tree species for arid and semi-arid regions

Botanical name	*Common Name*	*Economic importance*
Acacia albida	African Savana tree	Fodder, Firewood, tannin, gum
Acacia aneura	Mulga acacia	Fodder, firewood
Acacia nilotica Delile	Babul, Kikar	Fodder, tannin, gum, timber
Azadirachta indica	Neem	Timber construction farm implements seed oil, cake for manure, tannin, firewood, gum, medicinal

Table 2 : A few multipurpose tree species for humid tropical regions

Acacia auriculiformis		Small timber, firewood, charcoal, Tannin.
Albizia lebbeck	Siris	Fodder, firewood, charcoal, tannin, gum, timber, farm implements.
Anacardium occidentale	Cashew	Fruit, kernels, oil, Timber, firewood.
Anthocephalus chinensis	Kadam	Fodder, fruit, pulp-paper, firewood.
Areca catechu	Supari	Chewing gum, tannin, firewood, manure, mulching, pulp.
Artocarpus heterophyllkus Lamk.	Jackfruit	Fruit, vegetable, Fodder, packing boxes, plywood, furniture, ladders, farm implements, Firewood, medicinal.
Cordia dichotama Forst	Indian cherry	Fruits, fire-wood, fodder, tannin medicinal.

Table 3 : A few multipurpose tree species for temperate regions

Botanical name	*Common name*	*Economic importance*
Acacia dealbata Link	Lilvere wattle	Gum, flower are used in perfumery, tannin and dye, ornamental
Acacia decurrens Wild	Green wattle	Timber, firewood
Acer oblongum Wall	Maple	Fodder, firewood
Aesculus india Colebr.	Indian Horse	Timber
Bombax celba	Semal	Timber, firewood, fodder, silky floss, gum, medicinal
Cary Illionensis	Pican nut	Fruit and timber
Castanea sativa Mill	chestnut	Fruit, fodder and timber

13.6.9 Organic Farming

The organic farming implies

- Recycling farm wastes and residue to the native soil itself.
- Replenishing the nutrients depleted from the soil during crop growth.
- Encouraging the growth of soil micro-organisms which could regulate phased release of stored nutrients in the soil to the crop growth in right proportion.
- Maintaining soil health by balancing the soil moisture and soil aeration.
- Ensuring soil fertility by firmly binding the nutrient elements in the complex organic molecular.

Harmful effects of agricultural chemicals

- Residual toxicity is the major problem of agricultural chemicals. Many of the chemicals that are toxic to pests, are toxic to humans also and their admixture in the food may cause nerve and bone malformation in addition to blood clotting.
- Unlike the slow release of nitrogen from organic source, the chemical fertilizers like urea release nitrogen to the crops, abundantly within a short span of time and the excess ammonia can cause infant disease-methemoglobinemia. Even animals feeding on fodder supplied with excess urea can experience hair fall and skin disease.
- Amines produced due to high intake of chemical nitrogen may cause carcinogenic effects on human beings.
- Most of the pesticides and herbicides are nerve and respiratory poisons apart from upsetting Gastro-intestinal system of human beings. If these chemicals are not denatured before the food is being used for consumption, they are prone to affect human beings.
- The chemicals can also pollute the drinking water supply and cause bone abnormalities.
- The fumes emanated from the applied chemicals and fertilizers will cause bronchial asthma to the persons living nearby.
- Cotton garments produced from the cotton obtained from chemicals used fields cause skin lesions and irritations.
- Industries producing these chemicals and fertilizers pollute the atmosphere by releasing toxic fumes which are bound to affect the ozone cover.

How Organic farming is done ?

Composting of farm wastes and industrial wastes raising green manures as inter crops and insitu incorporation in the soil, biodegrading organic wastes such as coirpith, as a source of organic manure are some of the practical approaches for organic farming.

- Micro-organisms such as *Rhizobium, Azotobacter, Azospirillum, Blue Green Algae, Azolla mycorhiza* have ability to mobilize non-usable nutrients such as atmospheric nitrogen into usable form. Such biofertilizers will enrich the soil fertility.
- Vermicompost has gained popularity recently which refers to the compost prepared by earth worms. The symbiotic relationship between the plant & the soil born organisms, is an ecofriendly approach.

- The pests and diseases, affecting a crop must be checked by using biocontrol agents instead of chemical pesticides and fungicides. The wonder tree of Indian origin, neem has drawn the global attention for its pesticidal properties.

- The pests and diseases, affecting a crop must be checked by using biocontrol agents instead of chemical pesticides and fungicides. The wonder tree of Indian origin, neem has drawn the global attention for its pesticidal properties.

- Use of nuclear poly hedrosis virus (NPV), Bacteria, *Bacillus thuringiensis* in biocontrol have gained popularity recently. Release of egg parasities trichogramma species to control certain types of pests is another biocontrol method.
- Sex pheromones are used to attract major pests and they are trapped in light traps.

Advantage of organic farming

- Organic manure produce optimal conditions in the soil for high yields and good quality crops.
- They supply all the nutrients required by the plant (NPK, secondary) and micronutrients).
- They improve plant growth and physiological activities of plants.
- They improve the soil physical properties such as granulation and good tilth, giving good aeration, easy root penetration and improve's water holding capacity. The fibrous portion of the organic matter with its high carbon content promotes soil aggregation to improve the permeability and aeration of clay soils while its ability to absorb moisture helps in the granulation of sandy soils and improves their water holding capacity. The carbon in the organic matter is the source of energy for microbes, which help inaggregation.
- They improve the soil chemical properties such as supply and retention of soil nutrients and promote favourable chemical reactions.
- They reduce the need for purchased inputs.
- Most of the organic manures are wastes or by products which on accumulation may lead to pollution. By way of utilizing them for organic farming, pollution is minimized.
- Organic fertilizers are considered as complete plant food. Organic matter restores the pH of the soil which may become acid due to continuous application of chemical fertilizers.
- Organically grown crops are believed to provide more healthy and nutritionally superior food for man than those growth with commercial fertilizers.
- Organically grown plants are more resistant to disease and insects and hence only a few chemical sprays or other protective treatments are required.
- There is an increasing consumer demand for agricultural products, which are free of toxic chemical residues. In developed countries consumers are willing to pay more for organic foods.
- Organic farming helps to avoid chain reaction in the environment form chemical sprays and dusts.
- Organic farming helps to prevent environmental degradation and can be used to regenerate degraded areas.
- Since the basic aim is diversification of crops, much more secure income can be obtained than when they rely on only one crop or enterprise.
- In India many crops are produced as organic products such as; Basmati rice, pulses, tea, spices, oilseeds, fruits, vegetables, cereals and herbal medicines. Some other non edible products like organic cotton fiber, cosmetics, functional food products, body care products etc are also produced. India's production of organic products was 1.35 million metric tons in 2015-2016
- **Promotion of zero budget farming**, Govt. of India is promoting organic farming in the country through the dedicated schemes such a *Paramparagat Krishi Vikas Yojana* (PKVY) since 2015-16 and also through **Rashtriya Krishi Vikas Yojana (RKVY)**

Current Status Organic Farming: The current position of organic farming w.r.t. area covered across the country is 5.69 Million hectares under the schemes Paramparagat Krishi Vikas Yojana (PKVY), Mission Organic Value Chain Development for North Eastern Region (MOVCDNER) and National Programme of Organic Production (NPOP)

- Among all the states, **Madhya Pradesh** has covered largest area under organic certification followed by Himachal Pradesh and Rajasthan.

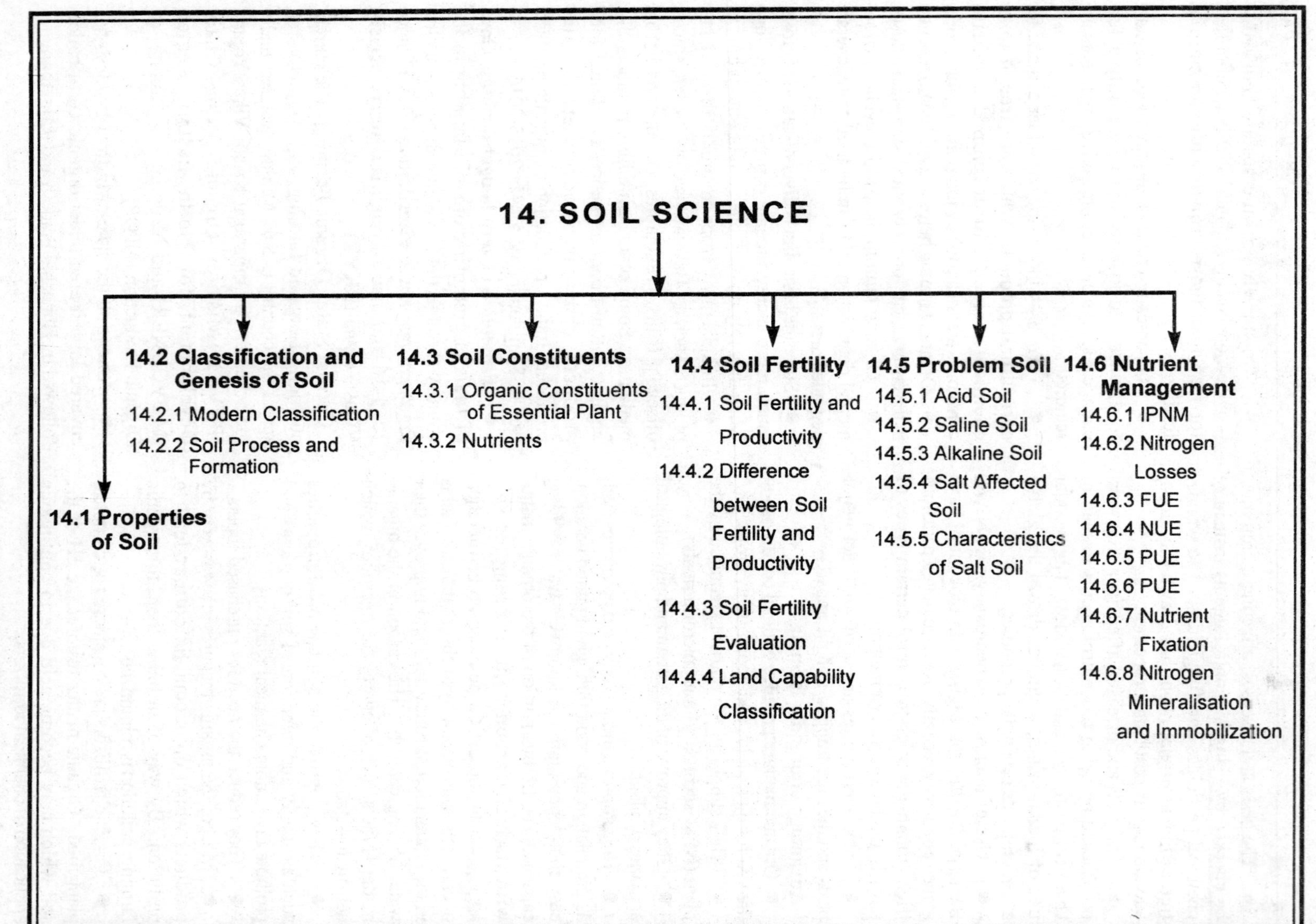
14. SOIL SCIENCE
14.1 Properties of Soil
14.2 Classification and Genesis of Soil
14.2.1 Modern Classification
14.2.2 Soil Process and Formation
14.3 Soil Constituents
14.3.1 Organic Constituents of Essential Plant
14.3.2 Nutrients
14.4 Soil Fertility
14.4.1 Soil Fertility and Productivity
14.4.2 Difference between Soil Fertility and Productivity
14.4.3 Soil Fertility Evaluation
14.4.4 Land Capability Classification
14.5 Problem Soil
14.5.1 Acid Soil
14.5.2 Saline Soil
14.5.3 Alkaline Soil
14.5.4 Salt Affected Soil
14.5.5 Characteristics of Salt Soil
14.6 Nutrient Management
14.6.1 IPNM
14.6.2 Nitrogen Losses
14.6.3 FUE
14.6.4 NUE
14.6.5 PUE
14.6.6 PUE
14.6.7 Nutrient Fixation
14.6.8 Nitrogen Mineralisation and Immobilization

14. Soil Science

Soils: Soil is an important layer of the earth's crust (outermost layer). It is the mixture of organic matter, minerals, gases, water and air and other organic organisms. It is the anchor to support vegetation which is the key to a food cycle in a ecosystem.

Important functions of soil :

1. It is a base /medium for plant growth (essential for life in our ecosystems).
2. As a means of storage, supply and purification of water for different uses.
3. It is modifier of earth's atmosphere and a sink for carbon dioxide.
4. It provides a natural habitat and basis for life on the earth.

Soil Science is the study of soil as a natural resource on the surface of the earth including soil formation, classification and mapping, physical, chemical, biological and fertility properties of soils, and management of soils. *Vasily Vasilievich Dokuchaev is regarded as father of soil science*, (A Russian scientist). He was first person to make wide geographical investigation of different soil types.

Two main branches of soil science are ; (*i*) Pedology, (*ii*) Edaphology

Pedalogy is the study of soil in its natural setting while

Edaphology is the study of soil in relation to soil dependent uses.

14.1 PROPERTIES OF SOIL

Physical	Chemical	Dynamic	Biological
1. Soil Texture	1. Soil Colloids	1. Soil Consistency	1. Macro-organisms
2. Soil Structure	2. Soil Reaction	2. Soil Compactions	2. Micro-organisms
3. Density of soil	3. Soil Organic Matters	3. Soil Strength	
4. Porosity of soil	4. Soil Acidity	4. Soil crusting	
5. Soil Temperature	5. Soil Alkalinity		
6. Soil colour	6. Neutrality		
7. Soil water			
8. Soil air			

Important Properties

Soil Texture

Can't be changed/altered.

1. Definition : It is the relative percentage of sand silt and clay in a soil.
2. Fundamental groups of textural classes
   - Sands : > 70% sand separate
   - Loams : Intermediate
   - Clays : > 35% clay separate
3. Determination of textural class
   - Feel Method
   - Laboratory Method

   (by USDA, based on the Mechanical Analysis)
4. Importance of soil Texture
   - State of Hydration and Exchangeable cations

Types of Structure

1. Plate like — Platy, Laminar
2. Prism like — Prismatic, Columnar
3. Block like — Angular, Sub-angular
4. Sphere like — Granular, Crumb

- Deflocculation Na^+
- Flocculation Ca^{2+}, Al^{3+}
- Range of Workability
- Heavy Soils
- Light Soils

Soil Structure

Can be changed/altered.

1. *Definition* : The mutual arrangement, orientation and organisation of particles in the soil.
2. *Types of structure* ;
3. Classes
   - Very fine
   - Fine
   - Medium
   - Coarse
   - Very coarse
4. *Grades*
   - Structure less
   - break
   - Moderate
   - Strong
5. *Factors affecting soil structure*
   (*a*) Climate
   (*b*) Organic matter
   (*c*) Adsorbed Cations
   (*d*) Tillage
   (*e*) Types of vegetation
   (*f*) Plant roots
   (*g*) Soil organisms
   (*h*) Crop rotation
   (*i*) Alternate wetting and drying

Density of Soil

1. *Definition* : Weight per unit volume of a substance.
2. *Types of Measurement*
   - Bulk Density
   - Particle Density

(A) ***Particle density/True density***

1. *Definition* : The weight per unit volume of the solid portion of soil.
2. $PD = \frac{\text{Mass of solids}}{\text{Volume of solids}}$
3. *Unit* : gm/cc.
4. *Characteristics* :
   - Depends on density of individual inorganic and organic constituents of soil.
   - Particle density decrease with increase in organic matter.
   - For normal soil particle density is 2.65 g/cc.

(B) ***Bulk Density***

1. *Definition* : The mass per unit volume of a dry soil.
2. $BD = \frac{\text{Weight of soil mass}}{\text{Soil volume}}$
3. *Unit* : g/cc
4. Characteristics :
   - Bulk density always smaller than particle density
   - Low Bulk density favourable physical condition.
   - Ideal Bulk density is 1.2 - 1.4 g/cc.
5. *Factors affecting* : Amount of pore space. Compactness of soil, texture, structure, organic matter.

Porosity of Soil

1. *Definition* : The percentage of soil volume occupied by pore space.
2. *Types of pore-space*

   Macro-pores :
   - Large sized pores
   - Found in between granules
   - Sandy soils

   Micro/Capillary Pores
   - Small sized pores
   - Found within granules
   - Clayey soils.

Soil Colour

1. Kinds of soil colour
   - *Lithochromic* : from Parent Material eg. Red Soil from Red Sandstone
   - *Pedochromic*: during soil formation e.g. Red soils from granite.

Factors affecting soil colour —

- Oxidation : Reduction.
- Alternate wetting and Drying.
- Iron compounds
- Silica lime and salts.

Soil Temperature

1. Sources of soil heat
   - Solar radiation
   - Bio-chemical reactions
   - Conduction
   - Precipitation, Exposure, vegetation
2. Loss of soil heat
   - Radiation
   - Conduction
   - Evaporation
   - Precipitation
3. Importance on plant growth and Nutrition
   (*i*) Germination of seeds
   (*ii*) Soil structure
   (*iii*) Decomposition of organic matter in soil.
   (*iv*) Absorption of water
   (*v*) Availability of Nutrients
4. Management of soil temperature
   (*i*) By regulating the amount of water
   - Drain
   - Add

   (*ii*) The colour of surface soil
   - Black : High temp.
   - White : Low temp.

   (*iii*) By providing different types of mulches on the soil surface.

Soil Water

It is a regulator of physical chemical and biological activities in soil.

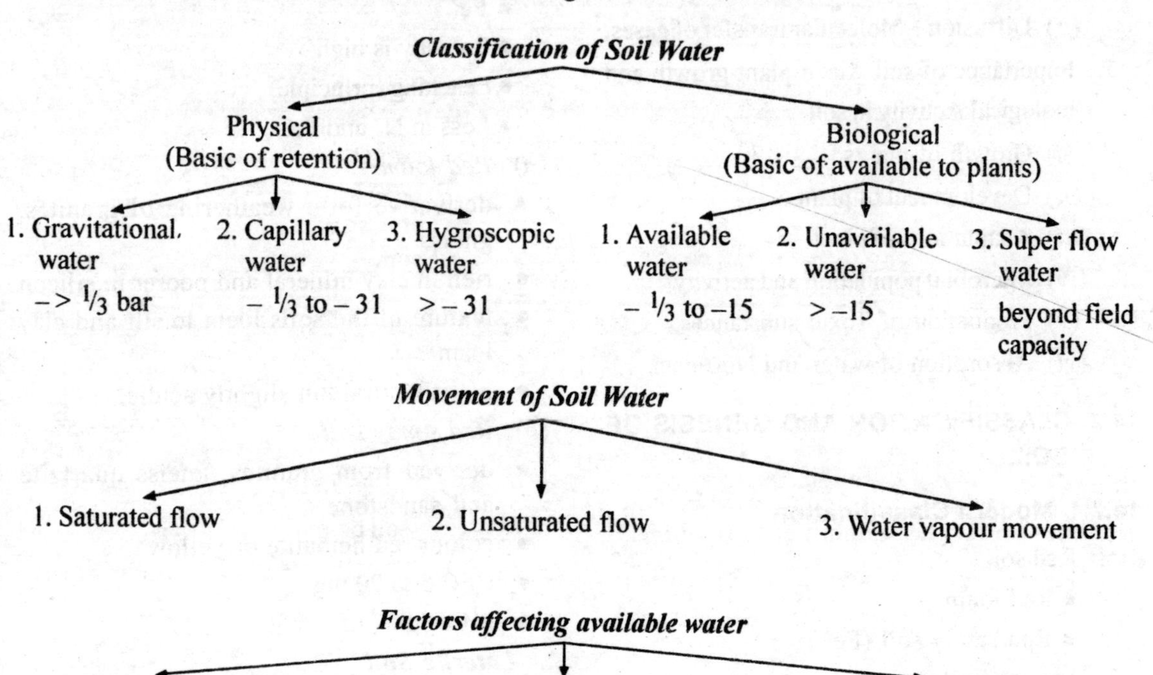

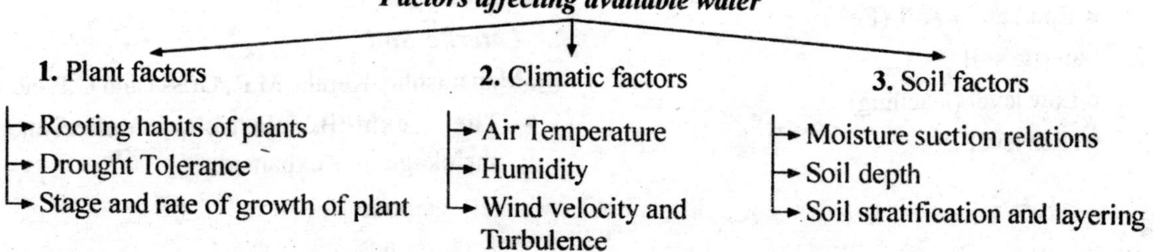

Soil Air

1. Definition : The exchange of CO_2 and O_2 gases between soil pore space and atmosphere.
2. Types of pores :
   - Inter crumb pores – between crumbs
   - Crumb pores – pores within crumbs.
3. Reasons for poor Aeration :
   - Excess soil moisture
   - Gaseous exchange
   - Water logged conditions
   - Saturated soil conditions
4. Mechanisms of exchange of gases
   (*i*) Mass flow – due to pressure differences between atmosphere and soil air.
   (*ii*) Diffusion – Molecular transfer of gases.
5. Importance of soil Air in plant growth and biological Activity in soil
   (*i*) Growth of plants
   (*ii*) Development of plant
   (*iii*) Adaptations of plant
   (*iv*) Microbial population and activity
   (*v*) Production of Toxic substances.
   (*vi*) Absorption of water and Nutrients.

14.2 CLASSIFICATION AND GENESIS OF SOIL

14.2.1 Modern Classification

1. Red soil
   - Red loam
   - Red sandy soil (Fe^{++})
2. Laterite soil
   - Low level (leaching)
   - High level
3. Black soil
   - Shallow black
   - Medium black
   - deep black soil
4. Alluvial soil
   - Coastal alluvium
   - Coastal sand
   - Deltaic alluvium
5. Desert soil
6. Tarai soil
7. Hill soil (brown)
8. Saline and Alkali soil
9. Peaty soils

1. *Red Soil*

Tamil Nadu, a part of A.P, M.P, Orissa, Bihar, U.P., W.B., Rajasthan

- Fe^{++}
- Porosity is high
- Leaching (principle)
- Less in N_2 and P_2O_5

(*a*) *Red loam*

- derivative form weathering of granites, gneiss
- rich in clay mineral and poorer in silicon
- texture of the soils loam to silt and clay loam
- pH is neutral but slightly acidic.

(*b*) *Red sandy soil*

- derived from granites, gneiss quartzite and sandstone
- colour red hematite or yellow
- CEC 5 to 20 mg
- pH acidic (4.5 - 6.5)

2. *Laterite Soil*

Maharashtra, Kerala, M.P., Orissa and U.P. etc

- They exhibit plasticity, cohesion, shrinkage and expansion.

- CEC 2 to 4 mg/100g of soil, poor water retention
- High organic matter poor in lime and Mg, deficient in P and K.
- p^H 5-6.

3. ***Black Soil***

Maharashtra, Gujarat, M.P, Rajasthan, U.P, A.P., Tamil Nadu, Karnataka

- Clay % of 40 - 60 %
- Clay - montmorlionite.
- pH - 7.5 to 8.5

Shallow black soil : (depth - 30 cm)

- Texture - silt loam to clay lime is usually present in fine grains
- Dark brown to dark yellow
- Structure is granular, blocky

Medium black soil : (30 to 100cm)

- Basaltic traps, Dharwar, Schists, basic granite, gneisses, hornblend.
- Moderately rich in organic matter and well drained
- Gypsum is available in upper surface.

Deep black soil :

- Derived from basalt
- % of clay - 40 to 60 %
- Lime is present in irregular nodules
- Alkaline
- Clay minerals

4. ***Alluvial soil*** ($\uparrow k \downarrow N \alpha P_2O_5$)

Rajasthan, U.P., Bihar, W.B., Gujarat, Assam, Orissa

Mainly formed deposition of silt over ages.

- High K less in N and P_2O_5

(*a*) *Coastal alluvium* :

- Structure - sand and silt, deep reddish brown to yellowish brown grey
- Derived from calcareous materials
- It is mainly for agriculture

(*b*) *Coastal sands* :

- Lack of profile development.
- Salinity is high.

(*c*) Delatic Alluvium

- It is formed by river deposition.

14.2.2 Soil Process and Formation

Soil Formation is a process of two distinct phases.

I. Weathering of Rocks and Minerals (Destructive Process)

II. Development of True Soil (Constructive Process)

I. Weathering : Process & transformation of solid rocks into soil

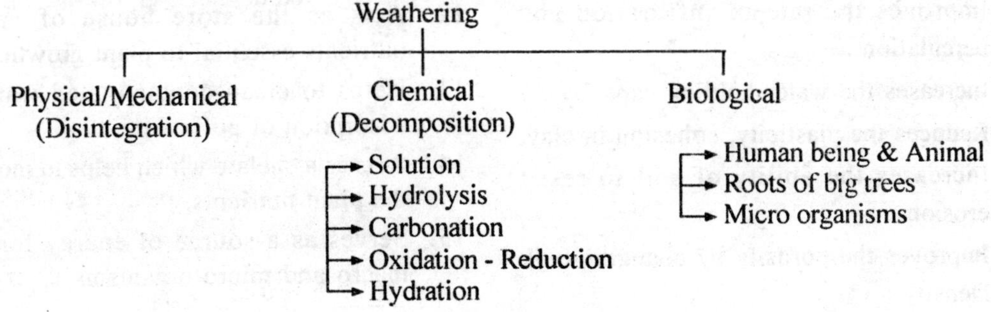

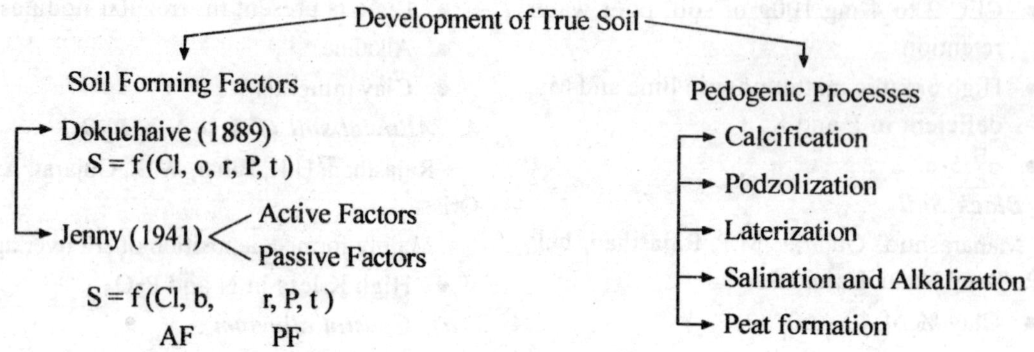

where, Cl - climate ; o - organism ; b - biosphere ;
r - relief ; P - parent material ; t - time.

Five Stages of Soil Development :

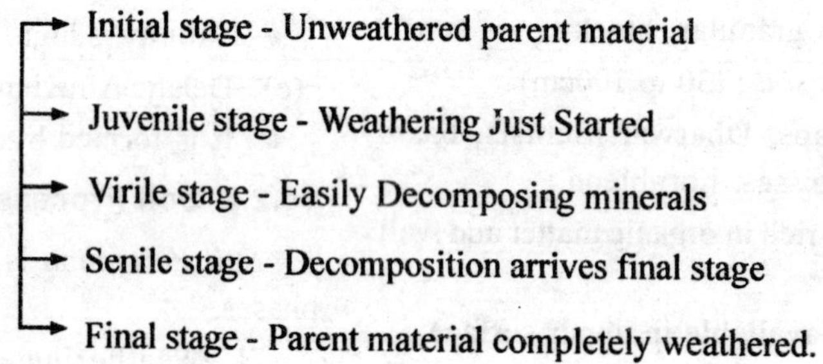

14.3 SOIL CONSTITUENTS

14.3.1 Organic Constituents

1. Influence the soil colour.
2. Improves the soil structure through aggregation.
3. Improves the rate of infiltration and perculation.
4. Increases the water holding capacity.
5. Reduces the plasticity, cohesion in clay.
6. Increases the ability of soil to resist erosion.
7. Improves the porosity by changing bulk Density.
8. Reduces the losses of surface soil by wind erosion.
9. Helps to reduce alkalinity of soil.
10. Acts as buffering agent.
11. Has solubiilzing effect on some mineral compounds present in the soil.
12. Acts as the store house of various nutrients essential to plant growth.
13. Helps to clean contaminated water by adsorption of pollutants.
14. Acts as a chelate which helps to mobilize the plant nutrients.
15. Serves as a source of energy for both macro and micro-organisms.

16. Interacts with pesticides.

Harmful Effects of Organic Matter

1. Serves as source of energy and carbon for many disease organisms.
2. Excessive application creates problems in mixing with soil thoroughly and abstructs easy planting.
3. Most of the toxins identified are short chain organic acids as Butyric, propionine etc.

C.N. Ratio (Carbon : Nitrogen Ratio)

1. Definition : The ratio of the weight of organic carbon to the weight of total nitrogen in a soil or organic material.
2. Ratio in soils :
   - Humus (10 : 1) is fairly constant in soils
   - Arable soils 5 : 1 to 15 : 1
   - Lower in warmer regions.
   - Narrow for sub soils than surface soils.
3. Ratio in plant and microbes :
   - Plant material is 20 : 1 (legumes 30 : 1)
   - FYM is 100 : 1
   - Microbes 4 : 1 to 9 : 1
4. Significance of C : N ratio :
   - The C : N ratio helps to conserve both the nitrogen and organic matter in the soil.
   - If organic residues with high C : N ratio is added, the N_2 is immobilized temporarily by micro organic *i.e.* ratio becomes narrow.
   - The converted N_2 is reconverted to simple inorganic forms.
5. Constancy of C : N ratio :
   - Decomposition process continues

     → C - lost as CO_2

     → N - lost as leaching/absorbed by plant
   - The percentage of total N_2 being depleted from soil equal to the percentage of carbon being lost.
   - At this point, ratio in top soil range from 10 : 1to 12 : 1 which is more or less constant.
6. C : N ratio and Organic matter
   - Organic carbon : Humus (1 : 1.7)
   - For most soils, the C:N ratio and organic matter is 20 : 1

14.3.2 Nutrients

1. *Nutrition* : Plant requires a supply of large number of substances from outside.
2. Total elements in plants → 60.
4. ***Essential Elements*** :

   CHO, NPK, Ca, Mg,

   Fe, Mn, Cu, Zn, Mo, B, Cl

(*a*) *Criteria* :

(*i*) A deficit of the elements make it impossible for a plant to complete it's vegetative and reproductive cycle.

(*ii*) Any one nutrient can't replaced by another.

(*iii*) Elements directly involve in metabolism.

5. ***Mineral Content of Plant***

(*i*) Water → 70%

(*ii*) Organic matter → 27%

(*iii*) Minerals → 3%.

6. ***Factors determine Nutrient availability***

(*i*) Soil characters like texture, structure moisture, pH. atmosphere and Salinity.

(*ii*) Interception and contact exchange.

(*iii*) Mass flow and diffusion.

(*iv*) Soil solution.

(*v*) Root growth and Morphology.

(*vi*) Root Exudation and Rhizophore.

(*vii*) Mycorrhizae.

Macronutrients

CHO – Not mineral, structural compound source - atmosphere, water.

2. *Nitrogen*

(*i*) *Source*

(*a*) Soil

(*b*) Biological N_2 fixation

(*c*) Industrial N_2 fixation

(*ii*) *Function* : Constituent protoplasm of chlorophyll and coenzyme.

(*iii*) *Forms of absorbtion*

(a) NO_3^-

(b) NH_4^+

(*iv*) *Deficiency Symptoms*

(*a*) Thin

(*b*) Pale yellow

(*c*) Protein decrease

(*d*) Root lengthening

(*e*) Prolonged dormancy

(*v*) *Disease*

(*a*) Starvation

(*vi*) Corrective

(*a*) NH_4SO_4

(*b*) CAN

(*c*) Urea

3. *Phosphorous*

Phosphorus is mobile in the plants and is constantly being recycled from the older parts to newer growing areas.

(*i*) *Function*

(*a*) root growth and ripening

(*b*) help translocating of CH_2O.

(*c*) Electron Transport System

(*ii*) *Forms of absorbtion*

(*a*) $H_2 PO_4^-$

(*b*) HPO_2^{-2}

(*iii*) *Deficiency symptom*

(*a*) Vein necrotic

(*b*) Tillering decrease

(*c*) Premature falling

(*d*) Stunting

(*iv*) Disease : Sickle leaf disease

3. *Potassium*

(*i*) *Source*

(*a*) Potash Feldspar

(*b*) Mica

(*c*) Glucomite

(*ii*) *Function*

(*a*) Cofactors for 40 Enzyme

(*b*) Stomatal regulation

(*iii*) *Deficiency symptom*

(*a*) Spot of dead tissue at tips.

(*b*) Margin of leaves (Interveinal chlorosis)

(*iv*) Disease

(*a*) Rosette

(*b*) Dieback

4. *Calcium*

(*i*) *Source*

(*a*) Chalk rock

(*b*) Ca Co_3

(*ii*) *Function* :

(*a*) Constituents of lamella and cell wall.

(*b*) Activator for many enzyme

(*iii*) *Deficiency* :

(*a*) Brown - scorching

(*b*) Terminal bud die

(*c*) Roots poorly developed.

(*iv*) *Disease* :

(*a*) Tip hooking, blosom end rot of tomato

5. *Magnesium*

(*i*) *Source* : Carbonate of Magnesium.

(*ii*) *Functions* : Chlorophyll Synthesis.

(*iii*) *Deficiency* :

(a) Mottled Chlorosis

(b) Withering (defoliation)

(iv) Disease : Sand-drown (Tobacco)

6. ***Sulphur***

(i) *Functions*

(a) Odour in Onion

(c) Nodule formation

(ii) *Deficiency*

(a) Leaf fall is rapid.

(b) Tea leaf turning into yellow

(iii) *Disease* : Akiochi (Rice)

7. ***Iron (Fe)***

(i) *Source* :

(a) Soil oxide

(b) Fe^{3+}.

(ii) *Functions* :

(a) N_2 Fixation

(b) Respiration

(iii) *Deficiency* :

(a) Interveinal chlorosis

(b) Complete bleaching

(c) Scorching of leaf margin

8. ***Manganese***

(i) *Functions* :

(a) Enzyme cofactor

(b) Formation of Chlorophyll

(ii) *Deficiency* :

(a) Dead spot scattered

(b) Grain formation decreased

(c) Root system become poor

(iii) *Disease* :

(a) Pahla blight– sugarcane

(b) Marsh spot of pea

9. ***Copper***

(i) *Function* :

(a) Compounds of plastocyanin

(c) Essential for Photosynthesis, respiration.

(ii) *Deficiency* :

(a) Vegetative growth is decreased.

(b) Whitening of terminal bud.

(c) Marginal chlorosis.

10. ***Boron***

(i) *Function* :

(a) CH_2O transport.

(b) Pollen gemination

(c) Water reaction

(d) N_2 Metabolism

(ii) *Deficiency* :

(a) Plant dwarf

(b) Dieback

(iii) *Disease* :

(a) Top sickness - Tobacco

(b) Hard fruit - citrus

(c) Hen chick - grape

11. ***Molybdenum***

(i) *Function* :

(a) Nitrate reductase

(b) N_2 Fixation

(ii) *Deficiency* :

(a) Light yellow and leaf blade fail to expand.

(b) Mottling of leaves.

(c) In oats, failure of grain formation.

(iii) *Disease* :

(a) Whiptail of Brassica

(b) Scald of legumes.

12. ***Chlorine***

(*i*) *Source* : Chlorides

(*ii*) *Function* :

(*a*) O_2 Evolution

(*b*) Photosynthesis

(*iii*) *Deficiency*

(*a*) Basipetal chlorosis

(*b*) Necrosis

13. ***Zinc***

(*i*) *Functions* :

(*a*) Enzyme alcohol dehydrase

(*b*) Co_2 evolution

(*c*) Chlorophyll synthesis

(*ii*) *Deficiency* :

(*a*) Leathery chlorosis

(*b*) Leaf margin sickle

(*iii*) *Disease* :

(*a*) Khaira – Paddy

(*b*) Rosette – Fruits

(*c*) White bud-maize

14.4 SOIL FERTILITY

14.4.1 Soil Fertility and Productivity

Soil Fertility : Soil fertility is the ability of the soil to provide all essential plant nutrients in available forms and in a suitable balance.

Soil productivity : The capability of soil to produce specified crop yield under well defined and specified systems of management of inputs and environmental conditions.

Factors Governing Soil Fertility

Parent material : Fertility of a soil depends on the chemical composition of parent material from which it derived.

Topography : Soils on the upper slope are less fertile than the soils on lower slope because high leaching and erosion on upper slope.

Climate : In tropical climate decomposition of organic matter is faster than temperate climate. Thus soils of tropical regions are less fertile when compared to temperate region.

Depth of Soil Profile

Deep soils are more fertile than the shallow soils and the roots are spread well enough in deep soils than the shallow soils.

Physical Condition of Soil

The soil texture and soil structure influence the soil fertility.

Artificial Factors

- Water logging
- Cropping system
- Toxic chemicals and pesticides in the soil.

14.4.2 Difference between Soil Fertility and Productivity

Soil Fertility	*Soil Productivity*
1. It is an index of available nutrient to plants	1. It is used to indicate crop yields.
2. Influenced by the physical, chemical and biological factors of the soil.	2. Depends upon fertility and location.
3. It is the function of available nutrients of the soil.	3. It is the function of soil fertility, management and climate.
4. All fertile soils are not productive.	4. All produtive soils are fertile.
5. It is an inherent property of the soil.	5. It is not the inherent property of the soil.

14.4.3 Soil Fertility Evaluation

It is the assessment of nutrient supplying capacity of the soil.

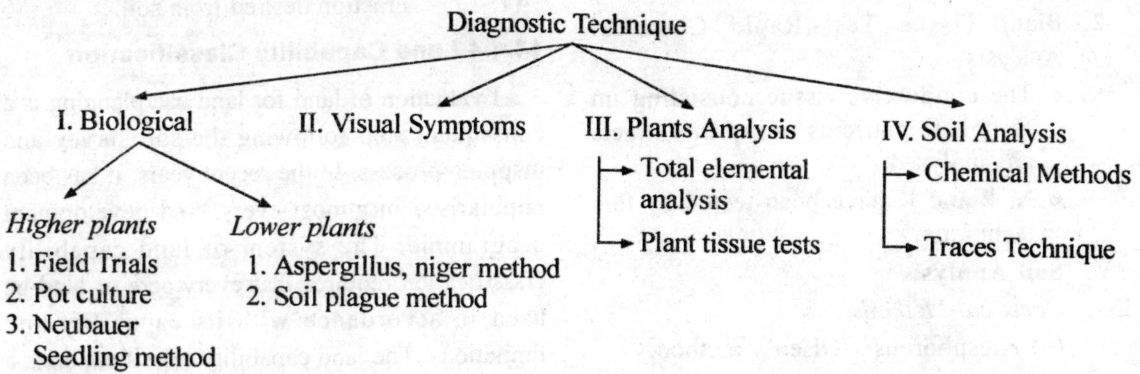

14.4.3 Soil Fertility Evaluation

It is assessment of nutrient supplying capacity of the soil

I. A. Biological Methods (using higher plants as indicator)

1. *Field Trials* : Direct way to assess the nutrient status in soil.
   - Trials with graded doses of nutrients will determine the exact requirement.
   - The effect is very much location specific.
   - Expensive and time consuming and growth condition can't be fully controlled.
2. *Pot Culture Methods* :
   - Same as field trials except that plants grown in pots with small volume of soil or under controlled field condition like greenhouse.
   - Limitations

     Preferential root growth

     Limited volume of soil

     Problems during packing of soil.
3. *Neubauer Seedling Method* :
   - Soil is exhausted of the available nutrients and entire plant is then analysed for the uptake.

II. B. Biological Methods (using Micro organism as indicator)

1. *Aspergillus Niger Method* :
   - For P, K and Mg in soils.
   - The amount of 'k' extracted in mycelium compares favourably with the content of exchangeable 'k' in soil under investigation.
2. *Soil Plaque Method* : by sockett and stewart for study of mineral deficiencies in soil.
   - If soil is deficient in 'k' or 'p' or both, the calories of the *Azotobactor* will not develop.
   - The intensity of growth of calories indicates the degree of deficiency.
3. Visual Symptoms
   - Indicator plants
   - Hidden hunger.

III. Plant Analysis Method

1. *Total Elemental Analysis* :
   - Leaf Analysis widely used.
   - Approach

     Visual Symptoms

     Deficiency

     Toxicity

     Critical nutrient level (CNL)

'CNL' is a range of concentration at which growth of plants is restricted in comparison with that of plant at a higher nutrient level.

2. Plant Tissue Tests/Rapid Chemical Analysis
   - The conductive tissue consisting un assimilate nutrients in sap or extracts are analysed.
   - N, P and K have been tested by this technique.

IV. Soil Analysis

1. *Chemical Methods* :
   (*a*) Phosphorous - Olsen's method,
   (*b*) Micro nutrient - Chelate DTPA
   (*b*) Gypsum requirement - Schoonover's method
2. *Tracer Technique* :
   - The plants of few supplied the phosphatic fertilizers with ^{32}P.
   - The harvested plants tested for total 'P' as well as ^{32}P.

$$A = B\left(\frac{1-Y}{Y}\right)$$

where, A - available 'P' in soil

B - amount of fertilizer 'P'

Y - fraction of 'P' in plant desired from fertilizer 'P'

$(1 - Y)$ - Fraction desired from soil.

14.4.4 Land Capability Classification

Evaluation of land for land use planning is a consequent step following the soil survey and mapping process. In the recent years, it has been popularised in almost every land development programme. The system of land capability classification requires that every acre of land be used in accordance with its capability and limitations. The land capability classification is a broad grouping of soils based on their limitations and also serves as a guide to assess suitability of the land for cultivation, grazing and forest plantation.

The land capability classification scheme developed by soil conservation service. United States Department of Agriculture (USDA).The classification scheme has four categories namely land suitability class, land capability classes (comprising eight classes). Land capability subclass and land capability units.

Table 1 : Land Capability Classification System

Land

Land use suitability (broad grouping)

Land suitable for cultivation				Land not suitable for cultivation			
Land capability class				Land capability class			
				Suitable for pasture and grazing			
Class I Very good land (with no limitation)	Class II Good land (with minor limitation)	Class III Moderately good land (with major limitation)	Class IV Fairly good land (with major limitation, occasional cultivation	Class V (with no limitation)	Class VI (with minor limitation)	Class VII (with major limitation)	Class VIII (Suitable for wild the and watershed)

Land capability subclass [grouped according to kind of limitation in each subclass *viz*., erosion (e), climate (c), excess of wetness (w), soil limitation (s)]

Land capability units

(Land management groups based on physical characteristics – land that is suited for essentially the same kind of management and same kind of soil and water conservation treatment)

14.5 PROBLEM SOILS

14.5.1 Acid Soil

Definition : The soil system proton (H^+ ions) donating capacity during its transition from a given state to a reference state.

Sources :

- Leaching due to heavy rain fall
- Acid fertilizer and soluble salts
- Humus and other organic acids
- Alumina silicate minerals
- Hydrous oxides - oxides of Fe and Al.

Classification of Acid Soils in India

- Laterite
- Lateritic red
- Mixed red, black and yellow
- Podsolic brown
- Foot hill soil
- Peat soil

Problems of Soil Acidity

(*i*) Toxic effect

(*a*) Acid toxicity : Acid anions and H^+ ions

(*b*) Toxicity of different nutrient elements

(*ii*) Acid Toxicity : The higher H^+ ion concentration is toxic to plants under strong acid conditions of soil.

(*iii*) Toxicity of different nutrient elements.

Fe and Mn → depends upon pH, organic matter and intensity of soil reduction increase organic matter in the soil the population of soil microbes increase the soil O_2 and results reduction of soil.

Toxicity of Al → problem in upland and lowland soils effect on crop by - Root growth restrict, affect cell division.

Restrict the absorption and translocation.

(*ii*) Nutrient Availability

(*a*) Non specific effect : Inhibition effect of root growth and thereby affects the nutrient availability.

(*b*) Specific Effect :

- Exchangeable bases - Ca, Mg
  (*i*) uptake ; (*ii*) Release
- Nutrient imbalances -
  Fe, Mn, Al → increase
  P, Mo, N, K, S → decrease

(*iii*) Microbial Activity

N - Fixation - Azotobacter sp

Symbiotic nitrogen fixation - Rhizobium sp

Fungi - grow well and cause disease

(*i*) root rot of tobacco

(*ii*) Blight of potato

Amelioration of Soil Acidity

Liming : Practicable feasible management practice

Lime requirement : Amount of lime added to increase the pH.

Principle

$$CaCO_3 + H_2O + CO_2 \rightarrow Ca(HCO_3)_2$$

$$Ca(HCO_3)_2 \rightarrow Ca^{2+} \downarrow + 2\,HCO_3^-$$

$$H^+ + HCO_3^- \rightarrow H_2CO_3 \rightleftharpoons H_2O + CO_2$$

Change of pH with the application of lime

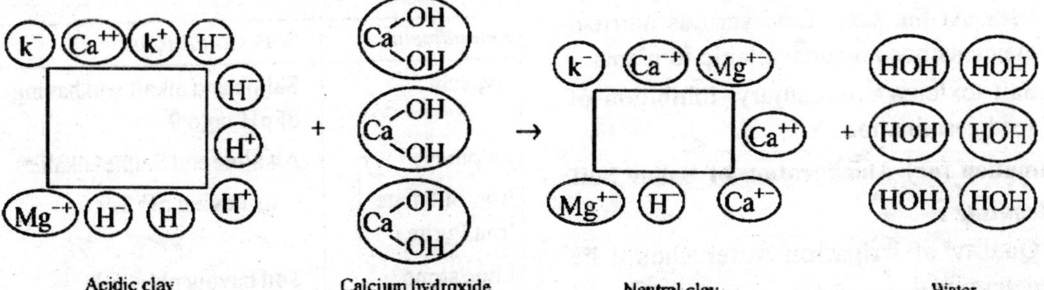

Factors Affecting Lime

- Moisture
- Temperature
- Amount of exchange Acidity

14.5.2 Saline Soil

- Contain Sufficiently Neutral Soluble Salts in chlorides and Sulphate of Na, Ca, Mg in the root zone adversely affect the crop growth.
- Contain small to appreciable quantity of Gypsum.
- Soluble carbonate always absent.
- Low permeability of soil cause poor drainage due to unfavourable soil structure and texture.
- Appearance of crop is patchy growth of crops with visible signs of salt injury (leaf tip burn/chlorosis).
- Presence of white salt crust on the surface so called white Alkali.

Saline soil interferes with the plant growth
or
Problems of salt affected soil

- Soils are usually barren but potentially productive
- Wilting coefficient of Saline soil is very high.
- Amount of available soil moisture is low.
- Absorption of water and Nutrients : Due to high salt concentration plants have to spent more energy to absorb water and to exclude salt from metabolically active sites. At the same time various nutrient elements become unavailable to plants.
- Salt toxicity - root injury, inhibition of seed germination.

Reclamation (or) Amelioration of Saline Soil

Pre-Requisite :

- Quality of irrigation water should be determined.
- Quality of soil
- Nature and distribution of salt in root zones.
- Level of subsoil water.

Technical requirements :

- Adequate drainage
- Availability of Sufficient water to meet the demand of crop and also leach the salt below the root zone in the soil.
- Perfect land levelling, good bunding for irrigation and advanced agronomic practices.
- Good quality of irrigation water.

Methods :

(A) *Eradication*

Scraping : Scrap the surface salt by mechanical means, limited success and temporary cure.

Flushing : Wash away the surface soil by flushing water over salt crusts.

Leaching : Effective way. The fraction of the irrigation water that must be leached through the root zone. Amount of salt leached depend on water Quantity depends on :

(*i*) Initial salt content
(*ii*) Final salt content desired
(*iii*) Extent of soil depth to be reclaimed under drainage.

(B) *Conversion*

In this conversion method different chemical amendments are used for changing part of the caustic alkali carbonates into sulphates and ultimately lost by leaching.

Amendments	*Soil Conditions*
Gypsum	Saline and alkali soil having of pH upto 9
Sulphur Iron Sulphate Iron Pyrite	Alkaline and Saline-alkali soils having pH 8-9
Lime stone	Soil having pH < 6-8

(C) *Control*

- Retardation of Evaporation to control saline soil
- Use of salt free irrigation water
- Avoid partial leaching of saline soil

14.5.3 Alkali/Sodic Soils

Exchangeable Sodium Cause :

- Soil dispersion
- Affecting soil physical properties
- Poor permeability of air and water
- Crust formation during dry
- Sticky during wet
- Low organic matter

Increased Soil pH 8.5 to 10

- Affect soil nutritional status
- From insoluble Ca carbonates
- Non-availability of Ca
- Accumulation of toxic elements - B, Mo, Na
- Reduce N-availability
- Reduce k-uptake
- Fe deficiency

Appearance

- Dispersed Soil Structure
- Water stagnation problem
- Hard crust formation at surfaces
- Dispersion of organic matter in soil (due to ↑ pH) and evaporation brings this dispersed organic matter to soil surface and give black coloured surface so called Black alkali/sodic.

Reclaiming Alkali soil

Pre requisites for reclaiming the Alkali soil.

Aim : To replace most of the exchangeable Na with Ca.

For quick results : Proceed cropping by chemical amendment followed by leaching of salts For delayed Results :

Include rice crop in the cropping sequence.

Along with this application of organic residues rice straw, farm yard manure etc.

Amendments and Practices

Direct Ca : Gypsum, $CaSO_4$

Indirect Ca : H_2SO_4, S (or) Iron Pyrite, $FeSO_4$

Soluble Sodium carbonate often present - so irrigate before gypsum application.

- Gypsum broadcasting at lower soil depths (10-12 cm)
- Gypsum should be of 2 mm sized.
- Organic matter with Gypsum give fast reclamation.

Proper Crop Selection

Rice followed by other sensitive crops even grasses. This can be followed by other field crops.

Water Management

Drainage : Lowering of water table from critical water table depth.

Irrigation : Apply irrigation in small depths and more frequently. Sprinkler irrigation is beneficial.

Cultural Practices :

- *Increase plant population.*
- *Tree planting technique* : Ridge planting.
- *Land levelling* - peripheral bunding in case of patchy growth.
- *Continuous cropping* - Rice and berseem (frequent irrigation crop).

New method : Precipitation of salt.

14.5.4 Salt Affected Soil

Salt affected soils are those soil, which have high amount of exchangeable soluble salt concentration or exchangeable sodium percentage variously called as saline soil/sodic soil.

Salt affected intrazonal soils are halomorphic soil.

Distribution of Salt Affected Soils in India

- Arid, Semiarid and Subhumid regions.
- Coastal areas
- In India 8.4 mn ha salt affected soils.

3 Main groups

- Saline Soil
- Alkali-Soil/Sodic-Soil
- Saline-Alkali Soil

Sources of Salts

- *Rock/Mineral weathering* - Find their way to sea.
- *Ocean/Sea water* - upliftment of marine water, sea intrusion.
- *Primary minerals.*
- *Arid and Semi Arid Climate* : Low Rainfall and high evaporation with in adequate drainage measures.
- *Irrigated Water* : Lack of drainage and leaching facilities increase water table and surface salt content in soil.
- *Salt Blown by Wind* : Arid regions near sea.
- *Excessive use of Basic Fertilizer* : Sodium nitrate, Basic Slag may develop Soil Alkalinity.
- *Change in Land use* : Deforestation, change in cropping/Fallow pattern practices cause rise in water table.

Characteristics of Salt affected Soil

	EC	ESP	pH	SAR	TSS
Saline	> 4 ds	< 15	< 8.5	< 13	> 0.1
Alkaline	< 4 ds	> 15	8.5-10	> 13	< 0.1
Saline-Alkali	> 4 ds	> 15	> 8.5	> 13	>0.1

14.5.5 Characters of Salt and Acid Soil

Characters	*Saline*	*Alkaline*			*Acidic*
		Non-saline alkali	*Saline alkali*	*Degraded alkali*	
1. Content	Na^+ – of soluble salt.	high exchangable Na	Na-soluble Salt	(H^+) (Na^+)	Exchangable Hydrogen
2. Exchangable Ca/Na	Exchangable Ca	Exchangable Na	–	–	H^+
3. Colour	White	Black	–	–	Red
4. Salt	SO_4,Cl, NO_3 of Na.	Na_2CO_3	–	$NaCO_3$-lower	
5. SAR (Sodium)	< 13	> 13	> 13	< 13	
6. ESR (Exchange sodium %)	< 15 of CEC	> 15	> 15	> 15 of CEC	
7. pH	< 8.5	8.5 – 10	> 8.5	8.5	less than 7.
8. EC (dsm^{-1})	> 4	< 4	> 4		
9. Physical condition	Floculation	Defloculation	Floculation	Compact Columnar Structure	
10. Organic matter	less than normal	very low	variable	low	
11. Total soluble salt content	0.1%	< 0.1%	> 0.1%	< 0.1%	

14.6 NUTRIENT MANAGEMENT:

14.6.1 Integrated Plant Nutrient Management (IPNM)

It is the intelligent use of optimum combination of organic, inorganic and biological nutrient sources in a specific crop, cropping system and climatic situation so as to achieve and sustain the optimum yield and to improve or maintain the soil's physical, biological and chemical properties.

Objectives

- To maintain soil fertility.
- Sustainable agriculture production.
- Improve farmer's profit.
- Increase availability of nutrient from all resources.
- Match demand and supply.
- Minimising loss of nutrients.
- Key principle : Balanced fertilizer.
- Aim of balanced fertilizer.
- Increase crop yield.
- Increase farm income.

Components

1. ***Chemicals :***

Urea, DAP to supply nitrogen, superphosphate and rock phosphate supply phosphorus, murate of potash and sulphate of potash supply potash. There are various complex fertilizers to supply combination of nutrients. Similarly there are many chemical compounds which supply various macro and micro nutrients in isolation or in combination.

2 ***Organic matter :***

Farmyard manure, vermicompost, oil cakes of different oilseeds like castor, neem etc. Incorporation of organic residues of proceeding crop in the fields, if time and soil conditions permit also serves as a source of organic nutrients for the succeeding crop.

3. ***Biological :***

(*a*) *Green manures* : cowpea, sunhemp, sesbania, etc. and green leaf manures like subabul, etc.

(*b*) Biofertilizers

Advantages

- Nutrients are easily and quickly available for the plants.
- Good for maintaining soil health
- Good for supplying micronutrient
- All sources can be generated on farm itself.
- It have advantage of being diversified into small units to meet the demands of the specific problems of location.
- Quality and shelf life of food products are increased.

Disadvantages

- Excessive use of fertilizers leads to imbalance in the soil pH.
- Most of the chemical fertilizers are high energy consuming.
- Chemical fertilizers pollute the environment.
- Organic sources are bulk in nature, low in nutrient content.
- Organic manures are having high C:N ratio and not properly composed can lead to temporary deficiency.

14.6.2 Nitrogen Losses

Nitrogen is a important macro nutrient for crop plants, which will be taken by the plants form soils.

This nutrient released in soils either by weathering or by decomposition of organic matter or by addition of fertilizers, induce N_2 losses from soils.

N_2 Losses

1. Leaching of loss/drainage loss of NO_3^-

   Depends on :

   Mobility : N_2 Mobile for leaching loss

   Region : Humid region N_2 loss is high than arid region.

   Soil Type : High loss in sandy soil than loam/clay.

   Land Nature : More loss in barren soil than crop covered land.

2. Run of loss like of NO_3^-.
3. Gaseous losses :

Takes place three ways :

(a) *Volatilization/Non-biological loss of NH_3:*

Volatilization : Volatilization is a gaseous loss of N in which inorganic N ions can be converted to gas and lost to the atmosphere.

Causes for Volatilization —

- Poor drainage : Volatilization occur in poorly drained soil. (e.g.) rice.
- Alkaline pH : Volatilization occurs in soil pH > 7.5.
- Soil pH coupled with form of fertilizer applied :

Acidic soil + NH_4^+ Fertilizer (Amm. Fertilizer) → No Volatilization (because soil pH don't increase)

Acid soil + Urea (NH_4 Forming Fertilizer) → Volatilization (because soil pH increase)

[Acid soil + $2H_2O^+$ + $CO(NH_2)^2$ → 2 NH_4^+ + HCO_3^-

$NH_4 \rightarrow NH_3 \uparrow$ (Volatializatoin)]

(b) *Chemical Decomposition of NO_2^- under acid condition.*

(i) Decomposition of $NH_4\ NO_2$

$$NH_4\ NO_2 \rightarrow 2\ H_2O + N_2 \uparrow$$

(ii) Van slyke reaction

$$R-NH_2 + HNO_2 \rightarrow R\text{-}OH + H_2O + N_2 \uparrow$$

(iii) Spontaneous decomposition of Nitrous acid.

$$3\ HNO_2 \rightarrow H_2O + 2\ NO \uparrow$$

$$NO + O \rightarrow NO_2 \uparrow$$

$$2\ HNO_2 \rightarrow NO + NO_2 \uparrow + H_2O$$

(c) *Microbial Denitrification : Liberation of N_2 and N_2O.*

- Denitrification : Nitrates are subjected to reduction in soils especially in those that are poorly drained or water logged and low in aeration with the help of micro organisms (or) Biological reduction of NO_3 to NO_2.
- Micro organism involved
  - *Pseudomonas denitrificans*
  - *Thiobacillus denitrificans*
- Process of denitrification in Soil :

$$\underset{\text{Nitrate}}{2\ NO_3^-} \xrightarrow{-2(0)} \underset{\text{(Nitrate)}}{2\ NO_2} \xrightarrow{-2(0)} \underset{\text{(Nitric oxide)}}{2\ NO} \longrightarrow$$

$$\underset{\text{(Nitrous oxide)}}{N_2O} \longrightarrow N_2 \uparrow \text{(Eliminated)}$$

Microganism removes O_2 act on e^- acceptor.

- Example : Rice ecosystem.

Causes of Denitrification —

- Water logged condition/anaerobic condition.

Significance of N_2 Losses

1. It maintains the N balances of N cycle.
2. N released during denitrification fixed by biological N fixation, makes available to plants.
3. In ecological point of view, leaching of NO_3 causes eutrophication, but denitrification prevents this eutrophication.

Way to Compensation

1. In case of leaching in alkali soil, add 25%, more N_2.
2. Denitrification can be reduced by addition of PO_4 residues.
3. Adequate drainage

Problems by N_2 Losses

- Eutrophication
- Atmospheric pollution

14.6.3 Fertilizers Use Efficiency (FUE)

What percentage of an applied fertilizer nutrient is utilized by the crops cropping system. This is termed as FUE.

Measures of FUE :

It can be measured by the percentage of added fertilizer nutrient recovered by the harvested portion of a crop (or) kgms of economic produce per kg of nutrient applied.

Problems of 'N' fertilizer - Leaching and volatilization.

Problems of 'P' fertilizer - Fixation, immobility and transformation.

14.6.4 Nitrogen Use Efficiency

- Mainly determined by various kinds of losses in the field, viz. volatilization in the form of ammonia, leaching and runoff, denitrification.
- N fertilizers are amide and ammoniacal forms.
- Converted into nitrate form, it becomes very much susceptible to loss by leaching with irrigation rain water
- Application of nitrogen fertilizer in split doses or as top dressing increases the efficiency of nitrogen use
- The NH^{4+} ions in ammonical fertilizers are adsorbed the soil clay. They may be utilised directly by certain crops eg rice, else they are transformed into nitrates by microbes and taken in this form.
- Higher the clay content of the soil and its exchange capacity the better is nitrogen use efficiency.

Urea Transformation in Soil

Volatilization loss

$$Co(NH)_2 \xrightarrow{urease} NH_4^+ \longrightarrow NH_3 \xrightarrow{Nitrosomonas} NO_3^-$$

$$\xrightarrow{Nitrobacter} NO_3^- \xrightarrow{Denitification} N_2O, NO, N_2$$

Plant uptake/leaching loss

Factors Determine Nutrient Uptake

- Crops, nature of root system
- Water
- Texture and pH soil
- Management practices (land preparation, choice of variety, timely sowing, optimum plant population, timely weed control, weed management, plant protection and balanced supply of essential plant nutrients)
- Agro climatic conditions

Practices to increase Nitrogen Use Efficiency

- Split application → apply 'N' in 2 or 3 installments to coincide with the peak period of nitrogen requirement of the crop.
- Sub surface application
- Pelleting with soil
- Incubation of urea with moist soil (1 : 6) for 2 or 3 days, resulting in the adsorption by soil clays of ammonia formed by the hydrolysis of urea.

Increase Efficiency through Slow Release Nitrogenous Fertilizers

1. Urea aldehyde condensation products *e.g.* urea form, oxamide, isobutytidene diurea.
2. Urea coated with sulphur, lac and neem. The coating is form a temporary barrier between urea granules and soil or soil water, thus reducing the rate of urea hydrolysis.
3. Blend Nitrogen fertilizer with a nitrification inhibitors :
   - non toxic to plants, soil micro-organisms, animal and fish.
   - block the conversion of $NH_4^+ \rightarrow NO_3^-$ by inhibiting Nitrosomonas activity.
   - not interfere with the transformation NO_2^- (nitrite) by nitrobactors.
   - be able to move with the fertilizer so that it will be distributed uniformly throughout the soil zone contacted by nitrogen fertilizer.
   - stable and long time inhibitory action
   - relatively inexpensive, so that it can be used as commercial basis.

   e.g. Nitrapyrin.
4. Judicious mix of manures and fertilizers.
5. Placement of urea super granules, made up of ordinary granules below the soil surface has been found to increase, to varying extent the NUE with regard to rice crop.

14.6.5 Phosphorous use Efficiency

- Fixation of phosphate is main problem.
- Water soluble phosphatic fertilizer after application to the soil react preferably with Fe and Al to form initial phosphate reaction products.
- Ca, mg carbonates, pH and water status of the soil control the nature of the reaction products.
- Efficiency of phosphatic fertilizers depends primarily upon the release of 'P' from the products rather than the fertilizer.
- Reaction products vary in their ability to release 'P'.

Measure to Increase Efficiency

- Minimum contact of the fertilizer with the soil to restrict phosphate fixation.
- Rise pH of acid soil by liming.
- Liming, deep placement and combined use of super phosphate and organic manures to increase the PUE.
- Rice to increase PUE → surface broadcast, followed by mixing during puddling.
- Wheat - Phosphate placement applied in seed furrow or drilling by just below the seeds.

- Crop rotation - to utilise 'P' direct as well as residual effect. Gram + rice rotation.
- In India application of 'P' Fertilizer with organic matter is beneficial. This increase the crop response and decrease fixation.
- Better utilisation of 'P' in acid soil mixed with Farm Yard manure.
- Sun hemp - Application of phosphate before ploughing best method to improve soil fertility.
- Crop may not use > 10% of 'P' if applied broad cast upto 30% efficiency when applied as concentrated band along the plant row.
- Clay soil have greater phosphate fixing capacity than sandy soil.

14.6.6 Potassium Use Efficiency

All potassium fertilizer are water soluble. Different 'k' fertilizer consist of 'k' in combination with chloride, sulphate, nitrate.

Both chloride and sulphate of k are soluble in water and on application to the soil then ionizes into k^+, Cl^- and SO_4^{2-} ions. The released k^+ ion from the fertilizer gets adsorbed on the soil colloids and also available to the plant through cation exchange reactions.

14.6.7 Nutrient Fixation

The process where by readily soluble plant nutrients are changed into less soluble form by reaction with inorganic or organic compounds of the soil restricting their mobility in the soil and thereby suffer a decrease in their availability to the plants.

Two kinds of fixation.

Cation fixation : N, K, Fe, Mn, Cu, Zn

Anion fixation : P, B, Mo

Phosphorous Fixation in the Soil

Phosphorous was fixed in the soil by the 3 general types of reactions.

1. Adsorption
2. Isomorphous replacement
3. Double Decomposition

(*i*) ***Adsorption***

(*a*) Physical Adsorption

(*b*) Chemical Adsorption

In Physical Adsorption phosphate is held in the soil solid surface and Chemical Adsorption phosphate penetrate more or less uniformly into the soil surface.

Adsorption of 'P' takes place on –

- on the surface of constant charge - crystalline clay minerals through cations.
- on the surface of variable charges like Fe^{3+}, Al oxides, organic matter.
- on the surface of kaolinite and Allophanes which have pH dependent charges on their crystal edges and surfaces.
- on the surface of organic matter which have pH dependent cations on their surface.

(*ii*) ***Isomorphous Replacement Reaction***

Phosphate is fixed by the Hydroxyl (OH^-) and silicate ions through attached to silicon and Al and are liable to either dissociate to give $(Si, Al)^+$ + OH^- (or), accept a proton (H^+ ion) and give rise to positively.

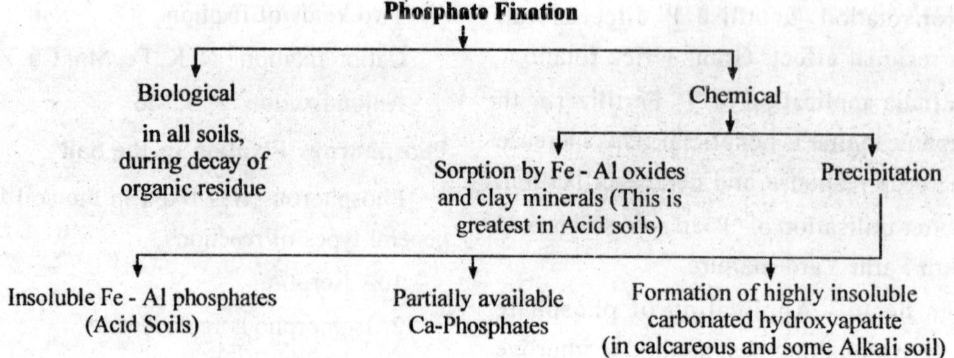

Charged clay, which then take part in the anion exchange with PO_4^-

$$(Si, Al) - OH \rightarrow -(Si, Al)^+ + OH^-$$

$$(Si, Al) - OH + H^+ \rightarrow (Si, Al) - OH_2^+$$

Another important mechanism for the phosphate fixation is a certain amount of silicate released from the tetrahedron.

But reaction of Fe, Al Hydroxides with the phosphate ions are the most significant for phosphate fixation in soils.

(*iii*) ***Double Decomposition Reaction***

This fixation (precipitation) is largely depend upon the pH of the system.

This phosphate fixation can be divided into two categories :

- Reaction involving Fe and Al - here P fixation can be reduced with increase in pH.
- Reaction involving Ca $(OH)_2$/Ca CO_3 - here P fixation can be increased by increasing pH.

Factors Affecting Phosphate Fixation

1. Nature and amount of Soil components
   a. Hydrous oxides of Fe, Al
   b. Type of clay
   c. Amount of clay
   d. Calcium carbonates
2. pH
3. Other ions
4. Organic matter
5. Temperature
6. Over liming.

14.6.8 Nitrogen Mineralisation and Immobilization

Nitrogen Mineralisation

It is a process of conversion of organic nitrogen to ammonia, which involves two reactions *viz.* aminization and ammonification.

Aminization

It is nothing but the process which converts the proteins into amimoacids and amines by the action of heterotrophic bacteria and fungi. Bacteria in alkaline and neutral soils. While fungi dominates in acid soils.

Proteins → Amino acids + Amines + urea + Energy

Ammonification

Here amines and amino acids are decomposed by other heterotrophs, there by releasing ammonia.

$$R-NH_2 + H_2O \rightarrow NH_3 + R\text{-}OH + \text{Energy}$$

$$NH_3 \xrightarrow{H_2O} NH_4 + OH^-$$

Fate of Ammonia

- It may be converted to NO_2^- and NO_3^- nitrification.
- Absorbed directly by higher plants.
- May be fixed in a biologically unavailable form.
- May be slowly released back to the atmosphere as N_2.

'N' Immobilization

It is the conversion of inorganic nitrogen to organic nitrogen and it is basically the reverse of nitrogen mineralisation. If decomposing organic matter contains low 'N' relative to 'c' microorganisms will immobilize inorganic nitrogen in the soil. Soil micro organism compete very effectively with plants for inorganic nitrogen during immobilization and plants may become readily deficient in nitrogen. For this nitrogen fertilizers are applied to compensate for immobilization and crop requirements.

Range in Quantity of N_2 Fixed by selected legumes

Legume	N fixed (lb/a/yr)	
	Range	Typical
Alfalfa	50-300	200
Beans	20-80	400
Birds foot Trefoil	50-100	70
Black gram	80-140	100
Chickpeas	20-100	50
Clovers (general)	50-300	150
Cluster beans	30-200	60
Common Bean	10-120	40
Cowpeas	60-120	90
Crimson clover	30-180	80
Faba beans	50-240	130
Field peas	50-180	100
Green gram	30-60	40
Groundnut	70-120	90
Hairy vetch	50-200	140
Kudzu	20-150	110
Ladino clover	80-200	150
Lentils	40-130	60
Leucaena	350-480	450
Lespedezas	30-120	85
Lucerne	180-250	220
Lupin	60-100	80
Mung bean	170-225	200
Peanuts	20-200	60
Pigeon Pea	140-220	200
Red clover	70-160	115
Soybean	20-260	100
Sweet clover	20-80	20
White clover	30-150	100
Winter peas	10-80	50

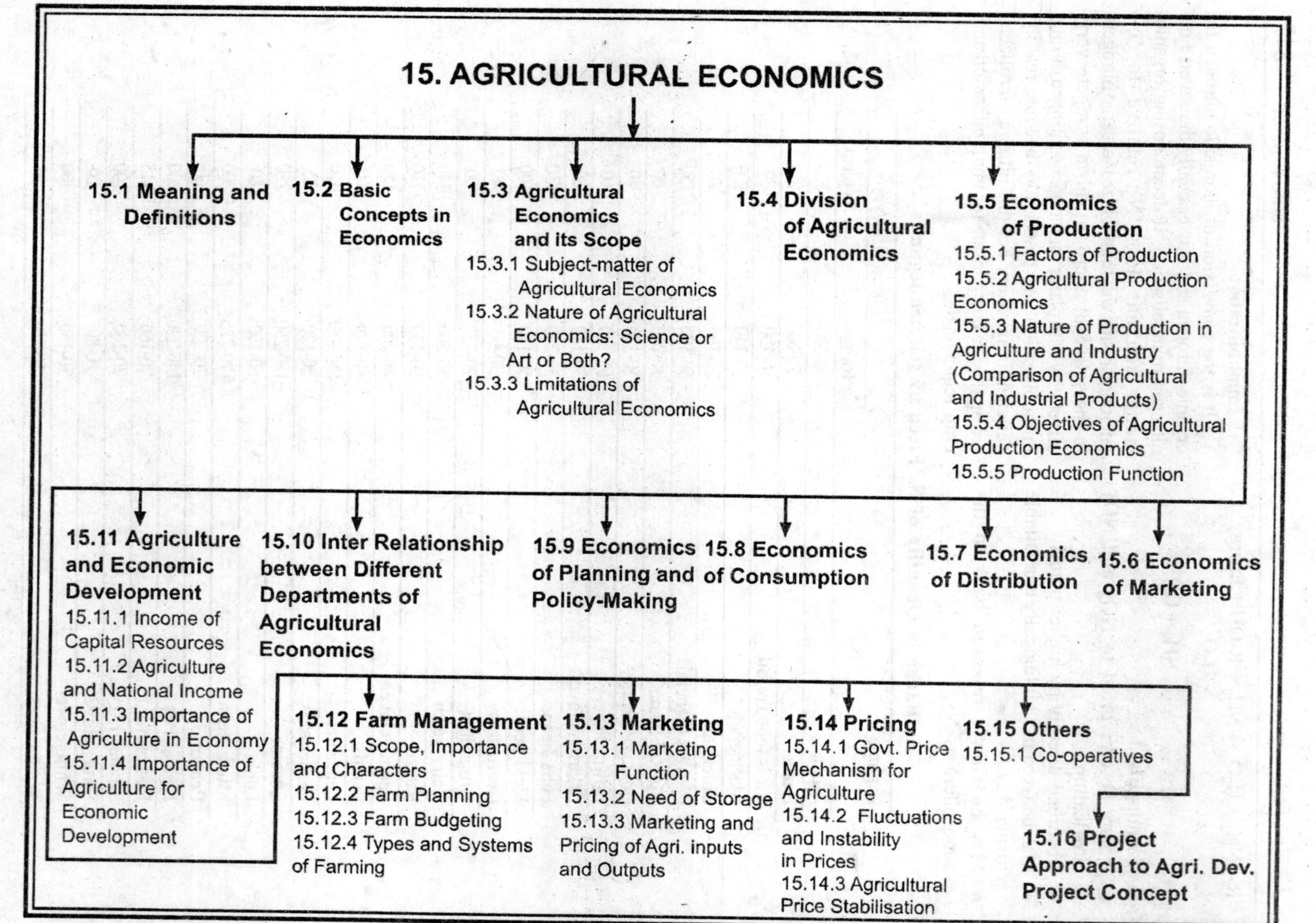
15. AGRICULTURAL ECONOMICS
15.1 Meaning and Definitions
15.2 Basic Concepts in Economics
15.3 Agricultural Economics and its Scope
15.3.1 Subject-matter of Agricultural Economics
15.3.2 Nature of Agricultural Economics: Science or Art or Both?
15.3.3 Limitations of Agricultural Economics
15.4 Division of Agricultural Economics
15.5 Economics of Production
15.5.1 Factors of Production
15.5.2 Agricultural Production Economics
15.5.3 Nature of Production in Agriculture and Industry (Comparison of Agricultural and Industrial Products)
15.5.4 Objectives of Agricultural Production Economics
15.5.5 Production Function
15.11 Agriculture and Economic Development
15.11.1 Income of Capital Resources
15.11.2 Agriculture and National Income
15.11.3 Importance of Agriculture in Economy
15.11.4 Importance of Agriculture for Economic Development
15.10 Inter Relationship between Different Departments of Agricultural Economics
15.9 Economics of Planning and Policy-Making
15.8 Economics of Consumption
15.7 Economics of Distribution
15.6 Economics of Marketing
15.12 Farm Management
15.12.1 Scope, Importance and Characters
15.12.2 Farm Planning
15.12.3 Farm Budgeting
15.12.4 Types and Systems of Farming
15.13 Marketing
15.13.1 Marketing Function
15.13.2 Need of Storage
15.13.3 Marketing and Pricing of Agri. inputs and Outputs
15.14 Pricing
15.14.1 Govt. Price Mechanism for Agriculture
15.14.2 Fluctuations and Instability in Prices
15.14.3 Agricultural Price Stabilisation
15.15 Others
15.15.1 Co-operatives
15.16 Project Approach to Agri. Dev. Project Concept

15. Agricultural Economics

15.1 MEANING AND DEFINITIONS

Agricultural economics is that branch of economics in which the studies are done pertaining to agricultural rules or problems of agriculturists. It is an applied field of economics in which principles of choice are applied in the use of source resources such as land, labour and management in farming and allied activities.

Some important definitions of Agricultural Economics are given as under:

"*Agricultural economics deals with the principles which underlie the farmer's problems of what to produce, how to produce it, what to sell and how to sell it- in order to secure for himself the largest net profit consistent with the best of society as a whole*". **— Taylor**

"*Agricultural economics is that branch of agricultural science which treats that manner of regulating the relations of different element comprising the resources of the farmer, whether it be the relations of each other, or to human beings in order to secure the greatest degree of prosperity to the enterprise*". **— Jouzier**

"*Agricultural economics as a social science is concerned with human behaviour during the process of producing, processing, distributing and consuming the products on farms and ranches*".

— Goodwin

"*Agricultural economics is the study of relationships arising from wealth-getting and wealth-using activity of a man in agriculture*".

— Hibbard

"*The science in which the principles and methods of economics are applied to the special conditions of agricultural industry*". **— Prof. Gray**

"*Agricultural economics is an applied social science dealing with how humans choose to use technical knowledge and the scarce productive resources such as land, labour, capital and management to produce food and to distribute it for consumption to various members of society overtime.* **— Prof. Gray**

15.2 BASIC CONCEPTS IN ECONOMICS

1. Utility: In popular parlance, utility means usefulness. In economics it means the capacity to satisfy a human desire (want) or serve a human purpose.

"Utility expresses a relationship between a subject and an object from the point of view of the problem of want - satisfaction". It is a purely subjective and relative concept. When we say that a certain object has utility, we simply mean that rightly or wrong, it is valued, that people desire to possess it. It follows then that liquor and opium have utility in the economic sense.

However, all utilities are not economic, *i.e.* are not wealth. Utilities of such things as air and light, which are given by the nature and unlimited quantities, however useful they may be are neverthless not wealth.

Utilities are classified into different kinds. A particular kind or type of utility for a commodity is found in a particular situation. The kinds of utility are 1. form utility 2. place utility 3. time utility and 4. possession utility.

2. Goods: The term is used for any commodity or service for which there is demand, without any consideration as to whether it is in any sense good or bad for person demanding it. In other words goods are anything that satisfies a human want.

3. Services: Services are nothing but consumer and producer goods which are non-material (intangible) and often consumed at the same time as they are produced *e.g.* services of teachers, singers, doctors or telephone operators, customer care officers etc.

4. Difference between goods and services: The difference between goods and services is that goods can be stored for future use but services cannot always be. For example, if the service of labourer are not used today, no correspondingly larger amount will be made available tomorrow.

In an economy some industries produce goods and other services. Service industries are often labour - intensive but they make a net contribution to the gross national product in largely the same way as goods do. Gross national product (GNP) is sum total of all goods and services produced in an economy during a particular period (usually an accounting year).

5. Classification of goods and services: Goods and services can be grouped in different ways. Fig. 1 shows a classification which is used by marketing men in which the major division is between consumer goods and services and industrial goods and services.

Consumer goods are sold to general public. **Industrial goods** are products which are not wanted for their own sake, but because they help firms to produce other goods.

Whether a good is classified as consumer or industrial may depends on the use to which it is put. For instance a thresher sold to a farmer is a consumer good, but when it is sold to a rice/wheat selling company it is an industrial good.

The other classification of goods are follows:

(i) Economic goods as free goods: This classification is based on supply.

Goods for which no price is charged and which are unlimited in supply are free goods.

Examples — air, water, sunshine, rainfall etc.

An economic good is any commodity or service which, directly by the individual or indirectly by commodity as a whole, must be paid for, in terms of money. In a barter economy (where money is not used as a medium of exchange) such goods are paid for in terms of other goods or labour service.

Usually when supply is less than demand a good becomes economic while supply is much more than demand the good becomes free. For example water which is a free good near canals and rivers for the consumers, become an economic good in water scarcity place. Similarly sand which is a free good in riverbeds becomes an economic good in places of house construction activities.

(ii) Consumer's and producer's good: This classification is based on consumption. Consumer goods are those from which consumer's directly

Table 15.1

Consumer goods and services			*Industrial goods and services*
A	*B*	*C*	
Convenience goods and services	(*a*) necessaries (*b*) luxuries	(*a*) once-for-all (*b*) durables	(*a*) Raw material (*b*) Intermediate goods (*c*) Capital goods
Shopping goods and services			

Fig. 15.1. Goods and services. A marketing classification

derive the satisfaction using the goods. Food, cosmetics, clothes, books, pens, etc. are examples of consumer goods.

Producer goods are those that help to produce other goods. They can be used by consumers or producers or both, because it depends upon how the good is used. From the consumer's point of view they give satisfaction indirectly. The examples are machines, factory, building, raw material etc.

***(iii) Durable goods*:** Based on durability goods are classified into monoperiod goods and polyperiod goods.

Those goods which are used only once to satisfy the need are called *monoperiod goods*. *Examples*— seeds, fertilizers.

Poly period or durable goods are those which are used again and again.

Examples— Machinery like tractors, agricultural equipments like spade, saw etc.

6. Commodity: In economics a commodity is synonymous with goods, or it may mean an article of trade (or exchange). Commodities (or goods) may be first, consumer goods like bread, butter etc. These are the goods that reach the consumers actually in the form they wish to have. Producer's good , on the other hand,are not wanted for their own sake but because they are creators of other goods. For example tractor is required to assist agricultural operations. Commodities are also classified as *primary goods* like raw materials, food articles, etc. and *manufactured goods* like an thresher, radio etc.

An important point to note is that not only are there many varieties of goods such as potatoes, cigarettes, fruits, motorcars etc, but economic importance of good does not depend solely on its physical characteristics, but on location and on its time of availability. Oil in Arab countries is not the same thing as oil in Delhi, or sand in desert is not the same thing as in urban areas, and orange in December (winter) is not the same thing as those in June (Summer). Hence it is difficult to define a commodity with precision. The assumption of perfect competition that a commodity is homogenous is not realistic. Thus, what distinguish a good from a commodity is the homogeneity of all the kinds in the case of latter. Thus a particular brand of cigarettes is a commodity but all cigarettes are goods.

7. Wealth: Economic goods are called wealth. A basic property of wealth is that it is a means of generating income, *i.e.* income is often regarded as the return on wealth. And in a world of scarcity, the primary objective of an individual, or society, is to maximize income. All economic goods are however not wealth. Goods to be wealth, must satisfy certain conditions.

***(i) It must possess utility*:** Goods which cannot satisfy human wants cannot be called wealth.

***(ii) It must be limited in supply*:** Thus free goods are not wealth. Only economic goods (or scarce means) constitute wealth, for they exist in limited quantities in relation to demand, require effort and are appropriated.

***(iii) It must have a money value*:** It should also be capable of being appropriated. That is some person (or persons) should be able to posses or enjoy it to the exclusion of others.

***(iv) It must be transferable and thus be external to man*:** It need not necessarily be possible to shift it bodily, but it must be possible to transfer its ownership from one person to another.

For example, a house cannot be broadly shifted, but the right to it can be transferred. Contrarily, health is not wealth in economic sense of the term, for it is non-transferable in as much as it is non-external also.

Types of Wealth

(*i*) ***Personal wealth***: The wealth of an individual is his stock of tangible or intangible possessions which have a market value. *Examples*— land, building, shares, bonds, good will of a business, copyrights etc.

(*ii*) ***Social wealth***: This consist of such things as publically-owned property - government schools, roads, libraries, parks etc.

(*iii*) ***National wealth***: It is an aggregate of all individuals' wealth and collective wealth of country. Rivers, mountains of country are also national wealth.

(*iv*) ***Cosmopolitan wealth***: The wealth belonging to world but not to one country. Rivers, oceans, mountains expanding over nations are examples of cosmopolitan wealth.

(*v*) ***Negative wealth***: Negative wealth is the exclusive debts owned by the individual.

8. Human capital: An individual's wealth include not only physical possessions such as house, shares, bank accounts etc. but also his business and professional connections, together with the value of particular skills he may possess.

Human (personal) capital denotes the skills, capabilities and abilities possessed by an individual, which permit him to earn an income.

Wealth and welfare: Wealth means a stock of goods (having extra value) possessed by an individual. Welfare implies the perception of well-being or the satisfaction derived by a man from possessing wealth. Welfare depends on a host of factors including wealth. A man cannot live without some amount of wealth. (*i*) But mere possession of wealth may not make a man happy.

Though wealth and welfare are closely connected with each other, they are not identical. For example

9. A man possessing: ₹ 1 lakh is not double happy compared to a man with ₹ 5 lakhs. Wealth may contribute to welfare, but it is not itself welfare. However certain things which are called wealth (like liquor or opium) not only do not contribute to welfare but may actually be ruinous to welfare in any fundamental sense of term.

Moreover, wealth alone cannot increase welfare. For example, air, sunshine etc being free, are not wealth but they are necessary for well-being (or welfare). Oppositely, if a business manager marries his female servant his wealth may remain unaffected but welfare may decline. For he may be socially may looked down upon. Lastly the actual production of wealth may involve much sacrifice of human welfare.

10. Economic agents: By an economic agent, we mean someone who performs a particular kind of economic function. Examples, a worker, an enterpreneur, an investor or a planner.

11. Economic unit: By economic unit we mean a group of individuals brought together for a common economic purpose such as a household, a firm, unit, or a government.

12. Firms: A firm is an economic unit responsible for the production and sale of goods and services. It may take various forms, an industrial, commercial, financial or agricultural etc.

13. Household: A consumer or a household is defined as a single person having an establishment for himself or a family consisting of a few individuals.

14. Income: In simple language, income can be defined as a flow of money or goods to an individual or group of firm or the economy over some time period.

It may originate from sale of productive services (as wages, interest, profit and rent).The concept of money is very important in Economics since " the sums of money that members of the household get from selling their labour or lending their savings or renting their land generally constitute the fund out of which they buy the goods for everybody's consumption".

15. Production: Production means creation of utility. A production encompasses creation of both goods and services in agriculture, industry, trade and various other sections of economy. The activities of men directed towards increasing the supply of goods may be called production. Says Credland "Production is the creation of goods and services to satisfy consumer demand".

16. Consumption: The starting point of all economic activity is the satisfaction of human wants, *i.e.* consumption. Adam Smith says "consumption is the sole end and purpose of all production and the interest of the producer ought to be attented to, only so far as it may be necessary for promoting that of the consumer".

17. Equilibrium: Equilibrium simply implies a position of rest or balance. For example, a market is in equilibrium if the quantity demanded at a particular price is exactly equal to the quantity supplied. When demand and supply for a commodity are equal a position of rest is obtained. Balance cannot be there if there is no equality between demand and supply.

18. Stocks and flows: In economics we often draw distinction between stock and flow variables. A stock variable has no dimension as such. Its value must be ascertained at some point in time. A flow variable is one which must be defined with a specific time interval. To illustrate, the number of units of goods X purchased over a time period of time is a flow, the inventories of good X held by producers at some point in time is a stock.

19. Value: The term 'value' is the central concept in economics. But it is ambiguous. It is often loosely used to signify utility. When we speak value - in use, we mean actual satisfaction which varies from man to man. Similarly intrinsic value means the inherent capacity possessed by an object yielding satisfaction. These two concepts are contrasted with the value that is current - the value that is expressed by actual rate of exchange.

Attributes of value:

1. Goods must possess utility.
2. Goods must be scarce, and
3. Goods must be transferable or marketable.

20. Price: Price is exchange value expressed in terms of money.

If we know the prices, of goods/objects, we also know their relative values.

'Value' and 'price' are often used as interchangeable terms. But this is not correct. If the prices of all things rise, it means that its value in terms of money has gone up, *i.e* value of money has gone down in terms of commodities.

21. Margin: In economics, nearly all decisions are made in terms of adding or subtracting marginal units - or as economist say, they are made at the margin.

15.3 AGRICULTURAL ECONOMICS AND ITS SCOPE

While discussing the scope of agricultural economics, we have to look into,

1. Subject- matter of agricultural economics.
2. Nature of agricultural economics.
3. Limitations of agricultural economics.

15.3.1 Subject-Matter of Agricultural Economics

Agricultural economics comprises all aspects of production, consumption, exchange and distribution. The agricultural economics deals with the problems of farmer as to-

1. What to be produced?
2. Auxiliary business to be done.
3. Quantity of production.
4. Time in which production quantity is to be sold and by whom?
5. What portion should be given to whom in his production?
6. What should be consumed?

Nowadays agriculture is carried on for marketing point of view and not for one's own subsistence. Hence he/she has to have a different regions and countries. Thus it is clear that the subject matter of agricultural economics is very exhaustive. This covers the individual farm organization and management to national and international strata farm organization and management. So it analyses in a micro and macroviews.

The important progress achieved in economics of agriculture is that there is a increased cooperation among nations.The aim of this co-operation is to make proper production and distribution of farm products. It is experienced that various nations have unequal distribution of agricultural resources. Hence more producing nations have responsibility towards less producing nations. The agricultural economics have worked on this behalf and the international commodity agreements can play an important role to distribute in best use of agricultural resources.

15.3.2 Nature of Agricultural Economics: Science or Art or Both?

In real science studies are done for cause and effect. Science is the systemized and generalized knowledge gained through observation and experiment. Analogous to science, an art is also systemized body of knowledge. It directs through a system of procedures to attain a given objective or goal. It tells how to do a thing.

Treating economics as a science, a given theory is formed through conduct of experiments, recording observations, analysis of data recorded, drawing of conclusions and finally testing them. In economics also the same procedure is followed to present any principle or theory. Hence economics is as good as any science. Only the question is regarding precision. The scientific experiments are conducted under laboratory conditions, while economic theories are subjected to several casual factors that influence human behaviour. The situation of controlled experiment in economics is not a possibility, since it deals with human behaviour, which is unpredictable. This indicates that the degree of precision in economics as a science is less, when compared with other science, but nevertheless economics is a science. As for example it shows in the field of agricultural production that in a given field if more doses of capital and labour is employed, each more doses diminish the production. Thus *law of diminishing production* applies to agriculture. But agriculture economist think over good or bad, right or wrong; he/she puts some ideals. As for example he/she tells that agricultural production in India is less than the other countries per hectare production which should be increased. Hence we can say it to be an ideal science.

Through it is a science, it is also an art. As an art, economics shows solution to the problems. It helps how to do a thing. The role of economics as an art can be found in any sphere of economic activity. Practical science or art tells us means to achieve the ideals. Agricultural economics studies means which can lead to increase in per hectare production.

Given a problem, the field of economics guides us to solve the same. Thus the field of agricultural economics has attributes of science and art. Agricultural economics therefore is a science as well as a art.

15.3.3 Limitations of Agricultural Economics

There are certain problems and limitations of agriculture which restrict the agricultural behaviour. We give here some of these:

(1) Crops are depending on favour of nature from sowing till ripening. In India crops are said to be dependent on monsoons. Due to limited scope of land, agriculture cannot be increased.

(2) Keeping in view climatic conditions, market demand, land and availability of capital the agriculturalists ascertain as to which crop they should be growing and how much land?

(3) Agriculturist cannot draw his capital invested in agriculture unlike in industries.

(4) Agricultural products are destructible and perishable in nature hence they cannot be stored for long time with the farmer.

(5) Agricultural crops need much space for storage and has high transportation costs.

(6) When all agricultural crops reach simultaneously, the rates come down and farmers has to bear the losses.

As agricultural business is scattered throughout the country, no division of labour is possible.

15.4 DIVISIONS OF AGRICULTURAL ECONOMICS

Agricultural economics can be divided into five major divisions. They are:

(1) Economics of production.

(2) Economics of distribution.

(3) Economics of consumption.

(4) Economics of planning and policy formulation.

15.5 ECONOMICS OF PRODUCTION

Production is a process where by some goods and services called inputs are transferred into other goods and services called *output*, which in turn creates utilities.

The inputs that are used in production of goods may be provided by nature and/or by other industries.

Characteristics of modern production:

1. Large scale production
2. Use of machinery
3. Immense use of capital
4. Division of labour
5. Competition and Combination
6. Privatization
7. Joint stock companies
8. Localization of industries
9. Unemployment and underemployment
10. Price and product differentiation
11. Advertisement
12. Speculation and
13. Risk

Production in agriculture is becoming complicated. In past it was only limited to way of life but now it has turned to be business hence an agriculturist it should know better- when to sow, how to sow and where to sow etc. Whatever farm - industry needs various inputs, their reciprocal relation should be in their knowledge. So that they should reach efficient production scale with reference to provided resources and costs. They should also know that which combination of individuals will be his profit and satisfaction.

It does not matter if agriculture is done for consumption or marketing. In both situations to reach maximum result has to be make optimal use of farm resources. They should also know the land, credit, marketing and taxation regulations. Then only they can influence government decisions in their favour. When they know farm management in individual way they can also deal with national and international problems. Atmosphere and institution management also come under the scope of agricultural economics.

Economics of production is not restricted to *factor- to-factor* and *factor-to-product* relationship and analysis but it extends to whole social and economic atmosphere in which agricultural work is being conducted. Other kind of economic system like land allotment, farm credit-labour, taxation, various government rules and regulations which influence agricultural production and sales are included in economics of production with the study of relationship of agriculturists.

15.5.1 Factors of Production

Productive resources required to produce a given product are called factors of production. These are 1. land 2. labour (human resource) 3. Capital and 4. Organization (enterprise).

1. Land: Land stands for all natural resources which yield income. According to **Marshall** land means " the material and forces which nature gives free for man's aid, in land and water, in air and light and heat".

Some of the characteristics of land are—

1. It is nature's gift and fixed in quantity.
2. It is undestructible and permanent.
3. It is immovable.
4. Geographical supply of land cannot be increased but economic supply can be increased by putting the land under intensive and higher use.
5. Land can be divided and sold like any other property.
6. Composition of land vary place to place *i.e.* land has variability in its structure and fertility.

2. Labour: Any work whether manual or mental which is undertaken for a monetary consideration is called labour. In the words of Marshal "any excretion of mind or body undergone partly or wholly with a view to earn some good other than pleasure derived directly from the work". Any work that is done for pleasure not come under labour.

Characteristics of labour:

1. Labour is inseparable from the human being (labourer) as it is a living thing.
2. Without the aid of labour, no land can be productive.
3. Labour has very weak bargaining power.
4. Change in prices effect labour supply and labour cost.
5. In the short run, labour supply cannot be adjusted with demand.
6. Supply of labour change with wages.
7. Labour does not last and is highly perishable commodity, in the sense the utilization of labour cannot be adjusted with demand.

Efficiency of labour: According to economists efficiency of labour means " the ability of labour by virtue of which it is productive". It indicates the qualitative and quantitative performance of the labourer. No doubt, it is a relative term, but when one person puts a better performance over others, we say that he/she is more efficient.

Division of labour: In modern production activity, the production of good is divided into several sub-process and each process is entrusted to a group of workers. Division of labour is meant to improve the efficiency of labourers.

3. Capital: Wealth which is used in producing further wealth, or which yields an income is called capital.

Forms of capital:

1. Fixed capital — land, buildings.
2. Working capital — Fertilizers, wages etc.
3. Sunk capital — Purchase of machinery and equipment.

Characteristics of capital:

1. It is productive.
2. It yields income.
3. It is prospective in the sense that we postpone the present use of it for future in anticipation of reward.
4. **Organization:** The role of organization or enterprise is co-ordinating and correlating the other factors of production process.

But for agricultural production only two factors namely labour and capital are important.

15.5.2 Agricultural Production Economics

Agricultural production economics is an applied field of science where in the principle of choice are applied to the use of capital, labour, land and management resources in the farming industry.

Agricultural production economics is concerned with the productivity of inputs. The agricultural production economists are concerned with any phenomena, which have a bearing on economic efficiency in the use of resources.

15.5.2.1 Basic terms in agricultural production Economics

Farm: Piece of land where crops and live stock enterprises are taken up under a single management and has specific boundaries.

Agricultural holding: The total area of land owned by individual or joint family whether cultivated out by family or rented out.

Family holding: It refers to a farm which yields a gross income of ₹ 1600 or a net income of ₹ 1200.

Operational holding: Total land area held under single management for the purpose of cultivation. It excludes any land leased out to another person.

Optimum holding: It refers to maximum size of holding which a family should possess. Three times the family holding is considered to be an adequate size of optimum holding.

Economic holding: It is defined as one which could provide a reasonable standard of living to the cultivators and give full employment for a family of nominal size.

Marginal farmer: Farmers owning less than 2.5 acres of dry land or 1.25 acres of wet land is called marginal farmer.

Small farmer: Farmer owning land holding ranging from 2.5 acres to 5 acres of dry land and below 2.5 acres of wet land is called small farmer.

Plant: It means group of technical units, *i.e.* 5 ha. farm or 1000 birds poultry or 10 animal dairy etc.

Farm firm: It is also known as *economic unit* which is run under one management. It represents an aggregation of resources for which costs and returns are worked out treating as unit. *Example*— A farm holding.

Resources: Anything that aids in production is called resource. *Examples*— seeds, fertilizers, feeds, veterinary medicines etc.

Resource service: The workdone by a person or a machine or livestock is called resource service. *Examples*— Services of labourer, machinery, farm implementation etc.

Fixed resources: Resources which remain unchanged irrespective of the level of production are called fixed resources. *Examples*— Land, machinery, implements etc.

Variable resources: Resources which change with the level of production are called variable resources. *Examples*— seeds, fertilizers, plant production chemicals feeds etc.

Flow resources: The resources which cannot be stored and should be used as and when they are available. For instance, if the services of a labourer available on a particular day are not used, then they are lost forever; similarly the services of machinery, farm building etc.

Stock resources: Stock resources are those which facilitate for their storage, when they are not used in one production period. *Examples* — seeds, fertilizers, feed etc.

Monoperiod resource: It is the resource which can be used only once in production Examples: seeds, fertilizers etc.

Polyperiod resource: It is the resource which is used in the production process over several period. *Examples*— Machinery, implements etc

Production period: It is the time period required for transformation of resources or inputs into products or output.

Farm entrepreneur: The person who organizes and operates the farm business and bears responsibility of the business.

Farm business manager: The person appointed by entrepreneur to manage and supervise the farm business and paid for his service.

Productivity: It denotes the efficiency with which various inputs are converted into products. It signifies the relationship between output and inputs. In simple terms output per unit of input is called efficiency. *Examples*—5 quintal of paddy/ha.

Efficiency: It means absence of wastage or using resources as effectively as possible to satisfy farmers' needs and goals.

15.5.3 Nature of Production in Agriculture and Industry (Comparison of Agricultural and Industrial Products)

The substantial difference of agricultural and industrial production is mentioned below in table. 15.2

Table 15.2

Agricultural Production	*Industrial Production*
1. Depends too heavily on seasonal and weather conditions.	1. Independent of weather conditions except supplies of certain raw material from industry.
2. Little control over the conditions of production.	2. Can be controlled.
3. Production is seasonal but consumption is regular so need storage.	3. Both production and consumption are almost regular. So not too much need of storage except few like crackers and woolen clothes etc.
4. It can be considered as industry & small scale.	4. Generally on large scale.
5. There is longer time between expenditure and return.	5. Shorter time between expenditure and return.
6. Agricultural products are joint products- Example: paddy + straw production.	6. Generally single product.
7. Products are bulky in nature and easily perishable.	7. Not all products are bulky and are not perishable in general.
8. Agricultural product cannot be offered in short time, with increase in market demand.	8. Can be offered in short time by increasing production rate/overtime to fulfill market demand.
9. Primarily concerned with basic necessities of life *i.e.* food crops.	9. Generally concerned with goods for comfort and luxuries.
10. Farming is slow business of low turnover.	10. Fast business having high turnover.
11. Demand of agricultural products is generally inelastic (or less elastic).	11. Demand of industrial products is generally elastic.
12. Agricultural labourers are generally disorganized and have low bargaining power.	12. Industrial labourers are generally organized and have high power.
13. Law of diminishing return begins to operate at early stage.	13. Law of diminishing return begins to operate at latter stage.
14. The variations in the farm products regarding appearance, size, shape, colour, length etc. are found due to availability of large number of crops; so poses difficulty in standardization and grading. Apart from this, quality variation bring in a wide variations in prices for the same commodity.	14. There is no such problem with industrial product, products are generally uniform in all respects. The production practices are standardized in industries. With the help of specialized machine it is possible to produce goods of uniform size and quality.
15. Larger proportion of land divided into several small and scattered holding.	15. There is no such case in industry as in industrial production economics of large scale production are not common feature in agriculture.

15.5.4 Objectives of Agricultural Production Economics

1. Determination and outlining of conditions giving optimum use of land, capital and management in the production of crops and live-stocks.
2. Determination of extent to which the existing use of resources deviates from optimum use.
3. Analysis of forces which condition production pattern and resource use.
4. Explanation of means and methods in getting the optimum use of resource from the existing one.

15.5.5 Production Function

Production function is a technical and mathematical relationship, describing the manner and extent to which a particular product depend upon the quantities of inputs or service of inputs used at a given level of technology and in a given period of time.

In the production relationship, output (Y) is a function of input (X) used in given by

$$Y = f(X)$$

15.5.5.1 Types of Production Function

1. **Continuous function:** The doses or levels of input can be split up into small units, which are used or done. Response of yield to different levels of inputs. *Example*- Fertilizer can be applied to a hectare of land in different quantities ranging from kilograms to quintals.
2. **Discrete production function:** Such functions are obtained for a input factor or work units, which are used or done in whole numbers such as one ploughing or a number of ploughings.
3. **Short run production function:** Production function which relates to factors and products where some resources are fixed.
4. **Long-run production function:** Those input- output relations which permit variations in the input of all factors.

The production function can be expressed in the forms—

(1) In the form of table

(2) Graphic (geometry) and

(3) Algebric forms.

Example — Production function for fertilizer input (X) and crop output (Y) in quintals.

1. In tabular form

Table 15.3

Fertilizer Input (X)kgs	*Crop output (Y) (quintal)*
0	7
5	9
10	13
15	17
20	21

2. Graphical form:

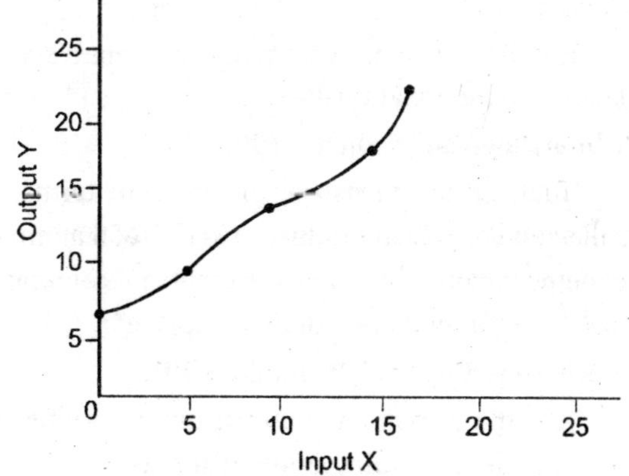

Fig. 15.1: Production function

3. Algebric form: The input-output relationship can be expressed by algebric equations. The relationship may be linear or non linear.

$Y = a + bX$ (linear)

$Y = a + bX + cX^2$ (nonlinear-quadratic)

$Y = a \times b$ (nonlinear exponential)

when a, b, c are constant coefficients.

15.5.5.2 Dependence of Production Function

Production function depends on the following factors.

1. Quantities of resources used.
2. Level of technical know-how of the producer.
3. Possible method in production.
4. Nature of firms organization.
5. Size of the firm.
6. Relative prices and combination of factors of production.

15.5.5.3 Factor Product Relationship

Factor product relationship is a basic production relationship between the input and output. This is mainly concerned with resource use and efficiency. It guides the producer in deciding as to how much to produce. The goal of this relationship is optimization of resources. The relationship is expressed by law of diminishing returns.

Certain concepts, which figure in analysis of factor product relationship are.

1. Total Physical Product (TPP)

Total output measured in physical terms is called total physical product. It is the total amount of output obtained by using different units of inputs measuring in units like quintals, kgs, etc.

2. Average Physical Product (APP)

It is the average amount of output produced by each corresponding units of inputs.

Average Physical Product (APP)

$$= \frac{\text{No. of units of output}}{\text{No. of units of input}} = \frac{Y}{X}$$

APP reflects the efficiency of variable input (technical efficiency).

3. Marginal Physical Product (MPP)

It is measured as the change in output which results from applying an additional unit of input.

$$\text{MPP} = \frac{\text{Additional output}}{\text{Additional input}} = \frac{\Delta \text{Y}}{\Delta \text{X}}$$

4. Elasticity of Production

It is a measure of responsiveness of output to an increase in input. It is defined as percentage increase in output as in relation to the percentage change in input.

Elasticity of Production (E_p)

$$= \frac{\text{Percentage change in output}}{\text{Percentage change in input}}$$

$$= \frac{\Delta Y}{Y} \times 100 \Big/ \frac{\Delta X}{X} \times 100$$

$$= \frac{\Delta Y}{Y} \times \frac{X}{\Delta X} = \left(\frac{\Delta Y}{\Delta X}\right) \times \left(\frac{X}{Y}\right)$$

$$= \text{MPP} \times \frac{1}{\text{APP}} = \frac{\text{MPP}}{\text{APP}}$$

5. Relationship Between TPP; MPP and APP

As long as MPP is increasing, TPP is increasing at an increasing rate. TPP goes on increasing at an increasing rate till the point of maximum MPP, TPP increases at a decreasing rate. When MPP becomes Zero, TPP attains the maximum. Negative MPP results in decreasing TPP, When TPP is increasing, MPP is positive, TPP is maximum MPP is Zero and when TPP declines MPP becomes negative.

6. Total Value Product (TVP)

Output measured in monetary terms is called total variable product. If it is from resource *X*.

$$. \text{TVP}_x = \text{TPP}_x \times \text{P}_{xy}$$

where P_y is the price of the price of product.

7. Average Value of Product (AVP)

It is total value product divided by amount of resource.

$$AVP = \frac{PVP_x}{X}$$

or $AVP = APP \times P_y$.

8. Marginal Value Product (MVP)

Marginal product measured in value terms is called marginal value product.

$$MVP = MPP \times P_y = \frac{\Delta Y}{\Delta X} \times P_y$$

Elasticity of Production, and Production Function

A production function with an elasticity of 1. ($E_p = 1$) throughout indicates constant returns.

A production function with elasticity of greater than 1 ($E_p > 1$) through out indicates increasing returns.

A production function with an elasticity less than 1 ($E_p < 1$) throughout indicates decreasing return.

15.5.5.4 Laws of Return

The nature of relationship between a single input and single output can be either the one or a combination of the types given below

1. Law of constant return.
2. Law of increasing return.
3. Law of decreasing return.

1. Law of Constant Return (Constant Margin Products): In constant returns, each additional unit of the variable input applied to fixed factor(s) produces an equal amount of additional product. The production function is a straight line (linear) having constant slope and expressed as

$$\frac{\Delta_{1r}}{\Delta_{1X}} = \frac{\Delta_{2r}}{\Delta_{2X}} = \frac{\Delta_{nr}}{\Delta_{nX}}$$

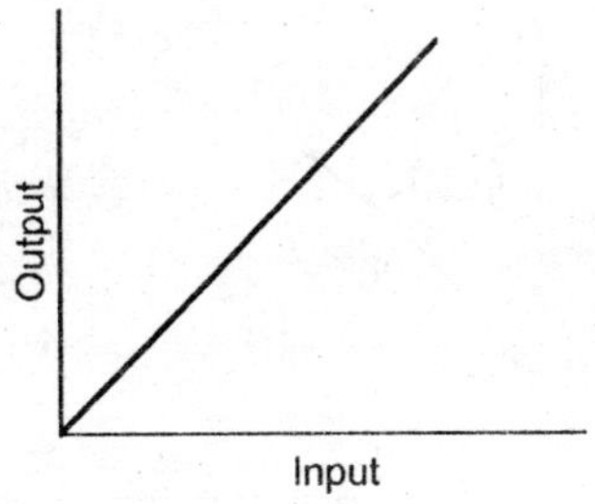

Fig. 15.2: Constant return to input use.

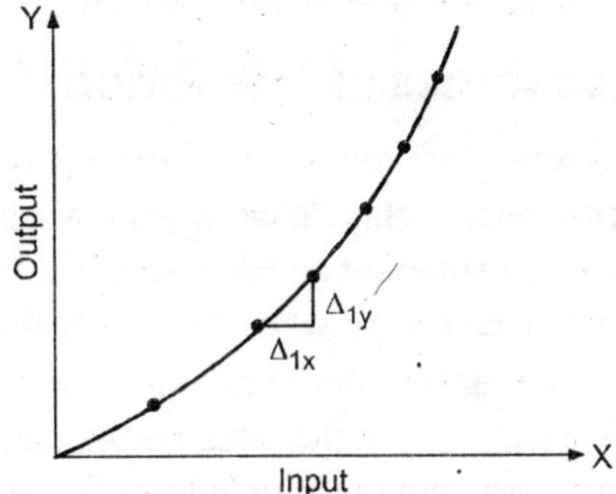

Fig. 15.3: Increasing returns to input use.

2. Laws of Increasing Return: (Increasing marginal productivity) In the law of increasing return, the addition of each successive unit of variable factor to the fixed factors in the production process, adds more to the total output than the previous unit.

The ratio of $\frac{\Delta Y}{\Delta X}$ will go on increasing as more and more units of input are added.

3. Law of Decreasing Return (Decreasing Marginal Productivity): The addition of each successive unit of the variable factor to the fixed factors in production process, adds less to the total output than the previous unit or each addition unit of input adds less to the total product than the previous unit did. This function exists in almost every practical situations of agriculture.

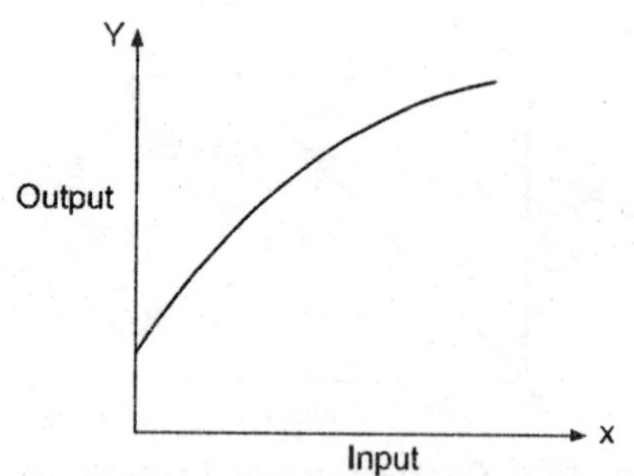

Fig. 15.4: Decreasing returns

15.6 ECONOMICS OF MARKETING

Efficiency of marketing depends on the appropriate relationship between various actively participating institutions and their services. Hence we have to decide as to when the crops are to be sold, and how many transportation vehicles will be needed, what will be the spread over marketing cost, what portion of the agriculture will get from the price taken from consumers and what will be profit margins of intermediaries. Economics of marketing also include marketing mediators profit, agricultural portion, price paid by consumer and besides influential government rules, export and import policies.

As agricultural production is not restricted to subsistence only and has become a business, technical progress and change in consumer preferences, it has made possible large scale economy and legislation of farm production has become a special job, hence with the economics of production, economics of marketing increase an importance now a days.

15.7 ECONOMICS OF DISTRIBUTION

This division concerns with the price distribution on resources of production. Without price no production work can be pushed further. Hence labours are paid wages, capitalists are paid interest and landlord is given rent. Profit is that price earned due to risk and uncertainty by the entrepreneur or person for his undo - mitable courage. This principle governs the distribution of wealth in society. The problem of distribution has gained importance now a days as there is increase in the difference of labour and income and distribution process is full of injustice. It is better to remember that general principles of distribution are also applicable to agriculture with few amendment.

15.8 ECONOMICS OF CONSUMPTION

This tells the agriculturists, the way to purview the price - parity between one's sales and purchases and proper adjustment of own family expenses. Thus it makes possible the attainment of welfare. There is the subject of terms of trade which is favourable to the farmer or not, is gaining importance as farmer's have their unions formed and government is paying more attention to the welfare of agriculturist community. The consumption of goods of the farmers are influenced by various taxations and as such their study also comes under agricultural economics. If the government is precautioned about the ill effects of taxation on agriculture they are totally left by the concerned government. Hence economic development is encouraged to need of proper use of farm income and family living.

15.9 ECONOMICS OF PLANNING AND POLICY-MAKING

Extra ordinance force is being applied to policies and practices relating to planning through out the world. Their is more necessity of planning for agriculture then for other spheres as everywhere agricultural economy is backward and consequently there is active force working for poverty and less development. Hence planning is only way out for this. For agricultural society planning is naturally needed to be quite different from industrial society as it has quite different priority order, procedure and urgency. Planning is limited to crops in agriculture and can be extended to whole agricultural management.

15.10 INTER RELATIONSHIP BETWEEN DIFFERENT DEPARTMENTS OF AGRICULTURAL ECONOMICS

All the division of agricultural economics are mutually related and dependent. Hence study of one cannot be separated from other. It should be borne in mind that the scope of agricultural economics is so vast that it is hard to bind them under these divisions.

Production and Consumption

Production has intimate relation with consumption. The interest in consumption shows the profitable conditions for the producer.In any country, level of consumption depends on limits on production capacity of the country.

Production and Marketing

Production is not perfect without marketing. With the extreme economic view point, production work is complete only when goods reach the consumers and this job is possible through efficient marketing. Again marketing has to bear in mind the nature of production as marketing is different for every good produced. Hence both division are related immediately.

Production and Distribution

Production is dependent on the process of distribution in economy up to great - extent. Availability of production resources is influenced much by distribution work which is not itself influenced by production. Every district should have project which will lead to accelerate the attempts for agricultural development. It is also suggested that successful agriculturists should be exchanged with different regions and states. Government of India has commenced a system of giving "National standard" to state having maximum production per acre.

It is great pleasure that planned development programs have cared more for agriculture recognization and much development is seen under agricultural conditions.

15.11 AGRICULTURE AND ECONOMIC DEVELOPMENT

While we talk about the problems of economic development of a country, we see at once to the prosperous society with a high living standard. We think that nation has a high per capita income. It must have an economy with various occupation and services and so different income resources. Still we should not overlook the high labour productivity, it will have no economic growth. As prosperity with high living standard can only be the effect of high productivity, so higher the percapita share of income, higher the production per worker. But we cannot have it for one sector only but overall production is increased and improvement is seen and thus we speak of industries, agriculture, manufacture and so on that we achieve a high labour productivity.

15.11.1 Income of Capital Resources

To increase labour productivity we require more of capital resources. We have limited land resources and cannot increase it. Only manpower increase can bring improvement in the resources, so we should use more capital on land. We invest in irrigation, prevent soil erosion, clear marshy land. Thus natural land resources are supplemented by man-made capital resources.

The capital resources should be applied to that sector from which we will be able to get economic surplus. We may think for a while that generally underdeveloped country, India though partly industrialised, still it is an agricultural country. Thus capital requirements are to be fulfilled by agriculture. It is only industry which has recently developed can provide capital resources. Hence only agriculture can provide the necessary economic surplus.

But if any how agriculture is not able to do this function, the economic growth is not possible. There must be great political pressure on agriculture and they should be brought under subsistence level to get economic surplus formation. Social measures should also be taken to let the farmers work hard and result much savings.

As for India, we see the quite low productivity in Indian agriculture. Indian soil is quite exhausted in productivity. It only gives a highly under-developed agriculture. Hence implementation of proper cultivating methods and inputs are necessary to make the soil to increase productivity. The second reason for low productivity is defective land tenure system and it has come still down due to present land reforms which have caused land owners and farmer to invest less.

15.11.2 Agriculture and National Income

The centuries before Christ, Cuto said that if there is delay in agriculture., every work is delayed. There is no doubt that before industrial revolution, most economic activities of people consisted of agriculture and handcrafts through out the world. India have achieved a high position. India is fastly related to agriculture since past. In reality, agriculture has influenced much through the centuries, Indian thoughts, aims and culture while the industrial country faced industrial revolution.

15.11.3 Importance of Agriculture in Economy

Present importance to agriculture is incomparable in Indian economy. Through its contribution is diminishing due to the pressure of industrialization. The agriculture occupation is the main source of economy. We get raw material from our agriculture for industries. The agriculture provides food and grass to the whole population and live-stock. Finances of centre and states are dependent on agriculture to large extent. In our exports agriculture produce and goods prepared from agriculture has 60% of the total export. Indian agriculture has an honorable place in international field. India has first place in the world for the production of tea and ground nuts. Rice, jute, sugar, sesamum, mustard and castor seed have the second place for India in the world.

In India agriculture is the main source of employment. More than 58% of total labour force of India is engaged in agriculture and dependent on it for there livelihood.

15.11.4 Importance of Agriculture for Economic Development

Though agriculture dominates the Indian economy but it is in backward status hence per capita income in agriculture is very less than that of industries. Many scholars have found agriculture itself liable for this low income. They put forth that while agriculture dominates in India, per - capita income cannot attain the desired or needed limit.

Thus agriculture dominance is responsible for economic backwardness which is not the cause but effect, though they are such intermingled mutually that we cannot point out as to which is the cause of effect. Besides it should be admitted that if we like future economic development of the country, agriculture must reach the prosperous stage. In economic development the inference of agriculture's important role can be had from the contribution it makes to other spheres.

15.12 FARM MANAGEMENT

Farm management is a science which deals with the proper combination and operation of production factors including land, labour and capital and the choice of crop and livestock enterprises to bring about a maximum and continuous return to the most elementary operation units of farming.

15.12.1 Scope, Importance and Characters

The term farm Management consists of two words, 'Farm' and 'Management'. 'Farm' is a modified piece of land held or operated as a unit for the production of agricultural products. 'Management' means the art of managing the farm. It involves the process of determining how the farm unit shall be organised and operated.

Management performs three unique and vital functions —

(*i*) **Management makes decisions:** To make decision means to decide upon each course of action ; to choose wisely among alternatives where such exist.

(*ii*) **Management acts on decisions:** After making decisions, the next step is to act, to work, the farm management selects the production level land technology and practices for products for producing and marketing the farm products. He must find and assemble all resources.

(*iii*) **Management takes Financial Responsibility:** The financial responsibility involves assembling. Using and maintaining the farm resources.

Farm management is concerned with decision making. The main decisions of agriculture business which will maximise the returns to the resources used in the farm business are:

1. **What to produce?** It decides which products to produce and shows the product-product relationship.
2. **How to produce?** It relates to choosing the most effective method of producing a given quantity of a particular product. It shows the factor-factor relationship.
3. **How much to produce? (Factor-Product-Relationship).** It relates to the problems in converting the several farm resources into the final farm product.
4. **Time relationship:** It shows the relationship between the time that farmer makes investment of capital and that later date when this investment returns the physical production.

Scope of Farm Management

The study of farm management comes under microeconomics. The subject of farm management covers selection, size and appraisal of farm, appraisal of farm resources, investment decisions, enterprise relationships, planning the crops, farm labour and livestock, farm power machinery and equipment, costs returns on individual enterprises, complete budgeting, risk and uncertainty and marketing of farm produce.

Nature of Farm Management

Farm management is both pure and applied. It is pure science. It is pure science because it deals with the collection, analysis and explanation of facts and the discovery of principles, it is an applied science because the ascertainment and solution of farm problems are within its scope. The study of farming from the point of view of art would involve a careful examination of the thought, process and skill used by farmers in farming. Thus farm management is both a science and an art.

Limitations

The limitations of the farm management are:

(*i*) *Small size of farm business*: Due to fragmentation and subdivision of holding the average size of operational holdings is very small.

(*ii*) *Inadequate capital*: It is the most serious problem in agriculture. The farmers do not get adequate and timely financial aid from the institutional agencies.

(*iii*) *Slow adoption of new technology*: The farmers do not get eager to adopt new technology, because it requires heavier inputs such as more fertilizers, timely irrigation, high yielding varieties of seeds and plant protection measures etc.

(*iv*) *Under-employment*: Under-employment reduces efficiency and productivity of rural people. Small size of operational holding, more family labour supply and seasonal nature of agricultural production are the source of the causes of under-employment in farming.

(*v*) *Managerial problems*: The illiterate farmers face difficult problems in managing the farm. Their managerial skill is very poor.

Economic Principles Applied to Farm Management

The economic principles applied to farm management have been developed directly or indirectly from the fundamental concepts and laws of economics. The economic science plays a very important part in all decisions of farm management. They are applicable to crop production, livestock production, machinery, farm buildings, labour and farm practices and production in general. The most important principles that help the farmers in making the choice and decisions to all phases of farming are:

1. *The Law of Diminishing Returns*

It is physical law of fundamental importance in organising and operating the farm business. It is an important guiding factor in farming to decide the level at which a farmer can increase the output. It is practical experience of every farmer that after a certain stage in cultivation, the application of successive doses of labour and capital to a given area of land does not bring about much return as the previous dose. Additional dose of labour and capital applied to the same place gives diminishing returns. This tendency is known as the law of Diminishing Returns.

Definitions:

"If the quantity of one productive service is increased by equal increments, with a quantity of other resource services held constant, the increment to total product may increase at first but will decrease after a certain point" — **E.O. Heady**

"An increase in capital and labour applied in the cultivation of the land causes in general, less than proportionate increase in the amount of product raised, unless it happen to coincide with an improvement in the art of agriculture."

— **Marshall**

Prof. Marshall has given three conditions which may delay the operation of the law in agriculture.

1. Improvement in the art (technique) of agriculture.
2. Increase in the efficiency or skill of the farmer.
3. Such precious doses of labour and capital applied to land that more than a proportional return is obtained from their increased application. The following table will illustrate principle of diminishing returns.

It is clear from the following table that as the units of inputs is increasing the marginal physical product increases upto 3rd unit of input, after that it becomes constant and from 4th unit the marginal physical product starts decreasing, it becomes negative on 7th unit of input. To find out optimum level of output it is essential to change the physical units of input and output. The economic point will be where marginal revenue and marginal cost are

Table 15.5

Units of input applied	*Addition to total inputs*	*TTP*	*MPP*	*APP*	*Cost of additional inputs @ Rs. 20 per unit per unit*	*Value of additional output @ Rs. 4*
0	1	0	0	0		
1	1	8	8	8	20	32
2	1	20	12	10	20	48
3	1	31	11	10.3	20	44
4	1	40	9	10	20	36
5	1	45	5	9	20	20
6	1	48	–3	8	20	–24
7	1	42	–6	6	20	–24
8		32	–10	4	20	–40

TPP = Total Physical Product, MPP = Marginal Physical Product, APP = Average Physical Product.

equal. In the above table 5th unit is the optimum level from the economic point of view.

Law of diminishing returns in general applied to the field of agriculture but its operation can be postpaned under the following condition.

(*i*) *Improved technology*: Technology components like high yielding varieties of seeds, irrigation, fertilizers, integrated pest management practice etc, would help in increasing the returns.

(*ii*) *New soils:* When virgin soils are brought under plough, they give more and more yield as their productivity is higher.

(*iii*) *Scarcity of capital*: Scarcity of capital limits the farmers in applying sufficient quantities of variable resources leading to the prevalence of increasing returns.

Reasons the operation of law of diminishing return

(*i*) Excessive dependence on agriculture

(*ii*) Less scope of division of labour

(*iii*) Less scope of mechanization

(*iv*) Cultivation of interior lands

(*v*) Continuous cultivation.

2. *Cost Principle*

This principle is of great importance in making decisions of farm business. Cost in general, refers to the expenses incurred on productive services and physical productivities are guided by the cost of various input factors. Mainly cost are of two types.

(*i*) *Fixed Cost:* Fixed costs present farming expenses of an overhead nature and do not change with the output. *e.g.*, interest on fixed capital, depreciation, revenue etc.

(*ii*) *Variable Cost:* It varies with the level of output. It includes changes of hired human labour, seed, fertilizer, irrigation, insecticides and pesticides etc. There are seven costs of production.

(*a*) *Total Variable Cost* (TVC): It is obtained by multiplying the amount of variable input by the price per unit in input.

(*b*) *Average Variable Cost* (AVC): It is computed by dividing total variable costs by the amount of output AVC varies with the level of production.

$$= \frac{TVC}{Output}$$

(*c*) *Marginal Cost* (MC). MC is the change in total cost as related to change in output.

$$= \frac{\Delta TC}{\Delta Y}$$

The marginal or added cost of each unit of output is important in determining how for we should push producing and how much of the various resources we should use.

(*d*) *Total Fixed Cost*. Total fixed cost shows the sum of expenditures which will be incurred irrespective of the output.

(*e*) *Average Fixed Cost*. It is the sum of total fixed cost and total variable cost.

TC = TFC + TVC

(*g*) *Average Total Cost*. It is worked out by dividing TC by output.

$$ATC = \frac{TC}{Output}$$

3. *The Law of Comparative Advantage*

The law of comparative advantage helps to explain regional specialization. Farmers tend to produce those items which give more income at a lowest relative cost. With the income, they buy the items needed for production and for living which are produced at lowest relative cost in other areas. In short, specialised and diversified farming are based on this principle.

4. *The Law of Equimarginal Returns*

Equi-means 'equal' and marginal means additional or incremental. The law states that profit will be greatest if each unit of labour, capital and land is used where it adds the most to the return. The table below illustrates this principle.

Table 15.6 Added or Marginal Returns to Capital on Three Enterprises

Amount of capital used (Rs.)	*Marginal returns*		
	Vegetable and crops	*Dairy*	*Poultry*
1000	1500	1400	1250
2000	1250	1250	1100
3000	1200	1000	1050
4000	1100	950	950
5000	1000	900	950
Total returns Form Rs. 5000.00	6050	5500	5250
Net profits	1050	500	250
Average Return/Rs.	1.21	1.10	1.05

In the above table, if we examine average return only, our advice to the farmer would be to produce only vegetable crops.

By investing ₹ 5000 he will get ₹ 1050 as net returns. But let us apply principle of equimarginal return for addition of resources. The farmer should apply ₹ 2000 on vegetable and crop, ₹ 2000 on dairy farming and ₹ 1000 on poultry keeping to get total returns of ₹ 6650 and net returns of ₹ 1650.00.

5. The Law of Substitution

The principle which helps selection from a number of alternatives is called law of Substitution. According to this law the least cost combination of inputs or practices is obtained when the inputs or practices replaced is equal to the value of the input or the practice added.

Procedure for finding out least cost combination —

1. Computation of Marginal Rate of Substitution (MRS) by dividing the number of units of the replaced input (x_1) by the number of units of added input (x_2).

$$\text{Substitution Ratio} = \frac{\text{No. of units of replaced input}}{\text{No. of units of added input}}$$

Symbolically it may be expressed as

$$\text{MRS} = \frac{\Delta X_1}{\Delta X_2}$$

where is X_1 being replaced by X_2.

2. Computation of price ratio by dividing the price of added input (Px_2) by the price of the replaced input (Px_1).

$$\text{Thus, price Ratio} = \frac{\text{Price of added input}}{\text{Price of replaced input}}$$

$$\text{Price Ratio} = \frac{Px_2}{Px_1}$$

where x_1 is replaced by x_2.

3. Calculation of least cost combination by

$$\text{equating (MRS)} = \frac{\Delta X_1}{\Delta X_2} = \frac{Px_2}{Px_1}$$

15.12.2 Farm Planning

Definition

It is a process of making decisions regarding the organisation and operation of a farm business.

Merits of Farm Planning

1. ***Income Improvements***

Income maximizations could be achieved from a given resources by reorganising present type of production activities as well as introducing changes in technology.

2. ***Educational Process***

Farm planning is an educational tool to bring about change in the outlook of the cultivators and the extension workers.

3. ***Desirable Organizational Changes***

Farm planning introduces desirable changes in farm operations and makes a farm a viable unit.

4. ***To Farmers***

- It helps him to examine carefully his existing resource situation and past experiences as a basis for decision.
- To make rational decisions on what to do.
- To find out the credit needs.
- To gives him an idea of the expected income after paying off his loans.

Characteristics of a Good Farm Plan

1. It should provide for efficient use of farm resources such as labour power and equipment.
2. Crop plan should have balanced combination of enterprises.
3. Avoid excessive risks.
4. Provides flexibility.
5. Utilize farmer's knowledge, training and experience and take account of the farmer's likes and dislikes.

15.12.3 Farm Budgeting

Definition

Farm budgeting is a method of analysing plans for the use of agricultural resources by the decision maker.

Types of Farm Budgeting

(a) *Partial Budgeting*

Partial Budgeting is a estimation of returns and expenses for a single enterprise or specific part of entire farm unit.

e.g. - Poultry unit (or) particular crop or one (or) few activities.

- It used as an aid in full budgeting.
- It used to estimate the effects or outcomes of possible adjustments in the farm business before such adjustments are actually made.

(b) *Enterprise Budgeting*

- It is used to estimate input required, costs involved, and expected returns from a particular enterprise.
- It is used to aid in selection of inputs and enterprises consistent with the resources available and to show combinations that will increase income from the farm business.

(c) *Complete (or) Total Budgeting*

- It involves estimation of returns and expenses for the entire farm as a single unit. Evaluates all enterprises of the farm *i.e.*, crops, livestock, labour, machinery, buildings, managerial ability to obtain an overall financial picture of a farm.
- Process by which the farmer readjusts his resources to his long run objectives and opportunities.
- Farmers make both long run and short run or annual budget plans.

Merits of Complete Budgeting

- Brings about progressive changes in income.
- Takes care of all factors affecting farm income.
- Resources and enterprises are considered simultaneously.

Steps in Farm Planning and Farm Budgeting

1. ***Inventory of limited resources***

Prepare a complete list of the farm resources which limit the size of the different farm enterprises.

2. ***Examine the Existing Farm Planning***

Full information on how each resource is being utilized and what are the outputs obtained from various enterprises adopted on the farm.

3. ***Locate the Weakness of the Present Plan***

Careful analysis of the resource use in the existing plan will throw up the imbalances. The various weaknesses in the existing plan will act as guidelines for bringing about improvements in the alternative plan.

Table 15.7: Differences between partial budgeting and complete budgeting

Partial Budgeting	*Complete Budgeting*
1. It considers a few alternatives.	1. In considers all alternatives.
2. Do not affect the organization vitally.	2. Affects the organization.
3. Easy and simple it requires less time and efforts.	3. It requires more time efforts and more data in accurate form.
4. It estimates only the changes in costs and returns associated with variations in organization and operation.	4. It involves estimating the results of a particular organisation and operation of a farm.

Technique of Farm Planning and Budgeting

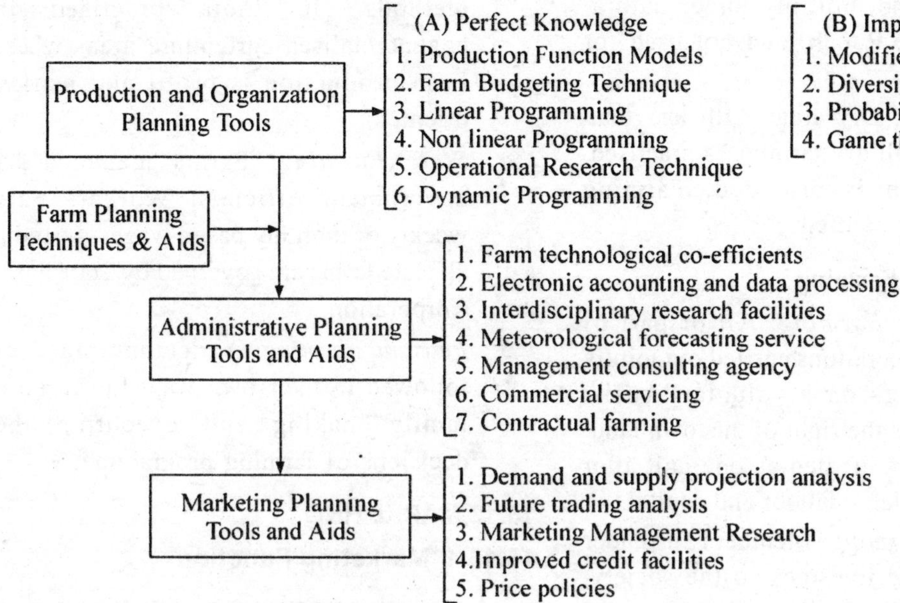

4. ***Prepare the alternative plan***

Within the framework of resource restrictions and keeping in view the weakness of the existing plan the alternative plans are developed.

5. ***Analysis of the alternative plans***

New plans are analysed for costs and returns and the optimum plan which promises the highest returns (to fixed resources) is selected.

6. ***Implementing the plan***

- There may be certain difficulties in implementing the plan, unless all the problems are properly anticipated.
- Good plan must provide for flexibility.

15.12.4 Types and Systems of Farming

Based on factors like crop and live stock raising and mode of social and economic planning, farming is classified into.

I. Types of Farming

When farms in a group are similar in the kinds and proportion of the crops and live stock that are produced and methods and practices followed in production.

II. Systems of Farming

Combination of products on a given farm and the methods or practices that are used in the production of the product.

I. Types of Farming

1. *Specilised Farming*: 50% or more of income derived from single source. In these areas where special market outlets and fairly uniform economic conditions for a long period are present.
2. *Diversified Farming* : No single product source income equals as much as 50% of the total Receipts. Farmers depend on several sources of income.
3. *Mixed Farming*: Combination of crop production with particular amount of livestock raising.

4. *Ranching*: Livestock grazes on natural vegetation and multiply under natural surroundings. Ranch land not used for cultivation.
5. *Dry Farming*: Dry areas with less than 20 inch rainfall dry farming is practised. No irrigation is provided. Farming predominantly rainfed.

II. Systems of the Farming

1. *Co-operative Farming*: All or part of agricultural operations carried out jointly by the farmers on a voluntary basis. Farmer retains the right of his own land. Land would be treated as one unit after pooling their land, labour and capital.
2. *Collective Farming*: Members surrender their land and livestock to the society. Members work together under management committee elected by themselves. Committee directs matters of allocation of work, distribution of income.
3. *Capital Farming*: Based on capitalistic methods. In India practised in commercialised agriculture areas where farm production is profit and market oriented.
4. *State Farming*: Farms managed by government officials. Workers paid weekly or monthly based wages. In India all state farms are governed by State Farm Corporation.
5. *Peasant Farming*: Agriculture practices followed in their own way. Entire farm family making and executing the decisions of farming programme.

15.13 MARKETING

15.13.1 Marketing Function

A marketing function is an act, operation on service by which the original producer and the final consumer are linked together.

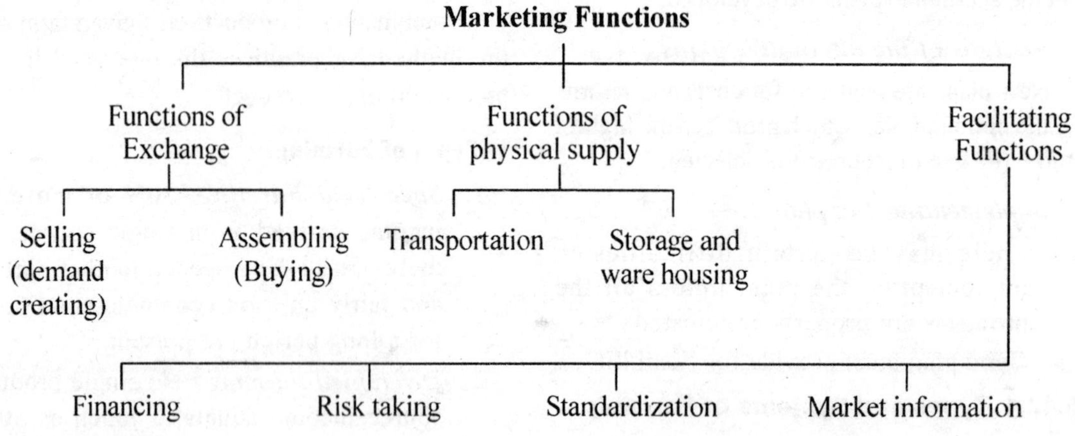

Broad Classification

1. Function of exchange
2. Function of physical supply
3. Facilitating functions

I. *Functions of Exchange*

Functions of exchange is a process of passing goods into the consumer hands. It includes buying, assembling and selling.

(*a*) *Buying*

- First step in the process of marketing.

(i) What to buy? (product)

(ii) When and how to buy? (Time)

(iii) From whom and where to buy? (Source)

(iv) On what terms and conditions and prices?

(b) *Assembling*

- Done after goods have been purchased.
- Involves creating and maintaining of stocks of goods

(c) *Selling*

- Right product available at right place
- In right quantity
- At right place
- At right time

II. *Functions of Physical Supply*

(a) *Transportation:* Through which commodity shifted from area of production to the markets.

Functions:

1. Helps to growth of industries
2. Increases demand for goods

(b) *Storage and Warehousing:* It involves holding and preserving goods, till they are needed for consumption.

15.13.2 Need of Storage

Agricultural product are seasonal, but are required for whole year consumption.

Grain Storage Structure

Small Scale S.S. :

(a) PAU bin- Capacity 1.5–15 quintal

(b) PUSA bin

(c) Hapur Tekka

Large Scale S.S. :

(a) CAP (Cover and Plinth): by FCI.

(b) Ware house- by FCI, CWC, SWC

Ware Housing

It is a scientific storage structure specially constructed for the protection of quantity and quality of stored products.

Functions:

1. Scientific storage
2. Financing
3. Price stabilization
4. Market intelligence

Types of Ware Houses

Based on ownership	Based on Commodity
1. Private	1. General
2. Public	2. Special commodity (cotton, wool, tobacco)
	3. Refrigerated ware house Temp. at 30-50°F less

Ware House in India

(a) National Co-operative Development and Ware Housing Board.

(b) Central Warehousing Corporation.

(c) State ware housing Corporation.

(d) Food Corporation of India.

Reasons for Poor Utilization

1. Lack of knowledge
2. Location
3. Complicated and time consuming procedure for deposit and withdrawing produce.

Processing

Activity which changes the form of a commodity

Farm Products $\xrightarrow{\text{Processing}}$ usable form

India's Position

- Primary processing
- 80% of value
- 12% food industry
- 33% in small scale tiny and cottage sectors.

Standardization

Determination of the standards to be established for different commodities.

Grading

Sorting of the unlike lots of the produce into different lots according to the quality specifications.

Advantages of Marketing Function —

1. It can make inter functional comparison of marketing cost.
2. Inter agency comparison of the cost of marketing function.
3. Inter commodity comparison of cost.

Disadvantages —

1. We don't know these functions are applied to specific business operations are not.
2. Numerous functions, so it is difficult to eliminate the unnecessary from necessary functions.

15.13.3 Marketing & Pricing of Agricultural Inputs and Outputs

Marketing of Agricultural Inputs

- No scope for extension of arable land, so production only through in crop productivity per unit of land.
- Judicious and efficient use of agro-inputs like seed, fertilizer, pesticide, irrigation, farm machines and credit play major role.
- This inputs are complementary through each others.
- There interrelation is so strong that one of them can be viewed in isolation.

1. ***Seed***
   1. In India 3 stage, Breeder, Foundation, Nuclear.
   2. 28 ICAR Projects, 24 companies were producing seed.
   3. 19 seed certification agencies and 86 seed testing labs.
   4. Private producing hybrid seed.
   5. Planting material from biotech (tissue culture)
   6. Critical Agricultural Inputs extra care should needed.
   7. State seed co-operative marketing federation state Agricultural Industrial Co. Providing seed to the farmer.

   Problem —

   - Very cumbersome and difficult certification.
   - In adequate and untimely available seed.

   Future strategies —

   1. Create efficient marketing and delivery system.
   2. Encourage greater participation of private, co-operative and NGOs.
   3. Encouraging, public private sector to increase infrastructure.

2. ***Fertilizer***

It contributes 50 to 60% of Incremental Agricultural Production.

Average fertilizer consumption - 77 kg/hectare

80% consumed in rice, wheat, sugarcane and cotton.

15.14 PRICING

15.14.1 Government Price Mechanism for Agriculture

Administered Prices

- It is the price fixed by the Govt. with the objectives of protecting farmers against a decline in prices during years of bumper production.
- Protecting consumers from excessive price increases during lean production years.
- Ensuring procurement for the maintenance of the buffer stock or operation of PDS.

MSP (Minimum Support Price):

MSP (Minimum Support Price): is a scheme of GOI to safeguard interest of farmers from middlemen. Under this scheme Government declarers the minimum support prices of various crops and agricultural produces and assures the farmers that their produce will be purchased at the MSP, thereby preventing the distress sale. FCI (Food Corporation of India) acts as the Nodal Agency for the GOI. *This scheme was first started by GOI in the year 1966-67 for wheat crop* and it now expanded to cover many more crops after the years in the wake of Green Revolution MSP and issue price are different terms, issue price is the price at which food grains are sold out from FCI. Before the harvesting during each Rabi/ Khariff season, GOI announces the MSP (Minimum Support Price) for procurement on the basis of recommendation of the CACP (Commission of Agricultural Costs and Prices).

Maximum or Ceiling Price

- It denotes upper price level of a commodity fixed by Govt.
- This provision prohibits a trader from charging a price higher than the maximum price.
- It is fixed by Govt. to protect the consumers from unwanted price increases.

Procurement Price

The GOI announces a set of prices for each agricultural marketing season, which guarantees to procure wheat and rice if market price falls below the announced price. The Commission for Agriculture Costs and Prices (CACP) is responsible for making price recommendations to the GOI on the basis of long run average costs of production of various crops, in a manner that covers costs of most farmers and reward efficient farmers. Procurement prices are meant to prevent distress sales of the produce in years of glut, and keep price for consumers at an affordable level. Over the time, support prices have become a guaranteed income support rather than a price risk hedge. Procurement prices are announced higher than minimum support prices because the government requires food grains to maintain buffer stocks.

Issue Price

- This is the price at which the commodity is made available to consumers at fair price shops.
- The issue price is always higher than the procurement price.
- The economic rationale of maintaining PDS lines in keeping issue prices lower than market retail price.

15.14.2 Fluctuations and Instability in Prices

Price Fluctuation

It refers to changes in prices of a commodity or group of commodities at a given market place or market areas.

Instability in Prices

It can be defined as the state in which prices continue to change with the passage of time at a given space.

Types of Instability

(*i*) *Short term instability*: The prices of a single commodity do not remains stable within a day, week or month.

(*ii*) Intra year instability (or) Inter Seasonal instability do not remain stable within a season.

(*iii*) *Inter year instability*: Prices of farm products never remain the same from one year to another.

(*iv*) *Long term instability* : Over the time, the economy may turn towards recession or depression and prices may start falling.

Factors Causing Price Fluctuations in Agricultural Commodities.

Short Term Fluctuations —

(*i*) Unexpected changes in market arrivals.

(*ii*) Unexpected changes in offtake from the markets.

(*iii*) Bottlenecks in the movement of commodities.

(*iv*) Announcement of relevant govt. policy.

Intra Year Fluctuations —

(*i*) Seasonal nature of production

(*ii*) Perishability of products

(*iii*) Seasonal nature of demand for certain products

(*iv*) Change in govt. policy on procurement

(*v*) Change in credit availability.

Inter Year Fluctuations—

(*i*) Fluctuations in production due to variation in Rainfall.

(*ii*) Fluctuations in supply due to floods and droughts etc.

(*iv*) Change in import-export policy.

(*v*) Change in public intervention policies like support or procurement prices.

Long Term Fluctuations

(*i*) Change in technology like evolution of high yielding, seeds and control measures for pest diseases.

(*ii*) Change in money supply.

(*iii*) Change in credit and interest rate policy.

(*iv*) Change in pay scales, wage rates.

(*v*) Change in sales tax, excise tax.

(*vi*) Change in policy towards hoarding, black marketing.

(*vii*) Extent of processing facilities available for farm products.

Effect of High Instability in Prices

General Effect on the Economy:

(*i*) A balance between supply and demand cannot be brought.

(*ii*) Consumers are not able to rationally allocate their scarce income/resources

(*iii*) Planning and mobilisation of resources for longterm investment projects become difficult.

(*iv*) It affects the income and living standards of the population.

Special Effect on Farmers and Farm Sector:

(*i*) Instability in farm product prices leads to instability in farm incomes.

(*ii*) Uncertain prices of those products which form the raw material for industries like cotton, oil seeds and sugarcane affect the costs, profitability in such industries and this in turn affects the growers of these crops.

(*iii*) Uncertain prices of farm products lead to uncertainty about the repayment capacity of farmer.

(*iv*) Instability in farm product prices affects the employment prospects of labour in the agriculture sector.

(*v*) High fluctuations in farm product prices makes it difficult to plan for exports or imports of commodities.

15.14.3 Agricultural Price Stabilisation

Agriculture price stabilisation means reduction in price fluctuations and regulation of price movements within a certain range.

Need for Stabilisation of Agriculture Prices

(*i*) Farm products account for a considerable part of the consumption expenditure of a majority of the population. Uncertainty in farm product prices lead to uncertainty in real incomes.

(*ii*) Prices of agriculture commodities have a high weightage in the general price index.

(*iii*) In the agriculture sector, farmers cannot reduce production when prices fall due to rigidities on account of seasonality and committed resources. So falling prices affect farmers adversely.

(*iv*) The efficiency of the farm sector and long term adjustments in resource allocation in and away from the farm sector depend on a relatively stable cost-price relationship.

Measures for Stablisation of Agriculture Prices

(A) *Non price Regulatory Measures*:

(*i*) Increasing and stabilising agriculture production.

(*ii*) Regulating imports and Exports.

(*iii*) Building up buffer stocks.

(*iv*) Ensuring procurement and public distribution.

(*v*) Expanding infrastructure.

(*vi*) Improving the efficiency of the marketing system.

(B) *Direct price control measures* :

(*i*) Fixation of Ceiling (Maximum) price

(*ii*) Fixation of Floor (Minimum) price.

(*iii*) Fixation of an administered price.

15.15 OTHERS

15.15.1 Co-operatives

The co-operative movement was started in India largely with a view to providing agriculturists funds for agricultural operations at low rates of interest and protect them from the clutches of money lenders.

They meet all types of credit *viz*., short term, medium term and long term credit requirements.

Co-operative Agricultural Credit Structure

Short term and Medium term credit (three tier):

1. State Co-operative Bank (SCB).
2. District Central Co-operative Bank (DCCB).
3. Primary Agricultural Credit Societies (PACS).

Primary Agricultural Co-operative Societies (PACS)

Functions

- To provide short term and medium term credit.
- To supply agricultural inputs like fertilizers and seeds.
- To provide marketing facilities for the sale of agricultural produce.
- To associate itself with economic and social welfare programmes of the village.

Sources of Funds

Share capital, deposits, reserve funds and loans borrowed from higher institutions or govt.

District Central co-operative Banks

Functions

- To maintain close and continuous contact with PACS and provide leadership and guidance to them.

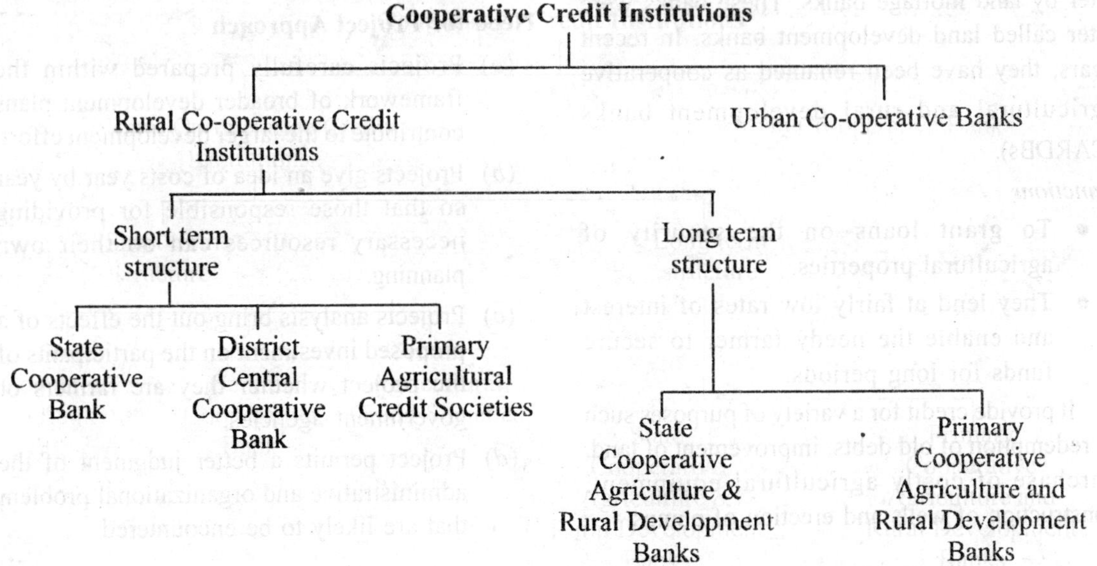

- To undertake non credit activities.
- To meet the credit requirement of member societies.

Sources of Funds

Share capital, deposit from public, borrowing from state cooperative banks, Govt. RBI, SBI and commercial banks.

State Co-operative Banks (SCB)

Functions

- Act as banker's bank to the district central co-operative banks.
- Uniform credit policies for the cooperative movement are formulated and executed by them.
- It also perform banking functions such as issuing drafts, cheques, collecting and discounting bills etc.

Sources of Funds

Share capital, reserve funds, deposit from members and non members, borrowings from NABARD, SBI and State Govt.

Long Term Rural Credit

The long term requirements of the farmers were traditionally met by the money lenders but later by land mortage banks. These banks were later called land development banks. In recent years, they have been renamed as cooperative agricultural and rural development banks (CARDBs).

Functions

- To grant loans on the security of agricultural properties.
- They lend at fairly low rates of interest and enable the needy farmer to secure funds for long periods.

It provide credit for a variety of purposes such as redemption of old debts, improvement of land, purchase of costly agricultural equipment, construction of wells and erection of pumps.

Sources of Funds

Share capital, deposits and issue of bonds. They are subscribed by commercial banks, SBI and RBI.

15.16 PROJECT APPROACH TO AGRICULTURAL DEVELOPMENT PROJECT CONCEPT

According to J.Price Gittiger, World Bank Economist, Project is the cutting edge of development. A project can be defined as the *investment activity in which the financial resources are expected to create capital assets that produce benefits over an extended period of time and which leads its planning, financing and implementation as a unit.* The essential features of a project are that is specific in objective, specific in location, specific in time. It has a set goal to achieve, the area or the region where it is to be implemented is predetermined and the time to terminate the commitment is fixed in advance. It is something which is measurable both in its major costs and return. It will have group activities which could be financed and a group can get benefits which could be identified and values are estimated. It will have a specific clientele which it is intended to reach.

Need for Project Approach

(*a*) Projects carefully prepared within the framework of broader development plans contribute to the larger development effort.

(*b*) Projects give an idea of costs year by year so that those responsible for providing necessary resources can do their own planning.

(*c*) Projects analysis bring out the effects of a proposed investment on the participants of the project whether they are farmers or government agencies.

(*d*) Project permits a better judgment of the administrative and organizational problem that are likely to be encountered.

(*e*) Project encourage conscious and systematic examination of alternatives.

(*f*) Projects help in building up a data base which is lacking in many developing countries.

Strategies for planning projects

Strategies in planning of agricultural projects should be guided by the National priorities, viz growth with social justice. Sometimes it might happen that the growth aspects are given prominence while the social justice aspects are relegated to the background. Consistent with the overall national objectives, project planning strategy should have the following priorities.

(*a*) Projects which have better distribution effects.

(*b*) Projects which increase benefits rapidly.

(*c*) Projects in areas which are relatively less developed.

(*d*) Projects which would benefit the most vulnerable sections of the community.

(*e*) Projects which have greater employment potential.

The best strategy to be adopted in agricultural project planning should be such that there is simultaneous development of those activities which would bring about the greatest benefit to the largest number of people.

Having decided the priorities, the next question would be to choose the projects. The strategy in deciding agricultural projects-single purpose or multipurpose projects. Planning and operation are easy in single purpose (as for *e.g.* dug well schemes, shallow wells schemes, schemes for the installation of pump sets etc.,) They are quick yielding and can be implemented by private bodies. The planning of multipurpose projects (major irrigation projects) on the other hand is complex and requires heavy financial involvement. The return from these projects is obtained after a long gestation period and hence generally only public bodies would be able to undertake these projects.

Project cycles

The effective phases in a project cycle:

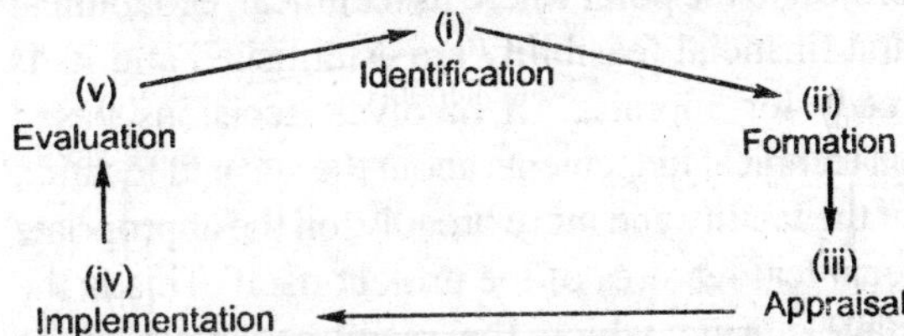

The reason why it is considered as a cycle is that each phase not only grows out of the proceeding one, but leads into the subsequent one and it is a self renewing cycle so that new projects grow out of old ones in continuous process.

Identification

It is the process of identifying, in a preliminary way, projects which are of high priority, which might be suitable for bank financing and which the borrower, the bank and the government are interested in considering. There are essentially three tests involved in identification of the project. These are;

(*a*) Whether the sector of the economy in which the project falls and the project itself are of high priority for development and recognized as such, in the Government's developmental plan.

(*b*) Whether, on prima facie grounds, the project seems to be feasible *i.e.*, whether a technical solution to the problem to which the project is addressed to, can be found at a cost commensurate with the benefits to be expected?

(*c*) Whether the government is willing to support the project by financial and other means, if required?

The following sources will help to identify new projects

(*a*) Governmental Five year plans.

(b) Existing Projects.

(c) Natural resources etc.

Formulation

This covers the steps necessary to bring a project to the point where its technical, economical and financial feasibility are established and its is ready for appraisal. It involves decisions based on technical judgements about the site and location of the facility and more broadly, on the appropriate technical features of the project itself. This is the stage during which the resources have to be judged and analysed. The alternatives are systematically to be explored. This is the stage wherein feasibility studies have to be conducted and preliminary investigations have to be made.

Appraisal

Appraisal covers the following aspects:

(a) Technical aspects

(b) Economic aspects

(c) Commercial aspects

(d) Financial aspects

(e) Managerial aspects

(f) Organizational aspects

The technical analysis will examines the possible technical conditions in the proposed agricultural project- the natural and physical resources, the availability and the potentials, the cropping pattern, the marketing arrangements etc.

The economic analysis covers the contribution of the project to the overall development of the economy and whether the project could contribute to the economy with justification for the utilization of the scarce resources.

The commercial analysis would deal with the arrangements for marketing the output produced by the project and the arrangement for supply of inputs needed to build and operate the project. The most important is that the demand and supply be ensured.

The financial analysis of a project covers all the financial effects of a proposed project on its beneficiaries. Farm budgets have to be prepared for the beneficiaries to determine the project's financial viability. The budget should determine the estimate year by year (gross receipts and expenditures), including the costs associated with the production and the loan repayments, to determine the net amount that will go to the family resources for its labour, management skills and capital. The farm budget, also becomes the basis for fixing the loan period and the repayment schedule.

The managerial aspects are crucial to good project design, implementation and operation. The following are to be considered.

(a) The managerial skills of the participating farmer;

(b) The availability of extension functionaries;

(c) The availability of technology;

(d) The availability of training programme for the farmers;

(e) The ability of the developmental departments to implement and monitor the project.

The organizational aspect deals with the authority of the organization and the linkages of the responsibility with the authority. The organization should have a sound system of accounting to ensure prompt disbursements under the project and to keep record of its progress.

Implementation

This is the most important step in the project cycle. A project has to be monitored during its implementation in order to ensure that it is completed as per the schedule and the expected benefits are realised by the beneficiaries. Problems may arise during the implementation of the project. The cost may go up and even the technologies may change. In order to have these settled, frequent monitoring is essential. Monitoring is the timely collection of data and analysis of the process and to find out whether the project is moving in the anticipated direction or not. If deviations occur, they have to be rectified.

Evaluation

The evaluation is the analysis study to determine the factors responsible for the success or failure of the project in order to learn how to plan better for the future. This is not only carried at the end but also in between while monitoring. The evaluation is always done on the objectives formulated for the project

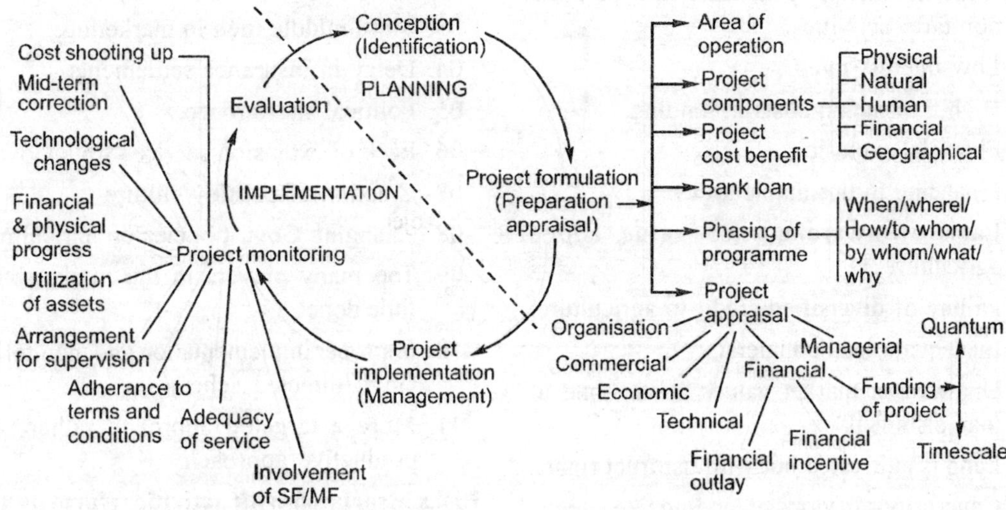

Fig. 15.5: Project approach to rural development project cycle

Risks and Risk Management in Agricultural financing

What is risk?

The exposure to the chance of injury or loss; a hazard/ threads or dangerous chance to any event.

What is risk mitigation?

Effort to reduce the severity, loss or injury to happen or happened.

Risks associated with borrowers

01. Interference of money lenders in funding during seasons.
02. Effect of ownership of lands and lending by bankers.
03. Unfavourable attitude of farmers towards agriculture.
04. Ignorance and illiteracy of farmers.
05. Low bargaining power of farmers- uncertainty of price.
06. Weak mind set of farmers leading to committing suicides.
07. Unaware of the concept of time value of money.
08. Sub-division and fragmentation of holding.
09. Unexpected events like death of the borrower.
10. Improper accounting standards.
11. Financial illiteracy
12. Distance from the place of production and place of sale.
13. Lack of awareness of value added agriculture.

14. Farmers are price takers and not price givers/ fixers.
15. Willful default.

Risks associated with bankers/ banks

01. Lack of knowledge of bankers on farm lending resulting in improper appraisal.
02. Under-financing/ Over-financing of farm/ non-farm activities.
03. Low interest rates.
04. High transaction costs in lending.
05. Financial exclusion.
06. Financing in unsuitable areas.
07. Lack of awareness of value added agriculture.
08. Failure of diversified credit to agriculture.
09. Inadequate loan collaterals.
10. Unaware of market values at the time of loan disbursal.
11. Land is cultivated but with usufruct rights.
12. Concessions in terms of lending like interest rates.
13. Lack of systems, experience and incentives to enforce loan repayments.
14. Lack of knowledge on project lending to agriculture.
15. System created is borrower dominated rather than depositors.
16. Cost over-run due to time over-run.
17. Foreign exchange risk.
18. Liquidity risk- mismatch between resources and requirements.
19. Fast lending.

Risks associated with uncontrollable events.

01. Unpredictable impact of weather/ climate- Vagaries of monsoon.
02. Effects due to pests and diseases– crops.
03. Effects due to diseases/ permanent total disability- livestock.
04. Effects due to the fire accidents.

Risks associated with others

01. Lack of proper planning for varying agro-climatic/ agri-export zones.
02. Improper price fixation, fluctuating *Minimum Support Price* and *Price concentrated risk.*
03. More middle men in marketing.
04. Delay in insurance settlements.
05. Political interference.
06. Lack of extension services from Govt. side
07. Creation of subsidy culture.
08. Changing Govt. policies on agriculture.
09. Too many players in the market but too little done.
10. Improper implementation and monitoring of Govt. directed schemes.
11. More a targeted approach rather than a productive approach.

Risks associated with activities (farm and non-farm)

01. Farming- still a traditional approach.
02. Uncertainty in the maturity pattern of crops.
03. Seasonality risk.
04. Poor yields/ productivity.
05. Yield uncertainty- reduction or increase in yield/ productivity.
06. Non-availability of inputs.
07. Non- availability of quality seeds and fertilizers.
08. Increase in cost of inputs.
09. Water of irrigation- a critical constraining factor.
10. Non-availability of agricultural labour during peak agriculture seasons.
11. Increase in cost of agricultural labour.
12. Migration of agricultural labour and professional shift.

13. Lack of good rural markets/ marketing and weak market infrastructure.
14. Continuous cultivation of land/ mono-crop leading to depletion of soil- mineral resources.
15. Excess application of fertilizers and pesticides.
16. High post harvest processing costs.
17. Post harvest losses.
18. Lack of storage structures.
19. Hoarding of produce leading to price hike.
20. Intermediary costs.
21. Price uncertainty.
22. Agri movable assets- high risk.
23. Pledging spurious jewels- frauds.

Self Help Groups

What is Self Help Group?

An SHG has an average size of about 15 people from a homogeneous class. They come together for addressing their common problems. They are encouraged to make voluntary thrift on a regular basis. They use pooled resources to make small interest bearing loans to their members. The process helps them imbibe the essentials of financial discipline in all of them. They also learn to handle resources of a size that is much beyond individual capacities of any of them. The SHG members begin to appreciate that resources are limited and have a cost. Once the groups show this mature financial behaviour, banks are encouraged to make loans to the SHG in certain multiples of the accumulated savings of the SHG. The bank loans are given without any collateral and at market interest rates. The groups continue to decide the terms of loans of their own members. Since the group's own accumulated savings are part and parcel of the aggregate loans made by the groups to their members, peer pressure ensure timely repayments.

Benefits of SHGs:

For members

- Discuss and help each other to solve common problems
- Collect and use own savings to make interest bearing small loans to each other.
- Learn basics of financial intermediation.
- Learn to appreciate other's needs and priorities their own needs'.
- Start handling resources of a size much beyond their individual capacities.
- Realise that resources are scarce and that they have a cost.
- Learn that repayment is not difficult, with regular savings habit.
- Use peer pressure as an effective substitute for collateral security.
- Win the confidence of the formal banking system through mature financial behaviour, leading to further access to need based funds.
- Learn to interact with the external environment in a meaningful way, leading to increased self esteem and confidence.

For Banks

- Benefits from reduced transaction costs through economies of scale.
- Learn to externalize credit supervision and servicing to the NGOs or the SHGs themselves.
- Benefits from mobilization of small savings through groups, gaining access to low-cost funds.
- Accept peer pressure within the SHGs as an excellent substitute for collateral securities, leading to more than 95% repayments.
- Get timely repayments leading to faster recycling of funds.

- Recognize SHGs as the appropriate medium for expansion of business of rural branches for wider coverage of clientele.
- Recognize the prospects of 'ripple effect' in quality among their clientele.
- Build goodwill among the rural clientele.
- Benefit from full refinance facility from NABARD for better fund management.

For NGOs

- Find SHGs as complimentary to their core functions.
- Use the synergy of social and economic programers for better impact on the poor.
- Deepen and widen the outreach to the poor through 'credit plus' approach.
- Gain recognition as socioeconomic change agents.
- Use the avenue for performing financially intermediation in underbanked areas.
- Act as meaningful agents between banks and the poor.
- Perform the role of propagators of innovative financial services delivery approaches.

SHG Bank Linkage Programme: Process as on 31st December 2005

Cumulative number of SHGs credit linked: 18,29,847

Cumulative bank loan disbursed: ₹ 83191 million

Cumulative refinance drawn by banks: ₹ 37414 million

Support from NABARD

General

- Conceptualized and introduced the pilot phase of SHG Bank Linkage Programme.
- Contributes to conductive policy framework.
- Value addition to the programme by developing and sharing different types of conceptual inputs for the stake holders.
- Making available on large scale capacity building inputs.
- Fund support for expendable and loan funding needs of participating agencies.
- Holds training consultations, supports stakeholders in training interventions.
- Supports banks to act as Self Help Promoting Institutions.

KISAN CREDIT CARD

→ **Eligibility**

→ All farmers.

→ **Purpose:**

→ To meet the short term credit requirements for cultivation of seasonal, annual, perennial crops.

→ The scale of finance is to be taken as indicatory cost and not mandatory. In case of high yielding varieties of irrigated crops, limit can be fixed upto 120% of scale of finance where there is tie-up arrangement with corporates when there is need to incur additional cost towards quality inputs.

→ Besides the above, KCC limit will be arrived at as given below:

→ Post-harvest/Household requirements of the farmer at 10% of the limit fixed subject to a maximum of ₹ 25000/- per farmer.

→ 10% towards maintenance of farm assets subject to a maximum of ₹ 25000/- per farmer.

Interest Rate:

Crop Loan and Term Loan (Repayment < 36 Months):

Upto 3 lakhs	@ 7%
₹ 3.00 to ₹ 5.00 lakhs	@ BPLR + 0.5 – 13.00
₹ 5.00 lakhs	@ BPLR+1.00%–13.50%

Other features of the scheme:

- The limit sanctioned is in the form of revolving credit. Hence, any number of debits/ credits may be allowed subject to the seasonal drawing limit.
- In case of natural calamities in the area declared, the amount released for the specified crop may be rephased/ rescheduled as term loan and further withdrawals upto the seasonal sublimits fixed may be permitted, if there are no other irregularities.
- Withdrawals in the account should be restricted to the seasonal sub-limit fixed for the respective seasons. Drawals/ repayment schedule have to be arrived at factoring in the nature of the crop cultivated.
- The loan amount released for one season should be adjusted within 2 months from the date of harvest of the crop.
- Renewal will be permitted only for satisfactorily operated accounts both in terms of drawal and adherence to seasonal sub-limits.
- Separate SB account need not be opened as the crop loan is given in the form of cash credit.
- Credit balance in the account will be entitled for interest on par with SB accounts.
- No service/ inspection fee to be levied for a loan limit upto ₹25000/. For loan amount exceeding the above, service charges will be levied.
- Chitta, adangal and other farm documents to be produced by the farmer once in 3 years.
- All eligible IBKC far rarners will be covered under Permanent Accident Insurance scheme.
- Annual premium under PAIS is ₹ 15 per farmer, of which ₹ 5/- will be borne by the borrower and the balance ₹ 10/- by the bank.
- Insurance of notified crops in notified areas under National Agri Insurance scheme is compulsory and the premium amount to be paid by the farmer at the time of availing credit under IBKC scheme.
- Disbursement can be in the form of cash. However, receipts for fertiliser, seeds should be produced by the farmer for ensuring end use.
- Tenants farmers are also eligible for availing IBKC.

Communication

Communication means transmitting and receiving information, signals or messages by means of gestures, words or symbols from one person to another. Communication is indispensable.

Communication is not what is transmitted but it is what is perceived. It is information coloured by the perceptions and intentions of both the sender and the recipient.

Basic Components are:-

1. What to say:	message
2. When to say:	timing
3. How to say:	medium
4. Who to say:	sender
5. Where to say:	receiver
6. Why to say:	reason

A successful communication should be CLEAR (evident to recipient), correct (free from faults), CANDID (sincere and open), CONCISE (brief in content), CONCRETE (real and specific). COMPLETE (full utility value) and COURTEOUS (soft and polite).

Seeing, thinking, understandings, reading, hearing, body posture, body movement, dressing and silence are ONE WAY communication.

Telling, listening, asking, selling, confrontation, face to face are TWO WAY communication.

Awareness in communicating with oneself. A man is known by his own words. When one communicates he transmits his feelings, attitudes, values, style of functioning, expectations, etc., peaking or writing does not call for extraordinary scholarship. It only calls for efforts and exercise. The idea of communication is no to impress but to express- to be understood well. Be brief. Brevity means precision and to the point.

Speaking and Writing

Use simple words, short words, popular words, mono syllable words, that are precise in meaning and rich in taste. Clarity improves results, Words compel more than mere numbers. One picture may replace 1000 words. Use one idea in one sentence. Use one topic in one paragraph. One cogent sentence is better than several paragraphs. One illustration is worth several pages of matter.

Grammar is more important than glamour.

Expressions of Avoid

Insult, to reader, offence to reader, irritation to reader, ambiguity in meaning, tactless messages, threats, circumlocution, padding and meaningless sentence, badly split words and faulty grammar are to be avoided in full.

Ambiguity

Express and do not impress. Ambiguity means more than one meaning is possible. It may be accidental, or intended to attract, or deliberate to cheat. Ambiguity is worse than faulty order of words, misplaced adjectives of pronouns or faulty sequence of tenses. Always look for dictionary meaning - never from a thesaurus which gives you synonyms with slight variations in meaning. Use correct spelling. Avoid confusing words, use exact word to mean what you mean.

Public Relations

You have a social status. People respect you, (It may be different with your staff). You must participate in social functions, clubs, meeting, official gathering. You must develop speaking skills. First you must be a good conversationalist. You must be interested, friendly, cheerful, good-humoured, animated but relaxed, courteous and tactful. Tact means thinking before you speak. Practise at home with family friends and strangers. Look for opportunities to speak out. Make friends with general public and with persons who are highly placed in society and government. Observe what they read, like and dislike. Listen to them and get hints on their habits. Ask personal questions in general. If people are shy, try to help them. Never argue or challenge them. You may win an argument but lose them permanently. First impression you make on them will have a lasting good impression. Never talk about their private, family affairs or criticise their close friends or relatives. Do not discuss about unpopular books or events or controversial topics, political issues, and faction. Look for the body gestures and expressions and signs. Gracefully leave their company when required.

Group Dynamics & Communication

Aggregation: Mere collection of people

Group: A number of people who are thought of being together.

A collection of individuals who have certain common qualities.

It is a social unit of two or more members all of whom engage, at some time or other, in interaction with one another. The members are dependent on one another for achieving a common goal.

The members of a group bear a psychological relationship to each other *i.e.*, Each member recognises the existence of the other member as an essential part of the group and his behaviour reflects their expectations. They influence one another and as they interact, the behaviour of one member will be modified by the actions of the other members.

According to the above definition, a group includes all words, symbols, gestures, which the members use communicate with each other. It includes the mutual and reciprocal influence between two or more members of the group.

The above definition gives 5 outstanding characters of a group in general:

Interaction of communication

Common goals/purposes

A set of norms

A set of roles

Inter-dependency

Small group: A small group is defined as any number of persons engaged in interaction with each other in a single face to face meeting or a series of meeting, in which each member receives some impression or perception of each other member. Usually 20 members are ideal for formation of a small group.

What for?

- If one wants to affect the interpersonal relations of member
- For a need of security, status, and self esteem, affiliation, power etc. For a goal achievement.
- To provide a synergy to the organisation (co-operation).
- To have a division of labour
- To resolve conflicts
- To complete a huge task in a short time.
- To share one's feelings.
- *Types of groups:*
- Formal and informal
- Committee
- Seminar
- Panel
- Symposium
- Conference
- Convention Forum
- Brainstorming
- Syndicate
- Workshop

Group dynamics: They are the influencing forces operating in groups that affect the work in a group. It describes various small details in a group.

The following are the factors that influence the group behaviour:

Group cohesiveness- the degree to which the group is attracted to and motivated to remain as a part of group.

- Time spent together
- Severity of initiation
- Group Size
- Internal and external threats
- Previous success
- Importance of the need to be together
- Homogeneity
- Division of labour
- Unbiased decisions
- Nature of communication
- Small group ecology -*e.g.* seating arrangement
- Status and respect with respect to power
- Leadership

- Group climate
- Conflict resolutions.

Group and its functions:

- **Task functions-** group and its work
- **Maintenance functions:** under-standing each other, resolving conflicts, rein-forcement of needs etc.,
- **Listening functions:** This depends on the acceptability amongst the members, awareness of the feelings of other, not making immediate judgments while listening, purpose and commitment of listening, avoiding distraction, use of words, etc.

Leadership and groups

Leader is a person recognised as being responsible for guiding the group through their tasks.

What is leadership?– the process of influencing others to take a desired action. Influence is the process of affecting the potential behaviour of others.

Different Leadership styles, and communication

Authority leadership- self styled, highly work oriented, more concerned about the tasks, and less about the people.

Paternalistic leadership- high work, oriented but has some concern for the people.

Participative leadership- likes the members to do their work in their own way, group oriented leaders.

Leadership qualities (functions of a leader):

- Willingness to accept the responsibilities
- Ability
- Ability to deal with people and their problems shows what is wrong.
- Strong bend towards action
- Good listeners, good initiator-sees man in worker.
- Taps all institutional resources
- Builds his own strength-commands respect
- Has good rapport
- Effective use of power and authority
- Enough self confidence
- Encourages good team work say "We" rather than "I".
- Tirelessly pursue
- Ready to face criticism
- Maintains cool temperature makes work a -makes work a gay
- Estimates situation anticipates reaction
- Has flexible strategies

Group and communication:

For sending messages effectively ....

- Use personal pronouns such as "I" and "my" by which you own your message
- Be complete and specific by communicating the frame of reference
- Be congruent in your verbal and non-verbal messages
- Repeat your message more than once using more than one channel of communication.
- Seek feedback about the way by which your message is received.
- Describe other members behaviour without evaluating and interpreting.

For receiving the messages effectively.....

- Paraphrase accurately and non-evaluatively (restate the words of the sender in your own words)

- Describe what you perceive to be the sender's feelings.
- State your interpretation of the sender's message

Patterns of communication among group members:

- Who talks, how often and how long
- Who talks to whom
- Who sparks off whom and in what ways

Seminar:

- Purpose is to share knowledge and to get the view of equally knowledgeable people
- One person presents a lead paper followed by group discussion.

Panel discussion:

- Purpose is to provide an interaction amongst the members
- Each member speaks on an announted topic.
- A panel discussion will be followed by a question-answer session

Symposium:

- A small group experts discuss different aspects of a problem for the benefit of the audience.
- Each speaker makes a prepared speech within the allotted time which is followed by a group discussion among the group members in which the audience also participate.

Conference:

- Purpose is to pool experience and opinions.
- It is a closed discussion
- A large gathering who confer to discuss a particular theme or exchange information or experience.
- It should result in a set of recommendations on the central theme.
- Participants have to register for attending the conference.

Convention

- Fellowship meeting of a closely linked fraternal group
- It is usually applied for professional gatherings.

Forum

- Public discussion wherein the audience participate with questions.

Brainstorming:

- It is a technique of generating ideas mostly in a problem solving exercise
- It accepts any type of suggestion.
- It is a group participative technique where in intense interaction takes place even on minute details.

Workshop

- It is a more work oriented group discussion with guidance and participating from all members.
- It is more skill oriented work.

Listening skills, communicating skills structure

However good the speaker or the skill he has in transmitting the message, it is effective only in the partnership with a good receiver. Hence listening is most essential feature of a communication process.

Hearing is a natural process which involves picking up catching sound vibrations but listening is a higher cognitive process which is under our control. It is making sense of what we hear. Listening requires paying attention, interpreting, and remembering sound stimuli which we can tune out, at any time, at will.

IMPORTANT KINDS OF LISTENING

01. *Critical listening:* lestening with an evaluation skill.
02. *Non-directive listening:* listening that allows the speaker emotionally express himself.
03. *Direction listening:* Listener leads the speaker by guiding the limits and direction of the conversation. The directing listener does not just listen; he or she takes control of the situation.
04. *Judgmental listening:* Listener reduces personal value judgments into the conversation and offers advice or makes statements regarding right or wrong conduct.
05. *Probing listening*: Listener asks lot of questions and inquisitive the point of frustrating the speaker.
06. *Soothing listening*: Listener's behaviour reassures the speaker who believes that conflicts should be avoided at all costs.
07. *Deliberative listening:* Listening for getting the content of the message.
08. *Empathic listening:* Listening for understanding the feeling context in which the communication takes place. Listener does not get inside the speaker to understand his communication.
09. *Active listening:* Listening without passing on judgment, but reflecting back what has been said to indicate the feelings of the speaker have been understood.

The most effective listener is the active listener because this person maintains the role of the true listener. The other types of listener are characterized by their attempts to influence or dominate the speaker.

What do we communicate by listening?

- The ideas we are interested in him.
- Feeling of importance for the message.
- Respect for the thoughts of the speaker.
- To inform that we are not interested in changing him or evaluating him.
- We want to understand the speaker
- We are ready to accept the problems.
- We accept it as a constructive behaviour.

Barrriers to active listening:

01. Drifting away from what the speaker says.
02. Counter arguments
03. Complete to impose in dialogue of their own and bigger anecdotes.
04. Filtering that part of the message which they do not readily fir in their frame of reference.
05. Distoring by interpreting in ways that belie the speaker's intensions.
06. Reacting by letting the person feeling about the speaker.
07. Assuming in the beginning itself that the subject is unimportant and uninteresting.
08. Day-dreaming.
09. Accepting the message which are consistent with their existing behaviour.
10. Allowing emotional words to block the message.

Points for effective/ good listening:

01. Positive thinking
02. Concentration
03. Determination
04. Mental alertness

05. Physical alertness
06. Intensity of communication.
07. Empathy
08. Acceptance
09. Willingness to take the responsibility.
10. Recognising the speaker's organization of ideas.
11. Relating one's ideas to our own knowledge.
12. Imaginative projection.
13. Receptive attitude
14. Positive relationship between the listener and the speaker.
15. Time of communication.
16. Up-right posture
17. Positive eye contact
18. Heard-behaviour-giving one's own experience only and not from other's experience.
19. Putting the speaker at ease
20. Removal of distractions.
21. Questioning the clarification-Questions beginning with any of the six servants of Kipling, demands that the speaker has done his own thinking and they are What, Who, How, Where, When & Why... The question for clarification have the tendency to unify the parties while the questions for justification on set the parties on opposite sides.

Interpersonal Communication

What it is?

It is the type of communication that take place between two persons under face to face situations directly.

Sender ⟺ Receiver

(Understand that the factors which affect the communication process and the tips for effective communication which we have seen in the earlier sessions will also hold good for the "effective interpersonal communication").

Supportive atmosphere for the interpersonal communication:

- Evaluation – Description
- Control - Problem orientation
- Strategy - Spontaneity
- Neutrality – Empathy
- Superiority – Equality
- Certainty – Receptivity

Responses that block interpersonal communications:

1. *Evaluation responses*: "you should.... " your duty....", "You are wrong ...", "you should know better ...." The speaker becomes too defensive.
2. *Topping response*: Phrases of "one up-manship". "when I was a child ...." "That is nothing, you should have seen .....", "you don't really mean that ...." , "the problem is that ...."
3. *Diagonising response*: Telling people what they feel or why they feel the way they do. "what you require is ..." "you don't really mean that ....", "the problem is that ...."
4. *Prying-question response*: Responses leading to interrogation. "why", "where", "when", "how", "what" "who".
5. *Warning, admonishing, commanding, response*: Produces resentment, resistance, rebellion attitude. "you better keep quite....", "if you don't ...", "you have to ...", "you will ...."
6. *Logical, lecturing responses*: Tends to make the other person inferior or defensive. 'Don't

you realize ....", "here is where you are wrong....", "the facts are....", "yes, but ...."

7. *Devaluation response*: Applying bandages in the wrong place to soothe the relationship. "It is not so bad ..." "don't worry ...", "you will get over it ...." "oh, you don't feel that way ....

Transactional Analysis

TA is a method of analyzing and evaluating the interpersonal communication. It was developed by Dr. Eric Berne. It involves the study of social transactions among the people. It leads to self awareness and better interpersonal relationship through better interpersonal communication.

Every, personality has three parts called PARENT, ADULT, & CHILD EGO STATES. Understand that these terms do not represent the physical age of person.

Parent ego state: Attitudes and behaviour learnt from external sources like sometimes nurturing or sometimes critical. In this state we think, feel and act as our parents did or any authority figure acted when we were small.

There are two ego states in this and they are: Nurturing parent and critical parent.

Nurturing parent .... "Yes, he deserves a lot of praise..."

Critical parent... "What can you expect from these people?"

Adult ego state: Deals objectively with reality. We gather information, reasoning it out, estimate the probabilities and make decisions. The response from the adult ego state will be analytical.

"What is the annual salary of this job? ...."

Child ego state: Impulses that naturally come to a child. The responses are usually emotionally charged.

"I wish I could take one day of and watch a cricket match ...."

Complementary transaction: parent- parent, adult – adult, child – child.

Crossed transaction..... Adult – adult to Parent- child/ adult – adult to child – parent.

Intrapersonal communication

It is the self communication. It is what goes on inside us we think, feel, value, react, imagine, dream etc. All sensory channels are very important. It depends on the perceptions. It is very important in decision making process for an individual and in an organization.

A good model for self disclosure in JOHRAI WINDOW. It helps to increase self awareness and effectiveness. It helps us to be more open, by allowing fresh thoughts to enter and our own ideas and thoughts to be share with others.

	Know to self	**Not known to self**
Known to others	I Public area	II Blind area
Not known to others	III Hidden area	IV Unknown area

An understanding of this very essential because it helps in good and effective communication in an organization.

Organizational Communication

A. What it is?

- It is the communication that takes place at various hierarchical levels in an organization.

B. Why is it essential?

- To build up human relations.
- To create good working atmosphere.
- To make members of organization to understand the objectives policies and problems of organization.
- To get opinions, ideas, suggestions and feedback from different hierarchical levels.

- To save labour and money
- To avoid wastage
- To solve conflicts
- To satisfy the human needs of recognition, self development a sense of belonging and sharing of feelings.

C. Flow of communication in an organization:

- Vertical communication
- Downward
- Upward
- Lateral/horizontal communication
- Diagonal communication
- Circular communication
- Serial transmission
- Rumoring

D. Flow of communication in an organization

- Vertical communication
- Downward communication

  Regulating and controlling through rules and regulations, norms of behaviour, duties and responsibilities.

Eg: Motivation *Eg*.: Training, instruction and order, advice and counseling, manuals of instruction etc.

Upward

Eg. Reports, Requests and Appeals, Representations, Complaints etc.

- **Lateral / horizontal communication**

Eg. Conferences, committees, circular, Group discussion etc.,

- Diagonal communication
- Circular communication
- Grapevine communication

In a business communication information about Product, Project, Market competition are of crucial importance.

Factors influencing organizational communication:

Source:

- Age
- Experience
- Secure credibility-safety factor, qualification, dynamism
- Receiver' source relationship
- Position occupied and power
- Capacity to convince
- Communication skills
- Attitude
- Knowledge on the message being communicated

Message:

- Source of message
- Nature of message ordered vs disordered, primacy vs recency
- Organization of the message
- Fear appeals
- One sided vs two sided communication
- Delivery of message- aggressive/ passive/ assertive / persuasive
- Perception about the message.

Receiver

- Age
- Experience
- Self esteem
- Anxiety
- Frame of reference
- Open/ closed mindedness
- Receiver/ source of relationship
- Attitude
- Cognitive needs

Channel:

- Cost of the channel
- Availability of the channel
- Time available for communication
- Knowledge about the use of channel
- Technical quality of the channel
- Uses of senses

Barriers in Communication Semantic Barriers

Words and Meanings:

- Vocabulary
- Meaning
- Lack of clarity
- Nature of message.
- Language of communication
- Structure of the message.
- Method of presentation.
- Communication skills.
- Incongruent verbal/ non-verbal messages.
- Selective perception
- Attitude towards messages.
- Metal make up (feelings and emotions)
- Situations of communication
- Improper listening.
- Knowledge variations.
- Skill variations.
- Defensiveness
- Cultural incompatibility.

Personal Barriers:

- Age
- Education
- Economic status
- Sex
- Experience
- Voice
- Mannerisms

Organizational Barriers:

- Hierarchical levels
- Specialization
- Office layout
- Pressure to complete a task
- Lack of responsibility
- Failure to use proper media
- Egocentism
- Red-tapism
- Multi layers in administration
- Duel subordination

Effective communication.

01. Understand people with whom you communicate, their needs, perceptions, feelings, situations and the total personality itself.
02. Trust persons involved in the communi-cation process.
03. Thinking is the basic element of communication process.
04. Be aware that the person with whom you are communicating is equally intelligent
05. Never be in tense while communicating with others.
06. Keep the receiver relaxed.
07. Select appropriate time to communicate appropriate message.
08. Have faith in the message communicated.
09. Communicate in a simple and local language.
10. Be a good and active listener
11. Never pressurize others to accepts your ideas.
12. Control your feeling and emotions while communicating with others.
13. Have a control on your tongue
14. Be open minded
15. Empathize while communicating with others.
16. Avoid physical barriers in communication.
17. Have a smile while taking to others.
18. Break the communication process when not liked by others.
19. Be assertive in thoughts and expressions.
20. Keep away your egoism that might bring personality clashes.
21. Practicing what one preaches increases the source credibility.
22. Beware that romours may boomerang.
23. Stop talking when not essential.

Avoiding pitfalls in communication

Define the problem

Determine the primary purpose of communication

Formulate the basic language

Tailor the message to the auidence

Establish the connection

Measure the result

15.17 THE AGRICULTURE EXPORT POLICY

The Agriculture Export Policy is framed with a focus on agriculture export oriented production, export promotion, better farmer realization and synchronization within policies and programmes of Government of India.

A new Draft Agriculture Export Policy was released recently by the Ministry of Commerce and Industry.

Objectives

- The 'National Agriculture Export Policy' is formulated in line with the vision to double the farmers' income by 2022.
- It is also towards the goal of increasing the share of agricultural exports from present about USD 30 billion to over USD 60 billion.
- Getting to one of the top 10 exporting countries of agricultural products is also one of the goals

Concerns addressed?

- **Trade** - India is today a leading global producer of foodgrain, dairy and several horticultural crops. But it holds a minuscule 2.2% share in global agri-exports.
- It is stuck at the lowest rung of the value chain, and India's farm exports are highly reliant on a handful of commodities.
- These include marine products, meat, rice and plantation crops.
- **Shortfalls** - Exports even in these items are frequently interrupted by self-imposed and arbitrary trade curbs.
- State-level curbs on movement of produce add to already high costs from fragmented farms and poor logistics.
- High rejection rates on consignments due to poor quality, antibiotic and pesticide residues & other phyto-sanitary grounds are major concerns.
- **Farmers**- Bumper crops continued to trigger a meltdown in food prices.
- Policy interventions such as e-Nam and the repeal of the APMC Acts by States have made scant progress.
- **Policy**-Domestic price and production volatility of certain agricultural commodities lead to using the existing policy for short-term goals.
- These include taming inflation, providing price support to farmers and protecting the domestic industry.
- These decisions may serve the immediate purpose of maintaining domestic price equilibrium.
- However, they end up distorting India's image in international trade as a long term and reliable supplier.
- It is thus imperative to frame a stable and predictable policy

Key recommendations?

- **Infrastructure** - The policy stresses on improving the infrastructure, and storage and exit point logistics to improve exports.
- It suggested a comprehensive need-gap analysis of existing export oriented infrastructure across the value chain for this.
- **R&D** - The policy emphasised promoting R&D activities for newproduct development for the upcoming markets.
- **Exports** - The policy aims to boost high value and value added agricultural exports, focusing on perishables.
- Dealing with sanitary and phytosanitary issues are also the priorities.
- Processed agricultural products and all kinds of organic products will not be brought under any kind of export restriction.
- **APMC**- Monopoly of the Agricultural Produce Market Committee is a long existing concern.
- The policy hinted at continuing the efforts with state governments to remove perishables from their APMC Acts.
- It also suggested better coordination between central ministries.
- **Mandi**-State governments would also be urged to standardize/ rationalize mandi taxes for largely exported agricultural products.
- Simplification or uniformity of mandi/ agricultural fee across states is required.
- This will empower the farmers, providing wider access to markets and enabling free trade across the country.
- **Products**- It is proposed that the agricultural export policy must focus on promotion of value added, indigenous and tribal products.
- Development of organic export zones/organic Food park with an integrated approach is suggested to help promote shipments.
- The draft policy considered working towards bringing in similar agencies in India.
- **Besides** the policy made a case for promoting contract farming as it would help in attracting investments.

16. EXTENSION EDUCATION

16.1 Agricultural Extension

16.1.1

16.1.2 Role of Extension Education in Agricultural Development

16.2 Extension Methods

16.2.2 Classification of Extension Teaching Methods

16.2.2 Evaluation of Extension Work

16.2.3 Training Programmes for Extension Workers

16.3 Others

16.3.1 Socio-economic Survey and Status of big, small marginal farmers and land-less agricultural labourers

16.3.2 Farm Mechanization

16.3.3 Programme-planning

16. Extension Education

16.1 AGRICULTURAL EXTENSION

16.1.1

The term extension was first used in the United States of America in the first decade of this century to connote the extension of knowledge from the Land Grant Colleges to the farmers through the process of informal education. In India, the terms community development and extension education became more popular with the launching of Community Development Projects in 1952 and with the establishment of the National Extension Service in 1953. Since then, Community Development has been regarded as a programme for an all-round development of the rural people, and extension education as the means to achieve this objective.

Extension education is an applied behavioural science, the knowledge of which is applied to bring about desirable changes in the behavioural complex of human beings usually through various strategies and programmes of change and by applying the latest scientific and technological innovations.

Extension education has now developed as a full-fledged discipline, having its own philosophy, objectives, principles, methods and techniques which must be understood by every extension worker and others connected with the rural development. It must be mentioned here that extension education, its principles, methods and techniques are applicable not only to agriculture but also to veterinary and animal husbandry, dairy science, home science, health, family planning, etc. Based upon its applications and use, various nomenclatures have been given to it, such as agricultural extension, veterinary and animal husbandry extension, dairy extension, home science extension, public health extension, and family planning extension.

Differences between formal education and extension education

It may, however, be mentioned here that when extension education is put into action for educating the rural people, it does not remain formal education. In that sense, there are several differences between the two. Some of these differences are :

Extension education	*Formal education*
1. The teacher starts with theory and works up to practicals.	1. The teacher (extension worker) starts with practicals and may take up theory later on.
2. Students study subjects.	2. Farmers study problems.
3. Students must adapt themselves to the fixed curriculum offered.	3. It has no fixed curriculum or course of study and the farmers help to formulate the curriculum.
4. Authority rests with the teacher.	4. Authority rests with the farmers.
5. Class attendance is compulsory.	5. Participation is voluntary.
6. Teacher instructs the students.	6. Teacher teaches and also learns from the farmers.
7. Teaching is only through instructors.	7. Teaching is also through local leaders.

Contd.

8. Teaching is mainly vertical.	8. Teaching is mainly horizontal.
9. The teacher has more or less homogeneous audience.	9. The teacher has a large and heterogeneous audience.
10. It is rigid.	10. It is flexible.
11. It has all pre-planned and pre-decided programmes.	11. It has freedom to develop programmes locally and they are based on the needs and expressed desires of the people.
12. It is more theoretical.	12. It is more practical and intended for immediate application in the solution of problems.

Objectives of extension education. The objectives of extension education are the expressions of the ends towards which our efforts are directed. Before starting any programme, its objectives must be clearly stated, so that one knows where to go and what is to be achieved. The fundamental objective of extension education is the *development of the people*. Which includes to raise the standard of living of rural area and farmers by helping them best use of natural resources in best way.

Agricultural extension in our country is primarily concerned with the following main objectives :

(1) The dissemination of useful and practical information relating to agriculture, including improved seeds, fertilizers, implements, pesticides, improved cultural, practices, dairying, poultry, nutrition, etc.;

(2) The practical application of useful knowledge to farm and home; and

(3) Thereby ultimately to improve all aspects of the life of the rural people within the framework of the national, economic and social policies involving the population as a whole.

Principles of extension education. The extension work is based upon some working principles and the knowledge of these principles is necessary for an extension worker. Some of these principles, as related to agricultural extension, are mentioned below.

1. *Principle of interest* and *need.* Extension work must be based *on* the needs and interests of the people. These needs and interests differ from individual to individual, from village to village, from block to block, and from state to state and, therefore, there cannot be one programme for all people.

2. *Principle of cultural difference.* Extension work is based on the cultural background of the people with whom the work is done. Improvement can only begin from the level of the people where they are. This means that the extension worker has to know the level of the knowledge, and the skills of the people, methods and tools used by them, their customs, traditions, beliefs, values, etc. before starting the extension programme.

3. *Principle of participation.* Extension helps people to help themselves. Good extension work is directed towards assisting rural families to work out their own problems rather than giving them ready-made solutions. Actual participation and experience of people in these programmes creates self-confidence in them and also they learn more by doing.

4. *Principle of adaptability.* People differ from each other, one group differs from another group and conditions also differ from place to place. An extension programme should be flexible, so that necessary changes can be made whenever needed, to meet the varying conditions.

5. *The grass roots principle of organization.* A group of rural people in local community should sponsor extension work. The programme should fit in with the local conditions. The aim of organizing the local group is to demonstrate the

value of the new practices or programmes so that more and more people would participate.

6. *The leadership principle.* Extension work is based on the full utilization of local leadership. The selection & training of local leaders to enable them to help to carry out extension work is essential to the success of the programme. People have more faith in local leaders and they should be used to put across a new idea so that it is accepted with the least resistance.

7. *The whole-family principle.* Extension work will have a better chance of success if the extension workers have a whole-family approach instead of piecemeal approach or separate and unintegrated approach. Extension work is, therefore, for the whole family, *i.e.* for male, female and the youth.

8. *Principle of co-operation.* Extension is a co-operative venture. It is a joint democratic enterprise in which rural people co-operate with their village, block and state officials to pursue a common cause.

9. *Principle of satisfaction.* The end-product of the effort of extension teaching is the satisfaction that comes to the farmer, his wife or youngsters as the result of solving a problem, meeting a need, acquiring a new skill or some other changes in behaviour. Satisfaction is the key to success in extension work. "A satisfied customer is the best advertisement".

10. *The evaluation principle.* Extension is based upon the methods of science, and it needs constant evaluation. The effectiveness of the work is measured in terms of the changes brought about in the knowledge, skill, attitude and adoption behaviour *of* the people but not merely in terms of achievement of physical targets.

16.1.2 Role of Extension Education in Agricultural Development

1. Making the adoption of new technology possible :
   - To help the farmer to accept new agricultural technology for obtaining higher yields.
   - Extension workers carry out field demonstrations about new technology.
2. Helps the farmers to visualize farming as a profitable occupation
   - Helps farmers to understand their own situation, resources and critical analysis helps them to take rational decisions for the best use of the available resources.
3. Trains the local leaders
4. Helps getting necessary inputs
5. Helps in decision making
6. Helps to change in attitude.
   - To bring about this change through education.
7. It helps in achieving cooperation of people.
8. It develops, strengthens and organise groups, institutes and people to achieve their aims.
9. Extension helps the planners, policy makers and administrators with local conditions and latest technologies suited to it.
10. Job of extension is to bridge the gap between the scientists on the one hand and the farmers on the other. This is done by taking new farm technology and innovations to the farmers door and helping him to adopt them.

It also brings the farmer's problems to the research stations for further investigations and to find out the required solutions.

16.2 EXTENSION METHODS

The methods are the tools and techniques used to create situations in which communication can take place between the rural people and the extension workers.

New knowledge, skills and practices are extended to them by arousing their interest.

16.2.1 Classification of Extension-teaching Methods

(A) *According to use and nature of contact :*

1. Individual contact methods	*2. Group-contact*	*3. Mass community contact*
• Face to face (or) person to person contact between the rural and extension workers • It is very effective in teaching new skills	• Group consists of 20-25 persons Groups are formed around common interest. • It involves face to face exchanged and discussed	• Large number of people are approached at shortest time to disseminate new information. • It is the quick method.
Examples :	*Examples :*	*Examples :*
1. Farm and Home visits	1. Method and Result demonstrations	1. Bulletins
2. Office calls	2. National demonstration	2. Leaflets
3. Telephone calls	3. Leader-training meetings	3. News papers
4. Personal letters	4. Conferences	4. Circular letters
	5. Discussion meetings	5. Radio
	6. Workshops	6. Television
	7. Field trips	7. Exhibitions
	8. Forum	8. Fairs
	9. Symposiums	9. Posters
		10. Campaigns

(B) *According to form :*

1. Written	*2. Spoken*	*3. Objective Visual*
1. Bulletins	1. General and Special meetings	1. Result demonstration
2. Leaflets	2. Farm and Home visits	2. Demonstration by posters
3. Folder	3. Official calls	3. Motion pictures (or) movies
4. News articles	4. Telephone calls	4. Charts
5. Personal letters	5. Radio	5. Slides
6. Circular letters		6. Film strips
		7. Models
		8. Exhibitions
		9. Posters
		10. Flannel graphs/charts
		11. Flash cards
		12. Puppet shows
		13. Photographs
		14. Black boards
		15. Cultural programmes
Suitable for literates only	Suitable for both Literates and illiterates	Suitable for both literates and illiterates

Table : Observation, Result Demonstration, Method demonstration (Comparison and Contrast)

Particulars	*Observation plot*	*Result Demonstration*	*Method Demonstration*
1. Purpose	To try a new practice recommended by research workers with a view to observe its value, suitability or otherwise, in a given area under farmer's condition	To show locally the worth or value of a recommended practice *i.e.* knowledge demonstration	To teach how to do job involving skill; (to teach 'doing' 'skills') *i.e.* skill demonstration
2. Conducted by	Farmer (Co-operator) under close supervision of extension worker.	Farmer (demonstrator) under the guidance of extension worker	Extension worker himself or local leader specially trained for the purpose.
3. For the benefit of	Extension worker to decide the suitability of a new practice to a given locality	The demonstrator as well as other farmers.	Person present at the demonstration.
4. Comparison	Essential Replications also necessary.	Essential (Not necessary to have replications in the same field).	Not essential
5. Maintenance of records	Absolutely necessary	Necessary	Not necessary
6. Time required	Substantial Period	Substantial	Relatively very little
7. Cost	Costly	Costly	Relatively cheap
8. Inter-relationship	Usually precedes results demonstration	Usually follows observation plots; may involve one or more method demonstrations.	Often paves the way for result demonstration.

For effective use of extension, methods, the appropriate method/combinations of methods should be selected. Farmers learn about new practices through several stages. These stages are (AIETA).

1. *Awareness stage* : Person comes to know about a new practice, but lacks the complete information.
2. *Interest stage* : He becomes interested in a new idea and wants to know more about it.
3. *Evaluation stage* : He mentally applies the new idea to his present situation and evaluates it.
4. *Trial stage* : He applies new idea/Practices in small scale to determine its utility under his condition.
5. *Adoption stage* : He decides to continue the full use of the practice.

16.2.2 Evaluation of Extension Work

In any organised system there is a need for stock taking in respect of goals and objectives.

Purpose

- To facilitate effective decision making
- To determine how well a desired behavioural changes have been taking place.

Types of Evaluation

1. *Casual every day evaluation :*
   Value judgements done everyday.
   Limitations —
   (*i*) Personal ideas used instead of standard measurements.
   (*ii*) Institution and personal bias cannot be eliminated.
   (*iii*) No systematic plan for arriving at conclusions.
2. *Self checking evaluation :*
   It makes conscious attempt to apply principles of evaluation *e.g.* checking the ordinary observations, talking with others, getting other people judgements.
3. *Do it yourself evaluation :*
   Apply principles of evaluation and are more systematically done. These require surveys or score cards.
4. *Extension studies :*
   More complex, use more scientific approach.
5. *Scientific approach :*
   Experimental studies, scientifically carried out to determine cause and effect relationship. It must be –

Analytical	:	Analyse the relationship of various factors.
Objective	:	Free of bias.
Impartial	:	Approach with an open mind and spirit of enquiry.

Procedure for Conducting Evaluation

1. *Need for the evaluation*
   (*i*) What extension project, problem, activity, job, method or situation do you want to evaluate ?
   (*ii*) Why evaluate it ?
   (*iii*) How can be results of the evaluation be used in your extension work ?
2. *Purpose of the evaluation*
   (*i*) An analysis of a situation.
   (*ii*) What question should be answered by the evaluation.
3. *Questions to be answered by the evaluation*
   (*i*) Analysis of a situation, clarify the kinds of information needed to answer the questions.
   (*ii*) If an evaluation of teaching, clarify the teaching objectives.
4. *Sources of information*
   (*i*) People
   - farmers, local leaders, club members, extension workers etc.
   - How will you draw a sample

   (*ii*) Recorded information - reports, census etc.
5. *Collecting the information*
   - How ? Interviews, mailed questionnaires, obseıvations etc.
   - Whom ? person making the evaluation, extension personal, local leaders etc.
6. *Interpreting, reporting and applying the finding*
   (*i*) Preparation of tables, charts and graphs
   (*ii*) Summary of findings.

Advantages of Evaluation

1. *Results*
   Evaluation helps us to determine the degree to which we are accomplishing that which we set out to do.
   (*i*) How effective extension work has been.
   (*ii*) How much of the plan of work has been accomplished.
   (*iii*) What improvements are desirable in plans for the years ahead.
2. *Programme Planning*
   It is the process of determining, developing and executing programmes. It is a continuous process, whereby farm people with the guidance and leadership of extension personnel attempt to determine, analyse and solve local problems.
   It gives a basis for adjusting the programme
   (*i*) How to judge the effectiveness of programme planning procedure.

(*ii*) What modifications in the planning procedures are desirable.

(*iii*) How is the present programme contributing to the long time programme.

3. *Evaluation Methods*

Evaluation serves as a check in teaching methods which methods are most effective in what situations and with what people.

4. *Reporting*

Evaluation provides a report to the public.

5. *Satisfaction*

Evaluation gives satisfaction to the local leaders and extension workers.

6. *Extension organisation*

(*i*) It help the specialists be distributed more systematically and efficiently to the workers.

(*ii*) It can help supervisors to give more effective guidance to extension work.

16.2.3 Training Programmes for Extension Workers

Definition

The process of a training employees or workers to gain effectiveness in their present or future work through the development of appropriate habits of thought and action, skills, knowledge and attitude.

Trainee —(knowledge skills, attitude, values / beliefs and understanding)→ Farmers, animal breeders, housewives

Importance of Training

- Communication of the desired knowledge
- Knowledge of the techniques, improvements in job efficiency
- Syllabus of trainee should be according to the needs of Extension agents
- Teaching of manual skills to improve efficiency.

Types of Training

1. Pre-service training
2. In-service training

Pre-service Training

- It is the training the extension worker receives before joining the actual job.
- It includes education at school level and college level.
- The extension worker receive specific training at the village level worker training centres.

In-service Training

It improves the ability of extension worker, keeps him abreast of the latest knowledge, gives him some special training in the new job he is required to do.

5 steps

1. Oriented towards the organisational setup.
2. Induction Training is given at the orientation centres where the extension works are given the knowledge of working and organisation of community development and extension service.
3. Short range courses organised by the state department of Agriculture through state Agriculture universities.
4. Periodical meetings and conferences are organized by state agriculture universities in collaboration with the state agriculture department for extension officers at district level or above.

Merits

- Training improves a person's skill and develops in him the desired attitude and values required for his work.
- Training helps to acquire occupational work skills and the latest knowledge.
- Makes him familiar with the objectives of the organization. He make this contribution in promoting the goals of his organization.

16.2.4 Lab to Land Programme

The LAB to LAND PROGRAMME (LLP) was launched by the ICAR in 1979 as a part of its Golden Jubilee celebration. The overall objective of the programme was to improve the economic condition of the small and marginal farmers and landless agricultural labourers, particularly scheduled castes and scheduled tribes, by transfer of improved technology developed by the agricultural universities, research institutes etc. The specific objectives of the lab to land programme.

1. Study and understand the background and resources of the selected farmers and landless agricultural labourers. To introduce low-cost relevant agricultural and allied technologies on their farms and homes for increasing their employment, production and income.
2. Assist the farmers to develop feasible farm plans keeping in view the availability of technologies, needs and resources of the farmers, and the resources which could be made available from external sources and agencies.
3. Guide and help the farmers in adopting improved technologies as per their farm plans, and demonstrate to them the economic viability of those technologies as well as methods of cultivation and farm management.
4. Organize training programmes and other extension activities, in relation to their adopted practices, and prepare them for active participation in agricultural development programmes of the State.
5. Make the farmers aware of the various opportunities and agencies which they could utilize to their economic advantage.
6. Develop functional relations and linkages with the scientists and institutions for future guidance.
7. Utilize this project as a feedback mechanism for the agricultural scientists and extension functionaries.

The programme was initiated with 75,000 farm families over the whole country. Major thrust in the programme was to introduce the most appropriate technologies that would help in the diversification of labour use and introduction of supplementary sources of income. The programme had been in operation in a number of phases. Normally a phase ran for a period of two years with a particular set of farmers.

16.2.5 Training and Visit System

By the middle of 1970s it was felt that extension services in the developing countries were suffering from a number of weaknesses, including the dissipation of extension workers energies on low priority tasks; the lack of a single, clear line of command; and a low level of agricultural knowledge and expertise among field level functionaries. As a means of reforming and strengthening the extension services, a reorganized system of agricultural extension, known as Training and Visit (T & V) system was introduced in India in 1974 with the World Bank assistance. It was presumed that transfer of technology through the 'contacting farmers' shall benefit all farmers.

In two and a half decades, T & V became the dominant method of restructuring the extension services in over sixty countries in Asia, Africa and Latin America. The system tries to achieve changes in production technologies used by the majority of farmers through assistance from well trained extension agents who have close links with agricultural research and supported by supply, service and marketing facilities.

T & V was regarded as an improved management system of agricultural extension and had the following *key features*–

1. *Professionalism.* Each extension agent is fully and continuously trained to handle one's particular responsibilities in a professional manner.
2. *Single line of command.* The extension service must be under a single line of technical and administrative command within the Ministry/Department of Agriculture.
3. *Concentration of effort.* All extension staff work only on agricultural extension. They are not responsible for any other activity not directly related to extension. In training sessions, attention is concentrated on important major points.
4. *Time-bound work.* Messages and skills be taught to farmers in a regular and timely fashion. The Village Extension Worker (VEW) must visit the farmers regularly on a fixed day, usually once each fortnight. All other extension staff must make timely and regular visits to the field. Recommendations for a specific area and for particular farming conditions for each fortnight periods are discussed and learned by Subject Matter Specialists (SMSs) at regular monthly workships; the recommendations are then presented to VEWs and Agricultural Extension Officers.
5. *Field and farmer orientation.* The contact with the farmers must be on a regular basis, on a schedule known to farmers, and with a large number of farmers representing all major farming and socio-economic types.
6. *Regular and continuous training.* Regular and continuous training of extension staff is required and discuss with them, to prepare specific production recommendations required by farmers for the coming fortnights and to upgrade and update their professional skills.

   Fortnightly training and monthly workshops, are the key means of bringing actual farmers' problems to the attention of research, of identifying research findings of immediate relevance to farmers.
7. *Linkages with research.* Problems faced by farmers that cannot be resolved by extension agents are passed on to researchers for either an immediate solution or investigation. Seasonal and monthly workshops, joint field visits, training of extension staff and formulation of production recommendations are some of the means by which linkages with research are maintained.

Organization of T & V

The T & V system has a hierarchical organization with one extension agent (VEW) for about 800 to 1000 farmers. The farmers are divided into eight groups of about equal size. Ten farmers in each group are selected as '*contact*' farmers. Extension message is to percolate to other farmers through these contact farmers. Eight VEWs are supervised by an AEO, and eight AEOs are supervised by a Sub-Divisional/District Extension Officer (SDEO/DEO), who are assisted by three to five Subject Matter Specialists (SMSs). Higher level supervision are provided by Zone Extension Officer (ZEO) and Director of Extension/Director of Agriculture of the State, with the assistance of SMSs.

16.3 OTHERS

16.3.1 Socio-economic survey and status of big, small and small farmers and land-less agricultural labourers

Introduction

When planning rural development of a country, improving the socio-economic condition of the rural population is important.

In an agricultural dominat economy, the landless labourers, marginal and small farmers are treated as target group, who should receive a high order of priority in the strategy of planning for rural development.

Agricultural is the dominant sector of the Indian economy.

- GDP contribution in 2.5%.
- Arable land is 143 in ha.
- Farm holding is about 89 million
- Average size of holding is 1.84 ha.
- The small and marginal farmers holdings constitute around 75% of the total holdings in our country.
- Small farmers have increased due to sub-division of medium sized holdings by their heirs.
- Marginal farmers have also grown in number when small farmers are shared by the heir.
- Large holdings have become medium sized holdings due to the operation of land ceiling act under land reforms.
- According to National sample survey, 63.0% of the holdings are < 2 ha and they occupy nearly 19% of the total cultivated land.
- Agricultural work force of 299 million in India (1931) which has grown to nearly 74% of this work force is agricultural.
- The landless labourer has no self-employment, they depend entirely on daily wages for their living. This category constitutes, the core of the employment problem in the rural sector.

Characteristics :

Small farmers and marginal farmers

- Poor land use
- Surplus family labour
- Under nutrition
- Higher debt
- Distress sale of farm produce

Economic Condition :

	Small farmers(%)	*Landless ha sources*
Work force	65.31%	55.72%
House hold engaged in agricultural	91.05%	98.90%
Literacy	37.45%	26.50%
Bullock power	13.43%	11.11%
Own bullock carts	2.2%	2.2%
Average gross income	Rs. 14,000	Rs. 3,400
Per capita	2,600	Rs. 765
An average consumption	Rs. 894	Rs. 623

Status (Social)

The status of small and marginal farmers varies only marginally and that of landless agricultural laboures is slightly less than the above two categories.

Extent of sub-division and fragmentation

Problems of small holdings :

1. Small size holding are many.
2. Many owners have small pieces of plots, so there is difficult to cultivate them together in a efficient manner.

Lack of capital surplus for investment :

Majority have less area (< 1 ha), so they are cultivate and get only less yield, so this is insufficient for them to create capital surplus investment in land development.

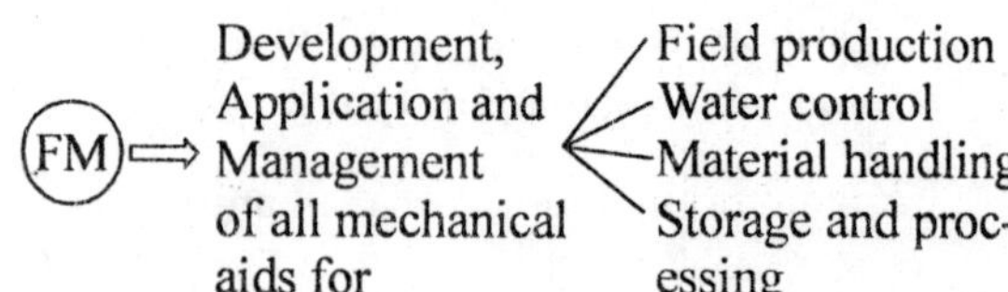

Status of Big farmers :

- They can mobilise surplus capital for investment in land development.
- More land under cultivation.
- Using modern equipments.
- Green revolution benefitted the big farmers than marginal and small farmers.

16.3.2 Farm Mechanisation

It is the application of engineering and technology in agricultural operations to do a job in better way to increase productivity.

It is a need based process which provides sufficient time gap for self adjustment of various inputs without causing sudden impact of changes.

Scope :

1. Increased irrigation facility
2. Introduction of HYV and new crops
3. Introduction of high dose of fertilizers and pesticides.
4. Multi cropping system and intensive cultivation.

Categories of Farmers on the Basis of Operational Holdings with Current Scenario

Size group	*Land ranges*	*Average area/ holding*	*% of holding holding to total 2010-11*	*% of holding holding to total 2015-16*
Marginal Farmer	Below 1.0 ha	0.39 ha	76.1 %	86.21 %
Small Farmer	1.0-2.0 ha	1.42 ha	17.9 %	13.22
Semi-Medium Farmer	2.0-4.0 ha	2.71 ha	10.0%	4.10%
Medium Farmer	4.0-10 ha	-	4.3 %	-
Large Farmer	More than 10 ha	-	0.70 %	0.57%

Benefits :

1. On time farm operation
2. Precision of operation
3. Enhancement of work and safety environment.
4. Reduction of less crops and food products.
5. Increased productivity of land.
6. Increased economic return to farmer.
7. Progress and prosperity in rural areas.
9. To perform difficult operations which cannot be done by animal and man power.
10. Proper utilisation of inputs like water, seed, fertilizer etc.

Limitations

1. Small land holdings (1.55 ha)
2. Low investing capacity of farmers.
3. Agriculture labour is easily available.
4. Adequate availability of draft animals.
5. Lack of suitable farm machine for different operations.
6. Lack of trained man power.
7. Lack of co-ordination between research organisation and manufacturer.
8. Inadequate quality control.
9. Lack of repair and service facilities for machines.

Future Needs of Farm Mechanisation

1. To develop multifunctional machines.
2. Development of combination machinery like minimum tillage, zero tillage, Till planter etc.
3. Better utilisation of solar, wind, energy and bio-fuels.
4. Efficient fuel utilisation - through better design and matching machinery.
5. Rain water harvesting and conservation.
6. Low cost micro irrigation and sprinklers using surface water needs to be developed and commercialised.

16.3.3 Programme-planning

The first step systematic attempt to promote rural development is to prepare useful programmes based on peoples needs. The development of such programmes, which harmonize with the local needs as the people see them and with the national interests with which the country as a whole is concerned, is an important responsibility of extension personnel at all levels-national, state, district, block and village.

Programme planning is the process of making decisions about the direction and intensity of extension-education efforts of extension-service to bring about social, economic and technological changes.

Principle of extension programme-planning. The planning of an extension programme is done on the basis of certain well recognized principles which should be clearly understood and followed by extension workers. The main principles are :

(*i*) The programme-planning should be based upon a careful analysis of a factual situation.

(*ii*) In a good programme-planning, problems for action are selected on the basis of recognized needs.

(*iii*) A good programme-planning determines objectives and solutions which are feasible and offer satisfaction.

(*iv*) The programme should be permanent and flexible to meet a long-term situation, short-time changes, and emergencies.

(*v*) A sound programme should have both balance and emphasis.

(*vi*) A good programme has a definite plan of work.

(*vii*) Programme-planning is a continuous process.

(*viii*) Programme-planning is a co-ordinating process.

(*ix*) Programme-planning should be educational and directed towards bringing about improvement in the ability of the people to solve their own problems individually and collectively.

(*x*) A good programme-planning provides for the evaluation of results.

The programme-planning process. The steps involved in this process are as follows :

1. *Collection of facts.* Sound plans are based on availability of relevant and reliable facts. This includes facts about the village people, physical conditions, existing farm and home practices, trends and outlook. Besides, other facts about customs, traditions, rural institutions, peoples' organizations operating in the area, etc. should be collected. The tools and techniques for collecting data include systematic observations, a questionnaire, interviews and surveys, existing governmental records, census reports, reports of the Planning Commission, Central Bureau of Statistics, and the past experiences of people.

2. *Analysis of the situation,* after collecting facts, they are analyzed and interpreted to find out the problems and needs of the people.

3. *Identification of problems.* As a result of the analysis of facts the important gaps between 'what is' and 'what should be' are identified & the problems leading to such a situation are located. These gaps represent the peoples needs.

4. *Determination of objectives.* Once the needs and problems of the people have been identified, they are stated in terms of objectives and goals. The objectives represent a forecast of the changes in the behaviour of the people and the situation to be brought about. The objectives may be long-term as well as short-term, and must be stated clearly.

5. *Developing the plan of work.* In order to achieve the stated objectives and goals, the means and methods attaining each objective are selected and the action plan, *i.e.* the calendar of activities is developed. It includes the technical content, who should do what, and the time-limit within the work will be completed. The plan of work may be seasonal, short-term, annual or long-term.

6. *Execution of the plan of work.* Once the action plan has been developed, arrangement for supplying the necessary inputs, credits, teaching aids, extension literature etc. has to be made and the specific action has to be initiated. The execution of the plan of work is to be done through extension methods for stimulating individuals and groups to think, act and participate effectively. People should be involved at every step to ensure the success of the programme.

7. *Evaluation.* It is done to measure the degree of success of the programme in terms of the objectives and goals set forth. This is basically done to determine the changes in the behaviour of the people as a result of the extension programme. The evaluation is done not only of the physical achievements but also of the methods and techniques used and of the other steps in the programme planning process, so that the strong and weak points may be identified and necessary changes made.

8. *Reconsideration.* The systematic and periodic evaluation of the programme will reveal the weak and strong points of the programme. Based on these points the programme is reconsidered and the necessary adjustments and changes are made in order to make it more meaningful and sound.

Programme-planning is not the end-product of extension activities but it is an educational tool for helping people to identify their own problems make timely judicious decisions.

From the above mentioned cycle, it is clear that the planning of an extension programme comprises a logical series of consecutive steps. The first 4 steps form the programme-phase. The steps 5-7 form the action-phase. The step 8, *i.e.* reconsideration, joins the 2 phases together, where it leads to the fact-collecting step, thus beginning once more the never ending or continuous process of planning the extension programme.

The procedure followed for planning extension programmes, including agricultural development programme, is as follows :

A state government indicates to the districts and blocks the tentative outlays proposed in their respective areas under the various heads and the financial and other assistance that may be expected. The blocks prepare their respective annual plans, taking into account the special local problems and the local contributions and the state finances available. These are then consolidated districtwise, on the basis of which the state plan is formulated. The plans are prepared by the *panchayat samities* for the blocks and the *zila parishads* for the districts where the *panchayati raj* is in force. Where this set-up has not yet been introduced, the plans are prepared by the block development committee and the district development committee for the block and the district respectively.

The outlay of the state plans are finally approved by the Planning commission.

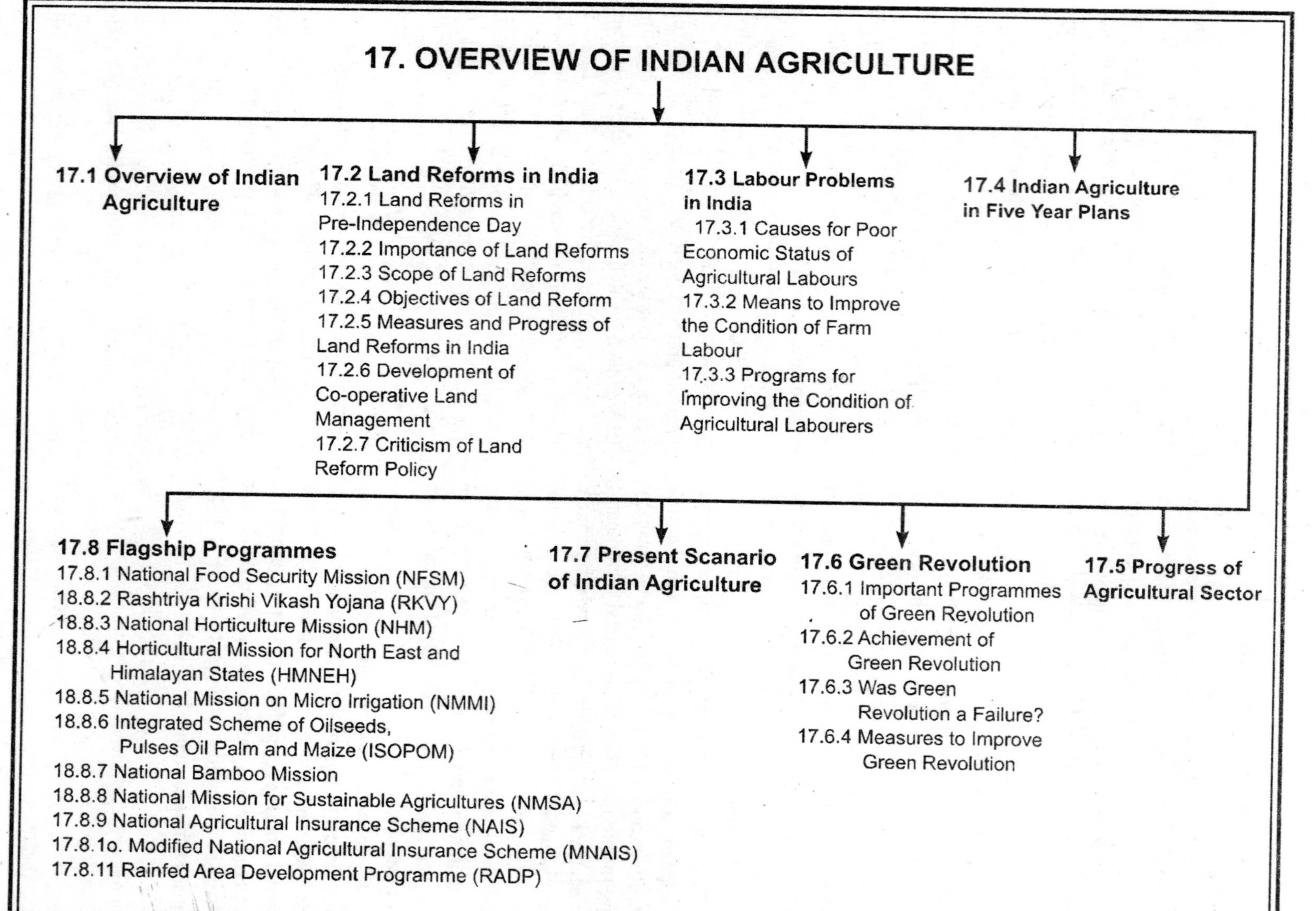
17. OVERVIEW OF INDIAN AGRICULTURE
17.1 Overview of Indian Agriculture
17.2 Land Reforms in India
17.2.1 Land Reforms in Pre-Independence Day
17.2.2 Importance of Land Reforms
17.2.3 Scope of Land Reforms
17.2.4 Objectives of Land Reform
17.2.5 Measures and Progress of Land Reforms in India
17.2.6 Development of Co-operative Land Management
17.2.7 Criticism of Land Reform Policy
17.3 Labour Problems in India
17.3.1 Causes for Poor Economic Status of Agricultural Labours
17.3.2 Means to Improve the Condition of Farm Labour
17.3.3 Programs for Improving the Condition of Agricultural Labourers
17.4 Indian Agriculture in Five Year Plans
17.5 Progress of Agricultural Sector
17.6 Green Revolution
17.6.1 Important Programmes of Green Revolution
17.6.2 Achievement of Green Revolution
17.6.3 Was Green Revolution a Failure?
17.6.4 Measures to Improve Green Revolution
17.7 Present Scanario of Indian Agriculture
17.8 Flagship Programmes
17.8.1 National Food Security Mission (NFSM)
18.8.2 Rashtriya Krishi Vikash Yojana (RKVY)
18.8.3 National Horticulture Mission (NHM)
18.8.4 Horticultural Mission for North East and Himalayan States (HMNEH)
18.8.5 National Mission on Micro Irrigation (NMMI)
18.8.6 Integrated Scheme of Oilseeds, Pulses Oil Palm and Maize (ISOPOM)
18.8.7 National Bamboo Mission
18.8.8 National Mission for Sustainable Agricultures (NMSA)
17.8.9 National Agricultural Insurance Scheme (NAIS)
17.8.1o. Modified National Agricultural Insurance Scheme (MNAIS)
17.8.11 Rainfed Area Development Programme (RADP)

17. Overview of Indian Agriculture

17.1 OVERVIEW OF INDIAN AGRICULTURE

India now being one of the fastest developing countries of world in many fields like industry, space research, education, information technology etc. Though the contribution of agriculture to overall GDP of India has fallen short from 30% in 1991 to less than 15% in 2011-12, a trend is expected in the development process of any economy, yet agriculture is the backbone of country. Although there are many development in the field of agriculture in India yet it is at slow pace due to many problems in the path. Here in this chapter we take a overview of Indian agriculture considering problems, reforms etc.

17.2 LAND REFORMS IN INDIA

17.2.1 Land Reforms in Pre-Independence Day

It was under the tenancy system that half the crops went to the owner of the land and remaining half was received by tenant. Then there were many intermediates. There were problems of rack renting, insecurity of tenure, lack of adequate incentive for investment on land because of uncertainly as to his continuity on the land.

Thus for developing economy of India, planners have prescribed the aims to make possible maximum production to increase the chances of employment so as to avail the concentration of economic power and maintain social Justice; Hence it was thought good to have revolutionary change in the social institution.

17.2.2 Importance of Land Reforms

Its importance lies in the fact that it has to eliminate all elements of exploitation and social injustice within agricultural system. Abolishment of rent - receiving tenures and the reforms under the tenancy are promulgated. These are the necessary steps like reorganization of agriculture, consolidation of holding, prevention of fragmentation, ceiling-fixing on holding and surplus land distribution to landless person, with co-operative farming.

Land reforms depend mostly on two factors. They are —

(1) Technological factors

(2) Institutional improvements.

Technological factors include means and method of farming like good seeds, fertilizers, new plough, tractors, harvest irrigation etc. All these tend to increase productivity. *Institutional improvements* include redistribution of land ownership, increase in the size of farms, security of lease hold, regulation of rent. The farmer become hopeless with the existence of feudal relations, small farm size, fragmentation, insecurity of leasehold, high rents. These all tend to weaken the capability to enjoy savings, investment and success in farming.

17.2.3 Scope of Land Reforms

The concept of land reforms is related to abolition of intermediate and bringing the tenants in direct touch with the state government. Tenancy reforms and regulation of rent create such atmosphere that farmers are hopeful to get the fruit of their labour.

Thus land reforms comprise

1. Abolition of intermediates
2. Tenancy reforms
3. Ceiling of land holding
4. Reognization of land.
5. Organization of co-operative farms.

17.2.4 Objectives of Land Reform

Emphasis has been given during five year plans on land reforms. The objectives of plans were to eradicate the following under the scheme.

(1) Existence of intermediates between the state and the farmer.

(2) Insecurity of land settlement by which farmers do not take any initiative to land reform.

(3) High rate of rents due to which farmers are not motivated to grow more and do not take any initiative to land reform.

(4) Small and scattered field due to which good systems could not be carried out.

(5) Unequal distribution of land so the some section are landless or having less fields so that agriculture is not profitable.

17.2.5 Measures and Progress of Land Reforms in India

First Phase (1936-55)

1. ***Abolition of intermediaries***: Zamindari has been abolished. The tillers have become the owners of land. The compensations have been given to inter-mediaries and waste lands are reclaimed for cultivation. Such lands have come under state governments or directly under Panchayats.
2. ***Tenancy reforms***: Tenants paid high rates without any security. The cultivator become the tenants of the land.
3. ***Regulation of rent***: The crop payment system has been changed to money-payment system.
4. ***Security of the tenure***: Ejectment has been staged. The tenants have been given permanent and heritable rights.
5. ***Conferment of ownership on tenants***: The tenants were given rights to purchase their holding in various states so that they may become owner cultivators.

Second Phase (1956-71)

Ceilings on holdings: Due to the vast number of landless labours, ceiling on land have been created so that available land through this source may be given to landless labourers through Panchayat of Labourers, Co-operative Societies.

Third Phase (1972-1995)

Consolidation of holdings: It is exchange of land of a compact mass for various separate parts. Thus a tenant will be substituted land at a place compactly and it will be an economic holding. To prevent fragmentation floor limit has been passed by legislation which leads to non-subdivision or transfer of holdings which will be reduced to unprofitable or uneconomic.

Co-operative farming: Under this the unit will be larger. Hence large scale operation benefit will be received, sources of receiving money, better implements, tractor, deep ploughs are all available. This would alter the social, economic and cultural life of people.

Benefits of Co-operative Farming :

1. Increase in output, 2. Better inputs possible, 3. Research is possible for farmers, 4. Carrying of off farm activities, 5. Close contact with Government.

Important Agricultural Co-operative Organization in India

1. National Cooperative Development Corporation (NCDC) established in 1963.
2. National Agricultural Co-operative Marketing Federation of India (NAFED).
3. Tribal Co-operative Marketing Development Federation of India Ltd. (TRIFED).
4. Indian Farmers Fertiliser Cooperative Ltd., (IFFCO).
5. Indian Farm Forestry Development Cooperative Ltd. (IFFDC)
6. Cooperative Rural Development Trust (CORDET).
7. Krishak Bharati Co-operative Ltd. (KRIBHCO).
8. National Bank for Agriculture and Rural Development (NABARD).

17.2.6 Development of Co-operative Land Management

Land problems could be solved by co-operative village management. It has following specialities.

1. The whole village agriculture land is taken as single unit.
2. Agriculture work is so organised that gives maximum benefit to village society.
3. Villages resources are well organized that it gives maximum benefit to village society.
4. Landless labourers are paid price for their work and those who are land owners are provided extra amount towards their rights on land.
5. Reorganised blocks are distributed by village management.
6. This system is promulgated while the two-thirds land owners or half the number of tenants of tilled land agree to it.

17.2.7 Criticism of Land Reform Policy

1. Reforms are in right direction but execution is defective.
2. Slowness of promulgation.
3. Weak economic condition of tenants.
4. Lack of uniformity in execution.

17.3 LABOUR PROBLEMS IN INDIA

Agricultural labour means the person who is not only engaged in the production of crops but also those who are kept as labourer or kind in agriculture occupations, like dairy and poultry business. The agricultural labour family means that family which has its most of income from farm labour.

Farm labourer is one who works for the wages in the form of crops and has no rights on land and has no risk or liability for the sowing the crops. He gets share in crops as wages and not as profit.

Kinds of Farm Labourers

According to "Congress Agrarian Reforms Committee", they can be categorised in the following three forms.

1. **Field workers:** They include ploughers, sowers and transfer sources of crops.
2. **Ordinary labourers:** They include clay diggers, fencing workers and clodders.
3. **Skilled workers:** They include carpenter, mistry, black-smith etc.

As per Labourer Intensive Survey Report there are two classes — (1) *Attached labours* are engaged for long terms with the farm-owners and cannot be free to work elsewhere till they reap the rewards to their labour (2) *Casual laboures* are for temporary work and free to work elsewhere and get wages at market rates.

Number of Agricultural Workers

Till the year 1971 farmers were 12.6 crore in number. The size of labour population is increasing over time. According to National Commission on Agriculture. (1991) 33.30 million household were agricultural labourers among 108.3 million rural household. Thus amounts to 30.75 percent.

As per census 2011 the number of agricultural labourers area and sex-wise is as follows:

u **Total agricultural labourers: 14,43,29,833**

Rural :	1,19,42,619
Urban:	63,88,688
Males:	8,27,40,351
Females:	6,15,89,482

Wages and Earnings of Agricultural Labours

It was pointed out by various reports and commissions the agricultural labourers in India did not receive notified minimum wages except in certain states of the country viz., Kerala, Punjab, Haryana and Uttar Pradesh. Even in these states females did not get sufficient remuneration of notified in the minimum wage act. Due to prevalence of Labour unions or organizations in certain areas wages are closer to minimum wages. But in the most of areas agricultural labourers were disorganized and have weak bargaining power.

Due to these factors there is a gap between actual wages and minimum wages fixed by the Government. Despite this the real wages of agricultural laboures have been increased with time and also the minimum wages decided by Government is also changing with time.

In the periodical revision of minimum wages, provisions was made to equal wages to all labourers irrespective of sex, cast and creed. The practice of serfdom was also abolished. Maximum agricultural labours are below the poverty line and caught in the vicious circle of utter poverty and misery.

17.3.1 Causes for Poor Economic Status of Agricultural Labours

1. **Social and economic status:** Most agricultural labourers belong to suppressed and depressed sections of society, who were neglected for ages together. These are the socially depressed and never has the courage to assert themselves.
2. Illiteracy and ignorancy.
3. Village labourer's do not have organizations and unions.
4. Less bargaining power.
5. Discontinuous work on farms *i.e.* they are seasonally employed.
6. Paucity of non-agriculture occupation in Indian villages.
7. Less demand of agricultural works.
8. Rural indebtedness.
9. Pledge of movable property and borrow of the amount at exorbitant rates of interest.
10. Some of them are bonded labours accepting lower wages.

17.3.2 Means to Improve the Condition of Farm Labour

National planning should act to the improvement of farm labour without that no improvement is possible. Some of the factors which could lead to the improvement of condition of farm labour are:

1. Education of labourers
2. Education of farmers
3. Waste land should be tilled for agriculture
4. Irrigation facilities
5. Intensive farming
6. Mechanised farming
7. High yield seeds
8. Multiple cropping
9. Cooperative rural management
10. Bhoodan yojana
11. Organization and village industries
12. Cheap transport facilities
13. Rural labour exchange
14. Farm labour organization
15. Establishment of advisory boards
16. Regulation of work hours
17. Improvement in working conditions
18. Safety of females and children
19. Vocational education
20. Technical education
21. Development of facilities of rural life
22. Establishment of social clinics
23. Arrangement of benefits of social security
24. Increase in rate of wages
25. Prescription of minimum wages
26. Organization of labour cooperative

17.3.3 Programs for Improving the Condition of Agricultural Labourers

1. National Rural Employment Program (NREP): Central Government had launched food for work programme on 1-4-1977 to help the rural people particularly agricultural labourers by generating additional employment to them and create durable states in the rural area. Later the same programme was renamed as National Rural Employment Programme (NREP) in October 1980. The programme was part and parcel of 6th five year plan and implemented on the basis of 50:50 share basis between the central and state governments.

The allocation of resources to the states and union territories was made on fixed formula based on 75 percent weightage to population of agricultural labourers and marginal farmers and 25 percent weightage for incidence of poverty in state.

2. Training of Rural Youth for Self-employment (TRYSEM): This national scheme was implemented in August 1979, with the prime aim of alleviating unemployment among rural youth. The main thrust on this scheme is directed against imparting necessary skills and made the new technology adopted by the youth for enabling them to take up suitable vocations of self employment. The target fixed for the scheme was 2 lakh rural youth under IRDP areas and non IRDP areas. Separate funds were provided for non-IRDP areas but not for IRDP areas. The programme was expected to cover all the blocks of the country in a phased manner. In effect TRYSEM is a part and parcel of IRDP programme. Training is given to the rural youth between 18 to 35 years for self employment avocations. Sixth plan had made a provision of Rs. 500 crore for strengthening TRYSEM.

3. **Small Farmers Development Agency (SFDA) and Marginal Farmers and Agricultural Labourers Development Agency (MFAL):** This project was started in the fourth-five year plan to improve the economic condition of rural poor and agricultural labourers. Initially 46 SFDAs and 41 MFALs were started to tackle the problem of small and marginal farmers and agricultural labours and benefit about 50,000 small farmers, 15,000 marginal farmers and 25000, agricultural labourers. The main thrust of this program was laid on the crop husbandry which included several intensive multiple cropping practices, water harvesting methods in rainfed areas such as land development soil conservation, development of minor irrigation, development of horticulture and adoption of potential technology for dry farming. These projects implemented animal husbandry programs *viz.* supply of much animals, poultry piggery, sheep and goat rearing, fisheries etc.

4. **Tribal Development Agency (TDA):** About 6 tribal development projects were implemented in 1971 and 1972 with a view to bring about economic development and accelerate pace of development in tribal areas of the country.

5. **Hill Area Development Agency (HADA):** It was started in fifth five year plan and being continued is some districts of state of Assam, Tamilnadu, Uttar Pradesh, West Bengal, Western ghat region constituting the states of Maharashtra, Karnataka, Tamilnadu, Kerala and Goa. The strategy of this programme is to develop economic status of agricultural labourers and rural poor by taking up the projects on watershed development, crop husbandry, development of horticulture, animal husbandry and infrastructure in rural areas.

6. **Drought Prone Area Programme (DPAP):** It was started in fourth five year plan. The main objectives of DPA pare.

- Development of small and marginal farmers and agricultural labourers.
- Development and management of irrigation sources.
- Restructuring the cropping pattern and pasture development.
- Soil, water conservation and afforestation, and
- Live stock development.

7. **Desert Development Programme (DDP):** This was launched in the country in the year 1977-78, based on recommendation of National Commission on Agriculture (NCA). About 132 blocks in country are being benefited in the states of Rajasthan, Haryana, Gujarat, Jammu and Kashmir and Himachal Pradesh. This programme aims at the integrated development of desert areas and increasing the productivity, income levels and employment opportunities of rural people.

8. **Minimum Wages Act for Agricultural Labourers:** This act was passed in 1948 in some states. Minimum wages in agriculture was not revised for a long period. There is a strong opinion which is contrary to the minimum wage act in rural areas. The basic argument is that at the peak agricultural seasons, labourers are demanding more wages than wages stipulated in the minimum wage act. Therefore, the enforcement of minimum wages has no meaning, but on the other hand, due

to slack season, due to excessive supply of labourer's and paucity of work, labourer are paid less wages than that indicated minimum wage act. The National Commission on Labour (NCL) gave due weightage to this view and strongly for continuance of minimum wage act on the ground of rights given to labourer's quality and growth in the economy. As regards the enforcement of minimum wage act, the commission suggested that Panchayat should be given adequate right to implement minimum wage acts. We also notice considerable disparities in the payment of wages across region, crops, states, sex, age etc. Labour Ministers' conference held in August 1981, decided that minimum wages fixed for labourers should not fall below poverty line. Wages should be properly linked with consumer price index. In pursuance of their recommendations, general minimum wages were fixed, taking into necessities like food, fuel and shelter in respect to workers.

17.4 INDIAN AGRICULTURE IN FIVE YEAR PLANS

Planning as an instrument of development was underlined by Sri M. Visveswarayya in the year 1934. But the first attempt regarding planning was made by *National Planning committee* under the chairmanship of Pandit Jawaharlal Nehru in 1938. However the committee could bring out the reports on planning in 1948 only. Mean while, the all India congress committee appointed the Economic Programme Committee with Pandit Jawahar Lal Nehru as the chairman in 1947. The first five year plan draft was issued by the planning commission.

The ***first five year plan*** was started in 1951 and ended in 1956. It gave importance to agriculture and industry. Community development schemes were started in such places which have more population and pressure and agricultural labours around. While less wages had assignation of minimum wages, occupancy rights of dwellings, organization of labour cooperatives and rehabilitation of land less farmers. This has expense of one crore rupees. Mechanized farms were established, Central Teacher Union has settled landless labours on the improved lands. Still the progress was very slow.

Second Five Year Plan (1956-61): Increased the cattle business with farm production, with expansion of cottage and small-scale industries and agricultural legislation. This increased chances of employment in rural economy, reallotment of land and facilities of education which in turn developed the status of farm labourer, efficiency, capability etc. to improve the status of the labourers. This scheme brought the expenditure of 90 crores of rupees. Approximately 20,000 families were settled on 1,00,000 acres of land.

Third five year plan (1961-66) took the cottage and small-scale industries, electrification of village, rural residence, facilities of drinking water, irrigation, increase in agricultural production, education etc. to improve the status of labourers. Programme for undertaking work projects in rural areas were carried out to give extra employment to 25 lakh labourers, while agricultural work was slack. Backward classes have the welfare work to the tune of 99.14 crores of rupees.

Fourth five year plan (1969-74) had two major objectives for the development of agricultural sector. The first objective aimed at annual growth of 5 percent in agriculture and the second objective was aimed at facilitating economic development of large number of rural population including small cultivators, dry land farmers, land less agricultural labourers etc. It included 88 project, 51 districts had development programme. Rural industrial projects were started to employ small farmers and field workers. National Agriculture Commission has recommended to give agricultural credit to small farmers and field workers.

Fifth five year plan (1974-79) emphasized on the projects started under the fourth plan and few new plans were started. Rural development scheme was started in 1971. During 1977-78, Krishi Vigrana Kendras and training centres were established for imparting advanced and potential agricultural technology to farmers. 78,000 gobar gas plants were also installed during this period. National seeds projects were also launched in the state of Andhra Pradesh, Maharashtra, Bihar and Karnataka.

Sixth five year plan (1980-85) aimed at achieving growth rate of 3.8 percent in agriculture. Wheat and rice production was keeping pace with their target, where as rest of the crops did not achieve the target. A centrally sponsored scheme of community nurseries of rice minkit, wheat, bajra, maize and ragi were intensified in order to create a clear impact on production of cereal crops. NABARD was established in July 1982.

Seventh five year plan (1985-90) prominence were given to dairy and fisheries development as a result their production increases substantially. The plan aimed at direct attack on problems of rural poverty, rural employment and regional imbalances in the development. Rural employment programme were launched during this plan.

Eighth five year plan (1992-97) aimed at the generation of surplus of food grains for export and attain self-sufficiency in pulses and oilseeds. The growth rate expected was 4 percent in respect of gross value of output and 3 percent with regard to value added products. The strategy that was spelt out to achieve the targeted production levels were laying emphasis on dry land farming, spreading benefits of green revolution to the eastern region and improving efficiency of irrigation.

Ninth five year plan (1997-2002) targeted annual growth rate of 4.7 percent. The targeted growth rate of agricultural crops output is expected to increase at 3.82 percent per annum and that of agriculture at 4.5 percent. Actual targeted growth rate for livestock was 6.59 percent, poultry 7.5 percent and fisheries 6.5 percent.

10th five year plan (2002-2007). In tenth plan GOI has proposed to eradicate poverty and unemployment and also double the per capita income in next 10 years. 10th plan expected to follow a regional approach rather than sectoral approach to bring down the regional disparities and inequality among the people. 10th plan could achieve a growth rate of 7.6% as against 5.5% of the 9th plans.

11th five year plan (2007-2012): To accelerate GDP Growth from 8% to 10%. Increase in Agricultural growth rate to 4% per year. Creation of 70 million new job opportunities and reduction in educated unemployment to below 5%. Provision of clean drinking water for all by 2009. Reduction in malnutrition among children 0-3 years age to half its present level, Reduction in anemia among women and girls by 50%. Increase in Forest cover by 5% points. Increase in energy efficiency by 20% points by 2016. The concept of inclusive growth was introduced. Total plan size proposed was 36.45 lakh crore more than double of 10th plan.

12th five year plan 2012-2017): Average growth rate set at 8.2%. Areas of main thrust are infrastructure health and education. Agriculture current plans growth rate at 3.3% compared to 2.4% during the 10th plan period .

The theme of the plan is "Faster, sustainable and more inclusive growth. Total plan size has been estimated at Rs. 47.7 lakh crore 135% more than 11th plan. 12th plan target one lakh MW in power generation.

17.5 PROGRESS OF AGRICULTURAL SECTOR

We can divide the progress of agricultural sector in four phases.

First phase (1947 to mid sixties): It witnessed tremendous agrarian reforms, institutional changes, development of major irrigation project and strengths of cooperative credit institution. This phase involved the land reforms. The community development programme, decentralized planning and the Intensive area development programmes were also initiated for regenerating Indian agriculture that had stagnated during the British period. Despite the institutional changes and development programmes by the government during this phase, India remained dependent upon foreign countries for food and feed for the rising population.

The **second phase (mid sixties to 1980)** adopted the new agricultural strategy which relied on high yielding variety of crops, multiple cropping, modern farm practices and spread of irrigation facilities. Agrarian reforms during this phase took back seat while research, extension, input supply, credit, marketing, price support and spread of technology were the prime concern of policy makers.

The **third phase (1980-1991)** in Indian agriculture witnessed the process of diversification which resulted into fast growth in non-food grains output like milk, poultry, fishery, vegetables, fruit etc. which accelerated the growth in agricultural GDP during 1980s. There has been a considerable increase in subsidies and support to agricultural sector during this period while public sector spending in agriculture for infra structure development started showing decline in real term but investment by farmers kept on moving on rising trends.

The **fourth phase** stated after initiation of economic reform process in 1991. Economic reform process involved deregulation, reduced government participation in economic activities and liberalization. Although there is no any direct reforms for agriculture but the sector was affected indirectly by devaluation of exchange rate, liberalization of external trade and disinvestment to industry. During this period opening up of domestic market due to new international trade accord and WTO was another change that affected a agriculture was launched by Indian government in July 2000. This raised new challenges among policy maker. Because of this, a New Agricultural Policy was launched by Indian Government in July 2000.

17.6 GREEN REVOLUTION

In the modern age, industrial revolution and wonders of science has given new form to traditional agriculture arrangement which is called Green Revolution.

The term green revolution has been used but it is not been adopted in the real sense to agriculture. Many persons have put their efforts and energies in discussing whether there is green revolution or not in the country. Those who are of the view that there is green revolution point out that during the present years capital inputs under fertilizer, insecticides, tractors, electric pumps and high productive quality of land have shown a heavy and high increase. They point out increased rate of production relating to few crops and field. But those who are of opposing views point out that we can judge green revolution from the figures of agricultural production. But as referred to these figures they do not indicate that there is green revolution.

The beginnings of the Green Revolution are often attributed to **Norman Borlaug,** an American scientist interested in agriculture. In the 1940s, he began conducting research in Mexico and developed new disease resistant high-yield varieties of wheat. By combining Borlaug's wheat varieties with new mechanized agricultural technologies, Mexico was able to produce more wheat than was needed by his own citizens, leading to exporter of wheat by 1960s. Due to success of Green Revolution in Mexico, its technologies spread worldwide in 1950s and 1960s. Countries all over the world in turn benefited from the Green Revolution work conducted by Borlaug and his research institution. India for example was on brink of mass famine in the early 1960s because of its rapidly growing population. Borlaug and the Ford foundation then implemented research there and they developed a new variety of rice IR8 that produced more grain per plant when grown with irrigation and fertilizers. Today India is one of the world's leading rice producer and IR 8 usage through out the Asia in decades following rice's development in India.

17.6.1 Important Programmes of Green Revolution

The following important programmes have been included under green revolution.

1. **Multiple crop programme:** To fulfill the fundamental objective of green revolution, it is necessary to produce multiple crops in a year 1970-71, multiple programme was carried on 38.12 lack hectares and in 1971-72, 58.06 lack hectare land. By the end of fifth five year plan it was used to convert 110 lack hectares of land under multiple crop programme.
2. **Standard seeds of more product:** Under green revolution most important place in green to authorized seeds. National seed corporation has authorized good variety of seed project in various states. With the help of World Bank, National Seed Project was started under which four seed corporation have been established in 1975-76 and during 1976-77 had come under establishment.
3. **Application of good chemical and manure:** To increase the agricultural production; the important elements required are chemical fertilizers. As compared to other countries India used less amount of manure on fields per hectare and per capita.
4. **Irrigation:** It was necessary to increase production in country so irrigation system was needed to enhanced. During 1951, the capacity to irrigate was 277 lakh hectare land, during 1973-74 it was increased to 450 lakh hectares. During the fourth five year plan irrigation was allowed to spend ₹ 511 crores and during fifth five year plan it was increased to ₹ 800 crores.
5. **Use of tractors in agriculture:** Better and more agricultural production system proved use of tractors more facilious. During 1970-71 demand for tractors was about 40,000. It reached to 1 lakh up to 1979-80
6. **Proper marketing price of agricultural yield:** Towards this government has established an agriculture price commission.
7. **Establishment of important commissions:** To fulfill all the programmes of green revolutions, government has established important commissions such as *National Seeds Corporation. National Co-operative Development Commissions, Agro Industries Corporations, Rural Electrification Corporation, Indian Food Corporations* etc.

17.6.2 Achievement of Green Revolution

The following achievements have been seen in India since green revolution.

1. **Maximum utilization of land:** Land is best utilized under the plan, *i.e.* small holding had produced more by rotation of different crops.
2. **Solution of food problems:** Green revolution has solved the problem of food in the country. Available figures indicate that agricultural production had reached 107.8 metric ton in 1970-71, while in 1977 it reached to 120 metric tons.

   The result of green revolution is that country is now able to overcome the problem of food grain shortages. The country is almost self-sufficient in food grains and it's no longer depends on import of food grains.

3. **Good plant variety:** The crops developed during the Green Revolution were high yield varieties meaning they were domesticated plants *breed* specifically to respond to fertilizers and produce an increased amount of grain per acre planted.
4. During the Green Revolution, plants that had the largest seeds were selected to create most production possible. After selectively breeding these plants, they evolved and all have the characteristics of larger seeds. These larger seeds then created more grain and a heavier ground weight.

17.6.3 Was Green Revolution a Failure?

It is said that green revolution failed and there are criticism against it.

We can say that green revolution is a failure because of the fact that limited agricultural field have been brought under the experimentation of wonder seeds or high yielding seeds. If we take the five main crops of rice, wheat, maize, jowar and bajra about which high yielding seeds were known, only 15% land under them was brought to the application of high yielding seeds. Remaining 85% part was brought under traditional farming. If we take all the crops, it will be seen that only 10% portion of land was brought under cultivation with wonder seeds while 90% land was under traditional farmer.

17.6.3.1 Main causes of failure of Green Revolution

1. Lack of education and training in farmers.
2. Lack of modern mechanization in agriculture.
3. Uneconomic size of economic holding.
4. Growth in unemployment.
5. Regional differences in agricultural production.

17.6.3.2 Criticism of the Green Revolution

1. Repeated cultivation of same crops leads to deficiency in a particular nutrient of soil which in turn reduces the soil productivity in the long term
2. Over utilization of fertilizer leads to degradation of soil fertility.
3. Continuous stagnation of water in the field increased the salinity of soil because hybrid variety need more amount of water.
4. It reduces the crop species diversity which in turn make crop more prone to disease and pests.

Biofortification

To tackle malnutrition effectively a new Scientific option is biofortification crops.

Biofortification is the idea of breeding crops to increase their nutritional value. This can be done either through conventional selective breeding, or through genetic engineering. Biofortification differs from ordinary fortification because it focuses on making plant foods more nutritious as the plants are growing, rather than having nutrients added to the foods when they are being processed. This is an improvement on ordinary fortification when it comes to providing nutrients for the rural poor, who rarely have access to commercially fortified foods. As such, biofortification is seen as an upcoming strategy for dealing with deficiencies of micronutrients in the developing world. In the case of iron, WHO estimated that biofortification could help curing the 2 billion people suffering from iron deficiency-induced anemia.

India will soon be the first country to commercially cultivate biofortified pearl millet, or bajra. The crop has been biofortified to improve its iron and zinc nutrients, and will be released in 2012 by HarvestPlus, a global alliance of research and implementing agencies.

To be distributed in Rajasthan, Maharashtra, Gujarat and Uttar Pradesh, the crop has been developed by HarvestPlus' sister concern International Crops Research Institute for the Semi-Arid Tropics (ICRISAT). "It targets hidden hunger of undernourished and malnourished people, especially the poor who are not able to meet micronutrient requirements from the food they eat," said Howarth E Bouis, director, HarvestPlus. He also said that varieties are high yielding. The price of nutrition's staple grain in retail markets will be equal to the price of todays lower nutrient grain varieties.

The Indian Council for Agricultural Research and the Indian Council of Medical Research (ICMR) plan to propose introduction of bio-fortification to the Planning Commission for the 12th Five Year Plan.

17.6.4 Measures to Improve Green Revolution

While the agriculture is not developed to the sufficient and proper modern ways, Green Revolution cannot achieve success. Hence it is quite essential to improve irrigation system, maximum use of electric power, better chemical fertilizer use. With these some other important measures are also essential to its success which are given below:

1. **Prevention of flood:** Flood should be prevented by erecting new dams and reservoirs to collect the flood water and use it for agricultural purposes. Thus floods will be decreased.
2. **Increase in the means of irrigation:** To irrigate the fields water ponds should be made to collect rain water and flowing village water so that the water will be collected there in and used for irrigation purposes when required.
3. **Use of better kind of manure:** To enrich the agriculture it is necessary to apply better and authentic manures.
4. **Application of insecticides:** To enforce the green revolution it is necessary to apply insecticides. These should be taxless item.
5. **Use of good quality authentic seeds:** To make the success of green revolution it is also important to use new variety of authentic seeds and then only. Indian agriculture can be progressive regarding quality and quantity.
6. **Use of modern agricultural instruments:** The last important suggestion of green revolution is that government should provide the agriculturists modern instruments for agriculture on easy payable instrument.

Conclusion

It must be borne in mind that solution of a problem can create an another problem. The use of fertilizers, insecticides are creating water pollution so proper planning should be made for the use of chemicals in agriculture so that they do not cause more environmental problems.

Despite of criticism though, the Green Revolution has forever changed the way agriculture is conducted worldwide, benefiting the people of many nations in need of food production.

17.7 PRESENT SCANARIO OF INDIAN AGRICULTURE

Agriculture as the largest private enterprise in India, is the life line of economy.

The total geographical area of country is 328.7 million hectare of which 140.3 million hectares is the net area, while 193.7 million hectares is the gross cropped area. The gross and net irrigated area is 85.8 million hectares and 60.9 million hectares respectively with a cropping intensity of 138 percent.

Statistically, the agricultural sector contributes to approximately.

- 14.6 percent of India's GDP (at 2004-05 prices) in 2011-12 and
- 10 percent (provisional) of total exports, besides.

- 58.2 percent of the workforce was provided employment in this section.
- 43 percent of India's geographical area is used for agricultural activity.

The overall target of GDP growth in country for eleventh plan is 9 percent per annum with an annual average growth rate 4 percent in agriculture. The strategy for accelerating agriculture to 4 percent per annum requires actions in terms of bringing technology to the farmers improving the efficiency of investments, increasing system support and rationalizing subsidies, diversifying, while also protecting food security concerns and fostering inclusiveness through a group approach, by which poor will get better access to land, credit and skills.

The annual growth rate of total GDP vis-a-vis growth rate in agriculture and allied sector from 2007-08 to 2011-12 (1st revision) is given in table 17.1.

Table 17.1: Growth in agriculture and allied sector GDP vis-a vis total GDP from 2007-08 to 2011-12 1st revision (at 2004-05 prices).

Year	Total GDP annual growth	Agriculture and allied sector (%) Actual growth
2007-08	9.3	5.8
2008-09	6.7	0.1
2009-10	8.6	0.8
2010-11	8.9	7.9
2011-12	6.7	5.0
2012-13	4.5	1.4
2013-14	4·9	4·6

Source: CSO (Central Statistics Office)

Gross capital formation in agriculture and allied sectors

Agriculture and allied sectors and their share in total GDP of the country during past years is presented in table 17.2.

Table.17.2: Gross capital formation (GCF) in agriculture and allied sector at 2004-05 (in ₹ crore).

Year	GDP	GCF	CCF as % of GDP
2005-06	594487	86604	14.56
2006-07	619190	92057	14.87
2007-08	655080	105741	16.14
2008-09	655689	127127	19.38
2009-10	660987	133162	20.14
2010-11	713477	131224	18.39
2011-12	739495	146578	19.82

Investment in agriculture

In recent years the share of Gross Capital Formation (GCF) of agriculture and allied sector in total GCF has hovered between 6 to 8 percent whereas it was 18 percent during the early 1980s.

As a result of initiatives taken by Government, the share of total investment in GCF in agriculture and allied sectors have been growing up in recent years. During 2004-05 to 2009-10, the total investment in agriculture increased in the range of 7.5% to 7.7% per annum. This trend continued in subsequent years. Actual expenditure during 10th Five year plan was ₹ 14992.36 crore against the approved out lay of ₹ 16093.32 crore. Out of total outlay of ₹ 66577.00 crore for seventh plan the anticipated expenditure up to 2010-11 would be ₹ 44413,01 crores.

17.8 FLAGSHIP PROGRAMMES

The department of Agriculture and Cooperation has implemented a number of Flagship Programmes some of which are discussed here.

17.8.1 National Food Security Mission (NFSM)

The National Development Council (NDC) in 2007 adopted a resolution to launch food security mission comprising rice, wheat and pulses to increase the production of rice by 10 million tones wheat by 8 million tones and pulses by 2 million tones by the end of eleventh five years plan (2011-12). Accordingly, a centrally, sponsored scheme

National Food Security Mission (NFSM) was launched.

The main objective of NFSM are:

1. To increase production of rice, wheat and pulses through area expansion and productivity enhancement in suitable manner in identified districts of country.
2. Restoring soil fertility and productivity at individual farm level, creation and employment opportunities.
3. Enhancing farm level economy (*i.e.* farms profits) to restore confidence among farmers.

NFSM is being implemented in 482 districts of 19 states viz. Andhra Pradesh, Bihar, Chattishgarh, Gujrat, Haryana, Jharkhand, Jammu and Kashmir, Karnataka, Odisa, Punjab, Rajasthan, Tamilnadu, Tripura, Uttar Pradesh and West Bengal. About 20 million hectares of rice area and 13 million hectares of wheat area is included in these districts that roughly constitute about 50% of cropped area for wheat and rice. For pulses additional 4.5 million ha (20% cropped area) would be created.

The National Food Security Mission has three Components

(*i*) National Food Security Mission-Rice (NFSM-Rice).

(*ii*) National Food Security Mission-Wheat (NFSM-Wheat)

(*iii*) National Food Security Mission-Pulses (NFSM-Pulses)

17.8.2 Rashtriya Krishi Vikash Yojana (RKVY)

RKVY is the biggest scheme in the agricultural sector presently which aims at achieving 4% annual growth in agriculture sector during Eleventh Plan Period, by ensuring holistic development of agriculture and allied sectors. The main objectives of the scheme are:

(*i*) To incentivise the states so as to increase public investment in agriculture and allied sectors.

(*ii*) To ensure the preparation of agriculture plans for the districts and states based on agro-climatic conditions, availability of technology and natural resources.

(*iii*) To provide flexibility and autonomy to states in the process of planning and executing agriculture and allied sector schemes.

(*iv*) To ensure the local needs are better reflected in agricultural plans of the states.

(*v*) To achieve the goal of reducing the yield gaps in important crops, through focused interventions.

(*vi*) To maximise returns to farmers in agriculture and allied sectors.

(*vii*) To bring about quantifiable change in production and productivity of various components of allied sectors by addressing them in holistic manner.

An outlay of ₹ 25,000 for Eleventh Five Year plan has been earmarked for RKVY. The outlays under RKVY for year 2010-11 have been substantially increased to ₹ 6722 crore, and ₹ 10350.00 crore is proposed for 2011-12. The allocation under RKVY for 2012-13 has been ₹ 9217 crore.

A total of 5768 projects were taken up by the states in the eleventh plan for which 3343 had been completed till December 2012.

Nine special programme/schemes implemented as sub-schemes of RKVY

1. Bringing green revolution in Eastern India
2. Integrated development of 60,000 pulses village in Rainfed Areas.
3. Special Programme on Oil Palm Areas Expansion (OPAE).
4. Vegetable Initiatives for Urban clusters.
5. Initiative for Nutritional Security Through Intense Millets Production (NSIMP).
6. National Mission for Protein Supplement (NMPS).
7. Accelerated Fodder Development Programme (AFDP)

8. Rained Area Development Programme (RADP)
9. Saffron Mission Revival of Cultivation in Jammu and Kashmir.

17.8.3 National Horticulture Mission (NHM)

The National Horticulture Mission was launched in 2005-06 for the holistic development of this sector in the country, daily ensuring vertical and horizontal linkages with active participation of stakeholders.

At present 371 districts in 18 states and 3 union territories have been covered under NHM. All horticulture crops such as fruits, species, flowers, medicinal and aromatic plant are included for area expansion.

Under the post harvest management component 153 pack houses, 157 cold storages, 3 controlled atmosphere storage, 1 refrigerated van, 101 mobile/primary processing units and 83 rural markets were set up and functional infrastructure for collection, grading etc., was created which helped in proper handling and marketing of horticulture produce.

17.8.4 Horticultural Mission for North East and Himalayan States (HMNEH)

The Technology Mission for North-Eastern States, including Sikkim aimed at holistic development of all the horticulture crops has now been renamed as *Horticulture Mission for North-East and Himalayan States* (NMNES). The major areas under the mission are setting up nurseries for production and distribution of quality planting materials, area expansion, rejuvenation of old/ unproductive orchards, precision farming, organic farming, creation of water resources cold storage units and market infrastructure development. During 2009-10, a budget allocation of which ₹ 325.72 crores have been utilised. The final allocation for 2010-11 it ₹ 432.40 crores, which includes ₹ 312.40 crores for North-Eastern states and ₹ 20.00 crore for Himalayan states. So far an amount of ₹ 268.98 crore has been released. The outlay approved for 2011-12 was ₹ 500.0 crores.

17.8.5 National Mission on Micro Irrigation (NMMI)

The main objective of micro irrigation scheme is to increase the area under efficient methods of irrigation, that is, drip and sprinkler for the optimal use of surface as well as ground water resources. NMMT scheme aims to improve water use efficiency which also leads to reduced requirement of fertilizer due to improvement in fertilizer use efficiency on account of fertigation. The savings in water can be used to augment area under irrigation and thereby increase productivity. Assistance is provided at the rate 60% of total cost of systems to small and marginal farmers and at the rate 50% to general farmers. Since 2009-10, an area of 8.22 lakh hectare have been covered- 4.00 lakh ha, under drip system and 4.22 lakh hectare under sprinkler system.

17.8.6 Integrated Scheme of Oilseeds, Pulses Oil Palm and Maize (ISOPOM)

In order to provide flexibility to States implementation, based on regionally differentiated approach in promoting crop diversification and to provide focused approach to programmes, centrally sponsored scheme of integrated scheme of oilseeds, pulses, oil palm and maize is being implemented since 1.4.2004. The scheme is being implemented in 14 major states for oilseeds and pulses, in 15 states for maize and in 9 states for oil palm.

17.8.7 National Bamboo Mission

The harness potential of bamboo crop in the country a centrally sponsored scheme "National Bamboo Mission" is under implementation in twenty seven states. Mission envisage promoting holistic growth of bamboo sector by adopting area based regionally differentiated strategy, under bamboo cultivation and improve marketing of bamboo and its product.

Under this mission steps have been taken to increase availability of quality planning material by supporting up of new nurseries/tissue culture units and strengthening existing ones.

17.8.8. National Mission for Sustainable Agricultures (NMSA)

NMSA seeks to transform indian agriculture into a climate resilient production system through suitable adaptation and mitigation in domain of both crops and animal husbandry. NMSA as a programmatic intervention focuses on promotion of location specific integrated/composite farming systems, resource conservation technologies, comprehensive soil health management, efficient on-farm water management and main streaming rainfed technologies. NMSA identify 10 key dimensions namely *seed and culture water, pest, nutrient, farming practices, credit, insurance, market, information* and *livehood diversification* for promoting suitable agriculture practices that cover both adaptation and mitigation through four functional areas, namely, (*i*) Research and development (*ii*) Technologies, (*iii*) Products and practices and (*iv*) Infrastructure and capacity building.

17.8.9 National Agricultural Insurance Scheme (NAIS)

General sector scheme namely, National Agricultural Insurance Scheme (NAIS) is being implemented since Rabi 1999-2000, as part of risk management in agriculture with intention of providing financial support to farmers in event of failure of crops due to natural calamities, pests and disease.

The scheme covers following crops.

1. Food crops (cereals, millets and pulses),
2. Oilseed (groundnut, sunflower, soybean, castor, *sesame*).
3. Annual commercial/annual horticultural crops (sugarcane, cotton, potato, onion, ginger, turmeric, banana, pineapple, jute etc.)

During last twenty crop season (*i.e.* from Rabi 1999-2000 to Rabbi 2009-10, 1762 lakhs farmers have been covered for an area of 2685 lakh hectares insuring total sum amounting ₹ 221307 crore.

17.8.10 Modified National Agricultural Insurance Scheme (MNAIS)

Modified NAIS has been approved for implementation on pilot basis in 50 district from Rabi 2010-11 season in the country. Administrative and Operational Guidelines for the scheme have already been issued on 17-09-2010 and 28-9-2010 respectively.

The major improvements made in MNAIS.

1. Actuarial premium will be paid for insuring the crops, hence the claims liability would be on the insurer; 2. The unit area of insurance for major crops in village panchayat; 3. Indemnity amount shall be payable for prevented sowing / planting risk and for post harvest losses due to cyclone; 5. On account payment up to 25% of likely claims would be released as advance for providing immediate relief of farmers; 6. Uniform seasonality discipline for loanee and non-loanee farmers; 7. More proficient basis for calculation of threshold yield and minimum indemnity level of 70% of instead of 60%; 8. Modifies NAIS with improved features will have two components that is compulsory and voluntary. Loanee farmers will be insured under 'voluntary category'. 9. Private sector insurers with adequate infrastructure and experience would also be allowed in implementation of MNAIS.

17.8.11 Rainfed Area Development Programme (RADP)

To ensure growth in the rainfed areas, GOI launched in the year 2011-12 a sub schemes under Rastriya Krishi Vikas Yojana "RAINFED AREA DEVELOPMENT PROGRAMME". This scheme aimes at improving quality of life of farmers especially small and marginal farmers by offering a complete package of activities to maximize farm returns to farmers. The scheme (RADD) focuses on Integrated Farming System (IFS) for enhancing productivity and miximizing risks associated with climate variability. This scheme has been piloted to 10 states in the year 2011-12 with an outlay of Rs. 2500 crore.

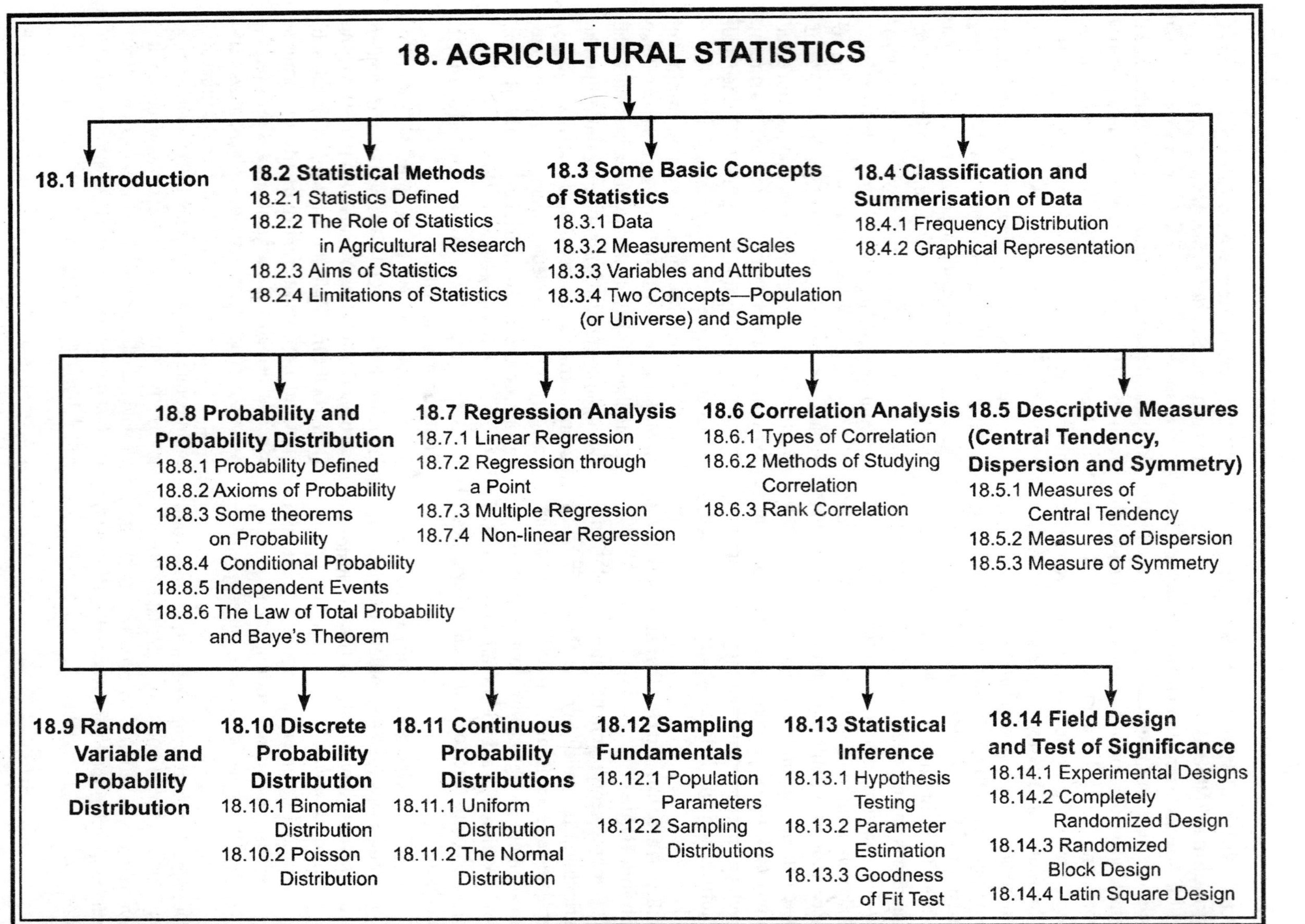
18. AGRICULTURAL STATISTICS
18.1 Introduction
18.2 Statistical Methods
18.2.1 Statistics Defined
18.2.2 The Role of Statistics in Agricultural Research
18.2.3 Aims of Statistics
18.2.4 Limitations of Statistics
18.3 Some Basic Concepts of Statistics
18.3.1 Data
18.3.2 Measurement Scales
18.3.3 Variables and Attributes
18.3.4 Two Concepts—Population (or Universe) and Sample
18.4 Classification and Summerisation of Data
18.4.1 Frequency Distribution
18.4.2 Graphical Representation
18.5 Descriptive Measures (Central Tendency, Dispersion and Symmetry)
18.5.1 Measures of Central Tendency
18.5.2 Measures of Dispersion
18.5.3 Measure of Symmetry
18.6 Correlation Analysis
18.6.1 Types of Correlation
18.6.2 Methods of Studying Correlation
18.6.3 Rank Correlation
18.7 Regression Analysis
18.7.1 Linear Regression
18.7.2 Regression through a Point
18.7.3 Multiple Regression
18.7.4 Non-linear Regression
18.8 Probability and Probability Distribution
18.8.1 Probability Defined
18.8.2 Axioms of Probability
18.8.3 Some theorems on Probability
18.8.4 Conditional Probability
18.8.5 Independent Events
18.8.6 The Law of Total Probability and Baye's Theorem
18.9 Random Variable and Probability Distribution
18.10 Discrete Probability Distribution
18.10.1 Binomial Distribution
18.10.2 Poisson Distribution
18.11 Continuous Probability Distributions
18.11.1 Uniform Distribution
18.11.2 The Normal Distribution
18.12 Sampling Fundamentals
18.12.1 Population Parameters
18.12.2 Sampling Distributions
18.13 Statistical Inference
18.13.1 Hypothesis Testing
18.13.2 Parameter Estimation
18.13.3 Goodness of Fit Test
18.14 Field Design and Test of Significance
18.14.1 Experimental Designs
18.14.2 Completely Randomized Design
18.14.3 Randomized Block Design
18.14.4 Latin Square Design

18. Agricultural Statistics

18.1 INTRODUCTION

The subject of statistics as it seems, is not a new discipline but it is as old as the human society itself. Its origin can be traced to old days when it was regarded as the 'Science of State-craft' and was the by-product of the administrative activity of the state. The term '*statistics*' seems to have been derived from the Latin word '*status*' or Italian word '*statista*' or the German word '*statistik*' each of which mean a political state. The science of state is said to have two main sources.

1. Government Records and
2. Mathematics.

The government records is the earliest foundation because all cultures with a recorded history had recorded statistics and the recording, as far as is known was done by agents of the government for government purposes.

Statistics is said to be branch of applied mathematics. The present body of statistical methods particularly those concerned with inferences about population from a sample, is based on mathematical theory of *probability* which marked a major step in intellectual history of the world.

Everything dealing with the collection of data, processing of data, analysis of data, and interpretation of numerical data belongs to the domain of statistics. In agriculture, this includes such diversified tasks as calculating the average length of the particular type of plant and the numbers involved in the field of agriculture evaluating the effectiveness of fertilizers, insecti-sides, etc.

The statistical principles are now applied to a wide variety of problems pertaining to biological, physical and social sciences as well as administration, business and industry. When the theory of statistics is applied to a specific scientific field, it is named after that science viz. Agricultural Statistics, Biometrics, economic statistics, engineering statistics. Though specific techniques of statistics may be emphasised in particular science, the basic statistical principles are same.

18.2 STATISTICAL METHODS

The word 'Statistics' came into existence during the middle of 18th century and now being used in the singular as well as plural sense. In the singular, it is the science which investigates the statistical methods and deals with number of applications. Statistics in the prural sense, are numerical statement of facts collected systematically with definite objective in view in any field of enquiry like other sciences.

Statistics may be classified in two categories (*i*) *Pure statistics* (*ii*) *Applied statistics*. Recently, statistics have a number of developments dealing with research related to pure and applied fields. In pure statistics, there is a mathematical derivation related to the formula, refined techniques. In applied statistics, there are a number of developments related to planning, economics, business, industrial sector, biology, water resources, medical science and agriculture etc. Now a days, it is very much essential to use statistics in every discipline without which the conclusions drawn will not be proved at global level.

18.2.1 Statistics Defined

There have been many definitions of term 'statistics' —indeed scholarly articles have carefully collected together hundreds of definitions. Some have defined statistics as *statistical data* (*plural sense*) whereas others as statistical method (singular sense). In general we can defined statistics as.

"Statistics is the study of collection, classification and analysis of numerical data for drawing valid conclusions and making reasonable decisions".

Some of the definitions of statistics in plural and singular sense given by some well known authors are as follows.

(i) Statistics as Numerical Data (i.e. plural sense).

* *"Statistics may be defined as the aggregate of facts affected to a marked extent by multiplicity of causes, numerically expressed, enumerated or estimated according to reasonable standard of accuracy, collected in a systematic manner, for a predetermined purpose and placed in relation to each other".* ***—Prof. Horac Secrist***

* *"Statistics are the classified facts representing the conditions of people in a state specially those facts which can be stated in number or in tables of numbers or in any tabular or classified arrangement"* ***–Webster***

* *"Statistics are numerical statement of facts in any department of equity placed in relation to each other".* ***–Bowley***

(ii) Statistics as Statistical Methods (i.e. singular sense).

* *"Statistics is the branch of scientific method which deals with the data obtained by counting or measuring the properties of natural phenomena".* ***–Kendall***

* *"Statistics refers to the body of technique or methodology which has been developed for the collection, presentation and analysis of quantitative data and for the use of such data in decision making".* ***–Nelter & Wasperman***

* *"Modern statistics refers to a body of methods and principles that have been developed to handle the collection, description, summarization and analysis of numerical data".* ***–Licon L. Chao***

Thus from various definitions we can say that

"Statistics are the numerical statements of facts capable of being analyzed and interpreted and the science of statistics is the study of principles and methods used in the collection analysis and interpretation of numerical data in any sphere of enquiry".

18.2.2 The Role of Statistics in Agricultural Research

Statistical methods are now used practically in all branches of agriculture. In fact, it was only due to need of statistical tools to handle problem in agricultural research, that was a tremendous development in statistical theory.

1. Statistical methods are used for efficient planning of experiments and for interpreting experimental data.
2. In agricultural research, efforts are made to know the hidden regularities of the some aspects of soil, crops and other related biological phenomena. It may be yield potentials of crop, effects of pest incidence on crop yields, effect of changes in climate on crops and pests, effect of fertilizer application on crop yields, effect of cross fertilization of different crop varieties etc.
3. Modern statistical methods and statistical data are being found increasingly useful in research in different field experiences about

crop yields with different types of fertilizers and different types of soils are very often designed and analyzed according to statistical methods.

4. Statistical methods are used for testing efficiency of chemicals and methods of treatment.
5. Agricultural economists use forecasting procedures in order to determine the future demand and supply of food. They are regression analysis in empirical estimation functional relationships between quantitative variable.

18.2.3 Aims of Statistics

There are three aims of statistics

1. To study the population.
2. To study the variation and its causes.
3. To study the methods of reduction of data.

18.2.4 Limitations of Statistics

The important limitations of statistics are as given below :

1. It does not study qualitative phenomena.
2. Statistical studies are true only on an average.
3. It does not study individuals.
4. It does not reveal the entire story.
5. It is liable to be misused.

18.3 SOME BASIC CONCEPTS OF STATISTICS

18.3.1 Data

Data means facts or information. In general, data can be defined as information collected for analysis and interpretations for a predetermined purpose.

Types of Data: Based on nature of characteristic observed, statistical data are classified as *attribute data* and *measurement data.* These data may be called as *primary data* or *secondary data* depending on the source of data.

"*Data originally collected in the process of investigations are known as primary data*". Primary data consists of figures collected at first hand in order to satisfy the purpose of a particular statistical enquiry. Primary data is generally collected by an authority conducting enquiry. The same authority makes analysis, interpretations and publishes the data.

"*Secondary data consists of figures which are already in existence, and which have been collected for some other purpose than the answering of question in hand*".

The secondary data are less expensive but it may not give all necessary information.

The observed data given as such is known as *row data.* When the observed data are grouped in groups or classes, they are known as *grouped data.*

18.3.2 Measurement Scales

In the process of data collection we make observation on objects of interest with respect to specified characteristics. These observations are made using *measurement scales*. Measurement scales having a constant interval size like metre and kilogram are called *equal interval* scales. Equal interval scales include area, volume, rate, length of time, degree of temperature and number of objects. We cannot use such scales to measure soil type, crop variables, colour of flowers, shape of seeds etc. They can only be as belonging to one group or the other. Thus when they give only a name to each observations, it is called *nominal scale*. When the categories are ordered or ranked like tall, short and medium plants the scale is referred as *ordinal scale*. Combining both form *non interval scale*

18.3.3 Variables and Attributes

An important distinction in technical language of statistics is that between *variables* and *attributes*. When a record is made of an actual measured quality characteristics, such as a dimension expressed in hundredths of millimeter, the quality is said to be expressed by *variables*. When a record shows only number of articles conforming and the failing to confirm to any specified requirements, it is said to be record by attributes. *Variable data* are measured on a continuous scale with individual measurements rounded to some designed number of decimal points. *Attribute data* involve counts of articles or counts of events. Thus a quantitative or qualitative characteristic that varies from observation to observation in the same group is called *variable*. In case of quantitative variables; observations are made using interval scales where as in case of qualitative variables nominal scales are used. *Conventionally, the quantitative variables are termed as variables and the qualitative variables are termed as attributes.* Thus, yield of a crop, available nitrogen in soil, daily temperature, number of leaves per plant, and number of eggs laid by insects are all variables. The crop varieties, soil types, shape of seeds, season and sex of insects are attributes.

The variables itself can be classified as *discrete* and *continuous* variable.

A variable which can assume only a countable number of real values is called *discrete variable*. Thus the variables not taking fractional values and only integral values are discrete.

The variables which can take all possible values (integral as well as fractional) between certain limits are said to be *continuous variable*. For example, The number of grains per panicle of paddy can be counted in full members only like 90, 92, 96 etc. Thus number of grains per panicle is a discrete variable. Similarly in the range of 8 kg to 10 kg yield of a crop the yield may be 8.004, 8.296, 9.756 etc. Hence yield is a continuous random variable.

18.3.4 Two Basic Concepts—Population (or Universe) and Sample

In a statistical investigation the interest usually lies in the assessment of the general magnitude and the study of variation with respect to one or more characteristics relating to individual belonging to a group. This groups of individual under study is called *population* or *universe*.

To emphasize population as the entire collection of units, we term the *population of units*. The unit is a single entity, usually an object or person, whose characteristics are of interest. The complete collection of units about which information is sought is called *population of units*.

A statistical population is the set of measurements (or records of same quality trait) corresponding to each units about which information is sought.

Instead of examining the entire group (*i.e.* population) which may be difficult or impossible to do, one may arrive at the idea of examining only a small part of this population called a sample.

A sample from a statistical population is the subset of measurements that are actually collected in course of an investigation.

The sample needs both to be representative of the population and to be large enough to contain information to answer the questions about the population that are crucial to the investigation.

For any population there is the value, for each unit, of a characteristics or variable of interest. For a given variable or characteristic of interest, we call the collection of values, evaluated for every unit in the population the *statistical population* or just the *population*. Here, we refer to the collection of units as the *population of units* when there is a need to differentiate it from the collection of values.

Generally, any statistical approach to learning about the population begins by taking a sample.

18.4 CLASSIFICATION AND SUMMERISATION OF DATA

Statistical data need to be collected to provide the vital information necessary to solve problems, once gathered, these data must be described and analyzed to produce summery information. After collection and editing of data, the first step toward further processing the same is classification. *Classification is the grouping of related facts into classes.*

Once the data have been collected we have a *mass or raw data*. Our mind cannot readily group the overall content of such a *mass data*. Hence we have to condense and summaries tables and graphs. Graphical presentation can often be most effective way to communication and information. Mass data may further be *condesed* in terms of simple numerical values like the *averages* and the measures of *dispersion*. All these methods are commonly known as *descriptive statistics*.

18.4.1 Frequency Distribution

The word 'frequency' is derived from how frequently a variable occurs, *i.e.* how many times a particular value of the variable is repeated. A tabular arrangement of the data which gives the frequency corresponding to each successive value of the variable is called *frequency distribution*, *i.e.* a frequency distribution refers to data classified on the basis of some variable that can be measured.

Depending on nature of variable the frequency is known as

(*i*) *Discrete frequency distribution* when variable involved is discrete and

(*ii*) *Continuous random variable* when variable involved is continuous.

For example: The number of rainy day in the months of year will represents the *discrete frequency distribution* while the quantity of precipitation (or rain) in some range (*viz a* cm to *b* cm) and number of days in the year represent the *continuous frequency distribution*.

The other classification of frequency distribution may be as

(*i*) *Simple or ungrouped frequency distribution*: In this frequency distribution, the data are arranged in ascending order of magnitude with their corresponding tally marks and frequency.

(*ii*) *Grouped frequency distribution*: In this distribution data is condensed by putting them into smaller groups or classes called *class intervals*.

(*iii*) *Relative Frequency distribution*:

(*iv*) *Cumulative frequency distribution*

(*v*) *Cumulative relative frequency distribution*: *less than cumulative relative frequency distribution*, and '*more than cumulative relative frequency distribution*.

18.4.1.1 Formation of discrete frequency distribution

The process of preparing of this type of distribution is very simple. We have to just count the number of times a particular value is repeated is called *frequency of that* class. In order to facilitate counting prepare a column of 'tallies'. In another column place all possible values of variables from lowest to highest. A frequency table is simply a table in which that data are grouped into classes and number of cases which fall in each class are recorded. The numbers in each class is referred as "frequencies" hence the term "frequency distribution,.

Example 18.1 : *In a survey of 35 families in a village, the number of mango tree per family was recorded and the following data obtained.*

1	0	2	3	4	5	6
7	2	3	4	0	2	5
8	4	5	12	6	3	2
7	6	5	3	3	7	8
9	7	9	4	5	4	4

Represent the data in discrete frequency distribution.

Solution:

Table 18.1

No. of mango tree	*Tallies*	*Frequency*
0	11	2
1	1	1
2	1111	4
3	~~1111~~	5
4	~~1111~~1	6
5	~~1111~~	5
6	111	3
7	1111	4
8	11	2
9	11	2
12	1	1
		Total 35

18.4.1.2 Formation of continuous frequency distribution

This type of classification is most popular in practice. The following technical terms are important when a continuous frequency distribution is formed or data are classified according to class intervals.

1. Class limit : The class limits are lowest and highest values that can be included in the class, For example, take the class 20-40. The lowest value or *lower limit* is 20 and the highest value (*upper limit*) is 40. The *lower limit* of class is the value below which there can be no item in the class. The upper limit of a class is the value above which no item can belong to the class.

2. Class interval : The difference between the upper and lower limits of the class is known as class interval. For example in the class 20-40, the class interval is 20 (*i.e.* 40–20). A sample formula to obtain the estimate of appropriate class interval *i.e.* *i* is

$$i = \frac{L - S}{k}$$

where L = largest term ; S = smallest term and k = the number of classes.

For deciding the number of classes *Sturges' Rule* can be opted. According to him number of classes can be determined by the formula

$$k = 1 + 3.222 \log N$$

where N = total number of observations.

It should be noted that since logarithm is used in the formula, the number of class shall generally be between 4 and 20 — it cannot be less than 4 even if N is less than 10 and if N is 10 lakh k will be $1 + (3.222 \times 6) = 20.9$.

Sturges suggested the following formula for determining the magnitude of class interval.

$$i = \frac{\text{Range}}{1 + 3.22 \log N}$$

Range is the difference of largest and smallest item.

3. Class frequency: The number of observation corresponding to particular class is known as class frequency.

4. Class Midpoint or Class Mark:

It is the value lying half way between the lower and upper limits of a class-interval.

Midpoint of a class

$$= \frac{\text{Upper limit} + \text{Lower limit}}{2}$$

Example 18.2 : *For the data given below, prepare a continuous frequency distribution table.*

64	*70*	*25*	*55*	*36*	*51*	*60*	*40*	*65*	*58*
33	*38*	*45*	*62*	*60*	*48*	*47*	*50*	*63*	*54*
76	*65*	*40*	*40*	*41*	*82*	*52*	*55*	*35*	*64*
30	*58*	*33*	*61*	*15*	*64*	*48*	*42*	*26*	*50*
20.	*55*	*42*	*53*	*50*	*48*	*46*	*45*	*18*	*9*

Solution : Here $N = 50$

$\therefore \quad k = 1 + 3.222 \times \log 100$

$= 1 + 3.222 \times 6\ 1.6989$

$= 1 + 5.47 = 6.47$ say 7

Class size $i = \dfrac{82-9}{7} = \dfrac{73}{7} = 10.43$

So we take $i = 10$

Table-18.2: Formulation of frequency Distribution

Marks	*Tally Bars*	*Frequency*
0-10	1	1
10-20	11	2
20-30	111	3
30-40	~~1111~~ 1	6
40-50	~~1111~~ ~~1111~~ 111	13
50-60	~~1111~~ ~~1111~~ 11	12
60-70	~~1111~~ ~~1111~~	10
70-80	11	2
80-90	1	1
		Total 50

5. Exclusive and inclusive methods of classification:

There are two methods of classifying the data according to class-interval namely

(*i*) '*exclusive*' method and

(*ii*) '*inclusive*' method.

Exclusive method : When class intervals are so fixed that the upper limit of one class is the lower limit of next class is known as 'exclusive' method of classification.

The following data are classified on this basis.

Table-18.3: Exclusive method

Height of plant (cm)	*No. of plants*
5-10	2
10-20	5
20-30	7
30-40	6
40-50	5
	Total 25

Inclusive method : Under the 'inclusive' method of classification, the upper limits is included in the class itself. The following example illustrates this method.

Table-18.4: Inclusive method

Height of plant (cm)	*No. of plants*
2-4	3
5-7	9
8-10	2
11-13	3
13-15	3
	Total 20

6. Cumulative frequency distribution tables

Sometimes it is useful to have cumulative frequency distribution. The *cumulative frequency* of a class is the total frequency up to and including the class. The table of cumulative frequency is called *comulative distribution table*. Again there are two types of cumulative frequency distribution. The cumulative frequency distribution of all values greater than or equal to the lower limit of each class is called *more than cumulative frequency distribution*. The cumulative frequency distribution of all values less than or equal to upper limit of each class is called *less than cumulative frequency distribution*. Whenever we refer to cumulative frequency distributions without qualification the "less than" type is implies.

The results of example 18.2 are presented as cumulative distributions in tables 18.5 and 18.6

Table 18.5: Less than cumulative frequency distribution for the data in table 18.2

Less than (<) class	***Cumulative frequency***
<10	1
<20	1+2=3
<30	3+3 =6
<40	6+6=12
<50	12+13=25
<60	25+12=37
<70	37+10=47
<80	47+2 =49
<90	49+1=50

Table 18.6: More than cumulative frequency distribution for data in table 18.2

Less than (<) class	***Cumulative frequency***
> 0	50
> 10	50 – 1 = 49
> 20	49 – 2 = 47
> 30	47 – 3 = 44
> 40	44 – 6 = 38
> 50	38 – 13 = 25
> 60	25 – 12 = 13
> 70	13 – 10 = 3
> 80	3 – 2 = 1

18.4.2 Graphical Representation

A data can be presented as bar chart or Pareto diagrams or dot diagram etc.

The most common form of graphical presentation of a frequency distribution is the *histogram*. Others two are *frequency polygon* and *cumulative frequency curve–ogive*.

18.4.2.1 Histogram

A histogram is a set of vertical bars whose areas are proportional to the frequencies represented. While constructing histogram the variable is always taken on *X*-axis and the frequencies on *Y*-axis. Each classes are then represented by a distance on that scale that is proportional to its class interval. The distance for each rectangle on *X*-axis shall remain the same in case of class-intervals are uniform throughout (Fig. 18.1). If they are different then vary.

The selection of class interval can effect the appearance of frequency histogram. If the class interval are very large, the table is compact but lacks detail. If the intervals are too small, the table may be bulky and not succinct enough. For the choice of class interval the following guidelines may be considered.

(*a*) *Brooks and Carruthers rough guides*: No. of Classes ≤ 5 log (sample size).

(*b*) *Charlier's rule of thumb*:

w = (maximum value – minimum value)/20, where w is the size of class interval. The number of classes is generally 15 to 25.

(*c*) According to *Struges*, the class interval can be estimated by $k = 1 + 3{\cdot}22 \log (N)$ where k is the number of classes and N is the number of observations.

18.4.2.2 Frequency polygon

A frequency polygon is a graph of frequency distribution. It has more than four sides. It is particularly more effective in comparing two or more frequency distribution.

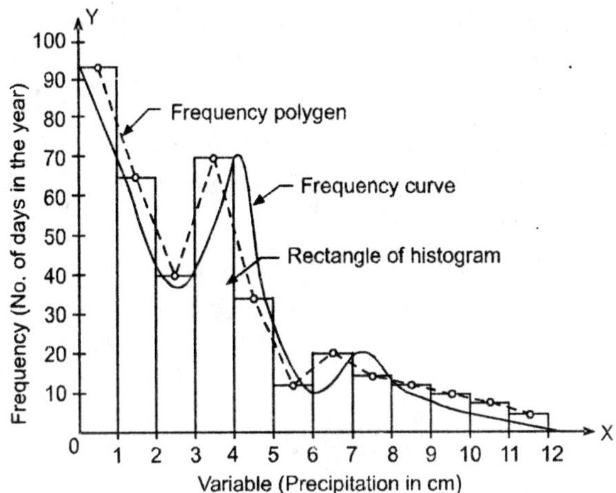

Fig. 18.1: Histogram, frequency polygon and frequency curve

For an ungrouped distribution, the frequency polygon is obtained by plotting points with abscissa as the variate values and the ordinate as the corresponding frequencies and joining the plotted points by means of straight lines. For a grouped frequency distribution, the abscissa of points are mid values of the class intervals. For equal class intervals the frequency polygon can be obtained by joining the middle points of the upper sides of the adjacent rectangles of the histogram by means of straight lines. If the class intervals are of small width, the polygon can be obtained by drawing a smooth freehand curve through the vertices of frequency polygon.

The frequency polygon so obtained should be extended to the base line (X-axis) at both the ends so that it meets the X-axis at the mid point of two hypothetical classes, *viz* the class before the first class and classes after last class each assumed to have zero frequency. (Fig. 18.1).

1. Simple Bar Diagram : If the classification is based on attributes and if attributes are to be compared with respect to single character we use simple bar diagram. For example, the area under different crops in a state, the food grain production in different years etc. can be represented by simple bar diagram. Simple bar diagram consist of vertical bars of equal width. The heights of these bars are proportional to the volume or magnitude of the attribute. All bars stand on same base line. The bars are separated from each other by equal intervals. The bars may be coloured or marked.

Example 18.3 : *The cropping pattern in Uttar Pradesh in the year 2012-13 was as follows. Draw the simple bar diagram for the given data.*

Crops	*Area (in 1000, hectare)*
Cereals	*5540*
Oil seeds	*2050*
Pulses	*760*
Cotton	*540*
Others	*1035*
	Total 9925

Solution : The simple bar diagram for the given data is shown in Fig. 18.2.

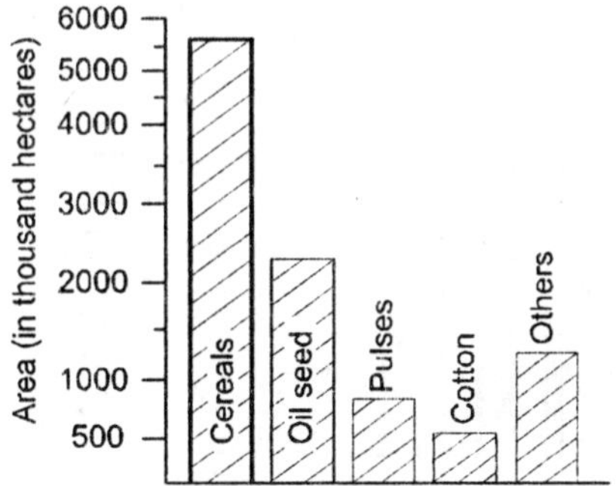

Fig. 18.2 Simple bar diagram showing the cropping pattern in U.P. in the year 2012-13.

2. Multiple Bar Diagram: If the data is classified by attributes and if two or more characters or groups are to be compared with in each attribute, we use multiple bar diagram. The multiple bar diagram is simply extension of simple bar diagram. For each attribute two or more bars representing separate characters or groups are to be placed side by side.

Each bar within an attribute will be marked or coloured differently in order to distinguish between them. Same type of marking or colouring should be done under each attribute, a foot note has to be given explaining the markings or colouring.

Example 18.4: *The cropping pattern in U.P, Bihar and West Bengal in the year 2012-13 was as follows. Present the data on multiple bar diagram.*

Crops	*Area (in thousand hectare)*		
	U.P	*Bihar*	*West Bengal*
Cereals	5040	3550	3400
Oil seeds	2050	1625	1950
Pulses	760	445	640
Cotton	540	215	260
Others	1035	540	615

Solution:

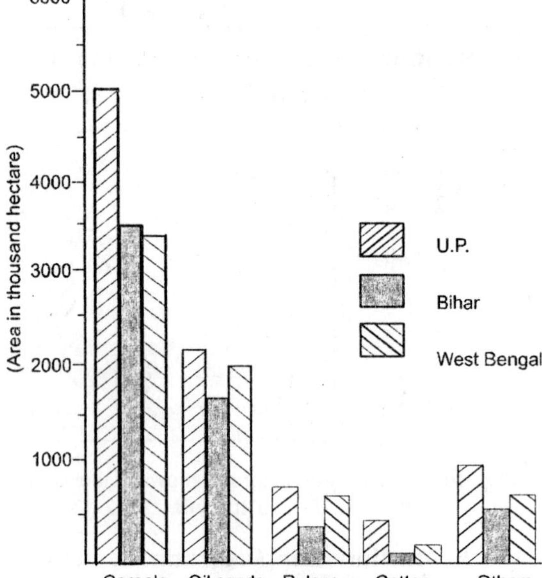

Fig. 18.3: Multiple bar diagram for cropping pattern in U.P., Bihar and W.B.

3. Component Bar Diagram: Sub-divided or component bar diagrams are useful for presenting several items of a variable graphically and enable us to make comparative study of different parts among themselves and study relationship between each part and a whole. The various components in each bar should be kept in the same order. To distinguish between different components a separate colour or shade is used for each component.

Example 18.5: *The distribution of bacterial population (10^5/g-soil) in the rhizosphere of certain ornamental plants in different physiological stages is given as follows*

Rhizoshpere	*Physiological stage*			
	Vegetative	*Flowering Stage*	*Seed setting stage*	*Total stage*
Clock's Comb	150	135	35	320
Saliva	140	65	55	260
Aster	60	80	38	178
Zinia	100	40	12	152
Total	**450**	**320**	**140**	

Present the data on component bar diagram

Solution:

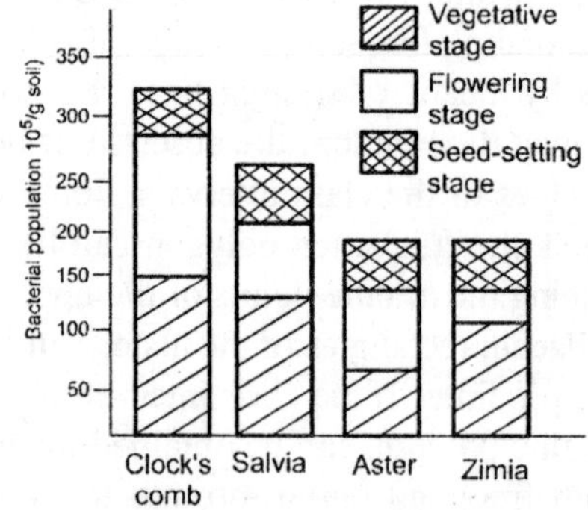

Fig. 18.4: Components bar diagram showing the distribution of bacterial population in the rhizoshpere of certain ornamental plants in different physiological stages.

4. Percentage Bar Diagram: Sub divided or component bar diagram presented graphically on percentage basis give *percentage bar diagram*. The total for each bar is taken as 100 and the value of each component or part is expressed as percentage of respective totals. In a percentage bar diagram, all the bars are of equal height and the various segments of the bar representing the different component will vary in height depending on their percentage value of the to the total.

Example: 18.6: *We reconsider the above problem (Example 18.5) to draw percentage bar diagram.*

Solution: Let us find the percentage for different stages as shown in Table 18.7.

Table 18.7

	Vegetative Stage	*Flowering stage*	*seed-setting stage*	*Total*
Cock's comb	$\frac{150}{320} \times 100 = 46.875$	42.1875	10.933	100
Saliva	$\frac{140}{260} \times 100 = 53.846$	25.0	21.154	100
Aster	$\frac{60}{178} \times 100 = 33.7$	44.94	21.342	100
Zinnia	$\frac{100}{152} \times 100 = 65.78$	26.312	7.89	100

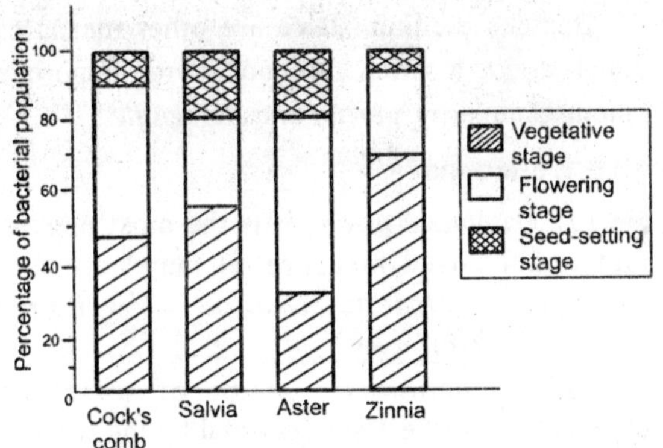

Fig. 18.5: Percentage bar diagram showing the distribution of bactorial population in the rhizoshpere of certain ornamental plants in different physiological stages.

5. Pie Diagram or Annular Diagram or Sub divided circular Diagram : A pie diagram is a circle subdivided into component sectors with areas proportional to the size of corresponding components. Pie diagrams are constructed on the percentage basis and not on absolute basis because a series of pie diagrams showing absolute figures would require that larger totals be represented by larger circiles which will involve many difficulties.

Steps in the construction of Pie diagram.

1. Construct each component as percentage of total.
2. Multiply this percentage by $3.6\left(\frac{360}{100}\right)$ to convert into degrees.
3. Starting with 12 O'clock position on the circile (clock wise) draw the largest component sector.
4. Draw other components in clock wise succession in descending order of magnitude except that for catch-all components like "All others" and miscellaneous which are shown last.
5. Use different shades or colours to distinguish between different colours.
6. Explain briefly the different components either with in the components in the figure or outside by arrows.

Example 18.7: *Consider the data in Example 18.3 and draw a pie diagram.*

Solution:

Table 18.8

Crops	*Area*	*Percentage area* $= \frac{A_i}{A} \times 100$	*Corresponding sector angle (in degree)* Q_i *= Percentage × 3.6*
Cereals	5540	55.8186	200.9469
Oil seed	2050	20.6549	74.3576
Pulses	760	7.65.74	27.5666
Cotton	540	5.4408	19.5868
Others	1035	10.4282	37.5415

The corresponding pie diagram's shown in Fig. 18.6.

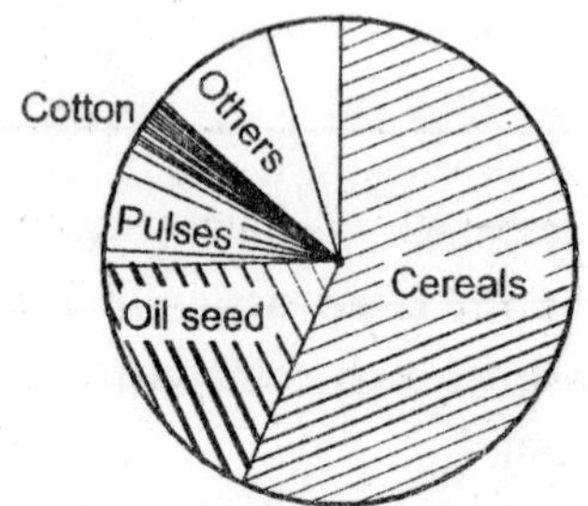

Fig. 18.6: Pie diagram showing the cropping pattern in U.P. in the year 2012-13.

18.5 DESCRIPTIVE MEASURES (CENTRAL TENDENCY, DISPERSION AND SYMMETRY)

18.5.1 Measures of Central Tendency

One of the most important objective of statistical analysis is to get one single value that describes the characteristic of entire mass of unwieldy data. Such a value is called an *central value or an average*. According to Croxton & Cowden. *"An average value is a single value within the range of the data that is used to represent all of the values in series"*.

Since an average is some where within the range of the data it is some times called *measure of central value*. The important measures are following:

Following are the different types of averages which are in common are :

1. ***Mean***
   (*a*) Arithmetic Mean (*b*) Geometric Mean
   (*c*) Harmonic Mean (*d*) Weighted Mean
2. ***Median***
3. ***Mode***

Besides median, there are other measures which divide a series into equal parts. Important among them are *Quartile* and *Percentile*.

18.5.1.1. Means

(*a*) *Arithmetic mean:* It is the most popular and widely used measure. It may be either (*i*) *simple arithmetic mean* or (*ii*) *weighted arithmetic mean.*

Simple arithmetic mean of a set of observations is there sum divided by the number of observations.

$$\therefore \quad \bar{x} = \frac{1}{n}\sum_{i=1}^{n} x_i \; (i = 1, 2... n)$$

Arithmetic mean of discrete series

(*i*) *Direct method*

$$\bar{x} = \frac{1}{N}\sum_{i=1}^{n} f_i x_i , \; N = \Sigma f_i \qquad (i = 1, 2, 3...n)$$

(*ii*) *Short cut method*

If data contain big figures we may use *short cut method*

$$\bar{x} = A + \frac{\Sigma f_i d_i}{\Sigma f_i}$$

where A is assumed mean (generally mid term in x_i) and $d = x_i - A$, f_i is respective frequency.

Arithmetic mean of continuous series

We find the midpoint of class mark (x_i) and than can use any of the following method.

(*i*) *Direct method*

$$\overline{x} = \sum_{i=1}^{n} f_i x_i$$

(*ii*) *Short cut method*

$$\overline{x} = A + \frac{\Sigma f_i d_i}{\Sigma f_i}$$

where $d_i = x_i - A$, A being assumed mean.

(*iii*) *Step deviation method*

$$\overline{x} = A + \frac{\Sigma f_i d_i'}{\Sigma f_i} \times i$$

where, A is assumed mean,

$$d_i = \frac{x_i - A}{i}$$ where i is the class size.

In *weighted arithmetic mean* relative importance of the different items are different. The weighted arithmetic mean is calculated as

$$\overline{x}_w = \sum_{i=1}^{n} \frac{w_i x_i}{\sum_{i=1}^{n} w}$$

where $\overline{x}_w$ represents the weighted arithmetic mean x_i represent the variable values and w_i represents the weights attached to variable values.

Merits of Arithmetic Mean

1. It is rigidly defined.
2. It based on all the observations.
3. It can be easily calculated.
4. Its algebraic treatment is specially easy and definitely possible.

Demerits

1. It is very much affected by the values at extremes and hence, often abnormal results are obtained due to some abnormal value.
2. If the items of the beginning and end of the series are not known, arithmetic mean can not be calculated.

Uses of A.M.

1. It is calculated to compare two or more series with respect to certain character.
2. It is used in calculating the standard deviation of the data.
3. It is calculated, for calculating the regression coefficient and the correlation coefficient.

(*b*) *Geometric mean:* Geometric mean of a set of n observations is the nth root of their product.

$$G = (x_1, x_2, x_3 \ldots\ldots x_n)^{1/n}$$

or $$G = \text{Antilog}\left(\frac{1}{N}\sum_{i=1}^{n} \log x_i\right)$$

It is generally used in rate of population growth and rate of interest and in construction of index number.

(*c*) *Harmonic mean:* Harmonic mean of a number of observations, none of which is zero, is the reciprocal of the arithmetic mean of the reciprocals of given values.

$$H = \frac{1}{\frac{1}{N}\sum_{i=1}\left(\frac{1}{x_i}\right)}$$

Example 18.8 : Find out the arithmetic mean of the following fifteen numerical quantities.

62, 85, 103, 107, 109, 110, 114, 115, 126, 151, 154, 161, 165, 170, 175.

Here $$\text{A.M} = \frac{\Sigma x}{n}$$

$$= \frac{62 + 85 + \ldots\ldots\ldots\ldots + 175}{15}$$

$$= \frac{1907}{15} = 127.133$$

Example 18.9: *Arithmetic mean for discrete frequency distribution (using direct method)*

Table 18.9

Daily wages (Rupees) (x_i)	*Frequency* (f_i)	$f_x x_i$	*Calculations*
2	2	4	Here $N = 100$
4	5	20	and $\Sigma f_i x_i = 1212$
6	8	48	Arithmetic Mean
8	10	80	$= \frac{\Sigma f_i x_i}{N}$
10	12	120	$= \frac{1212}{100}$
12	20	240	$= 12.12$
14	13	182	
16	15	240	
18	11	198	
20	4	80	
	$N = 100$	$\Sigma fx = 1212$	

Example 18.10 : *Calculation of arithmetic mean (using short cut method)*

Table 18.10

Class interval	*Frequency* (f_i)	*Mid value* (x_i)	$d_i = x_i - A = x_i - 25$	$f_i d_i$	*Calculations*
0-10	2	5	–20	–40	Here, $A = 25$
10-20	4	15	–10	–40	Arithmetic Mean
20-30	6	25	0	0	$= A + \frac{\Sigma f_i d_i}{N}$
30-40	8	35	10	80	$= 25 + \frac{200}{30}$
40-50	10	45	20	200	$= 25 + 6.666$
					$= 31.67$
	$N = 30$			$\Sigma f_i d_i = 200$	

18.5.1.2. Median

Median of a distribution is the value of the variable which divides it into two equal parts. It is the value which exceeds and is exceeded by the same number of observations *i.e.* it is the value such that the number of observations above it is equal to the number of observations below it. The median is thus a positional average.

Method of Computation —

(A) Computation from the raw data.

Median: When all the observations are arranged in ascending or descending order of magnitude, the middle one is known as the *median*.

If N is the total number of observations, the value of the $\left(\frac{n+1}{2}\right)^{th}$ item will be called *median* $\left(\frac{n+1}{2}\right)$, which stands for the serial order of the median is known as *median number*.

Example 18.11 : (When the number of item is odd).

Find the median of the following figures :

10, 11, 13, 11, 13, 10, 13, 11, 12, 15, 12

Here, median number $= \frac{11+1}{2} = 6$

Arranging the figures in the ascending order of magnitude, we get

10, 10, 11, 11, 11, 12, 12, 12, 13, 13, 15

Here 6th figure is 12.

Therefore, the median is 12.

Example 18.12: (When the number of item is even).

Find the median of the following figures :

10, 7, 11, 9, 8, 10, 11, 7, 9, 12

$$\text{Here, median number} = \frac{10+1}{2} = 5\frac{1}{2}$$

But, serial order of a term can never be fractional. In such a case there are two middle terms 5th and 6th. The mean of these two terms will be median.

Arranging the terms, we get

7, 7, 8, 9, 9, 10, 11, 11, 12

Here, 5th term = 9

and 6th term = 10

$$\therefore \quad \text{Median} = \frac{9+10}{2} = 9{\cdot}5$$

(B) Median for grouped data.

We first prepare the cumulative frequency table. Let the total number of observations be N. We determine the median class *i.e.*, the class in which $(N/2)^{th}$ observation lies, then we use the formula

$$\text{Median} = l + \left(\frac{N - c.f}{f}\right) \times i$$

where,

l — is the lower limit of the median class,

N — is the total frequency,

i — is the range (upper limit - lower limit) of the median class; or class interval.

$c.f$— is the cumulative frequency of the class just before the median class and f_1 is the frequency of the median class.

f – Frequency of medium class.

Merits of Median

1. It is easy to calculate.
2. It is not affected by the extreme values.
3. It can be located graphically.
4. It is best suited for open-end classes.

Demerits of Median

1. It can be located precisely, when the items are grouped.
2. It is not based on all the observations in the given data.
3. It may not be a true representative of a given data, when the items vary greatly in magnitude.
4. For finding out the median of a data, it is necessary to make an array in ascending or descending order. So when the number of item is quite large, this operation becomes tedious.

Example 18.13: *For the frequency distribution of weights of 190 Sorghum ear-head given below calculate the median.*

Weight of ear-head (in g)	*No. of ear-heads*
40-60	*6*
60-80	*28*
80-100	*35*
100-120	*55*
120-140	*30*
140-160	*5*
160-180	*12*
180-200	*9*
	N = 100

Solution: Frequency distribution of weights of 190 Sorghum ear-heads.

Table 18.11

Weight of ear-heads (ig g)	*No. of ear heads*	*Cumulative frequency*	
40 – 60	6	6	
60 – 80	28	34	
80 – 100	35	69 →	$\frac{N}{2} = \frac{190}{2} = 95$
100 – 120	55—f	124	*Median class*
120 – 140	30	154	
140 – 160	15	169	
160 – 180	12	181	
180 – 200	9	190	
Total	**190**		

$$\text{Median} = 100 + 100 + \frac{20}{55}(95 - 69)$$

$$= 100 + \frac{20}{55} \times 26 = 109.45\text{g}$$

18.5.1.3 Quartiles and other percentiles

In addition to the median, which divides a set of data into halves we can consider other division points. When a ordered data set is divided into quarters, the resulting division points are called *sample quartiles*. The First quartile, Q_1 is the value that has one fourth, or 25% of the observations below its value. The first quartile is also the sample 25th percentile $P_{0{\cdot}25}$. More generally, we define the sample 100 *pth* percentile as follows.

The sample 100 *pth* percentile is a value such that at least 100p% of observations are at or below this value and at least 100 (1 – *p*)% are at or above this value.

The following rule simplifies the calculation of sample percentiles.

Calculating the sample 100 pth percentile

1. Order the *n* observations from smallest to targets.
2. Determine the product *np*.

If *np* is not an integer, round it up to next integer and find corresponding ordered value.

If *np* is an integer, say *k*, calculate the mean of k*th* and (*k* + 1) *th* or desired observation.

In case of median, which is 50*th* percentile, this may not be uniquely define a percentile.

Sample Quartiles. When individual observations are ordered then

1. *First quartile,* Q_1 = 25 *th* percentile

$$Q_1 = \frac{(n+1)}{4} th \text{ item, when } n \text{ is odd}$$

$$= \frac{\frac{n}{4} th \text{ item} + \left(\frac{n}{4}+1\right) th \text{ item}}{2},$$

when *n* is even

2. *Second quartile or medium* Q_2 = 50 *th* percentile

$$Q_2 = \left(\frac{n+1}{2}\right) th \text{ item, when } n \text{ is odd}$$

$$= \frac{\left[\frac{n}{2} th \text{ item} + \left(\frac{n}{2}+1\right) th \text{ item}\right]}{2},$$

when *n* is even.

3. *Third quartile* $Q_3 = 75\ th$ percentile

$= \frac{3(n+1)}{4} th$ item, when n is odd

$= \frac{3n}{4} th$ item $+ 3\left(\frac{n}{4}+1\right) th$ item,

when n is even

4. *Interquartile range* : The amount of variation in the middle half of the data is described by the *Interquartile range = third quartile* (Q_3) – first quartile (Q_1).

Example 18.14. *From the given ordered data, find Q_1, Q_2, Q_3 $P_{0.95}$, range and interquartile range.*

6·2	7·7	8·3	9·0	9·4	9·8	10·5	10·7	11·7	11·2
11·8	12·3	12·8	13·2	13·3	13·5	13·9	14·4	14·5	14·7
15·2	15·5	15·8	15·9	16·2	16·7	16·9	17·0	17·3	17·5
17·6	17·9	18·0	18·0	18·1	18·1	18·4	18·5	18·7	19·0
19·1	19·2	19·3	19·4	19·4	20·0	20·1	20·1	20·4	20·5
20·8	20·9	21·4	21·6	21·9	22·3	22·5	22·7	22·7	22·9
23·0	23·5	23·7	23·9	24·1	24·3	24·6	24·6	24·8	25·7
25·9	*26·1*	*26·4*	*26·6*	*26·8*	*27·5*	*28·5*	*28·6*	*29·6*	*31·8*

Solution : We have $n = 80$

***First quartile Q_1* :** We have $np = 80 \times \frac{1}{4} = 20$, is even which is an integer, so we take the mean of 20th and 21st ordered observation.

$$Q_1 = \frac{14{\cdot}7 + 15{\cdot}2}{2} = 14{\cdot}95$$

***Second quartile (Q_2)* :** We have $np = 80 \times \frac{1}{2} = 40$ is even so the second quartile or median is the mean of 40th and 41st ordered observations.

$$Q_2 = \frac{19{\cdot}0 + 19{\cdot}1}{2} = 19{\cdot}05.$$

***Third quartile (Q_3)* :** We have np $= 8 \times \frac{3}{4} = 60$, is even so third quartile is the mean of 60th and 61st item.

$$Q_3 = \frac{22{\cdot}9 + 23{\cdot}0}{2} = 22{\cdot}95.$$

$P_{0{\cdot}95}$: We have $np = 0{\cdot}95 \times 80 = 76$ is even, so we take $P_{0{\cdot}95}$ as mean of 76th and 77th item.

$$P_{0{\cdot}95} = \frac{27{\cdot}5 + 28{\cdot}5}{2} = 28{\cdot}0$$

Range : We have, minimum = 6·2, and maximum = 31·8\

Range = maximum – minimum

= 31·8 – 6·2 = 25·6

Interquartile range : $Q_3 - Q_1$

= 22·5 – 14·95 = 8·0

18.5.1.4 Mode

This is the value of the variable which occurs most frequently or whose frequency is maximum. Also, if several samples are drawn from a population, the important value which appears repeatedly in all the samples is called the *mode.*

Mode for Individual Data

Example 18.15: The marks obtained in a test by 12 students in a class are given below :

37, 23, 16, 19, 34, 23, 5, 27, 36, 23, 20, 38

Find the modal mark.

Solution : Arranging the data in ascending order.

5, 16, 19, 20, 23, 23, 23, 27, 34, 36, 37, 38

Obviously, 23 occurs maximum number of times

$\therefore$ Mode = 23

Mode for Grouped Data

$$\text{Mode} = l + \frac{f_2}{f_1 + f_2} \times i$$

where,

l = lower limit of class preceding modal class

f_1 = Frequency class succeeding modal class

f_2 = Frequency of the next higher class

i = width of the class interval

Merits of the Mode

1. It is easily comprehensible.
2. It can be obtained simply by inspection. Even if it is to be computed, it can be computed very easily.
3. This is the value whose expectation is the greatest in whole of the statistical series.
4. Neither the extremes are needed in its computation nor it is affected by them.

Demerits

1. In various cases, there is no single and well defined mode.
2. Its computation is not based on all the observations.
3. It is unsuitable for any algebraic treatments.
4. When there are more than one mode in the series, it becomes difficult and takes much time to compute it.
5. In any set of observations, where, there are very small differences among the items of the same size, mode does not prove to be a good representative.

Use of mode

Its use is increasing everyday especially in the field of business.

Example 17.16 : *Find the mode for the data given in Example*

Solution:

Table 18.12: The frequency distribution of weight of 190 sorghum year heads.

Weight of ear-head (ing)	*No. of ear-heads*	
40-60	6	
60-80	28	
80-100	$35 - f_1$	
$l \rightarrow$ 100-120	55 –	Modal class
120-140	$30 - f_2$	
140-160	15	
160-180	12	
180-200	9	

$$i = 120 - 100 = 20$$

$$l = 100, f_1 = 35, f_2 = 30$$

$$\text{Mode} = 100 + \left(\frac{30}{35+30} \times 20\right) = 109.23\text{g}.$$

18.5.2 Measures of Dispersion

The measures that are constructed to indicate the spread of the data with respect to the average in a set of observations are called the measures of dispersion.

18.5.2.1 Range

Range is the simplest measure or dispersion. It is the difference between the highest and the lowest terms of a series of observations.

18.5.2.2 Mean deviation

If the deviations of all the observations from their mean are calculated, their algebraic sum will be zero. When this sum is always zero, it is impossible to get the average of these deviations.

To overcome this difficulty these deviations are added irrespective of plus or minus sign and then the average is calculated.

The deviations without any plus or minus sign are known as absolute deviations. The mean of these absolute deviations is called the mean deviation. The mean deviation may be calculated from mean or median.

For deviation from mean

$$M.D. = \frac{1}{\Sigma f_i} \cdot f_i \mid D_i \mid = \frac{1}{N} f_i \mid x_i - \bar{x} \mid$$

where, D_i = Deviation from the mean = $x_i - \overline{x}$

$|D_i|$ = Absolute deviation

N = Number of observations.

f_i = frequency.

Example 18.17 : *Calculate mean deviation from the following series.*

x	10	11	12	13	14
f	3	12	18	12	3

Solution:

Table 18.13: Calculation of mean deviation

x_i	f_i	$\lvert D_i \rvert$	$f_i \lvert D_i \rvert$
10	3	2	6
11	12	1	12
12	18	0	0
13	12	1	12
14	3	2	6
	N = 48		$\sum f_i \lvert D_i \rvert = 36$

Mean deviation from median

In this case we take deviation from median then it is called mean deviation from median.

$$\text{M.D. (median)} = \frac{1}{N} f_i \,|\, D_i \,|$$

where, $D_i = x_i - M$

Example 18.18 : *From a yield of large number of sesamum plant a variety of 100 plants were selected at random. The seed yields (g) per plant were recorded. The results are as follows:*

Seed yield (g)	2.5-3.5	3.5-4.5	4.5-5.5	5.5-6.5	6.5-7.5	7.5-8.5	8.5-9.5	9.5-10.5
No. of plants	4	6	10	26	24	15	10	5

Find the mean deviation (from mean) for the given data.

Solution :

Table 18.14: Frequency distribution of seed yield of 100 sesamum plants

Seed yield (x) (g)	*No. of plants* f_i	*Midclass* x_i	$f_i x_i$	$\lvert D_i \rvert = \lvert x_i - \overline{x} \rvert = \lvert x_i - 6.7 \rvert$	$f_i \lvert D_i \rvert$
2.5 – 3.5	4	3	12	3.7	14.8
3.5 – 4.5	6	4	24	2.7	16.2
4.5 – 5.5	10	5	50	1.7	17.0
5.5 – 6.5	26	6	156	0.7	18.2
6.5 – 7.5	24	7	168	0.3	7.2
7.5 – 8.5	15	8	120	1.3	19.5
8.5 – 9.5	10	9	90	2.3	23.0
9.5 – 10.5	5	10	50	3.3	16.5
Total	**100**		**670**		**132.4**

$$\bar{x} = \frac{\sum f_i x_i}{N} = \frac{670}{100} = 6.7$$

Mean deviation

$$\frac{\sum f_i |D_i|}{N} = \frac{\sum f_i |x_i - \bar{x}|}{N}$$

$$= \frac{132.4}{100} = 1.324 \text{ g}$$

18.5.2.3 Standard deviation

Its calculation is based on the deviations from the arithmetic mean.

In case of mean deviation the difficulty that the sum of the deviations from the arithmetic mean is always zero is solved by taking these deviations irrespective of plus or minus signs. But here, that difficulty is solved by squaring them and taking the square root of their average.

It is defined by the following formula

$$\text{S.D.} = \sqrt{\frac{\Sigma(x-\bar{x})^2}{N}}$$

where, x = An observation or variate value
m = Arithmetic mean of the population
N = Number of given observations.

For frequency distribution (Discrete series)

$$\text{S.D. } (\sigma) = \sqrt{\frac{\sum_{i=1}^{n} f_i x_i}{N}},$$

where $x_i = x_i - \bar{x}$

Alternatively,

$$\text{S.D. } (\sigma) = \sqrt{\frac{\sum d_i^2}{N} - \frac{\left(\sum d_i^2\right)^2}{N}}$$

where $d_i = x_i - A$ (A is assumed mean)

For continuous series

$$\text{S.D. } (\sigma) = i\sqrt{\frac{\sum f_i d_i^2}{N} - \left(\frac{\sum f_i d_i}{N}\right)^2}$$

where $d_i = \frac{x_i - A}{i}$

x_i is midpoint and A is assumed mean.

For sampling distribution we find sample standard eleviation

Sample Standard Deviation

$$s = i\sqrt{\frac{\sum(x-\bar{x})^2}{N-1}}$$

For practical purposes, the formula can be written as

$$s = \sqrt{\frac{\sum x_i^2 - \frac{\left(\sum x_i\right)^2}{N}}{N-1}}$$

For grouped data

$$s = \sqrt{\frac{\sum f_i x_i^2 - \left(\sum f_i x_i\right)^2}{N-1}}$$

Alternatively,

$$s = i\sqrt{\left[\frac{\sum f_i x_i^2 - \frac{\left(\sum f_i d_i\right)^2}{N}}{N-1}\right]}$$

Sample standard deviation (s) are used to estimate population standard deviation.

Example 18.19 : *Calculate the standard deviation from the data given below:*

Size of item	*Frequency*	*Size of item*	*Frequency*
3.5	3	7.5	85
4.5	7	8.5	32
5.5	22	9.5	8
6.5	60		

Solution : Calculation of standard deviation

Table 18.15

x_i	f_i	$d_i = x_i - 6.5$	$f_i d_i$	$f_i d_i^2$
3.2	3	–3	–9	27
4.5	7	–2	–14	28
5.5	22	–1	–22	22
6.5	60	0	0	0
7.5	85	1	85	85
8.5	32	2	64	128
9.5	8	3	24	72
	$N = 217$		$\sum f_i d_i = 128$	$\sum f_i d_i^2 = 362$

$$\sigma = \sqrt{\sum \frac{f_i d_i^2}{N} - \left(\frac{\sum f_i d_i}{N}\right)^2}$$

$$\therefore \quad \sigma = \sqrt{\frac{362}{217} - \left(\frac{128}{217}\right)^2} = \sqrt{1.668 - 0.349} = 1.149$$

Example 18.20 : *Find the standard deviation from the following data*

Class-interval	*0-10*	*10-20*	*20-30*	*30-40*	*40-50*	*50-60*	*60-70*
Frequency	*6*	*14*	*10*	*8*	*1*	*3*	*8*

Solution:

Table 18.16

Class interval	*Mid Value* x_i	*Frequency* f_i	$\frac{x_i - 45}{10} = d$	d_i^2	$f_i d_i$	$f_i d_i^2$
0–10	5	6	–4	16	–24	96
10–20	15	14	–3	9	–42	126
20–30	25	10	–2	4	–20	40
30–40	35	8	–1	1	–8	8
40–50	45	1	0	0	0	0
50–60	55	3	1	1	3	3
60–70	65	8	2	4	16	32
					$\Sigma f_i d_i = 75$	$\Sigma f_i d_i^2 = 305$

$$\text{S.D} (\sigma) = i\sqrt{\frac{\sum f_i d_i^2}{N} - \left(\frac{\sum f_i d_i}{N}\right)^2} = 10 \times \sqrt{\frac{305}{50} - \left(\frac{-75}{50}\right)^2} = 10 \times \sqrt{6.1 - 2.25} = 19.6$$

Example 18.21 : *For the given data given in example 18.18 Find sample standard deviation*

Solution

Table 18.17: Computation of sample standard deviation

Seed yield $(x_i)g$	*No. of plants* (f_i)	*Mid Class* x_i	$\frac{x_i - A}{i} = d_i$	$f_i d_i$	$f_i d_i^2$
2.5 – 3.5	4	3	–3	–12	36
3.5 – 4.5	6	4	–2	–12	24
4.5 – 5.5	10	5	–1	–10	10
5.5 – 6.5	26	6	0	0	0
6.5 – 7.5	24	7	1	24	24
7.5 – 8.5	15	8	2	30	60
8.5 – 9.5	10	9	3	30	90
9.5 – 10.5	5	10	4	20	80
Total	**100**			**70**	**324**

$$\sigma = \sqrt{\sum f_i d_i^2 - \left(\frac{\sum f_i d_i}{N}\right)^2} \times i = \sqrt{\frac{324 - \frac{(70)^2}{100}}{100 - 1}} \times 1 = 1.66\,g$$

Characteristics of S.D.

1. If all the variate values are the same, S.D. is equal to zero.
2. S.D. is least affected by fluctuations of sampling.
3. Its computation is based on all the observations.
4. It is affected by the change of scale, but not affected by the change of origin.

Uses

S.D. is used in computing statistical quantities like, regression coefficients, correlation coefficient, etc.

18.5.2.4 Variance

Variance is the square of the standard deviation variance $= (S.D.)^2$

18.5.2.5 Coefficient of variance

This is also a relative measure of dispersion. It is important on account of the widely used measure of central tendency and dispersion *i.e.*, Arithmetic Mean and *S.D.*

It is given by —

$$\text{C.V.} = \frac{\text{S.D.}}{\text{A.M.}} \times 100$$

18.5.3 Measure of Symmetry

It measures how the data is scattered about mean. If the data is exactly symmetrically displaced about the mean, the measure of asymmetry should be zero. For symmetrical distribution, mean = mode = median and the curve is bell shaped. (Fig. 18.7).

If the data to the right of the mean are more spread out from the mean then those on the left then, by convention, the asymmetry is positive and the distribution is said to be *positively skewed* but if data to the left of mean are more spread out from mean than those on the right then, asymmetry is negative and distribution is said to be *negatively skewed.*

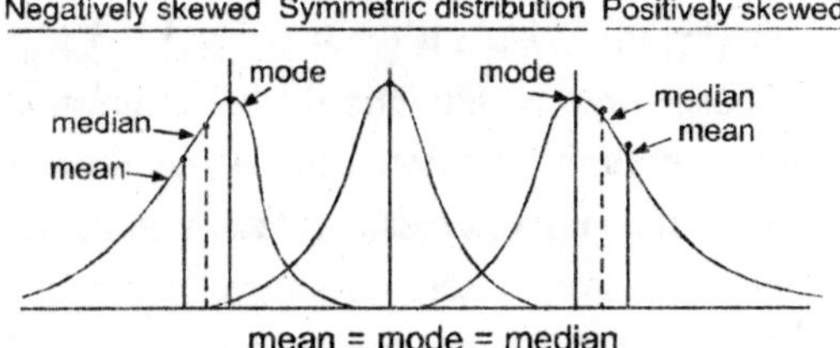

Fig. 18.7: Symmetrical and skewed distribution

Difference between dispersion and skewness : Dispersion is concerned with the *amount of variation* rather than with its direction. Skewness tells about the direction of the variation or departure from symmetry.

In fact, the measures of skewness are dependent upon the amount of dispersion. It may be noted that although skewness is an important characteristic for defining the precise pattern of a distribution, it is rarely calculated in business and economic series. Variation is by far is most important *characteristic of distribution.*

18.5.3.1 Moments and Kurtosis

1. **Moments.** Moment is a familiar mechanical term in physics which refers to the measure of a force with respect to its tendency to provide rotation. However, the term moment as used in physics has nothing to do with the moment used in statistics. The only analogy being that in statististics we talk of moment of random variable about *same point.* The moments in statistics are used to describe the various characteristics of frequency distribution like *central tendency, variation skewness* and *kurtosis.* It can be seen that the formula for a moment coefficient is identical with that for an arithmetic mean. This identity has led statisticians to speak of arithmetic mean as the "*first moment about the origin*".

Technically the mean is a moment coefficient and not a total moment, but in the case of frequency curves, with which mathematical statistics is concerned, the total frequency N is generally taken as unity, so that the total moment and moment coefficients are identical. In any case it has become customary in statistics to speak of the mean as first moment about the origin and the distinction between the total moment and moment coefficient is ignored. The concept of moment is also extended to higher powers.

The arithmetic mean of various powers of these deviations in any distribution is called *moments of the distribution.* If we take the mean of the first power of deviations, we get the first moment about mean the mean of square of deviations gives the second about mean the mean of the cubes of the deviations given is the third moment about the mean; and so on. The moments about mean is called '*central moments*' and are denoted by Greek letter μ (read as mu).

When the actual mean is in fractions it is difficult to calculate moments by applying the formula for central moments. In such case we can first compute moments about an arbitrary origin 'a' and then convert these moments into moments about actual mean. Moment about arbitrary origin, also called (raw moments) are denoted by symbol μ' to distinguish from moment about mean which are denoted by μ. Thus μ_1' would stand for the first moment about an arbitrary 'a' 'μ_2'' for the second moment about an arbitrary origin a and so on.

Formulae:

(*i*) *Central moments*: The rth moment of a variable x about the mean $\bar{x}$, usually denoted by μ_r is given by

$$\mu_r = \frac{1}{N}\sum f_i(x_i - \bar{x})^r, \quad r = 0, 1, 2, 3...n$$

In particular,

$$\mu_0 = \frac{1}{N}\sum f_i(x_i - \bar{x})^0 = \frac{1}{N} f_i = \frac{N}{N} = 1$$

and $\mu_1 = \frac{1}{N}\sum f_i(x_i - \bar{x}) = 0$

(Since sum of deviations about mean is zero).

Also $\mu_2 = \frac{1}{N}\sum f_i(x_i - \bar{x})^2 = \sigma^2$ (variance)

(*ii*) *Moment about arbitrary origin*: The *r*th moment of variable *x* about a point $x = a$ usually denoted by μ_r' is given by

$$\mu_r' = \frac{1}{N}\sum f_i(x_i - a)^r = \sigma^2, \ \sum f_i = N$$

Clearly

$$\mu_0' = \frac{1}{N}\sum f_i(x_i - a)^0 = \frac{N}{N} = 1$$

$$\mu_1' = \frac{\sum(X-a)}{N}, \ \mu_2' = \frac{\sum(X-a)^2}{N},$$

$$\mu_3' = \frac{\sum(X-a)^3}{N} \text{ and so on}$$

(*iii*) *Moments about the mean in terms of moments about any point:*

We have

$$\mu_r = \frac{1}{N}\sum f_i(x_i - \bar{x})^r$$

$$= \frac{1}{N}\sum f[(x_i - a) - (\bar{x} - a)]^r$$

$$= \frac{1}{N}\sum f_i[(x_i - d)^r,$$

where $x_i = x_i - a$, $d = \bar{x} - a$.

$$= \frac{1}{N}\left[\sum f_i x_i^r - {}^rC_1 d \sum f_i x_i^{r-1} + {}^rC_2 \sum f_i x_i^{r-r} .....\right]$$

$$= \mu_r' - {}^rC_1 d\, \mu'_{r-1} + {}^rC_2 d^2\, \mu''_{r-2}...$$

In particular, on putting $r = 2$, 3 and 4

$\mu_2 = \mu_2' - {\mu'_1}^2$

$\mu_3 = \mu_3' - 3\,\mu_2'\,\mu_1' + 2\,{\mu'_1}^3$

$\mu_4 = \mu_4' - 4\,\mu'_3\,\mu'_1 + 6\,\mu_1{}^1\,\mu_1{}^2 - 3\,{\mu'_1}^4$

These three results should be committed to memory. It should be noted that in each of these relations, the sum of coefficients of various terms on the right side is zero. Also each term on the right side is of the same dimension as the term on the left.

Moments about zero:

The moments about zero are often denoted by ν_1, ν_2, ν_3 etc. and are obtained as follows

$$\nu_1 = \frac{\sum fx}{N}, \quad \nu_2 = \frac{\sum fx^2}{N}$$

$$\nu_3 = \frac{\sum fx^3}{N}, \quad \nu_4 = \frac{\sum fx^4}{N}$$

Also, $\nu_1 = a + \mu'_1$ or mean.

$\nu_2 = \mu_2 + (\nu_1)^2$

$\nu_3 = \mu_3 + 3\nu\nu_2 - 2{\nu_1}^3$.

Purpose of moments: The concept of moment is of great significance in statistical work. With the help of moments we can measure the central tendency of a set of observation and there variability . Because of great convenience in obtaining measures of various characteristics of frequency distribution, the calculation of first four moments about the mean may well be made the first step in the analysis of frequency distribution.

Measure of skewness based on moments. A measure of skewness is obtained by making use of third moment about the mean. When the method of moment is applied β_1 is used as relative measure of skewness. β_1 is defined as

$$\beta_1 = \frac{\mu_3^2}{\mu_2^3} \quad \text{or} \quad \sqrt{\beta_1} = \frac{\mu_3}{\sqrt{\mu_2^3}}$$

This is also symbolised as γ_1. The value of β_1 should be zero for a perfectly symmetrical distribution. Greater the value of β_1, the more skewed is the distribution.

2. **Kurtosis:** Kurtosis in Greek means '*bulginess*'. In statistics kurtosis refers to the degree of flatness or peakedness in the region about the mode of frequency curve. The degree of kurtosis of a distribution is measured relative to the peakedness of normal curve. In other words, measures of kurtosis tell us the extent to which a distribution is more peaked or flat-topped than the normal curve. If a curve is more peaked than normal curve, it is called '*leptokurtic*'. In such a case the items are more closely related to bunched around the mode. On the other hand, if a curve is more flat topped than normal curve, it is called '*platykurtic*'. The normal curve itself is known as *mesokurtic* and in this case mean = mode = median. The condition of peakedness or flat-toppedness itself is known as kurtans or excess. The [Fig. 18.8] illustrates the shape of three different curves.

Kurtosis measuring the peakedness of a distribution is given by

$\beta_2 = \mu_4 / \mu_2^2$,

$\gamma_2 = \beta_2 - 3$ gives the excess of kurtosis.

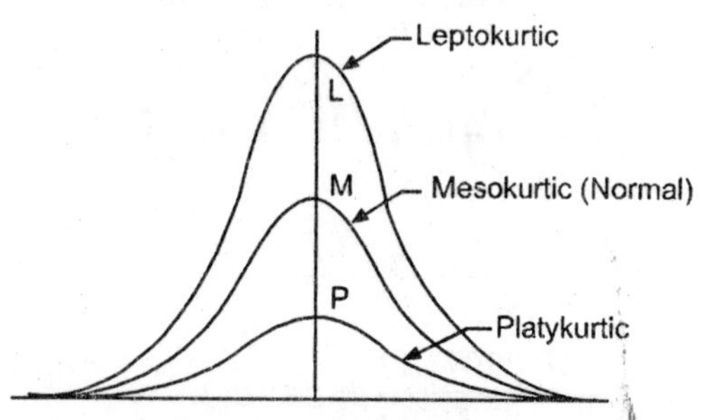

Fig. 18.8 : Leptokurtic, mesokurtic and platykurtic curve

For leptokurtic curve $\beta_2 > 3$

For platykurtic curve $\beta_2 < 3$ and for mesokurtic (normal curve)

$\beta_2 = 3$.

For a normal distribution $\gamma_2 = 0$, if γ_2 is positive, the curve is leptokurtic and if γ_2 is negative, the curve is platykurtic.

For a moderately skewed distributions. The averages has the following relationship.

Mode = 3 Median – 2 Mean

Example 18.22 : *Find the measure of skewness and kutosis for the following data:*

X	*3*	*4*	*5*	*6*	*7*	*8*	*9*	*10*
F	*4*	*6*	*10*	*26*	*24*	*15*	*10*	*5*

Solution:

Table 18.18: Computation of moments

x_i	f_i	$x_i - \bar{x} = d_i$	$f_i d_i$	$f_i d_i^2$	$f_i d_i^3$	$f_i d_i^4$
3	4	– 3.7	– 14.8	54.76	– 202.612	749.6644
4	6	– 2.7	– 16.2	43.74	– 118.130	318.8646
5	10	– 1.7	– 17.0	28.90	– 49.130	
6	26	– 0.7	– 18.2	12.74	– 8.918	83.5210
7	24	0.3	7.2	2.16	0.648	6.2426
8	15	1.3	19.5	25.35	32.955	0.1944
9	10	2.3	23.0	52.90	121.670	42.8415
10	5	3.3	16.5	54.45	179.685	592.9801
Total	**100**		**0**	**275.00**	**– 43.800**	**2074.1300**

μ_2 (varience)

$$= \frac{\sum f(x_i - \bar{x})^2}{N} = \frac{275}{100} = 2{\cdot}75$$

$$\mu_3 = \frac{\sum f(x_i - \bar{x})^2}{N} = \frac{-43.8}{100} = -0.438$$

$$\mu_4 = \frac{\sum f(x_i - \bar{x})^4}{N}$$

$$= \frac{2074.13}{100} = 20.7413$$

$$\beta_1 = \frac{(\mu_3)^2}{\mu_1^3} = \frac{(-0.438)^2}{(2.75)^3} = 0.0093$$

$$\beta_2 = \frac{\mu_4}{\mu_2^2} = \frac{20.7413}{(2.75)^2} = 2.7465$$

Since $\beta_1 = 0.0093$, so it is only slightly skewed. It is negatively skewed since μ_3 is negative.

The value of μ_3 is 2.7153, < 3. Hence the distribution is platykurtic.

18.6 CORRELATION ANALYSIS

In a bivariate distribution we may be interested to find out if any change in one variable affects a change in other variable? If one variable change with change in other variable then the two variables are said to be *correlated.*

Some of the definition of correlation/correlation analysis are given below:

1. "*Correlation is an analysis of variation between two or more variable*"

   —*A. M. Tuttle*

2. "*If two or more quantities vary in sympathy so that moments in one tend to be accompanied by corresponding moments in other(s) then they are said to be correlated*".

   —*L. R. Cornner*

3. "*Correlation analysis attempts to determine the 'degree of relations between variables.*"

   —*Croxton and Cowden*

18.6.1 Types of Correlation

Correlation is described or classified in several different ways. Three of the most important ways of classifying correlation are as follows:

(i) ***Positive and negative correlation***: If the two variables deviate in same direction, *i.e.* if the increase (or decrease) in one results in corresponding increase (or decrease) in other, correlation is said to be *direct* or *positive*. But if they constantly deviate in opposite direction, *i.e.* if increase (or decrease) in one results in corresponding decrease (or increase) in the other, correlation is said to be *diverse* or *negative.*

For example the correlation between no. of persons and use of water, is a kind of positive correlation while the correlation between volume of water vapour and pressure is an example of negative *correlation.*

Correlation is said to be *perfect* if the deviation in one variable is followed by corresponding and proportional deviation in other.

(ii) ***Simple, partial and multiple correlation***: When only two variable are studied it is a problem of *simple correlation.* Where three or more variables are studied it is a problem of either multiple or partial correlation.

In partial correlation more than two variables are recognised but only two variable influencing each other are considered. While in multiple all the recognised variables are considered.

(iii) ***Linear and Non-linear (curvilinear) correlation***: If the amount of change in one variable tends to be a constant ratio to the amount of change in the other variable then

correlation is said to be linear [Fig. 18.9(*a*)] but if the amount of change in one variable does not bear a constant ratio to the change in other variable then it is called *curvilinear* or *non linear correlation*. [Fig. 18.9 (*b*)]

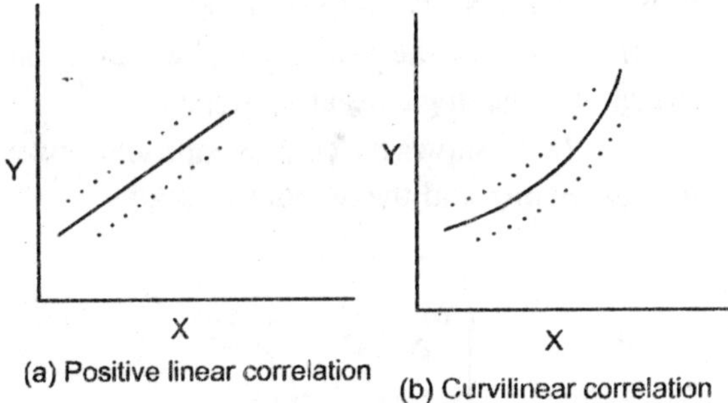

(a) Positive linear correlation (b) Curvilinear correlation

Fig. 18.9

18.6.2 Methods of Studying Correlation

The various methods of ascertaining whether two variables are correlated or not are.

(*i*) Scatter diagram method
(*ii*) Graphic method
(*iii*) Karl Pearson's coefficient of correlation.
(*iv*) Rank method
(*v*) Method of least squares

1. Scatter Diagram: It is the simplest way of diagrammatic representation of bivariate data. Thus for a bivariate distribution (x_i, y_1), i, = 1, 2,....n, if the values of variable X and Y are plotted along the X-axis and Y-axis respectively in the X-Y plane the diagram of dots so obtained is called *scatter diagram*.

From the scatter diagram, we can form a fairly good, though vague, idea whether the variable are correlated or not, *e.g.*, if the points are vary dense *i.e.* very close to each other, we should expect a fairly good amount of correlation between the variable and if points are widely scattered, a poor correlation is expected. This method is however not suitable if the number of observation is very large. Fig. 18.10 illustrates some scatter diagrams.

2. Karl Pearson's coefficient of correlation: As a measure of intensity or degree of correlationship between two variables, Karl Pearson a British Biometrician, developed a formula called correlation coefficient.

Correlation coefficient between two random variable X and Y, usually denoted by r is a numerical measure of linear relationship between them and is defined as

$$r = \frac{Cov.(x, y)}{\sigma_x, \sigma_y}$$

where, Cov, (x, y) is the covariance of X and Y and σ_x and σ_y variable are standard deviation of x and y respectively

If (x_i, y_i), i = 1, 2,..... n is the bivariable distribution, then

$Cov\ (x, y)$

$$= \frac{1}{n}\Sigma(x_i - \bar{x})(y_i - \bar{y})$$

$$= \frac{1}{n}\Sigma x_i y_i - \Sigma\bar{x}\ \Sigma\bar{y}$$

$$\sigma_x = \sqrt{\frac{1}{n}\Sigma(x_i - \bar{x}_i)^2} = \sqrt{\frac{1}{n}\Sigma x_i^2 - (\bar{x})^2}$$

and

$$\sigma_y = \sqrt{\frac{1}{n}\Sigma(y_i - \bar{y})^2} = \sqrt{\frac{1}{n}\Sigma y_i^2 - \bar{y}^2}$$

The correlation coefficient

$$r - \frac{\Sigma xy}{n\sigma_x \cdot \sigma_y} = \frac{\Sigma xy}{\sqrt{\Sigma x^2 \Sigma y^2}}$$

σ_x, and σ_y are standard deviation and n is the number of pairs of observations.

where $x = x_i - \bar{x}$ and $y = (y_i - \bar{y})$

The correlation coefficient is independent of change of origin and scale.

The value of correlation coefficient obtained from above formula lies between ± 1. Conclusions from values of r are as follows.

(*i*) when $r = +1$, perfect positive correlation.

(*ii*) when $r = -1$, perfect negative correlation.

(*iii*) $0 < r < 1$, negative correlation

(*iv*) $-1 < r < 0$, negative correlation and

(*v*) $r = 0$, no correlation

The magnitude of r describe the strength of a linear relation and its sign indicate the direction.

If $r = +1$, all pairs (x_i, y_i) lie exactly on a straight line having a positive slope.

If $r > 0$ then the pattern in the scatter diagram runs from lower left to upper right.

If $r < 0$ then the pattern in the scatter diagram runs from upper left to lower right.

If $r = -1$ then the pairs (x_i, y_i) lie exactly on a straight line having a negative slope.

Fig. 18.10 shows the correspondence between the values of r and the amount of scatter.

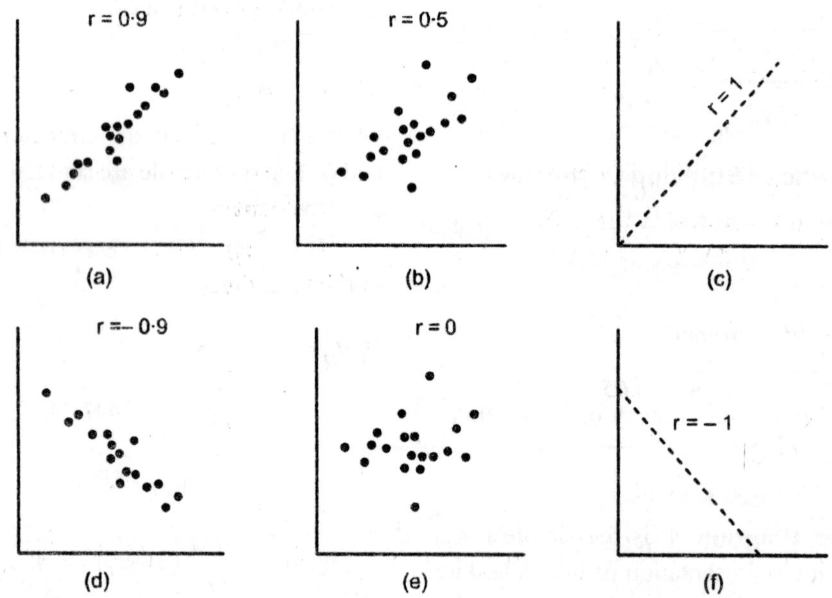

Fig. 18.10: Correspondence between the values of *r* and the amount of scatter.

Example 18.23. *Simultaneous observations of two random variables, x and y, are given below :*

x	1	2	3	4	5
y	2	3	4	6	5

Estimate (i) mean and standard deviation of x, (ii) mean and standard deviation of y, (iii) covariance of x and y, and (iv) correlation coefficient between x and y.

Solution: (*i*) *Mean and standard deviation of x*

Mean of x,

$$\bar{x} = \frac{1+2+3+4+5}{5} = \frac{15}{5} = 3 \cdot 0$$

Standard deviation, of x,

$$\therefore \quad \sigma_x^2 = \frac{\Sigma(x_i - \bar{x})^2}{n}$$

$$= \frac{(1-3)^2 + (2-3)^2 + (3-3)^2 + (4-3)^2 + (5-3)^2}{5}$$

$$\text{or} \quad \sigma_x^2 = \frac{10}{5} = 2$$

$$\therefore \quad \sigma_x = \sqrt{2} = 1 \cdot 414$$

(ii) Mean and standard deviation of y
Mean of y,

$$\bar{y} = \frac{2+3+4+6+5}{5} = \frac{20}{5} = 4$$

Standard deviation of y

$$\sigma_y^2 = \frac{(2-4)^2+(3-4)^2+(4-4)^2+(6-4)^2+(5-4)^2}{5}$$

$$= \frac{10}{5} = 2$$

$$\sigma_y = \sqrt{2} = 1 \cdot 414$$

(iii) Co-variance of x and y

The covariance of x and y is given by *Cov.* $(x, y) = \frac{1}{n}\Sigma(x-\bar{x})(y-\bar{y})$

So we prepare the table. 18.19

Table 18.19

x	y	$x-\bar{x}=x-3$	$y-\bar{y}=y-4$	$(x-\bar{x})(y-\bar{y})$
1	2	–2	–2	4
2	3	–1	–1	1
3	4	0	0	0
4	6	1	2	2
5	5	2	1	2
			$\Sigma(x-\bar{x})(y-\bar{y})$	$=9$

As $n = 5$

$$\therefore \text{ Cov. } (x, y) = \frac{1}{5}\times 9 = 1 \cdot 4$$

(iv) Now, coefficient of correlation

$$r = \frac{\text{Cov. }(x, y)}{\sigma_x \sigma_y} = \frac{1 \cdot 4}{\sqrt{2}\sqrt{2}} = 0 \cdot 7$$

Example 18.24 : *Find out the correlation coefficient between seed yield per plant and plant height of sesame using the following data.*

Plant height (in cm)	*94.2*	*85.4*	*76.1*	*95.7*	*85.5*	*74.4*	*83.5*	*94.5*	*76.9*	*81.3*
Seed yield Per plant (in g)	*8.39*	*6.05*	*4.67*	*7.21*	*7.56*	*5.22*	*8.98*	*2.39*	*5.17*	*10.08*

Solution:

Table 18.20

Plant height x (cm)	***Seed yield per plant y (in g)***	x^2	y^2	xy
94.2	5.22	8873.64	27.2484	491.724
85.4	9.98	7293.16	99.6004	852.292
76.1	2.39	5791.21	5.7121	181.879
95.7	5.17	9158.49	26.7289	494.769
85.5	10.08	7310.25	101.6064	861.84
75.4	8.39	5539.36	70.3921	624.216
83.5	6.05	6972.25	36.6025	505.175
94.5	4.67	8930.25	21.8089	441.315
76.9	7.21	59.13.61	51.9841	554.449
81.3	7.56	6609.69	57.1536	614.628
847.5	66.72	72391.91	498.8374	5622.287

Solution: We have $n = 10$

$$\therefore \quad r = \frac{n\sum xy - (\sum x)(\sum y)}{\sqrt{n\sum x^2 - (\sum x^2)}\sqrt{n\sum y^2 - (\sum y)^2}}$$

$$= \frac{10 \times 5622.287 - 847.5 \times 66.72}{\sqrt{10 \times 72391.91 - (847.5)^2}\sqrt{10 \times 498.8374 - (66.72)^2}}$$

$$= \frac{-322.33}{75.252 \times 23.17} = -0.1848$$

Hence there is negative correlation.

18.6.3 Rank Correlation

This method of finding out covan ability or lack of it between two variables was developed by British psychologist, Charts Edward Spearman. This measure is specially useful when quantitative measure for certain factors (such as evaluation of leaders, ability or the judgement of female beauty, beauty of flowers etc. cannot be fixed, but the individuals in the group can be arranged in order thereby obtaining for each individual a number indicating his (her) rank in group. *Spearman's rank correlation coefficient is defined as*

$$R = 1 - \frac{6\sum D^2}{n^3 - n}$$

where R denotes rank coefficient of correlation and D refers to difference of ranks between paired item in two series.

In case ranks are repeated

$$R = 1 \frac{6\left(\sum D^2 + \frac{1}{12}(m^3 - m) + \frac{1}{2}m^3 - m + ...\right)}{n^3 - n}$$

where m stands for the number of items with common rank, this value is added as many times the number of such groups.

Example 18.25 : *Calculate the Spear man's coefficient of rank correlation from the following data.*

x	53	98	95	81	75	61	59	55
y	47	25	32	37	30	40	39	45

Solution:

Table 18.21: Calculation of rank correlation coefficients

X	Y	R_1	R_2	$D^2 = (R_1 - R_2)^2$
53	47	1	8	49
98	25	8	1	49
95	32	7	3	16
81	37	6	4	4
75	30	5	2	9
61	40	4	6	4
59	39	3	5	4
55	45	2	7	25
				$\sum D^2 = 160$

We have $R = 1 - \frac{6\sum D^2}{n^3 - n}$, $\Sigma D^2 = 160$, $n = 8$

$$\therefore \quad R = 1 - \frac{6 \times 160}{8^3 - 8} = -0.905$$

Example 18.26 : *The following data relate the seed yield (x) and the number of capsules per plant (y) of 15 sesame plant.*

Find out the Spearman's rank correlation coefficient.

x	4	3	4	5	6	8	6	5	3	7	6	5	9	5	3
y	25	22	27	28	32	40	33	25	20	30	29	26	42	18	18

Solution: In X series, 3 occurs 3 times, 4 occurs 2 times and 6 occurs 3 times.

Starting with 3 (least) the three 3's occupy first 3 places 1, 2, 3 so rank of 3

$$= \frac{1+2+3}{2} = 2$$

4's occupy 4th and 5th positions so rank of 4 = $\frac{4+5}{2}$ = 4.5. Similarly other ranks can be found as shown in table 18.22.

Table 18.22: Computation of rank correlation

		Ranks		
X	Y	$R_1\,(X)$	$R_2\,(Y)$	$D^2 = (R_1 - R_2)^2$
4	25	4.5	5.5	1.00
3	22	2	4	4.00
4	27	4.5	8	12.25
5	28	7.5	9	2.25
6	32	11	12	1.00
8	40	14	14	0
6	33	11	13	4.00
5	25	7.5	5.5	4.00
3	20	2	3	1.00
7	30	13	11	4.00
6	29	11	10	1.00
5	26	7.5	7	0.25
9	42	15	15	0
5	42	15	1.5	36.00
3	18	2	1.5	0.25
				71.00

$$\therefore \quad R = 1 - \frac{6\left[71 + \frac{1}{12}(3^3 - 3) \times 2 + \frac{1}{12}(2^3 - 2) \times 3 + \frac{1}{12}(4^3 - 4)\right]}{15^3 - 15} = 0.8544$$

18.7 REGRESSION ANALYSIS

The term 'regression' literally means "stepping back toward the average.

Some of the definition of regression are

Regression analysis is the mathematical measure of the average relationship between two or more variables in terms of the original units of the data —*Blair*

"The term regression analysis refers to the method by which estimates are made of the values of a variable from a knowledge of the values of one or more other variables and to the measurement of the errors involved in this estimation process" —*Morris Hamberg*

"Regression alalysis attempts to establish the nature of relationship between the variables—that is, to study the functional relationship between the variables and there by provide a mechanism for prediction, or forecasting" —*Ya—Lu—Chou*

18.7.1 Linear Regression

If the variable in a bivariate distribution are related, we will find that the points in scatter diagram will cluster round some curve called the "*curve of regression*". If the curve is a straight line, it is called the *line of regression* and therefore, is said to be *linear regression.*

Linear regression is widely used to describe linear cause and effect relation between two variables. The objective is to predict a dependent variable in terms of a single independent variable. For example, let us first treat the case where the regression curve of y on x is linear. The linear regression equation in this case,

$$y_i = a + bx_i + \epsilon_i, \qquad i = 1, 2, ... ,n$$

where ϵ_i, a random variable, y_i is the ith value of dependent or regressed variable. The regression line crosses the y-axis at point 'a' and has a slope 'b'.

18.7.1.1 The line of regression

The line of regression is the line of the "*best fit*" and is obtained by the *principle of least squares.*

The line of regression is the line which gives the best estimate to the value of one variable for any specific value of other variable.

Let us suppose that in a bivariate distribution (x_i, y_i), $i = 1, 2, 3, ...n$ where, y is dependent variable and x is independent variable. Let the line of best fit (or regression line) be

$$y = a + bx \qquad (i)$$

The above equation represents a family of straight lines for different values of arbitrary constants 'a' and 'b'.

The problem is to determine 'a' and 'b' so that line is *line of best fit*. The paired observations are around the line as shown in scatter diagram Fig. 18.11.

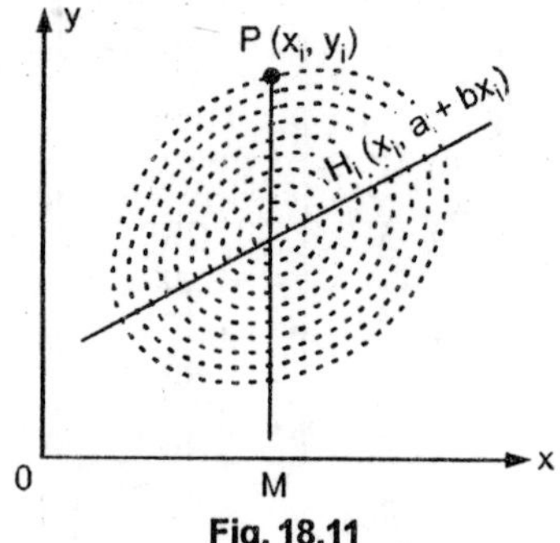

Fig. 18.11

The term 'best fit' is interpreted in accordandance with Legender's principle of *least squares which consists* in minimising the sum of square of the deviation of actual values of y from their estimated values as given by line of best fit.

Let P_i (x_i, y_i) be any general point in scatter diagram (Fig. 18.11). Draw $PM \perp$ to x-axis meeting the line at H_i. The abscissa of H_i is x_i and since H_i lies on the line, its ordinate is $a + bx_i$. Hence co-ordinate of H_i are $(x_i, a + bx_i)$.

Now $P_iH_i = P_iM - H_iM = y_i - (a + bx_i)$ is called the error of estimate (e_i) or the residual for y_i.

It may be positive or zero or negative depending upon P lies above, on or below the curve.

Let these residuals like P_iH_i be denoted by e_1, e_2, e_3,...., e_n.

Then according to *principle of least squares*, we have to determine a and b so that,

$$E = \sum_{i=1}^{n} e_i^2 = \sum_{i=1}^{n} (y_i - a - bx_i)^2 \qquad (ii)$$

e = error for individual data

E = Total error

If E is small, the fit is said to be good and if E is large the fit is poor.

For E to be minimum, the partial derivatives of E with respect to a and b must vanish separately.

i.e., $$\frac{\partial E}{\partial a} = 0$$

or $$-2\sum_{i=1}^{n}(y_i - a - bx_i) = 0$$

and $$\frac{\partial E}{\partial b} = 0$$

or $$-2b\,(y_i - a - bx_i) = 0 \qquad (iii)$$

Thus

$$\sum_{i=1}^{n} y_i = \Sigma a + b\sum_{i=1}^{n} x_i = na + b\sum_{i=1}^{n} x_i$$

and

$$\sum_{i=1}^{n} x_i y_i = a\sum_{i=1}^{n} x_i y_i = a\sum_{i=1}^{n} x_i + b\sum_{i=1}^{n} x_i^2 \quad (iv)$$

Equation (*iii*) and (*iv*) are known as normal equations and can be solved as simultaneous equation to get values of a and b. The values of a and b are next substituted in equation (*i*) to get the line of best fit.

The working procedure for line of *best fit* from given paired observations are as follows:

Step I: From normal equations

$$(i) \quad \sum_{i=1}^{n} y_i = na + b\sum_{i=1}^{n} x_i$$

$$(ii) \quad \sum_{i=1}^{n} x_i y_i = a\sum_{i=1}^{n} x_i + b\sum_{i=1}^{n} x_i^2$$

Step II: Solve the above two equations for a and b.

Step III: Substitute for a and b in the line $y = a + bx$ which will be required line of best fit.

18.7.1.2 Two types of regression lines

If we take the case of two variables x and y, we shall have two regression lines as the regression of x on y and regression of y on x. The regression line of y on x gives the most probable values of y for given value of x and the regression line x and y gives the most probable value of x for given value of y. However, when there is either perfect positive of perfect negative correlation between the two variables (*i.e.* $r = \pm 1$), the two regression line will coincide.

The regression line y on x is given by

$$y - \bar{y} = b_{yx}(x - \bar{x})$$

where b_{xy} is slope of the line of y on x and known as regression coefficient of y on x and is given by

$$b_{yx} = r = \frac{\sigma_y}{\sigma_x}$$

where σ_y and σ_y are standard deviations

Thus regression line of y on x, is given by

$$y - \bar{y} = r\frac{\sigma_y}{\sigma_x}(x - \bar{x})$$

The regression line of x on y is given by

$$x - \bar{x} = r\frac{\sigma_x}{\sigma_y}(y - \bar{y})$$

where b_{xy} is the slope of line x on y and known as regression coefficient of x on y,

It is given by

$$b_{xy} = r\frac{\sigma_x}{\sigma_{xy}}$$

∴ Thus equation of line of regression of X on Y is given by

$$x - \bar{x} = r\frac{\sigma_x}{\sigma_y}(x - \bar{y})$$

The two regression line passes through $(\bar{x}, \bar{y})$, hence intersect of $\bar{x}$, $\bar{y}$

15.7.1.3 Properties of Regression Coefficients

(*i*) *Correlation coefficient is the geometric mean between regression coefficients*

Proof: $b_{xy} \times b_{yx} = r\frac{\sigma_x}{\sigma_y} \times r\frac{\sigma_y}{\sigma_x} = r^2$

or $r = \pm\sqrt{b_{xy} \times b_{yx}}$

Clearly the two regression coefficients must have same sign and sign of r is taken same as that of b_{xy} and b_{yx}.

(*ii*) *Both the regression coefficients cannot be greater than unity.*

Proof : We have

$$r = \pm\sqrt{b_{xy} \times b_{yx}} \text{ but } |r| \le 1$$

$$\therefore \quad |b_{xy} \times b_{yx}| \le 1$$

Thus both b_{yx} and b_{xy} cannot be greater than one.

(*iii*) *Regression coefficients are independent of the change of origin but not the scale.*

(*iv*) *Angle between two regression lines*

$$\tan\theta = \left|\frac{b_{yx} - b_{xy}}{1 + b_{yx} \cdot b_{xy}}\right| = \left|\frac{r^2}{r}\right|\left(\frac{\sigma_x\sigma_y}{\sigma_x^2 + \sigma_y^2}\right)$$

$$= \frac{1 - r^2}{|r|}\left(\frac{\sigma_x\sigma_y}{\sigma_x^2 + \sigma_y^2}\right)\left(\because r^2 \le 1\right)$$

$$\therefore \quad \theta = \tan^{-1}\left\{\frac{1 - r^2}{|r|}\left(\frac{\sigma_x\sigma_y}{\sigma_x^2 + \sigma_y^2}\right)\right\}$$

Special cases

(*i*) If $r = 0$, $\tan\theta = \infty \Rightarrow \theta = \frac{\pi}{2}$, thus if the two variables are uncorrelated, the lines of regression become perpendicular to each other.

(*ii*) If $r = +1$, $\theta = 0$ or π. In this case two lines of regression either coincide or they are parallel to each other. But since both lines passes through $(\bar{x}, \bar{y})$ then can not be parallel. Hence in case of perfect correlation (positive or negative) the two lines of regression coincide. The above discussed properties are illustrated in Fig. 18.12.

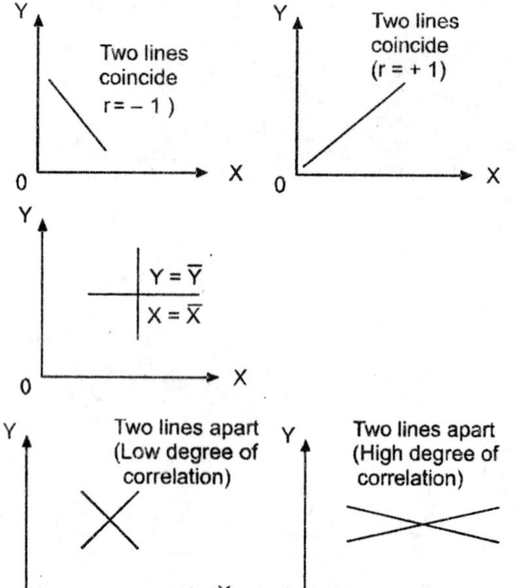

Fig. 18.12: Properties of regression lines

18.7.1.4 Criterian for best fit line relating two random variables

One of the main purposes of curve fitting is to estimate one of the variables (the dependent variable) from the other (the independent variable).

The regression equation y on x is given by $y = a + bx$.

A straight line fitted by least square should have following characteristics.

(*a*) It gives the best fit to the data.

(*b*) The deviation above the line is equal to deviation below the line.

(*c*) The straight line goes through the overall mean of the data.

The two parameters a and b can be found by solving simultaneously the following two equations.

18.7.1.5 Difference between correlation and regression analysis

(*i*) Whereas correlation coefficient is a measure of degree of covariability between X and Y the objective of regression analysis is to study the 'Nature of relationship between the variables so that we may be able to predict the value of one on the basis of another.

(*ii*) In correlation analysis the regression coefficient r_{xy} and r_{xy} are symmetric ($r_{xy} = r_{yx}$) but in regression analysis the regression coefficients b_{xy} and b_{yx} are not symmetric *i.e.* $b_{xy} \neq b_{yx}$.

(*iii*) Correlation coefficient is independent of change of scale and origin while regression coefficient are independent of change of origin but not the scale regression coefficient of correlation (r) takes the same sign as the regression coefficient (b_{xy} and b_{yx}).]

Example 18.27 : *From a paddy field, 36 plants were selected at random. The length of panicles (x) and number of grains per panicle (y) of selected plants were recorded. The results are given below. Fit a regression line of y and x.*

x	y	x	y	x	y
22.4	95	24.5	143	22.9	112
23.3	109	23.6	127	23.9	131
24.1	133	21.1	92	24.8	147
24.3	132	21.4	88	21.2	90
23.5	136	23.4	99	22.2	110
22.3	116	23.4	129	22.7	106
23.9	126	21.6	91	23.0	127
24.0	124	21.4	103	24.0	145
24.9	137	23.3	114	20.6	85
20.0	90	24.4	124	21.0	94
19.8	107	24.4	143	24.0	142
22.0	108	22.5	108	23.1	111

Solution:

From the given values of x and y, we have

$n = 36, \sum x = 822.9,$

$\sum x^2 = 18876.83, \ \bar{x} = 22.86$

$\sum y = 4174, \ \sum y^2 = 4962558,$

$\bar{y} = 115.94$

$\sum xy = 961834$

Now repression coefficient

$$b_{yx} = \frac{n\left(\sum xy\right) - \sum x \sum y}{n\sum x^2 - \left(\sum x\right)^2}$$

$$= \frac{36 \times 96183.4 - 822.9 \times 4174}{36 \times 18876.83 - (822.9)^2}$$

$= 11.5836$

$\therefore$ Regression of y on x

$y - \bar{y} = byx\ (x - \bar{x})$

$y - 115.94 = 11.5836\ (x - 22.86)$

$= 11.5836x - 264.80 + 115.94$

18.7.2 Regression through a point

If we want that the regression line must pass through a point (x_0, y_0), that is, if some condition on x and y must be satisfied, then we have the following modification. We desire that the line

$$y = a + bx$$

must pass through (x_0, y_0). This means y is fixed at y_0. Thus $y_0 = y_0$. This gives

$$y_0 = a + bx_0$$

On substituting a, becomes

$$y = y_0 + b\ (x - x_0)$$

which has only one unknown, namely b. We have

$$\text{SSE} = \Sigma\ \{y_i - y_0 - b\ (x_i - x_0)\}^2$$

To find β we minimise SSE by differentiating it with respect to β. This yields

$$b = \frac{\Sigma\ (x_i - x_0)\ (y_i - y_0)}{\Sigma\ (x_i - x_0)^2}$$

Note that the above will not give the best fit but will give the best fit out of those which pass through (x_0, y_0)..

18.7.3 Multiple Regression

In multiple regression, we deal with data consisting of $n(r + 1)$ – tuples $(x_{i1}, x_{i2}, ...x_{ir}, y_i)$ where the x_{ij}'s are given assumed to be known without error while the y_i's are values of random variables. Data of the this kind arises, for example, in studies designed to determine the effect of various climatic conditions on resistance of metal towards the corrosion, effect of humidity, etc.

As in case of one independent variable, we shall first deal where the regression equation is linear, namely, where for any given set of values, $x_1, x_2,.........$ and x_r, fo r independent variables, the mean of the distribution of y is given by

$$b_0 + b_1x_1 + b_2x_2 + ...+ b_r x_r$$

For two independent variables, this is the problem of fitting a plane to a set of n points with coordinates (x_{i1}, x_{i2}, y_i) as shown in Fig. 18.13.

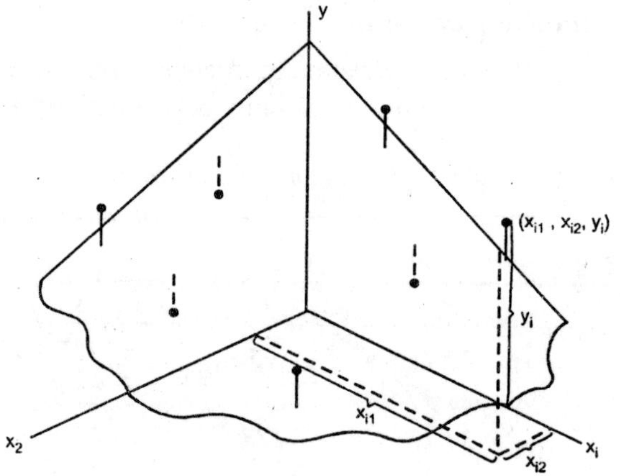

Fig. 18.13: Regression plane.

Using the methods of least squares we estimate the coefficient of β_0, β_1 and β_2, we minimize the sum of the squares of the vertical distances from the observations y_i to the plane. We minimize

$$\sum_{i=1}^{n}[y_i - (b_0 + b_1x_i + b_2x_{i2})]^2$$

The linear equation can be written as

$$y = a + b_1\ x_1 + b_2\ x_2$$

Normal equations for multiple regression with $r = 2$

$$\Sigma\ y = nb_0 + b_1\ \Sigma x_1 + b_2\ \Sigma x_2$$

$$\Sigma x_1 y = b_0\, \Sigma x_1 + b_1\, \Sigma x_1^2 + b_2\, \Sigma x_1 x_2$$
$$\Sigma x_2 y = b_0\, \Sigma x_2 + b_1\, \Sigma x_1 x_2 + b_2\, \Sigma x_2^2$$

In case of multiple correlation the correlation coefficient is given by

$$r^2 = \frac{b\sum(x_i - \bar{x})(y_i - \bar{y}) + c\sum(z_i - \bar{z})(y_i - y)}{\sum(y_1 - \bar{y})^2}$$

and the alternate formula is

$$r = \pm\sqrt{\frac{\{1 - \sum y_i - E(y_1)\}^2}{\sum(y_i - \bar{y})^2}}$$

18.7.4 Non-linear Regression

If the curve to be fitted contains non-linear terms in a, b etc, it is called non-linear regression. In non-linear regression the equations giving a, b etc. is not a linear system. So the solution becomes difficult. One case which is of frequent occurrence yields simple equations in these. This is the case of exponential regression. We discuss this below.

If the set of data consisting of n points (x_i, y_i) straightens out when $\ln y_i$ is plotted verses x_i, this shows that the regression of Y on x is exponential. This means that the mean $E(y)$ is given by ae^{bx}. Thus we take

$$E(y) = a\, e^{bx}, \qquad (i)$$

or

$$\ln E(y) = \ln a + bx \qquad (ii)$$

The relation Eq. (i) involves b non-linearly. The relation Eq. (ii) is linear in $\ln a$ and b.

Thus $E(z) = c + bx \qquad (iii)$

The relation Eq. (iii) is same as simple linear regression.

18.8 PROBABILITY AND PROBABILITY DISTRIBUTION

The probability of a given event is an expression of likelihood or chance of occurrence of an event. A probability is a number which ranges from zero to one, zero for an event which can not occur and one for an event certain to occur.

Definitions

1. ***Experiment and outcomes :***

Any operation on certain object or object which gives different results is called an 'Experiment' and the different possible results is called its 'outcomes'.

Example : Tossing of a coin is an experiment and appearance of head or tail are its outcomes.

2. ***Event***

One outcome or a group of outcomes associated with certain conditions. It is called '*Event*'.

Example : If one card is drawn from a pack of 52 playing cards, then drawing of : (*i*) Queen of heart; (*ii*) a card of spade; (*iii*) king of diamond etc., will be the examples of an event.

Sample Space: The set of all possible outcomes of a random experiment is called the sample space associated with it and it is generally denoted by S. Elements of sample space are called *sample points*. Total number of possible outcomes is called *exhaustive number of cases*.

3. ***Equally likely cases :***

The cases are said to be equally likely when we have no reason to expect any one more rather than the other.

Example : In a throw of an unbiased die any one out of the six faces may turn up so that all the six faces are equally likely.

4. ***Favourable Cases*** :

Out of all equally likely cases, those cases which fulfil the conditions of the 'Event' or the cases which entail the happening of the "Event" are known as favourable cases for that event.

Example : If a die is rolled and the event is $x > 4$, then the favourable cases are two *viz.*, $x = 5$ and $x = 6$, the remaining cases are unfavourable.

5. ***Exhaustive cases*** :

The total number of possible outcomes of a random experiment in a trial is known as the exhaustive number of cases.

Example : In throwing of a die the exhaustive number of cases is 6, since any one of the six faces marked with 1, 2, 3, 4, 5, 6 may come uppermost.

6. ***Mutually exclusive events*** :

Events are said to be mutually exclusive or incompatible if the occurrence of any one of them prevents the occurrence of all the other *i.e.*, if no two or more of them can occur simultaneously in the same trial.

7. ***Simple events*** :

The occurrence of a single event is known as the simple event.

8. ***Compound event*** :

Events obtained by combining together two or more elementary events are known as the compound events.

Example : In a throw of a die the event getting a multiple of 2, is a compound event because this event occurs if any one of the elementary events 2, 4 or 6 occurs.

9. ***Independent event*** :

Events are said to be independent if the happening of one event is not affected by the happening of others.

Example : If two cards are drawn from a well shuffled pack of 52 cards one after the other with replacement, then getting an ace in the first draw is independent to getting a king in the second draw but if the first card drawn in the first draw is not replaced, then second draw is dependent on the first draw.

18.8.1 Probability Defined

Broadly speaking there are four different schools of thought on concept of probability.

1. Classical (or priory or mathematical) Definition of Probability: If there are n-elementary events associated with a random experiment and m of them are favourable to an event A, then the probability of happening of A is given by

$$P(A) = \frac{m}{n},$$

The probability of an event can not be negative and cannot be greater than 1.

i.e. $$0 \leq P(A) \leq 1$$

The probability of non happening of A event

$$P(\bar{A}) = \frac{n-m}{n} = 1 - \frac{m}{n} = 1 - P(A)$$

Thus, $P(A) + P(\bar{A}) = 1$

In case $P(A) = 1$, then it is sure event and if $P(A) = 0$, it is impossible event.

Limitations of classical approach

This definition of classical probability breaks down in the following cases.

(*i*) If the various outcomes of random experiments are not equally likely or equally probable.

(*ii*) If the exhaustive number of outcomes of the random experiment is infinite or unknown.

2. Axiomatic Definition of Probability

If S be the sample space associated with a random experiment and let A be subset of S representing an event.

The probability of the event A is

$$P(A) = \frac{\text{Number of elementary events in } A}{\text{Number of elementary events is } S} = \frac{n(A)}{n(S)}$$

$$P(\bar{A}) = \frac{n(\bar{A})}{n(S)} = \frac{n(S) - n(A)}{n(S)} = 1 - P(A)$$

$\therefore$ $$P(A) + P(\bar{A}) = 1$$

3. Statistical or Empirical or Relative Frequency Definition of Probability

If an experiment is performed repeatedly under essentially homogenous and identical conditions, then the limiting value of the ratio of number of times the event to occur to the number of trials, as the number of trials becomes indefinitely large, is called probability of happening of event, it being assumed that limit is finite and unique.

Symbolically, if in n trials and event E happen m times, then the probability of the happening of E denoted by $P(E)$ is given by

$$P(E) = \lim_{n \to \infty} \frac{m}{n}$$

Limitations: (*i*) If an experiment is repeated a large number of times, the experimental conditions may not remain identical and homogeneous.

(*ii*) The limit may not attain a unique value however large the n may be.

4. Subjective approach of probability: The subjective probability is defined as the probability assigned to an event by an individual based on whatever evidence is available. The *subjective school of thought* is also known as the *personalistic school of probability*.

It may be pointed out that out of four interpretation of concept of probability, each has its own merits and one may use whichsoever approach is convenient and appropriate for the problems under consideration.

18.8.2 Axioms of Probability

A purely mathematical definition of the probability cannot give as the actual value of $P(A)$, the probability of occurrence of the event A and this must be considered as a function defined on all events.

If $P(A)$ is the probability function defined on a field B of the events if the following axioms hold

Axiom 1: (*Axiom of non-negativity*)

For each $A \subset B$, $P(A)$ is defined is real and $P(A) \geq 0$

Axiom 2: (*Axiom of certainty*)

For a sure event S in class B, $P(S) = 1$

Axiom 3: (*Axiom of additivity*)

If A_1 and A_2 are two mutually exclusive (or disjoint event) then

$$P(A_1 \cup A_2) = P(A_1) + P(A_2)$$

If $\{A_n\}$ is any finite or infinite sequence of disjoint events in B, then

$$P\left(\bigcup_{i=1}^{n} A_i\right) = \sum_{i=1}^{n} P(A_i)$$

18.8.2.1 Algebra of Events

Table 18.23: Algebra of Events

Event	*Set theoretic notation*
Not A	$\bar{A}$
A or B (at least one of A or B)	$A \cup B$
A and B	$A \cap B$
A but not B	$A \cap \bar{B}$
Neither A nor B	$\bar{A} \cap \bar{B} = \overline{A \cup B}$
At least one of A, B or C	$A \cup B \cup C$
Exactly one of A and B	
$(A \cap \bar{B}) \cup (\bar{A} \cap B)$	

Exhaustive events: For a random experiment E, a set of events is said to be exhaustive if one of them must necessarily occur every time the experiment is performed.

If $E_1, E_2, \ldots E_n$ are the subsets of a sample space S, and if

$E_1 \cup E_2 \cup E_3$ ..... $E_n = S$, then $E_1, E_2, ... E_n$ form a set of exhaustive events.

Further if $E_i \cap E_j = \phi$ for $i \neq J$, then event E_1 and E_2 are mutually exclusive and exhaustive events.

Odds of an Events: *Odds in favour of an event* $A = \frac{P(A)}{P(\bar{A})} = \frac{n(A)/n(S)}{n(\bar{A})/n(S)} = \frac{n(A)}{n(\bar{A})}$

$$= \frac{\text{No. of favourable out comes}}{\text{No. of unfavourable out comes}} = \frac{m}{n}$$

Total number of out comes, $n(S) = m + n$

Odd against the event A =

$$\frac{P(\bar{A})}{P(A)} = \frac{\text{No. of favourable out comes}}{\text{No. of unfavourable out comes}} = \frac{n}{m},$$

$n(S) = m + n.$

18.8.3 Some theorems on Probability

Theorem 1: The probability of impossible event is zero *i.e.* $P(\phi) = 0$

Theorem 2: If $A_1 \subset A_2$ then $P(A_1) \leq P(A_2)$ and $P(A_2 - A_1) = P(A_2) - P(A_1)$

Theorem 3: For every even A

$$0 < P(A) \leq 1$$

i.e. probability of an event lies between 0 and 1.

Theorem 4: If A' is the complement of A then $P(A')$ or $P(\bar{A}) = 1 P - (A)$

Example 18.28. *Two dice are thrown and the sum of numbers which came up on the dice is noted. Let us consider the following events.*

A: "The sum is even", B: "The sum is multiple of 3" C: "The sum is less than 4", D: "The sum is greater than 11".

Which pairs of these events are mutually exclusive?

Solution: We have

A = {(1, 1), (2, 2), (1, 3), (1,5), (2, 4), (2, 6), (3, 1), (3, 3), (3, 5), (4, 2), (4, 4), (4, 6), (5, 1), (5, 3), (5, 5), (6, 2), (5, 4), (6, 6)}

B = {(1, 2), (2, 1), (1, 5), (5, 1), (3, 3), (2, 4), (4, 2), (3, 6), (6, 3), (4, 5), (5, 4), (6, 6)}

C = {(1, 1), (2, 1), (1, 2)}

D = {6, 6}

Clearly

$A \cap B \neq \phi, A \cap C \neq \phi, A \cap D \neq \phi, B \cap D \neq \phi, C \cap D = \phi$

Thus C and D are mutually exclusive events.

Example 18.29. *A bag contains 50 tickets numbered 1, 2, 3, ... 50 of which five are drawn at random and arranged in ascending order of magnitude* ($x_1 < x_2 < x_3 < x_4 < x_5$). *Find the probability* $x_3 = 30$.

Solution: $n(S) = 50\ C_5$.

Since $x_3 = 30$, So x_1, x_2, x_2 must be small than 30 and x_4, x_5 must be greater than 30.

Number of numbers small than 30 = 29 and number greater than 30 = 20

∴ Required probability

$$= \frac{26_{C_2} \times 20_{C_2}}{50_{C_5}} = \frac{551}{15134}$$

18.8.3.1 Addition theorems on probability

Theorem 1: If A and B are two events associated with random experiment, then

$P(A \cup B) = P(A) + P(B) - P(A \cap B)$

Corollary:

In case A and B are mutually exclusive

Then

$P(A \cap B) = 0$ and $P(A \cap B) = P(A) + P(B)$.

Theorem 2: If A, B, C are three events associated with random experiment then

$P(A \cup B \cup C) = P(A) + P(B) + P(C) - P(A \cap B) - P(B \cap C) - P(C \cap A) + P(A \cap B \cap C)$

Corollary: If A, B, C are mutually exclusive events, then

$P(A \cap B) = P(B \cap C) = P(C \cap A) = P(A \cap B \cap C) = 0$

and $P(A \cup B \cup C) = P(A) + P(B) + P(C)$.

Theorem 3: Let A and B be two events associated with a random experiment.

Then

(*i*) $A(\bar{A} \cap B) = P(B) - P(A \cap B)$

(*ii*) $P(A \cap \bar{B}) = P(A) - P(A \cap B)$

Example 18.30. *A, B, C are three mutually exclusive and exhaustive events associated with a random experiment. Find P(A) it being given that* $P(B) = \frac{3}{2} P(A)$ *and* $P(C) = \frac{1}{2} P(B)$.

Solution: Let

$$P(A) = P \text{ then } P(B) = \frac{3}{2} P(A) = \frac{3}{2} P$$

$$P(C) = \frac{1}{2} P(B) = \frac{1}{2} \times \frac{3}{2} P(A) = \frac{3}{4} P$$

Since events A, B, C are mutually exclusive, $P(A) + P(B) + P(C) = P(S) = 1$.

$$\therefore \quad P + \frac{3}{2}P + \frac{3}{4}P = 1,$$

$$\text{or} \quad P = \frac{4}{13}$$

Example 18.31. *The probability that at least one of the events A and B occurs is 0·6. If A and B occur simultaneously with probability 0·2, then find* $P(\bar{A}) + P(\bar{B})$.

Solution: We have

P(at least one)

$= P(A \cup B) = 0{\cdot}6,\ P(A \cap B) = 0{\cdot}2$

Now $P(A \cup B)$

$= P(A) + P(B) - P(A \cap B)$

or $0{\cdot}6 = P(A) + P(B) - 0{\cdot}2$

$= 1 - P(\bar{A}) + 1 - P(\bar{B}) - 0{\cdot}2$

or $0{\cdot}6 = 1{\cdot}8 - \{P(\bar{A}) + P(\bar{B})\}$

or $P(\bar{A}) + P(\bar{B}) = 1{\cdot}8 - 0{\cdot}6 = 1{\cdot}2$

18.8.4 Conditional Probability

The probability of happening of an event A when the event B has already occurred is called conditional probability and is denoted by $P(A/B)$.

Similarly $P(B/A)$ means the probability of an event B when event A has already occurred.

$$P(A/B) = \frac{P(A \cap B)}{P(B)} = \frac{n(A \cap B)}{n(B)}$$

$$\text{Similarly, } P(B/A) = \frac{P(A \cap B)}{P(A)} = \frac{n(A \cap B)}{n(A)}$$

18.8.4.1 Multiplication theorems on probability

Theorem 1: If A and B are two events associated with a random experiment.

Then

$P(A \cap B) = P(A) \cdot P(B/A)$ if $P(A) \neq 0$

$P(A \cap B) = P(B) \cdot P(A/B)$ if $P(B) \neq 0$

Theorem 2. If $A_1, A_2 \ldots A_n$ are n events related to a random experiment then

$P(A_1 \cap A_2 \cap B_3 \ldots \cap A_n) = P(A_1) \cdot P(A_2/A_1) \cdot P(A_3/A_1 \cap A_2) \cdot \ldots\ldots P(A_n/A_1 \cap A_2 \ldots \cap A_n)$

Example 18.32. *A bag contains 5 white 7 red and 8 black balls. If four balls are drawn one by one without replacement, find the probability of getting all white balls.*

Solution: If A_1, A_2, A_3, A_4 denote the events of getting white balls in first, second, third and fourth draw respectively.

Then

Required probability

$= P(A_1 \cap A_2 \cap A_3 \cap A_4) = P(A) \cdot P(A_2/A_1) \cdot P(A_3/A_1 \cap A_2) P(A_4/A_1 \cap A_2 \cap A_3)$

$$= \frac{5}{20} \times \frac{4}{19} \times \frac{3}{18} \times \frac{2}{17} = \frac{1}{789}$$

18.8.5 Independent Events

Events are said to be independent, if occurrence or non occurrence of one does not affect the probability of the occurrence or non occurrence of the other.

In case of independent event $P(A/B) = P(A)$ and $P(B/A) = P(B)$.

18.8.5.1 Theorem on independent events

Theorem 1: If A and B are independent events associated with a random experiment, then

$$P(A \cap B) = P(A) \cdot P(B).$$

Corollary: If $A_1, A_2, A_3 \ldots. A_n$ are independent events associated with random experiment.

then,

$$P(A_1 \cap A_2\ A_3 \ldots. A_n) = P(A_1) \cdot P(A_2) \ldots P(A_n)$$

Theorem 2: If $A_1, A_2 \ldots A_n$ are n independent events associated with random experiment, then

$$P(A_1 \cup A_2 \cup A_3 \ldots. \cup A_n) = 1 - P(\bar{A}) \cdot P(\bar{A}_2) \ldots. P(A_n)$$

i.e. $\quad P(\text{at least one}) = 1 - P\ (\text{none})$

Example 18.33. *If event A and B are independent. Find P(B) if P(A) = 0·35 and P(A ∪ B) = 0·6.*

Solution: We have $P(A \cup B) = P(A) + P(B) - P(A \cap B)$

Since events are independent so $P(A \cap B) = P(A) \times P(B)$

$\therefore \quad 0{\cdot}6 = 0{\cdot}35 + P(B)\ (1 - 0{\cdot}35)$

or $\quad P(B) = \dfrac{0{\cdot}25}{0{\cdot}65} = \dfrac{5}{3}$

18.8.6 The Law of Total Probability and Baye's Theorem

Let S be the sample space and let $E_1, E_2, \ldots. E_n$ be n mutually exclusive and exhaustive events associated with a random experiment. If A is any event which occurs with E_1 or $E_2 \ldots.$ or E_n, then

$$P(A) = P(E_1) \cdot P(A/E_1) + P(E_2) \cdot P(A/E_2) + \ldots. + P(E_n) \cdot P(A/E_n).$$

Example 18.34. *Find the probability of drawing a one-rupee coin from a purse with two compartments one of which contains 3 fifty-paise coins and 2-one rupee coins and other contains 2 fifty-paise coins and 3 one-rupee coins.*

Solution: Let event $E_1 \rightarrow$ The first compartment of the purse is chosen.

$E_2 \rightarrow$ 2nd compartment of the purse is chosen.

$A \rightarrow$ The coin drawn is a one-rupee coin.

Then

$$P(E_1) = P(E_2) = \frac{1}{2},\ P(A/E_1) = \frac{2}{5},\ P(A/E_2) = \frac{3}{5}.$$

Thus by total law of probability

$$P(A) = P(E_1) \cdot P(A/E_1) + P(E_2) \cdot P(A/E_2) = \frac{1}{2} \times \frac{2}{5} + \frac{1}{2} \times \frac{3}{5} = \frac{1}{2}.$$

18.8.6.1 Baye's theorem

Let S be the sample space and let $E_1, E_2, E_3 \ldots. E_n$ be n mutually exclusive and exhaustive events associated with a random experiment. If A is any event which occurs with E_1 or E_2 or ..... E_n then.

$$P(E_i/A) = \frac{P(E_i) \cdot P(A/E_i)}{\sum_{i=1}^{n} P(E_i) \cdot P(A/E_i)},$$

$i = 1, 2, 3 \ldots\ldots n$

Example 18.35. *A man is known to speak truth 3 out of 4 times. He throws a die and reports that it is a six. Find the probability that it is actually a six.*

Solution: Let E_1 be the event six occurs on die and E_2 be the event that six does not occurs on die.

Then $P(E_1) = \frac{1}{6}$, $P(E_2) = \frac{5}{6}$.

If A be the event that man reports that it is a six.

Then

$P(A/E_1)$ = Probability that six has occurred and man reports the truth $= \frac{3}{4}$.

$P(A/E_2) = P$ (Man reports false, *i.e.* six has not occurred) $= 1 - \frac{3}{4} = \frac{1}{4}$.

$\therefore$ Required probability

$$= P(E_1/A)$$

$$= \frac{P(E_1)\cdot P(A/E_1)}{P(E_1)\cdot P(A/E_1) + E(E_2)\cdot P(A/E_2)}$$

$$= \frac{\frac{1}{6}\times\frac{3}{4}}{\frac{1}{6}\times\frac{3}{4}+\frac{5}{6}\times\frac{1}{4}} = \frac{3}{8}.$$

18.9 RANDOM VARIABLE AND PROBABILITY DISTRIBUTION

Random variable : The variable whose value is determined by outcome of a random experiment is called random variable. A random variable is also known as *chance variable* or *stochastic variable.*

A random variable may be thought of as a function defined over the elements of sample space. Random variables are denoted by capital letters X, Y, and so on to distinguish them from their possible values given in lower case.

Intuitively by a random variable we mean a real number X connected with the outcome of a random experiment.

For example in two tosses of a coin the number of heads may be 0, 1 or 2 which are random variable *i.e.* $X = 0, 1, 2$.

A random variable may be *discrete or continuous* and we have already disussed.

Probability distribution: Frequency distributions are broadly classified in two heads.

1. Observed frequency distribution and
2. Theoretical or expected distribution.

So far we have discussed the observed frequency distribution. Such distributions are based on observation and experiment.

The distribution as are expected on the basis of previous experience or theoretical considerations are known as '*theoretical distribution*' or '*probability distribution*'.

Like wise random variable the probability distribution may be discrete or continuous.

Distribution function: Let X be a random variable. The function F defined for all real x, by

$$F(x) = P(X \le x)$$

where x is any real number, *i.e* $-\infty < x < \infty$ is called the distribution function (d.f.) of random variable

In other words if

$$F(x) = P(X \le x) = \sum_{u \le x} f(u) = \int_{-\infty}^{x} f(x)dx$$

Then $F(x)$ is called distribution function of random variable X.

The distribution function is also called *commulative distribution function.*

The distribution function has following properties

(*i*) $F'(x) = f(x) \ge 0$ so that $F(x)$ is a non-decreasing function

(*ii*) $F(-\infty) = 0$; (*iii*) $F(\infty) = 1$

(*iv*) $P(a \le x \le b) = \int_{a}^{b} f(x)dx$

$$\int_{-\infty}^{b} f(x)dx - \int_{-\infty}^{a} f(x)dx = f(b) = f(a)$$

The behaviour of the random variable is completely described by the probability law which in turn may be characterised by either the *probability density function* or *commultive distribution function.*

18.10 DISCRETE PROBABILITY DISTRIBUTION.

It is generally used to those random events in which the outcome can be described as success or failure *i.e.*, there are two mutually exclusive events in an experiment. The Binominal or Poisson distribution can be used to find the probability of occurrence of an event r times in n successive years.

The use of discrete probability distributions is restricted generally to those random events in which the outcome can be described as success or failure *i.e.* there are only two mutually exclusive events in an experiment. Moreover, the successive trials are independent and the probability of success remain constant from trial to trial.

Two important discrete probability distributions are *Binomial distribution* and *Poisson distribution.*

Bernoulli trails

It is an experiment which has only two possible outcomes, success (S) and failure (F). Various outcomes of Bernoulli trails are: tossing a coin, firing a target, playing a game etc. Therefore, we can say that any chance mechanism whose outcomes can be grouped into two classes may be regarded as a Bernoulli trail.

18.10.1 Binomial Distribution

Let X be the random variable that equals the number of successes in n trails. To obtain probabilities concerning X, we proceed as follows:

If p and $(1 - p)$ are the probabilities of success and failure on any one trail, then the probability of getting n successes and $n - x$ failures in some specific order is $p^x (1 - p)^{n-x}$.

The number of ways in which we can select the x trails on which there is to be a success in ${}^nC_x$, the number of combination of x objects selected from a set of n objects, and is defined by

$p{\cdot}m{\cdot}f,\ f(x) = {}^nC_x p^x(1 - p)^{n-x}$

for $x = 0, 1, 2, ..., n$.

This probability distribution is called the binomial distribution because for $x = 0, 1, 2, 3,$ ...... and n, the values of the probabilities are the successive terms of binomial expansion of $[p + (1 - p)]^n$.

The term ${}^nC_x$ are referred as binomial coefficient. In some books, $(1-p)$ is referred as q, then

$$f(x) = {}^nC_x\ p^x q^{n-x};\ (x = 0, 1, 2, ..., n)$$

The sum of probabilities $= q^n + {}^nC_1\ p\,q^{n-1} + {}^nC_2 p^2 q^{n-2} + ... p^n = (q + p)^n = 1$

Properties of Binomial Distribution

(*i*) This distribution is specified by two parameters n and p.

(*ii*) Mean $(\mu) = np$,

Variance $= npq$,

Standard deviation $(\sigma) = \sqrt{npq}$,

$$\text{Skewness } (\gamma_1) = \frac{q-p}{\sqrt{npq}},$$

$$\text{Kurtosis } (\gamma_2) = \frac{1-6pq}{npq}$$

Example 18.36. *The maximum depth of rainfall at a station was 250 mm and return period of 50 years. Determine the probability of one day rainfall depth equal to or greater than 250 mm occurring (a) once in 20 successive years, and (b) two times in 15 successive years.*

Solution: Here, $P = \frac{1}{50} = 0{\cdot}20$

Case (*a*): Given ; $n = 20$, $x = 1$

Therefore,

$$P_{1,\,20} = {}^{n}C_x \cdot p^x (1-p)^{n-x}$$
$$= 20C_1 \cdot (0{\cdot}02)^1 \cdot (0{\cdot}98)^{20-1} = 0{\cdot}272$$

Case (*b*) = Given ; $n = 15$; $x = 2$

$$P_{2,15} = 15_{C_2} \cdot (0{\cdot}02)^2 (0{\cdot}98)^{15-2} = 0{\cdot}0292$$

18.10.2 Poisson Distribution

Let X be the random variable which takes non negative integer values only *i.e.*, $k = 0, 1, 2, ...$ Then the probability density function

$$P(X = k) = \frac{\lambda^k e^{-\lambda}}{k!} \quad \text{for } x = 0, 1, 2, \ldots\ldots\ldots..$$

where, $\lambda > 0$ is called the *Poisson distribution with parameter* λ. It is applicable to events in which p is very small and n is large. Since p is very small, these events are called rare events. It is also useful when it is possible to prescribe the number of times an event occurs but not the number of times it does not occur.

Under the condition that $p \to 0$ as $n \to \infty$, such that $np = \lambda > 0$, B binomial distribution approximates to poisson distribution.

For binomial distribution,

$$P(X = k) = n_{C_k} \left(\frac{\lambda}{n}\right)^k \left(1 - \frac{\lambda}{n}\right)^{n-k}$$

It can be shown that as $n \to \infty$.

$$P(X = k) = \frac{\lambda^k \cdot e^{-\lambda}}{k!}, \; k = 0, 1, 2, \ldots..$$

which is the Poisson distribution.

Here, λ = the mean of the distribution,

k = the specific value of the discrete random variable.

e = 2·718 (the base of natural logarithms)

The properties of Poisson distribution are;

Mean (μ) = λ, Variance = λ,

Thus for Poissons distribution Mean = Variance.

Standard deviation = $\sqrt{\lambda}$

Necessary conditions for application of Poisson distribution are:

(*i*) Discrete events,

(*ii*) Two events cannot coincide,

(*iii*) Average number of events per unit time is constant.

(*iv*) Independent event.

Poisson distribution is applicable in following cases:

→ number of rainy days in a given city,

→ number of droughts in a given time period.

→ probability of rare flood events.

Example 18.37. *If the probability that an individual suffers a bad reaction from injection of a given serum is 0·001, determine the probability that out of 2000 individual.*

(a) exactly 3, (b) more than 2, individuals will suffer a bad reaction

Solution: Let X denote the number of individuals suffering a bad reaction. X is Bernoulle distributed but sicne bad reactions are assumed to be rare events, we can suppose that X is Poisson distributed *i.e.*

$$P(X = x) = \frac{\lambda^x e^{-\lambda}}{x!} = 2000 \times 0{\cdot}001 = 2$$

where $\lambda = np$

$$(a)\; P(X = 3) = \frac{2^3 e^{-2}}{3!} = 0{\cdot}180$$

$(b)\; P(X > 2)$

$= 1 - P(X = 0) + P(X = 1) + P(X = 2)$

$$= 1 - \frac{2^0 e^{-2}}{0!} + \frac{2^1 e^{-2}}{1!} + \frac{2^2 e^{-2}}{2!} = 1 - 5e^{-2} = 0{\cdot}323$$

18.11 CONTINUOUS PROBABILITY DISTRIBUTIONS

If X is a continuous random variable the probability that X takes on any on particular value is generally zero. Therefore we cannot define a probability function in the same way as for discrete variable. In order to arrive at a probability distribution for a continuous random variable we note that the probability that the X lies between two values is meaningfull.

Let us postulate the existence of function $f(x)$ such that

(*i*) $f(x) \geq 0$

(*ii*) $\int_{-\infty}^{\infty} f(x)\,dx = 1$

where the second is a mathematical statement of the fact that a real values random variable must certainty lie between $-\infty$ and ∞. We often define probability that X lies between a and b by

$$P(a < X < b) = \int_a^b f(x)\,dx$$

A function $f(x)$ which satisfies the above requirements is called *probability function* of *probability distribution* of random variable but more often called a *probability density function* (pdf) or simply *density function* (*df*).

The distribution function $F(x)$ for a continuous random variable is given by

$$F(x) = P(X \leq x) =$$

$$P(-\infty < X < x) = \int_{-\infty}^{\infty} f(u)du$$

If $f(x)$ is the density function for a random variable X then can represent $y = f(x)$ graphically by a curve as in Fig. 18.14 (*a*).

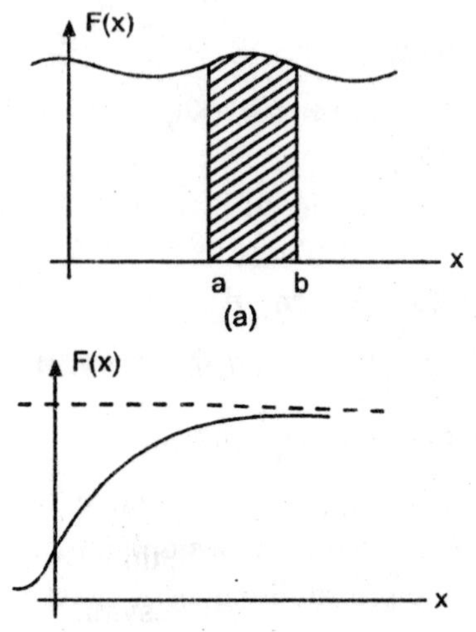

Fig. 18.14

Since $f(x) \geq 0$ the curve cannot fall below the x-axis. The entire area bounded by the curve and x-axis must be 1. Geometrically the probability that X is between a and b, *i.e.* $P(a < X < b)$, is then represented by the area shaded in 6.19 (*a*).

The distribution function $F(x) = P(X \leq x)$ is monotonically increasing function which increase from 0 to 1 and represented by curve as shown in Fig. 18.14 (*b*).

The commonly used continuous distributions are discussed below:

18.11.1Uniform Distribution

A random variable X is called a *uniform distribution* over $[a, b]$ if its *pdf* $f(x)$ is given by

$$f(x) = \frac{1}{b-a}, a \leq X \leq b$$

It is denoted by $U[a, b]$ and its expected value and variance are given by

$$E(X) = \frac{a+b}{2} \text{ and } V(X) = \frac{(b-a)^2}{12}.$$

The *cdf* of $U[a, b]$ is

$$F(x) = \int_a^x f(t)dt = \int_a^x \frac{1}{b-a}dt = \frac{x-a}{b-a}.$$

18.11.2 The Normal Distribution

The *normal distribution* also called the *normal probability distribution* happens to be most useful distribution for continuous variable. It was developed first by De Moivre as a limiting form of Binomial distribution.

One of the best known probability density function is that forming the similar bell-shaped curve for the normal distribution.

It is a bell shaped curve symmetric to mean both sides. The Fig. 18.15 show the shape of normal distribution (or normal probability density).

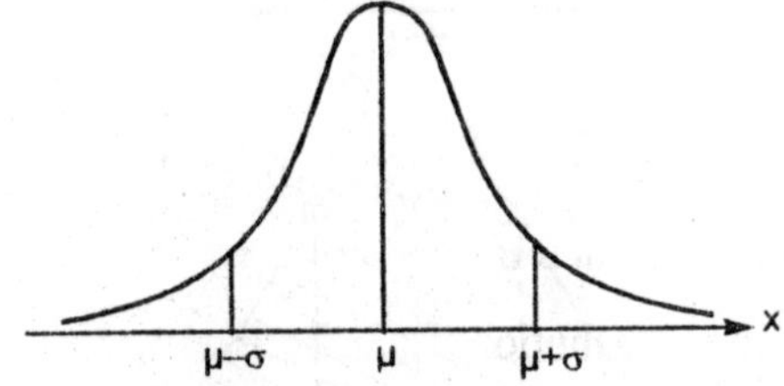

Fig. 18.15: Graph of normal probability density.

The equation of the normal probability density function (p.d.f) is given by

$$f(x\,;\,\mu, \sigma^2) = \frac{1}{\sqrt{2\pi}\,\sigma} e^{[-(x-\mu)^2/2\sigma^2]},$$

$$(-\infty < x < +\infty)$$

and $(-\infty < \mu < \infty), \sigma > 0$

where, μ called *mean* and σ^2 called *variance* are parametres.

The corresponding distribution function is given by

$$F(x) = P(X \leq x)$$

$$= \frac{1}{\sigma\sqrt{2\pi}} \int_{-\infty}^{x} e^{-(v-\mu)^2/2\sigma^2} dv$$

In this case we say that random variable X is normally distributed with mean μ and variance σ^2.

18.11.2.1 Standard normal distribution

Since the normal probability density cannot be integrated in closed form between every pair of limits a and b, probabilities relating to normal distributions are usually obtained from table 6. This table pertains to the standard normal distribution with $\mu = 0$ and $\sigma = 1$, and its entries are the value of standard normal variate.

The *p.d.f.*

$$f(z, 0, 1) = \frac{1}{\sqrt{2\pi}} e^{-\frac{1}{2}z^2}$$

where, $= \dfrac{x-\mu}{\sigma}$

This is often referred as *standard normal density function* or *distribution.* The corresponding distribution function is given by

$$F(z, 0, 1) = P(Z \leq z)$$

$$= \frac{1}{\sqrt{2\pi}} \int_{-\infty}^{z} e^{-u^2/2} du = \frac{1}{2} + \frac{1}{2\pi} \int_0^z e^{-u^2/2} du$$

We some times call the value of z of the *standarizated normal* variate Z the *standard score.*

The cumulative probabilities $F(z)$ correspond to the area under the standard normal density of the left of z as shown in Fig. 18.16.

To find the probability that a random variable having the standard normal distribution will take on a value between a and b, we use the equation

$$P(a < Z \leq b) = F(b) - F(a),$$

The area is shown in Fig. 18.17.

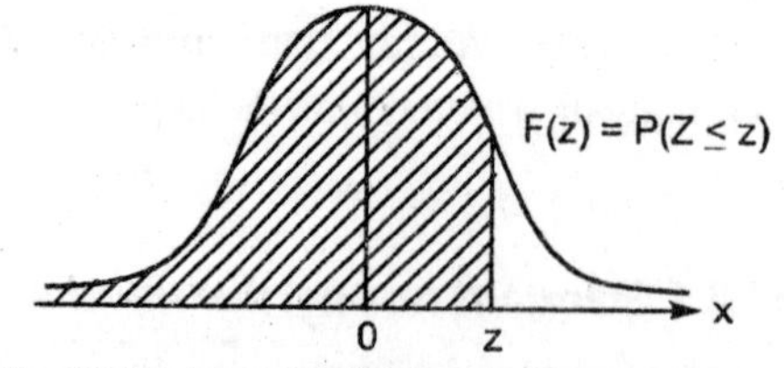

Fig. 18.16: The standard normal probabilities

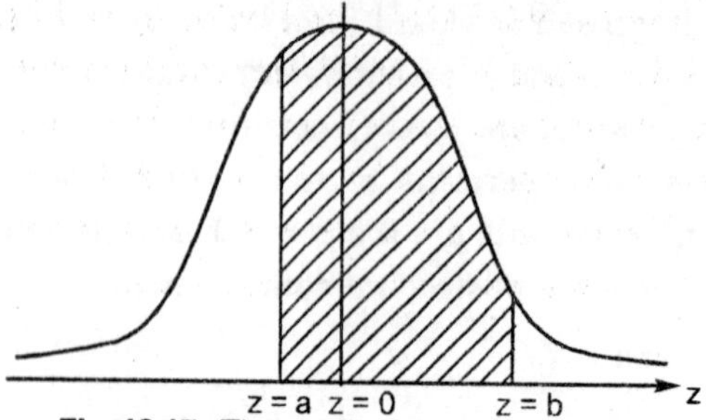

Fig. 18.17: The standard normal probability

The total area under standard normal curve is unity.

Example. 18.38. *If a random variable X is normally distributed with mean 113·4 and the variance 1·36, compute the probability that it assumes values between 112·1 and 115·4 and probability that its value is at least 113·0.*

Solution: Given

$$\bar{X} = 113{\cdot}4,\ \sigma^2 = 1{\cdot}36$$

$$\therefore \quad \sigma = \sqrt{1{\cdot}36} = 1{\cdot}166.$$

(*i*) *Probability that random variable X assumes the value between 112·1 and 115·4.*

$$z_1 = \frac{X_1 - \mu}{\sigma} = \frac{112{\cdot}1 - 113{\cdot}4}{1{\cdot}166} = -1{\cdot}114$$

$$z_2 = \frac{X_2 - \mu}{\sigma} = \frac{115{\cdot}4 - 113{\cdot}4}{1{\cdot}166} = 1{\cdot}792$$

Thus required probability

$$= P\,(112{\cdot}1 < X < 115{\cdot}4)$$

$$= P\,(-1{\cdot}114 < z < 1{\cdot}792)$$

Area under normal curve between

$$z = -1{\cdot}114 \text{ and } z = 1{\cdot}792$$

Area between $z = 0$ to $z = 1{\cdot}114$ + Area between

$z = 0,\ z = 1{\cdot}792$

$= 0{\cdot}3440 + 0{\cdot}4634 = 0{\cdot}8074$

Probability that the value is at least 113·0

For $X = 113{\cdot}0,\ z = \dfrac{113{\cdot}0 - 113{\cdot}4}{1{\cdot}166} = -0{\cdot}24$

∴ Required probability

$= P(X \geq 113) = P\,(z > -0{\cdot}24)$

$= 0{\cdot}5$ + Area curve between $z = 0$ and $z = -0{\cdot}24$

$= 0{\cdot}5$ + area of curve between $z = 0$ to $0{\cdot}24$

$= 0{\cdot}5 + 0{\cdot}0948 = 0{\cdot}5948$

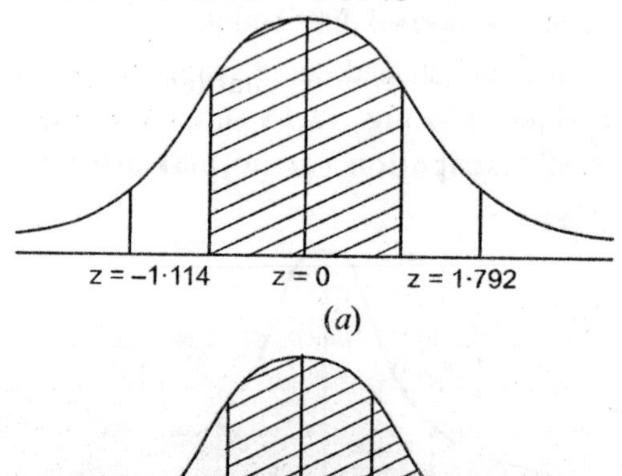

Fig. 18.18

18.12 SAMPLING FUNDAMENTALS

Sampling may be defined as the selection of some part of aggregate or totality on the basis of which a judgement or inference about aggregate or totality is made. In other words, it is the process of obtaining information about an entire population by examining only a part of it.

Sampling is used in practice for a variety of reasons such as

1. Sampling can save time and money. A sample study is usually less expensive than

a census study and produces result at a relatively faster speed.

2. Sampling may enable more accurate measurements for a sample study is generally conducted by trained and experienced investigators.
3. Sampling remains the only way when population contains infinitely many members.
5. Sampling remains the only choice when the test involves the destruction of the item under study.
6. Sampling usually enables to estimate sampling errors and, thus assists to obtain information concerning some characterstic of population.

18.12.1 Population Parameters

A population is considered to be known when we know the probability distribution $f(x)$ (probability function or density function) of the associated random variable X.

If, for example, X is normally distributed, we say that the population is *normally distributed* or that we have a *normal population*. Similarly, if X is binomially distributed, we say that the population is *binomially distributed* or that we have a *binomial population*.

There will be certain quantities that appear in $f(x)$, such as μ and σ in the case of the normal distribution or p in the case of the binomial distribution. Other quantities such as the median, moments, and skewness can then be determined in terms of these. All such quantities are often called *population parameters*. When we are given the population so that we know $f(x)$, then the population parameters are also known.

An important problem arises when the probability distribution $f(x)$ of the population is not known precisely, although we may have some idea of, or at least be able to make some hypothesis concerning, the general behaviour of $f(x)$. For example, we may have some reason to suppose that a particular population is normally distributed. In that case we may not know one or both of the values μ and σ and so we might wish to draw statistical inferences about them.

$P(X_1 = x_1, X_2 = x_2, \ldots\ldots\ldots\ldots, X_n = x_n) = f(x_1) \cdot f(x_2) \ldots\ldots f(x_n)$

Any quantity obtained from a sample for the purpose of estimating a population parameter is called a *sample statistic*, or briefly *statistic*. Mathematically, a sample statistic for a sample of size n can be defined as a function of the random variables, $X_1, \ldots\ldots\ldots X_n$ *i.e.*, $g(X_1, \ldots\ldots X_2)$. The function $g(X_1, \ldots\ldots\ldots X_n)$ is another random variable, whose values can be represented by $g(X_1, \ldots\ldots X_n)$. The word statistic is often used for the random variable or for its values, the particular sense being clear from the context.

18.12.2 Sampling Distributions

As we have seen, a sample statistic that is computed from $X_1, \ldots\ldots, X_n$ is a function of these random variables and is therefore, itself a random variable. The probability distribution, of a sample statistic is often called the *sampling distribution* of the statistics.

18.13 STATISTICAL INFERENCE

Statistical inference is that branch of statistics which is concerned with using probability concept that deal with uncertainty in decision making. The field of statistical inference has had a fruitful development since the later half of 19th century.

It refers to the process of selecting and using a sample statistic to draw inference about a population parameter based on a subset of it the sample drawn from the population. Statistical inference treats two different classes of problem.

1. **Hypothesis testing:** *i.e.* to test same hypothesis about parent population from which the sample is drawn.

Hypothesis testing begin with an assumption, called a hypothesis, that make about the population parameter. A hypothesis is the supposition made as a basis for reasoning. There can be several types of hypothesis.

For example, a coin may be tossed 200 times and we may get heads 80 times and tail 120 times. We may now interested in testing the hypothesis that the coin is unbiased. To take another example we may study the average weight of 100 students of a particular college and may get the result of 110 lb. We may now be interested in testing the hypothesis that the sample has been drawn from the population with 115 lb. Similarly, we may be interested in testing the hypothesis that the variables in the population are uncorrelated

2. **Estimation** *i.e.* to use the 'statistics' obtained from the sample as estimate of the unknown 'parameter' of the population from which the sample is drawn.

In both of the uses the particular problem at hand is structured in such as way that inference about relevant population can be made from population data.

18.13.1 Hypothesis Testing

There are many problems in which, rather than estimate the value of a parameter is available we must decide whether a statement concerning a parameter is true or false. This is a reason that we must test a hypothesis about a parameter. For example, the quality of water in the river violating the relevant standards.

The hypothesis tests are of two types:

(*i*) Parametric tests,

(*ii*) Non-parametric tests.

Parametric tests require the specification of the type of distribution of data. *e.g.*, *t*-test, linear regression test.

Non-parametric tests donot require specification of the distribution of data.

The main steps in conducting hyprothesis testing are as follows:

(*i*) We formulate a null hypothesis and an appropriate alternative hypothesis which we accept when the null hypothesis must be rejected.

(*ii*) We specify the probability of a type-*I* error; if possible desired or necessary, we may also specify the probabilities of type-*II* errors for particular alternatives.

(*iii*) Based on the sampling distribution of an appropriate statistic, we construct a criterian for testing the null hypothesis against a given alternative.

(*iv*) We calculate from the data the value of the statistic on which the decision is to be based.

(*v*) We decide whether to reject the null hypothesis or whether to fail to reject it.

The example of null hypothesis the distribution of flow at two stations are different.

After the hypothesis has been formulated, statistical analysis is carried out to either accept it or reject it. Table 18.24 produces four possible combinations.

Table 18.24

	Accept Hypothesis	***Reject Hypothesis***
Hypothesis is true	correct decision $(1-\alpha)$	Type-*I* error (α)
Hypothesis is false	Type-*II* error (β)	Correct decision $(1-\beta)$

If hypothesis is true and not rejected or false and rejected, the decision is in either case is correct. If hypothesis is true but rejected, it is rejected in error, and if hypothesis is false but not rejected, this is an error. The first of these errors is called a *type-I error* and is designated by the Greek letter α (alpha); the second is called a *type-II error* and is designated by the Greek letter β (beta).

The notation of the hypothesis:

H_1 : The alternative hypothesis is the claim we wish to establish.

H_0 : The null hypothesis is the negation of the claim.

Fig. 18.19 shows two sided test of hypothesis. The shaded area is termed as the area of rejection or the critical region.

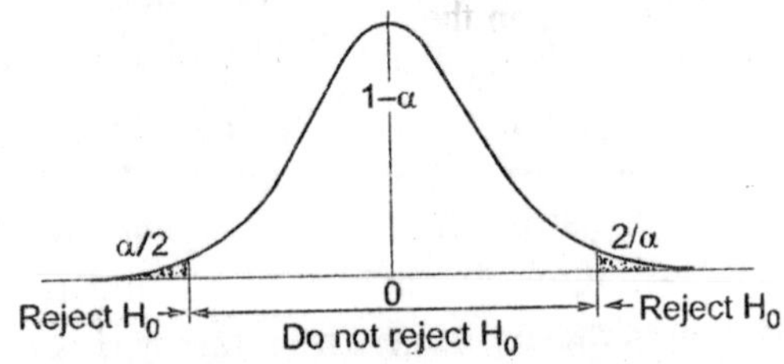

Fig. 18.19: Two-sided test of hypothesis

18.13.1.1 The *t*-test

The uncertainty in the estimates of parameters of a probability distribution, such as mean and standard deviation, can be studied using student's *t*-distribution. A random variable is said to follow student's *t*-distribution or simply *t*-distribution, if its p·d·f .is of the form

$$f(t) = k \cdot \left(1 + \frac{t^2}{n}\right)^{-\frac{(n+1)}{2}} \quad (-\infty < t < \infty)$$

where, k = is a constant.

n = number of degree of freedom.

t = a variable.

The percentage points of t distribution with n degrees of freedom are denoted by t_p, n or briefly t_p if the degree of freedom is understood.

Since the *t*-curve is symmetrical about zero [Fig. (18.20)].

$t_{1-p} = -t_p$ for example $t_{0\cdot 95} = -t_{0\cdot 05}$.

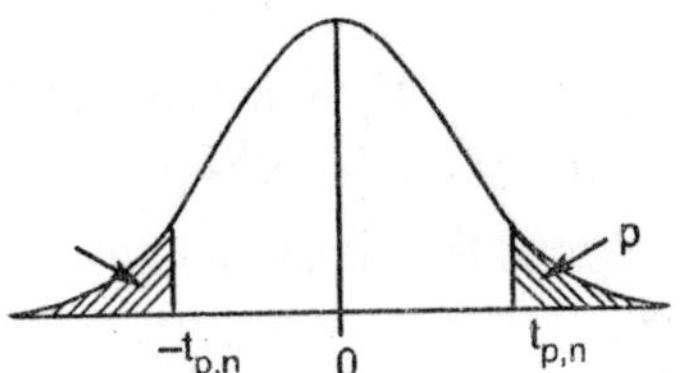

Fig. 18.20: t-distribution

18.13.1.2 Chi-square distribution

We speak of *goodness of fit* when we try to compare an observed frequency distribution with corresponding value of an expected, or theoretical distribution. The problem of testing the goodness of fit of theoretical distributions to sample distributions is essentially the same that of deciding whether there are significant differences between population and sample values. A important significance test for goodness of fit of theoretical distribution, the *Chi-square* test is described below. It enables us to find if the deviation of the experiment from theory is just by chance or is it really due to the inadequacy of the theory to fit the data.

Chi-square test

If O_i (i = 1, 2, ..., n) is a set of observed (experimental) frequencies and E_i (i = 1, 2, ...n)) is the corresponding set of expected (theoretical or hypothetical) frequency then *Karl Pearson's Chi-square* is given by

$$\chi^2 = \sum_{i=1}^{n}\left[\frac{(O_i - E_i)^2}{E_i}\right] \left(\sum_{i=1}^{n} O_i = \sum_{i=1}^{n} E_i\right)$$

with (n–1) d.f (degree of freedom).

χ^2–test is an approximate test for large values of n. For the validity of chi-square test of '*goodness of fit*' between theory and experiment following conditions must be satisfied.

(*i*) The sample observations should be independent.

(*ii*) Constants of cell frequencies, if any should be linear *e.g* $\Sigma O_i = \Sigma E_i$.

(*iii*) *N*, the total frequency should be reasonably large.

(*iv*) No theoretical cell frequency should be less than 5.

The goodness of fit test uses the Chi-square distribution to determine hypothesized probability distribution for a population provides a good fit. Acceptance or rejection of the hypothesized population distribution is based upon differences between observed frequencies (O_i) and expected frequencies (E_i) obtained under null hypothesis H_o.

Dicision result: Accept H_o if $\chi^2 < \chi_\alpha{}^2\ (n-1)$ and reject H_o if $\chi^2 > \chi^2_{\alpha\ (n-1)}$, where χ^2 is the calculated value of Chi-square for $(n-1)$ d.f and level of significance α.

Let, there be a sample size of *n* and values are taken from a normal population having a mean μ and standard deviation σ. The observations can be standardized using this relationship which is

$$Z = \frac{x-\mu}{\sigma}$$

A random variable *x* is said to follow chi-square (χ^2) distribution if its *p·d·f* is of the form.

$$f(x) = k \cdot e^{-x/2} \cdot x^{x/2-1}; \quad (0 < x < \infty)$$

where *k* is a constant. A variable which follows chi-square distribution is called a chi-square variate, and often denoted by the symbol χ^2. Fig. 18.21.

Characteristics:

(*i*) Mean = *n*, S.D. = $\sqrt{2n}$

where *n* is the number of degree of freedom (*d·f*).

(*ii*) The chi-square curve is positively skewed, and starting from 0 extends to infinity on the right, shown in Fig. (18.21)

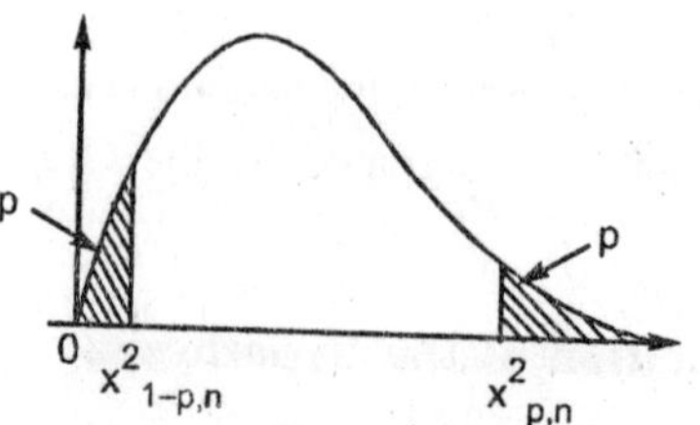

Fig. 18.21: χ^2-distribution

We speak of *goodness of fit* when we try to compare an observed frequency distribution with corresponding value of an expected, or theoretical distribution. The problem of testing the goodness of fit of theoretical distributions to sample distributions is essentially the same that of deciding whether there are significant differences between population and sample values. A important significance test for goodness of fit of theoretical distribution, the *Chi-square* test is described below. It enables us to find if the deviation of the experiment from theory is just by chance or is it really due to the inadequacy of the theory to fit the data.

18.13.2 Parameter Estimation

A statistic is called an *unbiased estimator* of a population parameter if the mean or expectation of the statistic is equal to the parameter. The corresponding value of the statistic is then called an *unbiased estimate* of the parameter.

If the sampling distribution of two statistics have the same mean, the statistic with the smaller variance is called a more *efficient estimator* of the mean. The corresponding value of the efficient statistic if then called an *efficient estimate*. Clearly one would in practice prefer to have estimates that are both efficient and unbiased, but this is not always possible.

For a normal population, the sampling distribution of the mean and median both have the same *mean*, namely, the *population mean*. However, the variance of the sampling distribution of mean is smaller than that of the sampling distribution medians. Therefore, the mean provides a more efficient estimate than the median.

Point Estimates and Interval Estimates

An estimate of a population parameter given by a simple number is called a *point estimate* of the parameter. An estimate of a population parameter given by two members between which the parameter may be considered to lie is called an *interval estimate* of the parameter.

For example If we say that a distance of 5·28 feet, we are giving a point estimate. If, on the other hand, we say that the distance is 5·28 ± 0·03 feet, *i.e.*, the distance lies between 5·25 and 5·31 feet, we are giving an interval estimate.

A statement of the error or precision of an estimate is often called its *reliability.*

18.13.2.1 Confidence interval estimates of population parameters

Let, μ_S and σ_S be the mean and standard deviation (standard error) of the sampling distribution of statistic. Then, if the sampling distribution of S is approximately normal (which as we have seen is true for many statistics if the sample size $n \geq 30$), we can expect to find S lying in the intervals $\mu_S - \sigma_S$ to $\mu_S + \sigma_S$, $\mu_S - 2\sigma_S$ to $\mu_S + 2\sigma_S$ or $\mu_S - 3\sigma_S$ to $\mu_S + 3\sigma_S$ about 68·27%, 95·45% and 99·73% of the time respectively.

Equivalently we can expect to find, or we can be confident of finding, μ_S in the intervals $S - \sigma_S + S + \sigma_S$, $S - 2\sigma_S$ to $S + 2\sigma_S$ or $S - 3\sigma_S$ to $S + 3\sigma_S$ about 68·27%, 95·45% and 99·73% of the time, respectively. Because of this, we call these respective intervals the 68·25%, 95·45% and 99·73%. Confidence intervals for estimating μ_S (*i.e.*, for estimating the population parameter, in the case of an unbiased S). The end numbers of these intervals ($S \pm \sigma_S$, $S \pm 2\sigma_S$, $S \pm 3\sigma_S$) are then called the 68·27%, 95·45%, and 99·73% *confidence limits*.

Similarly, $S \pm 1{\cdot}96\ \sigma_S$ and $S \pm 2{\cdot}58\ \sigma_S$ are 95% and 99% (0·95 and 0·99) confidence limits for μ_S. The percentage confidence is often called the *confidence level.* The numbers 1·96, 2·58 etc. in the coefficient limits are called *critical values*, and denoted by Z_c. From confidence levels we can find critical values and conversely.

In table 18.25 we give values of Z_c corresponding to various confidence levels used in practice. For confidence levels not presented in the table, the values of Z_c can be found from the normal curve area table.

Table 18.25

Confidence level	99·73%	99%	98%	96%	95·45%	95%	90%	80%	68·27%	50%
Z_c	3·00	2·58	2·33	2·05	2·00	1·96	1·645	1·28	1·00	0·6745

In cases where a statistic has a sampling distribution that is different from the normal distribution (such as Chi-square, t , or F); appropriate modifications to obtain confidence intervals have to be made.

18.13.2.2 Problems of parameter estimation

Sample values are used to estimate the parameters of a distribution function. Erros in sample data results inaccurate estimate. Each estimate of a parameter is a function of sample values which are observations of a random variable. Thus, the parameter estimate itself is a random variable having its own sampling distribution. A better estimate can be obtained by using a number of statistical properties. Some of the properties are discussed here.

1. Bias

BIAS can be defined as

$$\text{BIAS} = \frac{E(\hat{a}) - a}{a}$$

where $\hat{a}$ = estimate of a parameter of a.

For example, 600 samples of a given size generated then there are 600 values of parameter '*a*'. Thus, $E(a)$ is the average of the 600 estimates of '*a*' for a given sample size and is estimated as:-

$$E(\hat{a}) = \sum_{i=1}^{n} \frac{\hat{a}_i}{n}$$

where, n = no. of sample generated or the no. of values of a estimate.

If a_c is the estimate of the parameter '*a*' and $E(a_c) = a$, then a_c is known as unbiased estimate of '*a*'.

2. Consistency

Let, n = sample size,

a = the parameter,

a_c = consistent estimate of '*a*'.

The estimate a_c is known as consistent estimate of *a* if it converge to '*a*' with probability one as *n* tends to infinity.

Many unbiased estimates have variances of the type

$$\text{Var}(a_c) \cong \frac{c}{(n)^{0.5}}$$

where *c* is a constant.

If condition of consistency is satisfied then the probability density function of a_c would be more concentrated about *a*.

3. Efficiency

Let a_1 and a_2 are two estimates of '*a*', then the relative efficiency of a_1 with respect to a_2 is defined as

$$e = \frac{E(a_1 - \dot{a})^2}{E(a_2 - a)^2} \leq 1$$

i.e., $E(a_2 - a)^2 > E(a_1 - a)^2$

However, $e = 1$ means an efficient estimate. An estimate is said to be efficient if it is unbiased.

4. Sufficiency

Let a_1 and a_2 are two independent estimates of *a*. Now, a_1 is considered a sufficient estimate if the joint probability distribution of a_1 and a_2 has the property.

$$f(a_1, a_2) = f(a_1) \cdot f\left(\frac{a_2}{a_1}\right) = f(a_1) \cdot k(x_1, x_2, x_3, ..., x_n)$$

where, $f(a_1)$ = the distribution of a_1

$f\left(\frac{a_2}{a_1}\right)$ = the conditional distribution of a_2 given a_1.

$k(x_1, x_2, ..., x_n)$ = function of x_i.

In this case, a_1 is a sufficient estimate if a_2 does not produce any new information about *a*.

5. Standard Error

Standard error can be represented as, *S.E.* $= \frac{\sigma(\hat{a})}{a}$

where $\sigma(\hat{a})$ = standard deviation of *a*.

The standard deviation of *a* is computed as

$$\sigma(\hat{a}) = \left[\frac{1}{(n-1)} \sum_{i=1}^{n} \{\hat{a}_i - E(\hat{a}_i)\}^2\right]^{1/2}$$

where the summations are over *n* estimates $\hat{a}$ of *a*.

6. Root Mean Square Error (RMSE)

This is most frequently employed to check performance and is defined for parameter estimated as

$$\text{RMSE} = E[(\hat{a} - a)^2]^{1/2} / a$$

where $E[\cdot]$ = the expectation of $[\cdot]$. The relation between RMSE, BIAS, and SE as

$$\text{RMSE} = \left[\frac{(n-1)}{n}(\text{SE})^2 + (\text{BIAS})^2\right]^{1/2}$$

7. Robustness

A robust estimation is that which is resistant and efficient over a wide range of population fluctuations. According to the minimax criteria, both the average RMSE as well as maximum RMSE should be minimum.

8. Relative Mean Error

The relative mean error can be given by the following formula

$$\text{RME} = \frac{1}{N}\left(\sum_{i=1}^{N}\left[\frac{Q_0 - Q_c}{Q_0}\right]^2\right)^{1/2}$$

where, N = Sample size,

Q_0 = observed quantity of a given probability

Q_c = computed quantity of the same probability.

9. Relative Absolute Error (RAE)

RAE can be obtained from the following formula.

$$\text{RAE} = \frac{1}{N}\sum_{i=1}^{n}\left|\frac{Q_0 - Q_c}{Q_c}\right|$$

18.13.3 Goodness of Fit Test

The goodness of fit of a statistical model describes how well it fits a set of observations. Measures of goodness of fit typically summarize the discrepancy between observed values and the values expected under the model in question. Such measures can be used in statistical hypothesis testing, *e.g.* to test for normality of residuals, to test whether two samples are drawn from identical distributions (see Kolmogorov – Smirnov test), or whether outcome frequencies follow a specified distribution (see Pearson's chi-squared test). In the analysis of variance, one of the components into which the variance is partitioned may be a lack-of-fit sum of squares.

Example. 18.39. *In an engineering experiment, sample size 12, sample mean 5253·18 and sample variance 0·99. Test the hypothesis that the variance is equal to 2·50, against the alternative hypothesis that the variance is not equal to 2·50, at 90% confidence level.*

Solution: Null hypothesis

$$H_0 = \sigma^2 = \sigma_0^2 = 2{\cdot}50$$

Alternative hypothesis

$$H_1 = \sigma^2 \neq \sigma_0^2 = 2{\cdot}50$$

It is a two tailed problem. We apply chi-square test.

At 90% confidence level we will accept H_0 if

$$\chi^2_{0\cdot05} < \frac{ns^2}{\sigma_0^2} < \chi^2_{0\cdot95}$$

at degree of freedom $(n - 1) = 12 - 1 = 11$

i.e. $$5{\cdot}23 < \frac{ns^2}{\sigma_0^2} < 2{\cdot}0$$

We have $\chi^2 = \dfrac{ns^2}{\sigma_0^2}$

Here $\sigma^2 = 0{\cdot}9,\ \sigma_0^2 = 2{\cdot}5$

$n = 12$

$$\therefore \quad \chi^2 = \frac{12 \times 0{\cdot}99}{2{\cdot}5} = \frac{11{\cdot}88}{2{\cdot}5}$$

$$= 4{\cdot}752$$

Since calculated values does not lie between 5·23 and 21·0 so we reject the null hyphothesis and accept the alternative hypothesis. Hence at 90% confidence level it can be said that variance is not equal to 2·50.

Note: One may use $\chi^2 = \dfrac{(n-1)s^2}{\sigma_0^2}$ in that case $\chi^2 < 4{\cdot}752$. So if we also apply the above formula our null hypothesis is rejected.

18.14 FIELD DESIGN AND TEST OF SIGNIFICANCE

18.14.1 Experimental Designs

1. *Treatments* : The objects of comparison, which an experimenter has to try in the field for assessing their values are known

as treatments *e.g.* the varieties, manures, cultivation practices etc.

2. *Experimental Units*: Experimental material is divided into a number of ultimate units to which the treatments are applied and are known as experimental units.
3. *Experimental Material*: This is the name given to the material on which the experiment is performed *e.g.*, a field, soil, patients in a hospital etc.

Basic Principles of Field Experimentation

There are three basic principles of field experimentation which have been discussed below:

1. *Replication*: The allocation of the treatments under investigation is known as replication.

If the treatment is applied only once we have no means of knowing about the variation in the results of a treatment. Only when we repeat the application of the treatment several times we can estimate the experimental error. Only with the help of experimental error we can determine whether observed difference between treatment means are real or not. The standard error of mean is given by

$$SE(\bar{y}) = \frac{s}{\sqrt{r}},$$

where, s is the standard deviation and r is the number of observation (replication). That is, the standard error of the mean is proportional to the square root of the number of replications. Therefore, as the number of replications is increased the experimental error will be reduced. Thus the major functions of replications are : (1) to provide an estimate of experimental error and (2) to reduce the experimental error. The number of replication (r) required is given by the relation

$$r = \frac{2t^2 \cdot s^2}{d^2}$$

where

t = table value of t for a given level of significance and degrees of freedom.

d – the difference between the means of two treatments.

s^2 = error variance (taken from similar experiments conducted earlier), and

r = number of replications (to be computed)

2. *Randomization*: The allocation of the treatments to the different experimental units by a random process is known as randomization of treatments.

In other words if all the treatments have equal chances of being allocated to different experimental units, it is known as *randomization.* The actual procedure of randomization will vary according to the design we adopt.

3. *Local control*: The principle of making use of greater homogeneity in groups of experimental (*e.g.* blocks of a number of plots homogeneous within themselves) for reducing the experimental error as explained in the foregoing paragraph is known as local control. This is done by dividing the experimental material (*e.g.*, field) into homogeneous groups of experimental units (*e.g.*, blocks).

Uniformity Trial: In order to achieve local control we must have knowledge of experimental material. Had we made observations on experimental material previously, we may effect suitable grouping of the experimental units. When such knowledge is lacking, we, may carry out uniformity trial. The uniformity trial may be discussed with special reference to agricultural field experiments. It is conducted to know the nature of soil fertility gradient. The uniformity trial data may also be used to determine the size of the experimental unit.

Under, uniformity trial, a particular variety of crop is sown on the entire experimental field and uniformly managed throughout the growing season. At the time of harvest a substantial border is removed from all sides of the field. The remainder of the field is divided into small plots which are termed as *basic units*., The size of basic unit is decided by judgement, depending on crop. The produce from these basic units is harvested and recorded separately for each basic unit. Then the mean yield per basic unit is computed. The basic units giving yields above or below the specified percentage of the overall mean, say 5%, 10%, 15% etc., are marked on the plan of plots. Usually percentage are taken in such a way that will get 5 to 8 groups of experimental units. The similar units are then joined by lines to produce the contour map as shown if fig.18.22. Such map is called *fertility contour map* or *fertility gradient map* or *soil fertility map*. Using this map homogeneous experimental units can be grouped into blocks.

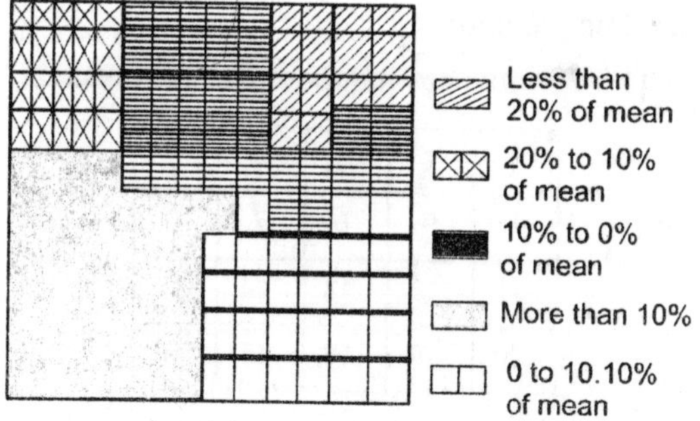

Fig. 18.22: Fertility contour map

Optimum Plot Size

There are two important methods that are used to determine the optimum plot size.

1. Maximum curvature methods : In this method the basic units of uniformity trials are combined to form new units. The new units are formed by combining columns, rows or both. Combination of columns and rows should be done is such a way that no column or row is left out. For each set of units, the coefficient of variation (C.V) is computed. A curve is plotted by taking plot size (in terms of basic units). The curve will be as shown in Fig. 18.23. The point at and which the curve takes a turn, that is point of *maximum curvature* is located by inspection. The value corresponding to the point of maximum curvature will be the optimum plot size.

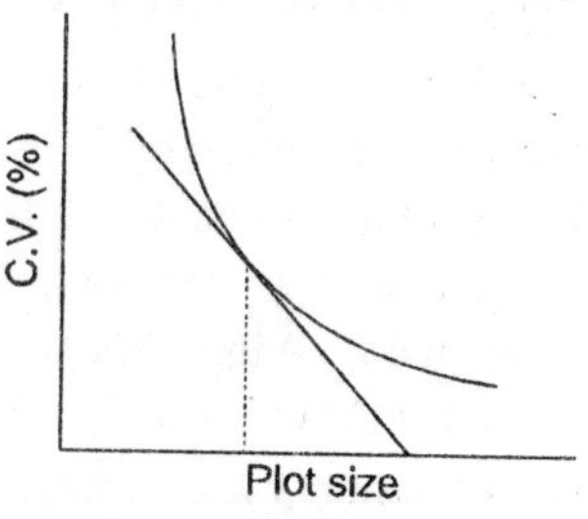

Fig. 18.23

2. Fairfield Smith's Variable Law: Fairfield smith developed an empirical model representing the relationship between plot size and variance of mean per plot.

This model is given by the relation

$$V_x = \frac{V_1}{x^b}$$

where,

x = the number of basic units in plot

V_x = the variance of mean per plot of X- units.

V_1 = The variance of mean per plot of one unit.

and

x^b = the regression coefficient

This equation may be written as

$$\log V_x = \log V_1 - b \log x.$$

The values of V_1 and b are determined by the principle of least squares.

Condition for Application of Designs

(*i*) C.R.D. (Completely Randomized Design) is apply when the experimental material is limited and homogeneous, such as the soil in the pot experiments.

(*ii*) R.B.D. is appropriate when the fertility gradient of the field is in one direction. It may be adopted upto 20 treatments without an appropriate loss of efficiency.

(*iii*) L.S.D. can replace the R.B.D., when the fertility gradient is in two directions instead of one. It may be adopted for number of treatments ranging from 5 to 8 or from 5 to 12 at the most.

18.14.2 Completely Randomized Design

This is the simplest type of design in which the whole experimental material is divided into a number of experimental units depending upon the number of treatments and the number of replications for each. After that the treatments are allotted to the units entirely by chance.

When the treatments are arranged randomly over the whole of a previously determined set of experimental units, the design is known as completely random mixed design. (CRD). In CRD the treatments are allotted to the experimental unit entirely by chance without making any effort to confine treatments to any portion of experimental material. Thus the randomization gives every experimental unit in the experimental material an equal probability of receiving the treatment.

Advantages:

1. The method of analysis remains simple when the results from some units are rejected.
2. This design is specially useful in small experiments where, the supply of experimental material is scarce and homogeneous.
3. In this design any number of treatments and replicates may be used.
4. The statistical analysis of the data is very easy.
5. The relative loss of information due to missing data in smaller than that with any other design.

Disadvantage : The major disadvantage CRD is that it is less accurate than other designs. When large number of treatments are included, a relatively large amount of experimental material must be used. Hence, the heterogeneity of experimental material will be increased. This will result in increased experimental error and reduced precision.

Layout of CRD : The placement of treatment on the experimental units along with arrangement of experimental units is known as lay out of an experiment.

Suppose that there are t treatments, namely T_1, T_2 ...... T_t. Further suppose that treatments are replicated t times each. If practical limitation dictate unequal replications should be made. We require, $t \times r = n$ experimental unit.

In case of unequal replications the number of experimental units required will be $r_1 + r_2$

$$t \text{ .....} + r_t = n$$

The entire experimental material is divided into n number of units. The units are numbered serially starting with one end and proceeding in a serpentine manner. Suppose that there are five treatments each with four replications. We need 20 experimental units.

The 20 units are numbered as follows.

1	2	3	4	5
10	9	8	7	6
11	12	13	14	15
20	19	18	17	16

Then n distinct three – digit random number are selected from random number table. The random numbers are written in order and ranked. The lowest random number is given rank 1 and the highest rank is allotted to largest random number. These ranks correspond to the unit number. The first set of runils are allocated to treatment T" the next r units are allocated to treatment T_2 and so on. This procedure is continued until all treatments have been applied.

For example, the selected number and their ranks are as follows:

Random number	*Rank*	*Treatment to be applied*
507	14	
962	19	
945	18	T_1
350	6	
225	4	
076	1	
984	20	T_2
375	7	
555	15	
233	5	
089	2	T_3
488	10	
382	9	
444	12	T_4
396	11	
754	16	
375	8	
494	13	
896	17	T5
098	3	

The final lay out is as follows:

The final layout is as follows

1	2	3	4	5
T_2	T_3	T_5	T_2	T_3
10	9	8	7	6
T_3	T_4	T_5	T_2	T_1
11	12	13	14	15
T_4	T_4	T_5	T_1	T_3
20	19	18	17	16
T_2	T_1	T_1	T_5	T_4

Application Conditions of C.R.D.

1. C.R.D. used when the experimental material is limited in quantity and homogeneous.
2. Where it is expected that some of the units will be destroyed.
3. In small experiments where the increased accuracy from the alternative design is not sufficient to exceed in importance the loss of error degree of freedom.

18.4.3 Randomized Block Design

In this design the whole experimental material is divided into homogeneous groups, each of which constitutes a single replication. Each of these groups is further divided into the number of experimental units which are equal in all respects. The treatments are applied to these units by any random process.

Some important points are given below about this design :

1. In R.B.D., the number of blocks must be equal to the number of replications fixed for each treatment.
2. The number of plots in each block should be equal to the number of treatments.
3. Randomization of treatments in each block should be afresh.
4. The experimental error within each block are to be kept as small as practically possible and the variation from block to block as great as possible. In this way all the treatments which are assigned to one block experience the same type of environmental effects and therefore comparable.

Lay out of RBD : The entire experimental material is divided into number of homogenous blocks. If there are *t* treatment bocks divided in *t* unit. The units in each block are numbered 1 to *t*. The treatments are also numbered conventionally.

By using random number tables, we select *t* distinct random numbers from 1 to *t*. These random numbers correspond to the treatment numbers. The first selected treatment is allotted to the first unit of a block, the second selected treatment to the second units, and so on, The randomization is done a fresh for each block in the same way.

Advantages of R.B.D.

1. When the material is heterogeneous the residual variance can be reduced by choosing blocks or plots such that the plots within any block are fairly similar.
2. This design allows of any number of treatments and any number of replications but when the number of treatments is very large the efficiency of error control decrease.
3. If a part of the experimental material is damaged by some agricultural disaster like insect, floods, water logging etc.
4. By means of grouping more accurate results are usually obtained than with the C.R.D.
5. Although a reduction in the number of replications leads to a larger standard error, yet it furnishes a result of some value at least.

Disadvantages:

When the number of treatments is increased, the block size will increase. If the block size is large it may be difficult to maintain homogeneity within blocks. Consequently the experimental error will be increased. Hence RBD may not be suitable for large number of treatment. But for this dis advantage, the RBD is a versatile design. It is most frequently used design in agricultural experiment.

18.4.4 Latin Square Design

In this design it is essential that each row and each column should contain the same application of treatments and hence, the variation between the row means and between the column means can be assessed and eliminated from the error increasing the precision of the estimates.

- The experimental design which simultaneously control the variation in two directions is known as Latin Square Design.
- The L.S.D. is very reliable to give precised results when the number of treatments is from 5 to 8 or at most 12. This design should not be used for less than 5 treatments.

The important points about L.S.D.:

1. No. of replications = No. of treatments
2. No. of rows = No. of column = No. of treatments
3. Randomization of treatments is done in such a way that each treatments occurs once and only once in each row and each column.

In LSD the number of rows and columns are equal, Hence the arrangement will form a square. It is identified as 5 × 5 latin squares, 6 × 6 latin squares etc according to the number of rows and columns. In LSD each row and each column is a complete block or column.

Lay out of LSD : The lay out of LSD starts with dividing the experimental material into number of rows and columns depending on number of treatments. A $t \times t$ standard square is then selected from Fisher and Yates statistical Table. The columns of the selected standard square are rearranged randomly by using random tables. Then keeping the first row of rearranged square as such, the remaining rows are randomized. The treatments are then allocated as in the final arrangement.

Advantages of L.S.D.

This design gives precision high enough to reduce the standard error to less than 1%.

L.S.D. control the variations due to two sources of variation.

Disadvantages of L.S.D.

- The analysis becomes very complicated when several plot yield are missing.
- An L.S.D. with two treatments is not possible.
- L.S.D. is less flexible in nature.

18.4.5 Split Plot Design (SPD)

- In this design, larger plots are taken for the factor which requires larger plots. Next each of the larger plots is split into smaller plots to accommodate the other factors. The different treatments are allotted at random to their respective plots. Such arrangement is called ***split plot design.***
- For example, experiments on irrigation, tillage, etc. requires larger areas. On the other hand, experiments on fertilizers, etc may not require larger areas. To accommodate factors which require different sizes of experimental plots in the same experiment, split plot design is used.

In split plot structure –

- the bigger plots are called primary plots/ main plots.
- smaller plots within the larger plots are called as sub plots.
- The factor levels allotted to the main plots are main plot treatments
- the factor levels allotted to sub plots are called as sub plot treatments.

Applications

SPD is used in following conditions –

- When all factors aren't important
- When factors of the different nature are to be tested in same experiment that is level of one factor require larger area as compared to other factor *eg.* Depth of ploughing and nitrogen levels, date of sowing and varieties, varieties and nitrogen level, irrigation level and varieties.
- When the levels of one factor produce larger differences as compared with the levels of other factor.
- When we want to study one factor with higher precision as compared to other factor and take it in sub plots.

Note: The number of error terms in a split plot design is 2

Strip plot Design (SPD)

- This design is also known as split block design.
- When there are two factors in an experiment and both the factors require large plot sizes it is difficult to carry out the experiment in split plot design. Also, the precision for measuring the interaction effect between the two factors is higher than that for measuring the main effect of either one of the two factors.

Note: The number of error terms in a split plot design is 3.

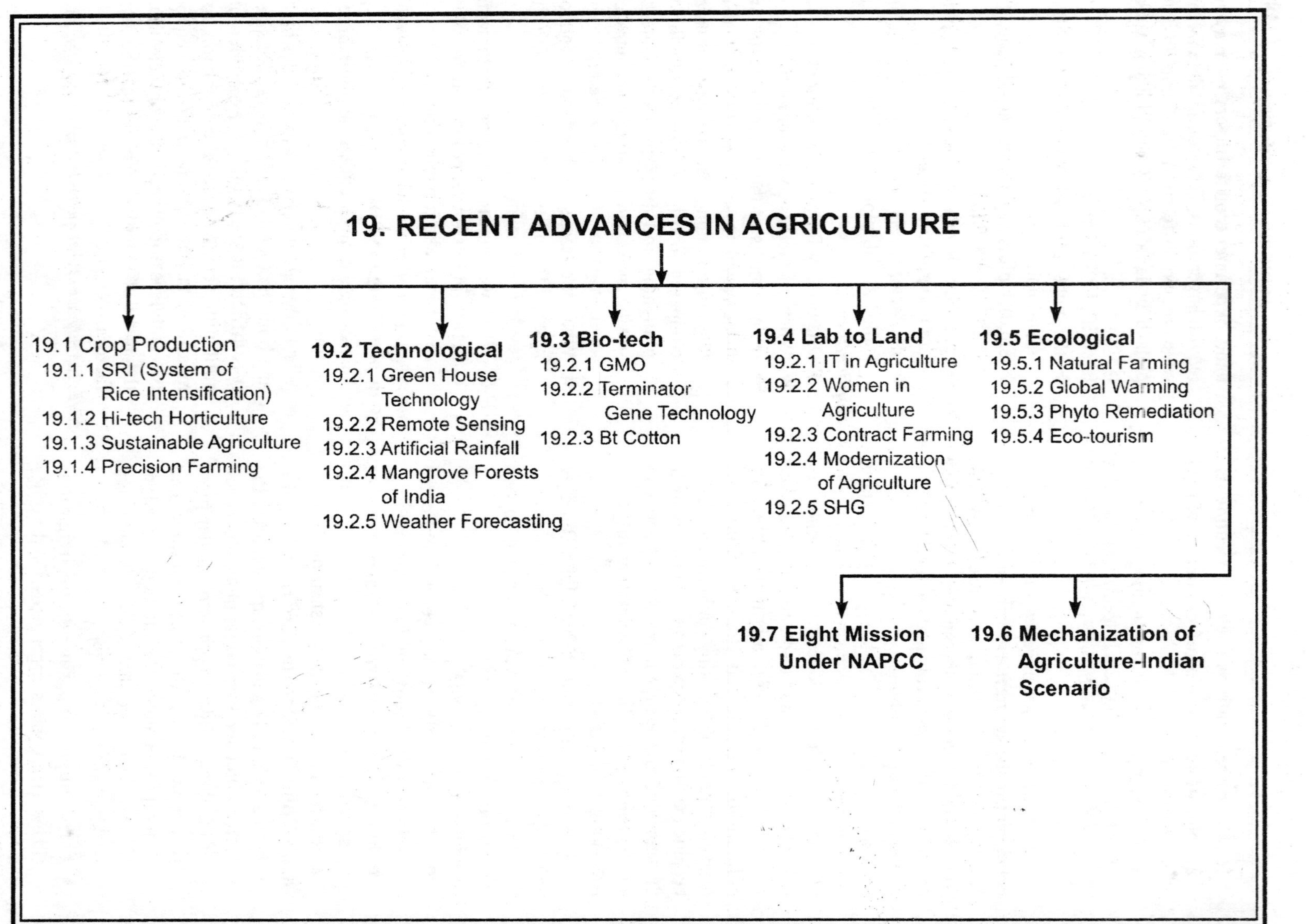
19. RECENT ADVANCES IN AGRICULTURE
19.1 Crop Production
19.1.1 SRI (System of Rice Intensification)
19.1.2 Hi-tech Horticulture
19.1.3 Sustainable Agriculture
19.1.4 Precision Farming
19.2 Technological
19.2.1 Green House Technology
19.2.2 Remote Sensing
19.2.3 Artificial Rainfall
19.2.4 Mangrove Forests of India
19.2.5 Weather Forecasting
19.3 Bio-tech
19.2.1 GMO
19.2.2 Terminator Gene Technology
19.2.3 Bt Cotton
19.4 Lab to Land
19.2.1 IT in Agriculture
19.2.2 Women in Agriculture
19.2.3 Contract Farming
19.2.4 Modernization of Agriculture
19.2.5 SHG
19.5 Ecological
19.5.1 Natural Farming
19.5.2 Global Warming
19.5.3 Phyto Remediation
19.5.4 Eco-tourism
19.7 Eight Mission Under NAPCC
19.6 Mechanization of Agriculture-Indian Scenario

19. Recent Advances in Agriculture

19.1 CROP PRODUCTION

19.1.1 System of Rice Intensification

Definition: System of rice intensification (SRI) concept is an alternate to the existing practises where fields are kept wet, not flooded with water. This system seems to be promising to overcome the shortage of water in irrigated rice The SRI appear to be a viable alternate for rice cultivation that saves expensive inputs, improves soil health besides optimum water-use efficiency.

Origin:

The system of rice intensification originated in Madagascar and was first synthesized in 1983 by Fr. Henri de Laulanie. Under the drought conditions of that year.

Significance:

There are three basic concepts constituting system of rice intensification. They are to be applied with adjustments in specific practices like spacing and timing as appropriate to local conditions. Transplanting the seedlings while still young, <15-days-old, *i.e.*, prior to the start of the 4th phyllochron of growth. This preserves plants potential for tillering and root growth that is reduced by transplanting later. Direct seeding is an option, however, since what is important is that plant roots should not be traumatized after they start their growth trajectory, on or about the 15th day. Careful transplanting promotes rapid resumption of growth.

Wide spacing between plants is maintained—with preferably just one plant/hill, and set them out in a square pattern, 25 cm × 25 cm or even wider if soil fertility is good. This gives room for profuse root and tiller growth, achieving 'the border effect' throughout the whole field. Another important aspect in SRI is to keep the soil both moist and aerated, at least during the vegetative growth period, so that roots have access to both oxygen and water. Under continuously hypoxic conditions, rice roots degenerate, with as many as 75% become dysfunctional by panicle initiation. Therefore, the aerobic conditions in SRI promote root growth substantially over submerged situations.

Specific practices include in addition to using young seedlings with still only two leaves, planted singly and widely spaced, frequent weeding, preferably with a rotating hoe that aerates the soil while it prevents weeds growth by churning them into the soil, application of small amounts of water daily, or alternate flooding and drying of fields for 3 to 6-day periods, and providing organic matter to the soil. Organic matter can be any decomposed biomass, including rice straw or weeds, manure or mulch can also be used. Green manures and cover crops are also useful. Correct implementation of the various, interdependent components (SRI practices) requires considerable field experience in which close observation and timeliness are crucial in minimizing labour requirements while maximizing total factor productivity.

Benefits associated with SRI

1. Greater root growth
2. Increased tillering
3. Increased grain filling
4. Improved grain quality and higher grain weight
5. Water savings
6. Less lodging
7. Reduced pest and disease attacks
8. Reduced or no need for chemical fertilizers
9. Seed savings
10. No need to purchase new seeds
11. Low cost of production
12. Increased factor productivity and profitability
13. Reduced risks

Conclusion

Rice yields all over the world have leveled out under the present system of flooded cultivation. We need to be looking for alternatives to existing practices with an open mind. SRI is still evolving and it is hoped that the scientific community will collaborate in further refining the SRI practices and working out the scientific reasons for the reported higher productivity. Planting young seedlings carefully and at wider spacing gives the plant more time and space for tillering and root growth. Careful water management keeping the field wet and not flooded gives better yield as it supports healthy root growth. This practice should be encouraged everywhere as the whole world is facing water shortages. Weeding rice fields with a rotary weeder helps by churning the soil and incorporating the weed biomass as it aerates the root zone. Therefore, seeing the overall benefits, the SRI needs to be evaluated further and refined to suit the local environments.

19.1.2 Hi-tech horticulture

Hi-tech technologies are which can be defined as those, which are modern, less environment dependent, capital intensive and have capacity to improve productivity and quality.

Thematic areas

1. Germplasm and crop improvement.
2. Hi-tech crop production of horticultural crops.
3. Protected cultivation.
4. Hi-tech plant protection.
5. Modern post harvest management.
6. Logistics and polices for Hi-tech horticulture.
7. Policy issues.

Overview of Hi-tech horticulture

1. Introduction

Horticulture crops occupy 13% of gross cropped area in the year of 2005-06. 37% of gross value of agriculture output and 50% of export earning. Horticulture today is not merely a means of diversification from an integeral part of food and nutritional security.

2. Genetic resource conservation of characteristics

1. In vitro collection and conservation of fruit like mango, jackfruit.
2. Tissue culture techniques.
3. Cryo preservation
4. gene bank.

3. Hi-tech propagation

3.1 *Micro propagation :*

1. Tissue culture units for propagation, Banana, papaya, cardamom, bamboo etc.

3.2 *Hybrid seed and seedling production in vegetable and flower crops :*

1. Use of male sterile

Flower : Marigold

Vegetable : Cabbage.

4. Hi-tech crop production

4.1 *High density planting*

1. *Mango* : Amarpali
2. *Apple* : Malling (m) and malling marton (Mm) root stock.

Growth retardants are used : CCC, B-9

4.2 *Integrated nutrient management* :

1. Fertilizer (slow release and coated)
2. Organic manuring (FYM, poultry etc.,)
3. Green manuring
4. Green leaf manuring
5. Composting of farm-town compost
6. Bio fertilizer
7. Vermiculture

4.3 *Micro irrigation*

1. Drip irrigation — root zone application
2. Sprinkler — for due effect.
3. Fertigation (Both water + fertilizer)
4. Chemigation
5. Herbigation

4.4 *Organic farming* :

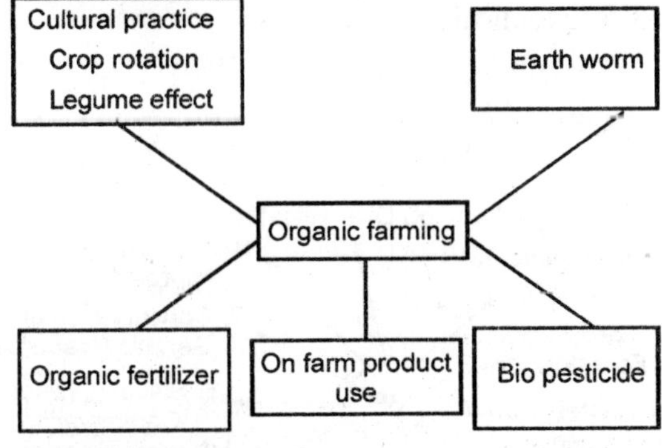

5. Protected cultivation

1. Green house cultivation
2. Low cost wire shade nets
3. Wrist chamber for propagation

6. Crop protection

1. Integrated pest management.
2. Pesticide application methods. Ultra low volume spraying.
3. Bio pesticide.
   - NPV
   - GV
   - Bt formulation
4. Pesticide residue management.
5. Molecular diagnostic.

7. Post harvest technology

7.1 *Grading* : (Maturity, colour, shape, texture, flavour)

7.2 *Packing and storage*

- modified atmosphere packaging
- Controlled atmosphere packaging
- Bio degradable film packaging

7.3 *Cold chain* : Cold chain is temperature management facility involving a number of equipment such as Pre-cooling unit, cold storage, humidity control,atmospheric storage, refrigerated and storage system.

7.4 *Irradiation* : It refers to physical means of exposing fresh or pre packed products to gamma rays.

1. *Flowers* : rose, gladiolus
2. *Vegetables* : Canned, persevered candy
3. *Fruits* : Mango, citrus

7.5 *Hi-tech processing* :

1. IQF—Individual quick freezing
2. Osmotic & Asecptic technologies.

7.6 *Mechanisation/Automation*

7.7 *Quality standards*

- Sanitary and phytosanitary measure (SP)
- Technical Barriers to trade (TBT)

8. Logistics and policies

- IPR issues.
- WTO concerns.

(1) Novelty

(2) distinctiveness

(3) Imitating

8.1 *Organisational support :*

1. National Horticultural mission.
2. Budget allocation to NHC (National Horticulture Commission).
3. NHB
4. APEDA

8.2 *Human Resource development*

- Advance training for student in Biotechnology micropropagation, green-house technology, seed production etc.

Training
- Horticulture engineering
- Land scape horticulture
- Processing (fruits & vegetables)

8.3 *Research thrust :*

- ICAR
- NBPGR
- CPRI
- CTCRI
- NATP project

9. Conclusions

1. Research in non core area to be taken up.
2. Various organisation like APEDA, NABARD, SFAC and FAO have to play a major pole.
3. Comprehensive, holistic approach towards integration of all technology leads to sustainable horticulture.

19.1.3 Sustainable agriculture

Sustainable agriculture is the successful management of agricultural resources to satisfy changing human needs, while maintaining or enhancing the environment quality and conserving natural resources.

It is also called as ecofarming, organic farming and biofarming.

Characteristics of sustainable agriculture

Maintaining a diversity of plant and animal species as a basis for ecological balance.

- Sustainable use of local resources.
- Minimal purchase of inputs.
- No adverse effects to the environment.

Need for sustainable agriculture

- For sustainable use of inputs.
- Low input for high return.
- To improve environment balance.
- To save natural resources.
- To improve quality of products.

Principle :

- Availability of Nutrients.
- Sustainable use.
- Conservation.

Dimensions :

1. Biologically feasible.
2. Ecologically stable.
3. Economically viable.
4. Socially acceptable.

Components

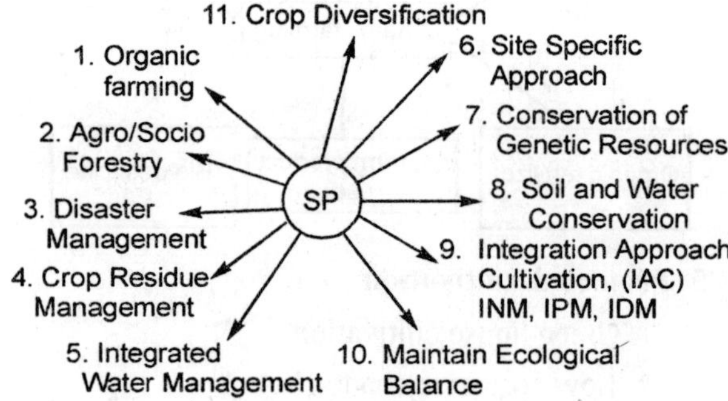

Advantages :

- Ecological balance.
- Low cost of cultivation.
- Nutritious food.
- Low input, high productive farming system.
- Resources are conserved for future.

Disadvantages :

Low yield in the short period.

Lack of timely used.

18.1.4 Precision Farming

1. Definition :

Precision farming is a site specific management approach in modern crop production for applying various input to the field in an economical and sound fashion.

2. Aim :

1. To identify and meet the real needs of variable parts of the field.

3. Used for :

Precise application of Agriculture inputs like

1. Seed 2 Fertilizer Pesticide 3. Irigation etc.

4. Components :

1. Yield monitoring 2. Yield mapping 3. Variable application of fertilizer 4. Salinity mapping 5. Guidance system 7. Record and analysis.

5. Requirements :

1. Field map 2. Variable rate of input 3. Differential Global Position System (DGPS).

6. Procedure :

1. Selection of inputs and resource 2. Proper study of GPS, GIS, range of sensor 3. Monitoring and collection of input data 4. Variable rate of application 5. Achievements by harvesting.

7. Advantage :

- Less time, money, input loss.

9. Disadvantage :

1. Input cost is high 2. Fragment condition of the field.

10. India's scenario :

1. Punjab, Haryana, it is possible because of 14 acre per capita 2. Other states by cooperative or contact farming is possible 3. It is in expanding stage.

19.2 TECHNOLOGICAL

19.2.1 Greenhouse Technology

Definition ; It is a controlled environment (protected) cultivation. It has been evaluated to create favourable micro-climate in which crop production can be made possible throughout the year /part of the year as required.

- It can be called as "*food factories*"
- It includes many of the followings :

(*i*) Bench /Beds.
(*ii*) Misters/pogger.
(*iii*) Drainage.
(*iv*) Light (Fluorescent/PAR lamps).
(*v*) Heating & cooling system.
(*vi*) Irrigation systems.
(*vii*) CO_2 enrichment.
(*viii*) Controlled system.

Benefits

(*i*) Raising crop any where, anytime, any crop.
(*ii*) Yield is maximum.
(*iii*) High Quality product (Exports).
(*iv*) Raising of seedlings.
(*v*) Crop productivity several times higher than open conditions.
(*vi*) Income generation for small farmers (Export).
(*vii*) Utilisation of wetland.

Types

- Thatch house; Making dark area. (Structures)

- Shade nets ; for filtration of sun light (entire nets).

Example

- **Vegetable** ; cauliflower, cabbage, tomato chillies, brinjal, cucumber.
- **Flowers;** Rose, orchids, chrysanthemum.
- **Fruits;** Strawberry.

Disadvantages

1. High labour requirement
2. Higher cost of production
3. Initial capital investment is high.

19.2.2. Remote sensing

Measurement or acquisition of information of some property of an object/phenomenon, by a recording device that is not in physical or inmate contact with the object/phenomenon under study.

Principle

Based on the reflected radiation energy from different areas on the land surface over a range of wavelength, the objects are distinguished.

Types of Sensors

Active : Sensing device directs/creates EMR.

Passive : Using other energy/natural energy sources (Sun).

Types of sensing

(1) *Airborne remote sensing*

- It is the method of remotesensing in which cameras or other devices fixed in an aircraft flying at fixes altitude are used to take photograhs of any sources on earth.

Types of aerial photographs.

(*i*) Infrared aerial photograph

(*ii*) Black & white aerial photography

- Particularly for mapping boundaries of water sources.

(*iii*) Colour aerial photography

- Disease detection.
- Identification of tree species in temperature zones.

Space/Satellite Remote Sensing

It is the method of remote sensing in which cameras, sensors and other devices attached to a satellite, orbitting round the earth take photographs of earth and its resources.

Stages of Remote Sensing :

(*i*) Origin of EMR.

(*ii*) Emission of energy from source to surface.

(*iii*) Interaction between energy in earth surface.

(*iv*) Transmission of reflected energy to the sensor.

(*v*) Detection of energy & converting into photographic image/electrical.

(*vi*) Recording the sensor.

(*vii*) Data processing.

(*viii*) Collection of special information.

(*ix*) Data processing & interpretation.

Basic concern

(*i*) Radiation from sun and other terrestrial Sources.

EMR spans a large spectrum of length from 10^{-10} m (α-ray) to 10^{6} m (long radiowaves).

In remote sensing useful regions are

(*a*) Visible (0·4 -0·7 μm)

(*b*) Infrared (0·7 - 3·0 μm)

(*c*) Thermal infrared (3-5 μm)

(*d*) Microwave regions (0·3-300 cm)

The sun is important source of EMR used in conventional optical remote sensing.

(*ii*) Effects of atmosphere

Electromagnetic radiation is scattered and absorbed by gases and particulate matters while passing through atmosphere.

Signatures concept

Any set of observable characteristic which is directly or indirectly related to the identification of an object and or its condition.

1. Spectral

Change in reflectance/emittance of objects as a function of wavelength. *Ex.* Forest.

2. Spatial

Arrangement of terrain features such as shape, size, texture of objects. *Ex.* Mountain

3. Temporal

Change of reflectivity/emissibility with time. *Eg.* Crop growth.

Table 1 : A brief account of remote sensing applications in agriculture

A. Scientific applications of relevance in agriculture	
Crop science including	Hydrology—surface run off ground water
All types of vegetation indices	Land surface climatology
Stress detection,	Soil science
Agro-climatology,	Terrestrial ecosystems
Agro-ecology,	Vegetation sciences
Global change,	Including forest science.
B. Management applications sensing in agriculture	
Agriculture	Yield prediction
Agro-ecology	Degraded land studies
Crop damage	Environmental impact
Crop condition	Studies of agricultural development,
Crop inventory-local,	urban development
Regional, national & global,	GIS—contributions to
Crop classification	regional agricultural
Live stock inventory	information systems
Yield estimation, Irrigated lands inventory	Land reclamation
Monitoring extent	Agricultural, Urban
Crop condition,	Soil pollution
Water utilisation	Vegetation
Regulation of water use	Range land management
Land evaluation	Range land condition
Land use	Range land classification
Land degradation	Range production
	Soil erosion
	Soil salinity

4. Polarisation

Changes in polarisation of radiation reflected or emitted by an object.

Analysis & generated data :

1. *Visual analysis*

   Eg : Colour, texture, shape size etc.

2. *Digital image processing techniques*

Computer analyse the shapes signature so as to associate each pixel with a particular feature of imaging.

Application of RS in Agriculture in INDIA

1. Crop production forecasting.
2. Land use
3. Mapping of waste lands
4. Soil mapping
5. Drought monitoring & its assessment
6. Monitoring of surface water bodies
7. Ground water exploration
8. Flood supply & damage assessment.

19.2.3 Cloud Seeding/Artificial Rainfall

Cloud Seeding

Cloud seeding is one of the tools to mitigate the affects of drought. It is defined as, "A process in which the precipitation is encouraged by injecting artificial condensation nuclei through aircrafts or suitable mechanism to induce rain from turbulent cloud".

The rain drops are about a million times heavier than cloud drops. So, rain develops only if the cloud droplets grow by some mechanism. These mechanisms are different for acid and warm clouds.

Seeding of Cold Clouds

This can be achieved by two ways.

1. **Dry Ice Seeding**

- Dry ice (solid carbon-dioxide) has certain specific features. It remains as it is at 80 °C and evaporates, but, does not melt.
- Dry ice is heavy and falls rapidly from top of cloud and has no persistent effects.

Steps involved are

1. Aircraft flies across the top of a cloud and 0.5-1.0 cm size dry ice pellets are released in a steady stream.
2. While falling through the cloud a sheet of ice crystals is formed.
3. From these ice crystals rain occurs.

This method is not economical as 250 kg of dry ice is required for seeding one cloud. To carry the heavy dry ice over the top of clouds special aircrafts are required, which is an expensive process.

2. **Silver Iodide Seeding**

Steps involved are

1. Minute crystals of silver iodide produced in the form of smoke acts as efficient ice-forming nuclei at temperatures below –5°C.
2. When these nuclei are produced from a flame on the ground generators, these particles are fine enough to diffuse with air currents.
3. Silver iodide is the most effective nucleating substance because, its atomic arrangement is similar to that of ice.
4. The time taken for silver iodide smoke released from ground generator to reach the super cooled clouds was often some hours, during which it would drift a long way and decay under the sun light.

5. So, the appropriate procedure for seeding cold clouds would be to release silver iodide smoke into super cooled cloud from an aircraft.

In seeding cold clouds silver iodide technique is more useful than dry ice technique, because

(*a*) Very much less of silver iodide is required per cloud.

(*b*) There is no necessity to fly to the top of the cloud, if area to be covered is large.

Seeding of Warm Clouds

1. **Water Drop Technique**

1. Coalescence process is mainly responsible for growth of rain drops in warm cloud.
2. The basic assumption is that the presence of comparatively large water droplets is necessary to initiate the coalescence process.
3. So, water droplets or large hygroscopic nuclei are introduced in the cloud.
4. Water drops of 25 mm are sprayed from air craft at the rate of 30 gallons per minute and rain occurs within a few hours.

2. **Common Salt Technique**

1. Common salt is a suitable seeding material for seeding warm clouds.
2. It is used either in the form of 10 per cent solution or solid.
3. A mixture of salt and soap avoid practical process.

19.2.4 Mangrove Forests of India

Bordering the Indian ocean and its vicinity, mangrove vegetation is found in the Sunderbans in the estuary of the Ganges, Brahmputra system, the Andaman Islands and the Irrawady delta. Small remnants of mangroves are also found in the estuaries of the Mahanadi, the Godavari and the Krishna on the east coast. The west coast mangroves can be divided into two forms - the open scrubby mangroves of the Kutch and Saurashtra coasts and the close forests extending from the mouth of the Narmada and Tapti southwards in the Karnataka and Kerala states.

(*a*) The coast line of Saurashtra and Kutch (Gujarat) : On sea coast *Avicennia officinalis* associated with *Aegiceras mayus, Pandanus tectorius*; on salt marshes along the coast Bruguiera gymnorhiza together with *Aeluropus villosus* grass; in salt marshes and tidal creeks *Rhizophora mucronata* and *R. conjugata* are common.

(*b*) The coast line of Maharastra from the mouth of the Narmada southwards: Coastal sea water and semi-fluid mud area has Avicennia officinalis, Acanthus ilicifolius, Rhizophoramucronata. Salt marshes of tidal creeks have Rhizophora mucronata, R. conjugata, Sonneratia apetala, S. acida, Excoecaria agallocha, Lumnitzera racemosa etc.

(*c*) The coast line of the Karnataka, Kerala and Tamilnadu : The mangrove forests have been divided into three zones. The coastal sea water and semi-fluid mud zone contain

Rhizophoramucronata, R. conugata, Kandelia rheedii, Bruguiera gymnorhiza, B. caryophylloides, Acanthus ilicifolius etc. Only the West coasts of Karnataka and Kerala contain *Ceriops candolleana* and *Bruguiera eriopetala.*

Salt marsh zone along the coast is dominated by *Lumnitzera racemosa, Excoecaria agallocha* and *Arthrocnemum indicum.* East coast of Tamil Nadu is more rich having species of *Avicennia, Salicornia, Scyphiphora* etc. In the salt marshes of tidal creeks *Aegiceras majus* and *Herritiera littoralis* occur dominantly in addion to *Sonneratia apetala, Avicennia officinalis* (in west coast of Karnataka and Kerala only). Towards the interior dry marsh lands of east coast of Tamil Nadu only, *Suaeda Mudiflora, S. maritima, S. Monoica* occur.

19.2.5 Weather forecasting

Within the year 2025 the world population would exceed 850 crores, but the production of Agricultural commodities remains uncertain. Agriculture has to meet this challenge, by increasing production in a sustainable way. Agriculture is still a gamble in a country like India as they lie near the equator.

The weather and climate near equator always remain dynamic and unstable. Hence the prediction of weather in these areas is highly difficult and not stable. Hence any strategy to increase agricultural production should take explicit amount of the complimentary of agrometeorology and development.

Already the area available under cultivation is getting reduced day by day due to urbanisation & other factors. Hardly there are lands to bring into cultivation. In this situation, increasing the production per unit area per unit time is essential for which the use of weather factors play a major role in our country. In India, crops on 67 percent of the 14.2 crores of arable land depends on rain. Therefore 44% of the food comes from dryland and 40% of the Indian population depend on the dryland agriculture. Dryland cultivation is mainly dependent on the prevailing weather parameters. 40% of food grains, 80% of maize, 95% of pearl millet and sorghum, 95% of pulses, 75% of oil seeds are grown in dryland areas.

The following aspects of weather concepts are essential to decide the sustainability of production and productivity in our country.

(*i*) the date of onset and termination of rainy season and variability.

(*ii*) the distribution of rainfall.

(*iii*) micro climate of the cropped area to decide the occurrence of pest and diseases etc., *i.e.,* air, temperature, relative humidity, dew fall, soil temperature wind speed of the cropped area.

By weather forecasting, decision can be taken in advance in agricultural operations and thus planning can be done.

The following table shows the emphasis and applicability of weather forcasting

Types	Validity & Accuracy	Forecast emphasis on	Applicability
1. Short range forecast	24-48 hours/ 70-80% accuracy	Temperature, wind, Velocity & direction, Sunshine duration time & amount of rain.	(*i*) Scheduling irrigation irrigation (*ii*) Adjusting time of Agrl. Operation.
2. Medium range forecast	upto 10 days/ 60-70% accuracy	Types of weather, sequence of rainy days, normal weather hazards	(*i*) Decide the sowing time (*ii*) Depth of sowing (*iii*) Deciding harvesting (*iv*) Time of spraying (*v*) Irrigation (*vi*) Decision of harvesting
3. Long range forecast.	30 days to one season/ 60% accuracy	Abnormality of temperature & precipitation	(*i*) To decide on soil moisture management. (*ii*) Selection of crop. (*iii*) Selection of suitable cropping pattern.

There are three types of weather forecasting:

Among all these types of weather forecast, medium range weather forecast is given much importance as their applicability can be materialised easily, as their validity and accuracy is satisfactory. In case of short range forecast, the validity is for only one or two days which is a small period, for the farmers to collect the recommendation and then to follow. In case of long range weather forecast the accuracy is not as good as medium range weather forecast.

19.3 BIO-TECH.

19.3.1 Genetically Modified Crops

Genetically modified crops (GM crops) or transgenics are expected to be the major players in food and nutritional security in future. Globally GM crop area has increased in a short span from 1.7 m ha during 1996 to 52.6 m ha (130 m acres) in 2001. The number of farmers that benefited from GM crops increased from 3.5 million in 2000 to an estimated 5.5 million in 2001. The six leading countries already growing GM crops are USA, Argentina, Canada, China, South Africa and Australia. The global market for GM product is projected to reach $ 8 billion by 2005 and $ 25 billion by 2010.

India has lagged much behind in commercializing the GM crops. In May 2002, three cotton hybrids have been cleared by Genetic Engineering Approval Committee (GEAC). The increase in yield is to the extent of 10-15% with IPM and a considerable saving in pesticides.

India has identified 12 major field crops (Rice, What, Maize, Sorghum, Cotton, Pigeon pea, Chick pea, Mung and Udad bean, Potato, Mustard, Soybean and Pearl millet) which need transformation for specific diseases like bacterial blight, rust, wilt, downy mildew and viruses and pests like shoot fly, pod borer and American boll worm.

Nutritional and food security are of utmost importance for country like India. There is no compelling scientific argument to demonstrate that GM crops are innately different from non-GM crops. But some novel GM crops may pose elevated risk to environment. Hence all cases of GM crops including novel genetic changes must be assessed on case-by-basis. GM rice varieties tolerant to drought, submergence, salinity and rich in micronutrients need to be commercialized. But gene can also be introduced for pest resistance. Golden rice in vitamin A is a major step towards nutritional security. It is worthy of note that rice, being a staple food, cannot be used as a bioreactor to produce drugs and pharmaceuticals.

Tomato, with its small genome, is an ideal crop for transformation. The increase in polyamine levels such as spermidine and spermine is associated with significant increase in concentration of antioxidant-lycopene with improvement in quality of juice and shelf life. Over expression (Antisense RNA technology) of S. adenosyl methionine decarboxylase gene (ySAmdc) increased the levels of polyamines during ripening. Thus ripening involves differentiation of chloroplasts to Chromoplasts which results in accumulation of beneficial nutrients such as β-carotene, vitamin E and lycopene.

Biopharming

Field crops such as corn and tobacco are programmed with recombinant DNA Technology to produce high-value-added pharmaceuticals. Biopharmed corn is modified to produce a vaccine against Escherichia coli (produces diarrhoea) Similarly, tomato and musk melon are addressed with r-DNA technology to produce edible vaccines. This is termed as biopharming where plants act as bioreactors and harvested to produce drugs and then purified. The potential of this technology is the economy of the process. It is much cheaper to produce the drugs from the plants. Nevertheless, the biopharmed crops should be distinct from food crops so that food security is not threatened.

Gene Flow

This phenomenon is universal. Crops and weeds have exchanged genes for centuries. But genetic engineering raised additional concerns due to introduction of novel genes. The fear is that the gene may lead to 'supra weeds' which may develop resistance to insects, diseases, herbicides and harsh growing conditions.

The risks of transgenic crops are :

(*a*) Cross pollination of GM crops with conventional varieties.

(*b*) Germination of volunteer GM seeds.

(*c*) Cross pollination of GM crops with crop cultivar distant for non GM crop or organic niche markets.

Now strategies are available for gene containment or gene flow. Refugia of 20% for cotton is already practised. Other strategies for containing gene flow are :

(*a*) Restricting flower opening and floral development (cleistogamy) in GM crops.

(*b*) Male sterility approach as in mustard.

(*c*) Seed sterility and apomixes (vegetative propagation).

(*d*) Destruction of volunteer seeds by chemical/physical means.

(*e*) Specific excision of transgene.

Recombiant systems have emerged as valuable tools for efficiently regulating the excision of transgene when their expression is no longer required.

Environmental risks

The fear is that GM crops might indirectly impact the environment. This might be associated with recombinant and novel combination of DNA passing into the environment.

19.3.2 Terminator Gene Technology

Definition

The technology that terminates the viability/ fertility of seed after a given time is known as terminator technology and the gene involved is popularly known as terminator gene.

The Terminator System in Cotton has Three Components :

1. A gene for a toxin that will kill the seed late in development, but that will not kill any other part of the plant.
2. A method for allowing a plant breeder to grow several generations of cotton plants, already genetically-engineered to contain the seed-specific toxin gene, without any seeds dying. This is required to produce enough seeds to sell for farmers to plant.
3. A method for activating the engineered seed-specific toxin gene after the farmer plants the seeds, so that the farmer's second generation will be killed. These three tasks are accomplished by engineering a series of genes, which are all transferred permanently to the plant, so that they are passed on via the normal reproduction of the plant.

Terminator Technology

The success of Terminator depends on its ability to make a lot of a toxin that will kill cells, and to confine that toxin to seeds.

To accomplish this, in the case of cotton, the plan is to take the promoter from a gene normally activated late in seed development in cotton, and to fuse that promoter to the coding sequence for a protein that will kill an embryo going through the last stages of development.

Problems with use of Terminator

It is likely that terminator will kill the seeds of neighbouring plants of the same species, under certain conditions. In any case, dead seeds, where they occur, would be a serious problem for the farmer whose fields are close to the Terminator crop. How many seeds die will depend on the degree of cross-pollination, and that is influenced by the species of plant, the variety of crop, weather conditions, how close the fields are to each other, and so on. They will contain the toxin and any other proteins engineered into the Terminator-protected variety. These new 'components' may make the seed unusable for certain purposes.

19.3.3. BT Cotton

Bacillus thuringiensis (Bt cotton) is generally modified varieties of cottons.

India, the world's third largest cotton producer, where the traders say Indian cotton yield is just 300 kg/ha less than half the global average of about 650 kg/ha whereas, India has the world's largest area under cotton cultivation covering nearly 9 million hectars.

Due to some reasons Indians are starving from low productivity of cotton *viz.*, low productivity of land, low and miscellaneous use of manures and fertilizers, applications of NPK, plant population, lack of irrigation, insects pests menaces, management practices and vagaries of monsoons etc.

Among these, the production loss due to insects is quite high. About 21 species of insects possess enmity against cotton of which 12 are so egregious. These insects are classified into two groups:

1. 'Sucking insects' viz., termites, aphids, whiteflies, hariquin bugs etc. affect flowering stage (40-50 days after sowing).

2. 'Bollworm complex' viz., spotted bollworm, green bollworm, pink bollworm etc., decrease 30-40% yield.

India spends nearly Rs. 2800 crores per year on pesticides to control the damage. Among which cotton uses more pesticides *i.e.*, Rs. 1600 crores than any other crop. Only 5% of the total cultivated area is occupied under cotton, which requires 55% of insecticides and pesticides per year. Obviously, it affects environment too. This also increases selection pressure in insect population, which influences the evolution of new resistant biotypes.

It is an urgent necessity to control this attack by developing new resistant host plants having resistant genes. This had not been developed earlier, but recently the biotechnology genes produce insecticidal crystal proteins from micro-organisms which are being transferred to the cultivated varieties of cotton. This cotton is known, as 'Transgenic Cotton'. In India, the research for developing transgenic cotton has been going or in Central Institute of Cotton Research at Nagpur (CICR), at the International Crops Research Institute for Semi-Arid Tropics (ICRISAT); several candidate genes are being evaluated for their biological efficacy against the sorghum shoot fly (*Atherigone soccate*), spotted stem borer (*Chilio partellus*), cotton bollworm which are the major crop pests in semi-arid tropics. Since last 4-5 years some work is in progress in different private sector companies too.

Fact behind BT Cotton

Bt is short for '*Bacillus thuringensis*' a gram positive, aerobic, sporulating omnipresent bacterium, first discovered in 1902, one of the most successful agent of biological control, synthesizes crystalline proteins coded by the Cry genes during sporulation. This BT gene (Cry 1AC) encodes a special type of toxin commonly known as 'Delta endo toxin'. Delta toxin kills bollworm only without any deleterious effect on living creatures. The effort was made towards transferring this gene in cultivars of cotton to produce transgenic cotton plants showing resistance against bollworm. There are several genes with different functions in various organisms. When we transfer a gene in a cultivated variety, it induces resistance against the insect in the similar fashion, like a child vaccinated against Polio. However, Bt was already known for the control of Bollworms. We were using same insecticides based on Bt such as Delfin, Dipeal-E, etc.

Mechanism of action of BT toxin

The mechanism of action of Bt cry proteins involves solublization of crystal in the insect midget protease, binding of the cry toxin to midget receptors and insertion of the cry toxin in to the apical membrane to create ion channels or pores, disturbing cellular osmotic balance and causing the cell to swell and lyse through a process that has been termed 'colloid-osmotic lysis', which results in death of insect.

History of Commercialized BT

Although Bt was described by Berliner in 1915 from the province of Thuringia in Germany, the earliest description was actually that of Ishiwata in 1902, from Japan. The name "*Bacillus Scotto*" was given to the bacteria which caused a diseased of silkworms, the name "Scotto" meaning limp. Berliner's isolation was from a disease Mediterranean flour moth, *Ephestia kubniella*, and he was able to show that the bacterium was toxic when the spores were fed to insects. Interest in

Bt was rekindled by Mattes in 1927 who was able to re-isolate a strain of Bt from Ephistia and subsequent field tests by Husz in 1928; with this isolate on the European corn borer gave promising results. This work eventually led to the first commercial product, Sporeine, which was first produced in France in 1938.

BT in plant protection

(*i*) BT as a biopesticide

Bt is now the most widely used biologically produced pest control agent. The use of biological pesticides in agriculture remains significantly behind that of synthetic chemical pesticides, several environmental and safety considerations favour the development of Bt Cry proteins that have been studied; they are not pathogenic to mammals, birds, amphibians, but very specific to the groups of insects pests against which they have activity. Since the Bt is not genetically engineered, there is no need to obtain specific biosafety permits to market Bt based products. It proves the possibilities of resistance development in insects, strategies to present such resistance and the need for providing an umbrella of integrated pest management (IPM) over the transgenic crops.

(*ii*) Novel BT biopesticide

Bt has evolved to produce large quantities of crystal proteins, making it a logical host for developing improved cry biopesticides. Natural isolates of Bt can produce target specified. On the other hand, certain combinations of cry proteins have been shown to exhibit synergise offers.

(*iii*) BT-transgenic plants

Delivery of Bt process through spray formulations, engineered Bt any other bacteria has certain limitations. The chemical sprays suffer from short life, mechanical method and inability to reach burrowing insects. Engineered bacteria vary after proliferating at a rate and quality not sufficient to kill the target insects. These disadvantages can be overcome if the cry proteins are expressed in the plant cells at levels sufficient enough to kill the larvae. The first transgenic plants using cry genes in crop plants like cotton, and potato conferred considerable protection against lepidopteran and coleopteran pests, respectively. Other crop species carrying various cry genes include Soybean, peanut, alfalfa, apple, white clover, broccoli, walnut, pear and sugarcane.

Some landmark events in the production of transgenic BT Cotton

March 10, 1995 : Department of Biotechnology (DBT) of the Government of India permits import of 100 gms of transgenic Cocker-312 variety of cotton seed cultivated in the United States by Mahyco. This variety contained the 'Cry 1 AC' gene from bacteriumn 'Bacillus thuringesis'.

April 1998 : Monsanto tie up. Gave permission for small trials of Bt cotton 100 gms per trial by Department of Biotechnology (DBT).

July 2000 : Mahyco allowed conducting large scale field trails including seed production at 40 sites in 6 States. The DBT set up committee to "independently" monitor and evaluate large scale yield trials.

July 18, 2001 : An open dialogue held between Monsanto and Ministry of Environment representatives and farmers to discuss Bt cotton with scientists,

June 19, 2001 : Genetic Engineering Approval Committee (GEAC) extends field trials of Bt cotton by another year. Mahyco conducts large-scale trial on 100 hectares in seven States.

Jan 23, 2002 : Dr. Manju Sharma, Secretary of DBT, declares that the round of Bt cotton trials was satisfactory and that is up to the GEAC and the Ministry of Environment to decide on a date of commercial release.

April, 5, 2002 : Mahyco had received approval from India's Genetic Engineering Approval Committee (GEAC) of the Environment and Forest Ministry for commercial cultivation of BT Cotton.

19.4 LAB TO LAND

19.4.1 Role of Information Technology in Agriculture

At present, benefits of this technology are available primarily to urban elite masses. Rural communities will continue to have difficulty in accessing crucial information due to their inability to understand so as to make timely decisions.

Meaning of IT

Information technology refers to a broad term comprising of new communication and computing technology. Computer hardware, software and Internet are keys to these systems that are designed, developed and managed by IT professionals.

Present scenario of IT

- The mobile internet is overtaking the wired internet.
- Networking is growing at a fantastic speed of 10-15% per month or about 200% a year.
- Specialists forecast that, in less than 10 years from now, Internet terminals would be as common as telephone instruments but much more useful.
- About 99% of the districts have Optical Fiber Network (OFN) providing desired bandwidth for Internet beings with computers such as keyboards, mouse devices, scanners. The advent of touch screen monitors that allow users to give input to computers by touching on the appropriate location of the monitor has made it possible to develop user-friendly interface for farmers which is easy, intuitive, circumvents language barrier and at the same time provides a relaxed environment to the users.

Output devices : Monitor screens, printers and plotters, data projectors support high resolution and good quality output.

Objectives of IT

- To put information close to the managers, scientists, teachers, extension workers and farmers.
- To improve the capacity of researchers, teachers and extension specialists to organize, store retrieve and exchange information.
- To evolve mechanism of information sharing.
- To strengthen national libraries and library's network through electronic access.
- To develop database for easy access and data based decision-making.

- To spread information among the farmers.

Unique features of IT

1. Access to the storehouse of information is easy.
2. Information is available instantaneously
3. Communication is interactive.
4. Information from any point in the globe is available.
5. Communication is dynamic and ever growing.

The key players for utilization of IT in agriculture

- The farmer: the actual person who can directly bring about an improvement in.
- Efficiency and productivity in agriculture.
- Various industries that provide inputs to agriculture.
- Various industries that deal with agriculture output.
- Institutions/organizations and NGOs working for the benefit of farmers.
- Agricultural universities and research centers.
- Central and State governments.

The key players listed here can make a big contribution to the economy with the assistance of IT.

The broad information inputs required by farmers

- **Awareness databases:** Those facilities to farmers for proper understanding of implications of the WTO on Indian agriculture
- **Decision support systems:** Information that facilitates farmers to make proper SWOT (strengths, weaknesses, opportunities and threats) analysis to take appropriate decisions.
- Information on new opportunities
- Monitoring systems for corrective measures

How IT helps in agricultural production

(*a*) As a tool for direct contribution to agricultural productivity.

(*b*) As an indirect tool for empowering farmers to take informed and quality decisions, which will have positive impact on the way agriculture and allied activities are conducted.

Information technology helps indirectly in the way of

- Precision agriculture
- Remote sensing
- Expert systems
- E-agribusiness

IT Centers for agricultural development

Agricultural Technology Information Centres (ATICs) provide

- Diagnostic services for soil testing, plant and livestock health
- Supply of research products such as seeds, planting materials, livestock breeds, poultry strains, fish seed, processed products etc., emerging from an institution for testing and adoption by various clientele
- Dissemination of information through literature, audio-visual aids and electronic media
- An opportunity to institutions for resource generation through sale of their technologies
- Support the district level ATMAs (Agricultural Technology Management Agencies) in technology dissemination.

Indian Society of Agricultural Information Technology (INSIT) & Mandates

- To mobilize farmers, scientists, institutions and organization
- To encourage teaching, research and extension activities
- To provide a forum for information exchange and dissemination
- To organize training programme

Information technology has a major role to play in all facets of Indian agriculture in addition to facilitating farmers in improving the efficiency and productivity of agriculture and allied activities; the potential of information technology lies in bringing about an overall qualitative improvement in life by providing timely and quality information inputs for decision making. It can be argued that the next revolution in urgent need of focus is IT revolution in field of Indian farming.

19.4.2 Role of Women in Agriculture

The women contributing to Agriculture production are usually called as Peasant Women (those who are working in the field) and Farm Women (those who are wives of farmers). The contribution of farm women to increased food production could be seen at terms of influencing farmers to (*i*) accept and adopt new technology to increase their farm income, (*ii*) modernise the farm through improved farm machinery, (*iii*) develop the farm with irrigation facilities, (*iv*) strength post harvest operations, (*v*) ensure safe storage of food grains, fruits, vegetables including food processing operations, (*vi*) timely marketing of produce to fetch attractive prices, (*vii*) encourage savings for investment on farm development, education of children etc.

It is evident that the farm women and peasant women are actively engaged directly/immediately in agriculture development process. The Government has taken up several activities to improve the professional skills and capabilites of these women.

With women the introduction of high yielding varieties programme and also technological advances in agriculture the gap between existing and desired skills has become wide. To narrow down the gap Farmers Training and Education Programme was started in 1966-67 under which 188 Farmer Training Centres (FTCs) were established. A lady demonstrator was appointed in each F.T.C. to take care of the peasant and farm women training and education.

A strategy to achieve the overall development of women in rural areas with special reference to women in agriculture include.

1. Formal and informal education of rural girls and women.
2. Training of rural women in farm and home activities.
3. Organisation of Rural Women Mandals, Rural Women Television Forums, Mass Campaigns etc.
4. Utilising interpersonal relations among rural Women to convince and motivate them for adoption of planned change.
5. Linking the planned development programmes for women with existing institutions *viz.* 188 Farmers Training Centres, 261 Krishi Vigyan Kendras, National Research Centre on Women in Agriculture etc.
6. Involvement of Women in Rural Institutions and all development programmes.
7. Education of Senior citizens in rural areas about role of Women in Development through film shows, poster, classes, discussions etc.

8. Modernisation of rural Social System through implementation of compulsory education for rural girls and adult education through functional literacy for rural women, intensive mass media facilities etc.
9. Liberalising the social and cultural values imposed on rural women.
10. Integrate the women groups in rural areas and form a network of rural women mandals in the country.
11. Create a voluntary movement of women for development of rural areas through assisting the change agents in implementation of developmental programmes.
12. Implementation of poverty eradication programmes designed for rural women.
13. Development of cooperative projects *viz.* fruit, vegetable and kitchen gardening, poultry rearing, bee-keeping, handicrafts, grocery stores, etc. minor construction projects-roofing, drainage channels etc. to help increase their income and improve the standard of living.

A special focus on development of peasant and farm women can be made through —

1. Creating a Women Extension Service Cadre in State Department of Agriculture to concentrate on specially identified clientele.
2. Identify and document the needs and problems of Women in Agriculture and accordingly design or refine the new or existing projects/programmes.
3. Promote environmentally and ecofriendly technologies for Women in Agriculture and encourage income generating enterprises to be undertaken by Women in Agriculture.
4. Incentives for outstanding contribution to boost the morale and encourage greater participation among other Women in Agriculture.
5. Telecast success stories/case studies to motivate the women for active participation in developmental programmes.
6. Modernise the implements/tools being used by women to reduce drudgery and to increased efficiency. In developing countries majority of women contribute to production of crops *i.e.* food production through manual/skilled labour advisory, extension service and also prepare and cook the food and their men and children thus it can be said that women feed the family, society, nation and the world.

19.4.3 Contract farming

In this era of globalization, the concept of contract farming has been acquiring a greater momentum. Contract farming is a system for the production and supply of agricultural products under forward contracts between cultivators or suppliers and buyers. Here to cultivator commits to provide an agricultural product of a specified type at a specified time and at a specified price that is required by the committed buyers.

The process of contract farming seems to attractive of implemented in its good sense. The main feature of contract farming is that the contractor supplies all the material inputs and technical advise required for cultivation to the cultivator. In turn the cultivator supplies the required land and labour.

In India the central government has drafted a model law on agriculture marketing to provide among other things, legal support to contract farming agreements. Some of the state government has entered into a contract farm for various crops.

19.4.4 Modernization of Agriculture

Agriculture is not an industry. It is an art and a way of life dealing with living beings. In the past years, it has undergone great changes all over the world. The important strategies of Agriculture are:

1. Increased food production
2. Saving the environment and attaining sustainability

There has been significant increase in the quantity of food produced and land conserved. This is the effect of modernization.

Drawbacks of Modernization

1. Reduce the number of jobs, people engaged.
2. Workers replaced.
3. Rural Culture put under pressure.
4. People migrate in search of work.
5. Local institutions weakened.
6. Resources once considered as precious, has now turned a waste to be disposed off.
7. Water, Soil contaminated.
8. Agriculture has become fossil fuel intensive.
9. Increased application of pesticide.
10. Agriculture has begun contributing to global warming.

Reasons for the rejection of Modern Methods by the poor farmers

1. High cost.
2. Poor extension.
3. Inappropriate incentives.
4. Inappropriate innovation.
5. General unawareness.
6. Local practices are better than modern methods.
7. Modern methods generate new problems.

Modernization of Agriculture in Indian Context

In India when the mechanization of bullock energy changed to power tiller use, about 8% increase in output required 43% increase in energy input.

For tractor 13% increase in yield cost 74% extra energy.

For wheat tractor 6% increase in yield cost 266% extra energy.

Biodiversity Loss due to Modernization of Agriculture

1. In India, once 30,000 varieties of rice were grown. But now 10 varieties cover 75% of the whole rice area. In the U.S. 65 types of vegetables lose 80-100% of the varieties of each since the turn of the 20th century. Of the 8207 varieties listed in 1903, only 607 are stored in seed forms. The greater the uniformity the greater the risk and danger of pest, risk and disease attack.
2. Loss of genetic diversity leads to future opportunity to raise adaptable crops and livestock.
3. A farmer prefers multi-cropping, inter-cropping, mixed-cropping but does not standardize his practices.
4. Farmers do not replace existing varieties of crops by a new variety.

Modernization of Agriculture and breakdown of rural communities

1. Number of jobs for local people gone.
2. Standardization has reduced the management skills.
3. Rural poverty increases.
4. Food processing jobs are no more localized. Far away central factories do that work.
5. Growing gap between the rich and the poor.

6. Landscape homogenized; farming system simplified.
7. Youth show less interest in labour intensive farming. Many farms have no successors.
8. Farm size increases, family farms vanishes.
9. Frustration, financial crisis, suicide on the increase among farmers.

19.4.5 Self Help Groups

Definition; It is a group of rural poor to have volunteer to organise themselves into a group for eradication of poverty of members".

3. Stages :

(*i*) Group formation.

(*ii*) Capital formation through revolving fund & skill development.

(*iii*) Income generation through economic activity.

Important features of SHG's;

- It consists of 10-20 persons.
- All members of group belongs to families of Below Poverty Live (BPL).
- There can be one person from one family.
- One person cannot be member of more than one group.
- Broad alliance with foreign countries.
- Both public and private investments should be encouraged.
- An efficient marketing system is available

Banking loans to SHG's

- Through NGO/Voluntary agency.
- NABARD promotes SHG's (improve rural poor).

Objectives

(*i*) Enabling member to become self dependent and Self-reliant.

(*ii*) Providing forum for members of discussing social and economic aspects.

(*iii*) Enhancing social status of members.

(*iv*) Providing patterns of members for exchange of idea.

19.5 ECOLOGICAL AGRICULTURE

19.5.1 Natural Farming:

Concept of Natural farming

Established by Masanobu Fukuoka. He wrote the famous book "*The One-Straw Revolution*".

Natural farming requires no machines, no chemicals and very little weeding. In other words natural farming is an attempt to develop a method of farming which could help to reverse the degenerative momentum of modern agriculture.

Principles of Natural farming

The following steps are important to make farming in natural way.

- No cultivation-stirring of the soil by plough
- No chemical fertilizers-addition of crop residues and organic manures
- No tillage-mulching with plant materials
- No dependence on chemicals-plant extracts can be used for plant protection

The main purpose of natural farming is to add ecological sustainability and agricultural development. According to Dr. M.S. Swaminathan, the basic principles of sustainability include, land, water energy, nutrient supply, pest management, post harvest systems and systems approach. These factors are closely linked together for sustainability.

Another important aspect in natural farming is preventive crop protection. The farmers should know agricultural ecosystems, like mixed farming and diversification, crop rotation, optimum planting season for all crops, manuring and plant health, choice of varieties and field hygiene. Since insects very often attack crops and other pests, the knowledge of above items is important. Use of pesticides had led to disturbances in the natural eco-system and poisonous chemicals up in ground water eventually having harmful environmental consequences. Hence, preventive crop protection principles are vital for natural farming.

Some ecologically friendly plants, *e.g.*, Neem, Pungam, Jatropha etc., help to promote ecological balance. The integrated farming systems can help in ensuring self-sufficiency in farm holding, increase and help in ecological balance by recycling of farm residue and wastes.

Natural farming is a traditional form of agricultural practice popularly associated with a back-to-nature movement. Hence, it is to make maximum possible use of natural resources, thereby avoiding the use of chemicals liable to kill or reduce the activity of beneficial soil organisms and avoiding direct feeding of land with soluble minerals.

19.5.2 Global warming A Grave concern under Indian Agricultural Scenario

Warming is accumulation of carbon dioxide, methane and other air pollutants collecting in the atmosphere like a thickening blanket, trapping the sun's heat and causing the planet to heat up. The green house effect or increasing world temperature is due to the accumulation of carbon dioxide as a result of combustion of fossil fuel methane is responsible for some of the depletion of the protective ozone layer. Methane arises largely from natural anaerobic ecosystems, rice paddies and ruminant animals.

Effects of Global warming

- Rapid melting of glaciers will cause floods in widespread damage to crops and livestock.
- Aberrant weather changes resulting in droughts and floods. Nearby two thirds of the kharif crop-paddy, oilseeds and pulses is at risk due to little or no rainfalls.
- Forests, farmer and cities will face troubles some new pests and more mosquito borne diseases.
- Disruption of habitats such as coral reefs and alpine meadows could drive many plant and animal species to extinction.
- Decreased crop fields and decreased arable land availability with subsequent starvation and malnutrition.
- Warm and weather conditions will increases the level of infectious diseases.
- The rising sea is expected to increase salinity across India's coastline.

United Sates of a America having only 4 percent worlds population, produce 25 percent of the carbon dioxide pollution from fossil-fuel burning (the largest share of any country). In fact, the United States emits more cabon dioxide than China, India and Japan combined.

CO_2 emissions across the globe (in billions of metric tonne)

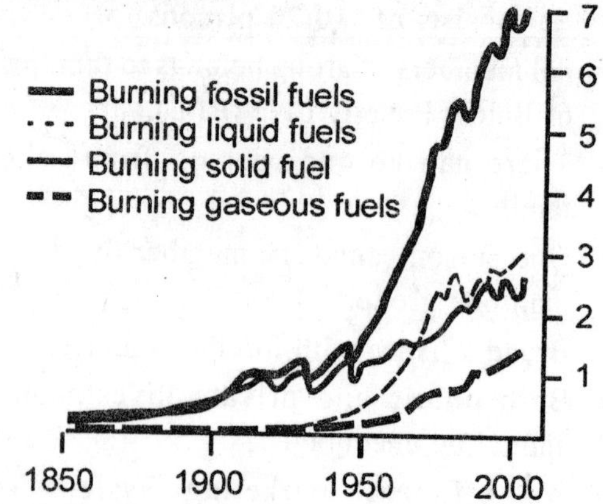

Remedial measures

- Maximum support should be given to research and development efforts that will enable fossil-derived fuels to be replaced by biomass-derived fuels.
- In the tropical countries, wetland rice may have to be discouraged in favour of dryland cereal production.

- In developing countries the production systems are much less dependent on inputs of fossil fuel.
- Emphasis must be given to crops that are most efficient in fixing carbon dioxide into biomass.
- Non renumerative species (especially pigs, poultry rabbits) should be promoted as meat producers, while meat production from ruminants should have a lower priority.

The consequences of global warming are dangerous. There should be a massive movement towards environmental friendly strategies that minimize production of carbon dioxide and methane.

19.5.3 Phytoremediation

Phytoremediation is the direct use of living plants for remediation of contaminated soil, sludges, sediments, or ground water through contaminant removal, degradation, or containment of the contaminant(s). Growing and, in some cases, harvesting plants on a contaminated site as a remediation method.

Significance of phytoremediation

Phytoremediation takes advantage of natural plant processes. It requires less equipment and labour than other methods since plants do most of the work. Trees and plants can make a site more attractive as well. The site can be cleaned up without removing polluted soil or pumping polluted ground water. This allows workers to avoid contact with harmful chemicals. Phytoremediation has been successfully tested in many locations.

The time it takes to clean up a site using phytoremediation depends on :

- Type and number of plants being used
- Type and amounts of harmful chemicals present
- Size and depth of the polluted area
- Type of soil and conditions present

These factors vary from site to site. Plants may have to be replaced if bad weather or animals destroy them. This adds time to the cleanup. Often it takes many years to clean up a site with phytoremediation, tree roots take in water and pollution from the ground polluted soil water enters tree where pollution is cleaned up.

Principal :

Phytoremediation works best at sites with low to medium amounts of pollutants. Plants remove harmful chemical from the ground when their roots take in water and nutrients from polluted soil, streams, and ground water. Plants can clean up chemicals as deep as their roots can grow. Tree roots grow deeper than smaller plants, so they are used to reach pollution deeper in the ground. Once inside the plant, chemicals can be :

- Stored in the roots, stems, or leaves.
- Changed into less harmful chemicals within the plant.
- Changed into gases that are released into the air as the plant transpires (breathes).

In addition to transporting nutrients, certain plants are capable of transporting environmental pollutants such as metals, radionuclides, chlorinated solvents, petroleum hydrocarbons, and ammunition wastes through biological means.

Contaminants and contamination levels

Phytoremediation is effective in the remediation of all types of soil contaminants; however it is more effective on lower concentration of contaminants.

Types of Vegetation Used

Some of the plants used in phytoremediation are:

- Alfalfa
- Hybrid Poplar Trees
- Blue-green Algae
- Duck Weed
- Arrowroot
- Sudan Grass
- Bermuda grass
- Alpine Bluegrass
- Yellow or White Water Lillies

Vetiver grass combined with applications of zeolite is being used to remove boron, cadmium and lead from indonesian soils contaminated by sludge from the textile industry.

Fate of contaminants

Once plants extract environmental contaminants, these contaminants are either broken down by the plant into non-hazardous substances or concentrated within plant tissues.

Concentration of contaminants in plant material gives rise to the possibility of animals and insects consuming the contaminated material thereby introducing contaminants into the food chain.

Pros and Cons of Phytoremediation

Advantages

Phytoremediation is cost effective.

It is suited to remediation of large areas of soil.

It is environmentally friendly.

Phytoremediation sites are more aesthetically pleasing.

Phytoremediation sites require low maintenance.

It involves no noisy and expensive equipment.

Disadvantages

Not as effective for sites with high contaminant concentration.

Phytoremediation is slower than conventional methods.

It does not work through the winter (Seasonally effective).

Risk to wildlife/food chain.

19.5.4 Carbon credits

The Kyoto Protocol Dec. 1997 had created a clean development mechanism under which polluting countries have voluntarily decided to bring down the level of carbon emission equivalent to the CO_2 emission at 1990. This is achieved by two means. Firstly, one can reduce the GHG by adopting new technology or improving upon the existing technology to reduce the carbon emission. If it is not possible to reduce emission, one has to pay for the excess emission by getting carbon credit from the developing countries who uses eco-friendly, new technology so that total emission of carbon into the atmosphere is reduced.

As per United Nations Framework Convention on Climate Change (UNFCCC), carbon credit is assigned to a firm or company or industry. This credit is depends upon how the rate of carbon emission is reduced from the accepted level.

Carbon credits are generated by enterprises in the developing world that shift to cleaner technologies that save on energy consumption and reduce greenhouse gas emissions. For each tonne of carbon dioxide emission reduced, the entity can get a carbon emission certificate which they can sell either immediately or through a futures market, just like any other commodity. The certificates are sold to entities in developed countries that have higher emission of carbon.

19.5.5 Carbon sequestration

Carbon sequestration can be defined as the capture and secure storage of carbon from the atmosphere on a long term basis. In other words, C sequestration can be defined as net removal of atmospheric CO_2 into long-lived pools such as terrestrial and oceans by photosynthetic pathway of plants and micro-organisms. The main objective of the carbon sequestration is to reduce the atmospheric CO_2 concentration to reduce the impact of global warming.

The balance between carbon additions of photosynthetic plant products to the soil and their losses via their subsequent decomposition and microbial respiration determines the amount of organic carbon present in the soil. Plant residues, manure, sewage sludge, and other organic carbon by-products are the major sources of carbon inputs in terrestrial ecosystem.

Natural carbon sequestration: It's simply the process by which nature has achieved a balance of **carbon** dioxide in our atmosphere suitable for sustaining life. Nature provided trees, the oceans, earth and the animals themselves as **carbon** sinks, or sponges.Main natural carbon sinks are **plants,** the **ocean** and **soil.**

19.5.6 Ecotourism

It is the responsible travel to natural areas that conserves the environment and improves the welfare of the local people. Eco tourism is a growing component of the larger tourism industry.

Benefits of Ecotourism

- It is an environment friendly and potentially more sustainable alternative to extractive activities. Such as farming, logging and mining.
- It offers local people the chance to escape from a cycle of poverty and by sharing their knowledge of the local terrain ecology with visitors.
- Business initiated through ecotourism ventures can become self sufficient with in a short span of time and can enhance the long term economic prospects of a community.
- It is an opportunity to preserve ecosystems and biological diversity.
- It can create jobs in remote regions.
- Protected areas and nature conservation provide many benefits to society including preservation of biodiversity.

Dangers of Ecotourism

- Tourism in a region that boast opportunities to view endangered species can cause the extinction of these species.
- A primary value of ecotourism is the employment opportunities it provides to local people. Yet many tourism operations are owned and run by corporations, providing no benefit to the locals.
- If ecotourism operations fail to live up to conservation ideals can have very serious environmental consequences.

The guiding principles for ecotourism

- Tourist activities must not degrade the resources.
- Tourism must respect the intrinsic value of natural resources.
- Tourist revenue must provide conservation, scientific or cultural benefits to the resources and local community.

Achieving the principle depends in large part upon the scale of the operation. Ecotourism can live up to the promises if it follows the principles, adequately monitor and protects its resources and ensure fair distribution of profits with in the community.

19.6 EIGHT MISSION UNDER NAPCC

Government of India has launched eight Missions as part of the National Action Plan on Climate Change (NAPCC) in specific areas *i.e.* Solar Energy, Enhanced Energy Efficiency, Sustainable Habitat, Water, Sustaining the Himalayan Eco-system, Green India, Sustainable Agriculture and Strategic knowledge for Climate Change which include assessment of the impact of climate change and actions needed to address climate change.

It is proposed, with the assistance of the World Bank to implement the Integrated Coastal Zone Management Plan Project which includes the mapping of a hazard line along the coastal areas of the country taking into account the sea level rise due to climate change and other parameters such as, shoreline change, tides and waves.

A survey conducted by the Organization for Economic Cooperation Development (OECD) in 2007 'Ranking Port Cities with High Exposure and Vulnerability to Climate Extremes' identifies Kolkata and Mumbai as "amongst the most vulnerable cities" to the rising sea level and storm surges due to climate change.

Although India does not have legal binding reduction commitments, a range of policies and programmes taken to respond to reducing emissions, inter-alia, include:

(*i*) Setting up of the Bureau of Energy Efficiency and encouraging energy conservation and introduction of energy efficiency labelling.
(*ii*) Promoting use of renewable energy
(*iii*) Power sector reforms and active renewable energy programme
(*iv*) Use of cleaner and lesser carbon intensive fuel for transport
(*v*) Fuel switching to cleaner energy
(*vi*) Afforestation and conservation of forests
(*vii*) Promotion of clean coal technologies
(*viii*) Reduction of gas flaring

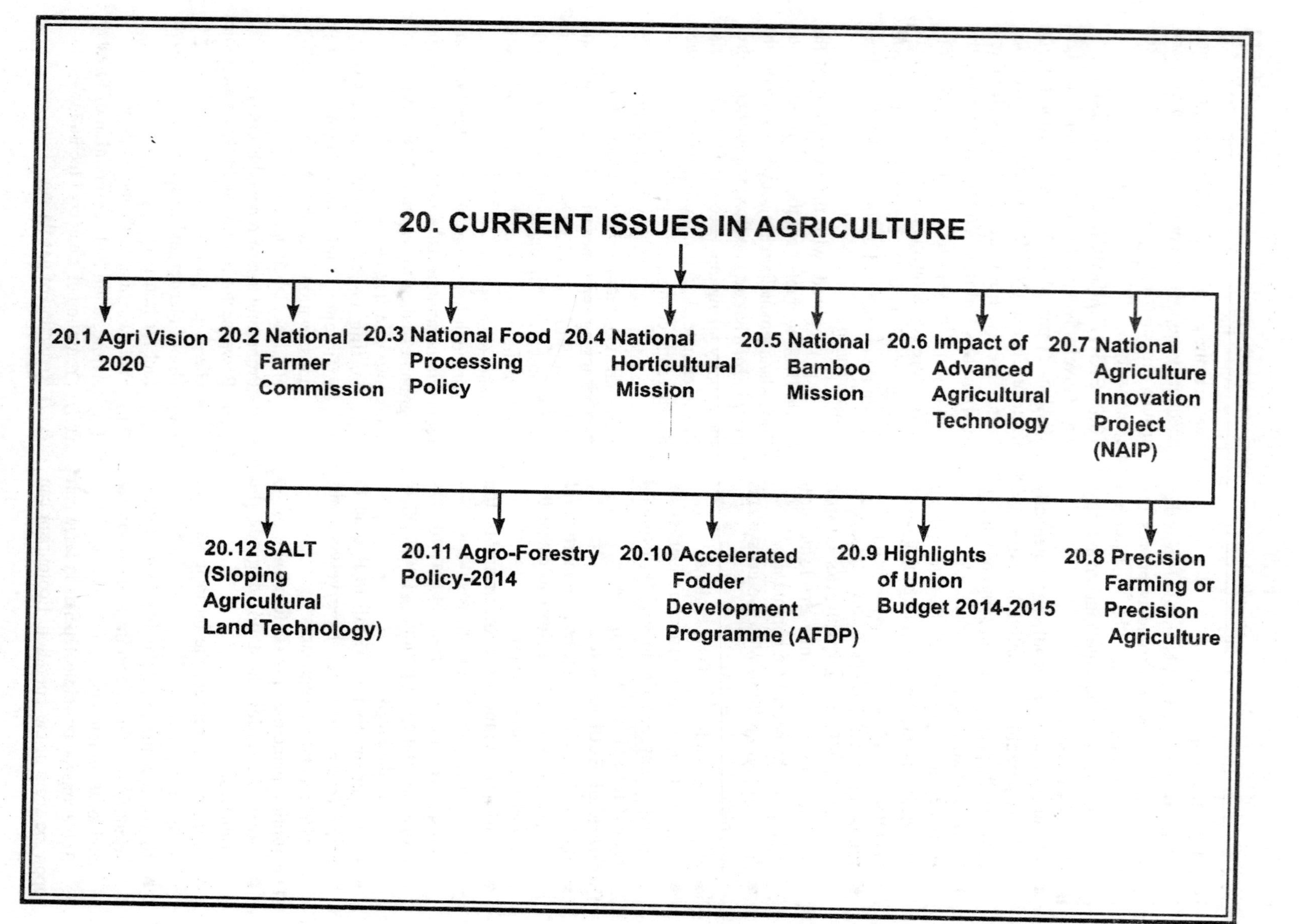
20. CURRENT ISSUES IN AGRICULTURE
20.1 Agri Vision 2020
20.2 National Farmer Commission
20.3 National Food Processing Policy
20.4 National Horticultural Mission
20.5 National Bamboo Mission
20.6 Impact of Advanced Agricultural Technology
20.7 National Agriculture Innovation Project (NAIP)
20.8 Precision Farming or Precision Agriculture
20.9 Highlights of Union Budget 2014-2015
20.10 Accelerated Fodder Development Programme (AFDP)
20.11 Agro-Forestry Policy-2014
20.12 SALT (Sloping Agricultural Land Technology)

20. Current Issues in Agriculture

20.1 AGRI VISION 2020

- Conservation, planned enhancement and utilization of agro-biodiversity.
- Enhancing productivity through evolution of high-yielding hybrids and varieties.
- Research on diversification, quality improvement, post-harvest technology, value-addition and export-oriented commodities.
- Sustaining enhanced productivity of irrigated agriculture and judicious development and use of energy, especially renewable sources of energy.
- Characterization and development of sustainable land-use models for rainfed agriculture in high-rainfall areas.
- Development of the Integrated Pest Management (IPM) and Integrated Nutrient Management System (INMS) approaches and systems for sustainable agriculture.
- Fostering excellence in relevant basic and strategic research.
- Generating research and technologies geared to promote equity among regions, sectors of society and gender.
- Strengthening social science, policy planning, agri-business and research monitoring mechanisms.
- Strengthening Agricultural Research Information System (ARIS).
- Promoting the Agricultural Human Resource Development (AHRD).
- Linking scientists with the farmers through Institute-Village-Linkage Programme (IVLP) as an innovative technology transfer model.
- Institutionalization and strengthening linkages/partnerships with CGIAR and under national and international agencies and research and development establishment, non-government organizations (NGOs), farmer organizations, private sector etc.
- Explore alternative or non-conventional plant resources for diversification of agriculture so as to enhance the food, fodder, fibre and fuelwood production.
- Provide solutions which are relevant, technically sound, economically viable, socio-culturally acceptable, eco-friendly and systems-responsive.
- Enhance the use of biofertilizers and biopesticides and biodegradable chemicals in agriculture.
- Strengthen the production base by harnessing modern scientific knowledge and translate it in the form of crop plants which are high-yielding, resistant to biotic and abiotic stresses, and efficient in input use.
- Improve skill and productivity of farm women with accent on social engineering and empowerment.

- State-of-the-art technology for post-harvest management and agro-processing.

20.2 NATIONAL FARMER'S COMMISSION

1. Chairman

Dr. M.S. Swaminathan

2. Objectives :

1. The Government to initiate policy interventions in the horticultural sector.
2. Implementation of sanitary and phytosanitary measures for facilitating exports and consumption.
3. Panel has estimated that current level of malnutrition reduce the national GDP by more than 3%.
4. It has suggested or 7-point national nutrition programme and linking horticulture to the national food for work and national rural employment schemes.

3. Six point action plan

1. Life cycle approach in the delivery of nutrition safety net programmes under the aegis of panchayati raj institutions. Need for a horizontal dimension to the vertically structured programmes.
2. Filling the gaps in the on-going programme, supplementary nutrition for adolescent girls, pregnant women and infants (within 2 years of age). This programme needs 2,50,000 tonne of foodgrain per year. Another 1,00,000 tonne of foodgrain for nutrition support of HIV^+ patients and those affected by tuberculosis.
3. Household entitlement cards to all eligible persons.
4. Integrated food guarantee programme-gender sensitive and enlarge scope of work in case of women.
5. Promote establishment and management of community food and fodder banks by self-help groups to widen food basket, alleviate food insecurity of both people and livestock.
6. Fight hidden hunger-supply, all antyodaya households with either vitamin-A or multiple fortified salt. This programme will require a budgetary provision of Rs. 200 crore.

20.3 NATIONAL FOOD PROCESSING POLICY

"Agriculture often suffers from varietal constraints. Oranges in Nagpur and mangoes in Malihabad in UP are instances where though there are gluts at harvest time, processing has not taken off due to want of processable varieties," the policy document said. In context the policy document suggested various steps to augment the sector.

- Changes in state laws for regulated markets for allowing private sector participation.
- Contract farming.
- Private sector participation in agricultural extension services.
- Procurement of farmers produces by large food retailers or markets.
- Improving post-harvest management, cleaning, grading, packaging and storage.
- Existing cold storage facility caters to less than 10% of the produces. It is mainly for potatoes and is rudimentary in nature. Hence there is a need for expanding cold storage facility for other perishable produces.
- Most of the limited controlled atmosphere storage facilities, technologies, protocols and machineries are imported. There is, therefore a need to upgrade manufacturing capabilities.
- Upgradation of research labs.
- Removal of conditionalities and caps on financing.
- Warehousing receipts should be made negotiable instruments also there should be a system of accreditation of warehouses.

- Warehousing receipts should be made negotiable instruments also there should be a system of accreditation of warehouses.
- Commodity markets should be stable and future markets more deepened. Banks, mutual funds and FIIs should be allowed in future trading and derivatives.
- Liberal funding from RIDF for infrastructure development for food processing sector. State should participate in joint ventures for infrastructure development.
- R & D on ethnic foods.
- Apex research institute should be set up to address R & D problems for the sector.
- Norms relating to Codex, HACCP, GMP and GHP should be met.
- Integrated food law and liberal labour laws.

20.4 NATIONAL HORTICULTURAL MISSION

National Horticultural Mission promoted by GoI. It was launched under the 10th Five year plan in the year 2005-06, 85% share by Central Government. and 15% share is contributed by State Governments.

Mission for Integrated Development of Horticulture (MIDH) (2013)

- Mission for Integrated Development of Horticulture (MIDH) is a Centrally Sponsored Scheme for the holistic growth of the horticulture sector covering fruits, vegetables, root & tuber crops, mushrooms, spices, flowers, aromatic plants, coconut, cashew, cocoa and bamboo. Under MIDH, Government of India (GOI) contributes 60%, of total outlay for developmental programmes in all the states except states in North East and Himalayas, 40% share is contributed by State Governments. In the case of North Eastern States and Himalayan States, GOI contributes 90%.

  In case of NHB (National Horticulture Board) Coconut Development Board, CIH (Central Institute of Horticulture Agencies. GoI contributes 100%. MIDH also provide technical advice and administrative support to State Governments/ State Horticulture Mission (SHMs) and Saffron Mission and other horticulture related activities RKVY/ NMSA.

- **MIDH Schemes :**
- **National Horticulture Mission (NHM)** is one of the sub schemes of Mission for Integrated Development of Horticulture (MIDH) which is being implemented by State Horticulture Missions (SHM) in selected districts of 18 States and four Union Territories.
- **Horticulture Mission for North East & Himalayan States (HMNEH)** is one of the sub schemes of Mission for Integrated Development of Horticulture (MIDH) which is being implemented by State Horticulture Missions (SHM) in the North Eastern States and Himalayan States.
- **National Horticulture Board (NHB)** is implementing various schemes under Mission for Integrated Development of Horticulture (MIDH) in all States and UTs.
- **Coconut Development Board (CDB)** is implementing various schemes under Mission for Integrated Development of Horticulture (MIDH) in all Coconut growing states in the country..
- **Central Institute for Horticulture (CIH)** was established at Medizipehima,

Nagaland in 2006-07 for providing technical back stopping through capacity building and training of farmers and Field functionaries in the North Eastern Region.CIH now one of the sub schemes of MIDH. However, CIH is not implementing any schemes directly

20.5 NANO UREA

Recently, the Indian Farmers Fertiliser Cooperative Limited (IFFCO) introduced the world's first Nano Urea Liquid for farmers across the world.

Objective was to enable Indian farmers to prosper through timely supply of reliable, high quality agricultural inputs and services in an environmentally sustainable manner and to undertake other activities to improve their welfare

- Developed by IFFCO
- 500 ml = 50 kg urea
- Size of the particles- 20-50 nm
- 4% N in nano urea

Central Public Sector Undertaking (CPSU) namely, NFL (National Fertilizers Ltd.) and RCF (Rashtriya Chemicals and Fertilizers Ltd.) have signed a MOU with IFFCO for transfer of Technology for setting up of nano urea plants at Nangal (Punjab) and Trombay (MH) respectively.

NANO-FERTILIZERS

Nano-sized fertilizers are **made using ammonium humate, urea, ammonia, peat, plant wastes, and other synthetic fertilizers.** A nano-sized nitrogen (N) fertilizer prepared by deposition of urea on calcium cyanamide is an example of nano-formulation.

Advantages of nanofertilizers over conventional mineral fertilizers:

- Nanofertilizers **feed the crop plants gradually in a controlled manner** in contradiction to rapid and spontaneous release of nutrients from chemical fertilizers.
- Nanofertilizers are **more efficacious in terms of nutrients absorption and utilization** owing to considerably lesser losses in the form of leaching and volatilization.
- Nanoparticles record **significantly higher uptake** owing to free passage from nano sized pores and by molecular transporters as well as root exudates.
- Nanoparticles also **utilize various ion channels** which lead to **higher nutrient uptake** by crop plants.
- Within the plant, nanoparticles may **pass through plasmodesmata** that results in effective delivery on nutrient to sink sites.
- Due to considerably small losses of nanofertilizers, these **can be applied in smaller amounts** in comparison to synthetic fertilizers which are being applied in greater quantities keeping in view their major chunk that gets lost owing to leaching and emission.
- Nanofertilizers offer the biggest benefit in terms of small losses which lead to **lower risk of environmental pollution.**
- Comparatively **higher solubility and diffusion** impart superiority to nanofertilizers over conventional synthetic fertilizers.

Limitations of nanofertilizers

- Nano fertilizers related **legislation and associated risk management** continue to remain the prime limitation in use.
- Another limiting factor is the **production and availability of nano fertilizers in required quantities.**
- The **higher cost of nano fertilizers** constitutes another hurdle in the way of promulgating them for crop production under varying pedo-climatic conditions across the globe.

- Another major limitation pertaining to nanofertilizers is the **lack of recognized formulation and standardization** which may lead to contrasting effects of the same nanomaterials under various pedoclimatic conditions.
- At present nano urea capacity is about 50 million bottles per year IFFCO has introduced innovative nano urea in the Indian market.

NANOTECHNOLOGY AND NANO UREA

Nanotechnology can increase agricultural production, and its applications include: (1) nanoformulations of agrochemicals for applying pesticides and fertilizers for crop improvement; (2) the application of nanosensors in crop protection for the identification of diseases and residues of agrochemicals; (3) nanodevices.

The government has approved the use of nanotechnology for plant nutrients, a decision which is expected to halve the use of urea and increase productivity by 15%. "The Central Fertiliser Committee, the regulatory body of fertiliser products, has approved the usage of nano nitrogen after a year-long field trial. Now nano nitrogen will soon be commercially available to farmers. "This will help cut our dependence on import".

20.6 IMPACT OF ADVANCED AGRICULTURAL TECHNOLOGY

Based on the case study conducted in Ludhiana district of Punjab. The following visible changes are noticed.

1. A real agricultural has taken place which has integrated the rural revolution.
2. Old traditions are disappearing, giving rise to new ways of living.
3. New schools are springing up.
4. Interior cultivation is practiced with higher level of inputs.
5. Rural banks advances are availed and some are taking up additional professions to augment the income from activities.
6. Huts and rural walled houses replaced by durable brick construction.
7. Many houses are provided with furniture, vehicles, TV, radio etc.
8. There is medical facility.
9. Mechanization of good proportion of farm operation.
10. There are many work opportunities available in agriculture.

CROP DIVERSIFICATION INDEX

The crop diversification index has a base value of one, which represents the percentage of total cultivated area devoted to a single crop. The current crop diversification index value demonstrates that there is a greater need for the diversification of crops in Telangana to achieve sustainable development. Telangana is the first state in the country to record its crop diversification patterns in the form of an index.

PM-KISAN is a scheme of the central government that was introduced in 2019. This scheme was launched to help farmers holding land with their financial needs. This year is the PM-KISAN scheme's third anniversary. Through Direct Benefit Transfer (DBT), Rs 6000/- is deposited into the bank accounts of farmers across the country in three equal installments every four months.

Accelerated Irrigation Benefits Programme (AIBP)

Central Govt., during 1996-97, launched an Accelerated Irrigation Benefits Programme (AIBP) to provide Central Loan Assistance (CLA) to major/ medium irrigation projects in the country, with the objective to accelerate the implementation of those projects which were beyond resource capability of the states or were in advanced stage of completion. While selecting the projects, special emphasis was to be given to Pre-fifth and Fifth Plan projects. Priorities were also given to those projects which were benefiting Tribal and Drought Prone Areas. However, under the revised AIBP Guidelines from the year 1999-2000 onwards, Central Loan Assistance under AIBP could also be extended to minor surface irrigation projects of special category states (N.E. States & Hilly States of H. P., Sikkim, J&K, Uttaranchal and projects benefiting KBK districts of Orissa).

20.7 NATIONAL AGRICULTURE INNOVATION PROJECT (NAIP)

Shri Sharad Pawar, then Minister of Agriculture launched NAIP on 26th July, 2006 with a total budget of 250 million US dollars (world bank will fund 200 million dollars as credit and Govt. of India will contribute 50 million dollars.

Objectives

1. To give the agricultural research and technology development system an explicit development and business perspective through innovative models.
2. To make the National Agricultural Research System a 'pluralistic' system where every organisation having stake in agricultural research: public, private or civil society, has to play a role.
3. Working in well defined partnership groups with clear common goals and understanding sharing responsibilities and benefits.
4. Funding through competition so that a wide choice of excellent innovative ideas come in from the stakeholders.
5. Work with focus, plan and time frames.

20.8 PRECISION FARMING OR PRECISION AGRICULTURE

It is an agricultural concept relying on the existance of in-field variability. It requires the use of new technology, such as global positioning (GPS), sensors, satellites or aerial images, and information management tools (GIS) to assess and understand variations. Collected information may to more and to more accurately predict crop yields. It seeks to avoid applying inflexible practices to a crop, regardless of local soil/ climate conditions, and may help to better assess local situations of disease or lodging.

Satellites allow farmers to easily survey their land. Global Poistioning Systems (GPS) monitor can find the location of a field to within one meter. It can then present a series of GIS maps that demonstrate which field are moist or dry, and where there is erosion of soil and other soil factors that stunt crop growth. The data can be used by the farmer to automatically regulate the machine application of fertilizer and pesticide.

The US. Department of Agriculture, NASA, and NOAA are among key agencies contributing to this revolution in large-scale agriculture. The goal is to improve farmers' profits and harvest yields while reducing the negative impacts of farming on the environment that come from over application of chemicals.

In India also this concept is gaining importance at present at Tamil Nadu Agricultural University, Coimbatore Tamil Nadu, a project Precision Farming is running under Director of Ext. Education to provide information about precision farming and its benefits to farmers and society.

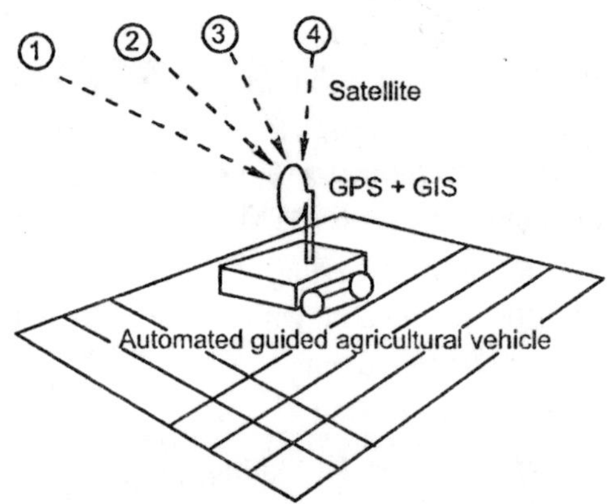

Fig. 20.1: Concept of precision farming

Precision farming is an agricultural system that has the potential of dramatically change agriculture scenario in 21st century. Precision agriculture is doing right thing at the right time. Knowing about the right thing involves data collection to a correct and viable decision.

Drone Rules, 2021

Ministry of Civil Aviation (MoCA) has released reorganized "Drone Rules, 2021" for public consultation. Draft rules were built on premise of trust, self-certification and non-intrusive monitoring. It will replace UAS Rules 2021 that was released on 12th March 2021. Draft Drone Rules 2021 As per new rules, there will be minimal human interface on digital sky platform and most permissions will be self-generated.

20.9 HIGHLIGHTS OF UNION BUDGET 2022-2023

Budget Focus

Entering Amrit Kaal, the 25 year long lead up to India @100, the budget provides impetus for growth along four priorities:

- PM GatiShakti
- Inclusive Development
- Productivity Enhancement & Investment, Sunrise opportunities, Energy Transition, and Climate Action.
- Financing of investments

PM GatiShakti

The seven engines that drive PM GatiShakti are Roads, Railways, Airports, Ports, Mass Transport, Waterways and Logistics Infrastructure.

PM GatiShkati National Master Plan

- The scope of PM GatiShakti National Master Plan will encompass the seven engines for economic transformation, seamless multimodal connectivity and logistics efficiency.
- The projects pertaining to these 7 engines in the National Infrastructure Pipeline will be aligned with PM GatiShakti framework.

Road Transport

- National Highways Network to be expanded by 25000 Km in 2022-23.
- Rs 20000 Crore to be mobilized for National Highways Network expansion.

Multimodal Logistics Parks

Contracts to be awarded through PPP mode in 2022-23 for implementation of Multimodal Logistics Parks at four locations.

Railways

- One Station One Product concept to help local businesses & supply chains.
- 2000 km of railway network to be brought under Kavach, the indigenous world class technology and capacity augmentation in 2022-23.
- 400 new generation Vande Bharat Trains to be manufactured during the next three years.

- 100 PM GatiShakti Cargo terminals for multimodal logistics to be developed during the next three years.

Parvatmala

- National Ropeways Development Program, Parvatmala to be taken up on PPP mode.
- Contracts to be awarded in 2022-23 for 8 ropeway projects of 60 Km length.

Inclusive Development

Agriculture

- Rs. 2.37 lakh crore direct payment to 1.63 crore farmers for procurement of wheat and paddy.
- Chemical free Natural farming to be promoted throughout the country. Initial focus is on farmer's lands in 5 Km wide corridors along river Ganga.
- NABARD to facilitate fund with blended capital to finance startups for agriculture & rural enterprise.
- 'Kisan Drones' for crop assessment, digitization of land records, spraying of insecticides and nutrients.

Ken Betwa project

- 1400 crore outlay for implementation of the Ken – Betwa link project.
- 9.08 lakh hectares of farmers' lands to receive irrigation benefits by Ken-Betwa link project.

MSME

- Udyam, e-shram, NCS and ASEEM portals to be interlinked.
- 130 lakh MSMEs provided additional credit under Emergency Credit Linked Guarantee Scheme (ECLGS)
- ECLGS to be extended up to March 2023.
- Guarantee cover under ECLGS to be expanded by Rs 50000 Crore to total cover of Rs 5 Lakh Crore.
- Rs 2 lakh Crore additional credit for Micro and Small Enterprises to be facilitated under the Credit Guarantee Trust for Micro and Small Enterprises (CGTMSE).
- Raising and Accelerating MSME performance (RAMP) programme with outlay of Rs 6000 Crore to be rolled out.

Skill Development

- Digital Ecosystem for Skilling and Livelihood (DESH-Stack e-portal) will be launched to empower citizens to skill, reskill or upskill through on-line training.
- Startups will be promoted to facilitate 'Drone Shakti' and for Drone-As-A-Service (DrAAS).

Education

- 'One class-One TV channel' programme of PM eVIDYA to be expanded to 200 TV channels.
- Virtual labs and skilling e-labs to be set up to promote critical thinking skills and simulated learning environment.
- High-quality e-content will be developed for delivery through Digital Teachers.
- Digital University for world-class quality universal education with personalised learning experience to be established.

Health

- An open platform for National Digital Health Ecosystem to be rolled out.
- 'National Tele Mental Health Programme' for quality mental health counselling and care services to be launched.

A network of 23 tele-mental health centres of excellence will be set up, with NIMHANS being the nodal centre and International Institute of

Information Technology-Bangalore (IIITB) providing technology support.

Saksham Anganwadi

- Integrated benefits to women and children through Mission Shakti, Mission Vatsalya, Saksham Anganwadi and Poshan 2.0.
- Two lakh anganwadis to be upgraded to Saksham Anganwadis.

Har Ghar, Nal Se Jal

Rs. 60,000 crore allocated to cover 3.8 crore households in 2022-23 under Har Ghar, Nal se Jal.

Housing for All

Rs. 48,000 crore allocated for completion of 80 lakh houses in 2022-23 under PM Awas Yojana.

Prime Minister's Development Initiative for North-East Region (PM-DevINE)

- New scheme PM-DevINE launched to fund infrastructure and social development projects in the North-East.
- An initial allocation of Rs. 1,500 crore made to enable livelihood activities for youth and women under the scheme.

Vibrant Villages Programme

Vibrant Villages Programme for development of Border villages with sparse population, limited connectivity and infrastructure on the northern border.

Banking

- 100 per cent of 1.5 lakh post offices to come on the core banking system.
- Scheduled Commercial Banks to set up 75 Digital Banking Units (DBUs) in 75 districts.

e-Passport

e-Passports with embedded chip and futuristic technology to be rolled out.

Urban Planning

- Modernization of building byelaws, Town Planning Schemes (TPS), and Transit Oriented Development (TOD) will be implemented.
- Battery swapping policy to be brought out for setting up charging stations at scale in urban areas.

Land Records Management

Unique Land Parcel Identification Number for IT-based management of land records.

Accelerated Corporate Exit

Centre for Processing Accelerated Corporate Exit (C-PACE) to be established for speedy winding-up of companies.

AVGC Promotion Task Force

An animation, visual effects, gaming, and comic (AVGC) promotion task force to be set-up to realize the potential of this sector.

Telecom Sector

Scheme for design-led manufacturing to be launched to build a strong ecosystem for 5G as part of the Production Linked Incentive Scheme.

Export Promotion

Special Economic Zones Act to be replaced with a new legislation to enable States to become partners in 'Development of Enterprise and Service Hubs'.

Sunrise Opportunities

Government contribution to be provided for R&D in Sunrise Opportunities like Artificial Intelligence, Geospatial Systems and Drones, Semiconductor and its eco-system, Space Economy, Genomics and Pharmaceuticals, Green Energy, and Clean Mobility Systems.

Energy Transition and Climate Action:

- Additional allocation of Rs. 19,500 crore for Production Linked Incentive for manufacture of high efficiency solar modules to meet the goal of 280 GW of installed solar power by 2030.

- Five to seven per cent biomass pellets to be co-fired in thermal power plants:
  - o CO_2 savings of 38 MMT annually,
  - o Extra income to farmers and job opportunities to locals,
  - o Help avoid stubble burning in agriculture fields.
- Four pilot projects to be set up for coal gasification and conversion of coal into chemicals for the industry
- Financial support to farmers belonging to Scheduled Castes and Scheduled Tribes, who want to take up agro-forestry.

Public Capital Investment:

- Public investment to continue to pump-prime private investment and demand in 2022-23.
- Outlay for capital expenditure stepped up sharply by 35.4% to Rs. 7.50 lakh crore in 2022-23 from Rs. 5.54 lakh crore in the current year.

  o Outlay in 2022-23 to be 2.9% of GDP.
- 'Effective Capital Expenditure' of Central Government estimated at Rs. 10.68 lakh crore in 2022-23, which is about 4.1% of GDP.

GIFT-IFSC

- World-class foreign universities and institutions to be allowed in the GIFT City.
- An International Arbitration Centre to be set up for timely settlement of disputes under international jurisprudence.

Mobilising Resources

- Data Centres and Energy Storage Systems to be given infrastructure status.
- Venture Capital and Private Equity invested more than Rs. 5.5 lakh crore last year facilitating one of the largest start-up and growth ecosystem. Measures to be taken to help scale up this investment.
- Blended funds to be promoted for sunrise sectors.
- Sovereign Green Bonds to be issued for mobilizing resources for green infrastructure.

Digital Rupee

Introduction of Digital Rupee by the Reserve Bank of India starting 2022-23.

Providing Greater Fiscal Space to States

- Enhanced outlay for 'Scheme for Financial Assistance to States for Capital Investment':From Rs. 10,000 crore in Budget Estimates to Rs. 15,000 crore in Revised Estimates for current year.
- Allocation of Rs. 1 lakh crore in 2022-23 to assist the states in catalysing overall investments in the economy: fifty-year interest free loans, over and above normal borrowings.
- In 2022-23, States will be allowed a fiscal deficit of 4% of GSDP, of which 0.5% will be tied to power sector reforms.

Fiscal Management

- Budget Estimates 2021-22: Rs. 34.83 lakh crore.
- Revised Estimates 2021-22: Rs. 37.70 lakh crore.
- Total expenditure in 2022-23 estimated at Rs. 39.45 lakh crore.
- Total receipts other than borrowings in 2022-23 estimated at Rs. 22.84 lakh crore.
- Fiscal deficit in current year: 6.9% of GDP (against 6.8% in Budget Estimates).
- Fiscal deficit in 2022-23 estimated at 6.4% of GDP.

Budget 2022 Should Focus upon Agri Infrastructure and Marketing through FPO and OFPO. Long-term growth of the agriculture sector is possible in agriculture only when the Indian agri-market will be competitive and remunerative for farmers. In this context, the Farmers' Producer Organisation (FPO) and Off-Farm Producer Organisation (OFPO) may play a bigger role in the coming days. In the last 7 years, NDA government has implemented numerous schemes to improve the agriculture sector. With a '*Beej Se Bazar Tak*' approach, government has brought remarkable changes in the agriculture sector. The budgetary allocation to the agriculture sector has gone up by more than 4 times in the last 7 years. In 2021-22, the department of agriculture was allocated 1,23,018 crore for three major schemes: *PM Kisan* (49%), *interest subvention on short-term credit to farmers* (15%) and *Pradhan Mantri Fasal Bima Yojana* (12%). Around 76% of the total allocated budget goes to these 3 major schemes. These three schemes have benefitted the farmers by reducing the input cost, risk and providing cheap credit for agriculture. Additionally, the government has also implemented the schemes like *PM Krishi Shinchayi Yojana*, 1 lakh crore *agriculture infrastructure fund* and umbrella scheme *PM SAMPADA* to promote food processing and agri-exports. In India, around 51 percent of the country's net sown area comes under the rainfed area. The livelihood of farmers in dryland areas remains vulnerable even in normal times. These farmers largely depend upon non-farm activities (40 percent of the total agricultural income). The government under *Atma Nirbhar Bharat Abhiyan* has taken various measures for making the dryland agricultural climate-resilient and allied activities profitable. Some of the important initiatives are promotion of *zero-budget natural farming, beekeeping, subsidy over solar pump, multi-tier cropping and providing short-term credit subvention to non-farm activities*. However, budgetary * With Department of Economic Analysis and Research (DEAR), NABARD, Mumbai. Usual disclaimers apply. 28th Jan 2022 7 provisions in some of these programmes are either inadequate or non-existence to address the challenges in dryland/rainfed areas. The upcoming budget focus upon promoting *FPOs* as the umbrella organisation, which should not just focus on collectivisation and input provider to members, but it should also aim at creating warehouses in the country, focusing on food processing and enhancing agri-exports. The Budget 2022 will also articulate agricultural stratagems that promote the non-farm sector through creating *OFPOs*. Promotion of new-age technologies in agriculture, encouraging data-driven solutions for *smart farming* and market connect, strengthening of FPOs and OFPOs and enabling investment in research and development (R&D) will provide necessary propulsion to agricultural growth. Union Budget 2022-23 have focus upon the following points to strengthen FPOs and OFPOs in the coming years:

1. Agro-Processing for farmers' prosperity FPOs have been doing a commendable job at reducing the input costs and facilitating the increase in the price farmers receive for their produce. However, the existing over 10,000 FPOs along with the target of forming new 10,000 FPOs by 2023-24 under the new Government of India's scheme needs to look beyond just aggregation of inputs and produce. The next step forward should be to handhold these FPOs through their transition to the small-scale processing unit. Indian secondary food processing is at 13-14% of agricultural output much less than China (23%) and USA (65%), this couples with huge wastage of fruits and vegetables (16%) in the country. It is

expected that this Union Budget will provide separate outlay for FPOs to undertake processing along with separate funding for their training and promoting the marketing of their processed product.

2. Harnessing FPOs potential to export Giant strides have been made in agri export since independence. India broke into the top 10 list of agricultural produce exporters in 2019 with a sizeable share in the export of rice, cotton, soybeans and meat, according to a World Trade Organization (WTO) report. However, this share is still not commensurate with our share in world production and we hold immense potential to compete with leading exporting countries of the world, having been endowed with multiple agro-climatic zones. FPOs as a novel idea hold immense potential in connecting our farmers directly to the world markets. Recently, the *Fresh Vegetables and Fruits Exporters Association* (*VAFA*), Mumbai has signed an MOU with FPOs from Varanasi for sourcing fresh vegetables and fruits for export to the international market. The time has come to connect our FPOs directly to clients overseas and not through a third party. It is expected that this Union Budget will have provision to provide technical support to FPOs to root out the problems faced by FPOs in procuring necessary 28th Jan 2022 7 certification and complying with Sanitary and Phyto Sanitary (SPS) requirements. Interest rate subvention, credit guarantee and incentive to undertake exports will do well to augment exports by FPOs.

3. The Impetus to Warehousing Infrastructure The government has taken a welcome step by integrating the *e-National Agriculture Market* (*e-NAM*) with *FPOs and warehouses* to provide end-to-end connectivity for farmers. Geotagging of Warehouses will also go long way in facilitating farmers in locating nearby warehouses and utilising its services. The time is opportune for government to enhance the FPOs access to warehouse infrastructure by providing additional incentives to them through higher interest subvention. At least one warehouse should be developed in each taluka/block to enable farmers' access to post-harvest pledge financing and increase cash flow with the farmers.

4. Realising the potential of the Off-Farm sector Indian landholding is highly fragmented with small and marginal farmers constituting around 86% of the total landholding with a share of around 46% of the area. There is 2.4 percent of the total agricultural households are landless labourers. In such a scenario, we cannot envision inclusive and sustainable agriculture growth without focussing on the allied sector. Collectivisation of handloom weavers, artisans into the *Off-Farm Producer Organisation (OFPO)* can prove a game-changer in alleviating many from poverty and supporting India's diverse handloom heritage. Grant support for the formation of OFPO and providing handholding to empower our weavers to market their products to the final consumer. In a long journey of eventually empowering our weavers to market their creations overseas themselves, this shall prove to be a game-changer step. Besides the efforts to boost growth in agriculture sector, the Government of India should increase spending on agricultural infrastructure. In previous budgets, greater importance was given to cash-based schemes as compared to spending on infrastructure. The creation of Agriculture infrastructure Fund (AIF) is a welcome move in that direction but target-based infrastructure needs to be created for sustainable and long-term growth of the agriculture sector. A better storage capacity itself will resolve most of the farmer's despairs. The COVID-19 pandemic has hit agricultural operations hard. Several reports have highlighted

that farm operation mainly suffered due to infrastructure bottlenecks, for instance, supply chain disruptions, lack of storage and marketing infrastructure. In all those reports, one of the important reasons pointed out was the lack of public infrastructure. Hence, it is the time when government should focus on developing the infrastructure for agriculture and allied sector. A four-pronged approach *i.e. promoting climate-resilient agriculture, converting FPOs and OFPOs in umbrella organisation, building agri-infrastructure and promoting 28th Jan 2022 7 allied sectors* should be the priority area for the budget 2022. This approach will help in achieving the dual goal of raising incomes for smallholder farmers and strengthening the sector's competitiveness.

NATIONAL NURSERY PORTAL (14th April, 2021)

Agriculture Minister launches a 'National Nursery Portal' to connect nursery operators with consumers.

The National Nursery Portal, developed by the National Horticulture Board, will help consumers to connect with nurseries and get information about the availability and prices of quality planting material, an official statement said.

It will also help nursery operators in assessing market demand for their products.

Minister said that the youth could become entrepreneurs in the horticulture sector and contribute to economic growth.

FOOD LEGUMES RESEARCH PLATFORM (FLRP)

The Union Cabinet has approved the establishment of Food Legumes Research Platform (FLRP) by International Center for Agricultural Research in Dry Areas (ICARDA) at Amlaha, Sehore, Madhya Pradesh with satellite hubs in West Bengal (for pulses) and Rajasthan (for natural resource management). (24th Feb., 2016)

The Cabinet also approved signing of lease deed with MP government for the land provided by them (70.99 hectare, 175.42 acre) at Amlaha Farm, Sehore on land rent of Re 1 per acre per year for 30 years on lease and to further lease it to ICARDA to establish the FLRP in MP. "The establishment of FLRP will enable India to harness the best of international science in meeting the emerging food security challenges. India would be able to rapidly and effectively absorb the research output achieved in the country by FLRP,". A major international R&D institution will make India an even bigger centre for agricultural research in the world and this, in turn, will attract further research and development investment in the country.

JAL SHAKTI ABHIYAN LAUNCHED BY GOI

- Union Jal Shakti Minister recently launches Jal Shakti Abhiyan.
- The Jal Shakti Abhiyan is a **campaign for water conservation** and water security in the country.
- The campaign will run through citizen participation in two phases, firstly, during the monsoon season, from 1st July, 2019 to 15th Sep., 2019 and then from 1st October, 2019 to 30th November, 2019 for states receiving the North East retreating monsoons.
- It is a collaborative effort of various ministries of the Government of India and state governments, being **coordinated by the Department of Drinking Water and Sanitation (DDWS)** that comes under the Jal Shakti Ministry.
- The focus of the campaign is on water stressed districts and blocks. The teams of officers from the central government will visit and work with district administration

in **1592 water stressed blocks in 256 districts,** to ensure five important water conservation interventions.

- The **five important water conservation interventions** are:
  - Water conservation and rainwater harvesting,
  - Renovation of traditional and other water bodies/tanks,
  - Reuse of water and recharging of structures,
  - Watershed development and
  - Intensive afforestation.
- The water conservation interventions will also be supplemented with special interventions including the development of block and district water conservation plans, promotion of efficient water use for irrigation and better choice of crops through Krishi Vigyan Kendras.

20.10 ACCELERATED FODDER DEVELOPMENT PROGRAMME (AFDP)

A special scheme "Accelerated Fodder Development Programme (AFDP)" was launched during 2011-12 as a sub-scheme of RKVY and the scheme has been extended to 6 States in the year 2013-14 with release of Rs.80 crore.

AFDP envisages accelerating production of fodder through promotion of integrated technologies and processes for enhancing the availability of fodder throughout the year. It will also help in mitigating shortage of fodder caused due to natural calamities like droughts and floods. During 2013-14, 1,163 clusters (each of 250-500 ha) were taken up under the programme and about 14.65 lakh ha area has been brought under various dual purpose/ fodder crops.

20.11 AGRO-FORESTRY POLICY – 2014:

Based on the recommendation of National Advisory Council (NAC), Agroforestry Policy has been formulated by GOI. The aim of the policy is to promote land use system which integrates trees and shrubs on farmlands and rural landscapes to enhance productivity, profitability, diversity and ecosystem sustainability. The Cabinet has approved the National Agro-forestry Policy, 2014 on 6th February, 2014. India became the world's first country to adopt a comprehensive agroforestry policy.

20.12 SVAMITVA SCHEME

SVAMITVA Scheme is a Central Sector scheme launched by Hon'ble Prime Minister of India on National Panchayat Day *i.e.* 24th April 2020. The Ministry of Panchayati Raj (MoPR) is the Nodal Ministry for implementation of the scheme. In the States, the Revenue Department / Land Records Department will be the Nodal Department and shall carry out the scheme with support of State Panchayati Raj Department. Survey of India shall work as the technology partner for implementation.

The scheme aims to provide an integrated property validation solution for rural India. The demarcation of rural abadi areas would be done using Drone Surveying technology. This would provide the 'record of rights' to village household owners possessing houses in inhabited rural areas in villages which, in turn, would enable them to use their property as a financial asset for taking loans and other financial benefits from Bank

The scheme seeks to achieve the following objectives:

i) To bring financial stability to the citizens in rural India by enabling them to use their property as a financial asset for taking loans and other financial benefits.

ii) Creation of accurate land records for rural planning.

iii) Determination of property tax, which would accrue to the GPs directly in States where it is devolved or else, add to the State exchequer.

iv) Creation of survey infrastructure and GIS maps that can be leveraged by any department for their use.

v) To support in preparation of better-quality Gram Panchayat Development Plan (GPDP) by making use of GIS maps.

vi) To reduce property related disputes and legal cases

NUTRIENT BASED SUBSIDY (NBS)

NBS (Nutrient Based Subsidy) for non-urea fertilizers raised by GOI (July 2019). The government hiked the subsidy on non-urea fertilisers to make available farm nutrients at affordable prices to farmers, a move that would cost the exchequer Rs 22,875.50 crore in this fiscal.

A decision in this regard was taken at the meeting of Cabinet Committee on Economic Affairs (CCEA) headed by Prime Minister Narendra Modi.

"The CCEA approves NBS rates for Phosphatic and Potassic (P&K) fertilizers for the year 2019-20; expected expenditure during 2019-20 to be Rs 22,875.50 crore," This will help in promoting balanced use of fertilisers.

The nutrient based subsidy (NBS) rate for sulphur increased by 31 per cent to Rs. 3.562 per kg from Rs. 2.722 in 2018-19 while keeping that for nitrogen, phosphorus and potassium unchanged.

The NBS subsidy for nitrogen, phosphorus and potassium would remain at Rs. 18.901, Rs. 15.216 and Rs. 11.124 per kg, respectively, an official statement said here.

20.13 WELFARE PROGRAMMES AND SCHEMES

1. Swachh Bharat Abhiyan: A national campaign by Indian Government launched officially on 2nd Oct., 2014 at Rajghat by Prime Minister Narendra Modi. India's biggest cleanliness drive covering 4,041 towns, to clean the streets, roads and infrastructure of the country. This campaign aims to achieve vision of clean India by 2nd Oct. 2019. (150th Birth Day of Mahatma Gandhi).

2. Pradhan Mantri Sansad Adarsh Gram Yojana

- The Saansad Adarsh Gram Yojana was launched for the development of model villages.
- Under the Yojana, Members of Parliament (MPs) will be responsible for developing the socio-economic and physical infrastructure of three villages each by 2019, and a total of eight villages each by 2024.
- The first Adarsh Gram must be developed by 2016, and two more by 2019. From 2019 to 2024, five more Adarsh Grams must be developed by each MP, one each year.
- This implies that a total of 6,433 Adarsh Grams, of the 2,65,000 gram panchayats, will be created by 2024.

Main Objectives of PMSAGY

- The development of model villages, called Adarsh Grams, through the implementation of existing schemes, and certain new initiatives to be designed for the local context, which may vary from village to village.
- Creating models of local development which can be replicated in other villages.

3. Pradhan Mantri Suraksha Bima Yojana: It is a low-cost personal accident cover to increase the insurance penetration level for poor in the country. Scheme was announced by Finance Minister in budget 2015-2016. This insurance policy covers death or disablement of the policy holder caused due to accident or any accidental injuries. This insurance cover is available for a premium of just Rs. 12 per year to people in the age group of 18-70 years. The scheme is offered by all public sector general insurance companies. This scheme ensures social security through convenient modes like auto debit facility.

4. Pradhan Mantri Jeevan Jyoti Bima Yojana: The scheme will be a one year cover, renewable from year to year. Insurance Scheme offering life insurance cover for death due to any reason.

Scope of coverage: All savings bank account holders in the age 18 to 50 years in Bank will be entitled to join. In case of multiple saving bank accounts held by an individual in one or different banks, the person would be eligible to join the scheme through one saving bank account only.

Benefits: Rs. 2 lakhs is payable on member's death due to any reason. If the scheme member dies during insurable membership, subject to policy being in force and all due premiums, Service Tax and other levies (if any) having been paid and subject to any restrictions or qualifications referred to in these Clauses, the amount specified as the Sum Assured for such Scheme Member shall become due to the Nominee of the Scheme Member. No maturity benefit is payable under the policy. No surrender benefit is payable under the policy.

5. Nai Manzil: A scheme launched by Dr. Najma Heptullah, Union Minister for minorities helping minority youth without a formal school-leaving certificate to obtain one and find better employment opportunities.

6. Shyama Prasad Mukherjee Urban Mission for integrated project based infrastructure in rural areas of the country. Mission aims to create 300 Urban growth clusters over the next 3 years.

7. Deen Dayal Upadhaya Gram Jyoti Yojna for feeder separation to augment power supply to Rural areas.

8. Pradhan Mantri Mudra Yojana: Micro Units Development and Refinance Agency (MUDRA) Bank is a new institution being set up by Government of India for development and refinancing activities relating to micro units. The purpose of MUDRA is to provide funding to the non corporate small business sector. Loans worth about Rs 1 lakh crore have been sanctioned to small entrepreneurs under the Pradhan Mantri MUDRA Yojana. Finance Minister announced this yojana during presenting the Union Budget for F.Y. 2016.

9. Pradhan Mantri Fasal Bima Yojana (PMFBY): Provide insurance cover to rabi and kharif crops and financial support to farmers in case of damage of crops and in order to make crop insurance simpler and cheaper for the farmers and to provide them with better insurance services, a Central Sector Scheme of Pradhan Mantri Fasal Bima Yojana (PMFSY) was launched by the Government of India replacing NAIS and MNAIS. Under the new scheme, farmers will have to pay a uniform premium of two per cent for all kharif crops and 1.5 per cent for all rabi crops. The scheme will be implemented from the kharif season from year, *i.e.* 2016.

New Guidelines for PMFBY for benefit of Farmers in the Country.

- Provision of penalties/Incentives for states, Insurance companies and Banks;
- 12% interest rate to be paid to the farmer by the insurance companies for delay in settlement of claim beyond two months of

prescribed cut off date. State Govt. similarly have to pay 12% interest for delay in release of State share of subsidy beyond three months of prescribe cut off date/ submission of request by Insurance company.

- More time is given to farmer to file his claim of 72 hours from 48 hours through any stakeholder or directly on the portal.

Wild Animals attack coverage under PMFBY

- The Centre has allowed Telangana government to provide add-on risk covered under the Pradhan Mantri Fasal Bima Yojana (PMFBY) due to two wild animals attack i.e wild boar and monkey attacks which is perceived to be substantial and identifiable.
- The risk coverage would be on the same as in case of regular crop insurance i.e the loss of yield against the notified indemnity level of the threshold yield.
- The indemnity level fixed for the State is 80%.
- The threshold yield is calculated based on the average yield of past seven years excluding a maximum of two calamity years as notified by the State Government.
- The risk cover for wild animal attack on crops would insure the farmers against income loss, save costs for measures such as electric-fencing for the standing crop and prevents accidents of electrocution due to such fencing.

10. Pradhan Mantri Jan Dhan Yojana: Pradhan Mantri Jan-Dhan Yojana (PMJDY) is National Mission for Financial Inclusion to ensure access to financial services, namely, Banking/ Savings & Deposit Accounts, Remittance, Credit, Insurance, Pension in an affordable manner.

Account can be opened in any bank branch or Business Correspondent (Bank Mitre) outlet. PMJDY accounts are being opened with Zero balance. However, if the account-holder wishes to get cheque book, he/she will have to fulfill minimum balance criteria.

Special Benefits under PMJDY Scheme

- Interest on deposit.
- Accidental insurance cover of Rs.1.00 lac
- No minimum balance required.
- Life insurance cover of Rs.30,000/-
- Easy Transfer of money across India

11. Sukhanya samridi account: Sukanya Samriddhi Account is another welcome step from Govt of India. Honorable Prime Minister of India, Narendra Modi launched Sukanya Samriddhi Account "A Small Savings Scheme" on 22nd January, 2015. It is part of "Beti Bachao – Beti Padhao" initiative of Government of India (GOI) also known as BBB Sukanya Samriddhi Account, government is trying to give a social message that girl child is not a financial burden if parents of a girl child secure their future through proper financial planning.

Objective

- Highest Interest Rate among all Small Savings Schemes offered by government of India
- Sukanya Samriddhi account will offer interest rate of 9.1% for financial year *i.e.* FY 2014-15. It is highest among all Small Savings Schemes.

12. Pradhan Mantri Jeevan Jyoti Bima Yojana (PMJJBY)

- Age of the Insured Bank account holders aged between 18 and 50 years are eligible to apply for this scheme. So, if you are aged more than 50 years, you are not eligible to enroll yourself for this scheme.

- Premium Amount – Less than 1Rupees a day or an annual premium of Rs. 330 is what you need to pay to get a life cover of Rs. 2 lacs.
- No matter what your age is, the premium is fixed at Rs. 330 for a life cover of Rs. 2 lacs.
- This annual premium of Rs. 330 has been fixed for the first three years from June 1, 2015 to May 31, 2018, after which it will again be reviewed based on the insurers' annual claims experience.

13. Krishi Amdani Beema Yojana

- To give an impetus to the dying agricultural practice
- There is 14 crore hectares of agricultural land in India, of which only 44 % in under irrigation
- Pradhan Mantri Gram Sinchai Yojana would be introduced so that more agricultural land is irrigated.
- Talking about the difficulty of small and marginal farmers he said that most of them were leaving the agricultural practice because of the uncertainty over the produce and returns.
- Krishi Amdani Beema Yojana so that the farmers don't bear any financial burden if their produce gets destroyed due to unexpected weather or for any other reason.

14. Pradhan Mantri Gram Sinchai Yojana -

- The programme will be supervised and monitored by an Inter-Ministerial National Steering Committee (NSC) will be constituted under the Chairmanship of Prime Minister with Union Ministers from concerned Ministries.
- A National Executive Committee (NEC) will be constituted under the Chairmanship of Vice Chairman, NITI Aayog to oversee programme implementation, allocation of resources, inter ministerial coordination, monitoring & performance assessment, addressing administrative issues etc.

Major objective of PMGSY

- Ensure water supply to farmers round the year.
- Basic contours of the agri-irrigation programme would be on lines of the PMGSY, under which each irrigation project would be selected for releasing funds by the state government after seeking nod of the concerned District Panchayat.
- Importance in the wake of poor implementation of various irrigation projects in some states despite release of central funds by different ministries under various schemes for several years.
- Besides, water is necessary for farmers as country's 50 per cent of the agriculture land is rainfed.

Pradhan Mantri Krishi Sichai Yojna: It is a national mission to improve farm productivity and ensures better utilization of the water resources in India. CCEA approved this schemes in 2015-16 with an outlay of 50000 crores for period of 5 years (2015 to 2019-20). The integrated watershed programme was subsumed into the current PMKSY on 26th Oct., 2015. Objectives of this scheme is to (1) Enhance the adoption of precision irrigation and other water saving technologies (More crop per drop) (2) Enhance recharge of aquifers and introduce sustainable water conservation practices. GoI has extended this scheme for 2021-26 for benefit of farmers of the country.

15. Soil Health Card Scheme: Farming as an activity contributes nearly 1/6th of our Gross Domestic Product and a majority of our population is dependent on it for their livelihood. Deteriorating soil health has been a cause of concern and that has been leading to sub optimal utilization of farming resources. Imbalanced use of fertilizers, low addition of organic matter and non-replacement of depleted micro and secondary nutrients over the years, has resulted in nutrient deficiencies and decrease in soil fertility in some parts of the country.

About Soil Health Card

- Soil health needs to be assessed at regular intervals so as to ensure that farmers apply the required nutrients while taking advantages of the nutrients already present in the soil.
- Government has launched a scheme to provide every farmer a Soil Health Card in a Mission mode.
- The card will carry crop wise recommendations of nutrients/fertilizers required for farms, making it possible for farmers to improve productivity by using appropriate inputs.
- A Soil Health Card is used to assess the current status of soil health and, when used over time, to determine changes in soil health that are affected by land management.

Benefits of Soil Health Card

- The budget allotted Rs.100 crore for issuing cards and an additional Rs.56 crore to set up 100 mobile soil testing laboratories across the country.
- The soil health card details existing nutrient status of the soil and crop-wise recommendations of nutrients and fertilizers required, making it easier for farmers to improve productivity by using appropriate inputs.
- The agriculture ministry released Rs.86 crore for the soil health card scheme.
- In comparison, between 2007-08 and April 2014, the ministry had spent Rs.112 crore on the scheme. The flagship scheme of the ministry has been sanctioned Rs.568 crore for the next three years.
- Imbalanced use of fertilizer is has led to declining crop productivity in the country, further fuelled by a skewed fertilizer policy where urea is heavily subsided, leading to overuse.
- While urea consumption increased from 59% to 66% of total consumption in 2012-13 over 2010-11, per hectare consumption of fertilizer declined from 140kg to 128kg over the same period, noted the Economic Survey of 2014.
- Current trends in farm output reveal that the marginal productivity of soil in relation to the application of fertilizers is declining, it reported.

CENTRALLY SPONSORED SCHEMES IN AGRICULTURE

Facitities And Services Provided By Centrally Sponsored schemes In Agriculture

Kisan Call Centres

In order to provide an easy access to agricultural information for decision support, the Government of India has taken up to major initiative of launching a network of Kisan Call Centres for farmers on 21st January, 2004. The Call Centres are accessible on Toll Free Telephone No. 1551 and are presently located in Delhi, Bangalore, Chandigarh, Chennai, Hyderabad, Kanpur, Kalkata and Mumbai. The Centres are manned by agricultural graduates to answer queries related to weather, seeds, fertilizer and similar issues. These Call Centres provide advice in the local language and operate on all seven days in a week. While, online answers are available during the day shift, the answers are given 24 hours of the day. Beyond office hours, the queries are recorded and responded to by post.

Kisan Satellite channel

The Prime Minister of India launched a Kisan satellite TV channel on 2lst Januury, 2004 so that well tested and easily adoptable agricultural technologies and technologies which are in the pipeline generated by the Agricultural Research institutions and State Agricultural Universities are easily transferred to the farming community and other stoke holders through o very vital medium *i.e.* Doordarshan. At present, the national telecast for kisan channel is through the Indira Gandhi National Open University cable network. The progromme is of one hour duration and transmitted from 6-7 AM,8-9,AM and 5.30-6.30 PM and 7.30-8.30 PM. The programme likely to be further exponded to two hours to be repeated four times in near future. Similarly, from February 2004, 96 FM Stations of the All India Radio and 12 Doordarshan Kendras in the country (Hisar, Varanasi, Muzaffarpur, Indore, Rajkot, Jalpaiguri, Sambalpur, Shillong, Dibrugarh, Gulbarga, Daltanganj and Viajyawara) will also have one hour agricultural broadcast/telecast

National Agricultural Insurance Scheme (NAIS)

Objectives :

- To provide insurance coverage and financial support to the farmers in the event of prevented sowing & failure of any of the notified crop as a result of natural calamities, pests & diseases.
- To encourage the farmers to adopt progressive farming practices, high value in-puts and better technology in agriculture.
- To help stabilize farm incomes, particularly in disaster years.

16. Swasth Dhara Khet Hara slogan (Healthy Earth Green Farm) given in 2015 for improving the fertility of agricultural land which has been increased by adopting scientific techniques and about 52 new technologies have been developed in last 3 years, on soil testing, organic farming packages and 12 mini-labs made accessible to farmers to test soil parameters. Through advanced technologies and new varieties of paddy, mustard and wheat are developed which can be irrigated by the saline waters. Vulnerability Atlas of India has been made available to 572 rural districts of country and 75 ideal watershed also developed to provide irrigation water.

17. National Livestock Mission: (Livestock Insurance Scheme) It is a scheme to protect livestock losses due to untimely death of animals. For first time all the districts and animals are covered under livestock insurance scheme. All indigenous and hybrid mulch animals, cattle (Horses, donkeys, mull, camel, pony and male cattle and buffaloes and other livestocks (goat, sheep, pig and rabbit) can be insured under this scheme.

18. Mission Organic Value Chain Development for North Eastern Region (MOVCDNER). This scheme of GOI from 2015 to 2018 was started on 11th January 2016. So far 2321 farmers groups and 8 farmers produce organizations have been made under this programmes.

19. Mera Gaon Mera Gaurav: The "Mera Gaon Mera Gaurav" has been conceptualized in which scientists of ICAR and Agricultural

Universities will identify villages in the vicinity of the Institutions for providing advisories and consultations to farmers for increasing farm productivity and production.

An innovative initiative "Mera Gaon Mera Gaurav" has been planned to promote the direct interface of scientists with the farmers to hasten the lab to land process. The objective of this scheme is to provide farmers with required information, knowledge and advisories on regular basis by adopting villages. The scheme started in 2015-2016 with 10,712 villages. The number now increases to 13,500 villages in 2016-17

Work expected from Scientists in a Village:

1. To identify a village and strengthen interface with farmers
2. To periodically update farmers about agricultural activities through phone and mobile messages.
3. To provide technology handout as per the agro-ecological conditions of the village.
4. To provide information to farmers about agricultural inputs, seed, fertilizer, chemical, agricultural machinery, climate, market, etc.
5. To educate farmers through newspapers, community radio, etc.
6. To create awareness among farmers about the programmes being implemented by various organizations and institutions working at local level *e.g.* voluntary organizations, farmers' organisation, ATMA, other Govt. departments.
7. To make farmers aware of the sensitive issues of national importance such as: Swachh Bharat Abhiyaan, climate change, water conservation, soil fertility, etc.
8. To organise farmer's meet by visiting the selected villages as per need and facilitate the participation of specialists of the concerned institutes.
9. To identify technical problems at village level and make use of those in prospective research programmes.
10. To generate technical, social and economic data related to a village and to submit quarterly report of work done.

20. Farmers FIRST: The objectives of the 'Farmer FIRST' initiative is to move beyond the production and productivity and to privilege the complex, diverse & risk prone realities of the farmers through enhancing farmers-scientists contact with multi stakeholders participation for technology development and application. The project has been conceptualized with focus on Enriching Farmers –Scientist interface; Technology Assemblage, Application and feedback; Partnership and Institutional Building and Content Mobilization.

21. ARYA: The ICAR has initiated a program on "Attracting and Retaining Youth in Agriculture (ARYA) in selected districts through KVKs with an objective for entrepreneurial development of Youth in Rural Areas to take up various Agriculture, allied and service sector enterprises for sustainable income and gainful employment. The identified youth are trained on entrepreneurship development skills by providing a basket of options to start agriculture ventures for self employment.

About NABARD

- The importance of institutional credit in boosting rural economy has been clear to the Government of India right from its early stages of planning.
- Reserve Bank of India (RBI) at the insistence of the Government of India, constituted a Committee to Review the Arrangements For Institutional Credit for Agriculture and Rural Development (CRAFICARD) to look into these very critical aspects.

- The Committee was formed on 30 March 1979, under the Chairmanship of Shri B. Sivaraman, former member of Planning Commission, Government of India.
- National Bank for Agriculture and Rural Development (NABARD) was approved by the Parliament through Act 61 of 1981.
- NABARD came into existence on 12 July 1982 by transferring the agricultural credit functions of RBI and refinance functions of the then Agricultural Refinance and Development Corporation (ARDC).

  Set up with an initial capital of Rs.100 crore, its' paid up capital stood at Rs. 5,000 crore as on 31 March 2016

Rates of Interest Charged by NABARD for Refinance to Banks under Various Schemes (28 February 2021)

S. No.	Particulars	Interest Rate (%)
1.	**Short Term refinance assistance:**	
a.	State Cooperative Banks for financing crop loans	4.50
b.	RRBs for financing crop loans	4.50
c.	DCCBs directly financing for crop loans	4.50
d.	Commercial Banks./RRBs in respect of their finance to PACS towards crop loans	4.50
e.	ST – Additional SAO/ST(Others)/ ST (SAO) -SCARDBs (Annual Product)	5.50
f.	StCBs/RRBs - Conversion of Short Term- crop loans into Medium Term loans (3% less than the interest rate charged by banks to ultimate beneficiaries subject to minimum interest rate at 8.10%)	8.10
2.	**Long Term refinance assistance**	
a.	LT loan to State Govt. for contribution to share capital of coop. credit institutions	5.8
b.	**RRBs/ StCBs**	
	- Refinance for 5 years and above	6.40
	- Refinance for 3 years to less than 5 years	6.00
	- Refinance for 18 months to less than 3 years	5.50
	SCARDBs	
	- Refinance for 5 years and above	6.45
	Refinance for 3 years to less than 5 years	6.05
	Refinance for 18 months to less than 3 years	5.55
3.	**Direct Lending**	
a.	Warehouse Infrastructure Fund (WIF)	
	State Governments	Bank rate – 1.50%
	State Government Corporation (with State Govt. Guarantee)	Bank rate – 1.50%
	Entities promoted by State / Central Government owned/assisted (without State Govt. Guarantee)	PLR + risk premium
	Other entities	PLR + risk premium
	* Current Bank Rate – 4.25%	
	* Current PLR–7.75%	
	Food Processing Fund (FPF)	
	State Governments	Bank rate – 1.50%
	Entities promoted by State Governments (with Govt. Guarantee)	Bank rate – 1.50%
	Entities promoted by State Governments(without Govt. Guarantee)	PLR + risk premium
	Other entities	PLR + risk premium
	* Current Bank Rate – 4.25%	
	* Current PLR–5.80%	

ICT AND AGRICULTURE

Across the developing world agriculture sustains the majority of rural livelihoods. Ever since people have grown crops, raised the live-stocks, and caught fish, they have sought information from one another or from some source to sustain or survive. These information include, 1) most effective planning strategy on steep slopes, 2) where to get and buy improved and certified seeds, 3) who is paying the highest price in the market, 4) where to get cheap and easy irrigation water etc. Over the time weather patterns and soil conditions change. Epidemics of pests and diseases come and go . Updated information allow farmers to cope with and benefits from these changes, practices and input use and real time information about the situation.

As agriculture is location specific and farmers need accurate and reliable weather forecasts and advice on agriculture in time. Providing such knowledge and information is a challenging task.

Harnessing the rapid growth of internet and associated technologies such as mobile phone is critical o helping the farmers obtain the information they need and to promoting transformational agricultural development. Interest is especially for creating the access to extension services provided by universities or govt agencies, improving marketing of outputs, and arranging logistics

National e-governance plan in Agriculture (NeGP-A)

- To promote rapid development of agriculture through use of ICT for ensuring timely access to agriculture related information for the farmers of the country
- To integrate various activities in the agriculture value chain so that farmers would able to make proper and timely use of the available information

GST on Agricultural products 2022-23

- 12 per cent GST on bio-medical waste treatment facilities
- 12 per cent GST on solar water heater instead of the previous 5 per cent
- 18 per cent GST on tetra packs, and cheques – lose or in book form
- 5 per cent GST on pre-packed, labelled food items like atta, paneer, and curd

Fertilisers an important element of agriculture was previously taxed at 6% (1% Excise + 5% VAT). In the GST regime, the tax on fertilisers has been increased to 12%. The same impact is on Tractors. Wavier on the manufacture of Tractors is removed and GST of 12% has been imposed.

Single packages of food items like cereals, pulses and flour weighing up to 25 kg will be considered as 'prepackaged and labelled', and liable to 5% GST. Other items such as curd, lassi, and puffed rice too would attract GST at the rate of 5% when pre-packaged and labelled.

Tetra Pak (or aseptic packaging paper) used for packaging liquid beverages or dairy products will now attract 18% GST instead of 12%.

18 % Goods and Services Tax (GST) will be applicable on bank cheque book/loose leaf cheques and 12 % on maps, atlas, and globes.

- The rate on machines for cleaning, sorting or grading seed, and grain pulses, machinery used in milling industry or for the working of cereals etc, 'pawan chakki', or air-based atta chakki, wet grinder goes up from 5% to 18%.
- Machines for cleaning, sorting or grading eggs, fruit or other agricultural produce and its parts, milking machines and dairy machinery will have rates going up from 12% to 18%. the food subsidy borne by FCI.

GST is a single tax on the supply of goods and services, right from the manufacturer to the consumer. Credits of input taxes paid at each stage will be available in the subsequent stage of value addition, which makes GST essentially a tax only on value addition at each stage. The final consumer will thus bear only the GST charged by the last dealer in the supply chain, with set-off benefits at all the previous stages.

INDIA'S EXPORT IN 2020-21 WITH TREMENDOUS GROWTH

India has consistently maintained trade surplus in the agricultural products over the years. India's agricultural and allied exports during 2019-20 were Rs. 2.52 lakh Crores and imports were Rs. 1.47 lakh Crores. Even, during the difficult time of Pandemic, India took care not to disturb the world food supply chain and continued to export. The export of Agri and allied commodities during Apr,2020 - Feb,2021 were Rs. 2.74 lakh Crore as compared to Rs. 2.31 Crore in the same period last year indicating an increase of 18.49%.

The commodities which posted significant positive growth in exports were wheat, other Cereals,Rice (other than Basmati), Soya meal, Spices, Sugar, Raw Cotton, Fresh Vegetables, Processed Vegetables, and Alcoholic Beverages etc.

Wheat and other cereals posted huge growth over last year, *i.e.*, increasing from Rs. 425 Crore to Rs. 3283 Crore and Rs. 1318 Crore to Rs. 4542 Crore, respectively. On specific demand from countries, NAFED has exported 50,000 MT wheat to Afghanistan and 40,000 MT wheat to Lebanon under G2G arrangement. India has witnessed tremendous growth of 727 % for Wheat export.

Country has witnessed significant growth of 132% in export of (Non-Basmati) Rice. Export of Non-Basmati Rice has gone up from Rs 13,030 crores in 2019-20 to Rs 30,277 crores in 2020-21.This increase in exports is on account of multiple factors, mainly being India capturing new markets namely, Timor-Leste, Papua New Guinea, Brazil, Chile, and Puerto Rico. Exportswere also made to Togo, Senegal, Malaysia, Madagascar, Iraq, Bangladesh, Mozambique, Vietnam, Tanzania Rep and Madagascar.

India also enhanced export of Soya meals by 132%. Soya meal has gone up from Rs 3087 crores in 2019-20 to Rs 7224 crores in 2020-21.

Other commodities of Agri& Allied basket witnessing significant increase in export during April, 2020 to February, 2021 as compared to corresponding period during 2019-20, have been Spices (Rs 26257 crore vs Rs 23562 crore; growth 11.44%), Sugar(Rs 17072 crore vs Rs 12226 crore; growth 39.64%), Raw Cotton(Rs 11373 crore vs Rs 6771 crore; growth 67.96%), Fresh Vegetable (Rs4780 crore vs Rs 4067 crore; growth 17.54%) and Processed Vegetables (Rs 2846 crore vs Rs 1994 crore; growth 42.69%)etc.

The imports of Agri and allied commodities during April, 2020 – Feb, 2021 were Rs. 141034.25 Crore as compared to Rs. 137014.39 Crore in the same period last year witnessing a slight increase of 2.93%.

Despite COVID-19, balance of trade in agriculture has favorably increased during April, 2020 – Feb, 2021 to Rs. 132,579.69 Crore as against Rs. 93,907.76 Crore during the same period in 2019-20.

PM KUSUM YOJNA

The Central Government has announced the Kisan Urja Suraksha evam Utthaan Mahabhiyan or KUSUM scheme which aims at furthering the production of solar power in India and also gives the benefits of solar farming to farmers. The Union Budget 2018-19 has set aside a sum of Rs.48000 crores for this program for ten years.

In March 2021, the Central Government introduced modifications to an existing component of the PM-KUSUM scheme – a farmer income support and a de-dieseling scheme – so as to channelise its focus on solarising agricultural feeders instead of pumps. This move would obviate the need for farmers to replace every existing pump in a village with a solar pump.

Kusum Scheme Benefits

- It shall enable the decentralisation of the production of solar power.
- Transmission losses of DISCOMS will be under control.

- Subsidy burden on DISCOMS in the sector of agriculture will be decreased to a large extent.
- This will give farmers the chance to sell to the grid the extra power that is generated by the solar plants that are stationed on their barren lands.
- It will provide a fillip to the emerging green economy in India.
- The scheme also has direct employment generation potential. As per available studies, around 24.50 job-years are created per MW of small capacity solar installation. Therefore, besides increasing self-employment, the scheme is likely to generate employment opportunities equivalent to 7.55 lakh job-years for skilled and unskilled workers.
- The program will also aid in the de-dieseling of the agricultural sector in India. This implies that the existing diesel pumps will be replaced.
- Other benefits that will accrue to the farmers due to the implementation of this program include conservation of water, water security as well as energy efficiency.

SAMPDA

New name **SAMPDA** given to revamp the National Mission on Food processing to launch with an outlay of Rs.6000 crore.

The objective of **SAMPADA** is to supplement agriculture, modernize processing and decrease agri-waste.

SAMPADA is an umbrella scheme incorporating ongoing schemes of the Ministry like Mega Food Parks, Integrated Cold Chain and Value Addition Infrastructure, Food Safety and Quality Assurance Infrastructure, etc. and also new schemes like Infrastructure for Agro-processing Clusters, Creation of Backward and Forward Linkages, Creation / Expansion of Food Processing & Preservation Capacities.

NANOTECHNOLOGY

Modern agricultural practices have triggered the process of agricultural pollution. This process can cause the degradation of eco-systems, land, and environment owing to the modern-day by-products of agriculture. The substantial use of chemical fertilizers, pesticides, and, contaminated water for irrigation cause further damage to agriculture. The current scenario of the agriculture and food sector has therefore become unsustainable. Nanotechnology has provided innovative and resourceful frontiers to the agriculture sector by contributing practical applications in conventional agricultural ways and practices. There is a large possibility that agri-nanotechnology can have a significant impact on the sustainable agriculture and crop growth. Recent research has shown the potential of nanotechnology in improving the agriculture sector by enhancing the efficiency of agricultural inputs and providing solutions to agricultural problems for improving food productivity and security. The prospective use of nanoscale agrochemicals such as nanofertilizers, nanopesticides, nanosensors, and nanoformulations in agriculture has transformed traditional agro-practices, making them more sustainable and efficient. However, the application of these nano-products in real field situations raises concern about nanomaterial safety, exposure levels, and toxicological repercussions to the environment and human health. The present review gives an insight into recent advancements in nanotechnology-based agrochemicals that have revolutionized the agriculture sector. Further, the implementation barriers related to the nanomaterial use in agriculture, their commercialization potential, and the need for policy regulations to assess possible nano-agricultural risks

21. TABLES

1. LAND USE PATTERN IN INDIA
2. LAND HOLDING PATTERN
3. AREA AND PRODUCTION OF MAJOR HORTICULTURE CROPS
4. MAJOR INDIAN STATES CULTIVATING SPECIFIC FRUITS
5. FRUITS AND PROCESSING PRODUCTS
6. FERTILIZER CONSUMPTION IN NUTRIENT DURING 2005- 06 TO 2009- 10.
7. FOOD GRAIN PRODUCTION
8. INTERNATIONAL CLASSIFICATION OF SOILS (ACCORDING TO TEXTURE)
9. PRODUCTION AND PER CAPITAL AVAILABILITY OF MILK
10. COMMERCIAL CROP PRODUCTION
11. NATIONAL FLOWERS OF THE DIFFERENT COUNTRIES
12. PROJECT FOOD GRAIN PRODUCTION AND CONSUMPTION
13. MAXIMUM ROOT ZONE DEPTH OF VARIOUS CROPS
14. SOIL MOISTURE RANGE AVAILABILITY FOR DIFFERENT CROPS
15. SCHEMES OF GOVERNMENT OF INDIA
16. LIVESTOCK POPULATION IN INDIA
17. ALL INDIA PRODUCTION OF MILK, EGGS AND WOOL
18. FISH PRODUCTION IN INDIA
19. PLAN-WISE SHARE OF PUBLIC SECTOR OUTLAYS AND EXPENDITURE UNDER AGRICULTURE AND ALLIED ACTIVITIES

21. Tables

Gross value added by economic activity at current basic prices (Agriculture, Forestry and Fishing) (2021-22):- **396,4312 crore rupees**

Gross value added by economic activity at constant basic prices (2011-12) Agriculture, Forestry and Fishing (2021-22):- **211,5048 crore rupees**

Sectoral share in gross value added at percentage at current basic prices (Agriculture, Forestry and Fishing) 2021-22:- **18.6%.**

Sectoral share in gross value added at percentage at constant basic prices (2011-12) :- **3.3%.**

Gross capital formation in Agriculture and allied sector (2020-21) at current prices:– **5,25,2309 crore**

Gross capital formation in Agriculture and allied sector (2020-21) at constant (2011-12) prices:– **4,11,9206 crore**

Share of gross capital formation in 2020-21 at current prices:– **15.9%**, a t constant prices:– (2011-12):– **13.5%**

1. LAND USE PATTERN IN INDIA

	Land Particulars	*Area (m.ha)*
1.	Total geographical area	328·73
2.	Forest	72·01 (23.40%) (As per India State Forest Report)
3.	Barren land and unculturable land	17·28
4.	Area under non agricultural uses	26·45
5.	Permanent pastures and grazing land	10·24
6.	Culturable wasteland	12·66
7.	Current fallow land	14·27
8.	Net area sown	140·0
9.	Total cropped area	194·39
10.	Net Irrigated area	71·55
11.	Area sown more than once	57.97
12.	Area under Micro-irrigation (2018)	7·73
13.	Area under Sprinkler Irrigation	2·45
14.	Gross-cropped area (2018-19)	197.32
15.	Reported Area for Land Utilization	307.79
16.	Average Rainfall	1190 mm

2. LAND HOLDING PATTERN

1.	Marginal holding	Less than 1 ha.
2.	Small holding	1 – 2 ha.
3.	Medium holding	4 – 10 ha.
4.	Large holding	More than 10 ha.

State 1st in production of various crops/ products in India

Crop	*States*
Total food grain	U.P. > Panjab
Cereals	U.P. > M.P.
Rice	West Bengal > U.P.
Wheat	Uttar Pradesh > Panjab
Jowar / Sorghum	Karnataka
Maize	Karnataka
Bajara	Rajasthan
Gram	M.P. > Raj.
Pulses	Raj. > Madhya Pradesh
Oil Seeds	Madhya Pradesh > Raj.
Gram	Madhya Pradesh > U.P.
Ground nut	Gujarat > T.N.
Rapeseed and Mustard	Rajasthan > M.P.
Sugarcane	U.P. > Maharashtra
Cotton bales	Maharashtra > Gujarat
Tobacco	Gujarat > A.P.
Soyabean	M.P. > Maharashtra
Rubber	Kerala
Potato	U.P.
Onion	Maharastra> Karnataka
Coffee	Karnataka
Tea	Assam
Sunflower	Karnataka > A.P.
Jute	West Bengal > Assam
Small millets	M.P.
Livestock Population	U.P. > Raj.
Egg Production	A.P.

3. FIRST ADVANCE ESTIMATES OF AREA & PRODUCTION OF HORTICULTUE CROPS (2021-22)

Area in '000 Ha, Production in '000 MT

Crops	2019-20		2020-21 Final		2021-22 (First Advance Estimate)	
Fruits	**Area**	**Production**	**Area**	**Production**	**Area**	**Production**
Almond	9	11	10	11	10	11
Aonla/Gooseberry	97	1164	100	1197	100	1206
Apple	310	2814	313	2276	313	2437
Banana	897	32597	924	33062	880	32454
Ber	54	606	53	580	53	570
Citrus						
(i) Lime/Lemon	322	3687	327	3548	322	3517
(ii) Mandarin	454	6136	477	6219	473	6265
(iii) Sweet Orange(Mosambi)	217	4040	217	3988	220	3894
(iv) Other Citrus	81	704	77	489	76	475
Citrus Total (i to iv)	**1075**	**14568**	**1097**	**14245**	**1091**	**14150**
Custardapple	50	395	47	407	45	387
Grapes	150	3181	155	3358	162	3490
Guava	292	4361	308	4582	307	4516
Jackfruit	187	1739	188	1893	187	1877
Kiwi	5	14	5	16	5	17
Litchi	97	726	98	720	98	724
Mango	2294	20317	2317	20386	2339	20336
Muskmelon	61	1368	75	1478	70	1510
Papaya	142	5780	146	5540	149	5744
Passion Fruit	12	56	11	70	11	57
Peach	19	119	18	111	18	108
Pear	43	290	42	276	42	279
Picanut	3	0	1	0	1	0
Pineapple	106	1732	106	1799	106	1808
Plum	23	88	24	84	23	78
Pomegranate	283	3186	288	3271	282	3216
Sapota	84	906	79	822	80	834
Strawberry	1	9	3	20	3	14
Walnut	108	286	109	281	109	282
Watermelon	116	3157	119	3254	120	3225
Other Fruits	258	2610	294	2741	359	3592
Total Fruits	**6774**	**102080**	**6930**	**102481**	**6967**	**102924**
Vegetables						
Beans	239	2269	261	2595	256	2520
Bittergourd	107	1268	109	1330	107	1334
Bottlegourd	189	3106	193	3171	192	3143
Brinjal	744	12682	749	12874	744	12768
Cabbage	401	9272	412	9560	413	9606
Capsicum	34	534	37	563	38	563
Carrot	105	1828	108	1885	110	1910
Cauliflower	467	8941	473	9225	473	9283
Cucumber	112	1656	117	1652	116	1608
Chillies (Green)	387	4119	411	4363	400	4221
Elephant Foot Yam	31	724	38	874	35	819
Mushroom		211		243		259

Okra/Ladyfinger	521	6355	531	6466	523	6416
Onion	1431	26091	1624	26641	1914	31129
Parwal/Pointedgourd	59	754	62	725	61	741
Peas	568	5848	567	5846	549	5680
Potato	2051	48562	2203	56173	2208	53603
Radish	207	3184	207	3263	209	3347
Pumpkin/Sitaphal/Kaddu	104	2183	106	2205	106	2218
Sweet Potato	108	1141	106	1121	106	1119
Tapioca	173	6060	183	6941	134	4742
Tomato	818	20550	845	21181	831	20300
Other Vegetables	1453	20945	1517	21550	1541	22555
Total Vegetables	**10310**	**188284**	**10859**	**200445**	**11065**	**199882**
Aromatics and Medicinal	**641**	**734**	**653**	**825**	**650**	**767**
Flowers Cut		676		828		791
Flowers Loose	323	2323	322	2152	267	2095
Total Flowers	**323**	**3000**	**322**	**2980**	**267**	**2886**
Honey		**120**		**125**		**125**
Plantation Crops						
Arecanut	747	1382	794	1563	774	1393
Cashewnut	1125	703	1159	738	1166	774
Cocoa	98	26	103	27	102	28
Coconut	2173	14006	2199	14301	2229	13657
Total Plantation	**4143**	**16116**	**4255**	**16629**	**4271**	**15852**
Spices#						
Ajwain	41	29	42	36	40	28
Cardamom	85	21	85	34	84	26
Chillies (Dried)	626	1835	702	2049	678	1874
Cinnamon/Tejpata	3	8	2	5	3	6
Celery,Dill & Poppy	33	35	25	31	25	30
Clove	2	1	2	1	2	1
Coriander	529	701	656	891	640	811
Cumin	1276	912	1087	795	1037	725
Fenugreek	126	182	156	241	169	252
Fennel	83	140	83	137	82	137
Garlic	352	2925	392	3190	393	3208
Ginger	178	1868	205	2225	193	2219
Nutmeg	24	16	24	16	24	15
Black Pepper	259	104	309	141	284	92
Vanilla	0	0	0	0	0	0
Saffron	4	0	4	0	4	0
Tamarind	44	165	42	156	46	174
Turmeric	296	1153	293	1124	306	1176
Mint (Mentha)	328	44	347	46	336	39
Total Spices	**4291**	**10137**	**4457**	**11117**	**4344**	**10816**
Total	**26482**	**320471**	**27476**	**334603**	**27563**	**333251**

Horticulture

In the world India's Position : 1st –Coconut, Arecanut, Cashew, Ginger, Turmeric, Black pepper
2nd – Fruits and vegetables.

Vegetables

Overall India ranks IInd. 1st – Cauliflower 2nd – Onion 3rd – Cabbage.

Plantation Sector

Tea : India is the largest Producer and Consumer.

Tea and Coffe Production

Year	*Tea (million kg)*		*Coffee (lakh tonnes*	
	Production	*Export*	*Production*	*Export*
2007-08	900·0	-	2·62	2·19
2008-09	900·0	-	2·62	2·04
2009-10	900·0	-	3.00	-
2010-11	1000·0	-	3.00	-
2011-12	1000·0	-	3.00	-
2013-14	1208.78	-	3.48	-
2014-15	1197.00	-	3.27	
2015-16	1233.00	-	3.48	
2016-17	1283.00	-	3.20	
2020-21	1257.52	-	3.54	

India produces two types of coffee: Arabica and Robusta. Arabica has high market value. Highest Coffee state is Karnataka. India is among 10 coffee producing country with about 3% of the global output.

Rubber

India rank 3rd in the World in Natural Rubber production after Indonesia and Brazil.

The total production of Natural Rubber during 2018-19 is provisionally estimated at 648000 tonnes.

4. MAJOR INDIAN STATES CULTIVATING SPECIFIC FRUITS

Sr. No.	*Name of the fruit*	*States*
1.	Mango	Uttar Pradesh, Maharashtra, Gujarat, Bihar, Andhra Pradesh, West Bengal
2.	Banana	Maharashtra, Tamil Nadu, Karnataka, Bihar, West Bengal, Gujarat
3.	Citrus	Maharashtra, Tamil Nadu, Punjab, Karnataka, Assam
4.	Apple	Himachal Pradesh, Jammu-Kashmir, Uttar Pradesh
5.	Litchi	Bihar, West Bengal, Uttar Pradesh
6.	Sapota	Maharashtra, Gujarat, Karnataka
7.	Guava	Uttar Pradesh, Madhya Pradesh, Bihar, Maharashtra
8.	Pineapple	Karnataka, Kerala, Orissa, Bihar, North-East State
9.	Papaya	Kerala, Bihar, Karnataka, Uttar Pradesh, Tamil Nadu, Maharashtra
10.	Pomegranate	Maharashtra, Karnataka, Andhra Pradesh

5. FRUITS AND PROCESSED PRODUCTS

Fruits	*Processed products*
Mango	Juice, Jam, Squash, Syrup, Pickle, Mango, Powder, Toffee, Canned Mango.
Guava	Jam, Jelly, Pickle, Powder, Puree, Peru Khand.
Orange	Juice, Jam, Squash, Syrup.
Custard Apple	Jam, Puree, Powder.
Grapes	Bedana (Kishmish), Jam, Juice, Syrup, Squash and Wine.
Jamun	Juice, Squash, Syrup, Jam, Jelly, Toffee, Seed Powder.
Ber	Ber candy, Ber powder, Ber dates, Ber squash, Syrup, Jam and Dried ber.
Papaya	Jam, Marmalade, Tutti-frutti (candy), Pickle, Wafers (Papad), Chocolate, Cannel papaya, Fridge dried papaya.
Pomegranate	Pomegranate Juice, Jelly, Syrup, Squash, Nectar, Anar rub, Anar crush, Dried pomegranate (Anar Dana), Powder, Pomegranate wine.
Chikku (Sapota)	Chikku Juice (RTS), Squash, Syrup, Jam, Toffee, Candy, Dried Fruit Scrap and Powder, Milk Shake, Ice cream.
Banana	Banana Powder, Banana Flour, Banana Puree, Banana Chips, Banana Jam and Banana Jelly, Banana vinegar, Sweet coat banana.
Aonla (Amla)	Juice, Syrup, Squash, Sweet Syruped Anola, Pickle, Dried Anola and in Ayurvedic Medicine.

Fertilizers

India is 2rd largest producer and consumer of fertilizer.

1st China and 2nd is India and 3rd USA.

Fertilizer production in FY 2020-21

- 43.48 million tonnes

Total consumption 2020-21 (upto kharif-2020)

- 32.853 million tonnes

India's average fertilizer consumption in 2021

- 137·15 kg/ha.

Highest fertilizer consumption state

- Telengana
- Panjab

Lowest fertilizer consumption state

- Arunanchal Pradesh, Nagaland, Sikkim

Area under certified organic farming (2021): 4.3 mha

Biopesticides usage is on rise in India with Maharashtra being highest consumer in 2019-20 total biopesticides consumption in India is 8,847 MT

Agriculture Machinery

Highest average sale of Tractors

- Uttar Pradesh.

Highest average sale of Power Tillers

- West Bengal

External Farm Inputs

Manual, animal, natural mechanical, electrical, chemical, hybrid seeds, tissue culture, bio-technology, digital technology, etc.

Livestock, Poultry and Fisheries

Total Livestock population- 535.82 million As per census (2019)

Milk Production (2019-20)

- **198.4 million tonnes**

India ranks 1st (Highest in world)

Per capita availability – 406 g/day.

Egg Production (2021)

- 122 Billion

India ranks 4th

Fisheries (2021)

- 14·2 Million tonnes

India ranks 2nd

2nd largest producer of inland fish.

Wool

- 40·4 million kg (2018-19)
- Negative growth rate in 2018-19

Honey

- 1,33,00 tonnes (2021)

6. FERTILIZER CONSUMPTION IN NUTRIENT TERMS DURING 2010-11 TO 2020-21 IN LAKH TONES

Product	*2010-11*	*2012-13*	*2018-19*	*2019-20*	*2020-21*
Nitrogenous (N)	165.6	168.2	176.7	191.0	204.0
Phosphatic (P)	80.5	66.5	69.1	76.6	89.7
Potastic (K)	35.1	20.62	26.80	26.07	31.5
Total (N+P+K)	281.2	255.4	272.22	293.7	325.35
Per Ha. Consumption (kg)	144.14	–			

7. AREA SOWN IN INDIA (Mha)

Area sown under Rabi 2020-21 (As on 24th Dec., 2020)

(Area in Lakh ha.)

Crop	Area Sown in 2019-20	Area sown in 2020-21	% Change
Wheat	297.39	313.24	5.33
Rice	13.47	12.49	–7.28
Pulses	141.64	149.29	5.40
Coarse Cereals	46.55	43.44	–6.68
Oilseeds	74.19	79.47	7.12
Total	**573.23**	**597.92**	**4.31**

Source: Ministry of Agriculture and Farmers Welfare, GOI.

Area sown under Kharif-2020 (as on 18th Sept., 2020

(Area in Lakh ha.)

Sr. No.	Crop	Area Sown in 2019-20	Area sown in 2020-21	% Change (2020-21)
1.	Rice	378.71	406.97	7.46
2.	Pulses	132.34	138.62	4.75
3.	Coarse Cereals	178.65	182.17	1.97
4.	Oilseeds	178.65	196.78	10.49
5.	Sugarcane	51.78	62.65	1.68
6.	Jute & Mesta	6.86	6.97	1.60
7.	Cotton	127.09	129.47	1.87
	Total	**1053.52**	**1113.63**	**5.71**

Source: Ministry of Agriculture & Farmers, GoI.

8. INTERNATIONAL CLASSIFICATION OF SOILS (ACCORDING TO TEXTURE)

Parts by weight of the Fractions in %

Kind of Soil	*Parts by Weight of the Fractions in %*					
	Clay Slit **< 2 μ**	***Slit*** **2.20 μ**	***Clay & Silt*** **< 20 μ**	***Total Sand*** **20-2,000 μ**	***Fine Sand*** **20-200 μ**	***Coarse Sand*** **200-2,000 μ**
1. Loamy Coarse Sand	–	–	< 15	> 85	–	> 45
2. Loamy Fine Sand	–	–	< 15	> 85	> 40	–
3. Coarse Sandy Loam	< 15	–	15-35	65-85	–	> 45
4. Fine Sandy Loam	–	–	15-35	65-85	> 40	–
5. Loam	–	< 45	> 35	< 65	–	–
6. Silty Loam	–	> 45	–	–	–	–
7. Sandy Clay Loam	–	< 20	< 45	>55	–	–
8. Clay Loam	15-25	< 20	< 45	> 55	–	–
9. Silty-clay Loam	> 45	> 60	< 40	–	–	–
10. Sandy Clay	–	< 20	< 45	> 55	–	–
11. Loamy Clay	25-45	<45	> 45	< 55	–	–
12. Silty Clay	–	> 45	> 70	< 30	–	–
13. Heavy Clay	> 45	–	–	–	–	–

Meaning of symbol ; 1μ = 1/1000 mm > = greater than < = less than

IMPORTANT BREEDS

Cattle–Gir, Kankrej, and Ongole Friesian etc.
Buffaloes – Zaffarbadi, Mahsana, Murrah and Surathi.
Improved Breeds- Mehsana and Godavari
Pig (exotic) – White yorkshire, Hampshire and Lindrace.

AGRICULTURAL TRADE (India's Position)

Exports

Agricultural crops	*Position*
Rice (Basmati)	I
Cashew	II
Wheat and Spice	III
Tea	IV
Tobacco	V

Agricultural products	*Position*
Marine products	I
Rice	II
Oil meals	III
Cashew	IV
Wheat and Spice	V

Imports

Agricultural products	*Position*
Vegetable oil	I
Cashew	III
Pulses	II
Sugar	IV
Fruits and nuts	V

9. PRODUCTION AND PER CAPITA AVAILABLITY OF MILK

Year	*Per capita (grams/day)*	*Milk proction (MT)*
1990- 91	176	53·9
2000- 01	220	80·6
2005- 06	241	97·1
2006- 07	246	100·9
2007- 08	252	104·8
2008- 09	258	108·5
2009-10	263	112·5
2010-11	281	127.9
2013-14	307	137.2
2015-16	337	155.5
2021	437	210.00

10. COMMERICAL CROP PRODUCTION (Million Tonnes)

Crop	*2011-12*	*2012-13*	*2014-15*	*2015-16*	*2016-17**	**2021-22 (Ad. Est.)**
Ground nut	6.96	4.75	5.93	5.37	7.05	8.24
Mustard and Rapseed	6.60	7.82	6.28	6.80	7.91	11.46
Soya bean	12.21	14.68	10.37	8.57	14.12	13.31
Total oil seeds	29.80	31.01	27.51	25.25	33.59	37.14
Cotton*	35.20	34.00	34.81	30.0	32.51	34.06
Jute** and Mesta**	11.40	11.30	11.12	10.52	10.06	9.16
Sugarcane	357.67	352.0	362.3	348.4	310.0	414.04

* Million bales of 170 kg each; ** million bales of 180 kg each; *2nd advanced estimate

11. NATIONAL FLOWERS OF THE DIFFERENT COUNTRIES.

Country	*Flower*
Australia	Golden Wattle
Argentina	Cattleya orchid
Austria	Edelweiss
Bangladesh	White Water Lily
Belgium	Azalea, Red or corn poppy
Bermuda	Blue-eyed grass
Bolivia	Inca magic flower
Brazil	Cattleya orchid
Bulgaria	Rose
Canada	Maple leaf, Rose as symbol of England, Fleur-de-lis for France Ceylon Wesak orchid
Chile	Copihue
China	Narcissus, Peony, Plum flower,
Colombia	Mayflower orchid (or Christmas orchid)
Costa Rica	Flor de San Sebastian Cuba Ginger lily
Czech Republic	Rose (traditional), Carnation (popular), Linden tree
Denmark	Red Clover
Ecuador	White nun orchid, Rose Egypt Lotus or Water Lily
England	Red tudor rose
Estonia	Blue Cornflower
Ethiopia	Calla
Finland	Lily of the valley
France	Lily
Germany	None
Greece	Guernsey lily
Holland	Tulip
Hong Kong	Bauhinia Blakeana
Hungary	Tulip
India	Lotus (Nelumbo nucifera) Indonesia Melati jasmine, Moon orchid Iran Rose
Iraq	Rose
Ireland	Shamrock
Italy	Poppy or White Lily Jamaica Lignum Vitae (Guiacum officinale)
Japan	Chrysanthemum
Jordan	Black Iris
Korea	Rose of Sharon
Laos	Rice
Latvia	Oxeye Daisy

Country	*Flower*
Libya	Pomegranate Blossom Lithuania Rue (Ruta)
Luxembourg	Rose
Malagasy	Traveler's tree Republic
Malaysia	Hibiscus
Maldives	Rose
Mexico	Dahlia
Monaco	Carnation
Nepal	Rhododendron
Netherlands	Calendula (official), Orange tulip (popular)
New Zealand	Kowhai
Nicaragua	Sacuanjoche or May Flower Northern Ireland Shamrock
Norway	Heather
Pakistan	Jasmine
Panama	Dove orchid (or Holy Ghost orchid)
Paraguay	Jasmine-of-the-Paraguay, Orange blossom
Persia	Rose
Peru	Inca magic flower
Philippines	Sampagita Jasmine
Poland	Cornflower
Portugal	Lavender
Russia	Chamomile, Daisy
Rumania	Dog rose
San Marino	Cyclamen
Scotland	Thistle
Singapore	Orchid: Vanda Miss Joaquim
Slovenia	Carnation
South Africa	King Protea
South Korea	Hibiscus
Spain	Red carnation, Pomegranate
Sweden	Twinflower
Switzerland	Edelweiss
Turkey	Rose (traditional), Tulip (popular)
Uruguay	Ceibo, Cockspur coral Vatican City Easter lily
Venezuela	Orchid
Wales	Leek and Daffodil
Yemen	Arabian coffee
Yugoslavia	Lily of the Valley
United States	Rose
Zimbabwe	Flame Lily

12. TOTAL PULSES: ALL-INDIA AREA, PRODUCTION AND YIELD ALONGWITH COVERAGE UNDER IRRIGATION

Year	*Area (Mha)*	*Production (Mt)*	*Yield (kg./hect.)*	*Area Under Irrigation (%)*
2000-01	20.35	11.08	544	12.64
2001-02	22.01	13.37	607	13.32
2002-03	20.50	11.13	543	13.97
2003-04	23.46	14.91	635	13.65
2004-05	22.76	13.13	577	14.04
2005-06	22.39	13.38	598	14.55
2006-07	23.19	14.20	612	15.45
2007-08	23.63	14.76	625	15.89
2008-09	22.09	14.57	659	16.47
2009-10	23.28	14.66	630	16.21
2010-11	26.40	18.24	691	14.92
2011-12	24.46	17.09	699	16.13
2012-13	23.26	18.34	789	18.51
2013-14	25.21	19.25	764	19.72
2014-15	23.55	17.15	728	19.86
2015-16	24.91	16.32	655	19.26
2016-17	29.45	23.13	786	18.83
2017-18	29.81	25.42	853	22.87
2018-19	29.16	22.08	757	23.20
2019-20	27.99	23.03	823	
2020-21*	28.83	25.72	892	

Source: Directorate of Economics & Statistics, DA&FW   * 4th Advance Estimates

13. MAXIMUM ROOT ZONE DEPTH OF VARIOUS CROPS

Crop	*Root Zone Depth (mm)*	*Crop*	*Root Zone Depth (mm)*
Wheat	500-750	Groundnut	600- 800
Barley	600-750	Cotton	900- 2000
Maize	400- 600	Tabacco	200- 750
Sorghum	1000- 1500	Mustard	500- 1500
Pearlmillet	300- 450	Toria	300- 500
Italian millet	300- 450	Sun flower	500- 1500
Crowfoot millet	300- 450	Safflower	500- 2500
Jute	450- 500	Seasamo	350- 370
Sugarcane	500- 750	Castor	1000- 4000
Gourds	400- 600	Cole crops	200- 500
Cucumber	400- 600	Berseem	500- 3000

14. SOIL MOISTURE RANGE AVAILABILITY FOR DIFFERENT CROPS

Crop	*Range (per cent)*	*Crop*	*Ragne (per cent)*
Wheat Tall	100-50	Tomato	100- 70
Dwarf	100-60	Cabbage and Cauliflower	100- 50
Barely	100- 40	Radish, Turnip, and Beetroot	100-75
Sorghum and other millets	100- 40	Turmeric, Ginger	100-60
Oilseeds	100- 50	Chillies	100- 50
Cotton	100- 20 (50- 60)	Fruit Trees	100- 75
Jute	100- 70	Papaya	100- 80
Jute	100- 50	Banana	100- 80
Sugarcane	100- 50	Maize local	100- 50
Tabacco cigar	100- 70	Hybrid and Composites	100- 25 to 30
Hookah	100- 70	Jute	100- 65 to 70
Bidi	100-60	Mustard	100- 50 to 75
Forage crops	100-50		
Vegetables	100- 70 to 80		

15. TOXINS PRODUCED IN PLANTS

Plants	*Toxins*
Cotton	Gossypol
Lathyrus	Neurotoxin-ß-oxalyl-L-a,ß-diaminopropionic acid (ODAP)
Rapeseed & Mustard	Erucic acid- High erucic content is associated with myocardial fibrosis in adults and lipidosis in children.
Soybean	Goitrogen
Lucerne/Alfalfa	Saponin and plant estrogen
Groundnut	Aflatoxin-High moisture, Fungi (Aspergillus flavus & A.Niger)
Sorghum	HCN glucosides/Prussic acid/ Dhurin
Sweet clover	Coumarin
Potato	Steroidal alkaloid
Yam	Alkaloids

Plants	*Toxins*
Mango	Resin/ Urushiol
Field pea	Antivitamin E factor
Safflower, Sunflower	Polyphenolics
Pigeon pea / French bean	Tripsin inhibitors
French bean	Heamaggutinin
White clover	Glucosides
Cucurbits	Cucurbitacins
Cassava	CN Glucosides
Pearl millet, Napiergrass	Oxalic acid
Gram leaf	Malic & Oxalic acid
Cole crops	Sinigrin
Celery	Appin
Tomato	Tomatine
Brinjal	Salasodine
Pumpkin	Choline esterase

16. LIVESTOCK POPULATION IN INDIA

(Million Numbers)

Species	*1951*	*1956*	*1961*	*1966*	*1972*	*1977*	*1982*	*1987*	*1992*	*1997*	*2003*	*2007*	*2012*	*2019*
Cattle	155.3	158.7	175.6	176.2	178.3	180.0	192.5	199.7	204.6	198.9	185.2	199.1	190.9	192.5
Adult Female Cattle	54.4	47.3	51.0	51.8	53.4	54.6	59.2	62.1	64.4	64.4	64.5	73.0	76.7	81.4
Buffalo	43.4	44.9	51.2	53.0	57.4	62.0	69.8	76.0	84.2	89.9	97.9	105.3	108.7	109.9
Adult Female Buffalo	21.0	21.7	24.3	25.4	28.6	31.3	32.5	39.1	43.8	46.8	51.0	54.5	56.6	55.0
Total Bovines	198.7	203.6	226.8	229.2	235.7	242.0	262.2	275.7	288.8	288.8	283.1	304.4	299.6	302.3
Sheep	39.1	39.3	40.2	42.4	40.0	41.0	48.8	45.7	50.8	57.5	61.5	71.6	65.1	74.3
Goat	47.2	55.4	60.9	64.6	67.5	75.6	95.3	110.2	115.3	122.7	124.4	140.5	135.2	148.9
Horses and Ponies	1.5	1.5	1.3	1.1	0.9	0.9	0.9	0.8	0.8	0.8	0.8	0.6	0.6	0.3
Camels	0.6	0.8	0.9	1.0	1.1	1.1	1.1	1.0	1.0	0.9	0.6	0.5	0.4	0.3
Pigs	4.4	4.9	5.2	5.0	6.9	7.6	10.1	10.6	12.8	13.3	13.5	11.1	10.3	9.1
Mules	0.1	0.0	0.1	0.1	0.1	0.1	0.1	0.2	0.2	0.2	0.2	0.1	0.2	0.1
Donkeys	1.3	1.1	1.1	1.1	1.0	1.0	1.0	1.0	1.0	0.9	0.7	0.4	0.3	0.1
Yak	NC	NC	0.0	0.0	0.0	0.1	0.1	0.0	0.1	0.1	0.1	0.1	0.1	0.1
Mithun	NA	NA	NA	NA	NA	NA	NA	NA	0.2	0.2	0.3	0.3	0.3	0.4
Total Livestock	292.9	306.6	336.5	344.5	353.2	369.4	419.6	445.2	470.9	485.4	485.0	529.7	512.1	535.8
Poultry *	73.5	94.8	114.2	115.4	138.5	159.2	207.7	275.3	307.1	347.6	489.0	648.8	729.2	851.8

Source : Livestock Censuses, Department of Animal Husbandry, Dairying & Fisheries, Ministry of Agriculture & Farmers' Welfare, GoI

17. ALL INDIA PRODUCTION OF MILK, EGGS AND WOOL

Year	*Milk (Million Tonnes)*	*Eggs (Billion Nos.)*	*Wool (Million Kgs.)*	*Meat (Million Tonnes)*
2009-10	116.4	60,267	43.1	4.5
2010-11	121.8	63,024	43.0	4.9
2011-12	127.8	66,449	44.7	5.5
2012-13	132.4	69,731	46.1	5.9
2013-14	137.7	73,438	47.9	6.2
2014-15	146.3	78,484	48.1	6.7
2015-16	155.5	82,929	43.6	7.0
2016-17	165.4	88,137	43.5	7.4
2017-18	176.3	95,217	41.5	7.7
2018-19	187.7	1,03,318	40.4	8.1
2019-20	198.4	1,14,383	36.8	8.6
2020-21*	209.9	1,22,049	36.9	8.8

"-" Not received/not available "*" Provisional Figures

Source: State/UT Animal Husbandry Departments

18. FISH PRODUCTION IN INDIA

Year	*Fish production (In Lakh Tonnes)*			*Annual Average Growth Rate (Percent)*		
	Marine	*Inland*	*Total*	*Marine*	*Inland*	*All India*
2009-10	31.04	48.94	79.98	4.23	5.52	5.02
2010-11	32.5	49.81	82.31	4.7	1.78	2.91
2011-12	33.72	52.94	86.66	3.75	6.28	5.28
2012-13	33.21	57.19	90.4	-1.51	8.03	4.32
2013-14	34.43	61.36	95.79	3.67	7.29	5.96
2014-15	35.69	66.91	102.6	3.66	9.04	7.11
2015-16	36	71.62	107.62	0.87	7.04	4.89
2016-17	36.25	78.06	114.31	1.14	8.63	6.12
2017-18	37.56	89.48	127.04	3.61	14.62	11.13
2018-19	38.53	97.2	135.73	2.58	8.62	6.84
2019-20	37.27	104.37	141.64	-3.2	7.37	4.35

Source: Department of Fisheries, States Government / UTs Administration

19. PLAN-WISE SHARE OF PUBLIC SECTOR OUTLAYS AND EXPENDITURE UNDER AGRICULTURE AND ALLIED ACTIVITIES

(Rs. Crores)

Five Year Plan / Annual Plan	*Percentage share of Agriculture and Allied Activities to Total*	
	Plan Outlays	*Actual Expenditure*
Ninth Plan (1997-2002)	4.4	4.0
Tenth Plan (2002-07)	3.9	3.8
Eleventh Plan (2007-12)	3.7	4.4
Annual Plan (2007-08)	3.2	4.2
Annual Plan (2008-09)	4.0	4.3
Annual Plan (2009-10)	3.6	4.1
Annual Plan (2010-11)	4.0	4.9
Annual Plan (2011-12)	4.2	4.5
Twelfth Plan (2012-17)*	4.7	N.A
Annual Plan (2012-13)	4.5	5.4
Annual Plan (2013-14)	4.7	4.8
Annual Plan (2014-15)#	2.4	2.4
Annual Plan (2015-16)#	2.0	1.9
Annual Plan (2016-17)	NA	NA

Source: Economic Survey and Budget Documents, Ministry of Finance

1. Figures include Outlays and Expenditure by Centre, States and Union Territories.
2. Actual Central Expenditure figures are including IEBR from 2009-10 onwards

* 12th Plan projections (At current prices).

For Centre only as figures for States & UTs are not yet available.

20. MAXIMUM PERMISSIBLE LIMITS OF POLLUTANTS IN THE AIR SET BY WHO

($\mu g/m^3$)

Pullutants	Health end point	Guideline average	Average time
Sulphur dioxide	Changes in lung function in asthmatics	500	10 minutes
	Exacerbations of respiratory symptoms in	125	24 hours
	sensitive individuals	50	1 year
Lead	Critical level of Pb in blood <25 μg Pb/l	0.5	1 year
Carbon monoxide	Critical level of COHb <2.5 %	10000	15 minutes
		60000	30 minutes
		30000	1 hour
		10000	8 hours
Nitrogen dioxide	Slight changes in lung function in asthmatics	200	1 hour
		40	1 year
Ozone	Respiratory function responses	120	8 hours

Source : *TERI Energy Data Directory & Yearbook, 2002-03,* Tata Energy Research Institute, New Delhi.

21. COMPOSITION OF AIR

Components	Concentration (ppm by volume)
Nitrogen (N_2)	78.009
Oxygen (O_2)	20.95
Argon (Ar)	0.93
Water vapour (H_2O)	–
Carbon dioxide (CO_2)	0.032
Neon (Ne)	0.0018
Krypton (Kr)	0.0001
Methane (CH_4)	0.00015
Helium (He)	0.00052
Ozone (O_3)	0.000002
Xenon (Xe)	0.000008
Nitrous dioxide (N_2O)	0.00002
Carbon monoxide (CO)	0.00001
Hydrogen (H_2)	0.00005
Ammonia (NH_3)	0.0000006
Nitrogen dioxide (NO_2)	0.0000001
Nitric oxide (NO)	0.00000006
Sulphur dioxide (SO_2)	0.00000002
Hydrogen sulphide (H_2S)	0.00000002

22. LIST OF PESTICIDES WHICH ARE BANNED REFUSED REGISTRATION AND RESTRICTED USE

(As on 01.07.2021)

I. PESTICIDES / FORMULATIONS BANNED IN INDIA.

A. Pesticides Banned for manufacture, import and use.

1. Alachlor (Vide S.O. 3951(E), dated 08.08.2018)
2. Aldicarb (vide S.O. 682(E) dated 17th July 2001)
3. Aldrin
4. Benzene Hexachloride
5. Benomyl (vide S.O 3951(E) dated 8th August, 2018)
6. Calcium Cyanide
7. Carbaryl (vide S.O 3951(E) dated 8th August, 2018)
8. Chlorbenzilate (vide S.O. 682(E) dated 17th July 2001)
9. Chlordane
10. Chlorofenvinphos
11. Copper Acetoarsenite
12. Diazinon (vide S.O 3951(E) dated 8th August, 2018)
13. Dibromochloropropane (DBCP) (vide S.O. 569 (E) dated 25th July 1989)
14. Dichlorovos (Vide S.O. 3951(E), dated 08.08.2018)
15. Dieldrin (vide S.O. 682(E) dated 17th July 2001)
16. Endosulfan (vide ad-Interim order of the Supreme Court of India in the Writ Petition (Civil) No. 213 of 2011 dated 13th May, 2011 and finally disposed of dated 10th January, 2017)
17. Endrin
18. Ethyl Mercury Chloride
19. Ethyl Parathion
20. Ethylene Dibromide (EDB) (vide S.O. 682(E) dated 17th July 2001)
21. Fenarimol (vide S.O 3951(E) dated 8th August, 2018)
22. Fenthion (vide S.O 3951(E) dated 8th August, 2018)
23. Heptachlor
24. Lindane (Gamma-HCH)
25. Linuron (vide S.O 3951(E) dated 8th August, 2018)
26. Maleic Hydrazide (vide S.O. 682(E) dated 17th July 2001)
27. Menazon
28. Methoxy Ethyl Mercury Chloride (vide S.O 3951(E) dated 8th August, 2018)
29. Methyl Parathion (vide S.O 3951(E) dated 8th August, 2018)
30. Metoxuron
31. Nitrofen
32. Paraquat Dimethyl Sulphate
33. Pentachloro Nitrobenzene (PCNB) (vide S.O. 569 (E) dated 25th July 1989)
34. Pentachlorophenol
35. Phenyl Mercury Acetate
36. Phorate (Vide S.O. 3951(E), dated 08.08.2018)
37. Phosphamidon (Vide S.O. 3951(E), dated 08.08.2018)
38. Sodium Cyanide (banned for Insecticidal purpose only vide S.O 3951(E) dated 8th August, 2018)*
39. Sodium Methane Arsonate 40. Tetradifon
41. Thiometon (vide S.O 3951(E) dated 8th August, 2018)
42. Toxaphene (Camphechlor) (vide S.O. 569 (E) dated 25th July 1989)
43. Triazophos (Vide S.O. 3951(E), dated 08.08.2018)
44. Tridemorph (vide S.O 3951(E) dated 8th August, 2018)
45. Trichloro acetic acid (TCA) (vide S.O. 682 (E) dated 17th July 2001)
46. Trichlorfon(Vide S.O. 3951(E), dated 08.08.2018)

23. EXTREME EARTH

Earth has been experiencing a surge in natural disasters, climate related extreme events and high intensity typhoons world-wide in the last three decades.

1. Warmer by the day: Year 2016 was the warmest year on the record. Earth has warmed up by 1.1 °C over pre-industrial levels and about 0.06 °C above the previous record set in 2015.

   Earth's average temperature hit a record level for the third consecutive year in 2016.
2. Becoming a gas chamber: The atmospheric concentrations of carbon-di-oxide, methane, an nitrous oxide hit a new high in the millennia.
3. Sea level is rising: Sea level rise caused when ice melts and warmer water expands, continues and appears to be accelerating.
4. The average ocean level was 70 mm higher in the year 2015 than 1993 water mark.

   Great barrier reef also underwent unprecedented second straight year of bleaching due to warming of seas. (Bleaching of coral reefs is the loss of endosymbiotic algae from the coral, either through expulsion or loss of algal pigmentation. Above average temperature is the leading cause of coral bleaching.
5. In the year 2017 Antarctic sea ice hits its lower recorded level.

24. CARDINAL TEMPERATURE FOR THE GERMINATION OF SEEDS OF CERTAIN CROPS

Crop	*Cardinal temperature(°F)**		
	Minimum	*Optimum*	*Maximum*
Rice	50-93	86-90	97-101
Maize	46-50	89-95	104-111
Sorghum	46-50	89-95	104
Wheat	39-40	77	86-90
Barley	39-40	68	82-86
Oats	39-41	77	86
Rye	34-36	77	86
Peas	34-36	86	95
Lentil	39-41	86	97
Sugarbeet	39-41	77	82-86
Carrot	39-41	77	86
Pumpkin	54	89-93	104
Melon	54-59	95	104
Alfalfa	34	86	99
Flax	35-37	77	86
Hemp	34-36	95	113
Vetch	34-36	86	95

* $\frac{C}{100} = \frac{F-32}{100} = \frac{K-273}{100}$

25. NET AVAILABILITY OF FOODGRAINS PER DAY IN INDIA

(gram/capita/day)

Year	Rice	Wheat	Other cereals	Cereal	Gram	Pulses	Food grains
1981	197.8	129.6	89.9	417.3	13.4	37.5	454.8
1986	212.0	151.0	70.7	433.7	16.2	43.8	477.5
1991	221.7	166.8	80.0	468.5	13.4	41.6	510.1
1996	204.4	176.0	62.0	442.5	11.3	32.7	475.2
2001	190.5	135.8	56.2	386.2	8.0	30.0	416.2
2002	228.7	166.6	63.4	458.7	10.7	35.4	494.1
2003	181.4	180.4	46.7	408.5	8.5	29.1	437.6
2009	188.4	154.7	63.9	407.0	12.9	37.0	444.0
2010	182.0	168.2	51.4	401.7	13.5	39.4	437.1
2011	181.6	163.6	65.5	410.7	14.5	43.0	468.2
2012	190.2	158.4	60.0	408.6	13.5	41.7	463.8
2013	197.4	183.1	52.7	433.2	15.3	43.3	491.9
2014	198.0	183.0	61.8	442.9	16.3	46.4	489.3
2015	186.0	168.0	77.7	421.4	14.0	43.8	465.1
2016	184.2	199.7	71.6	443.7	13.3	43.0	486.8
2017	183.0	182.7	80.6	434.0	17.3*	54.7	488.7
2018(P)	189.0	176.4	85.5	438.4	N.A.	54.5	484.3

Note : *Provisional.

Source : *Agricultural Statistics at a Glance 2017, Directorate of Economics & Statistics, Ministry of Agriculture & Farmers Welfare, Govt. of India. (Website: http://eands.dacnet.nic.in)*

26. PRODUCTION & PRODUCTIVITY IN AGRICULTURE DURING PAST 4 DECADES

(Area: million ha, Production & consumption: million t, Yield: kg/ha)

Crops		1960-61	1970-71	1980-81	1990-91	2000-01	2002-03	2010-11	2011-12*
All foodgrains	Area	115.58	124.32	126.67	127.84	121.05	111.50	126.67	125.03
	Production	82.02	108.42	129.59	176.39	196.81	174.19	244.49	257.44
	Yield	710	872	1023	1380	1626	1562	1930	2059
Rice	Production	34.58	42.22	53.63	74.29	84.98	72.65	95.98	104.32
	Yield	1013	1123	1336	1740	1901	1804	2239	2372
Wheat	Production	11.00	23.83	36.31	55.14	69.68	65.10	86.87	93.90
	Yield	851	1307	1630	2281	2708	2619	2988	3140
Oilseeds	Production	6.98	9.63	9.37	18.61	18.44	15.06	32.48	30.01
	Yield	507	579	532	771	810	710	1193	1135
Sugarcane	Production	110.00	126.37	154.25	241.05	295.96	281.57	342.38	357.67
	Yield	45549	48322	57844	65395	68577	64562	70091	70317
Pulses	Production	12.70	11.82	10.63	14.26	11.08	11.14	18.24	17.21
	Yield	539	524	473	578	544	556	691	694
Coarse cereals	Production	23.74	30.55	29.02	32.70	31.08	25.30	43.40	42.01
	Yield	528	665	695	900	1027	962	1531	1593
Milk	Production	20.00	N.A.	31.60	53.90	80.80	87.3	121.80	N.A.
Fish	Production	1.16	1.76	2.44	3.84	5.66	6.20	8.42	N.A.
Irrigated area		27.98	38.19	49.70	62.47	75.14	N.A.	N.A.	N.A.
Fertiliser consumption		0.29	2.17	5.52	12.55	16.70	16.09	28.12	27.74
Per capita availability of food grains (gms/day)		449.60	455.01	410.40	472.60	454.40	492.20	438.6	462.9

Note : N.A. Not available, * #Provisional, 4th Advance estimates.

Source : *Agricultural Statistics at a Glance, 2012*, Directorate of Economics and Statistics, Ministry of Agriculture, Govt. of India.

27. NUTRI/COARSE CEREALS: ALL-INDIA AREA, PRODUCTION AND YIELD

Year	*Area (Mha)*	*Production (Mt.)*	*Yield (kg/hect.)*
1961-62	44.73	23.22	519
1971-72	43.57	24.60	564
1981-82	42.45	31.09	733
1991-92	33.42	25.99	778
2001-02	29.52	33.38	1131
2011-12	26.42	42.01	1590
2020-21*	23.83	51.15	2146

Source: Directorate of Economics & Statistics, DA&FW * 4th Advance Estimates

28. NINE OILSEEDS: ALL-INDIA AREA, PRODUCTION AND YIELD

Year	*Area (Mha)*	*Production (Mt.)*	*Yield (kg/hect.)*
1950-51	10.73	5.16	481
1961-62	14.77	7.28	493
1971-72	17.27	9.08	526
1981-82	18.91	12.08	639
1991-92	25.89	18.60	719
2001-02	22.64	20.66	913
2011-12	26.31	29.80	1133
2020-21*	28.79	36.10	1254

Source: Directorate of Economics & Statistics, DA&FW * 4th Advance Estimates.

Note: Data for 1950-51 to 1969-70 relate to total of five major oilseeds viz. groundnut, 'castorseed, sesamum, rapeseed & mustard and linseed. After 1969-70 data also ' included 'Soybean, Sunflower, Nigerseed and Saflower

29. COTTON: ALL-INDIA AREA, PRODUCTION AND YIELD ALONGWITH COVERAGE UNDER IRRIGATION

Year	*Area (Mha)*	*Production (Mt.*	*Yield (kg./hect.)*	*Area Under Irrigation (%)*
1961-62	7.98	4.85	103	13.00
1970-71	7.61	4.76	106	17.34
1980-81	7.82	7.01	152	27.28
1991-92	7.66	9.71	216	33.33
2001-02	9.13	10.00	186	33.82
2011-12	12.18	35.20	491	35.90
2020-21*	13.01	35.38	462	

Source: Directorate of Economics & Statistics, DA&FW * 4th Advance Estimates.

30. FOODGRAINS: ALL-INDIA AREA, PRODUCTION AND YIELD ALONGWITH COVERAGE UNDER IRRIGATION

Year	*Area (Mha)*	*Production (Mt.)*	*Yield (kg/hect.)*	*Area Under Irrigation (%)*
1986-87	127.20	143.42	1128	32.64
1987-88	119.69	140.35	1173	33.48
1988-89	127.67	169.92	1331	34.42
1989-90	126.77	171.04	1349	34.99
1990-91	127.84	176.39	1380	35.07
1991-92	121.87	168.38	1382	37.37
1992-93	123.15	179.48	1457	37.42
1993-94	122.75	184.26	1501	38.66
1994-95	123.86	191.50	1546	39.61
1995-96	121.01	180.42	1491	40.13
1996-97	123.58	199.34	1614	41.71
1997-98	124.07	192.26	1552	41.71
1998-99	125.17	203.61	1627	43.31
1999-00	123.10	209.80	1704	44.63
2000-01	121.05	196.81	1626	43.70
2001-02	122.78	212.85	1734	43.58
2002-03	113.86	174.77	1535	43.42
2003-04	123.45	213.19	1727	42.60
2004-05	120.08	198.36	1652	44.59
2005-06	121.60	208.60	1715	45.70
2006-07	123.71	217.28	1756	47.18
2007-08	124.07	230.78	1860	47.28
2008-09	122.83	234.47	1909	48.47
2009-10	121.33	218.11	1798	47.84
2010-11	126.67	244.49	1930	48.06
2011-12	124.75	259.29	2078	49.82
2012-13	120.78	257.13	2129	51.30
2013-14	125.05	265.05	2120	51.92
2014-15	124.30	252.03	2028	53.06
2015-16	123.22	251.54	2041	52.66
2016-17	129.23	275.11	2129	52.16
2017-18	127.52	285.01	2235	53.00
2018-19	124.78	285.21	2286	54.32
2019-20	126.99	297.50	2343	
2020-21*	129.34	308.65	2386	

Source: Directorate of Economics & Statistics, DA&FW

* 4th Advance Estimates.

31. SCHEMES OF GOVERNMENT OF INDIA

Government at both Central and State levels encourages participation of the NGO's in developmental and welfare activities. It is a bird's eye view of the agriculture Government schemes.

I. Schemes of Ministry of Agriculture (Department of Agriculture and Co-operation)

1) Intensive Cotton Development Programme (ICDP) under Mini- Mission-II of Technology Mission on Cotton.
2) Integrated Scheme of Oilseeds, Pulses, Oil Palm and Maize (ISOPOM)
3) National project on Development and Use of Biofertilizers
4) Strengthening of Agriculture Extension Services
5) Small Farmer's Agri-Business Consortium
6) Natural Disaster Management Programme
7) Watershed Development Project in Shifting Cultivation Area (WDPSCA) Additional Central Assistance to State Plan Scheme
8) Scheme for Strengthening and Modernization of Pest Management Approach in India

II. Schemes of Ministry of Agriculture (National Horticulture Board Schemes)

9) Development of Commercial Horticulture through Production and Post-Harvest Management
10) Capital Investment Subsidy for Construction/Expansion/Modernization of Cold Storages/ Storages of Horticulture Product
11) Technology Development and Transfer for Promotion of Horticulture
12) Information Service for Horticulture Crops
13) Horticulture Promotion Service

III. Schemes of Ministry of Agro and Rural Industries

14) Scheme of Fund for Regeneration of Traditional Industries (SFURTI)
15) Programme for Promotion of V.I Cluster – Rural Industry Service Centre (RISC)
16) Prime Minister's Rozgar Yojana (PMRY)
17) Rural Employment Generation Programme (REGP)

IV. Schemes of Ministry of Food Processing Industries

18) Scheme for Human Resource Development
19) Scheme for Technology Upgradation/ Expansion/Establishment/ Modernization of Food Processing Industries
20) Scheme for Quality Assurance, Codex Standards and Research and Development
21) Scheme for Backward and Forward Integration and other Promotional Activities
22) SAMPADA: Umbrella scheme incorporating ongoing schemes of the Ministry for Mega Food Parks. Integrated Cold Chain, Infrastructure for Agro-processing Clusters. Food Safety and Quality Assurance Infrastructure.

Agriculture Terms Defined

Abiotic stress: Conditions or factors which can cause harmful effects to plants such as soil conditions, drought or extreme temperatures.

Achene: A small, dry seeded fruit that does not open to release the seed.

Aerosol: Very finely divided particles (of solution) in the air.

Agro ecosystem: The land or area used for crops, pasture and livestock, the adjacent uncultivated land that supports other vegetation and wildlife, and the associated atmosphere the underlying soils, groundwater, and drainage networks.

Alley cropping: Planting of crops in strips with rows of trees or shrubs on each side.

Allochemicals: They are non-nutritional chemicals, produced by an organism of one species and affect the growth, health behavior or population biology of individuals of another species.

Amino Acids: These are fundamental building blocks of proteins. There are 20 naturally occurring amino acids in animals and around 100 or more found in plants.

Anabolism: The synthesis of complex molecules from simpler ones together with the storage energy. All biosynthetic reactions in a living organisms.

Annual: Plant that grows, sets seed and dies in one growing season.

Antibiotic: A substance produced by microorganism which inhibit the growth of a poisonous or toxic chemical.

Arable land: Land that can be cultivated to grow crops

Aridity: A measure of dryness or water deficiency. An index based on water deficiency over water need.

Azotobacter: Azotobacter is defined as a genus of free living soil microbes that are responsible for nitrogen fixing in the atmosphere. These microbes play a very significant role by synthesizing some biologically active substances which helps in stimulating plant growth. Azotobacter are mainly used in agriculture, most importantly in nitrogen biofertilizers such as azotobacterin.

Benthos: It is the organisms living in the benthic zone or at the bottom of an aquatic ecosystem.

Binomial system: The system of dual latin names for plants and animals devised by Swedish Naturalist Linnaeus and now almost universally applied to living organism.

Biochemical pathways: Various steps involved in any bioconversion process.

Biocide: A general poison or toxicant used for killing.

Biodiversity: The variability among living organisms from all sources and the ecological complexes of which they are part and include diversity within species or between species and of ecosystems.

Bioeconomics: The relationship between pest numbers and economic losses.

Biofertilizers: These are organisms that enrich the nutrient quality of soil. Main sources of biofertilizers are bacteria, fungi and cyanobacteria (blue green algae).

Biofuel: Gas (as methane) or liquid fuel (ethyl alcohol) made from plant material (biomass).

Biological control: It is a pest management technique in which the natural enemies are introduced, encouraged, multiplied by artificial means and disseminated by man.

Biome: terrestrial regions characterized by certain types of vegetation and other forms of life.

Bioremediation: Process of using microorganism to degrade or consume to remove pollutants from environment.

Biotic stress: A stress that occurs in plants as a result of damage done by other living organisms, such as bacteria, viruses and fungi, parasites, weeds, insects and other native or cultivated plants.

Carbohydrates: They are the compounds either aldehydes or ketones having an empirical formula (CH_2O), *i.e.* they are composed of carbon, O_2 and H.

Carbon credit: A form of permit which allows a country or organization to produce a certain amount of carbon emissions and which can be traded if the full allowance in not utilized.

Carrying capacity: It is the maximum population density in a given environment which will support for a sustained period.

Chemical control: The use of chemicals to kill, deter or in any way suppress pest population.

Chemosynthesis: Process of carbohydrate synthesis in which the organisms use chemical reactions to obtain energy from inorganic compounds. *Example* bacteria of the genus Nitrosomon as oxidise ammonia to nitrite.

Chitin: A polymer of the nitrogenous polysaccharide n-acetyl-D-glucosamine that occurs in the exoskeleton of arthropods.

Chromatid: A thread of DNA formed during cell division when a chromosome splits along its length.

Chromosomes: Thread or rod like structure in eukaryotes found in nucleus; made up of nucleoprotein and contain all the nuclear genes. The paired, self-replicating genetic structures of cells that contain the cellular DNA; the nucleotide sequence of the DNA encodes the linear array of genes.

Cistron: A unit function in DNA. One DNA cistron specifies one polypeptide chain in protein synthesis.

Class: A division of animal kingdom below a phylum but above an order.

Classification: It is the arrangement of organism into groups on the basis of their relationship.

Clone: All individuals derived by vegetative propagation from a single original individual.

Coenzymes: These are organic molecules that are required by certain enzymes to carry out catalysis. They bind to the active site of the enzyme and participate in catalysis but are not considered substrates of the reaction. They often function as intermediate carriers of electrons, specific atoms or functional groups that are transferred in the overall reaction. An example is the role of NAD in the transfer of electrons in certain coupled oxidation reduction reactions. Some examples are: Nicotine adenine dinucelotide (NAD), nicotine adenine dinucelotide phosphate (NADP), flavine adenine dinucelotide (FAD), Thiamine pyrophosphate (thiamine vit. B_1) and biotin.

Cofactor: It is a non-protein chemical compound that is bound (either tightly or loosely) to a protein and is required for the protein's biological activity. These proteins are commonly enzymes and cofactors can be considered as "helper molecules/ions" that assist in biochemical transformations. Cofactors are often classified as inorganic substances that are required for, or increase the rate of, catalysis. Some examples are Zn^{++}, Fe^{+++} or Fe^{++}, Cu^{++} or Cu^{+}, K^{+} and Mg^{++}.

Condensation: Joining of monomers by covalent bonding with the accompanying loss of water molecules.

Coniferous trees: Cone bearing trees, mostly evergreens, that have needle-shaped or scale like leaves.

Contour farming: Ploughing and planting across the changing slope, rather than in straight lines to help retain water and reduce soil erosion.

Crust solid outer zones of the earth. It consists of oceanic crust and continental crust.

Cutin: A waxy water-repellent substance in the cuticle of plants, consisting of highly polymerized esters of fatty acids.

Dioxins: Family of 75 chlorinated hydrocarbon compounds formed by unwanted by-products in chemical reactions involving chlorine and hydrogen, usually at high temperature.

Dosage: Quantity of pesticide applied per individual or per unit area or per unit volume or per unit weight.

Drift: Spray or dust carried by natural air currents beyond the target area.

Ecological footprint: Amount of biologically productive land and water needed to supply a population with the renewable resources it uses and to absorb or dispose of the wastes from such resource use. It measures the average environmental impact of populations in different countries and areas.

Ecological tipping point: Point in the development of an environmental problem where a threshold level is reached, causing an irreversible shift in the behaiour of a natural system.

Ecosystem: The interacting system of the living organisms in a area and their physical environment.

Ecotone: The communities which are considered to be transitional between two biomes.

Ectoparasite: A parasite which lives externally on the body of its host.

Embryogenesis: Embryogenesis is a process whereby a plant is formed from one cell in a tissue, or callus by the normal zygotic pathway. Process is initiated by hormones or in the absence of hormones (*e.g.*, carrot, brome grass).

Enzymes: These are macromolecular biological catalysts. They help in catalyzing the biochemical and biological reactions by binding to the substrate at the substrate.

Explant: A plant organ or piece of tissue used to initiate a tissue culture.

Fallow land: It is the land which is kept for one or more growing seasons to kill the weeds or making the soil richer with nutrients and moisture.

Famine: Widespread malnutrition and starvation in a particular area because of a shortage of food, usually caused by drought, war, flood, earthquake, or other catastrophic events that disrupt food production and distribution.

Fecundity: Capacity to produce offspring, power of a species to multiply rapidly.

Fermentation: It is the process of energy production in a cell under anaerobic conditions without oxygen.

Fidelity: It is the accuracy with which a population estimate follows actual numbers in a population.

Fishprint: Area of ocean needed to sustain the consumption of an average person, a nation, or the world.

Food web: Complex network of many interconnected food chains and feeding relationships.

Fumigant: Pesticide exhibiting toxicity in the vapour phase.

Functional agro-biodiversity: Those elements of biodiversity on the scale of agricultural fields or landscapes; which provide ecosystem services that support sustainable agricultural production and can also deliver benefits to regional and global environment and to public at large.

Fuzz: Non separable part of fiber attached to cotton seeds.

Gene duplication: Genetic alteration in which a segment of DNA is repeated. Duplications may appear anywhere, but where the duplicated segment is adjacent to the original one is called *tandem duplication*.

Gene: A chromosomal segment that codes for a single functional polypeptide chain or RNA molecule.

Genetic drift: It is the change in the relative frequency in which a gene variant occurs in a population due to random sampling and chance rather than natural selection. In genetic drift, a population experiences a change in the frequency of a given allele; prompted by random luck rather than a need for adaptation.

Gibberellins: A group of plant hormones responsible for stimulation of the growth of leaves and shoots in general and internodal length of genetically dwarf plants in particular.

Gossypol ($C_{30}H_{30}O_8$): It is a polyphenol derived from the cotton plant. It acts as a inhibitor for several dehydrogenase enzymes. It is a yellow pigment.

Green revolution: Popular term for the introduction of scientifically bred or selected varieties of grain (including rice, wheat, and corn) that, with adequate inputs of fertilizer and water, can greatly increase crop yields.

Gregarious: The habit of living together in groups.

Guttation: Loss of water in the liquid state from uninjured parts of plants is known as guttation. It usually occurs from tips and margins of leaves during night or early morning when there is high atmospheric humidity.

Habitat: The place where plants and animals live.

Herbarium: It is a collection of preserved plant specimens. There specimens may be whole plants or plant parts; these are usually in dry form, mounted on sheet or may be kept in

Hormones: Organic substance secreted by plants and animals that functions in the regulation of physiological activities and in maintaining homeostasis.

Homokaryon: A bi or multinucleate cell having nuclei of the same kind in a common cytoplasm.

Humus: Slightly soluble residue of undigested or partially decomposed organic material in topsoil. This material helps retain water and water-soluble nutrients, which can be taken up by plant roots.

Hybrid: Offsprings received from two genetically dissimilar individuals, strains, purelines etc.

Hydrolysis: A process that occurs when a large molecule is broken down into smaller parts by the addition of water.

Inert: A material having no biological action.

Inoculum: A biological material (like a virus or toxin or immune serum) that in injected to increase immunity to a particular disease.

Inter cropping: Growing two or more different crops at the same time on a plot. Example a carbohydrate-rich grain that depletes soil nitrogen and protein rich legume that adds nitrogen to the soil may be intercropped.

Jhum cultivation: Also called slash and burn agriculture, is the process of growing crops by first clearing the land of trees and vegetation and burning them there after. The burnt soil contains potash which increases the nutrients contents of the soil for better yield. This is practiced in the states like Arunachal Pradesh, Meghalaya, Mizoram and Nagaland in India.

Kairomone: It is a chemical released by an organism which is beneficial to a member of another species, but not the emitter.

Key pest: The serious and persistant pests which attack a crop.

Leaching: In agriculture, leaching refers to the loss of water soluble nutrients from the soil due to rain or irrigation.

Leaf miner: An insect which lives in and feeds upon the cells between the upper and lower epidermis.

Lef area index: Ratio of leaf surface to soil surface in relation to utilization of solar energy for photosynthesis.

Lenticel: It is a porous tissue consisting of cells with large intercellular space in the periderm of the secondary thickened organs and the bark of woody stems and roots of dicotyledonous flowering plants allow gas exchange between the atmosphere and internal tissues.

Lichen: These are composite organisms consisting of a association of a fungus with a photosynthetic partner usually either a green alga (commonly *Trebouxia*) or (commonly *Nostoc*). The morphology, physiology and biochemistry of lichens are very different from those of the isolated fungus and alga in culture. Lichens occur in some of the most extreme environments on Earth, hot deserts, rocky coasts and toxic slag heaps.

Lipids: Compounds composed of C, O_2 and H which are insoluble in water. These are a class of organic compounds that tend to be soluble in non-polar solvents such as alcohol includes fats and oils. Fat and oils are used as long term storage compounds in plants and animals.

Loams: Soils containing a mixture of clay, sand, silt and humans. Good for growing most crops.

Lysis: Process of cell disintegration or membrane rupturing.

Mechanical control: Any means of pest control which involves mechanically trapping or killing insects or providing barriers to prevent insects gaining access to plants or other material.

Microbial control: The use of pathogenic micro-organisms to control pest species and is free from harmful residue, non-phytotoxic to crops at higher doses and harmless to beneficial insects.

Micronutrients: Chemical elements that organism need in smaller trace amounts to live, grow or reproduce like Ferrous, copper, chlorine and Iodine, zinc etc.

Microtubules: Submicroscopic, hollow tubes of protein that function throughout the cytoplasm to provide structural support and enable movement.

Molecular farming: The synthesis of certain chemicals through transgenic plants is known as molecular farming. The genes confirming quality characters are being transferred to get increased synthesis of some chemicals *viz.* Mannital dehydrogenage gene from *Escherichia coli* is transferred to tobacco for increased production of mannital.

Monoculture: Cultivation of a single crop, usually on a large area of land.

Mycorrhiza: This term refers to symbiotic association between the fungus and the roots of higher plants. Some of these fungi are present on the root surface only whereas others enter into roots as well. These fungus solubilizes phosphorus, produce plant growth promoting substances and protect host plant from soil pathogens. Mycorrhiza is classified into two types (*i*) ectotrophic and (*ii*) endotrophic

Mycotoxin: A deadly poison produced by fungi.

Nitrogenase: Nitrogenases are enzymes that are produced by certain bacteria, such as cyanobacteria (blue-green algae). These enzymes are responsible for the reduction of nitrogen (N_2) to ammonia (NH_3).

Nucleic acid: Complex molecules that store and transfer information within a cell. They are constructed of funda-mental monomers known as nucleotides.

Nucleoli: Nuclear structures composed of completed or partially completed ribosomes and the specific parts of chromosomes that contain the information for their construction.

Nucleoplasm: The liquid matrix of the nucleus composed of a mixture of water and the molecules used in the construction of the rest of the nuclear structure.

Nucleotide: A fundamental subunit of nucleic acid constructed of a phosphate group, sugar, and an organic nitrogenous base.

Nymph: The juvenile stage of an insect which undergoes incomplete metamorphosis.

Oogenesis: The specific name given to the gametegenesis process that leads to the formation of eggs.

Organelles: Cellular structures that perform specific functions in the cell.

Orthologues: Homologous proteins that perform the same function in different species.

Parasitoid: Parasitodids is a special kind of predotor that often about the same size as its host, kills its host and require only one host for development into free living adult *e.g.* Braconid wasps.

Perennial: Crops that can live for more than 2 years.

Persistence: The term applied to chemicals that remain active for a long period of time after application.

Pesticide: A chemical which kills pest organisms by virtue of its toxicity.

Pheromone: A chemical substance produced by an insect or other animal, which conveys a message to other members of the same species and usually results in a behavioural response.

Photoperiodism: It is the ability of plants to detect and respond to the relative length of day and night to which plant is exposed.

Phytoalexins: Phenolic compounds produced by plants when they become diseased or injured by insects; enable plants, once damaged, to resist further attack by insects.

Phytochrome: A pigment which is a protein molecule, located within the plasma membranes of the leaf cells that is extremely sensitive to certain forms of light.

Plasmids: Small and circular DNA molecule which has self replicating property in the host cell or an extrachromosal genetic plant found in bacterial cell and can replicate independently of chromosomal DNA.

Polyculture: Complex form of inter cropping in which a large number of different plants maturing at different times are planted together.

Polymerase chain reactions (PCR): A method for amplifying a DNA base sequence using a heat stable polymerase and two primers, one complementary to the (+) strand at one end of the sequence to be amplified and other complementary (–) strand at the other end. The faithfulness of reproduction of the sequence is to the fiedility of the polymerase. PCR was developed by Karl Mullis in the year 1980.

Protandrous plants: Where male reproductive organism (anthers) mature before the female reproductive organism (stigma) examples are; Maize, sunflower

Proteins: Proteins are large biomolecules or macromolecules consisting of one or more long chains of amino acids residues. They are one of the building blocks of the body tissues and can serve as a fuel source. There are about 20 different amino acids that occur naturally in proteins.

Protogynous plants: The plant flower in which female reproductive organism (stigma) mature before the males ones (anthers) thereby ensuring that self fertilization does not occur . Examples are *plantain* and *figwort, plearlmillet.*

Protoplast is a plant, bacterial or fungal cell that had its cell wall completely or partially removed using either mechanical or enzymatic means.

Quarantine: All operations associated with the prevention of unwanted organisms into a territory or their exportation from it.

Residue: Amount of pesticide remaining in or on plant tissues or in soil at harvest time.

Rhizobium: It is a symbiotic bacterium, and occurs in the root nodules of legumes. Rhizobium fixes atmospheric nitrogen in the presence of red pigment called leghae-moglobin. Rhizobium was first time isolated by Biejerinck in the year 1888.

Salinity: Amount of various salts dissolved

Salinization: Accumulation of salts in soil that can eventually make the soil unable to support plant growth.

Scavenger: Organism that feeds on dead organisms that were killed by other organisms or died naturally. Examples include vultures, flies, and crows.

Smother crop: A smother crop is a thick, rapidly growing crop that is used to suppress or stop the growth of weeds which have better root systems that help them compete with weeds for water and as a result, the root systems of weeds get weak. Meanwhile, the dense top growth of the smother crop suppresses the top growth of weeds. Effectively, smother crops successfully compete with weeds for vital resources (growth, space, water, light) and inhibit their germination and growth.

Strip cropping: Planting regular crops and close-growing plants such as hay or nitrogen-fixing legumes in alternating rows or bands to help to reduce depletion of soil nutrients.

Succulent plants: Plants such as desert cacti that survive in dry climates by having no leaves, thus reducing the loss of scarce water. They store water and use sunlight to produce the food they need in the thick, fleshy tissue of their green stems and branches.

Sustainable agriculture: Agriculture and agrobased food system that are economically viable and meet society's need for safe and nutritious food while conserving or enhancing natural resources and the environment for sustainable life for future generations.

Symbiosis: Association of two dissimilar organisms to their mutual advantage.

Synecology: The branch of ecology that deals with the structure and development of entire ecological communities and the interrelationship of the plants and animals with them. It is the study of entire ecosystem.

Synomone: It is a chemical released by an organism which benefits both the sender and receiver.

Tolerance: a mechanism of plant resistance to pests whereby the plant is able to grow and produce despite pest injury.

Transgenic plants: These are genetically modified plants which have been created by mobilizing genes of interest from one species to another species by means of biotechnology.

Trap crop: Crop of plants grown especially to attack insect pests.

Turnover number: The number of molecules of substrate that a single molecule of enzyme can react with in a given unit of time.

Vector: An insect or other animal which transmits viruses or other pathogens directly or indirectly.

Vernation: Arrangement and folding of leaves in a bud before its opens.

Virgin soil: The soil, which has not been used for cultivation.

Viruses: Infectious disease causing agents, smaller than bacteria, which always require intact host cells for replication and which contain either DNA or RNA as their genetic component.

Wetland: Land that is covered all or part of the time with saltwater or freshwater, excluding streams, lakes, and the open ocean.

Windrower: A machine used for the purpose of crop harvesting.

Xerophyte: Plant adapted to survive long periods without moisture. *Example-* Cacti.

Xylan: Polysaccharide found in the cell walls of plants, where it forms the bulk of the hemicelluloses component.

Yellowing: It is the result of diminishing rate of chlorophyll synthesis from plants.

Zeatin: Naturally occurring cytokinin derived from adenine which plays a role in the growth and development of plants.

Glossary (Economic Terms)

Acquisition: The acquiring or control of one corporation by another. In "unfriendly" takeover attempts, the potential buying company may inducements to stockholders.

Active Market: Term used by stock exchange which specifies the particular stock or share that deals in frequent and regular transactions. It helps the buyers to obtain reasonably large amounts any time.

Administered Prices: When the prices of an item or a commodity are decided by the government or any other agency and not on the basis of demand and supply, such types of prices are called *Administered Prices*.

Amalgamation: It means 'merger'. As and when necessity arises two or more companies are merged into a large organisation. The old firms completely lose their identity when the merger takes place.

Amortization: Payment of a loan in installments by the borrower. It is usually done in an agreed period and every installment includes a part of the total loan plus the interest

Appreciation: Appreciation means an increase in the value of something *e.g.* stock of raw materials or manufactured goods. It also includes an increase in the traded value of currency. It is an increase in the value of assets over a particular time period.

Asset: Anything which has a 'money value' owned by an individual or a firm is an asset. It is of *three* types:

(*i*) *Tangible asset:* All physical assets such as land, machinery, building, consumer durables (refrigerator, car, T.V., Radio, etc.) etc. (*the assets which are in the material form*).

(*ii*) *Intangible assets:* All non-physical/ immaterial assets such as brand names, good will, credit-cxworthiness, knowledge, know how, etc.

(*iii*) *Financial assets:* All financially valid valuables other than tangibles and intangibles such as currencies, bank deposits, bonds, securities, shares, etc.

Assets: Everything a corporation or an organisation owns or that is due to it cash, investments, money due it, materials and inventories, which are called current assets; buildings and machinery, which are known as fixed assets; and patents and goodwill, called intangible assets.

Autarchy: It means self-sufficiency and self-reliance of an economy. Autarchy is an indicator of self-sufficiency. It means that the country itself can satisfy the needs of its population without making imports from other countries.

Balance of Payment: It is the difference between country's payments and receipts from other countries during a year. In other words the balance of payment shows the relationship between the one country's total payment to all other countries and its total receipts from them. Balance of payment not only includes visible export and imports but also invisible trade like shipping, banking, insurance, tourism, royalty, payments of interest on foreign debts.

Balance of payments: A balance sheet of an economy showing its total external transactions with the world calculated on the principles of accounting is an annual concept.

Balance of Trade: It refers to the relationship between the values of country's imports and its export, *i.e.* the visible balance. Balance of trade refers to the total of country's export commodities and total value of imports commodities. Thus, balance of trade includes only visible trade *i.e.* movement of goods (exports and imports of goods). Balance of trade is part of Balance of Payment statement.

Balance Sheet: It is a statement showing the assets and liabilities of a business at certain date. Balance sheet helps in estimating the real financial situation of a firm.

Balanced budget: The annual financial statement (*i.e.*, the budget) of a government which has the total expenditures equal to the taxes and other receipts. Most governments, in practice run unbalanced budgets, *i.e.*, deficit budgets or surplus budgets either the expenditures being higher or lower than the taxes and the other receipts, respectively. It is done to regulate the economic activities.

Bank: Bank is a financial institution. It accepts funds on current account and savings accounts. It also lends money. The bank pays the cheques drawn by customers against current or savings bank account. The bank is a trader that deals in money and credit.

Bankruptcy: It is a situation in which a person is unable to discharge his debt obligations.

Basis Point: One gradation on a 10-0 point scale representing 1%; used especially in expressing variations in the yields of bonds. Fixed income yields vary often and slightly within one percent and the basis point scale easily expresses these changes in hundredths of 1%. For example, the difference between 12.83% and 12.88% is 5 basis points.

Bill of Exchange: A unconditional order in writing addressed by one person to another requiring the address to pay on demand or at a fixed future time a certain sum of money to the order of the specified person or to the bearer.

Birth Rate: Birth Rate (or Curde Birth Rate) is number of the births per thousand of the population during a period, usually a year. Only live births are included in the calculation of birth rate.

Black Money: It is unaccounted money which is concealed from tax authorities. All illegal economic activities are dealt with this black money. Hawala market has deep roots with this black money. Black money creates parallel economy. It puts an adverse pressure on equitable distribution of wealth and income in the economy.

Blue Chip: It is the most reliable industrial shares on a stock exchange. It is concerned with such equity shares whose purchase is extremely safe. It is a safe investment. It does not involve any risk.

Book Value: An accounting term. Book value of a stock is determined from a company's records, by adding all assets then deducting all debts and other liabilities, plus the liquidation price of any preferred issues. The sum arrived at is divided by the number of common shares outstanding and the result is book vlaue per common share. Book value of the assets of a company or a security may have little relationship to market value.

Boom: The point at which price and employment are the maximum. The trade is also at its highest point and beyond this no upward movement is possible.

Bounty: It is a subsidy paid by the government to exporters. It reduces the price of exportable goods and hence act as incentive to enhance exports.

Brain-Drain: It means the drift of intellectuals of a country to another country. Scientists, doctors and technology experts generally go to other prominent countries of the world to better their lot and earn huge sums of money. This Brain-Drain deprives a country of its genius and capabilities.

Buffer Stocks: These are the stocks (generally of primary goods) accumulated by government agency when supply is plentiful. These stocks are released at times of shortage of supply. In India Food Corporation of India (FCI) accumulates foodgrains as buffer stocks.

Bull Market: It is a market where the speculators buy shares or commodities in anticipation of rising prices. This market enables the speculators to resale such shares and make a profit. The opposite of Bear Market.

Bullion: It is gold or silver having a specific degree of purity. Generally it is in the form of gold or silver bars.

Buyer's market: A short period of market situation in which there is excess supply of goods/services forcing price fall to the advantage of the buyers.

Buyer's market: When the market is favourable to buyer's market. This situation occurs when there is a change from boom to recession *i.e.* demand is less than supply.

Callable: A bond issue, all or part of which may be redeemed by the issuing firm, institution or organisation under specified conditions before maturity. The term also applies to preferred shares that may be redeemed by the issuing organisation.

Capital adequacy ratio: A regulation on commercial banks, co-operative banks and the non-banking financial companies to maintain a certain amount of capital in relation to their assets (*i.e.*, loans and investments) as a cushion against probable losses in their investments and loans.

Capital Market: A market for long term loans. Capital market is the market which gives medium term and long term loans. It is different from money market which deals only in short term loans.

Capital: Capital is one of the three main factors of production (*labour* and *natural resources* are the other two), classified into *physical capital* (*i.e.*, factories, machines, office, etc.) and *human capital* (*i.e.*, training, skill, etc.).

Capital: Capital: Capital is one of the three main factors of production (*labour* and *natural resources* are the other two), classified into *physical capital* (*i.e.*, factories, machines, office, etc.) and *human capital* (*i.e.*, training, skill, and knowledge etc.). In a joint stock company, the capital has various specific terms showing different forms of the *share.*

(*i*) *Authorised capital:* This is the amount of share capital fixed in the Memorandum of Association (MoA) and the article of association of a company as required by the Companies Act. This is also known as the *Nominal* or *Registered capital.*

(*ii*) *Paid-up capital:* The part of the authorised capital of a company that has actually been paid up by the shareholders. A difference may arise because all shares authorised may not have been issued or the issued shares have been only partly paid-up by then.

(*iii*) *Subscribed capital:* The capital that has actually been paid by the shareholders (as they might have committed more than this to contribute). It means, the subscribed capital is the actually realised paid-up capital (paid-up capital is subscribed capital plus credit/due on the shareholders).

(*iv*) *Issued capital:* The amount of the capital which has been sought by a company to be raised by the issue of shares (cannot exceed the authorised capital).

(*v*) *Called-up capital:* The amount of share capital the shareholders have been *called* to pay to date under the phased payment terms. It is usually equal to the 'paid-up capital' of the company except where some shareholders have failed to pay their due installments (known as *calls in arrears*).

Capitalism: The economic system based on free enterprise and private profit. Capitalism is a economic system in which all means of production are owned by private individuals. Self-profit motive is the guiding feature for all the economic activities under capitalism. Under pure capitalist system economic conditions are regulated solely by free market forces. This system is based on 'Laissez-faire system' *i.e.* no state intervention. Sovereignty of consumer prevails in this system.

Cash flow: Reported net income of a corporation plus amounts charged off for depreciation, depletion, amortization, and extraordinary charges to reserves, which are bookkeeping deductions and not paid out in actual rupees and paise.

Cash Reserve Ratio (CRR): It refers to that portion of banker's total cash reserves which they are statutorily required to hold with the R.B.I. The commercial banks are required to keep a certain amount of cash reserves at the central bank *i.e.* RBI. This percentage amount is called CRR. It influences the commercial bank's volume of credit because variation in CRR affects the liquidity position of the banks and hence their ability to lend.

Cheque: Cheque is an order in writing issued by the drawer to a bank. If the customer has sufficient amount in his account, the cheque is paid by the bank. Cheques are used in place of cash money.

Collateral : Securities or other property pledged by a borrower to secure repayment of a loan.

Conglomerate: A company or an organisation that has diversified its operations usually by acquiring enterprises in widely varied industries.

Consortium: An *ad hoc* grouping of firms, governments, etc. brought together to undertake a particular project by pooling their resources and skills for success of major construction projects, loans, etc.

Corporate sustainibility index: The Bombay Stock Exchange (BSE) has proposed to come out with a corporate sustainability index (CSI) a possible new stock exchange which will be created for developing *trust marks* to denote a corporate's sustainability achievements. This will be the first such index in Asia.

Cost Price Index (CPI): It is used for measuring cost of living and it covers large number of commodities than Wholesale Price

Index (WPI) which is used for measuring rate of inflation.

Coupon bond: Bond with interest coupons attached. The coupons are clipped as they come due and presented by the holder for payment of interest.

Current assets: Those assets of a company that are reasonably expected to be realized in cash, sold or consumed during one year. These include cash, Government bonds, receivables and money due usually within one year, as well as inventories.

Debentures: A document which enlists the terms or conditions of a loan. The debentures are used by corporate sector (companies). The debenture holders are to be paid a fixed annual rate of interest and they have the first claim on the assets of a company as creditors.

Decentralisation: Decentralisation means the establishment of various units of the same industry at different places. Large scale organisation or industry can not be run at one particular place or territory. In order to increase the efficiency of the industry, various units at different places are located.

Deflation: It is a fall in the general price level over a particular period of time. It is opposite to inflation.

Depression: It is just opposite to "boom". It implies a state of economy when lack of demand result in heavy unemployment and stagnation in economy.

Direct cost: The direct material and labour cost of a product proportionally varies with the total output.

Direct Tax: It is a tax whose burden cannot be shifted *i.e.* the burden of direct tax is borne by the person on whom it is initially fixed, *e.g.* personal income tax, social security tax paid by employees, death tax etc.

Dividend: It is earnings on stocks paid to shareholders.

Dumping: Exporting a good at a price lower than its price in the domestic market. To neutralise the effects of dumping the importing country may impose a *surcharge* on such imports which is known as the *antidumping* duty.

Dumping: It means selling goods in international market at a price which is lower than that in domestic or home market.

e-Business: Using computers and the Internet to link both the *internal* operations (*i.e.*, transactions and communications between the various departments/divisions of the business firm) and its *external* operations (*i.e.*, all its dealings with the suppliers, customers, etc.).

e-Commerce: Method of buying and selling goods and services over the Internet a kind of direct marketing *i.e.* without the help of any middle arrangement of sales.

Economies of scale: The long-run reduction in average/unit cost that occurs as the scale of the firm's output increases. The opposite situation is known as *diseconomiess of scale.*

Elasticity of demand: The responsiveness of demand of commodity to the change in its price is known as elasticity of demand.

Engel's law: Ernest Engel, the 19th century German statistician, analysed the budget data of working families and established a relationship between the families income and expenditure. According to the Law "When a family's income increases the percentage of its income spent on food decreases."

Environmental audit: Assessment of the environmental impact of a firm/public body through its activities. This is done with an major objective to reduce or eliminate the pollution.

Free and open market: A market in which supply and demand are freely expressed in terms of price. Contrasts with a controlled market in which supply, demand and price may all be regulated.

Free goods: The goods which are in abundance (*as air and water*) and are not considered as scarce economic goods. As such goods have *zero supply price* and they will be used in large volumes resulting into rising environmental pollution (point should be noted that today air and water may not be considered as the typical free goods, at least the 'pure air' and 'pure water').

Free Trade: It implies absence of any protective tariffs or trade barriers by any economy with respect to export and import.

Free trade: The international trade among an agreed-upon group of countries without any barriers (such as tariffs, quotas, forex controls, etc.), promoted with the objective of securing international specialisation and an edge in their foreign trade.

Funded debt: usually interest-bearing bonds or debentures of a company. Could include long-term bank loans. Does not include short-term loans, preferred or common stock.

Gini coefficient: An inequality indicator in an economy. The coefficient varies from 'zero' to 'one'. A 'zero' Gini coefficient indicates a situation of perfect equality (*i.e.*, every household earning the same level of income) while a 'one' signifies a situation of absolute inequality (*i.e.*, a single household earning the entire income in an economy).

Golden handshake: A payment (usually generous) made by a company to its employees for quitting the job prior to their service.

Gross National Product (GNP): Gross Domestic Product plus net factor income from abroad is equal to Gross National Product.

Hoarding: An act of unproductive stocking of *money* or *goods.*

Hypothecation: The pledging of securities as collateral-for example, to secure the debit balance in a margin account.

Indenture: A written agreement under which bonds and debentures are issued, setting forth maturity date, interest rate and other terms.

Indifference curve: A curve on the graph showing the alternative combinations of two products, each giving the same utility/satisfaction.

Inflation: It is sustained increase in general price level over a particular period of time. It reduces the purchasing power of money.

Insolvency: The situation when the liabilities of an individual or a firm to creditors exceeds its assets–inability to pay the liabilities from the assets. Also known as *bankruptcy.*

International Monetary Fund (IMF): A multinational institution set up in 1945. It started working as a independent organisation in 1947. It seeks to maintain cooperative and orderly currency arrangements between member countries with the aim of promoting increased international trade and BOP equilibrium.

Inventory: The stocks of finished goods, goods under the production process and raw materials held by a firm.

IRA: Individual retirement account. A pension plan with tax advantages. IRAs permit investment through intermediaries like mutual funds, insurance companies and banks, or directly in stocks and bonds through stockbrokers.

Isocost curve: A curve on the graph showing the varying combinations of factors of production (*i.e.*, labour, capital etc.) that can be used to produce a given quantity of a product with a given technology.

Keogh plan: Tax-advantaged personal retirement program that can be established by a self-employed individual.

Laissez-faire: Literally it means 'to let people do as they choose'. It is an economic doctrine which emphasizes the superiority of 'free' trade and 'free' markets over state's interference in economic affairs. It is a French orgin of which British variation is 'Laissez-faire'.

Liabilities: All the claims against a corporation. Liabilities include accounts, wages and salaries payable; dividends declared payable; accrued taxes payable; and fixed or long-term liabilities, such as mortgage bonds, debentures and bank loans.

Liquid asset: The monetary asset that can be used directly as payment.

Liquidation: The process of converting securities or other property into cash. The dissolution of a company, with cash remaining after sale of its assets and payment of all indebtedness being distributed to the shareholders.

Liquidation: A process of 'winding up' a joint-stock company as a legal entity.

Liquidity trap: A situation when the interest rate is so low that people prefer to hold money rather than invest it. In such situations investors do not go to increase investment even if the interest rates on loans are decreased. J. M. Keynes suggested for increased government expenditure or reduction in taxes to fight such a situation.

Marginal unity: The increase in satisfaction/utility a consumer derives from the use/consumption of one *additional* unit of a product in a particular time period–it goes on decreasing, *i.e.*, the *diminishing marginal* utility.

Market Value: The market value of an equity share is the price at which it is traded in the market. This price can be easily established for a company that is listed on the stock market and actively traded (For a company that is listed on the stock market but traded very infrequently, it is difficult to obtain a reliable market quotation. For a company that is not listed on the stock market, one can merely conjecture as to what its market price would be if it were traded).

Microcredit: Smaller credit/loan to small and needy borrowers who are outside the reach of commercial banks, for the purpose of undertaking productive activities in any area.

MODVAT: The modified system of value added taxation is based on the idea of tax final products and not inputs that go into production.

Monopoly: It is a type of market structure having one seller and many buyers. There is a lack of substitute products and entry of new firms into market is not possible.

MoU: The concept of Memorandum of Understanding (MoU) was introduced in 1988. The main objective of MoU is to reduce the quantity of control and increase the quality of accountability. The emphasis in on achieving the negotiated and agreed objectives rather than interfering in the day-to-day affairs.

NASDAQ: An automated information network that provides brokers and dealers with price quotations on securities trade over-the-counter. Nasdaq is an acronym for National Association of Securities Dealers Automated Quotations.

Net asset value: Usually used in connection with investment companies to mean net asset value per share. An investment company computes its assets daily, or even twice daily, by totaling the market value of all securities owned. All liabilities are deducted, and the balance is divided by the number.

Net Domestic Product (NDP): The money value of a nation's annual output of goods and service, less capital consumption (depreciation) experienced in producing that output.

Net National Product (NNP): Net National Product is equal to Net Domestic Product Plus Net factor income from abroad.

Net worth : Net worth for a company is its total assets minus total liabilities.

Octroi: It is an internal tariff system among different region of a country.

Overbought: An opinion as to price levels. May refer to a security that has had a sharp rise or to the market as a whole after a period of vigorous buying which, it may be argued,has left prices "too high".

Oversold: The reverse of overbought. A single security or a market which, it is believed, has declined to an unreasonable level.

Per Capital Income: It implies income per person. It is obtained by dividing national income of country by its population.

Population trap: A situation of population growth rate greater than the achievable economic growth rate. This makes it difficult to alleviate poverty; government is suggested to implement population control measures.

Portfolio: Holdings of securities by an individual or institution. A portfolio may contain bonds, preferred stocks, common stocks and other securities.

Poverty Line: The poverty line has been fixed by the planning commission on the basis of an average daily intake of 2400 calories per person is rural areas and 2100 calories per capita in urban areas. In monetary terms the poverty line is commented to be Rs. 76 per month in rural and Rs. 88 in urban areas in terms of 1979-80 prices.

Poverty trap: A situation where an unemployed getting unemployment allowance is not encouraged to seek work/employment because his/her after-tax earnings as employed is less than the benefits as unemployed also known as the *unemployment trap.*

Preferred stock: A class of stock with a claim on the company's earnings before payment may be made on the common stock and usually entitled to priority over common stock if the company liquidates. usually entitled to dividends at a specified rate-when declared by the board of directors and before payment of a dividend on the common stock-depending upon the terms of the issue.

Recession: A economic cycle characterised by a modest downturn in the level of economic activity means fall up of demand.

Redemption price: The price at which a bond may be redeemed before maturity, at the option of the issuing company. Redemption value also applies to the price the company must pay to call in certain types of preferred stock.

Refinancing: Same as refunding. New securities are sold by a company and the money is used to retire existing securities. The object may be to save interest costs, extend the maturity of the loan, or both.

Reflation: It is an increase in the level of National Income and Output. Reflation is often deliberately brought about by the authorities in order to secure full employment and to increase the rate of economic growth.

Rule of thumb: A rough-and-ready decision-making aid that provides an acceptably accurate approximate solution to a problem acceptable to all.

Scheduled bank: It is a bank included in the second schedule of RBI. It has a minimum cash reserve of "Rs. 5 lakh".

SEBI: It was set up in 1988 by the Government of India to regulate the operations in stock market of India. The SEBI stands for Securities and Exchange Board of India

Seller's Market: It is market situation which exists for a short time period. During this period there is an excess demand for good and services at current prices which forces price up to the advantage of the seller.

Sensex: The Stock Exchange Sensitive Index (popularly referred to as the SENSEX) reflects the weighted arithmetic average of the price relative of a group of share included in the index of sensitive shares.

Share Capital: Money raised by issuing of shares is called Share Capital.

Shares: These are the equal portions of the capital of a listed company. Shares in a company do not carry fixed rate of interest. The holders of the ordinary shares carry the residual risk of the business; they rank after debenture holders and preference shareholders for the payment of dividends and they are liable for losses, although this liability is limited to the value of the shares and to the limit of guarantee given by them. Preference shares are such shares of a company on which interest is paid before any others, and owners have prior right to repayment of capital if company is wound up.

Sinking fund: Money regularly set aside by a company to redeem its bonds, debentures or preferred stock from time to time as specified in the indenture or charter.

Social costs: The costs borne by the society at large resulting from the economic activities by the firms pollution being a prominent example or costs borne by society after bandhs, closure or burning of public properties.

Speculator: One who is willing to assume a relatively large risk in the hope of gain.

Standard deviation: A statistical technique to measure how far a variable moves over time away from its mean (average) value.

Stock exchange: An organized marketplace for securities featured by the centralization of supply and demand for the transaction of orders by member brokers for institutional and individual investors.

Sunk costs: The costs in commercial activities that have been incurred and cannot be reversed. The cost on advertisement, research and development, etc. are examples of such costs.

Technological unemployment: Unemployment which results from the automation of the production activities (*i.e. machines replacing men*).

Third market: Trading of stock exchange-listed securities in the over-the-counter market by non-exchange member brokers.

Treasury stock: Stock issued by a company but later reacquired. It may be held in the company's treasury indefinitely, reissued to the public or retired. Treasury stock receives no dividends and has no vote while held by the company.

Variable annuity: A life insurance policy where the annuity premium a set amount of dollars) is immediately turned into units of a portfolio of stocks. Upon retirement, the policy holder is paid according to accumulated limits. Its objective is to preserve, through stock investment, the purchasing value of the annuity which otherwise is subject to erosion through inflation.

VAT: It seeks to tax the value added at every stage of manufacturing and sale with a provision of refunding the amount of VAT already paid at earlier stages to avoid double taxation.

Yield: Also known as return. The dividends or interest paid by a company expressed as s percentage of the current price.

Zero tilling: A relatively new farm production process, is a one-time operation in which a small drill places the seed and the fertiliser in a small furrow, saving the farmer a lot of time and other resources. At first utilised in Haryana in 1999–2000, by now it has spread to the other wheat growing states like. Punjab, Uttar Pradesh, Uttarakhand, and Bihar particularly. The technique gives comparatively higher yield (by over 5 per cent) than the conventional wheat farming.

Index

D

Q

R

S

COMMITTEES AND COMMISSIONS

NEW COMMITTEES FORMED DURING 2016-2017

1. ***Ashok Dalwai Committee:*** Blueprint for Doubling Farmers Income by 2022.
2. ***T. Haque Expert Group***: For Land leasing
3. ***Deepak Nayak Committee***: Revival of Godavari River
4. ***Arvind Subramanyam Committee***: For tackling of pulse shortage and to review MSP and bonus for pulses
5. ***Madhav Chitale Committee***: To prepare Guidelines for desiltation of river ganga
6. ***Kaddiyam Srihari***: To look into issues of girl education in the country
7. ***S. Ayyapan***: To draft National Policy on marine fishery.
8. ***A. K. Sinha Expert Committee***: Study the silt in the rivers
9. ***Pradeep Kumar Sinha:*** GST council.
10. ***Akhilesh Ranjan:*** To examine the business models for e-commerce and to identify the direct tax issue in relation to e-commerce transactions and provide suggestions.
11. ***Rajnath Singh:*** For Central Assistance to Karnataka, Puducherry and Arunachal Pradesh affected from drough and floods.
12. ***Sudarshan Committee:*** To study regulatory issue is relating to financial technology and digital banking in India.

SOME OTHER IMPORTANT COMMITTEES

1. ***Khusro Committee* (1986)**: For Agricultural Credit System
2. ***Purshotam Das Committee***: For Agricultural Finance and Co-operative societies.
3. ***Abhijit Sen Committee* (2002)**: For long term food security
4. ***Shankar Lal Gori Committee***: Agricultural Marketing
5. ***UK Sharma Committee***: Role of NABARD in Rural Regional banks
6. ***Malhotra Committee***: Insurance Reforms
7. ***Bhanu Pratap Singh Committee***: Agriculture
8. ***Chakravarty Committee***: Banking sector reforms

NATIONAL AND OTHER COMMISSIONS

1.	First Irrigation Commission	–1901
2.	Royal Commission on Agriculture	–1926
3.	Central Water commission	–1945
4.	Finance Commission	–1951
5.	Central Forestry Commission	–1965
6.	Khadi and Village Industry Commission	–1957
7.	Commission for Agricultural Cost and Prices	–1965
8.	Second Irrigation commission	–1972
9.	Agricultural Commission	–1976
10.	National Flood Commission	–1976
11.	National Commission for Women	–1992
12.	National Commission on Backward Classes	–1993
13.	National Commission on Cattle	–2002
14.	Commission for Enterprise in Unorganized Sector	–2004
15.	National Commission on Farmers	–2004
16.	National Statistical Commission	–2005
17.	National Forest Commission (Review of Forest Policy)	–2013